Principles and Practices for Petroleum Contaminated Soils

Edited by

Edward J. Calabrese
Paul T. Kostecki

CRC Press is an imprint of the
Taylor & Francis Group, an **informa** business

CRC Press
Taylor & Francis Group
6000 Broken Sound Parkway NW, Suite 300
Boca Raton, FL 33487-2742

First issued in paperback 2019

ISBN-13: 978-0-367-45014-4 (pbk)
ISBN-13: 978-0-87371-394-8 (hbk)

Visit the Taylor & Francis Web site at
http://www.taylorandfrancis.com

and the CRC Press Web site at
http://www.crcpress.com

To the Memory of Ed Lewis

Edward J. Calabrese is a board certified toxicologist who is professor of toxicology at the University of Massachusetts School of Public Health, Amherst. Dr. Calabrese has researched extensively in the area of host factors affecting susceptibility to pollutants, and is the author of more than 270 papers in scholarly journals, as well as 18 books, including *Principles of Animal Extrapolation; Nutrition and Environmental Health*, Vols. I and II; *Ecogenetics; Safe Drinking Water Act: Amendments, Regulations and Standards; Petroleum Contaminated Soils*, Vols. 1, 2, and 3; *Ozone Risk Communication and Management; Hydrocarbon Contaminated Soils*, Vols. 1 and 2; *Hydrocarbon Contaminated Soils and Groundwater*, Vols. 1 and 2; *Multiple Chemical Interactions; Air Toxics and Risk Assessment; Alcohol Interactions with Drugs and Chemicals, Regulating Drinking Water Quality, Biological Effects of Low Level Exposures to Chemicals and Radiation*, and *Contaminated Soils: Diesel Fuel Contamination.* His most recent book is *Risk Assessment and Environmental Fate Methodologies.* He has been a member of the U.S. National Academy of Sciences and NATO Countries Safe Drinking Water committees, and of the Board of Scientific Counselors for the Agency for Toxic Substances and Disease Registry (ATSDR). Dr. Calabrese also serves as Chairman of the International Society of Regulatory Toxicology and Pharmacology's Council for Health and Environmental Safety of Soils (CHESS) and Director of the Northeast Regional Environmental Public Health Center at the University of Massachusetts.

Paul T. Kostecki, Associate Director, Northeast Regional Environmental Public Health Center, School of Public Health, University of Massachusetts at Amherst, received his PhD from the School of Natural Resources at the University of Michigan in 1980. He has been involved with risk assessment and risk management research for contaminated soils for the last eight years, and is coauthor of *Remedial Technologies for Leaking Underground Storage Tanks*, coeditor of *Soils Contaminated by Petroleum Products* and *Petroleum Contaminated Soils*, Vols. 1, 2, and 3, and of *Hydrocarbon Contaminated Soils*, Vols. 1 and 2, and *Hydrocarbon Contaminated Soils and Groundwater*, Vols. 1 and 2, of *Contaminated Soils: Diesel Fuel Contamination*, and coeditor of *Risk Assessment and Environmental Fate Methodologies*. Dr. Kostecki's yearly conferences on hydrocarbon contaminated soils draw hundreds of researchers and regulatory scientists to present and discuss state-of-the-art solutions to the multidisciplinary problems surrounding this issue. Dr. Kostecki also serves as Managing Director for the International Society of Regulatory Toxicology and Pharmacology's Council for Health and Environmental Safety of Soils (CHESS), as Executive Director of the newly formed Association for the Environmental Health of Soils (AEHS), and as Editorial Advisor to the *Journal of Soil Contamination* and *SOILS* magazines.

PREFACE

The past decade has witnessed a considerable regulatory and public health interest in the problems associated with petroleum contaminated soils. During this time a spate of research papers has been published that has sought to offer unique insights and sound guidance in the multifaceted aspects of assessing the environmental and public health aspects of the petroleum contaminated soil domain. Consequently, the field has experienced a rapid expansion of the available literature on petroleum contaminated soils, and it has become difficult to place the vast amount of literature in perspective, while it is easy to become somewhat confused and a bit overwhelmed. In response to these circumstances, we decided that a book on *Principles and Practices* would be appropriate in order to not only bring order to the field but also because the field has been rapidly maturing and stable perspectives are emerging. This conclusion is derived from our experience in helping to organize and direct national conferences and workshops each year on hydrocarbon contaminated soil, compiling an annual bibliography of soil remediation references, and helping direct technical input for *SOILS* magazine, as well as our specific research in the field.

At first it was thought that we would consider the publication of the "best" articles published in recent conference proceedings. However, this notion was quickly put aside and all authors were requested to update and expand their efforts. While this added an additional nine months to the publishing time, the improvement in the final product was substantial, making the book well worth the wait.

Principles and Practices is designed to be a substantial and broadly based reference addressing each major technical element of the problem, including site assessment, analysis, remediation and risk assessment. It will be useful for a broad range of interested parties including graduate students and new employees, and for molding a specific orientation to the field as well as to the research and regulating communities. It likewise can serve as a ready reference on a nearby shelf.

The contributing authors are generally recognized on the national and international levels as experts and maintain strong, active research programs. We believe that the careful selection of authors based on technical knowledge and experience, along with the choice of the critical topical areas will help assure that this book will serve its goal as a strong and stable reference in the critical area of petroleum contaminated soils.

CONTENTS

Part I: Analysis & Testing

Part II: Environmental Fate and Modeling

Part III: Remediation

Part IV: Health Assessment

Principles and Practices for Petroleum Contaminated Soils

CHAPTER 1

Analysis of Petroleum Contaminated Soil and Water: An Overview

Thomas L. Potter, Mass Spectrometry Facility, Massachusetts Agricultural Experiment Station, University of Massachusetts, Amherst

INTRODUCTION

Spills and leaks of petroleum products including gasoline, diesel fuel, and lubricating and heating oil often result in the contamination of soil and water. Analyses required to evaluate the extent of such releases and the threat they present to public health and the environment take a variety of forms. Analytical objectives are also diverse and often poorly specified. They range from a simple assessment of "presence or absence" to determination of the concentration of certain toxic substances these products contain. The least specific and most general analytical approach to the problem involves some form of "total petroleum hydrocarbon" (TPH) measurement. In contrast are analyses which are focused on selected target compounds.

Methods[1-4] which are commonly used for petroleum contaminated soil and water analysis were developed by the U.S. Environmental Protection Agency (EPA). Various modifications of well-known EPA techniques such as "Modified Method 8015"[5] also see wide application. It should be emphasized that the EPA methods have their origin in techniques developed by the agency for compliance monitoring in certain regulatory programs. The methods were not specifically developed for the analysis of petroleum contaminated soil and water, nor have they been systematically evaluated for this purpose.

In this review, the "state of the art" of petroleum contaminated soil and water analysis is discussed with emphasis on the EPA Methods.[1-5] Examples demonstrate that the methods can be used effectively but problems do arise. The need for method development is emphasized.

THE PROBLEM DEFINED: PETROLEUM PRODUCT CHEMISTRY

The source material for nearly all petroleum products is crude oil. Initial processing involves distillation into a series of fractions characterized by distillation temperature ranges and pressures. In general, the lighter fractions (lower boiling) represent gasoline-range material. The intermediate or middle distillate fractions represent feedstock for diesel and jet fuels and "light" heating oils. The residuum in this process serves as heavy fuel oils or other products. The trend from gasolines to the residual fuels is from the highly volatile to the nonvolatile, recognizing that in this case volatility is functionally defined.

Beyond distillation, numerous refinery processes are utilized to optimize the yield of certain products and to achieve desired product characteristics. The result is that some products may have little resemblance to the distillate fractions obtained in the initial crude oil processing. Gasoline is probably the best example. This product is blended from numerous refinery streams, and various additives are used to meet engine performance criteria.

Regardless of production modes or producers, most products are exceptionally complex materials with a wide range of physical and chemical properties. Gasoline, diesel fuel, and related products may contain hundreds or even thousands of individual constituents with boiling point distributions on the order of hundreds of degrees Celsius. Further, several chemical classes are usually represented, including paraffins, olefins, aromatics, heteroaromatics, and polar hydrocarbons containing oxygen, nitrogen, and sulfur. In turn, each class of compounds is characterized by various homologous series within which structural, enantiomeric, and other types of isomerism are exhibited. The higher alkyl substituted homologs also predominate.[6]

Another significant characteristic of the products is that their composition is variable. This is primarily in terms of the relative amounts of the various hydrocarbons the products contain. Relative product composition may also change dramatically after release into the environment. Processes responsible include volatilization, dissolution, and biotic and abiotic degradation. Each process influences to greater or lesser degree certain compounds or groups of compounds, and the rates of change are a function of environmental conditions.

These factors and others make petroleum product residue analysis in soil and water a formidable analytical challenge. They require that analytical methods be broad in scope. Where target compound analyses are involved there is also need for very high degrees of analytical selectivity and specificity. Compounds must be able to be detected in the presence of numerous potential interferences, and considering the toxicity of many petroleum constituents, high sensitivity is needed. Detection limits in the 1 to 10 micrograms per liter per component must be routinely achieved.

"TOTAL PETROLEUM HYDROCARBONS"

In light of the physical and chemical complexity of petroleum products, the analytical process is often reduced to the measurement of indicator parameters. Measurements of this type focus on determination of TPH. EPA methods include Method 413.1: "Oil and Grease" and Method 418.1: "Total Recoverable Petroleum Hydrocarbons."[1] These methods are similar to other well-known techniques.

Methods Description

Methods 418.1 and 413.1 specify the extraction of hydrocarbon residues from solids and water using an organic solvent, trichlorotrifluoroethane. After extraction the sample is discarded and the solvent treated with silica gel to remove interfering "humic" materials. Solvent concentration using rotary thin film evaporation and other techniques follows.

In Method 418.1, measurement of the "total hydrocarbon" content is performed using an infrared spectrophotometer set at 2930 cm^{-1}. Petroleum hydrocarbons, namely n-paraffins, exhibit a strong adsorbtion band at this wavelength. This is due to the presence of $-CH^2-$ groups in the molecules. Total hydrocarbon concentration is expressed relative to the detector response to a standard mixture containing a fixed ratio of aromatic and paraffinic hydrocarbons, or to a petroleum product reference sample.

With Method 413.1, the total hydrocarbon content in solvent extracts may be determined gravimetrically. In this case the solvent is completely evaporated and the residue weighed.

Applications

A distinct advantage of the TPH approach is that instrumentation costs are modest and extensive technical training of analysts is not required. This translates to low cost. Excellent measurement precision (i.e., reproducibility) can also be obtained. Unfortunately, there are some important limitations. Measurement accuracy may vary widely, depending on the products involved and the extent of post-release chemical changes (weathering) which may have occurred. Even more significant is the uncertainty that the use of data obtained from analyses of this type may introduce into risk assessment and management schemes.

Specific problems relate to the fact that a significant portion of the more volatile compounds in gasoline and light fuel oil may be lost in the solvent concentration step. This is especially so with gravimetric techniques. With residual fuels and other "heavy distillates," low recoveries often result for another reason. This is because many of their constituents are poorly soluble in trichlorotrifluoromethane and are not effectively extracted.

Another problem, at least with the infrared procedures, is in the selection of standards. The relative response of an infrared spectrophotometer to a hydrocarbon mixture is a function of the relative amounts of aromatic and aliphatic hydrocarbons it contains and the wavelength setting. At 2930 cm^{-1}, the wavelength specified in Method 418.1, detector response is only obtained for compounds which have a -CH^2- group. In short, the method has very poor sensitivity for aromatics.

A related issue is that hydrocarbon mixtures used for instrument calibration have constant composition (in terms of aromatics/paraffins content), whereas the relative composition of petroleum products and their residues are highly variable. This may introduce substantial uncertainty into measurements. Attempts have been made to compensate by using samples of petroleum products as standards and in some cases by artificially weathering them.[7] However, there has been no systematic evaluation of the relative effect of this approach on method precision and accuracy. Some improvement is expected, but the choice of product "standards" and the extent to which laboratory "weathering" should be carried out are complex variables. Various data show that the aromatics content of products may vary by at least a factor of two and after release the relative composition of residues is highly variable, depending on numerous environmental factors.[8,9]

"TARGET COMPOUND" METHODS: THE 500, 600, 8000 SERIES METHODS

At the opposite extreme of indicator parameter monitoring is the direct measurement of specific constituents in petroleum contaminated soil and water. The U.S. EPA 500, 600, and 8000 series methods are the most widely used methods for this purpose.[2-5] Each of these methods has an associated list of target compounds for which it was specifically developed and evaluated.

Methods Description

With the passage of amendments to the Clean Water Act in 1972 and the subsequent consent decree settlement in 1976, the EPA developed a list of compounds termed the "Priority Pollutants." The agency also responded to the need for analytical procedures which could be used to detect these compounds as residues in water and wastewater by development of the 600 series methods.[10] Use of these methods is now nearly universal in public and private sector laboratories.

Related methods developed by the agency for drinking water and solid waste analysis include the EPA 500 and 8000 series methods, respectively.[3-5] In most cases, there are few, if any, conceptual or procedural differences between corresponding methods in these and the 600 series.

A common feature of the methods is that nearly all utilize gas liquid chromatography (GC), and for the most part "packed column" technology is specified. The methods also depend on highly selective detectors. GC/MS techniques in which gas chromatography columns are interfaced to mass spectrometers are emphasized.[10] GC/MS instruments are well known for their excellent sensitivity and ability to specifically detect organic compounds. An alternate detector used with methods which target monoaromatic hydrocarbons is the photoionization detector (PID). It has a relatively high selectivity for aromatics over aliphatic hydrocarbons.[11]

Only two sets of methods (610 and 8100, 8015), use the flame ionization detector (FID). This detector gives nearly universal response to hydrocarbons[12] and offers no selectivity. Identifications are based on chromatographic separations alone.

A key aspect of all the methods is that they may be categorized as either a "volatile" or "semivolatile" method, depending on the relative volatility of their target compounds. The "volatile/semivolatile" approach was taken in the development of the methods so that chromatographic separation, sample preconcentration, and injection techniques could be optimized.

A functional definition of the "volatiles" is those compounds which can be effectively recovered from soil or water using purge-and-trap techniques at room temperature. This involves purging of the sample with an inert gas at room temperature and trapping volatile compounds stripped from the sample with a porous polymer adsorbent. The trapped compounds are desorbed directly into the inlet of a gas chromatograph by rapidly heating the trap after the column carrier gas has been diverted to flow through it.

Alternate chromatographic and sample preconcentration techniques were developed for the higher boiling "semivolatile" compounds. These methods involve liquid/liquid and liquid/solid solvent extraction with accompanying pH adjustment for recovery of "acidic" and "base/neutral" compounds, and as such are often termed the "extractables." Solvent extracts are concentrated and aliquots injected directly into gas chromatographs.

Applications

The "volatiles" methods (602, 502, 502.2, 503.1, 8020, 524.1, 624, 8240, 8260) are routinely used where gasoline is involved. With other products such as diesel fuel, kerosene, and #2 fuel oil, the "semivolatiles" methods (610, 625, 8250, 8270) are relied upon. This reflects the relative volatility of the products.

An example of the use of the "volatiles" approach to gasoline contaminated water is shown in Table 1. In a laboratory experiment, water was equilibrated with an unleaded gasoline. The water was then analyzed using conditions equivalent to EPA Method 8240. Compounds detected included benzene, toluene, ethylbenzene, and xylenes (BTEX). These compounds are target analytes in this and related methods and are found at relatively high concentration in most gasoline.

Table 1. "Volatiles" Analysis of Water Equilibrated with an Unleaded Regular Gasoline[a]

Compound	Concentration (milligrams per liter)
"Target Compounds"	
Benzene	29.5
Toluene	42.6
Ethyl-benzene	2.4
Xylene isomers[3]	14.7
"Nontarget Compounds"[b]	
Methyl tert-butyl ether	116.0
1,3 pentadiene	0.1
2-methyl-2-butene	0.1
2-methyl-1-butene	0.1
methyl pentadiene isomer	0.1
2-methyl-thiophene	0.1

[a]Analysis using analytical conditions equivalent to U.S. EPA Method 8240.[3]
[b]Tentative identification (except for MTBE) based on tabulated spectra in Reference 17. The reported concentration is an approximate result.

BTEX also has a relatively high aqueous solubility and, at least in the case of benzene, is considered very toxic. The maximum contaminant level (MCL) for this compound in drinking water is only 5 micrograms per liter.[13]

Based on these results, it is clear that use of this "volatiles" method in investigation of gasoline contamination is a reasonable approach. Application of various "semivolatiles" methods where diesel fuels and related products are involved is similarly effective in that key compounds like naphthalene and phenanthrene are targeted. Nevertheless, both types of applications encounter numerous problems. Troublesome issues include detection of "nontarget" compounds, the limited range of the methods relative to the product composition, and poor chromatographic resolution with the packed and capillary columns specified.

Returning to Table 1, note that among the compounds detected, the one present at highest concentration was a nontarget compound. Indeed, the concentration of methyl tert-butyl ether (MTBE) exceeded the sum of the concentrations of all other compounds combined. MTBE does not appear on any of the EPA target compound lists and unless identification and quantitation of nontargets are specifically requested, the presence of this compound in contaminated samples may be overlooked.

In fact, "false negative" results for MTBE may be quite common. This is alarming, considering that it is now the octane booster of choice in unleaded gasolines and is used at relatively high concentration.[14] It also is apparently transported in groundwater at much faster rates than other gasoline hydrocarbons.[15] In addition, the compound is perceived to be relatively toxic. Interim drinking water standards set in various states are in the 5 to 100 micrograms per liter range.[16]

One of the reasons why MTBE may not be detected is that the potential for this compound to occur as a groundwater contaminant is not widely appreciated.

A closely related factor is economics. In turn, this is often the reason why the packed column GC/PID volatiles methods (502, 602, 8020) are used instead of the more complex GC/MS methods (524, 624, 8240). In most commercial laboratories, per unit charges for GC/PID analyses are typically less than half that of corresponding GC/MS methods. Both types of methods have BTEX in common as target compounds and can be effectively applied to their measurement in petroleum contaminated soil and water. The problem is that the GC/PID methods have limited capability to specifically detect MTBE and other nontarget compounds.

MTBE is perhaps the best example of a significant nontarget compound. Others which are of concern include methanol, tert-butyl alcohol, and various polar aromatics.

A related case history involves a site where discharges from an underground gasoline tank leak had resulted in soil and groundwater contamination. In response, the tank was removed, free product recovered, and a "packed tower" air-stripping device installed to treat contaminated groundwater. Over several years of operation the influent of the "packed tower" was routinely monitored for "volatiles" using U.S. EPA Methods 602 and 624. When the concentration of benzene, toluene, and related compounds in the groundwater had fallen below detection limits of 2 micrograms per liter, it was concluded that the established objectives of the remediation program had been met. The water treatment equipment was dismantled and removed from the site.

Several years later, a sample from a monitoring well on the site confirmed that the concentration of the "volatile" hydrocarbons in the groundwater was close to the detection limit (Figure 1a). However, this was not the case for the "extractable" hydrocarbons (Figure 1b). The total concentration of these compounds was in the parts per million range. The initial response to the data was that a recent fuel oil leak was responsible. This was a reasonable conclusion based on the "volatiles"/"extractables" analytical scheme. However, it did not consider that many of the more water-soluble compounds found in middle distillate fuel oils are represented in the "heavier ends" of gasolines. Ultimately, it was determined that there had been no recent releases of fuel oils on the site. The compounds that had been detected in the groundwater were gasoline residues from the original spill.

In this case what had occurred was that the more soluble "volatiles" (BTEX) were leached relatively rapidly from contaminated soil at the site. Left behind was a significant portion of the "heavier ends," which apparently continues to be leached into groundwater. Further remediation is under consideration, and it is not surprising that there are several legal and financial complications.

This situation developed in part because it was not recognized that while the "volatiles" methods used are effective procedures for BTEX monitoring, they are not applicable to the entire range of compounds found in gasolines. The heavier ends of these products are in the "semivolatile" range. Typically, the volatile hydrocarbons in gasoline contaminated water exceed the concentration of the semivolatile compounds by an order of magnitude. But, as the product is weathered, the relative distribution can change dramatically.

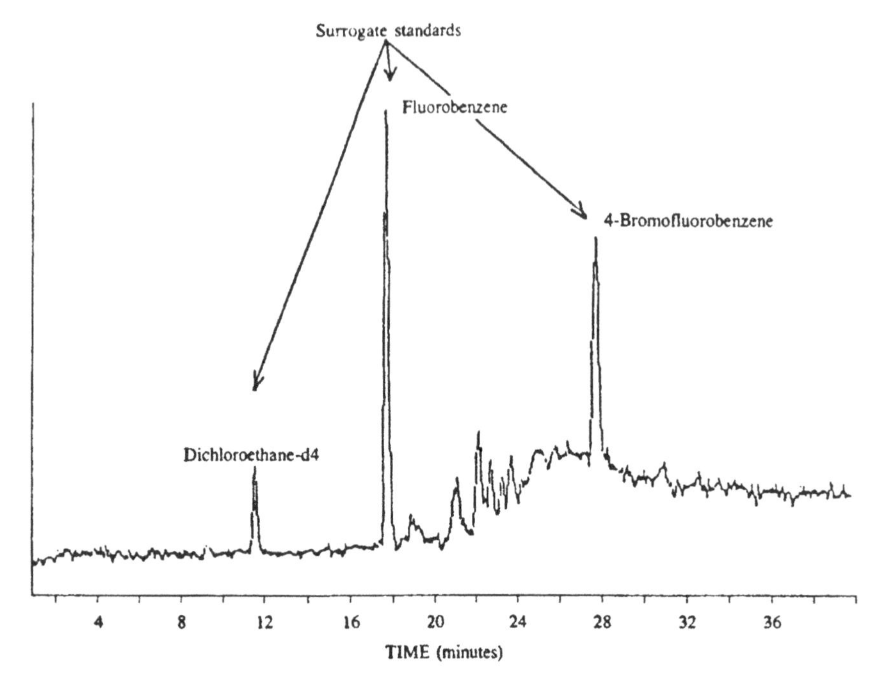

Figure 1a. Total ion current chromatogram from the "volatiles" analysis, U.S. EPA Method 8240 (Reference 3).

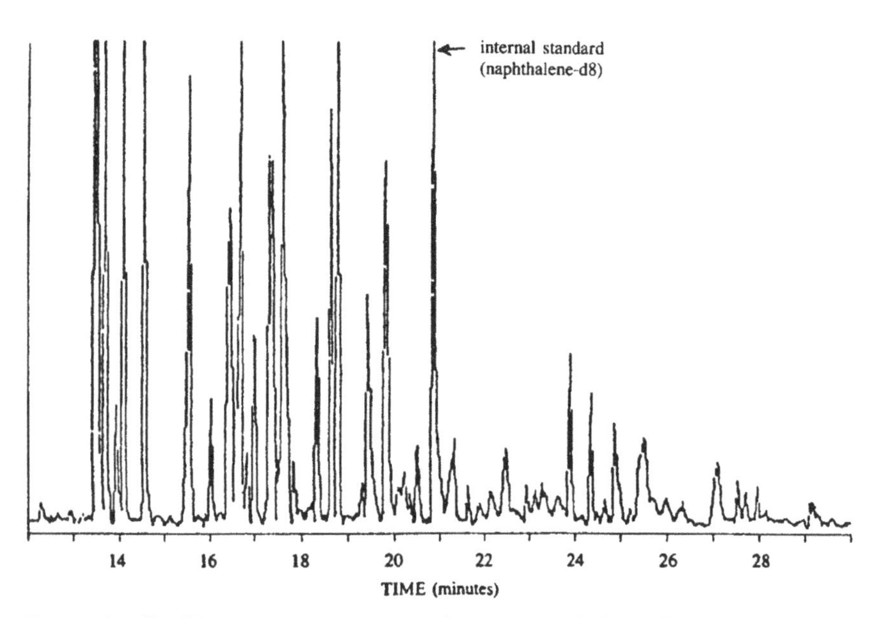

Figure 1b. Total ion current chromatogram from the "semivolatiles" analysis, U.S. EPA Method 8270 (Reference 3).

The corollary to this is that the "semivolatiles" methods are not applicable to the entire range of compounds which occur in middle distillate products like diesel fuel. In fact, the data presented in Table 2 show that significant quantities of "volatiles" (BTEX) can be leached from diesel and even from residual fuels.

Table 2. Benzene, Toluene, Ethyl-Benzene, and Xylenes (BTEX) Concentration in Water Equilibrated with Various Petroleum Products

	Concentration (milligrams per liter)			
Product	Benzene	Toluene	Ethyl-Benzene	Xylenes
Gasoline	29.5	42.6	2.4	14.7
Diesel fuel	.13	.41	.18	.70
#6 Fuel oil[a]	.01	.03	.007	.05
Drinking water standards[a]	.005 (MCL)	2.0 (MCLG)	.66 (MCLG)	.44 (MCLG)

[a]#6 Fuel oil data and drinking water standards from References 18 and 13, respectively.

The picture that emerges is that most petroleum products are not distinct entities, but rather represent a continuum over broad ranges. This is shown by the total ion chromatograms of a gasoline and three middle distillate fuel oils shown in Figure 2. Considering this situation, it is probably appropriate in many circumstances to analyze samples for both "volatiles" and "semivolatiles," regardless of the product type. The solution, however, is not as simple as it may sound. It is limited by chromatography problems, especially with the packed column "volatiles" methods. The "heavier ends" in gasoline and most of the compounds in middle distillate fuels elute very slowly from the GC columns used in these methods. The result is that post-analysis column conditioning is usually required. This can substantially increase analysis time.

With the "semivolatiles" methods, chromatography is also a problem, even when high resolution capillary columns are used. In this case it is because the very large number of compounds these products contain cannot be resolved on a single chromatography column. The "humps" observed in the total ion current chromatograms of the middle distillate products shown in Figure 2 are indicative of this. What they represent are an "envelope" of unresolved compounds. The poor resolution makes identification of target compounds which elute in this region very difficult.

To some degree, the "humps" and their accompanying resolution problems are alleviated when analyzing dissolved hydrocarbons in water. This is one of the reasons why acceptable results were obtained in the analysis of the water-soluble fraction of the diesel fuel described in Table 2. In effect, what occurs when a product dissolves in water is a group-type separation, with the aqueous phase strongly favoring the lower molecular weight aromatics. Contaminated soils, however, typically retain the complexity of the products, and analytical problems are correspondingly difficult.

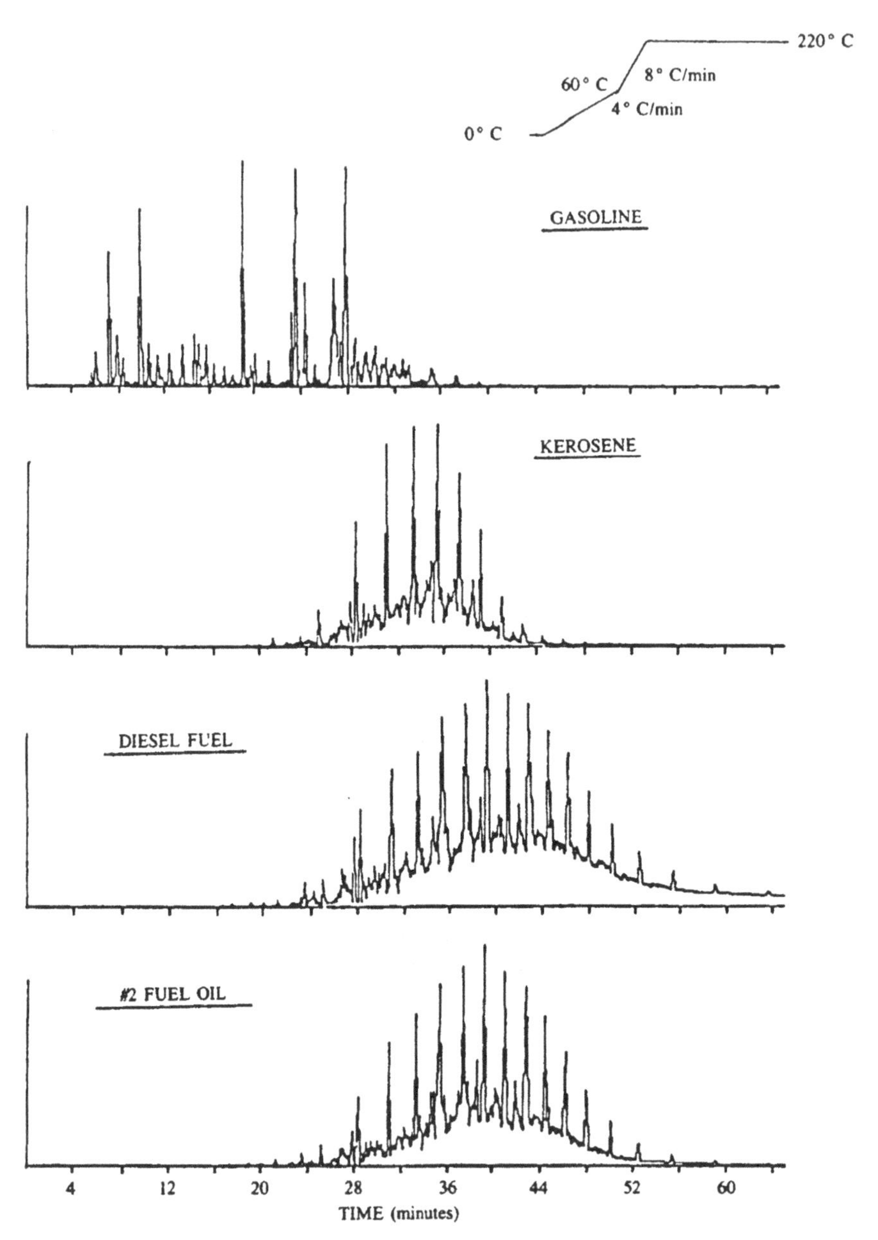

Figure 2. Total ion current chromatograms of four petroleum products on a 60 meter DB-1 (J + W Scientific) fused silica capillary column.

One other point regarding the chromatography problems associated with petroleum products can be made by examining Figure 2. Notably, chromatograms of "heavy distillate" products like motor lubricating oils and residual fuels are not presented. Their omission from the figure is not an oversight. These products are mixtures of very "high boiling" hydrocarbons. Their analysis using gas chromatographic techniques is difficult at best. "Humpograms" with little separation of constituents are typical.

METHOD DEVELOPMENT NEEDS

There are clearly many potential problems in the application of the EPA methods to petroleum contaminated soil and water. In part this is because the methods were not specifically developed, nor have they been systematically evaluated for this purpose. The modification of existing methods and/or the development of new methods is needed.

With the TPH methods, identification of solvents or solvent mixtures which more effectively extract higher molecular weight aromatics is necessary. However, problems will remain with low recoveries of the "lighter" hydrocarbons and with standardization, especially with infrared based methods. There is no simple way of predicting the relative amount of aromatic or aliphatic hydrocarbons which will occur in a product or its residues recovered from soil and water.

Alternative approaches to TPH measurements such as "Modified Method 8015"[5] appear superior. The method specifies use of an FID, a nearly universal and nonspecific hydrocarbon detector. Yet, this method is not without limitations. Quantitative results may vary widely due to poor chromatographic separation and relatively low hydrocarbon recoveries in solvent extracts.

This leads to the conclusion that fundamental changes in the TPH analytical approach are needed. One possibility for contaminated soils is "flash thermal desorbtion" (FTD) combined with GC/FID or an FID alone. With FTD, solid samples are heated very rapidly, resulting in desorbtion of hydrocarbons into the inlet of a GC/FID or FID. In many ways it is analogous to traditional purge-and-trap methods used for water analysis. FTD has been shown to be applicable to the analysis of a wide range of hydrocarbons (up to C30) in solid samples and has been used extensively in petroleum exploration.[19] A distinct advantage of FTD is ease of sample preparation.

For the target compound methods, the recent advances in capillary gas chromatography, information processing, and coupled and multidimensional chromatographic techniques are expected to play an important role. The value of large diameter capillary columns with thick films has already been recognized.[20] They can accept high desorbtion flows from purge-and-trap devices and provide enhanced resolution over packed columns. The capillary columns also offer chromatography conditions which are applicable to a much broader range of compounds.

To obtain the full value of these columns in the analysis of petroleum contaminated soil and water, advances in the purge-and-trap methodology are necessary. Studies have shown that desorbtion kinetics and other factors may limit the chromatographic improvement that capillary columns offer.[21] Traps are needed which have high breakthrough volumes for broad ranges of hydrocarbons and which allow quantitative and rapid thermal desorbtion. It is realistic to expect that heated purge-and-trap techniques can be developed which would allow recovery of the higher boiling compounds in gasoline and the various middle distillate products without sacrificing recovery of the more volatile compounds. Another improvement with the non-GC/MS volatiles methods would be use of coupled detectors. For example, the advantages of operating PID and FID in series have been shown.[11] With this approach the number of compounds detected in a single analysis can be greatly expanded and the detection of nontarget compounds facilitated.

Capillary columns have been in use for a relatively long period of time with the semivolatiles methods. These columns offer many advantages over packed columns when applied to compounds in this category. This includes higher resolution and inertness, and they can be directly coupled to mass spectrometers. Significant limitations remain. This is symbolized by the aforementioned "hump" in the chromatograms of middle distillate products (Figure 2).

Complete separation of all the compounds in highly complex mixtures such as diesel fuel, kerosene, and related products is beyond the limits of a single gas chromatography column.[22] What is needed is group type separations prior to gas chromatography. That is, separation of the compounds into more homogeneous groups by liquid chromatography. Unfortunately, group-type separations are time-consuming and labor-intensive. Advances in combined microbore liquid chromatography/high resolution gas chromatography offer the potential for performing this "on-line."[23]

One final point that requires emphasis is that target compound lists need to be reexamined to more comprehensively address the contaminants which may occur in petroleum contaminated soil and water under varying conditions. This must be done with the recognition that numerous factors determine information and, ultimately, analytical needs. Experiences with MTBE should tell us that petroleum product chemistry is not constant. As sources of crude oil, other raw materials, and economic factors change, significant changes in product composition are expected. Analytical methods must change accordingly.

ACKNOWLEDGMENTS

Preparation of this document was supported in part by funds provided by the Massachusetts Agricultural Experiment Station and the Massachusetts Department of Environmental Protection.

REFERENCES

1. "Methods for the Chemical Analysis of Water and Wastes,"EPA-600/4-79-020, U.S. Environmental Protection Agency, 1979.
2. "Guidelines Establishing Test Procedures for the Analysis of Pollutants Under the Clean Water Act; Final Rule and Interim Rule," U.S. Environmental Protection Agency, 1984. 40 CFR Part 136. *Federal Register* 49(209):1–210.

3. "Test Methods for Evaluating Solid Waste," 3rd ed., Doc. No. SW-846, U.S. Environmental Protection Agency, 1986.
4. "Methods for the Determination of Organic Compounds in Drinking Water," EPA-600/4-88/039, U.S. Environmental Protection Agency, 1988.
5. "Leaking Underground Fuel Tank Manual: Guidelines for Site Assessment, Cleanup, and Underground Storage Tank Closure," State of California Leaking Underground Fuel Tank Task Force, California Water Resources Control Board, Sacramento, CA, 1988.
6. Speight, J. G., "The Chemistry and Technology of Petroleum," (New York: Marcel Dekker Inc., 1980).
7. DeAngelis, D. "Quantitative Determination of Hydrocarbons in Soil," in M. Kane, Ed. *Manual of Sampling and Analytical Methods for Petroleum Hydrocarbons in Groundwater and Soil,* Publication 4449, American Petroleum Institute, Health and Environmental Sciences Department, Washington, DC, 1987.
8. Ury, G. B., "Automated Gas Chromatographic Analysis of Gasolines for Hydrocarbon Types, *Anal. Chem.* 53: 481-485 (1981).
9. Edgerton, S. A., R. W. Coutant, and M. V. Henley, "Hydrocarbon Fuel Dispersion in Water," *Chem.* 16(7):1475-1487 (1987).
10. Telliard, W. A., M. B. Rubin, and D. R. Rushneck, "Control of Pollutants in Wastewater," *J. Chromatog. Sci.* 25:322-327 (1987).
11. Driscoll, J. N., and M. Duffy, "Photoionization Detector: A Versatile Tool for Environmental Analysis," *Chromatography* 2(4):21-27 (1987).
12. Tong, H. Y., and F. W. Karasek, "Flame Ionization Detector Response Factors for Compound Classes in Quantitative Analysis of Complex Organic Mixtures," *Anal. Chem.* 56:2124-2128 (1984).
13. "Drinking Water Regulations and Health Advisories," U.S. Environmental Protection Agency, Office of Drinking Water, Washington, DC, 1990.
14. Anderson, E., "MTBE Strengthens Hold on Octane Booster Market," *Chem. Eng. News*, October 13, 1986, p.8.
15. Garrett, P., M. Moreau, and J. Lowry, "Methyl Tertiary Butyl Ether as a Ground Water Contaminant," in *Proceedings of Third National Conference on Petroleum Hydrocarbons and Groundwater*, National Water Well Association, Dublin, OH, 1986.
16. Personal communications: Maine Department of Environmental Protection, Connecticut Department of Environmental Protection and Florida Department of Environmental Management.
17. *Eight Peak Index of Mass Spectra*, 3rd ed., Royal Society of Chemistry, The University, Nottingham, UK, 1983.
18. Burchette, G., "Number Six Fuel Oil and Ground Water," *Ground Water Monitoring Review* 6:32 (1986).
19. Crist, W. A., J. Ellis, J. de Leeuw, and P. A. Schenck, "Flash Thermal Desorbtion as an Alternative to Solvent Extraction for the Determination of C8 to C35 Hydrocarbons in Oil Shales," *Anal. Chem.* 58: 258-261 (1985).
20. Reding, R., "Chromatographic Monitoring Methods for Organic Contaminants Under the Safe Drinking Water Act," *J. Chromatog. Sci.* 25: 338-344 (1987).
21. Mosesman, N. H., L. M. Sidsky, and S. D. Corman, "Factors Influencing Capillary Analyses of Volatile Pollutants," *J. Chromatog. Sci.* 25:351-355 (1987).
22. Pitzer, E. W., "Contributions of Stereoisomerism to Peak Shapes of Branched Paraffins in the High-Resolution Gas Chromatographic Analyses of Jet Propulsion Fuels," *J. Chromatog. Sci.* 26:223-227 (1988).

23. Duquet, D., C. Dewaele, and M. Verzele, "Coupling Micro-LC and Capillary-GC as a Powerful Tool for the Analysis of Complex Mixtures," *J. High Res. Chromatog. and Chromatog. Commun.* 11: 252–256 (1988).

CHAPTER 2

Fingerprinting Petroleum Products: Unleaded Gasolines

Thomas L. Potter, Mass Spectrometry Facility, Massachusetts Agricultural Experiment Station, University of Massachusetts, Amherst

At most petroleum storage tank facilities there are at least several tanks onsite containing the same or different types of products. It is also common for tanks or pipelines to be located on nearby properties. Thus, when a product release is detected there may be many potential sources, including both on- and offsite structures. Under these circumstances source identification is essential for the selection and implementation of appropriate corrective actions and ultimately for assigning responsibility. To this end, chemical analysis of products may play a critical role. Ideally, through chemical fingerprinting the release can be traced directly to its source.

A variety of analytical schemes have been published for petroleum product fingerprinting,[1-3] although few have focused on gasolines. Those that have, typically begin with a simulated distillation analysis and proceed as necessary to more complex analyses involving identification of additives such as octane boosters and hydrocarbon profiling by gas chromatography and/or mass spectrometry.[4-6] In general, methods and procedures based on high resolution gas chromatography-mass spectrometry (HRGC/MS) have shown the greatest success.[6,7]

In this chapter, HRGC/MS analysis of a set of unleaded gasolines is discussed with respect to product fingerprinting. It is shown that computer-based pattern recognition techniques like cluster analysis can facilitate data analysis. Nevertheless, uncertainty in results may be relatively large and the burden of proof beyond the limits of data that can reasonably be obtained.

ANALYSIS

Instrumentation and analytical conditions were based on U.S. Environmental Protection Agency (EPA) Method 8270.[8] This HRGC/MS technique represents the "state of the art" for trace organic analysis. Modifications were that injections were made using the "split" mode, toluene-d8 was used as an internal standard, and alternate chromatographic conditions were utilized.

Use of split injection permitted direct injection of the products into the gas chromatograph without the use of dilution solvents. By avoiding dilution solvents, the entire chromatographic profile of each product was obtained. Figure 1 provides a typical total ion current chromatogram.

The alternate internal standard, toluene-d8, was selected since it has physical-chemical properties which are similar to many gasoline constituents and can be

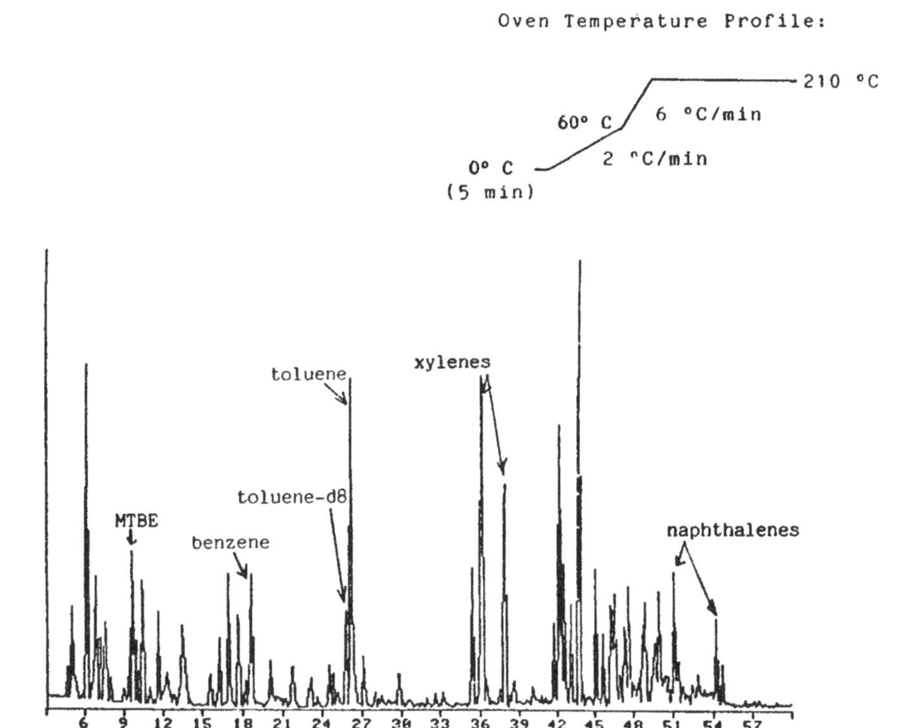

Figure 1. Total ion current chromatogram of an unleaded regular gasoline.

quantified in gasolines under the analytical conditions described. The toluene-d8 was spiked directly into each gasoline sample at the rate of 0.5% by volume. Note that all quantitation was based on relative response to the internal standard and a series of relative response factors determined from the replicate analyses of a standard hydrocarbon mixture. This approach substantially improved the precision of concentration measurements (Table 1).

Table 1. Coefficient of Variation for the Concentration of Selected Compounds in a Gasoline Using Internal and External Standardization[a, b]

	Coefficient of Variation (%)[c]	
Compound	**Internal**	**External**
Methyl-tert-butyl ether	4.7	13.1
Benzene	3.3	14.4
Pentane, 2,2,4-trimethyl	1.9	15.7
Toluene	1.1	18.2
Ethyl-benzene	4.2	19.6
Xylenes, para and meta	3.2	18.8
Xylene, ortho	4.6	15.7
Naphthalene	3.1	16.6

[a]Data from three replicate analyses of an unleaded regular gasoline (product M).
[b]Analysis by HRGC/MS with split injection and quantitation based on the internal standard, toluene-d8.
[c]Coefficient of variation is an expression of the standard deviation divided by the mean and reported as relative percent.

Differences in chromatographic conditions were in the GC oven temperature profile. A lower initial temperature and slower oven temperature program rates were used than are specified in method 8270. These conditions were selected to allow better separation of the constituents in the products. For example, the alternate conditions permitted complete resolution of the octane booster, methyl tert-butyl ether (MTBE), from various five carbon alkanes. This improved MTBE quantitation by eliminating interferences from the heavy isotope ions (C13) of these compounds.

The alternate chromatographic conditions did not, however, permit complete separation of all constituents. This is reflected in the fact that isomer specific identifications of the C8-alkanes (Table 4) were not attempted. Such identifications were beyond the limits of the analytical conditions.

RESULTS AND INTERPRETATIONS

The sample set was a series of six regular unleaded gasolines purchased at service stations operated in the Amherst, Massachusetts area. In addition, a "weathered" sample was prepared by 50% evaporation of one of the products. Concentration

Table 2. Concentration of Selected Compounds in Six Unleaded Regular Gasolines[a]

Compound	Concentration (Grams per Liter)					
	Product					
	M	Ge	A	T	S	G
MTBE[b]	23.0	50.9	36.0	7.2	13.6	<.1
Cyclohexane	1.1	1.3	2.5	1.3	1.7	1.9
Isooctane[c]	35.4	44.1	56.1	64.2	64.5	26.6
Benzene	12.7	9.6	14.2	9.4	20.2	28.7
Toluene	47.8	36.8	41.6	37.5	68.0	68.2
Ethyl-benzene	15.8	13.3	10.4	14.2	17.5	16.2
Xylenes	67.0	60.7	58.9	63.2	78.4	67.3
Naphthalene	6.7	8.0	7.1	8.7	6.2	2.8

[a]Analysis by HRGC/MS with split injection and quantitation based on the internal standard, toluene-d8.
[b]MTBE is the octane booster, methyl tert-butyl ether.
[c]Isooctane is the common name for 2,2,4-trimethyl pentane.

Table 3. Concentration of Selected Compounds in an Artificially Weathered Unleaded Regular Gasoline[a]

Compound	Concentration (grams per liter)	
	Product (T)	50% Evaporated (T50)
MTBE[b]	7.2	0.8
Benzene	9.4	4.9
Cyclohexane	1.3	2.9
Isooctane[c]	64.2	80.7
Toluene	37.5	59.6
Ethyl benzene	14.2	24.6
Xylenes	63.3	128
Naphthalene	8.7	24.5

[a]Analysis by HRGC/MS with split injection and quantitation based on the internal standard, toluene-d8.
[b]MTBE is the octane booster, methyl tert-butyl ether.
[c]Isooctane is the common name for 2,2,4-trimethyl pentane.

Table 4. Concentration of C8-Alkanes in Five Unleaded Regular Gasolines and an Artificially Weathered Sample[a]

Peak No.[b]	Concentration (grams per liter)					
	Products					
	M	Ge	A	T	S	T50[c]
1	60.9	101	128	147	148	178
2	9.4	10.5	11.2	15.3	15.8	23
3	9	10.8	11.3	14.7	13.7	20
4	13.9	23.5	31	34.2	43.4	54
5	25.7	27.5	24.1	26.9	26.8	39
6	21.1	22.5	19.3	22.6	22.7	33

[a]Analysis by HRGC/MS with split injection and quantitation based on the internal standard, toluene-d8.
[b]Peak 1 was identified as 2,2,4-trimethyl pentane.
[c]T50 was prepared by 50% evaporation of product T.

data for selected constituents in each sample has been summarized in Tables 2, 3, and 4.

Additives Analysis

The utility of additives analysis is that presence or absence of a unique compound may allow a definitive product identification. To this end, the data summarized in Table 2 show that five out of the six products contained the octane booster, MTBE. Within the sample set the product that did not contain this compound was easily distinguished from the others. The converse is that among the products which contained MTBE, detection of this additive provided little information regarding product source. Note that no other unique compounds were detected at a detection limit of .01% by weight.

This is not to say that distinctive additives were absent. The problem is that most additives are used at low product concentrations and that "a priori" knowledge of additive concentration and structure is required before methods can be developed for their analysis. In general, this information is proprietary and not available to the analyst.

Hydrocarbon Profiling

In the absence of unique additives, fingerprinting usually turns to hydrocarbon profiling. This approach involves identifying a unique pattern for each product based on the relative concentrations of various constituents. Techniques for data interpretation range from simple visual comparison of chromatograms to computer-based pattern recognition techniques. An example of the latter is cluster analysis.

The basic concept of cluster analysis and related techniques is that a product can be represented as a point in a space whose dimensionality is equivalent to the number of number parameters for which concentration data is available.[9] Within this space, the distance between two points is a relative measure of similarity, and points that are close to each other are said to group or "cluster." Operationally, what is accomplished through computation is the generation of alternate variables which allow the data to be represented in fewer dimensions while retaining most of the inherent variability within the data set.

Results of the cluster analysis of the Table 2 data are presented graphically in Figure 2 (only data for the products which contained MTBE were used). Computations were performed on a Zenith 386 workstation using the software package, "Spring-stat."[10]

That the points on this two-dimensional plot are well separated is indicative of the unique chemical character of each product. It also indicates that the subset of gasoline hydrocarbons on which the cluster analysis was based effectively describes the chemical variability of these products. These compounds are representative of at least four major refinery streams which are combined in varying proportions to yield products meeting performance specifications.[11]

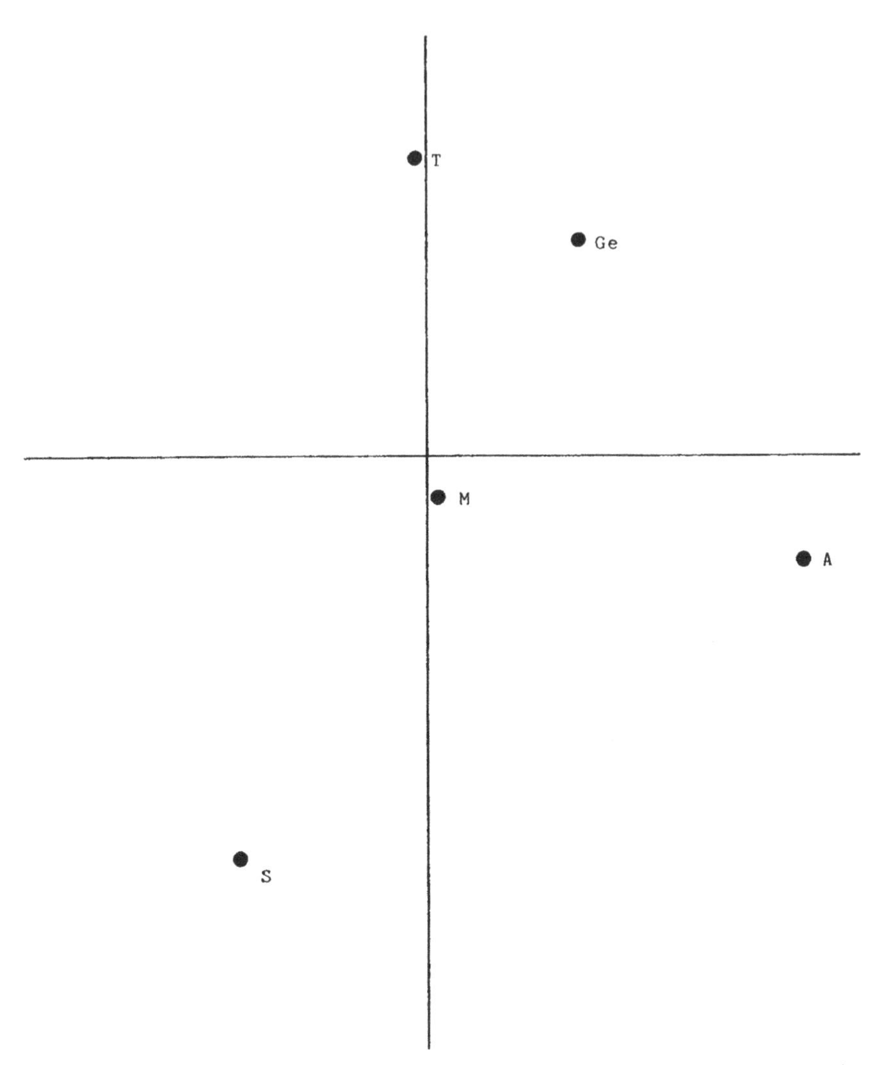

Figure 2. Factor score plot based on the concentrations of nine principal constituents in five gasolines.

The expectation is that by collecting similar concentration data for an unknown, and performing a cluster analysis with the reference sample data (Table 2), a product similarity index can be derived. The more similar the composition of an unknown and product, the closer they will cluster, and the more likely their source is related.

Unfortunately, even with the high precision measurements obtained in this study, this approach was not found to be an effective predictor of sample source for the artificially weathered sample (Table 3). In fact, Figure 3 shows that T50 failed

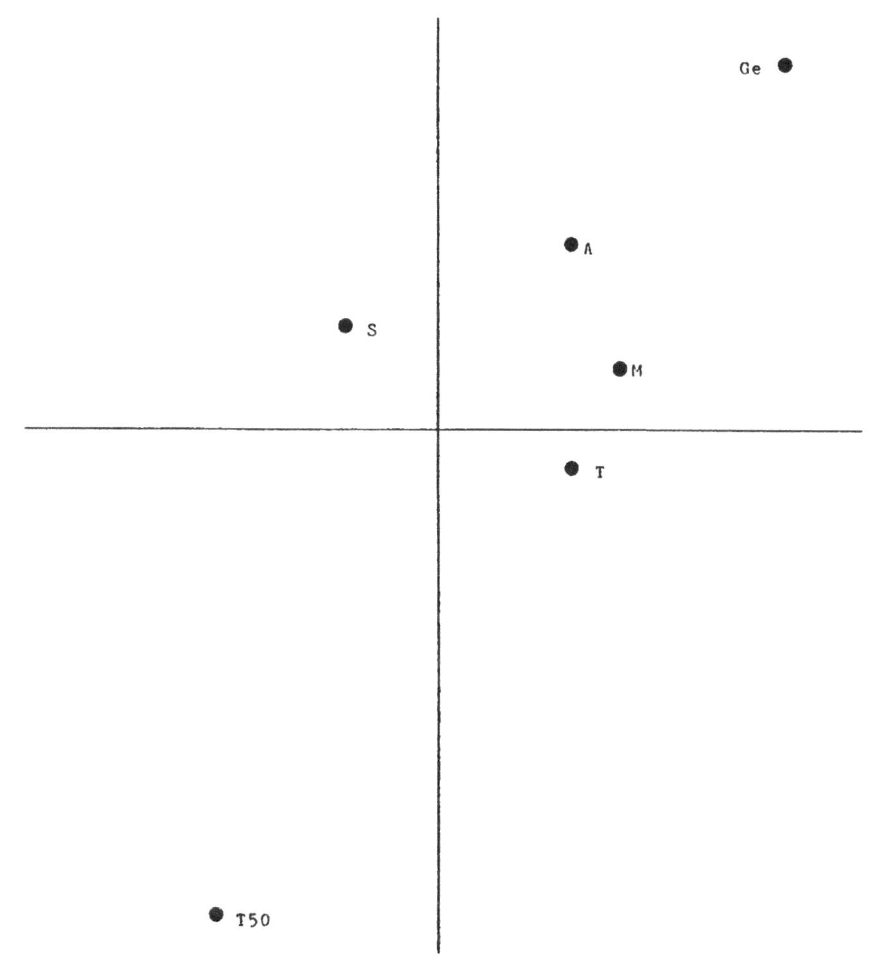

Figure 3. Factor score plot based on the concentrations of nine principal constituents in five gasolines and an artificially weathered product.

to cluster with any of the products, and that product T (its source) was more like the other products than it is T50.

The breakdown in this product matching system occurred because the Table 2 compounds evaporate from gasolines at different rates. For example, the Table 3 data show that after 50% evaporation, the product concentration of MTBE and benzene had decreased, whereas the concentration of isoooctane and other compounds had increased.

Given this behavior, alternative cluster analyses based on the relative concentrations of various isomers were attempted. Isomeric groups evaluated included C2-benzenes, C1-naphthalenes, C3-benzenes, and C8-alkanes. Each of these groups are well represented in gasolines, and their concentration in the product phase tends to increase as weathering advances. The rationale for examining isomers is that aqueous solubilities and vapor pressures of isomers exhibit relatively small differences.[12]

In short, it was found that the only isomeric group which had potential for fingerprinting was the C8-alkanes. Indeed, with the C8-alkane data, the weathered and unweathered samples, T50 and T, clustered closely (Figure 4). The limitation is that only two clusters were observed for the five products; thus, an exact match was beyond the limits of the data.

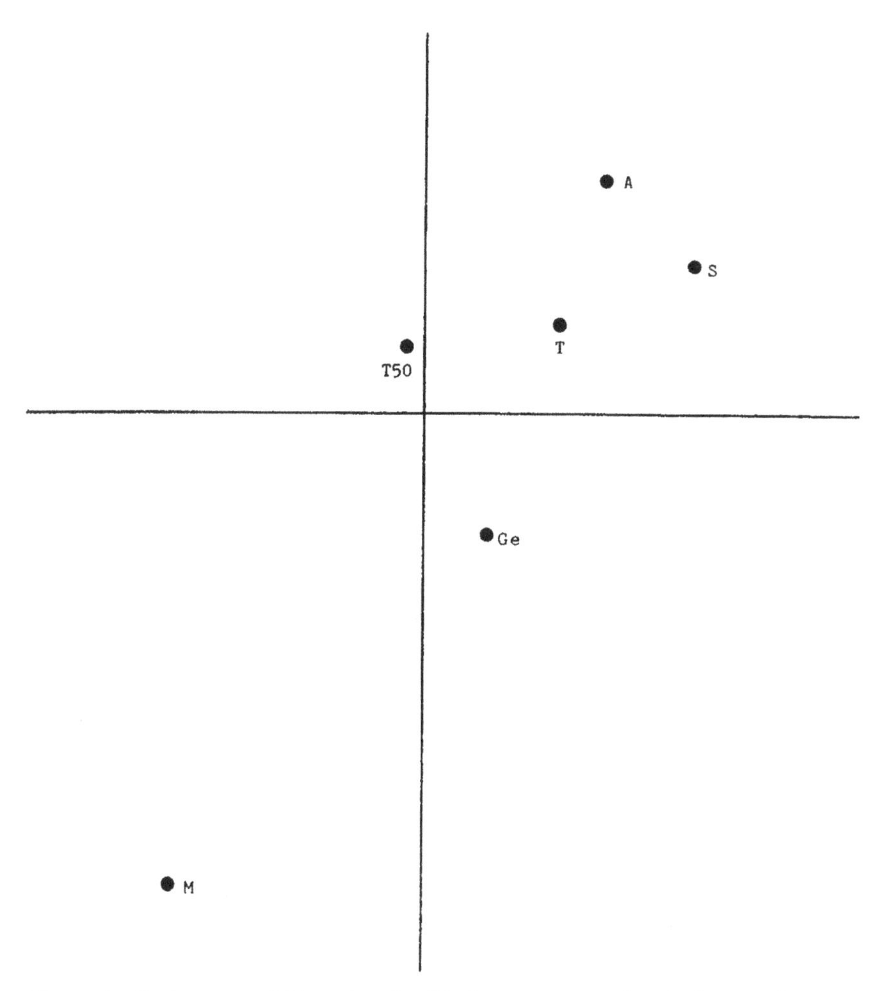

Figure 4. Factor score plot based on the C8-alkane concentrations in five gasolines and an artificially weathered product.

What was found for the other isomers was that the relative distributions of isomers within a group were nearly identical. This is reflected in the fact that gasolines are highly engineered products. Their composition is a function of refinery and blending processes which promote a "leveling" effect among products.[11] With less refined products the relative distributions of various isomers is usually more distinctive. Calculation of isomeric concentration ratios is one of the principal tools of crude oil fingerprinting.[2,3]

CONCLUSIONS

The examples cited illustrate that unleaded gasolines can be fingerprinted through additives analysis and hydrocarbon profiling, but there are significant limitations. The occurrence of (or the ability to detect) unique additives in unleaded products is apparently rare, and weathering can obscure much of a product's unique chemical character.

Other obstacles to effective gasoline fingerprinting is that, unlike the situation described in this work, product reference samples which can be directly related to a release are rarely if ever available, and product composition is not constant. Product mixing (co-mingling) in pipeline and other distribution systems, product switching (i.e., trading among refiners and blenders) are equally if not more important problems.[11,13]

This serves to emphasize that uncertainty in fingerprinting results may be large. The work described in this chapter was both expensive and time-consuming, and "state of the art" analytical technology was utilized. Yet a definitive match could not be obtained for the weathered sample. Simply, the burden of proof may be beyond the limits of data which can reasonably be obtained.

ACKNOWLEDGMENTS

Financial support for this work was provided by the Massachusetts Agricultural Experiment Station, University of Massachusetts, Amherst, MA.

REFERENCES

1. ASTM Standard Method, D3327-78, Comparison of Waterborne Petroleum Oils by Gas Chromatography, 1978 *Annual Book of ASTM Standards*, Part 31, American Society for Testing Materials, Philadelphia, PA., 1978.
2. Butt, J. A., D. F. Duckworth, and S. G. Perry, eds. *Characterization of Spilled Oil Samples* (New York: John Wiley & Sons, 1986).
3. Bentz, A., "Who Spilled the Oil," *Anal. Chem.* 50: 655A (1978).
4. Rygle, K., "Methods for 'Free' Product Analysis," in M. Kane, ed., *Manual of*

Sampling and Analytical Methods for Petroleum Hydrocarbons in Ground Water and Soil. Publication 841-44490, American Petroleum Institute, Washington, DC (1987).

5. Youngless, T. L., J. T. Swansiger, D.A. Danner, and M. Greco. "Mass Spectral Characterization of Petroleum Dyes, Tracers, and Additives," *Anal. Chem.* 57:1894–1902 (1985).
6. Sleck, L. W., "Fingerprinting and Partial Quantification of Complex Hydrocarbon Mixtures by Chemical Ionization Mass Spectrometry," *Anal. Chem.* 51:128–132 (1979).
7. Sutton, D. L., "Component Analysis by High Resolution Gas Chromatography," in M. Kane, ed., *Manual of Sampling and Analytical Methods for Petroleum Hydrocarbons in Groundwater and Soil,* Publication 841-44490, American Petroleum Institute, Washington, DC (1987).
8. "Test Methods for Evaluating Solid Waste," 3rd ed., U.S. Environmental Protection Agency, Doc.: SW-846, 1987.
9. Hopke, P., "An Introduction to Multivariate Analysis of Environmental Data," in B. Natusch and P. Hopke, eds., *Analytical Aspects of Environmental Chemistry* (New York: John Wiley & Sons, 1983).
10. *SpringStat,* version 2.2, Spring Systems Inc., Chicago, Illinois.
11. Whitmore, I., "Identification of Spilled Hydrocarbons," in M. Kane, ed., *Manual of Sampling and Analytical Methods for Petroleum Hydrocarbons in Groundwater and Soil,* Publication 841-44490, American Petroleum Institute, Washington, DC (1987).
12. Mackay, D., and W. Y. Shiu. "A Critical Review of Henry's Law Constants for Chemicals of Environmental Interest," *J. Phys. Chem. Ref. Data* 10:1175–1199 (1981).
13. Anderson, E. V. "Fuel Ethanol Wins Two Regulatory Decisions," *Chem. Eng. News* Feb. 2, 1987, pp. 14–15.

CHAPTER 3

Differentiation of Crude Oil and Refined Petroleum Products in Soil

Ann L. Baugh, Unocal Corporation, Los Angeles, California
Jon R. Lovegreen, Applied Geosciences Inc., Irvine, California

As part of the societal recycling of depleted oil fields, assessing whether a petroleum hydrocarbon detected in soil samples is a crude oil or a refined petroleum product has increased in importance. Neither federal nor California regulations list soils containing crude oil to be hazardous wastes.[1,2] Soils containing refined petroleum products typically require mitigation even though regulations do not define these soils as a hazardous waste, especially in the case of leaking underground fuel tanks.[3] The differentiation of crude oil from refined petroleum products has important economic impacts. Crude oil can be left in place provided it can be demonstrated that it is not hazardous[2] and that there is a low likelihood of migration to the water table.[3] However, for refined petroleum products, gasoline may be required to be mitigated to levels as low as 10 parts per million (ppm) and diesel fuel to 100 ppm.[3] Benzene and toluene may be required to be mitigated to 0.3 ppm, and total xylenes and ethylbenzene to 1 ppm.[3]

CHEMICAL ANALYSES

Time and economic constraints of a typical site investigation call for use of data provided by standard analyses available from state-certified and Environmental Protection Agency (EPA) contract laboratories. The approach for differentiating between crude oil and refined petroleum products reported here emerged in the course of site investigations[4-8] associated with the redevelopment of depleted oil field properties in southern California. Use was made of analyses routinely performed in the course of the site investigation. In Table 1, a list is given of the analyses that have been found to be helpful in making judgments as to the likelihood that the origin of a petroleum hydrocarbon is from crude oil or refined petroleum product.

Table 1. Chemical Analyses Used for Differentiation of Crude Oil and Refined Petroleum Products in Soil

1. Total Recoverable Petroleum Hydrocarbons (TRPH), EPA Method No. 418.1[9]
2. Total Petroleum Hydrocarbons (TPH), EPA Method No. 8015 (Modified)[3]
3. Benzene, Toluene, Total Xylene and Ethylbenzene (BTXE), EPA Method No. 8020[10]
4. Volatile Organic Compounds (VOCs), EPA Method No. 8240[10]
5. Semi-Volatile Organic Compounds (Semi-VOCs), EPA Method No. 8270[10]
6. Organic Lead, Inductively Coupled Plasma/Mass Spectrometry (ICP/MS)[11]

The total recoverable petroleum hydrocarbon (TRPH) analysis, EPA Method No. 418.1, is an infrared spectrophotometric procedure[9] in which an extract using fluorocarbon-113 (1,1,2-trichloro-trifluoroethane) is quantified using the C-H stretch band at about 2930 cm-1 against a standard containing n-hexadecane ($C_{16}H_{34}$), a straight chain hydrocarbon; isooctane (C_8H_{18}), a branched hydrocarbon; and chlorobenzene (C_6H_5Cl), an aromatic hydrocarbon. These compounds are representative of those found in a mineral oil or a light fuel,[9] but do not include the heavier fractions (C_{20}+) found in crude oil. Hence, the TRPH analysis provides only an estimate of the concentration of a crude oil in soil. It is described as a useful survey tool in the Leaking Underground Fuel Tank (LUFT) Field Manual[3] because of the low cost. Also, the loss of about half of any gasoline present during the extraction process is reported in the analytical procedure.[9] Nyer and Skladany[12] report 25% or more variability. The result of a TRPH analysis is a single concentration, reported generally in ppm.

The total petroleum hydrocarbon (TPH) analysis, which is routinely referred to in California as EPA Method No. 8015 (modified), in reality is a method described in the LUFT Field Manual[3] and developed by the California Department of Health Services (DOHS). This method emerged in the development of analyses for gasoline and diesel fuel from leaking underground fuel tanks. It resembles the EPA method in that it uses a temperature programmable gas chromatograph (GC) with a flame ionization detector (FID) and a similar column. The DOHS method[3] describes the GC operating conditions as follows:

"Column temperature is set at 40 degrees Celsius at the time of injection, held for four minutes, and programmed at 10 degrees Celsius per minute to a final temperature of 265 degrees Celsius for 10 minutes."

In California, the method used from laboratory to laboratory when requesting EPA Method No. 8015 (modified) is not standardized, and can vary to some degree, especially in the temperature programming. DOHS is attempting to standardize this and recommends that laboratories be given the procedure as described in the LUFT Field Manual.[3]

When TPH analyses are being performed on soils for the purpose of differentiating crude oil and refined petroleum products, the extraction procedure is used. The solvent suggested in the DOHS procedure is carbon disulfide; although other solvents such as ethyl acetate or methylene chloride may be used, provided the solvent can extract the petroleum hydrocarbons, and does not interfere with the resulting gas chromatogram.[3] When the extraction method is used, the detection limit for TPH is normally 10 ppm for both gasoline and diesel.[3] The results are reported as ppm for generally no more than two of the four refined petroleum products in this boiling range, i.e., gasoline, mineral spirits, kerosene and diesel fuel. Peaks above the diesel range are reported as $C_{20}-C_{30}$ hydrocarbons. Some laboratories report only one number as TPH in ppm, which is calculated using only one standard, generally gasoline or diesel. When the analysis is being performed for the purpose of a leaking underground fuel tank, often the contents of the tank are known and this method is appropriate. However, for our purposes in differentiating crude oil and refined petroleum products, the use of two standards with a report of gasoline, diesel fuel, or $C_{20}-C_{30}$ segments is preferred. The $C_{20}-C_{30}$ segment is usually calculated using the diesel standard for comparison. This practice is likely to produce slightly low results for the $C_{20}-C_{30}$ fraction because the instrument response per unit concentration of a heavier hydrocarbon is probably lower than that of diesel fuel.

The EPA Method No. 8020 analysis is for aromatic volatile organic compounds (VOCs) and provides data on chlorobenzene and dichlorobenzenes as well as benzene, toluene, xylene and ethylbenzene (BTXE). A purge-and-trap method is used with soils for sample injection into the GC. A temperature programmable GC is used with a photoionization detector (PID). Due to the fact that samples can be contaminated by the diffusion of volatile organics, especially chlorofluorocarbons and methylene chloride, through the sample container septum during shipment and storage, a field sample blank prepared from reagent water and carried through sampling and subsequent storage and handling is recommended to serve as a check.[10] With soil samples, the detection limit for BTXE can be as low as 1 part per billion (ppb). This will vary depending on actual concentration in the sample and interference from other compounds and can reach 250 ppb.[10] The results are reported in ppm or ppb, as appropriate.

The EPA Method No. 8240 is used to determine VOCs in a variety of solid waste matrices, including soil. The method is applicable to most VOCs that have a boiling point below 120°C and vapor pressures of a few millimeters of mercury at 25°C and that are insoluble or slightly soluble in water. More than 50 compounds are reported to be analyzed by this method,[10] including BTXE. When this analysis is performed on a sample, EPA Method No. 8020 can usually be omitted. The method consists of a purge-and-trap process for injection of the volatiles into the GC; a temperature programmable GC; a scanning, electron impact mass spectrometer (MS); and a computerized data system.

The detection limits for BTXE with this method is generally 5 ppb for soils with low concentrations, and increases to over 600 ppb for high concentrations. Samples require the same precautions for contamination by other VOCs as discussed above for EPA Method No. 8020.[10] The VOCs other than BTXE detected by this method are for the most part not petroleum hydrocarbons. However, some laboratories report mixtures of compounds such as C_5-C_{11} aliphatic and alicyclic hydrocarbons and C_9-C_{10} alkylbenzenes, which are constituents of gasoline and the gasoline fraction of crude oil. The results of an EPA Method No. 8240 analysis are reported in a computer printout from the mass spectrometer data system with chemical abstract number, compound name, concentration in either ppb or ppm, as appropriate, and detection limit. All chemicals routinely analyzed are listed, whether present or not; so typically, a data sheet will contain a large number of NDs (not detected).

The EPA Method No. 8270 is similar to 8240 in that it is a gas chromatography/mass spectrometry (GC/MS) method. It is used to determine the concentration of semivolatile organic compounds (semi-VOCs) in extracts prepared from all types of solid waste matrices, soils, and groundwater. The method is applicable to most neutral, acidic, and basic organic compounds that are soluble in methylene chloride and are capable of being eluted without derivatization as sharp peaks from a GC fused-silica capillary column with a slightly polar silicone. Such compounds include polynuclear aromatic hydrocarbons, chlorinated hydrocarbons, pesticides, esters, aldehydes, ethers, ketones, anilines, pyridines, quinolines, aromatic nitro compounds, and phenols. The EPA[10] lists more than 100 compounds as routinely detected by this method, which includes semi-VOCs found in crude oil. As with EPA Method No. 8240, the results of an EPA Method No. 8270 analysis are reported in computer printout from the mass spectrometer data system with chemical abstract number, compound name, concentration in either ppb or ppm, as appropriate, and detection limit. Also, it includes typically more than 50 standard compounds with ND (not detected) reported for most compounds.

The analysis for organic lead does not appear to be standardized. The LUFT Field Manual[3] describes a DOHS procedure using a xylene extraction followed by a flame atomic absorption spectroscopic (AA) method. A similar method using inductively coupled plasma/mass spectrometry (ICP/MS) has been developed[11] which reports total organic lead. The ICP/MS method was used for the analyses reported in the case studies.

CRITERIA USED FOR THE DIFFERENTIATION

The criteria used for the differentiation of crude oil and refined petroleum products in soil samples are listed in Table 2. The primary criterion is the interpretation of gas chromatograms obtained from the TPH analysis. In this analysis, the GC scans eluted compounds in the temperature range from 50°C to 300°C at a heating rate of 10° per minute and held at 300°C for five minutes, resulting in a total run time of 30 minutes. The chromatograms are visually

Table 2. Criteria Used for the Differentiation of Crude Oil and Refined Petroleum Products in Soil

1. Appearance of TPH gas chromatogram
2. Comparison of TRPH and TPH
3. BTXE concentrations
4. Other VOCs present
5. Semi-VOCs present
6. Presence of organic lead

compared to standards, then calculated and reported in generally no more than two of four refined petroleum products in this boiling range, i.e., gasoline, mineral spirits, kerosene and/or diesel fuel. Peaks above the diesel range are generally reported as $C_{20}-C_{30}$ hydrocarbons. The peaks that occur in the last five minutes of the scans are from the higher boiling point tars and asphaltenes in the sample. Their occurrence is a primary indication of a crude oil. In Figures 1, 2, and 3, the TPH gas chromatograms of the standards for gasoline, kerosene, and diesel fuels are given, respectively. The number printed above the peaks are retention times in minutes. The occurrence of peaks for each standard is confined to a particular section of the chromatogram representing the boiling fraction of the crude from which it was refined. Gasoline elutes from the beginning of the scan to approximately 10 minutes, kerosene from approximately 5 to 16 minutes, and diesel fuel from approximately 7 to 22 minutes. Also, the chromatograms of the refined products have a bell-shaped appearance. In Figure 4, a representative chromatogram is given of crude oil collected from operating wells in the oil field in which all of the case studies are located.

The laboratory results for the TRPH, TPH, and BTXE analysis of the four crude oil samples are given in Table 3. The occurrence of peaks for the crude oil is relatively uniform throughout the chromatogram from 2 to 29 minutes. This uniform occurrence of peaks throughout the chromatogram is also a primary indication of crude oil, despite the fact that the laboratory analysis for TPH will report the different segments as gasoline, kerosene, or diesel fuel fractions. In summary, the uniform occurrence of peaks throughout the 30-minute scan, plus the occurrence of peaks in the last five minutes, are primary indications of a crude oil, rather than a refined product, whereas the occurrence of segmented peaks, especially with a bell-shaped appearance, is an indication of a refined petroleum product.

A second criterion used to assess whether petroleum hydrocarbons detected in soil samples are refined products or crude oil is the comparison of the TPH results with the results of analyses for TRPH. The TPH analysis measures hydrocarbon content to a maximum of about C_{30} with the ability to quantify using standards terminating with the carbon range of diesel fuel, which is generally up to approximately C_{23}.[3] The TRPH analysis is reported[9] to lose approximately half of the gasoline present during the extraction process, and extracts the higher molecular weight fuels and oils. Hence, the TPH analysis is biased to the lower

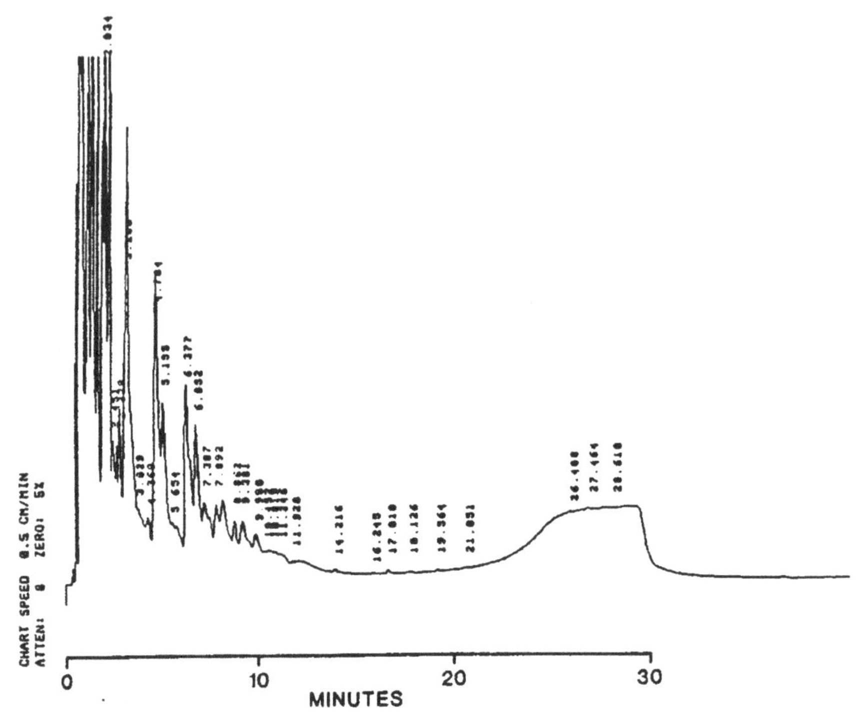

Figure 1. Standard TPH gas chromatogram for gasoline.

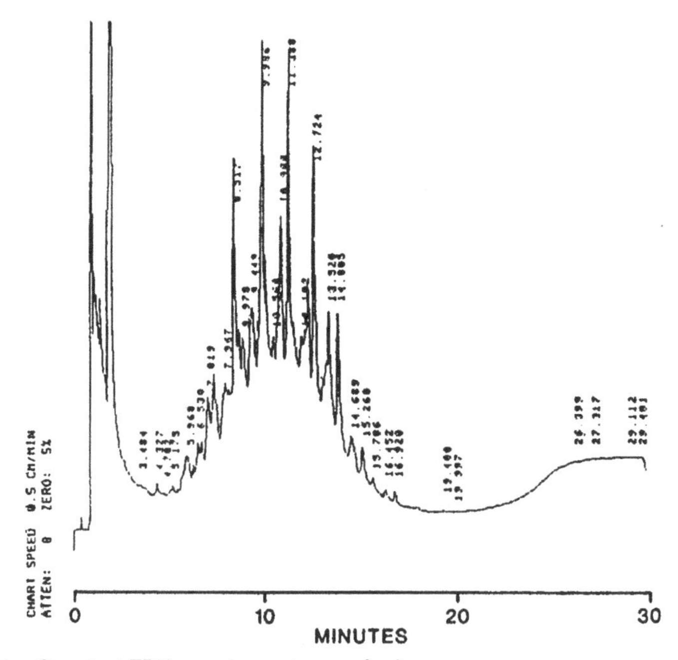

Figure 2. Standard TPH gas chromatogram for kerosene.

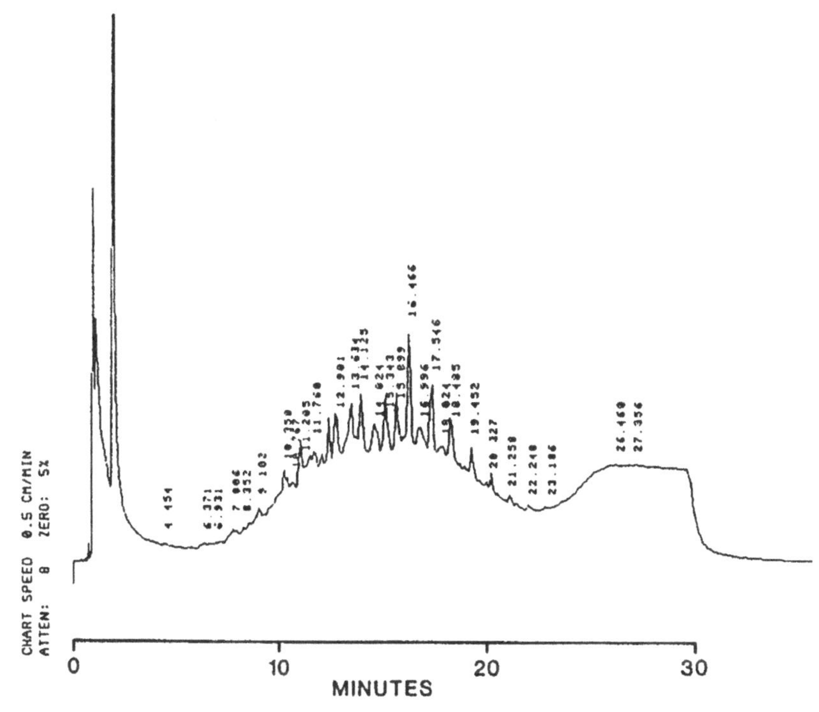

Figure 3. Standard TPH gas chromatogram for diesel fuel.

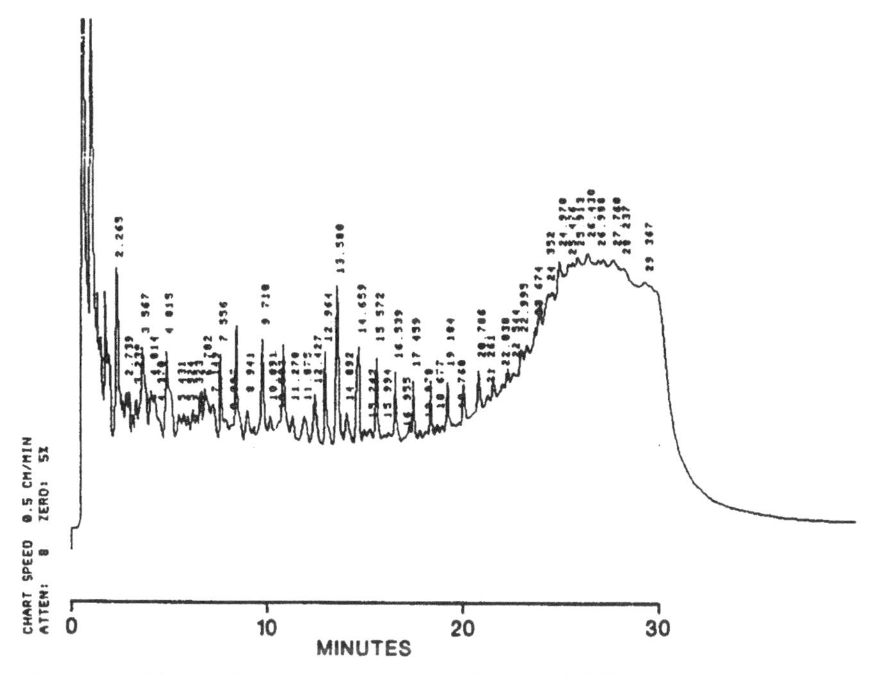

Figure 4. TPH gas chromatogram for crude oil sample 0-639B.

Table 3. Crude Oil Sample Analysis (Percent)

Analysis	O-620N	O-634D	O-639B	O-645C
TRPH	99	83	47	87
TPH	48.1	45.5	44.8	45.4
Gasoline	28.8	24.8	24.7	23.6
Diesel	10.4	9.1	10.6	14.2
$C_{20} - C_{30}$	14.9	11.6	9.5	7.6
Benzene	0.24	0.14	NA[a]	NA
Toluene	0.54	0.33	NA	NA
Total xylenes	0.16	0.10	NA	NA
Ethylbenzene	0.83	0.59	NA	NA

[a]Indicates not analyzed.

molecular weight petroleum hydrocarbons, while the TRPH analysis is biased to the higher molecular weight compounds. The comparison of TRPH with TPH concentrations indicates the relative distribution of low and high hydrocarbon fractions. When the TRPH concentration is much larger than the TPH concentration, the presence of a crude oil is indicated, whereas a higher TPH concentration is a clear indication of a refined petroleum product.

The variability of the TRPH analysis can be anticipated to limit the application of this criterion. In the analysis of four crude soil samples collected from operating wells in Table 3, the TRPH results range from 47% to 99%, while the ratio of the TRPH and TPH concentrations for three samples is approximately 2:1 and greater than 1:1 for the fourth sample.

A third criterion that can be used to assess the presence of gasoline is the results of analyses for BTXE, as provided in either EPA Method No. 8020 or 8240. BTXEs are natural constituents of crude oil, and typically are found in concentrations of a few percent.[13] In Table 3, the percent by volume of BTXE is given for two crude oil samples collected from operating oil wells on the site in Case Study I. The concentrations for BTXE in Table 3 are quite similar to the concentrations reported for a representative petroleum of the API Research Project 6.[13]

BTXEs are primarily found in the gasoline fraction of a crude oil and also are added to gasoline in the refining process. Historically, BTXEs constitute 15% to 25% of regular or unleaded gasolines, and as high as 40% for premium gasolines.[14] Since BTXEs are in the boiling range of the gasoline fraction of petroleum, they tend not to be found in diesel fuels, although higher boiling aromatic compounds may be found in low concentrations.

The VOCs analysis using EPA Method No. 8240 can detect, in addition to BTXEs, other VOCs such as short chain aliphatic and alicyclic hydrocarbons, which can provide backup data for the assessment process.

The semi-VOCs analysis provided by EPA Method No. 8270 detects higher molecular weight compounds that provide backup data which is especially useful for the indication of the presence of crude oils. In Table 4, a list is given of some of the compounds detected by the semi-VOCs analysis that would indicate the presence of a crude oil and not a gasoline or diesel fuel.

Table 4. Semivolatile Organic Compounds in Crude Oil[a]

Benz(a)anthracene
Chrysene
Dimethylnaphthalene
1-Methylnaphthalene
Fluoranthene
2-Methylnapthalene
Napthalene
Phenanthrene
Pyrene

[a]Kirk and Othmer.[15]

An additional indicator of gasoline as a refined product is the presence of organic tetraethyl or tetramethyl lead. Organic lead has, historically, been added to gasoline at concentrations up to 800 ppm. It can be analyzed for using ICP/MS.[11] The LUFT Field Manual[3] describes a DOHS method for organic lead using flame AA spectroscopy.

CASE STUDY I

One of the first sites[6] on which the differentiation criteria were used was located in the north-central portion of a large southern California oil field. The 20-acre site was the location for a proposed business park. The site was bound on the north, west, and south by major city streets, and on the west by railroad tracks. At the time of the site investigation, there were 18 active/idle and 19 abandoned oil wells within the site boundaries. Numerous oil-field-related pipelines traversed the site.

In the immediate site vicinity (the area within approximately one mile of the site), past and present primary land use included oil field development and chemical manufacturing/refining. Commercial development had been taking place since the early 1980s, resulting in the conversion of land that had been used historically for oil field and industrial purposes to office, warehouse, and light manufacturing facilities.

Oil-field-related potential source areas onsite that were judged to warrant investigation for the presence of crude oil or other potentially hazardous waste included former sump areas, former aboveground storage tank areas, and the abandoned, active and idle oil wells. A site schematic is given in Figure 5 that indicates the location of these potential source areas, along with the approximate location of trenches and borings used for the characterization of the potential source areas. Sumps and aboveground storage tanks (ABT) are designated with letters A through T as follows:

Sumps: A, B, C, D, H, K, L, N, Q, R, S, T (12 total).

Aboveground storage tanks: E, F, G, I, J, M, O, P (8 total).

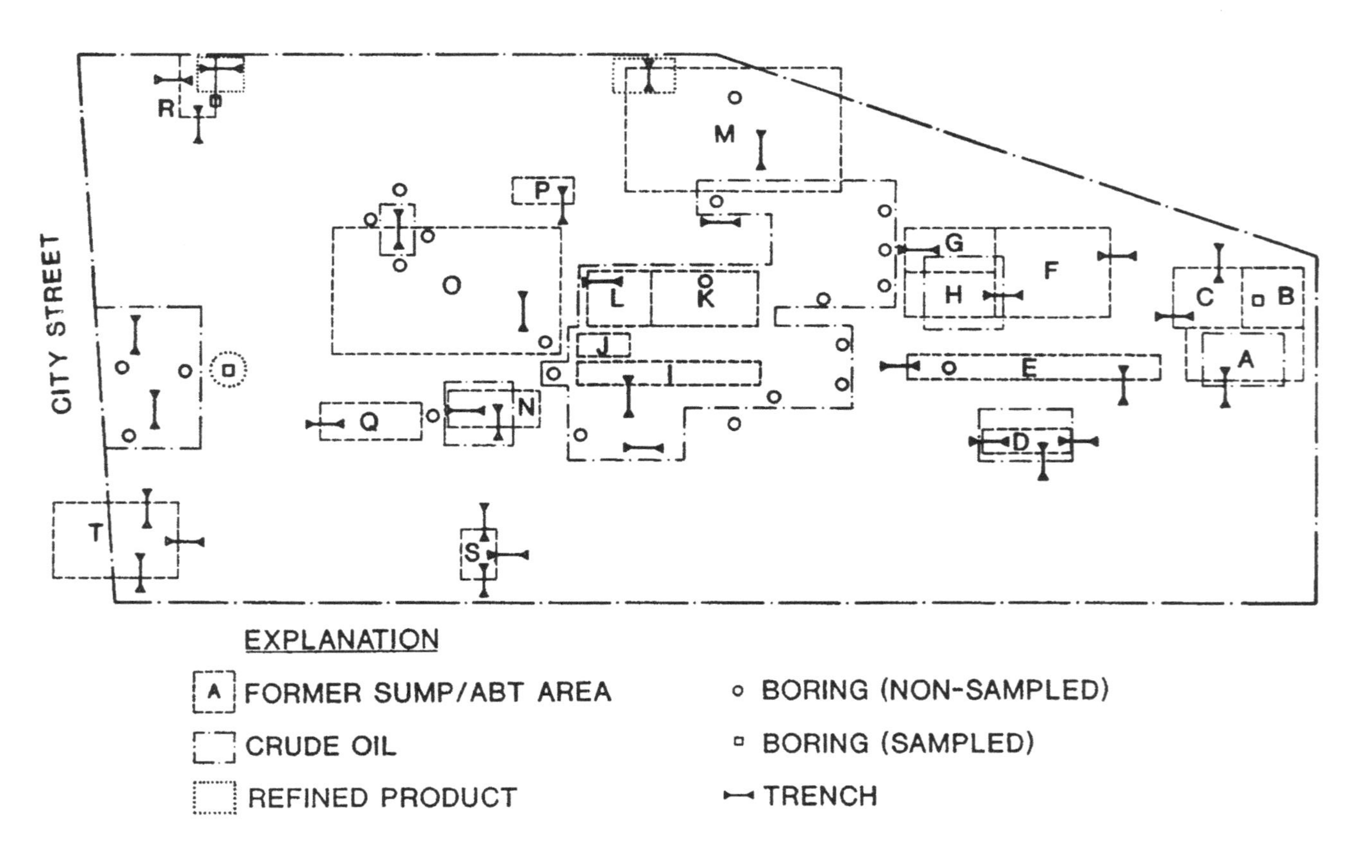

Figure 5. Site I schematic with petroleum hydrocarbon degraded areas.

The areas that were judged to be degraded with crude oil and refined petroleum product as a result of the site investigation are also shown in Figure 5. Most of the petroleum hydrocarbons detected at this site were assessed to be crude oil, with a number of areas exceeding 10,000 ppm TRPH for which remediation may be required. Three areas were assessed to contain refined petroleum product: Sump R, as primarily diesel fuel; Tank M, as primarily gasoline; and a background area, as a mixture of gasoline and diesel. Representative laboratory results for source areas assessed as crude oil and refined petroleum products are provided in Table 5. In Table 6, the evaluation of these data in terms of the criteria described above are given to exemplify the application of this approach. The assessment of crude oil for the majority of the samples is based on the appearance of the chromatograms and on the comparison of the TRPH and TPH results. The chromatograms for Sump H and Tank O, which were judged to be crude oil, are provided in Figure 6. These chromatograms are typical of those that are judged to be crude oil. They have a uniform occurrence of peaks throughout the scan, including peaks in the last five minutes of the run. While the intensity of the peaks in the diesel range is greater than the other areas, it is not interpreted to represent diesel fuel. In aged crude oil soil samples, the gasoline fraction is lost through volatilization and biodegradation. The C_{20} to C_{30} fraction of crude oil has a lower concentration than the diesel fraction and the instrument response to this fraction is also reduced. The diesel fuel standard in Figure 3 can be seen to have a much more dramatic bell-shape than in these chromatograms. The interpretation of these chromatograms as a crude oil was made with confidence.

The chromatograms for Sump R and Tank M are given in Figure 7. Each of these chromatograms was interpreted to be a refined petroleum product with a background of crude oil. For each of the soil samples, the TPH result was larger than the TRPH concentration as given in Table 5. The chromatogram for Sump R area exhibits an intense segmented, bell-shaped scan from about 2 minutes to 12 minutes, with scattered weak peaks throughout the balance of the run. These weak peaks are believed to indicate a background concentration of crude oil underlying the refined petroleum product. The TPH concentration was 14,700 ppm, primarily in the gasoline range, versus a TRPH concentration of 1,300 ppm. The chromatogram for Tank M area also has a prominent segmented, bell-shaped portion from approximately 7 minutes to 20 minutes, primarily in the diesel range. In addition, the chromatogram has weak peaks throughout the balance of the run which are interpreted to be a background of crude oil. The TPH concentration was 1,730 ppm versus a TRPH concentration of 1,100 ppm.

In this case, the appearance of the gas chromatograms and the comparison of TPH and TRPH provided consistent evidence for the presence or absence of refined petroleum products. The judgments were made with confidence. While reliance on the analyses for BTXE, other VOCs, semi-VOCs, or organic lead was not used, the results for these analyses are consistent with the interpretations.

Table 5. Case Study I: Laboratory Results (ppm)

Analysis	Sump H	Tank O	Sump R	Tank M
TRPH	104,000	12,000	1,300	2,100
TPH	44,000	7,350	14,700	3,620
Gasoline	4,000	370	14,700	ND
Diesel	36,000	6,300	ND	3,500
$C_{20}-C_{30}$	4,000	680	ND	120
Benzene	ND[a]	12	ND	ND
Toluene	ND	29	18	ND
Total xylenes	4	27	130	ND
Ethylbenzene	1	3	19	ND
Organic lead	ND	NA[b]	NA	ND
Other VOCs			NA	ND
C_5-C_{11} Aliphatic and alicyclic hydrocarbons	1,000	900		
C_9-C_{10} Alkylbenzenes	300	200		
Semi-VOCs			NA	
C_8-C_{35} Hydrocarbon matrix	300,000	40,000		10,000
Benzo (B&K) fluoranthenes	ND	1		
Dimethylnaphthalenes	600	90		
Fluorene	19	3		
1-Methylnaphthalene	100	20		
2-Methylnaphthalene	230	35		
Naphthalene	92	12		
Phenanthrene	56	8		
Pyrene	ND	1		

[a]Indicates not detected.
[b]Indicates not analyzed.

Table 6. Case Study I: Assessment of the Origin of Petroleum Hydrocarbons in Soil

Criterion	Sump H	Tank O	Sump R	Tank M
1. Appearance of TPH gas chromatogram				
a. Peaks last 5 min.	YES	YES	YES	YES
b. Segmented, bell-shaped peaks	NO	NO	YES	YES
c. Uniform occurrence of peaks	YES	YES	NO	YES
2. TRPH>>TPH[a]	YES	YES	NO	NO
3. BTXE present	YES	YES	YES	NO
4. Other VOCs present	YES	YES	NA[b]	YES
5. Semi-VOCs present	YES	YES	NA	NO
6. Organic lead present	NO	NA	NA	NO
JUDGMENT	Crude oil	Crude oil	Refined product, crude oil background	Refined product, crude oil background

[a]Indicates concentration of total recoverable petroleum hydrocarbons (TRPH) is much greater than concentration of total petroleum hydrocarbons (TPH) detected by EPA Method Nos. 418.1 and 8015 (modified), respectively.
[b]Indicates not analyzed.

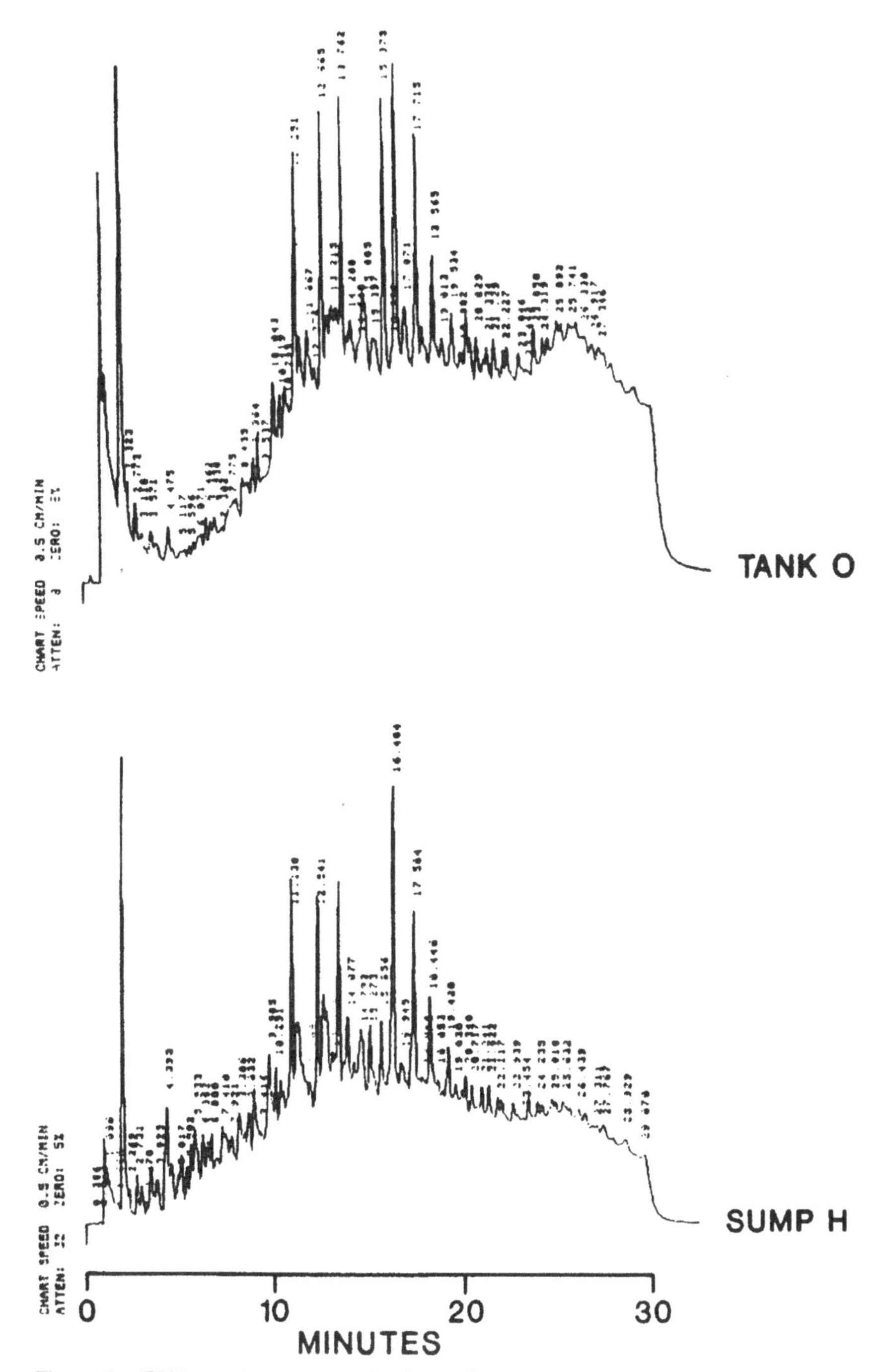

Figure 6. TPH gas chromatogram for Site I, Sump H, and Tank O areas, judged to be crude oil.

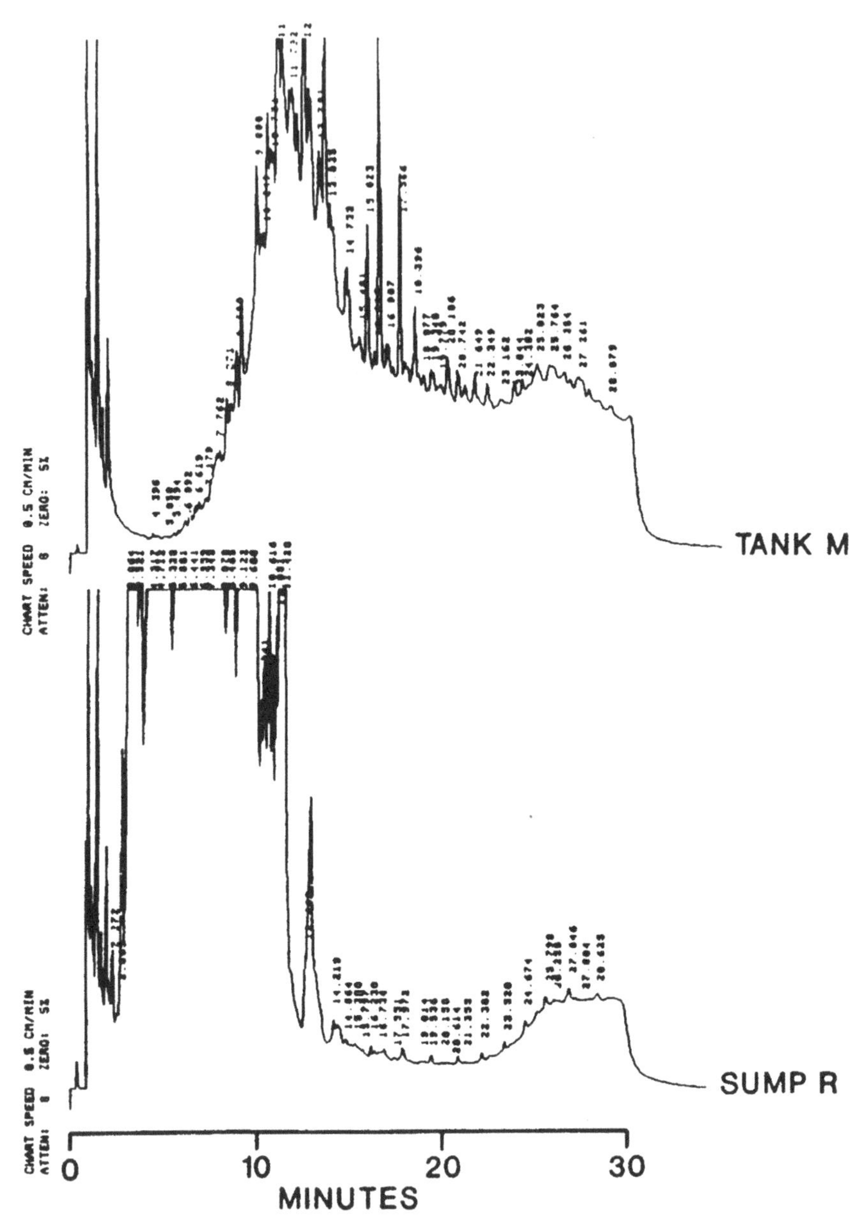

Figure 7. TPH gas chromatogram for Site I, Sump R, and Tank M areas, judged to be refined petroleum product and crude oil background.

CASE STUDY II

Site II was a 2.5-acre parcel located in the north-central portion of a southern California oil field. The site was bound on the south by a city street, on the north by an operating oil field, and on the east and west by light industrial buildings. The immediate site vicinity, the area within approximately a one-mile radius of the site, consisted largely of operating oil fields and some industrial properties. The objective of this investigation was to assess the nature, extent, and potential migration of petroleum hydrocarbons reported to be in the subsurface from a previous investigation.

A schematic of Site II is given in Figure 8 with the location of borings from the previous investigation, as well as from the existing investigation shown. Also, the areas judged to have petroleum degraded soils are indicated in Figure 8. In Table 7, the results of chemical analyses used for the differentiation of crude oil and refined petroleum products are given. The chromatograms of two of the soil samples used for the evaluation of the appearance criteria are provided in Figures 9 and 10. The assessment of the origin of the petroleum hydrocarbons detected, including the judgment reached for each sample, is described in Table 8. The Area I sample was interpreted to be crude oil, while the Area II sample was judged to be refined petroleum product (gasoline). The chromatogram for Area I clearly exhibits a uniform occurrence of peaks throughout the scan, as well as peaks in the last five minutes of the run, for indications of the presence of predominately crude oil. The chromatogram for Area II, which was judged to be gasoline, exhibits peaks in the first 7 to 10 minutes of the run, with a bell-shaped appearance and no peaks in the balance of the scan.

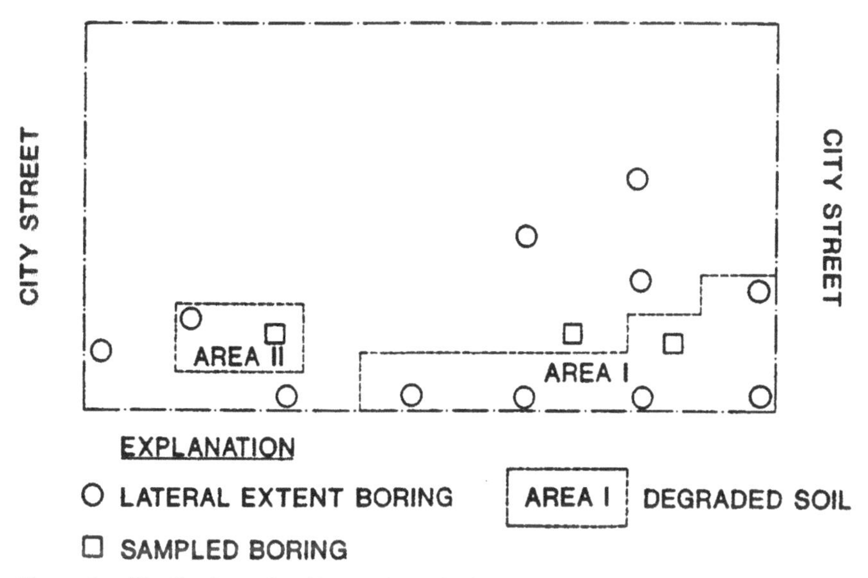

Figure 8. Site II schematic with petroleum hydrocarbon degraded areas.

Table 7. Case Study II: Laboratory Results (ppm)

Analysis	Area I	Area II
TRPH	9,000	4,300
TPH	790	3,500
Gasoline	ND[a]	3,500
Diesel	630	ND
$C_{20}-C_{30}$	160	ND
Benzene	ND	1.4
Toluene	0.3	44
Total xylenes	2.7	93
Ethylbenzene	0.9	30
Organic lead	ND	ND

[a]Indicates not detected.

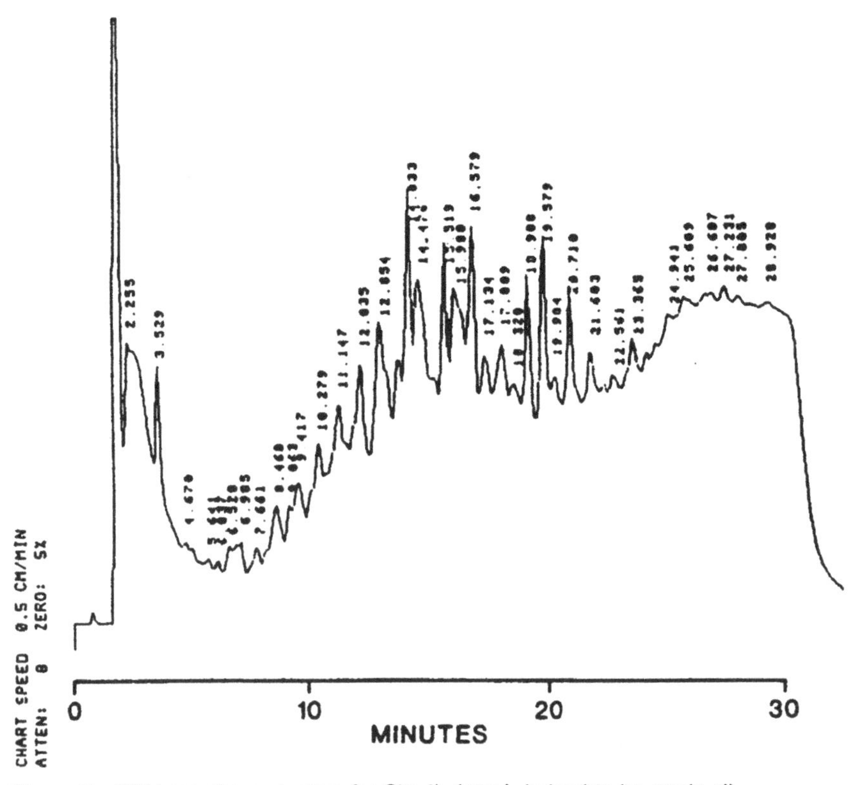

Figure 9. TPH gas chromatogram for Site II, Area I, judged to be crude oil.

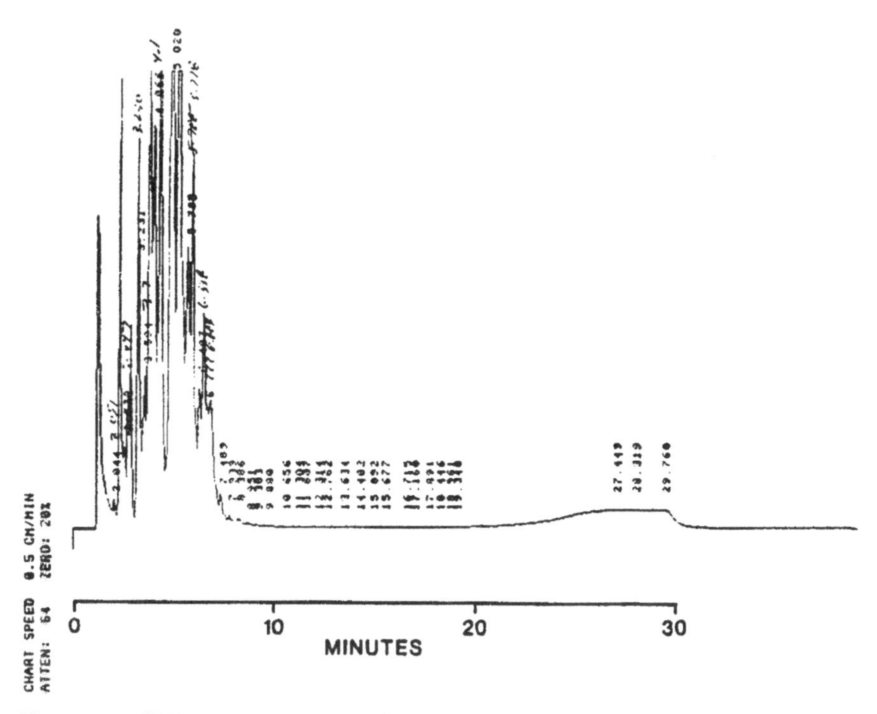

Figure 10. TPH gas chromatogram for Site II, Area II, judged to be refined petroleum product.

Table 8. Case Study II: Assessment of the Origin of Petroleum Hydrocarbons in Soil

Criterion	Area I	Area II
1. Appearance of TPH gas chromatogram		
a. Peaks last 5 min.	YES	NO
b. Segmented, bell-shaped peaks	NO	YES
c. Uniform occurrence of peaks	YES	NO
2. TRPH> >TPH[a]	YES	NO
3. BTXE present	YES	YES
4. Other VOCs present	NA[b]	NA
5. Semi-VOCs present	NA	NA
6. Organic lead present	NO	NO
JUDGMENT	Crude oil	Gasoline

[a]Indicates concentration of total recoverable petroleum hydrocarbons (TRPH) is much greater than concentration of total petroleum hydrocarbons (TPH) detected by EPA Method Nos. 418.1 and 8015 (modified), respectively.

[b]Indicates not analyzed.

Based on these judgments and the results from a previous investigation, it was recommended that the degraded soil in both Areas I and II be excavated and remediated to a depth of 5 feet. Also, the soil in Area 2 that was judged to be degraded with gasoline should be studied for the feasibility of mitigating with an in situ vapor extraction system.

CASE STUDY III

Site III provides an example in which the application of the criteria resulted in conflicting assessments. Site III was a four-acre parcel located in the south-central portion of a large southern California oil field. The site was bound on the east and south by major city streets, on the north by a landscaping business, and on the west by light industry. In the immediate site vicinity, historical and present primary land use included oil field development and refining. Limited industrial development had been taking place since the early 1980s, converting the depleted oil field properties to office, warehouse, and light manufacturing facilities. The objective of the effort was to characterize the extent and degree of soil degradation and to assess the origin of the petroleum hydrocarbons

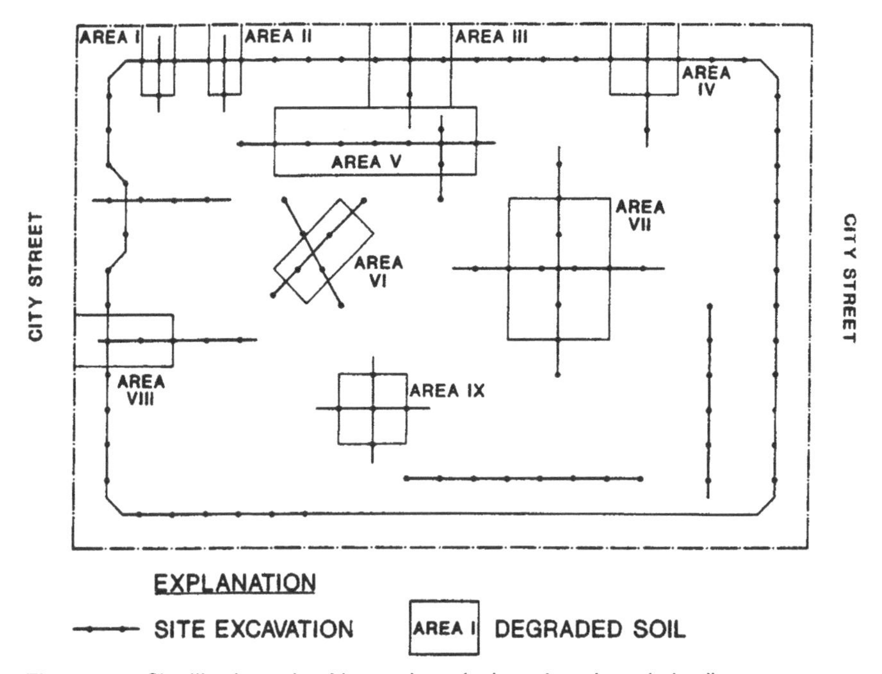

Figure 11. Site III schematic with petroleum hydrocarbon degraded soil areas.

detected in terms of crude oil or refined petroleum products. In Figure 11, a schematic of the site is given showing potential source areas, test pits, and the perimeter excavation used for characterization and sampling. The areas assessed to be degraded are also indicated.

In the investigation, 1 perimeter trench and 18 test pits were excavated. The degraded areas included three pipeline leaks along the northern property boundary, Area III along the northern boundary, and a portion of Area VIII along the western boundary. In Table 9, laboratory results used to assess the origin of petroleum hydrocarbons in terms of crude oil or refined petroleum products are given for three typical soil samples at the site. In Figures 12 and 13, the TPH gas chromatograms are given for the three representative samples. In Table 10, the assessment of the data and the gas chromatograms are given. Each of the chromatograms exhibits peaks in the last five minutes of the run, a uniform occurrence of peaks throughout the 30-minute run, and no segmented, bell-shaped features. All of these characteristics are consistent with an assessment of crude oil. However, only the Area V sample had a TRPH result greater than the TPH analysis and the presence of a refined product cannot be ruled out with confidence. The other criteria provide no assistance, because of no analysis or nondetect. The detection of phenanthrene in the semi-VOC analysis for Area V sample does confirm the assessment of crude oil for this sample, but does not help to rule out a refined product. The BTXE also detected for the Area V sample adds strength to the crude oil assessment because the refined petroleum product, in this case, is judged to be diesel fuel which generally contains little or no BTXE. The final judgment for this site reported the presence of predominantly crude oil, with the possibility of some overlapping diesel fuel. This assessment is likely to require remediation to a much lower level than might otherwise have been needed if an assessment of crude oil only could have been justified on the basis of the appearance of the chromatograms.

Table 9. Case Study III: Laboratory Results (ppm)

Analysis	Area V	Area IX	Area II
TRPH	7,600	3,500	15,000
TPH	6,600	26,800	17,700
Gasoline	1,000	7,400	1,300
Diesel	2,900	13,000	10,000
$C_{20} - C_{30}$	2,700	6,400	6,400
Benzene	ND[a]	0.080	0.081
Toluene	0.029	9.3	12
Total xylenes	0.120	4.7	6.1
Ethylbenzene	ND	4.0	7.2
Organic lead	ND	NA[b]	NA

[a]Indicates not detected.
[b]Indicates not analyzed.

Table 10. Case Study III: Assessment of the Origin of Petroleum Hydrocarbons in Soil

Criterion	Area V	Area IX	Area II
1. Appearance of TPH gas chromatogram			
a. Peaks last 5 min.	YES	YES	YES
b. Segmented, bell-shaped peaks	NO	NO	NO
c. Uniform occurrence of peaks	YES	YES	YES
2. TRPH>>TPH[a]	YES	NO	NO
3. BTXE present	YES	YES	NA
4. Other VOCs present	NA[b]	NA	NA
5. Semi-VOCs present	YES[c]	NA	NA
6. Organic lead present	ND[d]	NA	NA
JUDGMENT	Crude oil	Crude oil and diesel fuel	Crude oil and diesel fuel

[a]Indicates concentration of total recoverable petroleum hydrocarbons (TRPH) is much greater than concentration of total petroleum hydrocarbons (TPH) detected by EPA Method Nos. 418.1 and 8015 (modified), respectively.
[b]Indicates not analyzed.
[c]Phenanthrene detected at 0.640 ppm.
[d]Indicates not detected.

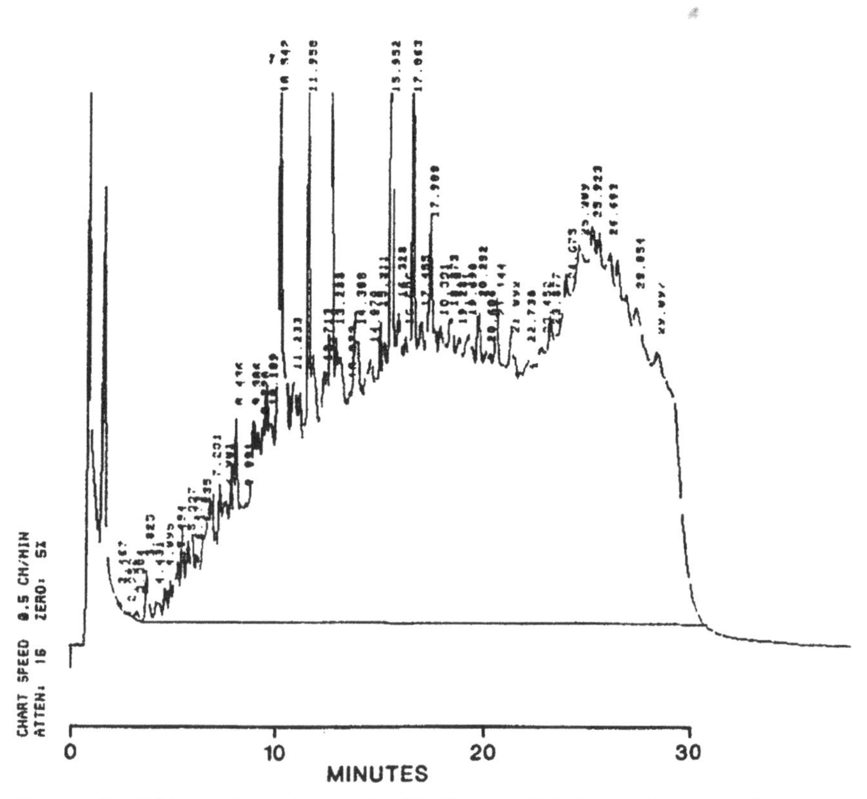

Figure 12. TPH gas chromatogram for Site III, Area V, judged to be crude oil.

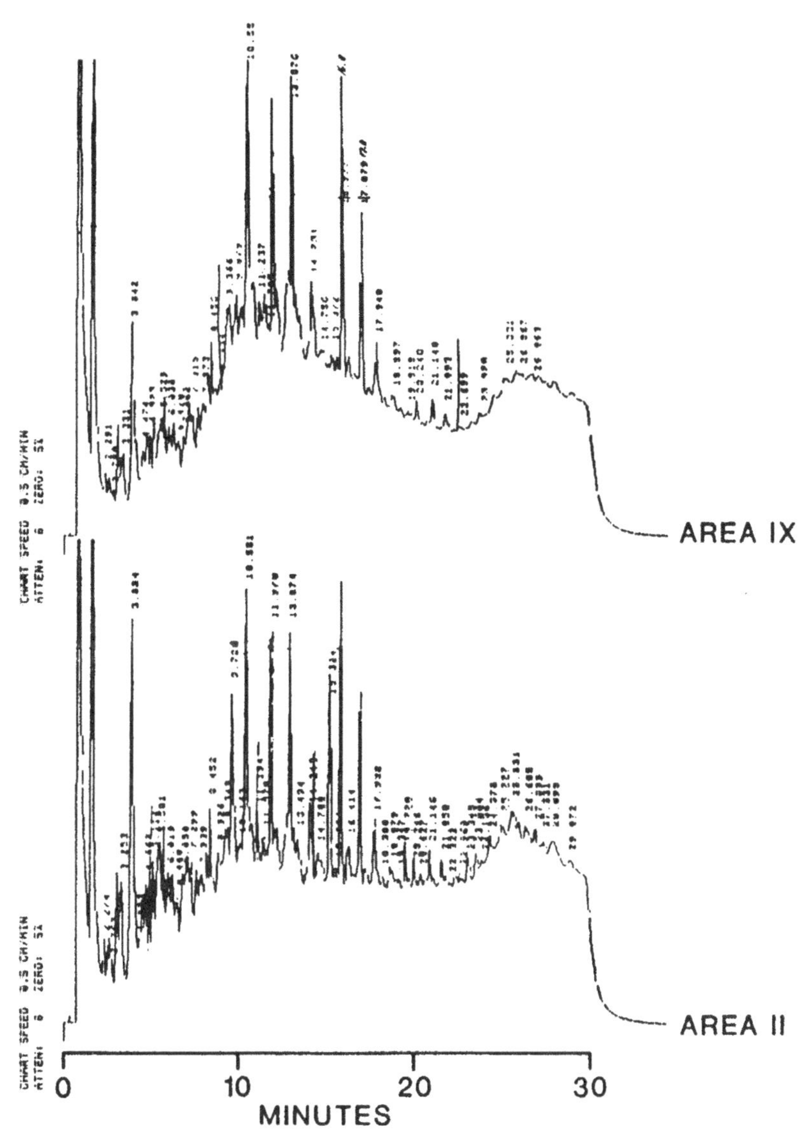

Figure 13. TPH gas chromatogram for Site III, Areas II and IX, judged to be crude oil and refined petroleum product.

LIMITATIONS AND FUTURE WORK

The assessment of the origin of petroleum hydrocarbons using the criteria described above has, in a number of cases, provided regulators with the evidence needed to justify reduced mitigation measures based on the absence of refined petroleum product. To that end, these criteria have served in reducing the cost of remediation. However, the reliance on TRPH is a recognized limitation of this approach because it is more of a survey tool than a reliable site characterization analysis. Future improvements in the analysis for the concentration of crude oil in soils could be made on two bases. First, a better understanding of the ability of EPA Method No. 418.1 to measure crude oil could improve the interpretation of the comparison of this TRPH with the TPH result. Retaining the use of the TRPH analysis has the advantage of being a simple, inexpensive method. Second, more reliable improvement would be to find another analysis, perhaps one that is routine in the petroleum industry, to substitute for the TRPH analysis.

When investigating a site which historically has been used as an oil field, the question of the presence of crude oil is not generally the issue. The issue to resolve is the absence of a refined petroleum product. A further development in the TPH analysis that included standards for crude oil and for mixtures of crude oil and diesel fuel might provide the evidence needed. If reliable standards were found, convincing evidence for the absence of diesel fuels might be possible using the appearance of TPH gas chromatograms without a comparison to TRPH.

Senn and Johnson[16] have discussed the use of capillary column GC to estimate the degree of degradation and relative age of petroleum hydrocarbons in soil and groundwater. Here, use is made of ratios of pristane and phytane, C_{17} and C_{18} alicyclic hydrocarbons, respectively, which biodegrade more slowly than the C_{17} and C_{18} aliphatic hydrocarbons. Therefore, larger C_{17} to pristane and C_{18} to phytane peak ratios indicate less degradation between samples. They also report that the presence of light-end peaks also indicate that a gasoline product in a sample is relatively fresh.

In our case studies, the occurrence of light-end peaks is rare. Hence, our concern centers primarily on assessing the absence of diesel fuel when crude oil is believed to be present. Diesel fuels are reported[3] to have predominate carbon chain lengths of C_{16} and C_{17} in the C_{10} to C_{23} mixture. If this predominance of C_{16} and C_{17} does not occur in the crude, this feature could be very helpful in providing evidence for the presence or absence of diesel fuels. Even with a more expensive analysis to provide this data, the reduced costs in remediation if the absence of a refined product is accepted by a regulatory agency would overshadow the cost of the analysis.

REFERENCES

1. CFR (Code of Federal Regulations), Identification and Listing of Hazardous Waste: 40 CFR 261, Revised 1 July 1987.
2. CCR (California Code of Regulations), Title 22, Division 4, Environmental Health, Chapter 30, Minimum Standards for Management of Hazardous or Extremely Hazardous Waste, 1988.
3. SWRCB (State Water Resources Control Board), *Leaking Underground Fuel Tank Manual: Guidelines for Site Assessment, Cleanup, and Underground Storage Tank*

Closure State of California Leaking Underground Fuel Tank Task Force, Sacramento, California, revised 5 April 1989, 121 pp.
4. Applied Geosciences Inc. "Report on the Phase II Site Characterization of a 4.3-Acre Property," unpublished report prepared for a confidential client, 2 June 1988(a), 33 pp.
5. Applied Geosciences Inc. "Report on the Results of an Assessment of Oily Soil Adjacent to Two Oil Wells," unpublished report prepared for a confidential client, 6 July 1988(b), 22 pp.
6. Applied Geosciences Inc. "Phase II Toxic Hazard Investigation of a Planned Business Park," unpublished report prepared for a confidential client, 16 August 1988(c), 49 pp.
7. Applied Geosciences Inc. "Oily Soil Assessment," unpublished report prepared for a confidential client, 14 December 1988(d), 7 pp.
8. Applied Geosciences Inc. "Phase II Site Characterization Report," unpublished draft report in preparation for a confidential client, in preparation September 1989.
9. U.S. EPA (Environmental Protection Agency), Methods for Chemical Analysis of Water and Wastes, EPA 600/4-79-020, revised March 1983, U.S. EPA Environmental Monitoring Laboratory, Cincinnati, Ohio.
10. U.S. EPA, Test Methods for Evaluating Solid Waste; Physical/Chemical Methods, SW-846, Third Edition, Office of Solid Waste and Emergency Response, U.S. EPA, Washington, D.C., 1986.
11. Northington, J. D., Technical Director, West Coast Analytical Service, Inc., Santa Fe Springs, California, oral communication, 1988.
12. Nyer, E. K., and G. J. Skladany. "Relating the Physical and Chemical Properties of Petroleum Hydrocarbons to Soil and Aquifer Remediation," *Ground Water Monitoring Review* Winter 1989, pp 54–60.
13. Rossini, F. D. "Hydrocarbons in Petroleum," *J. Chem. Ed.* 37:554–561 (1960).
14. Guard, H. E., J. Ng, and R. B. Laughlin, Jr., Characterization of Gasoline, Diesel Fuels and Their Water Soluble Fractions, Naval Biosciences Laboratory, Naval Supply Center, Oakland, California, September 1983.
15. Kirk, R. E., and D. F. Othmer, Eds., *Encyclopedia of Chemical Technology* (New York: John Wiley & Sons, 1983).
16. Senn, R. B., and M. S. Johnson. Interpretation of Gas Chromatography Data as a Tool in Subsurface Hydrocarbon Investigations, in *Proceedings of the NWWA/API Conference on Petroleum Hydrocarbons and Organic Chemicals in Ground Water—Prevention, Detection and Restoration,* Houston,Texas, 13–15 November 1985, National Water Well Association, Dublin, Ohio, pp. 331–357.

CHAPTER 4

Onsite Analytical Screening of Gasoline Contaminated Media Using a Jar Headspace Procedure

John Fitzgerald, Division of Hazardous Waste, Massachusetts Department of Environmental Protection, Woburn, Massachusetts

INTRODUCTION

Portable analytical instrumentation is now widely used to investigate and document conditions of environmental contamination by volatile organic compounds, including media contamination by complex mixtures such as gasoline and other light petroleum distillates. In addition to providing substantial cost-savings over traditional laboratory procedures, the utilization of such equipment provides immediate onsite data for use in assessment or remedial response actions.

A majority of portable analytical units currently in use contain or consist of a photoionization detector (PID) or a flame ionization detector (FID). Most units are designed or operated solely as PID or FID response meters, without chromatographic separation, and provide quantitative data on "total organic vapor" (TOV) concentration.

Due to simplicity and expediency of operation, virtually all portable field units are designed to analyze gaseous samples at ambient temperatures. Multimedia analysis of soil and water samples is accomplished by a "headspace" technique, which involves the mass transfer or partitioning of volatile contaminants from an aqueous or bulk soil matrix to an overlying, confined gaseous phase.

Despite widespread usage, very little information has been published on the applications and limitations of such field procedures.

Without qualitative (chromatographic) definition, TOV headspace methodologies are intrinsically limited to the analysis of samples containing a known contaminant or contaminant mixture. Even within this limited universe, additional questions arise on the interpretation, accuracy, and significance of resultant headspace data. Indeed, aside from the broader interpretative issues, the evaluation and even comparison of TOV headspace data has traditionally been hampered by the lack of a standard, universally accepted procedure.

It is the intent of this author to begin discussions on this subject. Research was conducted to investigate the use and utility of a PID and an FID meter for the field headspace analysis of soil and water samples contaminated with unleaded gasoline. A series of laboratory and field experiments were conducted in order to:

(1) evaluate instrument operation and response characteristics
(2) evaluate the rate, extent, and chemistry of jar headspace development
(3) evaluate the effects of various physical and environmental factors on headspace development and instrument detection and response, and
(4) develop a preliminary correlation of "total organic vapor" headspace measurements with headspace and media concentrations of targeted gasoline constituents of concern, including benzene, toluene, ethylbenzene, and xylenes (BTEX).

BACKGROUND

In order to correlate or interpret TOV headspace screening data for a complex mixture such as gasoline, it is first necessary to gain an understanding of: (1) the chemistry and environmental fate of gasoline, (2) the volatilization process, and (3) the operational mechanics and selectivity of portable PID and FID instrumentation. While a detailed discussion of these three items is beyond the scope of this chapter, pertinent elements are briefly outlined in the following paragraphs.

The Chemistry and Fate of Gasoline

Industry specifications for gasoline products are based upon physical and performance-orientated criteria, and not upon a designated chemical formulation.[1] In addition to a variety of additives, refined gasolines may contain any combination of petroleum hydrocarbons from C_2 through C_{13}. While formulations are variable, Domask has reported, excluding additives, 42 specific constituents generally comprise about 75% of unleaded blends.[2]

With the notable exception of certain water-soluble and/or toxic additives, the monoaromatic compounds benzene, toluene, ethylbenzene, and xylenes (BTEX) are generally considered the most environmentally significant components of gasoline. The BTEX fraction typically constitutes approximately 15% of unleaded gasolines.[2] Because of its relatively high water-solubility, volatility, and toxicity, benzene, which generally comprises between 1% and 3% of gasoline, is normally targeted as the individual gasoline constituent of greatest concern.

In addition to the BTEX compounds, however, dozens of other volatile gasoline constituents will elicit a response on typical PID and FID field units. In this regard, three categories or component groupings may be designated within a gasoline formulation, as graphically illustrated in Figure 1, which depicts a (packed-column) chromatogram of an unleaded blend used during experimentation.

Peak elutions within Figure 1 are generally reflective of the vapor pressures of the individual (or co-eluting) gasoline components. The "first third" grouping represents a number of relatively low-boiling-point hydrocarbons, predominated by normal paraffins, isoparaffins, cycloparaffins, olefins, and cyclo-olefins. The "second third" grouping is composed primarily of the BTEX compounds. Finally, the "last third" components are comprised of heavier molecular weight (greater than 110), high-boiling-point hydrocarbons, and are predominately alkyl-aromatic compounds.

Once released to the environment, the chemical composition of gasoline residuals will be altered by environmental fate processes. In general, within the context of Figure 1, fate processes tend to result in a shift of relative component groupings from left to right: initial volatilization of "first third" hydrocarbons, with concurrent and subsequent leaching, volatilization, and biodegradation of the "second third" (BTEX) compounds. The specific extent and combination of fate processes, temporally and spatially, will dictate the media concentration and distribution of gasoline constituents, which in turn will dictate the extent and chemistry of headspace development and portable PID or FID unit response.

Volatilization and Headspace Development

While volatilization on a macro-environmental scale can significantly influence gasoline composition and sample chemistry, short-term volatilization on a micro-environmental scale is of principal concern in the analytical headspace screening of sample media.

Theoretical and empirical relationships used to describe and model volatilization are based upon mass transfer principles. Graphical depictions of the volatilization process for aqueous and soil systems are presented in Figures 2 and 3, respectively.

For aqueous systems, the "two-layer film" theory of mass transfer resistance is generally utilized to describe and model volatilization. "Henry's Law," an empirical relationship concerning the partitioning of volatile chemicals between the aqueous and vapor phases, is widely used to predict the extent of compound volatilization. According to this relationship, at equilibrium, the concentration of a gas dissolved in a liquid is directly proportional to the partial pressure of that gas above the liquid; the constant of proportionality being the Henry's Law constant.[3]

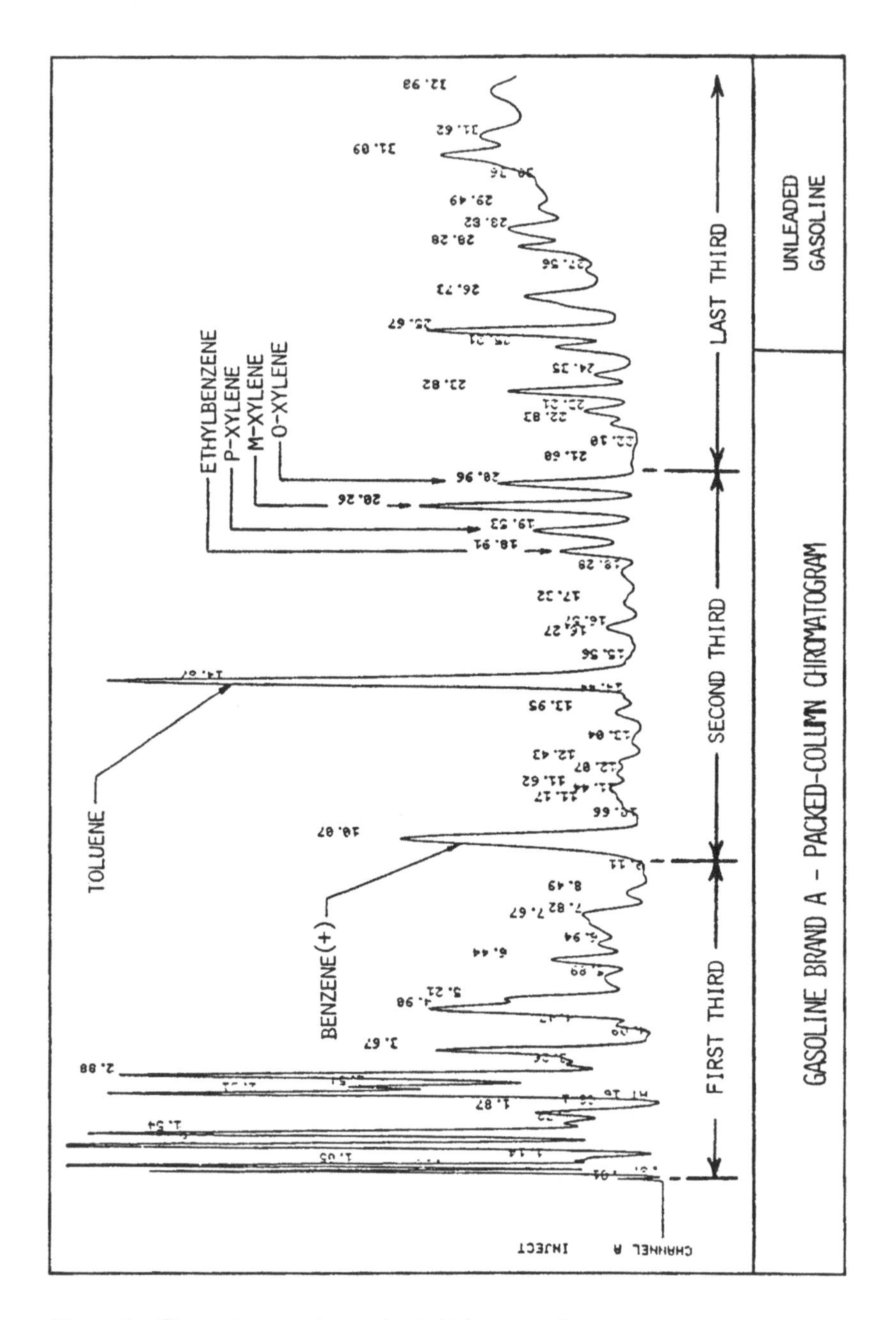

Figure 1. Chromatogram of an unleaded blend gasoline.

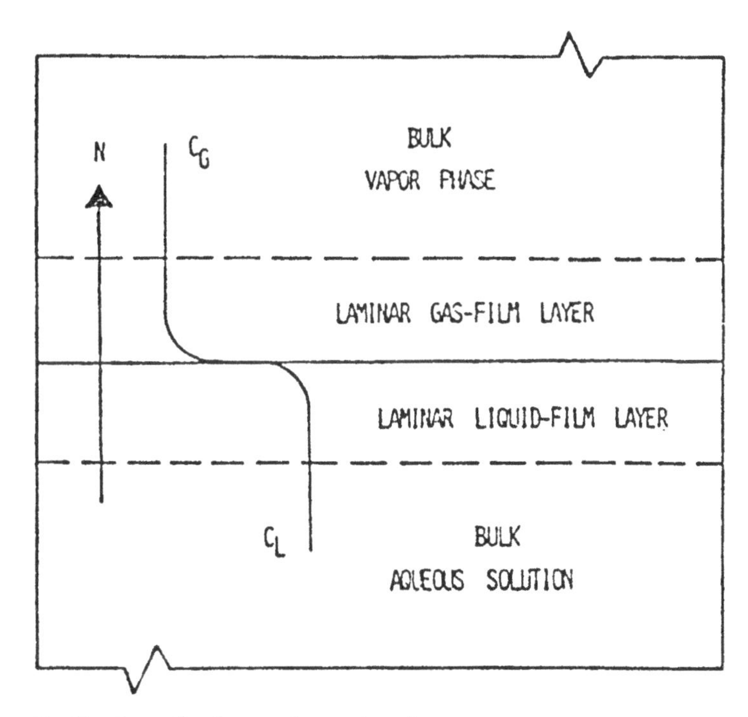

Figure 2. Two-layer film theory of mass transfer.

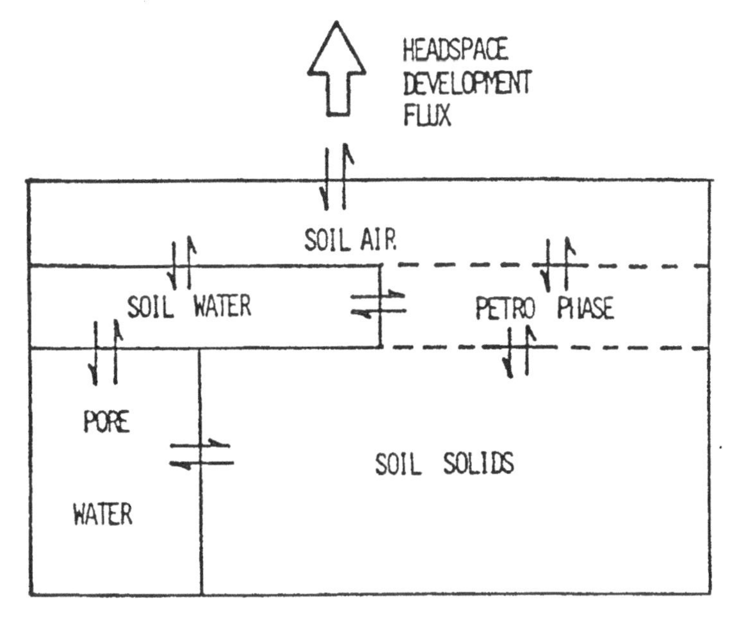

Figure 3. Contaminant partitioning and volatilization within an elemental soil block.

Volatilization from a bulk soil matrix is considerably more complex than volatilization from an aqueous solution, as the former includes solid, liquid, and gaseous compartments. From an analytical perspective, jar headspace development above a soil sample involves mass-transfer from the solid, immobile liquid, and/or mobile liquid(s) into the soil pore-gas, with subsequent diffusion or advection to the soil surface, and final diffusion/advection through the overlying containerized headspace. In moist soils without free (immiscible) gasoline, constituents volatilize from the soil/water interface into the soil pore-gas. The concentration at the interface is maintained by the liquid diffusion of the contaminant from adsorption sites on the soil solids.[3]

The equilibrium state of contaminant partitioning within an elemental soil block (based upon Jury, et al., in Lyman, et al.[3]) may be expressed by the following equation:

$$C(t) = \gamma Cs + \theta Cw + nCa \qquad (1)$$

where:

$C(t)$ = total contaminant concentration ($\mu g/cm^3$)
γ = soil bulk density (g/cm^3)
Cs = contaminant sorbed onto soil ($\mu g/g$)
θ = volumetric soil water content (cm^3/cm^3)
Cw = contaminant concentration in soil water ($\mu g/cm^3$)
n = volumetric soil air content (cm^3/cm^3)
Ca = contaminant concentration in soil air ($\mu g/cm^3$)

Substituting a Freundlich isotherm relationship for Cs, assuming adsorption is controlled by the organic carbon content of the soil, and substituting a Henry's Law relationship for Ca, it is possible to estimate the equilibrium state of a contaminant in terms of the soil water contaminant concentration, Cw:

$$C(t) = \gamma(Koc)(\%Soc)(Cw)/100 + \theta Cw + (n)(H')(Cw) \qquad (2)$$

where:

Koc = soil (organic carbon) partition coefficient ($\mu g/g$) / ($\mu g/cm^3$)
$\%Soc$ = percentage of soil organic carbon (%)
H' = dimensionless Henry's Law constant

Since gasoline is a mixture of many hydrocarbons, partitioning relationships between individual compounds would be difficult to predict or model. Nevertheless, for perspective, applying the relationships described in Equation 2 to a typical soil sample ($n = 0.2$, $\theta = 0.1$, $\gamma = 1.58$) containing (only) benzene at a contaminant concentration of 1 $\mu g/g$, the equilibrium partitioning between soil solids, water, and air at a low, medium, and high organic carbon content (%Soc) would be as shown in Table 1.

Significantly, mass percentage distribution of benzene (and the remainder of the "second third" BTEX compounds) favors the solid and water compartments. While this is even more true of the "last third" constituents, the high vapor pressures and Henry's Law constant of "first third" compounds result in substantially higher gaseous-phase partitioning. Though the relative mass partitioning of benzene soil gas may seem minor, it is important to note that even at a soil organic carbon content of 5%, the predicted equilibrium gaseous phase concentration of 0.02 $\mu g/cm^3$ represents a volumetric gaseous concentration of 6.2 ppm (20 °C and 1 atm), well within the detection limits of portable field PID and FID units.

Table 1. Equilibrium Partitioning in a Soil Sample.

Soc %	Cs $\mu g/cm^3$	Cs %mass	Cw $\mu g/cm^3$	Cw %mass	Ca $\mu g/cm^3$	Ca %mass
0.2	0.693	43.9%	0.624	39.5%	0.26	16.6%
2.0	1.400	84.5%	0.127	8.0%	0.05	3.2%
5.0	1.504	93.2%	0.005	3.4%	0.02	1.3%

Portable Field Instrumentation

In order to produce meaningful screening data, it is essential that the operating principles and mechanics of field PID and FID units be understood and that the manufacturers' recommendations on operation, maintenance, and calibration be instituted. In this regard, several items are worth noting:

(1) The difference in detector response and selectivity between PID and FID systems is significant. While FID response is relatively uniform for most volatile gasoline hydrocarbons, PID response increases with degree of compound unsaturation, resulting in an approximately 5 times greater response to benzene than hexane using a 10.2 eV lamp. Selectivity of this nature may be useful in focusing on the BTEX monoaromatic compounds within a media sample.

(2) Instrument flow rate and response characteristics vary. These parameters are especially relevant when attempting to analyze headspaces in small sampling containers.

(3) Normally, maximum jar headspace response is obtained 2 to 5 seconds following sampling initiation. Meter response and maximum deflection is difficult to discern on a unit with digital readout, unless a "maximum hold" feature is available.

(4) All field units are susceptible to environmental and temperature effects. Moreover, PID response may be significantly reduced under conditions of elevated humidity.

(5) Unlike PID units, FID systems will respond to methane, which may cause significantly higher TOV headspace concentrations in certain media samples and/or analytical settings.

EXPERIMENTAL PROCEDURES

Experimentation was conducted in three areas: (1) evaluation of portable unit operation, calibration, and jar headspace response characteristics, (2) aqueous headspace development and analysis, and (3) soil headspace development and analysis. Media experiments in areas (2) and (3) centered on the examination of a field headspace procedure comprised of the following:

(a) placement of a soil or water sample in a glass container (generally to one-half jar capacity)
(b) application of a sheet of aluminum foil to seal the glass jar
(c) allowance of a period of time for static or dynamic headspace development
(d) subsequent analysis of the jar headspace via the insertion of a PID or FID unit sampling probe through the aluminum foil seal, recording the maximum meter response as the "total organic vapor" headspace determination.

For increased relevance to "field" objectives and limitations, experimental evaluation focused on short-term development techniques under conditions and restrictions which may apply at a field site.

In all areas of experimentation, correlative headspace data was obtained by comparison of "total organic vapor" meter responses with the chromatographic analysis of a 5 mL split-sample. For comparative purposes, both PID and FID field units were calibrated to respond "as benzene" (i.e. assume a benzene response factor).

Media experiments were performed on water and soil samples contained in 5 oz (140 mL) through 16 oz (480 mL) glass jars in which basic headspace development parameters were varied, including:

- shape/volume of jar
- depth of sample
- headspace development temperature
- headspace development time
- static vs dynamic headspace development
- gasoline concentration

Aqueous samples were formulated by adding measured quantities of a well-characterized ("fresh" and evaporatively weathered) unleaded gasoline stock to distilled, organic-free water. Soil headspace analyses were conducted on silica sand samples contaminated with unleaded gasoline stock and subjected to various degrees of leaching and passive aeration. Limited analyses of actual field samples were also conducted.

In total, 183 aqueous samples and 128 soil samples were analyzed by an H-Nu Systems and/or Century Systems unit, with split-sample chromatographic analyses performed on 16 of the aqueous samples and 24 of the soil samples.

Physical and environmental variations within discrete experimental setups allowed a relative evaluation of the "mechanical" aspect of headspace development (i.e. sample size, agitation, temperature, etc.). Moreover, viewed externally with the results of other compatible setups, data obtained from individual and collective experimentation allowed a limited evaluation of the overall extent and kinetics of headspace development, headspace chemistry, and instrument response.

Two field units were evaluated during the course of this research project. The PID unit was an H-Nu Systems Portable Photoionization Analyzer, Model PI-101, with a 10.2 eV lamp probe. The FID unit was a Century Systems Portable Organic Vapor Analyzer, Model OVA-128, utilized on a direct FID response mode.

Chromatographic identification and quantification of gasoline and headspace vapor samples was accomplished on a Hewlett-Packard (HP) Model 5790 gas chromatograph equipped with either a packed or capillary column and a flame ionization detector. The HP gas chromatograph was interfaced with a Spectra-Physics Model SP 4270 Computing Integrator for FID signal processing. Chromatographic operating procedures were based upon a modification of a methodology developed by Bruell and Hoag at the University of Connecticut.[4]

EXPERIMENTAL RESULTS AND CONCLUSIONS

Experimental observations, results, and conclusions are summarized below:

Portable Instrumentation

(1) Suppression of H-Nu PI-101 (PID) meter response was observed under conditions of elevated atmospheric humidity. For benzene, response reduction was as high as 40% under certain conditions; suppression of other BTEX compounds may not be as pronounced.

This finding is consistent with information published by Barsky et al.[5] and Chelton et al.[6] Although not ionizable with a 10.2 eV lamp source, water vapor molecules apparently inhibit PID response (a) by absorbing emitted ultraviolet radiation, thereby decreasing photon intensity to ionizable sample compounds, and/or (b) by colliding and reacting with photoionized species.

Water vapor effects were not noted on the Century Systems OVA (FID) unit.

(2) H-Nu PI-101 response suppression was also noted on gasoline-vapor concentrations above 150 ppm (v/v) Total Organic Vapors (TOV) as benzene. Although apparently independent of the water vapor

condition, this phenomenon may be attributable to similar mechanisms of ultraviolet absorption and/or collisional deactivation.

(3) With both units calibrated to respond "as benzene," the H-Nu PI-101 to Century OVA (or PID/FID) response ratio ranged from 0.56 for "fresh" gasoline vapor to 1.35 for (evaporatively) weathered gasoline stock vapor. This finding is graphically displayed in Figures 4 and 5 and is consistent with detector selectivities and gasoline constituent distribution.

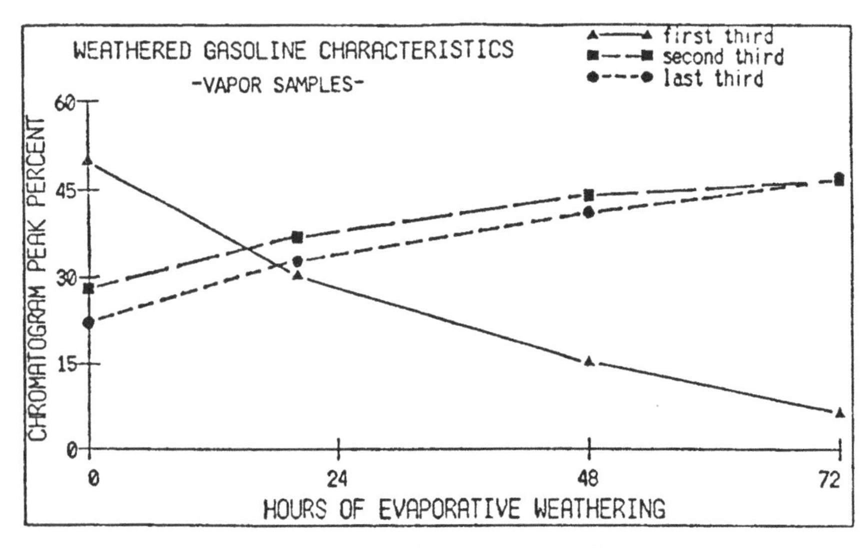

Figure 4. Weathered gasoline characteristics—vapor samples.

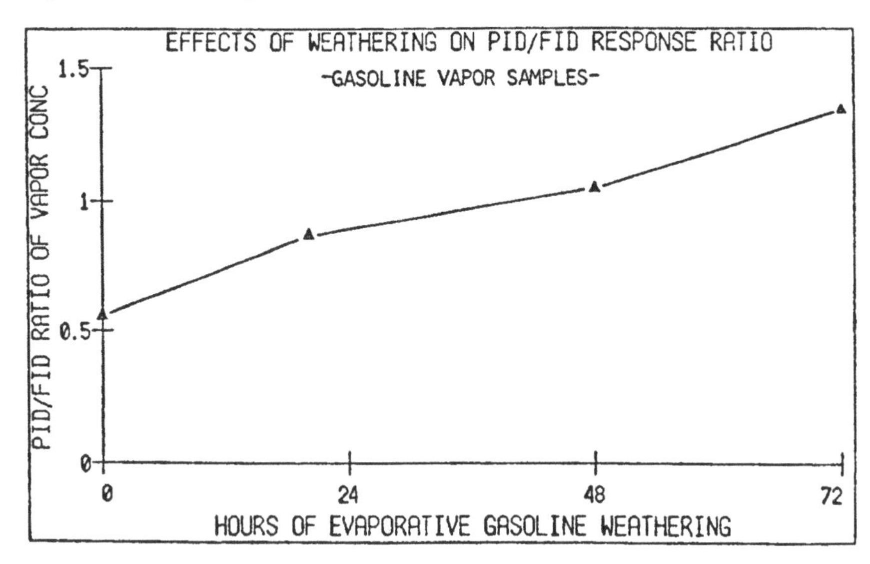

Figure 5. Effects of weathering on PID/FID response ratio.

Jar Headspace Analytical Screening Technique

(1) Using the "probe insertion" jar headspace sampling technique previously described, maximum H-Nu and Century OVA instrument response was generally obtained 2 to 5 seconds following probe insertion, as illustrated in Figures 6 and 7. This response value was recorded as the jar headspace sample concentration. Headspace concentrations above 150 ppm (v/v) TOV as benzene would on occasion produce erratic or significantly nonlinear meter responses. Accordingly, all data comparisons were limited to samples yielding less than this value.

(2) Using the "probe insertion" method, H-Nu meter response (accuracy) was observed to be approximately 50% of the true headspace vapor concentration in one-half filled 9 oz and 16 oz sample jars. OVA Model 128 response (accuracy) was approximately 65% for 16 oz jars, and 55% for 9 oz jars. Significantly lower response levels were noted in 5 oz sample jars, indicating insufficient sample headspace volume.

(3) In a comparison of 16 oz and 9 oz sampling jars subjected to the "probe insertion" technique, overall analytical screening reproducibility (precision) was superior in 16 oz jars. Using the H-Nu unit, the Relative Standard Deviation (RSD) of replicate sample headspace analyses was approximately 5% for 16 oz jars and 9% for 9 oz jars. For the OVA unit, based on a limited database, the RSD for 16 oz jars was approximately 5% to 10%, and approximately 10% to 15% for 9 oz jars. For comparative purposes, all sample jars were filled to one-half capacity.

(4) Jar agitation ("dynamic development") was generally observed to significantly increase short-term headspace development, especially in aqueous samples.

(5) The temperature of soil and water samples and the ambient temperature of headspace development did not significantly affect short-term dynamic headspace development over the range of temperatures examined. A majority of the media samples evaluated were cooled to about 12°C (to simulate ground/groundwater temperatures in Massachusetts) and developed at ambient temperatures which ranged from 20 to 25°C. Selected samples, however, were cooled to as low as 4°C and/or developed in ambient temperatures as low as −12°C. While effects on headspace development were minor, it is stressed that ambient temperature effects on the operation and response of field units may have a significantly more pronounced impact.

(6) The shape of sample jars did not significantly affect analytical results (headspace development and subsequent instrument response) of dynamically developed media samples.

(7) While a majority of experimental setups consisted of jars filled with soil or water samples to one-half capacity, a limited database obtained from selected samples indicated that short-term dynamic development results were lower (particularly in aqueous samples) in 1/4 filled jars and somewhat higher in jars filled to 3/4 depth. Due to decreased headspace volumes, however, the reproducibility (precision) of 3/4 filled jar samples were lower than in half-filled setups.

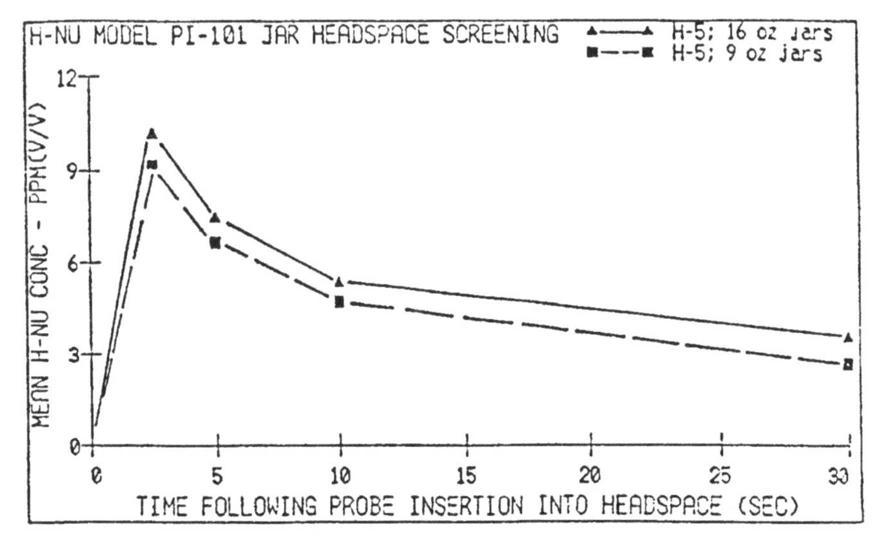

Figure 6. H-Nu Model PI-101 jar headspace screening.

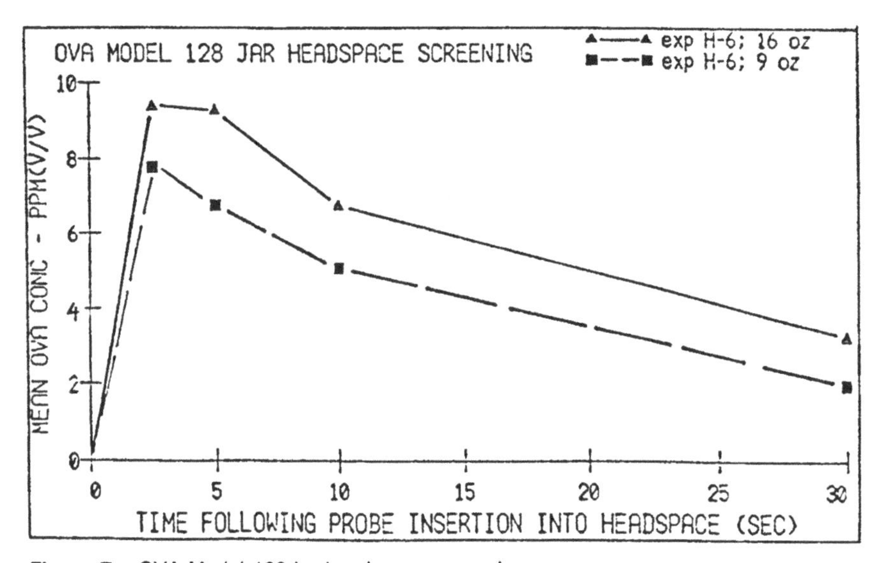

Figure 7. OVA Model 128 jar headspace screening.

Aqueous Headspace Evaluation

(1) In short-term evaluations (headspace development less than 30 min), headspace partitioning in agitated jars was one order of magnitude greater than in statically developed jars. Generally, over 90% of ultimate headspace development was achieved following 30 sec of vigorous agitation over a 5 to 10 min development period (see Figures 8 and 9).

(2) Aqueous headspace development was not significantly different between 9 oz and 16 oz sample jars.

(3) In "fresh" gasoline aqueous samples, during short-term dynamic development, benzene partitioning into the jar headspace ranged from 53% to 95% of the predicted Henry's Law (single component) equilibrium condition, with a mean value of 70% and a RSD of 18%. Total BTEX partitioning ranged from 25% to 56%, with a mean value of 40% and RSD of 24%. In (evaporatively) "weathered" aqueous stock samples, BTEX partitioning increased to 50% of the predicted equilibrium condition.

(4) Using the "probe insertion" technique, the BTEX headspace concentration comprised about 33% of the H-Nu meter TOV headspace concentration in "fresh" gasoline stock and about 50% in the "weathered" gasoline aqueous stock. The BTEX headspace concentration comprised about 20% of the Century OVA FID response in "fresh" stock and about 65% in "weathered" aqueous gasoline stock. This observation is consistent with the increased relative concentration of unsaturated ("second and last third") compounds in the more weathered samples and intrinsic PID/FID detector selectivity.

Soil Headspace Evaluation

(1) Experimental observation confirmed that headspace development and chemistry above gasoline-contaminated soils is subject to substantial variability, influenced by physical/chemical and moisture conditions of the soil and by the mechanism(s) of gasoline contamination and weathering. Due to this parameter variability and (relatively) limited scope and extent of experimentation, only preliminary conclusions may be drawn from examined data.

(2) Jar agitation significantly increased short-term headspace development over static conditions, with a 10% to 25% increase observed between 10 ppm (v/v) and 100 ppm (v/v) TOV and a 20% to 100% increase observed at concentrations less than 10 ppm (v/v) TOV.

(3) Contrary to aqueous experiments, headspace development values were consistently higher in 16 oz jars compared to 9 oz jars. Overall,

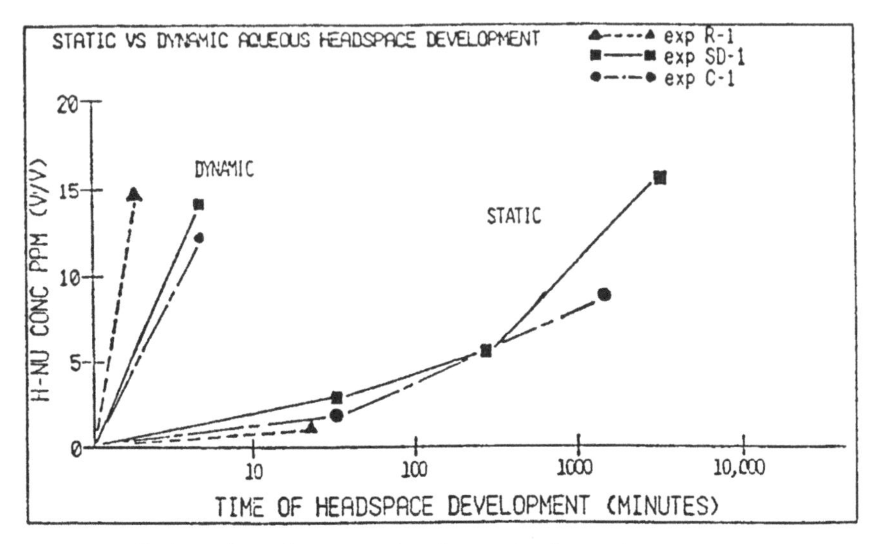

Figure 8. Static vs dynamic aqueous headspace development.

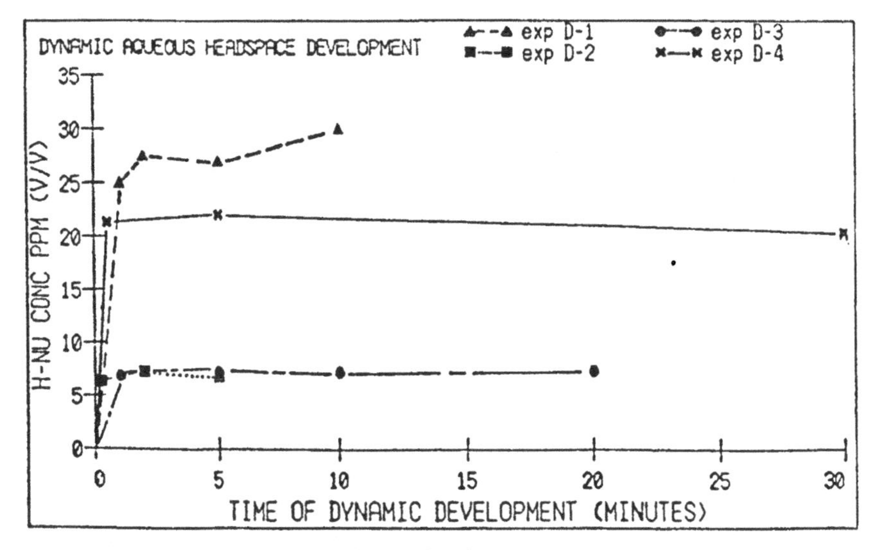

Figure 9. Dynamic aqueous headspace development.

the mean difference was approximately 14%, although differences were more pronounced at lower headspace concentrations (less than 10 ppm v/v TOV) where 16 oz jars were 20% to 50% higher than 9 oz jars.

(4) The observed relationship between "total organic vapor" PID and FID jar headspace measurements and volumetric headspace vapor

concentrations of benzene and BTEX constituents is graphically depicted in Figures 10 through 13.

(5) Based upon a limited database, short-term dynamic development of (moist) sandy soils yielded between 50% and 100% of the ultimate jar headspace development, and produced headspace benzene concentrations of 25% to 50% of the predicted equilibrium benzene condition. The total organic vapor (TOV) concentration in ppm (v/v) as benzene via the jar headspace procedure was seen to be 2 to 3 orders of magnitude greater than the soil benzene concentration in μg/g. As a preliminary conclusion, a (moist) sandy gasoline-contaminated soil sample producing a jar headspace concentration of 100 ppm (v/v) TOV as benzene would be expected to contain benzene in the bulk soil matrix in the range of 0.1 to 1.0 μg/g. Additional data is needed, however, to better define this relationship.

(6) Data analysis suggest the PID/FID response ratio may be a useful indicator of headspace and soil chemistry. In seven samples evaluated, the PID/FID response ratio "as benzene" varied from 0.42 to 2.9, the lower value representing less "weathered" gasoline contamination.

(7) Empirical observations and mathematical predictions indicate that the presence of even minute quantities of immiscible-phase gasoline within a soil sample will produce a TOV headspace concentration above 150 ppm (v/v), which is believed to be the upper limit of (relatively) linear PID response in contaminated soil samples.

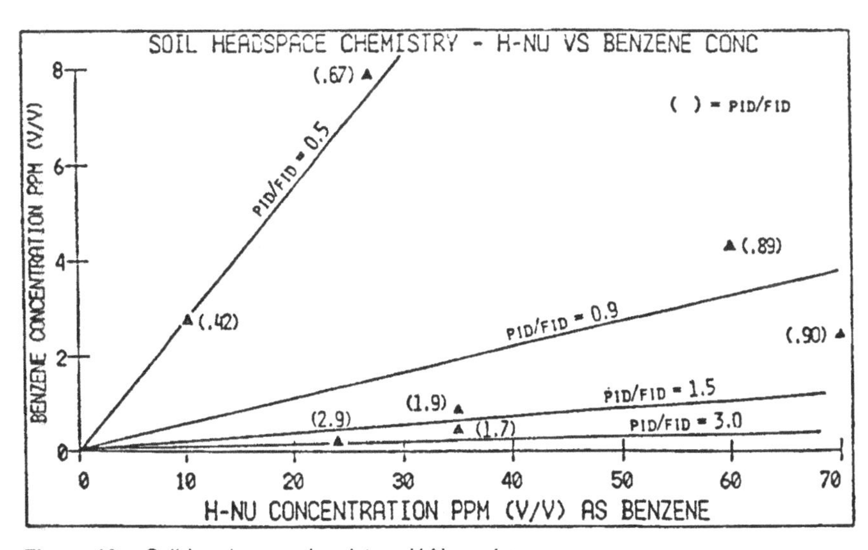

Figure 10. Soil headspace chemistry—H-Nu vs benzene.

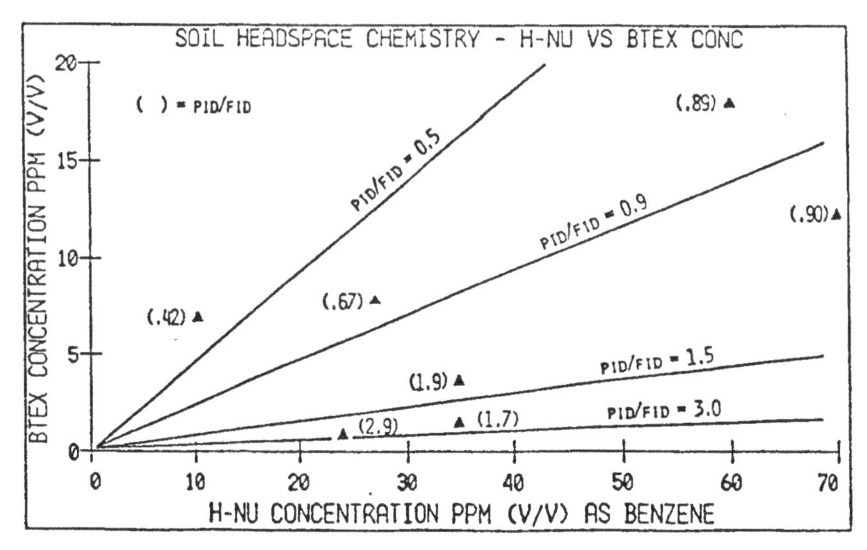

Figure 11. Soil headspace chemistry—H-Nu vs BTEX.

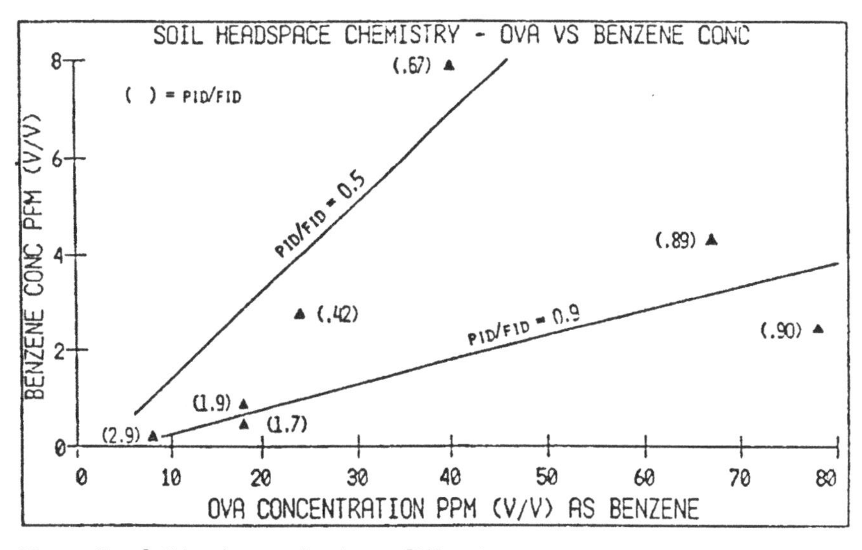

Figure 12. Soil headspace chemistry—OVA vs benzene.

CONCLUSIONS

Analytical and interpretative limitations exist in the utilization of portable PID and FID response meters. Historically, improper and/or inconsistent use of these investigative tools has perhaps limited their acceptance and field utility. Nevertheless, certain, defined applications appear possible and desirable, particularly in

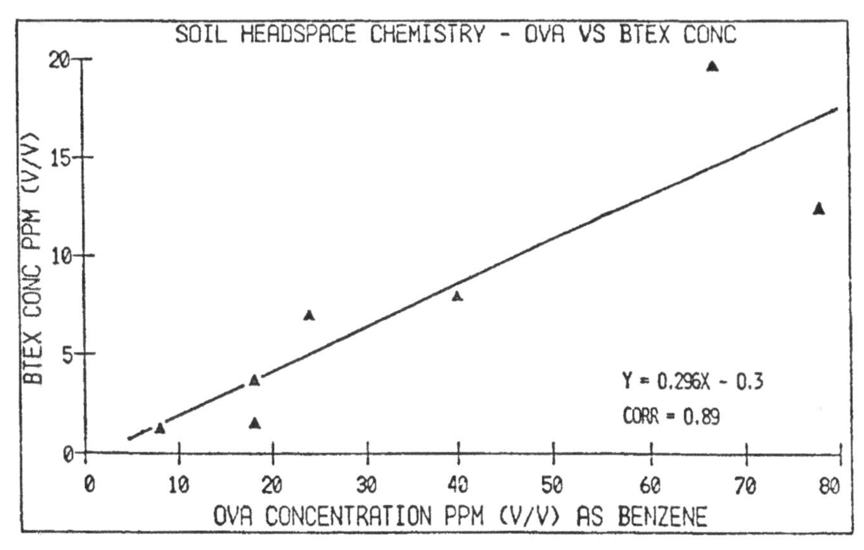

Figure 13. Soil headspace chemistry—OVA vs BTEX.

the areas of remedial response field evaluation and documentation.

Preliminary data has been obtained to aid in the interpretation of analytical headspace screening data. Additional investigation is necessary.

Based upon an evaluation of field headspace development and "probe insertion" sampling parameters, the following procedural elements are recommended:

(1) Utilize glass jars between 9 oz and 16 oz in total capacity; 16 oz jars are recommended, particularly for field instrumentation with sampling flowrates greater than 500 mL/min.
(2) Half-fill sample jars. Seal with a clean sheet of aluminum foil.
(3) Vigorously agitate sample jar for at least 30 sec over a 5 to 10 min headspace development period.
(4) Quickly insert the PID or FID sampling probe through the aluminum foil seal and record the maximum meter response as the "total organic vapor" headspace concentration. Maximum response should occur between 2 and 5 sec when using a 9 oz or 16 oz jar. Erratic meter responses should be discounted. PID measurements above 150 ppm (v/v) TOV as benzene may be significantly nonlinear.

REFERENCES

1. Price, D., and J. Chen. "Chemical-Physical Parameters and Processes Effecting Petroleum Fuel Migration," Draft Report, General Software Corporation, U.S. EPA Contract No. 68-02-3970, September 25, 1985.

2. Domask, W. G., "Introduction to Petroleum Hydrocarbons, Chemistry and Composition in Relation to Petroleum-Derived Fuels and Solvents," *Proceedings of the Workshop on Kidney Effects of Hydrocarbons*, (Boston, MA: American Petroleum Institute, 1983), pp. 5–33.
3. Lyman, W. J., W. F. Reehl, and D. H. Rosenblatt. *Handbook of Chemical Property Estimation Methods* (New York: McGraw-Hill Book Company, 1982).
4. Bruell, C. J., and G. E. Hoag. "Capillary and Packed-Column Gas Chromatography of Gasoline Hydrocarbons," in *Proceedings of Petroleum Hydrocarbons and Organic Chemicals in Groundwater*, National Water Well Association, 1984.
5. Barksy, J. B., S. S. Que Hee, and C. S. Clark. "An Evaluation of the Response of Some Portable, Direct-Reading 10.2 eV and 11.2 eV Photoionization Detectors, and a Flame Ionization Gas Chromatograph for Organic Vapors in High Humidity Atmospheres," *J. Am. Ind. Hyg. Assoc.*, 46:9–14 (1985).
6. Chelton, C. F., N. Zakraysek, G. M. Lautner, and R. G. Confer. "Evaluation of the Performance of the Bacharach TLV Sniffer and H-Nu Photoionization Gas Analyzer to Common Hydrocarbon Solvents," *J. Am. Ind. Hyg. Assoc.* 44:710–715.

CHAPTER 5

Relationships Between Chemical Screening Methodologies for Petroleum Contaminated Soils: Theory and Practice

S. A. Denahan, B. J. Denahan, W. G. Elliott, W. A. Tucker, M. G. Winslow, Environmental Science & Engineering, Inc., Gainesville, Florida
S. R. Boyes, GeoSolutions Inc., Gainesville, Florida

The contamination of the environment by petroleum products from surface spills and leaking underground storage tanks (USTs) is a leading environmental problem today. Identification, assessment, and remediation of petroleum contaminated sites is a major task facing civilized society. Thousands of these sites exist in the country. Traditional assessment technology for the definition of the spatial extent of contamination involves the collection and laboratory analysis of samples of the various environmental media. The economic pressures engendered by the extremely large number of petroleum sites to be dealt with has spawned the development of several screening methodologies, which can provide a cost-effective alternative to the traditional sampling and laboratory analysis.

This chapter discusses one of the most useful of these methodologies, the measurement of vapor phase petroleum compound concentrations in the vadose zone. Because most petroleum products have a significant component of volatile compounds in their makeup, they are well suited to vapor phase analysis. Whenever petroleum products are released into the environment a vapor phase component will be present. Most commonly, a three phase equilibrium is established, with petroleum compounds present as vapor phase, liquid phase, and also adsorbed to the solid phase, mainly organic soil components. Several different methodologies have been employed to sample and analyze these vapor phase components. These methodologies can be divided into two broad categories: in situ methods and head space methods.

The in situ methods provide a mechanism for collecting a sample of the vapor phase components in situ without the need for the collection of a soil sample. There are two primary variations of this technique (active and passive), with the principal differences being how the sample of the vapor phase material is obtained. In the active technique, a sample of vapor is extracted from the vadose zone by means of a hollow probe and a vacuum pump. In the passive technique, a collection device containing an adsorbent medium is buried in the soil for a specified period of time and then exhumed with whatever vapor phase components have been adsorbed. The passive technique has several apparent disadvantages in that it requires two and possibly more field efforts, and the chemical data are not acquired in real time. Additionally the passive technique relies on the natural flux of vapor phase components to move to the collector, and the rate of this natural movement is often very slow and also variable, depending on soil permeability, moisture content, temperature, and atmospheric pressure. Only active methods will be discussed in this chapter.

The head space methodologies require the collection of a soil sample from the zone of suspected contamination. This sample is then placed into a closed container (various types are used) and the vapor phase components concentration in the head space above the soil is measured directly using a portable instrument. Certain states have mandated the use of this methodology for the field determination of the extent of petroleum contaminated soils.

INSTRUMENTATION

Two principal types of portable analytical instruments are commercially available for the analysis of the vapor phase components of petroleum contaminated soils, flame ionization detectors (FIDs), and photo ionization detectors (PIDs). There are significant differences between the responses of these two types of instruments. These differences and the practical implications of them are one of the principal topics of discussion in this chapter.

Flame Ionization Detectors (FID)

The Flame Ionization Detector (FID) in today's commercially available portable screening instruments is a nonspecific detector designed to measure total organic compounds present in the vapor phase. Vapor phase components are drawn through a probe into an ionization chamber containing an oxygen (air)-hydrogen flame where chem-ionization of organic molecules occurs:

$$CH + O- > CHO^+ + e^-$$

The ions and electrons pass between electrodes to which a voltage is applied, decreasing the resistance and causing a current to flow in the external circuit. The FID should respond to all organic molecules containing an ionizable carbon atom. Clean air components, H_2O and CS_2 should yield no effective response. The FID's response to different organic molecules differs only in the total number of ionizable carbon atoms in the molecules. A substituent (e.g., Cl) on an individual carbon atom in a molecule generally reduces the ionization compared to hydrogen. When measuring petroleum hydrocarbon contamination in the vapor phase, the detector response will, in general, increase with increasing carbon number. The detector response should also be approximately the same for alkanes, alkenes, and aromatics with the same number of effective (ionizable) carbon atoms. The screening FID responds quantitatively only to the individual compound to which it is calibrated. Response to an unknown compound or mixture of compounds, as in the case of petroleum hydrocarbons, can only be nonqualitative and semiquantitative in nature.

Photoionization Detectors (PID)

The Photoionization Detectors (PID) in today's commercially available portable instruments that are designed to screen vapor phase contaminants is a nonspecific detector that measures total ionizable compounds, both organic and inorganic. Unlike the FID, the PID is a nondestructive detector (i.e., molecules exit the detector chemically unaltered from their original state). Vapor phase components are drawn through a probe into an ionization chamber which is separated by a window from an ultraviolet (UV) lamp. The UV lamp emits photons with a specific energy (several interchangeable lamps of varying intensities are available for most commercial instruments). The photons are transmitted through the window into the ionization chamber. Molecules that have lower ionization potentials than the energy of the radiated photons will absorb a photon and become ionized:

$$AB + \text{photon}- > AB^+ + e^-$$

The ionization chamber also contains two electrodes to which a voltage is applied. The ionized molecules and electrons flow into the electric field and the resulting increase in current is proportional to the concentration of ionized molecules in the vapor phase. The PID does not respond to the components of clean air or H_2O because they have ionization potentials higher than the energy of any commercially available UV lamp (8.4, 9.5, 10.2, 10.6, and 11.7 eV). Standard lamps that are most commonly used in field screening (10.2 and 10.6 eV) of petroleum hydrocarbon contaminants will respond well to numerous inorganic gases such as CS_2, H_2S, pH_3, and NH_3. All compounds manifest a unique ionization potential, but all ionizable compounds are not detected by the PID equivalently. In general, the lower the ionization potential of the compound, the greater the PID's response to it. Within a homologous series, the detector response will generally increase with increasing carbon number and is different between homologous series. Although most petroleum hydrocarbon compounds will be detected by the PID equipped with a standard lamp (10.2 or 10.6 eV), the response increases with increasing unsaturation of the molecule; benzene > cyclohexene > cyclohexane and for different types of petroleum hydrocarbons sensitivity increases as follows: aromatics > alkenes > alkanes, and cyclic compounds > noncyclic compounds. Like the FID, the screening PID responds quantitatively only to the individual compound to which it is calibrated. Response to an unknown or unknowns can only be nonqualitative and semiquantitative.

FIELD SAMPLE COLLECTION PROCEDURES

This section describes the methods used to collect field data using various detection techniques. The spacing between sample stations should be chosen with consideration of the potential size of the contaminated zone to be detected. The interval between soil gas stations can be greater for potentially larger contaminated areas and conversely smaller for potentially small areas.

In Situ Soil Gas Measurements

To obtain in situ samples from the vadose zone, a soil gas probe, consisting of a hollow steel tube, is driven into the soils to a desired depth. The soil gas probe is evacuated, and a representative sample is collected. The desired depth of the soil gas probe should be determined based on the physical soil characteristics of the study area. The tip of the soil gas probe should be driven into the soil far enough to isolate the sampling port from ambient air. In sandy soils the sand tends to close around the soil gas probe. In silty or clayey soils the hole will tend to remain open; therefore, care must be used to assure a tight seal. All soil gas probes for a given study should be driven to approximately the same depth below ground surface. A depth of 3 to 5 ft. below ground surface is generally sufficient. The probe can be driven with a slide hammer or a pneumatic hammer.

At each sample location, a soil gas probe constructed of 0.5 in. diameter galvanized steel tubing should be driven to its sampling depth and capped. For sampling, the cap is removed, and the soil gas probe is attached by Teflon® tubing to a 1 liter Tedlar® purge bag. The purge bag is placed inside a laboratory desiccator; a vacuum pump is then connected to the desiccator and used to evacuate the space inside the desiccator surrounding the bag. This negative pressure causes the soil gas to be drawn into the purge bag as it fills the evacuated volume. Prior to sampling, approximately two liters of gas should be evacuated to purge the ambient air from the sampling train and to draw in gases from the soil. One bag should be dedicated to purging operations.

After purging the sampling train, a new labelled bag is placed in the desiccator and used to collect the sample for analysis. Once the sample is collected, the sample bag is removed from the desiccator, and the contents are analyzed using one of several techniques previously described. If more than one method of analysis is employed, a single sample bag of gas should be used to supply soil gas to all instruments for a given sampling location. In this way, all data from a specific soil gas probe will be generated from a single sample of a specific concentration.

Gas to be analyzed by field gas chromatograph (GC) is removed from the sample bag using a gas-tight syringe and injected into the GC for quantification. Tygon® tubing is used to convey the sample from the capture bag to the other field instruments such as the FID or PID. This is done by connecting the Tygon tubing directly from the sample bag to the field instrument and allowing the instrument to draw in the soil gas. The instrument reading is then entered in the field notebook.

Dynamic trapping, another method of sampling soil gases for analysis, relies on adsorption of the contaminant on a medium of activated charcoal. Prior to sampling, each charcoal adsorption tube is thermally desorbed and stored in double-walled containers to prevent contamination. The soil gas probes are installed and evacuated. Sampling is accomplished by connecting a thermal desorption tube in-line to the soil gas probe and vacuum pump. Soil gas is evacuated through the tube for 15 minutes at a rate of 100 cubic centimeters per minute, for a total of 1500 cubic centimeters of gas sampled. Quality assurance samples may be collected by pumping ambient air through the thermal desorption tubes. After sampling is completed, each tube is sealed in a double-walled container, transported to the laboratory and thermally desorbed, then analyzed using EPA Method 602 for volatile aromatic compounds.

Head Space Measurements

Head space measurements are obtained by collecting a soil sample and placing it in a sealable container without aerating the sample. The container is then sealed and the temperature of the contents allowed to equilibrate for a minimum of five minutes. After the sample equilibrates, the gases in the headspace of the container are analyzed. If a GC is used, the sample can be removed from the container using a syringe and injected into the GC for analysis. If a FID or PID is used, consideration should be given to the technique used to remove the sample from the container, the type of container used (rigid or flexible) and the instrument being used (destructive or nondestructive). If a rigid container is used and the sampling port of the instrument is inserted into the sample container, only the initial moments of the instrument reading will reflect the true concentration of the contaminants in the soil gas. This occurs because as the instrument pumps the gases out of the container the sample is diluted.

There are several ways to avoid compromising the analytical results. If a nondestructive instrument such as a photoionization detector (PID) is used, the soil gas can be recirculated back into the sample container by use of the vent port. A Tygon® tube can be attached to the vent port of the instrument and inserted back into the container. The sample will enter the sampling port, circulate past the ionizing lamp and out the vent port back into the jar. This method does not introduce air into the system, nor does it destroy the sample. If the field instrument uses a flame ionization detector (FID), the sample is destroyed, and recirculating the gases through the instrument will not provide satisfactory results. In this case, a flexible container should be used. The container will collapse as the gas is removed, maintaining the integrity of the sample.

Ambient Screening

Ambient screening is the simplest method of collecting soil gas data on soils as they are obtained from a bore hole. Utilizing this method provides real time information concerning vertical distribution of the contaminant in the vadose zone. Although this method is very quick and inexpensive it often can prove difficult to exactly reproduce the results. The screening is usually conducted as follows, as soils are removed from the bore hole, a field screening instrument such as a PID is moved along the soil sample. Any positive response above the background is recorded, along with the sampling depth interval. To increase the likelihood of detecting any contaminants, the soil sample is sometimes scored or split open to increase the rate of release of volatile soil gases. This screening must be completed immediately after the soils are obtained to assure the best possible results.

Several environmental factors, given below, influence the results of this screening technique and should be considered while collecting and interpreting data:

1. If soil temperatures have not equilibrated, variations in ambient temperatures will influence the results.
2. Variations in wind speed can skew the data; obviously high ambient wind speeds will tend to disperse the volatiles coming from the soil samples.
3. Sample moisture content will influence the response of some instruments.

Collection of Samples for Laboratory Analysis

The collection of representative samples for laboratory analysis of volatile compounds requires preparation and planning prior to sampling. Care should be taken throughout the sampling effort to minimize aeration of the soils. Sampling for volatile compounds consists of taking an undisturbed sample, and quickly isolating the sample from the ambient conditions. The sample should be sealed into a container with minimum or no headspace, chilled, and taken to a laboratory for analysis. This often presents a problem in the field, as packing a soil sample into a rigid container with no headspace requires considerable manipulation of the soil, with the resulting loss of volatiles.

THEORETICAL CONSIDERATIONS

Uncontaminated soil is a three phase system of solids, liquid, and air. Soil solids are predominantly minerals such as silica or calcite that do not interact chemically with petroleum hydrocarbons but also include organic matter, a phase that preferentially absorbs petroleum hydrocarbons. As a result, it is sometimes useful to think of soils as comprised of an inert mineral phase, organic matter, water, and air.

Units of Measurements

In the assessment of multimedia contaminant fate and transport, confusion regarding the units of measurement and appropriate methods of conversion to permit intercomparison of contaminant concentrations occasionally obscures more substantial phenomena. The following discussion is provided to clarify such issues and to facilitate communication and understanding.

Contaminant concentrations in soils are most commonly expressed on a mass/mass basis. In laboratory analysis for volatiles, a soil sample at "field moisture" is weighed and then extracted to determine the total mass of contaminant that had been present in soil, liquid, and vapor phases at the time of collection. A corollary portion from the same sample is weighed before and after drying to determine its moisture content. Then the total mass of contaminant that had been present in the bulk sample (all three phases) is referenced to the mass of a dry soil (bulk soil minus water).

On a mass/mass basis this concentration may be expressed as $\mu g/g$ or mg/kg, or ppm by weight.

Contaminant concentrations in water are commonly reported on a mass/volume basis, as in mg/L. The metric mass and volume scales are linked via the specific gravity of water: 1 L of water has a mass of approximately 1 kg. Consequently, the concentration expressed in mg/L, is numerically equivalent to a concentration in mg/kg, so either of these units may be referred to as ppm.

Furthermore, the bulk density of most soils is between 1.5 and 2.0 kg/L; not so different from the specific gravity of water. As a result it is not too misleading to compare soil concentrations and water concentrations without rigorously considering the difference between mass/mass versus mass/volume as routinely reported.

This comfortable situation does not prevail when considering soil gas concentration data. The standard reporting convention for vapor phase concentrations is in volume/volume, which is precisely equivalent to moles/mole based on the Ideal Gas Law. One ppm of a contaminant in air would represent, for example, one $\mu L/L$ or one micromole of contaminant per mole of air.

Air is predominantly a mixture of nitrogen (79%) and oxygen (21%), and behaves like a gas with a molecular weight of 29. (This is similar to the determination of the atomic weight of an element based on the fractional abundance of its isotopes.) A mole of a gas at standard temperature and pressure occupies a volume of 22.4 liters. As a result the density of air is 29 grams per 22400 milliliters or

1.29×10^{-3} gm/mL: air is approximately 800 times lighter than water and more than 1000 times lighter than bulk soil.

To convert from contaminant concentrations in air, routinely expressed as ppm by volume, to a mass per volume basis, one must consider the molecular weight of the contaminant. Benzene has a molecular weight of 78. At STP, then pure benzene vapor has a density of 3.5×10^{-3} gm/mL (78/22400). If the concentration of benzene in air is 1 ppm, then 1 L of air contains 3.5 μg of benzene. By contrast, if 1 L of water contained 3.5 μg of benzene, the concentration might be expressed as 3.5 ppb.

Equilibrium Partitioning in the Three Phase System

Organic contaminants are readily exchanged between soil phases, and tend to partition between phases according to the chemical potential gradients across phase boundaries. Water and air move relatively slowly through soil pores, allowing time for contaminants to partition between phases, establishing geochemical equilibrium. There are complications in predicting or estimating what this equilibrium distribution will be in real soils: organic compounds dissolved in the soil solution such as humic and fulvic acids alter the nature of the aqueous solution that is often simplistically assumed to be pure water. The solid phase conceptualized as soil organic carbon is very complex and may differ from one soil to the next. Nonetheless, a simplified model has proven to be useful in soil contaminant data interpretation and transport modeling. This model, presented concisely by Jury et al.[1] and applied to petroleum hydrocarbons by Tucker et al.,[2] considers soil to consist, geochemically, of water, air, and organic carbon. The contaminant is assumed to distribute itself among these phases so as to establish chemical equilibrium. Furthermore, this equilibrium condition is assumed to be the same in all soils. The relative amounts of contaminants sorbed to solids, dissolved in water, or in vapor form depends solely on the contaminant physical chemical properties and the relative amounts of each of these phases in the particular soils.

This equilibrium condition is well defined based on soil organic carbon content, moisture content, bulk density or porosity, and the organic carbon adsorption coefficient and Henry's law constant of the contaminant. With knowledge of the contaminant concentration in bulk soil, or any single phase such as soil gas, the contaminant concentration in other phases can be estimated. The simplified model is expressed as follows:

$$C_s = KC_w \quad (1)$$

$$C_a = HC_w \quad (2)$$

where:

C_s is the concentration of contaminant on solid phases (mass/mass, dry weight)
K is the soil adsorption coefficient (vol/mass)
Cw is the concentration in the soil moisture (mass/vol)
Ca is the concentration in the soil air (mass/vol) and H is the Henry's law constant expressed in dimensionless concentration units.

The soil adsorption coefficient is given by K = Koc foc where Koc is the partition coefficient between soil organic carbon and water (vol/mass) and foc is the organic carbon content of the soil (mass/mass).

Based on this equilibrium partitioning relationship, the concentration in the bulk soil is comprised of contribution from different phases. Concentration in bulk soils are referenced to the dry mass of the soil. As shown by Jury et al.,[1] the concentration in bulk soil can be related directly to the concentration in any single phase. For this application the relationship to soil air is most relevant.

$$C_{st} = (Kocfoc/H + nw/\ bH + na/\ b)\ Ca \tag{3}$$

where:

$$nw + na = n \tag{4}$$

nw is the water filled porosity (vol/vol)
na is the air filled porosity (vol/vol)
n is the total porosity (vol/vol)
b is the soil bulk density (dry)
C_{st} is the contaminant concentration in bulk soil according to the standard reporting convention, i.e. referenced to dry weight.

A few other standard soils relationships assist in application of this equation;

$$b\ (1 - n)\ 2.65 \tag{5}$$

assuming the solid phase is predominantly silica or calcium carbonate. If soil moisture 0 is expressed as mass of moisture divided by dry mass of soil, then

$$nw = 0\ b \tag{6}$$

An example of how this equation may be applied is as follows. A soil gas sample has been taken from a sandy soil with a moisture content, 0 = 8%, foc = 0.6%, and b = 1.59.

These soil properties imply

$$n = 1 - \frac{1.59}{2.65} \quad \text{from Equation 5}$$
$$= 0.40$$
$$nw = 0.13 \quad \text{from Equation 6}$$
$$na = 0.27 \quad \text{from Equation 4}$$

Assume the soil gas contained 500 ppm, benzene. Referring to the section on Units of Measurement, the benzene vapor concentration, is 1.75 mg/L. Assume Koc is 100 and H is 0.235. We would, consequently, expect bulk soil sample from the same location to contain

C_{st} = [100(0.006)/0.235 + 0.13/1.59 (0.235) + 0.27/1.59] 1.75 mg/L = 5.4 mg/kg = 5.4 ppm = benzene by substituting the corresponding value in Equation 3.

Although a direct relationship between soil gas and bulk soil concentration is expected, the numerical relationship is complex and not a direct comparison. Soil properties and contaminant properties will affect the comparison. Generally soil concentrations of petroleum hydrocarbons, expressed in ppm (mass/mass) will be substantially less than soil gas concentrations in ppm (volume/volume).

Aging of Petroleum Products in the Environment

Petroleum products are complex mixtures of many petroleum hydrocarbons. Specific products tend to predominantly contain compounds with similar boiling points, as dictated by the distillation processes used in refining. Upon release of a petroleum product to the environment, certain fractions or groupings of similar compounds may quickly be lost via volatilization, dissolution/leaching, or biodegradation. This leaves behind a residue of more persistent fractions of the original mixture, or weathered product. A very important fate process affecting many refined products is volatilization. In fact, the prominence of this process accounts for the success of the soil gas monitoring techniques.

For example, a representative gasoline mixture may contain 70% paraffinic compounds and 30% aromatics. Over time the proportion of aromatics may increase since the lower molecular weight paraffins are extremely volatile and may not persist. The weathering proceeds, the aromatics and higher molecular weight paraffinic constituents are expected to persist and increase in fractional abundance relative to their contribution to fresh product. The weathering process is very complex and not readily predictable. Nonetheless, the interpretation of vapor monitoring must recognize that as product weathers in the ground, the vapor "signal" may fade, and the relationship between soil gas and bulk soil contaminant concentration may steadily change. Furthermore, monitoring devices that have differential sensitivities to compounds of different molecular weight or structure (aromatic versus paraffinic) may exhibit variable relative sensitivities to fresh product versus weathered product.

THE REAL WORLD: WHAT WORKS AND WHAT DOESN'T

FID vs PID Detectors

In this section we will present some real data collected in the course of site assessment and remediation activities at several sites in Florida. We will examine the relationships among the various types of data and explain those relationships in terms of the theory presented previously.

Table 1. Comparison of Field VOC Detection Methods for Soils Containing Adsorbed Gasoline Surrounding Subsurface Storage Tanks

		Central Florida Site 1		
ID	**PID (ppm)**	**OVA (ppm)**	**GC (ppm)**	**Method SW-846 8020**
26	471	>1000	3477	
27	187	300	245	
28	65	86	101	
29	501	>1000	2869	
30	490	>1000	10400	
31	547	>1000	10227	
32	469	>1000	6101	
33	458	>1000	13370	BTEX = <20 ug/kg
		North Florida Site		
ID	**PID (ppm)**	**OVA (ppm)**	**GC (ppm)**	**Method SW-846 8020**
2	122	160	320	
3	443	>1000	11630	
4	389	>1000	18800	
6	411	>1000	9297	
7	380	680	800	
8	460	>1000	6497	BTEX = 336 ug/kg
9	4	5	4.1	
10	280	400	322	
11	9	6		
		Central Florida Site 2		
ID	**PID (ppm)**		**GC (ppm)**	
1	1.6		0.6	
2	47		39.8	
3	1.3		0.3	
4	528		10480	
7	41		38	
8	71		110	
9	6.7		3.5	
10	1.8		0.4	
11	68		54	
12	131		151	
13	432		13539	
14	140		513	

PID = Photovac TIP®—calibrated to 100 ppm isobutylene.
OVA = Foxburough OVA®—calibrated to 100 ppm methane.
GC = FID & PID serial detectors, GC run isothermally at 75°C and calibrated with 100 ppm methane.
Method SW-846 8020 = Laboratory Analysis.

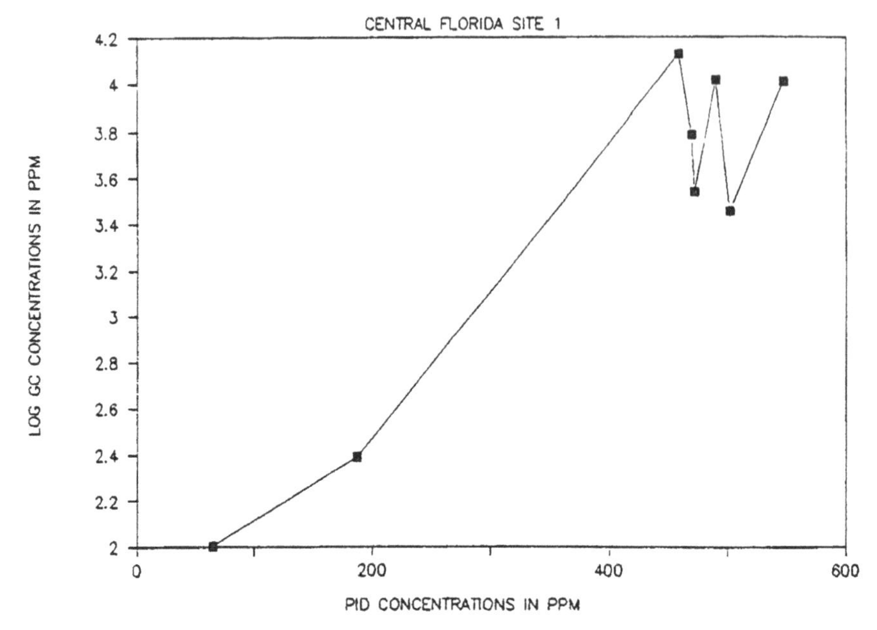

Figure 1. PID vs LOG GC concentrations (central Florida site 1).

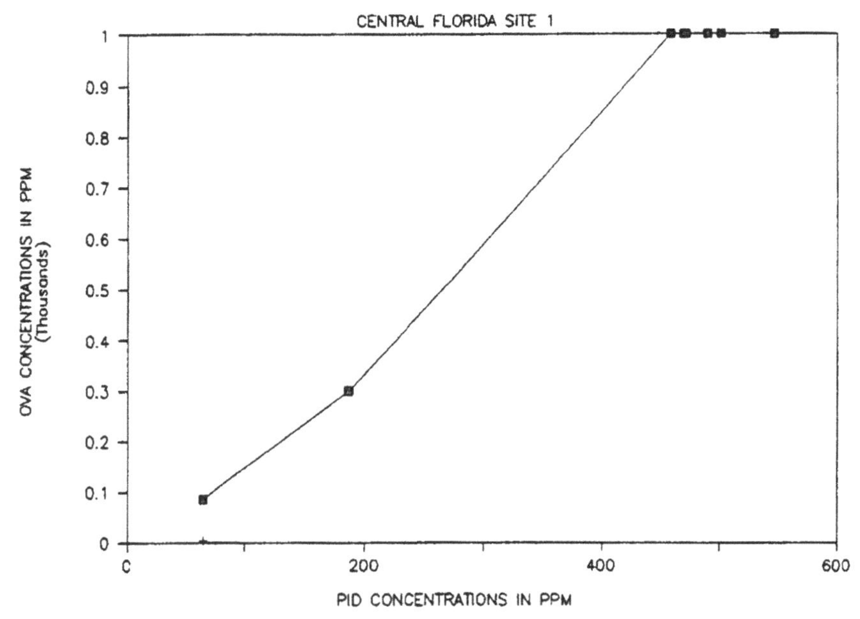

Figure 2. PID vs OVA concentrations (central Florida site 1).

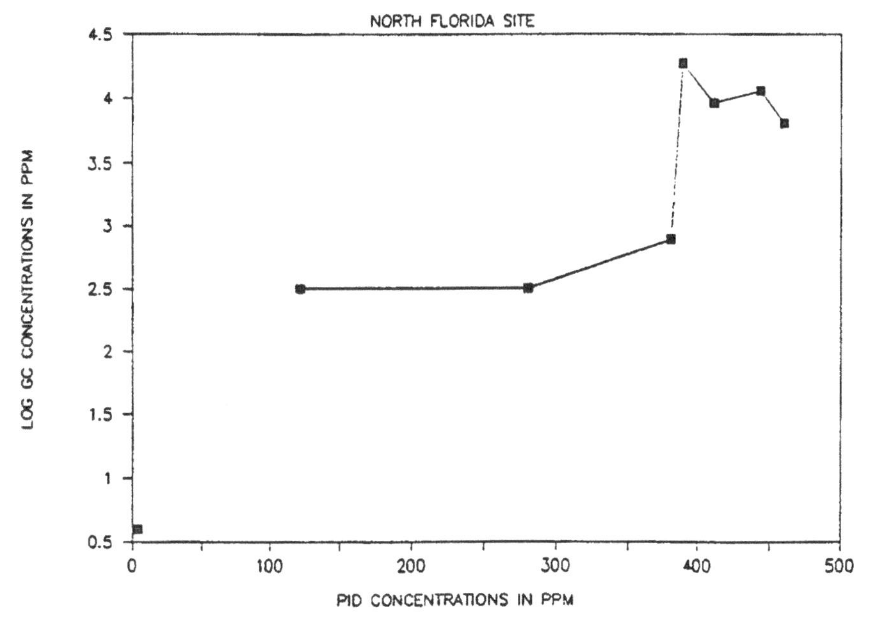

Figure 3. PID vs LOG GC concentrations (north Florida site).

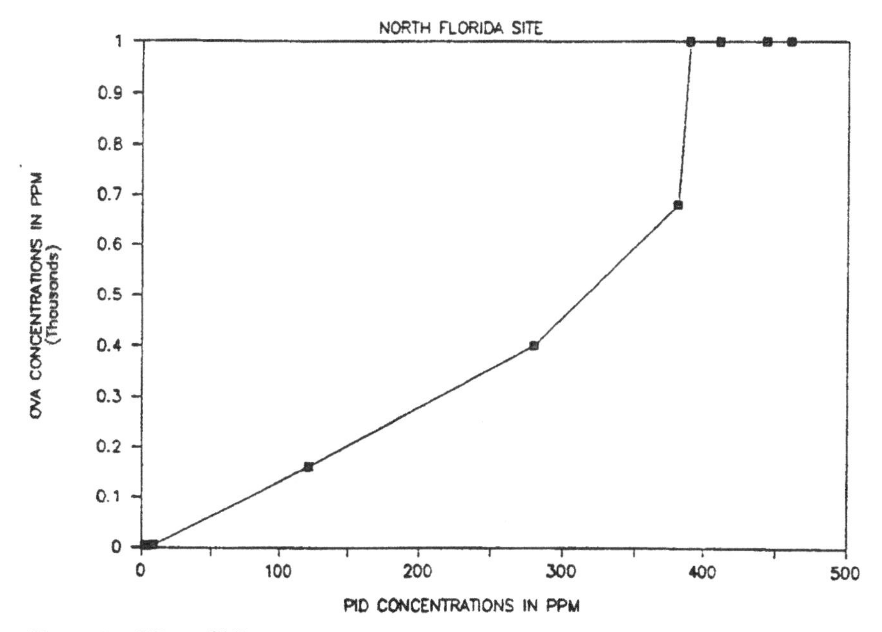

Figure 4. PID vs OVA concentrations (north Florida site).

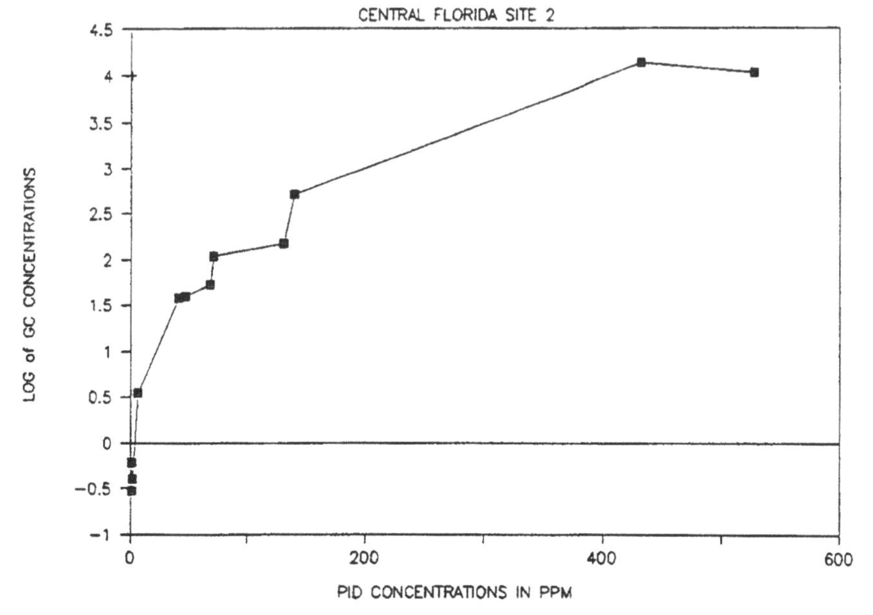

Figure 5. PID vs LOG GC concentrations (central Florida site 2).

Table 1 and Figures 1 through 5 provide a comparison of soil quality data developed during the excavation of contaminated soils from three former gas stations in Florida. The samples were collected following the removal of underground gasoline storage tanks at each of the facilities. The table compares the results of three different field methods (Photovac TIP I, a PID, Foxborough OVA, an FID and gas chromatography, using a PID and FID in series) used to quantify relative total contaminant levels in each sample. Two of the samples subsequently were analyzed for benzene, toluene, ethylbenzene, and xylenes (BTEX) by EPA method SW-846-8020 in the laboratory.

The Foxborough OVA (FID) was calibrated to 100 ppm methane, and the Photovac TIP I (PID) was calibrated to 100 ppm isobutylene. Both were calibrated in accordance with the methods outlined in their respective operating manuals.

The gas chromatograph was run isothermally at 75°C and calibrated to the peak height of a 100-ppm methane standard. Duplicates and controls were run for at least 10% of the analyzed samples. Calibration was verified after analysis was completed. Serial detection by both flame- and photo-ionization detectors on the GC allowed for the exclusion of methane from the calculated results.

Sandy soils containing adsorbed gasoline were excavated from around the subsurface tanks. The source of hydrocarbons detected in the material originated from tank overfill that had occurred over a period of more than 20 years. The adsorbed hydrocarbons in the material had been subjected to weathering by volatilization, biodegradation, and solution into infiltrating water. The samples were collected and placed in clean, 16-ounce, wide-mouth glass jars that were capped with an aluminum foil septum. The samples' temperature was reduced to approximately 20 × C. After approximately five minutes, the septum was perforated, allowing sampling of the head space by the three methods (GC, OVA, and PID) to determine relative concentrations of vapor phase hydrocarbons.

The results presented in Table 1 indicate variability between methods of field analysis. The variation is attributable to the detection method and the relative suite of contaminants (weathered, adsorbed gasoline) present in the samples. Each method appears to indicate a relative difference in concentration between samples; therefore, each can be used in the field to discriminate successfully between soils that contain adsorbed hydrocarbons and those that do not. Inspection of Figures 1 through 5 shows that for each site the variation between the responses of the FID and PID instrument is systematic. Considering the calibration procedures employed, one would have to expect nonequivalence of response. The FID was calibrated to methane, a single carbon compound to which the detector is not very sensitive. The PID, on the other hand, was calibrated to isobutylene, a compound which is about in the middle of the sensitivity range for the instrument. This calibration procedure has the effect of making the FID significantly more sensitive than the PID, even though the PID is generally more sensitive to petroleum compounds.

Additional factors to consider when choosing the PID or FID for screening include the portability of the instrument. The PID and FID are both portable, with rechargeable nickel-cadmium batteries powered with A/C power options. However, the FID utilizes a hydrogen flame detector running from compressed hydrogen gas; if the FID is required for extended field efforts, potentially hazardous compressed hydrogen gas must be kept onsite.

The ability of an instrument to detect methane is another consideration. The FID has the ability to detect methane in soil gas vapor, while the PID does not. Since methane occurs in both gasoline and natural soils, the possibility exists that misleading soil gas "hits" actually attributable to natural conditions may be measured using the FID.

The FID's destructive analysis of samples is a factor which is undesirable where repeat analysis of a sample or simultaneous analysis by multiple techniques is required. In these instances a PID would be the preferred instrumentation.

Soil Gas Methods vs Direct Chemical Analysis

Both the in situ and headspace soil gas screening methods are important screening tools for assessing soil contamination quickly and inexpensively. Both methods also permit onsite analytical results, allowing continuous adjustment of the scope and focus of the investigation while it is ongoing. However, a measured soil gas vapor concentration typically is not interpreted to represent the true analytical concentration of petroleum in soils.

The question of whether the soil gas techniques are as effective as direct chemical analysis of soil samples at detecting low levels of soil petroleum contamination was investigated by comparing headspace and/or in situ measurements for specific samples/locations with results of chemical analyses on selected soil samples.

The data presented in Table 1 and Figure 6 indicate that the correlation between soil gas screening and direct chemical analysis does exist, but is not a direct relationship; this was expected due to the differences in instrument response, sensitivity, calibration procedures, environmental conditions, and other factors discussed earlier which influence the measured soil gas vapor concentration. The relationship between soil gas concentration and laboratory analytical concentration appears to have a lower threshold, as shown in Figure 6. Soil gas headspace concentrations measured at around 300 ppm or less were below the EPA Method 8020 total VOA (TVOA) detection limit for soil samples analyzed. When soil gas vapor concentrations are above around 300 ppm, the direct chemical analysis also detected TVOAs.

This situation indicates that soil gas techniques are very good at detecting petroleum components in the vapor phase at low concentrations, below the saturation

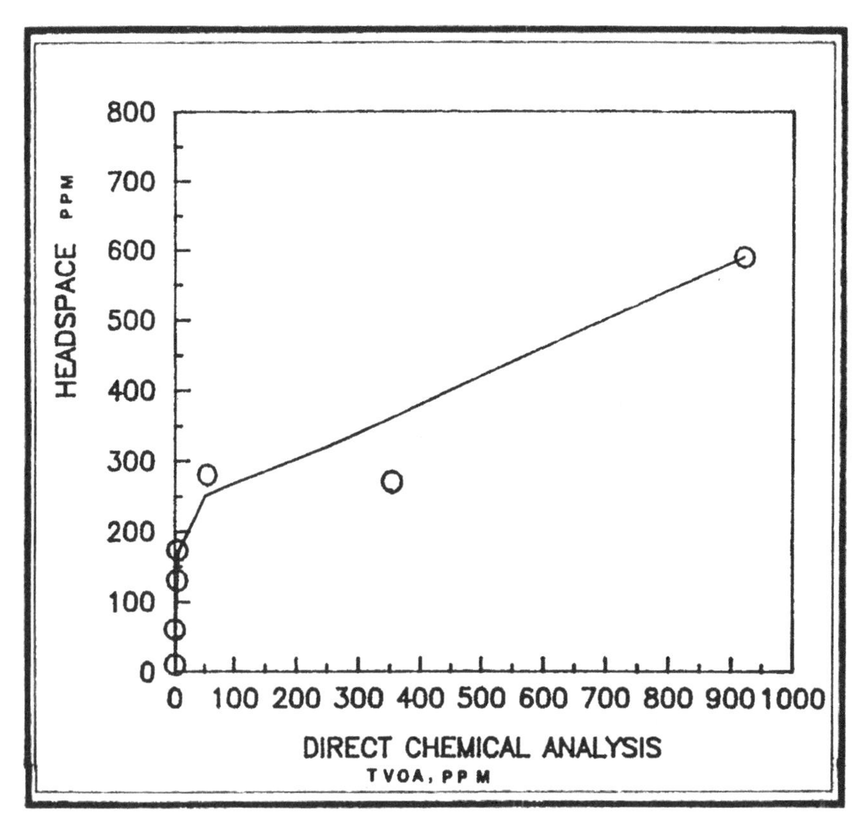

Figure 6. Correlation between headspace and direct chemical analysis.

point required for residual liquid to be detected by direct chemical extraction and analysis. When volatile petroleum components are present as vapor in the soil pores, soil gas techniques can be used effectively to track this vapor plume back to the spill or leak site (source). If the spill has reached phase equilibrium, the soil gas plume may extend well beyond the zone of soil saturation.

For this application, soil gas techniques prove to be more sensitive and much faster than direct laboratory soil analysis for determining whether petroleum products are present.

The Florida Case

The Florida Department of Environmental Regulation (FDER) has incorporated into its regulations a headspace screening procedure. The intent of this is to provide a quick procedure which may be used in the field to determine the extent of severe petroleum contamination in soils. The definition of "excessively contaminated soil" in the regulation is specified as 500 ppm. The procedure is described only briefly in the regulation and has required clarification through the issuance of a guidance document. Calibration procedures are not specified in either the regulation or the subsequent guidance, other than to follow manufacturers' recommendations. The regulation recommends the use of an FID instrument, but allows the use of a PID if the response can be demonstrated to be equivalent. As we have seen from the preceding discussion, instrument response is dependent on many factors, and the variety of equivalences should surprise no one.

REFERENCES

1. Jury, W. A., W. F. Spencer, and W. J. Farmer. "Behavior Assessment Model for Trace Organics in Soil: I. Model Description," *Journal of Environmental Quality,* 12(4):558–564 (1983).
2. Tucker, W. A., C. T. Huang, J. M. Bral, and R. E. Dickinson. "Development and Validation of the Underground Leak Transport Assessment Model," *Proceedings of Petroleum Hydrocarbons and Organic Contaminants in Ground Water: Prevention, Detection, and Restoration,* National Water Well Association, 1986.

CHAPTER 6

Electrochemiluminescent Sensing of Petroleum Contamination at Trace Levels

Brian G. Dixon, John Sanford, and **Brian W. Swift,** Cape Cod Research, Inc., East Falmouth, Massachusetts

INTRODUCTION

The longtime threat of water pollution has become an increasingly common and unfortunate reality in recent years. Juxtaposed to massive oil spills, as epitomized by that of the *Exxon Valdez*, are large numbers of small spills, especially of petroleum products, but also of chlorinated solvents and other chemicals. At this stage it is clear that an unsettlingly high number of these spill sites pose potential health threats to the public, and that remediation will be required. The first step in the remediation process is to determine the extent of the pollution hazard via detection. In recent years the groundwater pollution detection limits, required by the Environmental Protection Agency (EPA) and various state agencies, have continuously dropped.

The next generation of instrumentation, which will be needed very soon, will be required to accurately measure contaminants in the low, parts per billion, or even parts per trillion, range. Another demand, which significantly compounds the difficulty of the problem, is that it is abundantly clear that this new generation of devices will have to be durable and portable so that they can be used for in situ measurements in the field. This is not a trivial technological exercise. Under the best of conditions, it is hard to measure parts per billion of a compound in any carrier. Add to this the potential interference of a multitude of other species which may be present, and at much higher concentrations, and the scope of the problem becomes even clearer.

As sobering as these facts are, recent research has shown that it may be possible to accomplish this formidable task. In particular, work that is closely related to that described herein has demonstrated the potential for the remote laser-induced fluorescence assay for groundwater contaminants.[1-5] Various laser-fiber optic systems have been designed and constructed that are capable of detecting trace levels of contaminants in groundwater. From a practical standpoint these systems all suffer from the fatal flaw that they involve lasers. In terms of use in the field, lasers are inherently delicate since they require precisely aligned optics and shock-sensitive equipment. Lasers are also, in general, expensive. Most laser systems are dependent upon the use of excitation light in the ultraviolet region of the spectrum which can rapidly degrade the glass of fiber optic cables. This requires the use of quartz fibers which are much more expensive and brittle. Currently available fiber optic-laser systems also suffer from high attenuation problems.

Another approach is the use of bioluminescent organisms which are quite sensitive to the presence of pollutants of interest, such as benzene and naphthalene.[6-8] This, too, is an approach worth investigating but has some inherent drawbacks. First is the necessity of dealing with biological processes, which are delicate. Another drawback is that these biodetectors yield qualitative, not quantitative results.

The state of the art suggests that a technique which will allow for the detection of groundwater pollutants at very low concentration levels, but using a rugged device design, is appropriate. An electrochemiluminescence-based system as described in this research effort will prove to be less costly and complex, and is a technique that lends itself to deployment for remote sensing.

RESEARCH OBJECTIVES AND FUNDAMENTALS

The primary objective of this research was to develop an electrochemiluminescence (ECL) optrode (by definition an optrode is an optical electrode) system that would be capable of detecting aromatic hydrocarbons in water at concentrations in the low, parts per billion range. A second objective of this research was to use the feasibility studies to design a device which would be durable, dependable, easily transportable, and relatively low in cost. Other objectives were to establish the feasibility for using the novel ECL technique to qualitatively and quantitatively measure the concentrations of aromatic hydrocarbons in water.

ECL is a blend of the two distinct disciplines of photochemistry and electrochemistry. In effect it combines all of the rich and useful features of photochemistry, such as the phenomena of fluorescence and phosphorescence, with the radical cation/anion and electron transfer capabilities inherent in electrochemistry. In essence, ECL involves the use of electrodes and the application of small voltages to create photochemical excited states.

Aqueous ECL System

Both organic solvent and aqueous based luminescent systems are well known. Although much less studied, aqueous luminescent systems are, in fact, far more common. Examples of such systems include the firefly, luminescent bacteria, and a multitude of other bioluminescent organisms.[9-11] These naturally occurring luminescent systems are also almost invariably quite efficient as well. For example, the quantum yield of the firefly luminescence approaches 100%, a yield that is unmatched in synthetic, chemiluminescent systems. In general, water and oxygen quench excited states very efficiently, which has meant that the vast majority of laboratory studies have been carried out in rigorously dried organic solvents and in the absence of oxygen.[12-14] Over the last 10 years or so, research into aqueous based transition metal complexes in luminescence applications has been intense primarily because of their promise for solar energy harvesting applications. In an ongoing series of excellent studies, Bard has demonstrated the capabilities of ruthenium bipyridyl complexes to be sensitive ECL agents in aqueous systems.[15-18] The most studied of these ruthenium complexes is tris (2, 2'-bipyridyl) ruthenium (II) chloride [hereafter referred to as $Ru(bipyr)_3$], which is a crystalline, orange, water-soluble complex that possesses a rich and varied chemistry and the octahedral structure shown below:

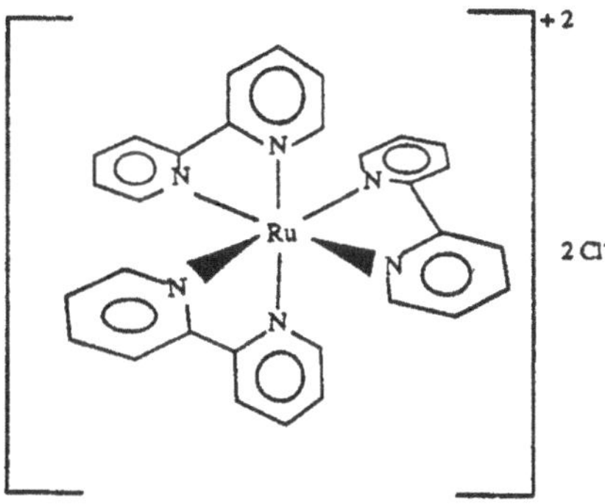

Among the interesting properties of this complex is the fact that it gives efficient ECL in aqueous solution. This is clearly an advantage for the application of detecting petroleum in soils or water. The ECL photochemistry of this compound is complex, but basically involves emission from a d-π^* excited state. Of immediate interest here is the fact that its ECL emission occurs in the visible range of the electromagnetic spectrum, specifically ~500 to 700 nm with a maximum at ~600 nm. This means that silicon-based photodetectors and nonquartz fiber optic cables can be used which are much less costly and more readily available. Of equal importance is the fact that the subject complex can be immobilized within a polymer or onto its surface, and thus the required ECL carried out in the solid state. The mechanism of the luminescence involves a reversible charge-transfer excited state complex between the Ru^{+2} and the bipyridyl ligands. The quantum efficiency of this process is dependent upon, among other factors, the nature of the solvation of the complex. This phenomena is used to great advantage in the described work.

Photomultiplier Detection System and ECL Studies

Both homogeneous and solid state ECL laboratory experiments were carried out with the $Ru(bipyr)_3$ complex under model conditions to establish the scope of the technology. Benzene was used in all of the experimental studies as a model compound for aromatic hydrocarbon pollutants that are present at fuel spill sites. In all of the experiments to be discussed, the benzene solutions were carefully prepared daily for a given set of experiments. All of the experiments were reproduced at least in duplicate. Control experiments were also performed with the various components individually to assure that the observed phenomena were real.

General Laboratory Techniques

Sensitive photodetection experiments were carried out with a photomultiplier-monochromator apparatus model (Model 77250 and Model 77762, respectively, the Oriel Corp.). This instrument consisted of a sensitive photomultiplier tube whose operational detection range is from 200 nm to 800 nm, a monochromator, and attached fiber optic cables. A polypropylene mold was machined to hold the quartz cell in which the electrochemical experiments were actually carried out. This mold also was modified so that the end of the fiber optic cable could be placed flush to the side of this quartz cell.

Experiments were carried out using the tris (2, 2'-bipyridyl) ruthenium (II) chloride in the presence and absence of trace levels of benzene. These experiments were carefully performed to assure that extraneous impurities would not be introduced into the test solutions. All glassware was first acid-washed, rinsed with deionized water, and thoroughly dried prior to use. New and carefully cleaned syringes were used as well. In the experiments about to be described, the stock solutions were prepared immediately prior to the performance of the experiments.

Effect of Voltage

There is little doubt that the experimentally observed ECL was due to the $Ru(bipyr)_3$ complex. The mechanism of this ECL involves a charge transfer ($d-\pi^*$) transition with the emitting excited state being a triplet. The crucial electron transfer occurs at a voltage of $\sim$0.95V vs SCE. Virtually no luminescence was detectable below this voltage while, at or above it, an increase of up to 40% in total intensity was observed. In addition, this emission corresponded to the well-known characteristic spectrum expected for this complex. The appropriate control experiments were carried out to ensure that Ru(bipyr)$_3$ ECL was actually being observed. These experiments included running all of the components of the stock solution independently and together, but in the absence of the Ru(bipyr)$_3$, at 0 and 1.000V, and confirming that no luminescence changes were observed.

Homogeneous (Aqueous) Phase ECL Studies

Figure 1 is composed of two plots, 1a and 1b. Figure 1a is a plot of the luminescence of a 1×10^{-4} M Ru(bipyr)$_3$ solution as a function of voltage applied to the solution and in the absence of benzene. The baseline corresponds to zero applied voltage. As can be seen, at an applied voltage of 0.950V there is an immediate response due to Ru(bipyr)$_3$ electrochemiluminescence. This response was found to be very reproducible and quite fast. Removal of the applied voltage resulted in an immediate drop in intensity to the background light level. Figure 1b is a plot of the luminescence after the injection of enough stock benzene solution to yield a final benzene concentration of 25 ppb. It should be noted that the volume of such an injection is insignificant in terms of the total volume of solution. As can be seen, there was a significant increase in the total detected luminescence. This result was found to be very reproducible and the electrode response was very quick as well.

Ruthenium Complex Emission Spectrum

An ECL spectrum of the Ru(bipyr)$_3$ complex was obtained by using the monochromator to scan from ~550 to 800 nm. Similar experiments were then carried out in the presence of benzene at various concentrations.

Figure 2 contains representative results showing the effect of benzene concentration upon the electrochemiluminescence spectrum of Ru(bipyr)$_3$ in water. This figure compares ECL in the absence of benzene and then in the presence of 24 and 48 ppb. The results contained in Figure 2 are interesting and very promising for a number of reasons. First, the observed Ru(bipyr)$_3$ ECL increases significantly as the benzene concentration goes up. In principle, this will allow for the quantitative as well as qualitative determination of a pollutant's presence. Of equal importance is the fact that the change in intensity with benzene is not linear over the whole spectral range. This is important since it should be possible to simultaneously monitor intensity changes at different wavelengths and thereby increase both the sensitivity and signal to noise ratio of the detecting device.

Solid State ECL Studies

The above described homogeneous phase work was then extended to the solid state. Bard and Anson have shown that it is possible to immobilize ruthenium bipyridyl complexes within polymer matrices and carry out solid state ECL.[18-21]

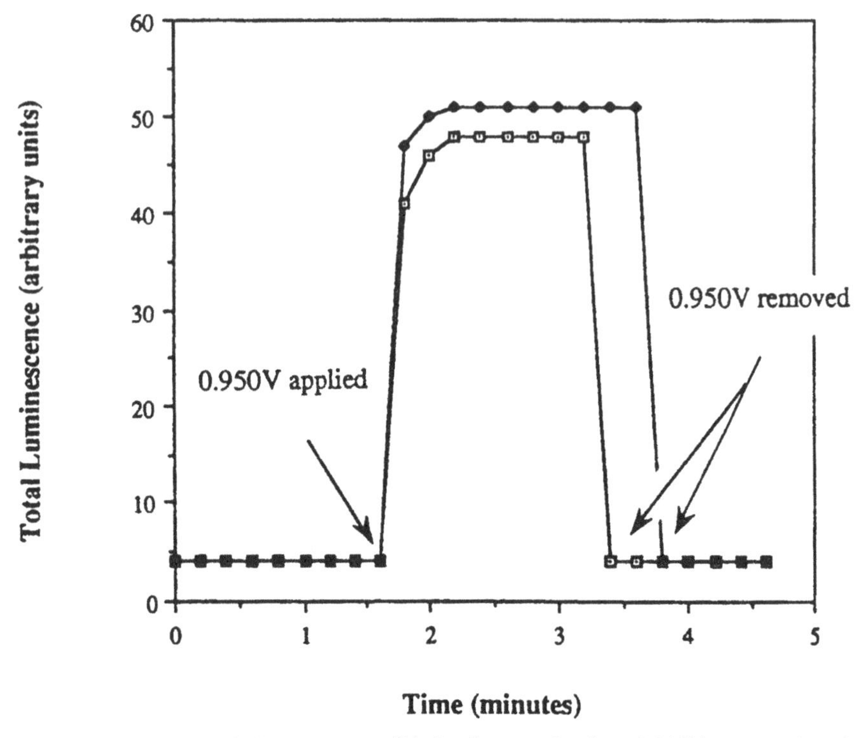

Conditions: $[Ru(bipyr)_3Cl_2]$ = 1×10^{-4}M; [sodium oxalate] = 0.025M prepared and added to the quartz cell; all solutions prepared in deionized water. Benzene was added from a stock solution. Applied voltage of 0.950V at ~1.6 minutes.

White squares = no benzene
Solid black diamonds = with 25 parts per billion benzene added

Figure 1. Electrogenerated chemiluminescence of the $Ru(bipyr)_3Cl_2$ with and without benzene in water.

In the current work, parallel experiments were carried out wherein the platinum working electrode was first coated with a thin layer of Nafion®, a perfluorinated polymer possessing pendant sulfonic acid groups. This polymer modified electrode was further modified by immersion in a 5×10^{-3}M solution of $Ru(bipyr)_3$ in 0.1M H_2SO_4. This process immobilizes the $Ru(bipyr)_3$ within the polymer via bonding to the sulfonic acid groups.

The modified electrode was placed into the cell so that its flat surface faced toward the end of the fiber optic cable. ECL experiments were then carried out as described above for the homogeneous work. Figure 3 is the solid state counterpart to Figure 1 and includes plots 3a and 3b. Figure 3a shows the results of these solid state scans in the absence of benzene. A comparison of these results with those of Figure 1a yields a couple of interesting conclusions. First, the

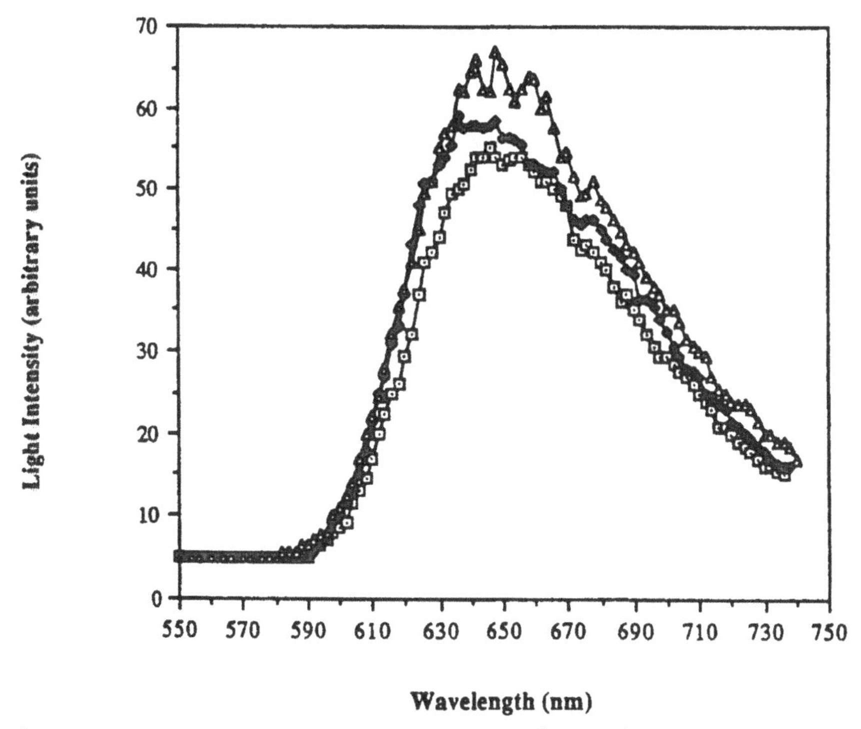

Conditions: In deionized water; $[Ru(bipyr)_3] = 1 \times 10^{-3}M$; $[Na_2C_2O_4] = 0.2M$; $[NaCl] = 0.1M$; applied voltage = 1.3V.

White squares = without benzene
Black diamonds = with 24 parts per billion benzene
Triangles = with 48 parts per billion of benzene

Figure 2. Effect of benzene concentration upon Ru(bipyr) ECL.

response of the electrode to the applied voltage, which is immediate in the solution experiments, is slower in the solid state. The decay of the luminescence upon removal of the voltage is also slow in the latter state. These results are not unexpected. Since the Nafion-$Ru(bipyr)_3$ composite is not an especially good electrical or ionic conductor, its electrochemical response is expected to be much slower than the solution phase case. In addition, the restriction of the mobility of the ruthenium complex in the solid state means that the various light-quenching mechanisms are drastically reduced. This allows the excited state $Ru(bipyr)_3$ species to have significantly longer lifetimes, and the observed luminescence will decay more slowly upon removal of the applied voltage, as is observed.

Figure 3b is the result of a luminescence scan run immediately after the lower run but in the presence of 25 ppb of benzene. As in the solution phase instance, a significant increase in the total luminescence was observed. This result has now been extended to parts per trillion concentrations of benzene as well.

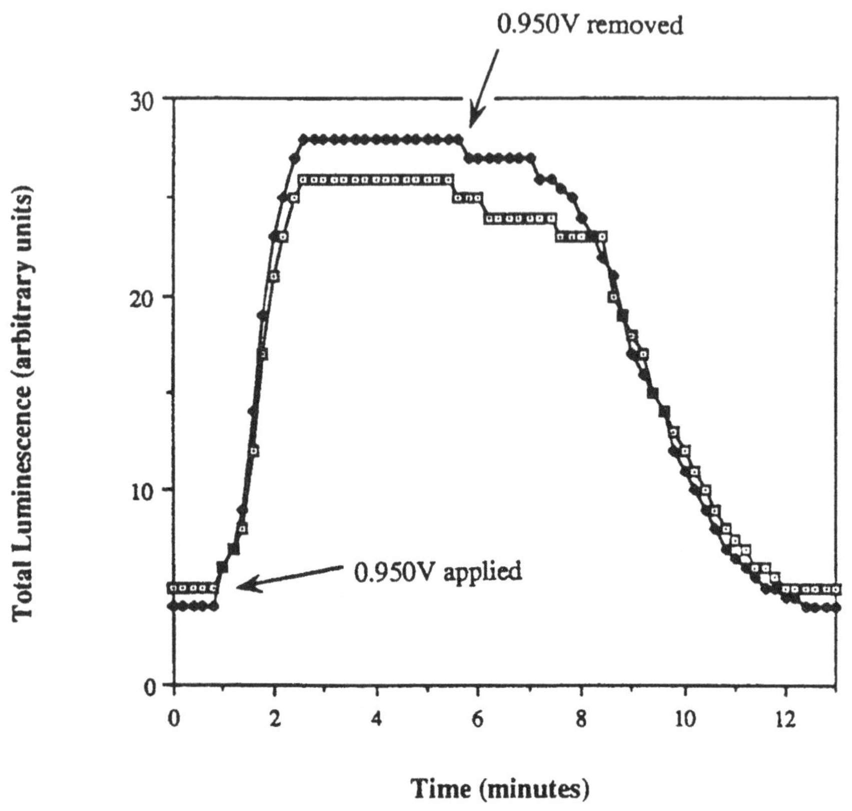

Conditions: [$Ru(bipyr)_3Cl_2$] immobilized by ion exchange onto the Nafion® coated, platinum working electrode. H_2O solution of [sodium oxalate] = 0.025M prepared and added to the quartz cell; all solutions prepared in deionized water. Benzene was added from a stock solution. Applied voltage of 0.950V at ~1 minute in both cases.

White squares = no benzene
Solid black diamonds = with 25 parts per billion benzene added

Figure 3. Solid state electrogenerated chemiluminescence of the $Ru(bipyr)_3Cl_2$ in Nafion® with and without benzene in water.

Clay Modified Electrode Studies

Recent investigations have extended the solid state studies to electrodes that are modified with thin layers of clay, into which the ruthenium complex is immobilized. Bard has pioneered the area of clay modified electrodes as well.[23] A recent series of interesting studies has involved a calcium montmorillonite clay hereafter coded STx-1. This clay was purchased from the Source Clay Minerals Repository (University of Missouri, Columbia, MO), and was purified as described previously.[23]

Platinum electrode substrates were used in these studies as well, exactly as described above for the Nafion® case. A number of different electrode-coating techniques were evaluated, and to a large degree, the thinner the clay layer was, the better the end product electrode turned out to be. In general, ~5–30 μL of a colloidal clay solution was dispersed via a syringe to each side of the flag to coat the electrode, which was dried at 22° to 25°C for 24 hr. The clay modification to the electrode gave the latter a dull, iridescent sheen. The clay modified electrode was then immersed (usually for an exposure time of 30 min) into an aqueous solution of 1×10^{-3}M $Ru(bipyr)_3$, to ion exchange the ECL complex into the clay, and then washed with copious quantities of water. After this exposure, the yellow-orange color due to the $Ru(bipyr)_3$ was evident.

Figure 4 contains the results of a sequential set of experiments with the STx-1 clay-modified electrodes and combinations of the model contaminants benzene and ethylene glycol. An important point is that the sequence of addition of these contaminants was varied, and plain water was added in a number of instances to assure that the addition sequence, in and of itself, did not determine whether

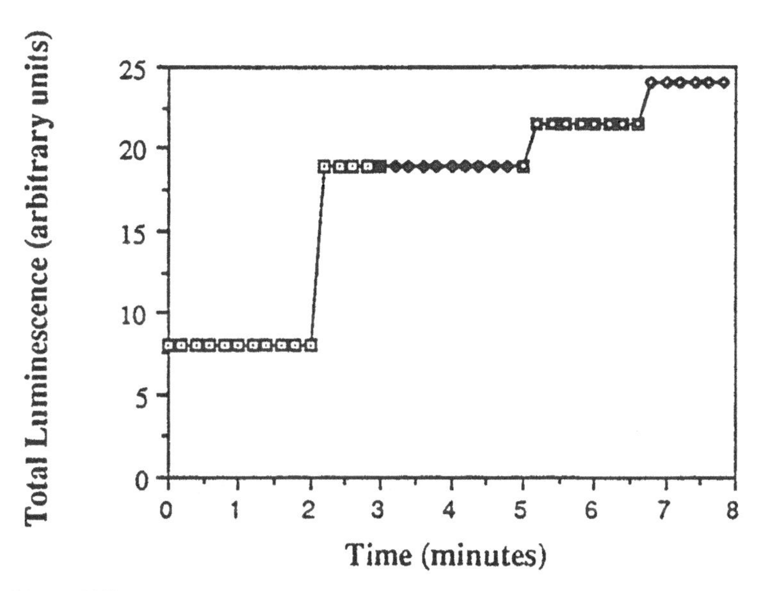

Conditions: H_2O solution of [sodium oxalate] = 0.025M; [NaCl] = 1×10^{-3}M; all solutions prepared in deionized water. Benzene and trichloroethane were added from freshly prepared stock solutions. Applied voltage of 1.0V.

White squares = steady state luminescence
Solid black diamonds = addition of 24 ppb Ethylene Glycol
Black squares = first addition of 24 ppb Benzene
White diamonds = second addition of 24 ppb Benzene (to give a total benzene concentration of 48 ppb)

Figure 4. Solid state electrogenerated chemiluminescence of $Ru(bipyr)_3Cl_2$/STx-1 clay. Effects of sequential ethylene glycol-benzene(1)–benzene(2) additions.

there was a response. More specifically, Figure 4 shows the results on the luminescence of the sequential addition of 5 μL ethylene glycol stock solution (to yield a final concentration of 24 ppb), then 5 μL of benzene stock (also to yield a final concentration of 24 ppb), and then another 5 μL of benzene stock (also to yield a final concentration of 24 ppb), and then another 5 μL benzene stock. As can be seen, only for the benzene additions was an effect observed. In no instance was a response obtained from the addition of either plain water or ethylene glycol. Benzene did, however, yield a response consistent with the results shown in Figure 3 above. Most importantly, it has been demonstrated that aromatic hydrocarbons, such as benzene and toluene, can be selectively detected in the presence of hydroxyl-containing species, such as ethylene glycol (and, of course, water).

Portable (Field) Device Design

The results of the research to date indicate that it will be feasible to use electrochemiluminescence as a technique for the detection of petroleum contaminants at ppb concentrations in the environment. The key to the success of this technique, however, is the design and construction of a portable device for field use. As currently envisioned, this solid state device will consist of two sections as detailed in Figure 5. A small PVC pipe containing the optrode sensor and light detector electronics can be lowered underground into a test well, if desired. A larger instrumentation package containing the control electronics for the system will sit on the surface. The two modules will be electrically joined together by a multiconductor shielded cable. The cable will have a connector on the sensor package end to allow for easy replacement of the optrode if it should fail. The PVC pipe containing the sensor will have a number of holes drilled in it to allow groundwater to freely pass over the optrode. The optrode itself will consist of a bundle of glass optical fibers coated with indium-tin oxide as the working electrode and the $Ru(bipyr)_3$ compound. The indium-tin oxide electrode is optically transparent, thus excellent light transmission characteristics are to be anticipated. Although it is not required to immobilize the ruthenium complex upon the fiber optic cable, it is attractive to do so since the maximum light sensitivity can be achieved using this arrangement. The electrochemical cell will be completed by a length of ruthenized titanium wire which will serve as the counter electrode, and a piece of silver silver-chloride wire which will be used as the reference electrode. The light generated by the optrode will be optically coupled to detector modules which will convert the light into an electrical voltage. The detector modules and their associated support electronics will be potted into polyurethane at the end of the PVC pipe. The output voltage signal produced by the detectors will be sent to the instrumentation package on the surface.

The heart of the portable solid state device will be a single board computer. This device will serve as the control unit and data logger for the system. A block diagram of the complete system is shown in Figure 6. The chosen computer is an extremely versatile device that squeezes a large number of functions into a

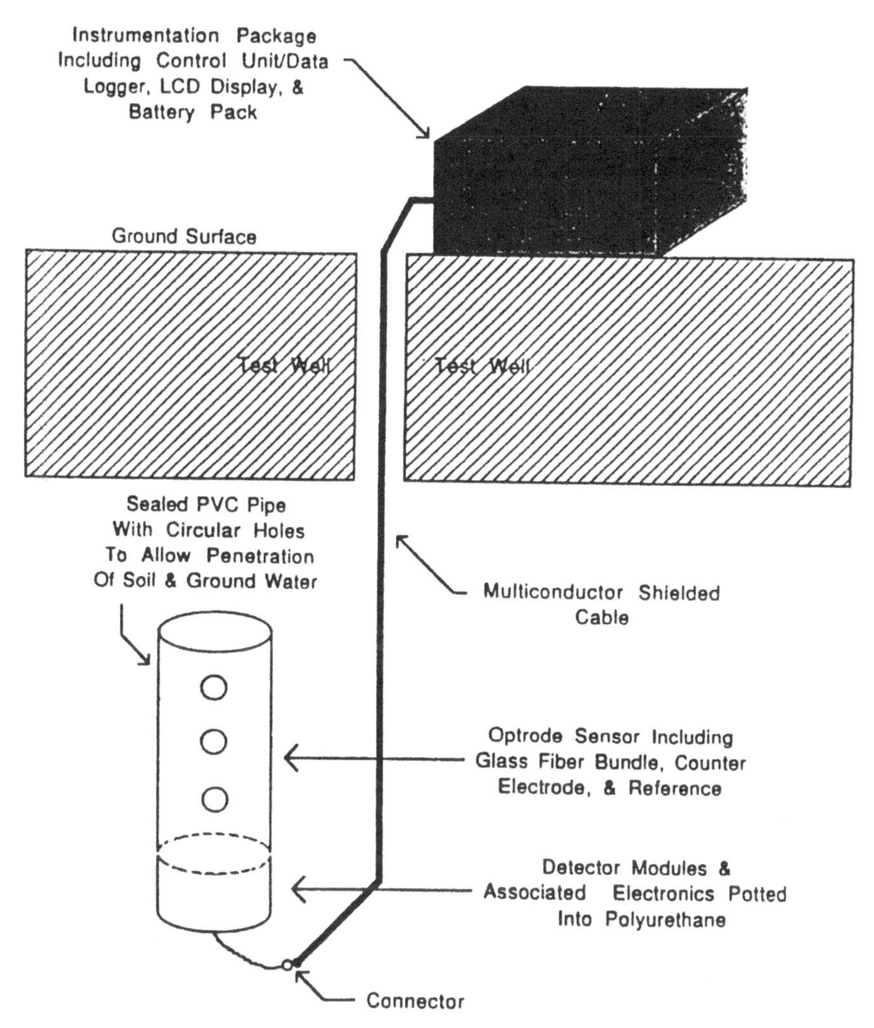

Figure 5. Physical layout of optrode system.

2.25″ by 3.725″ by 0.75″ space. The computer is programmable in a hybrid version of BASIC, or for greater speed, with a compiled version of BASIC. It is also possible to directly program the 8 bit Hitachi 6303Y CPU with assembly language code for minimum program space and maximum speed. Up to 28 kbytes are normally available for program and data storage, and an additional add-on memory board allows expansion to 512 kbytes. The computer draws very little power, requiring only 120 mW for normal operation and 24 mW in a standby mode. This will allow the portable device to operate for several months at a time with only a medium-sized 12-volt sealed lead acid battery. A solar panel could be used to trickle-charge the battery to allow the system to run for much longer periods. The computer also includes a backup lithium battery mounted on a board

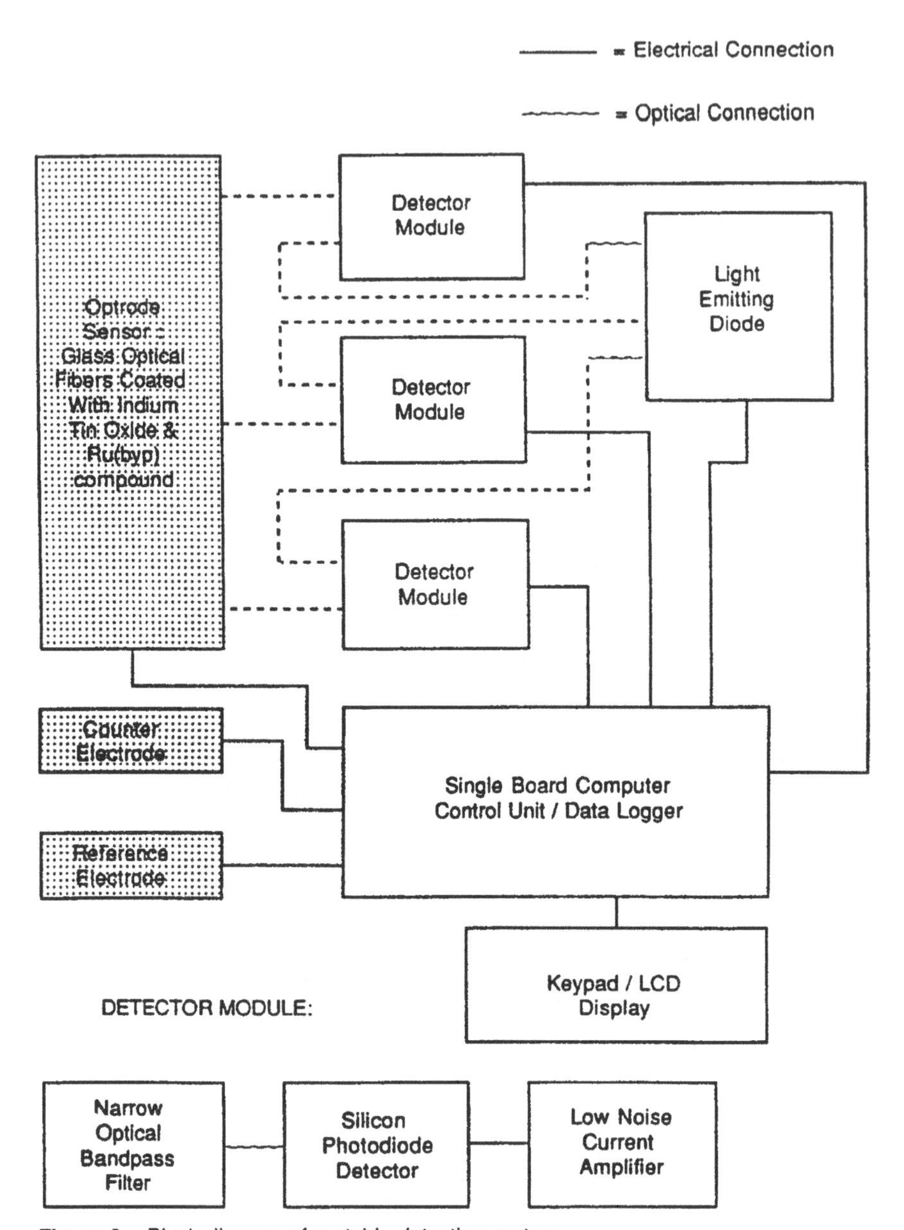

Figure 6. Block diagram of portable detection system.

to preserve the program memory and data if the main power should fail. Other features of the computer include a 10-bit, 11-channel analog to digital converter and 16 digital input/output lines for control functions. Also, it generates precision time intervals to allow for accurate sequencing of events ranging from

10 ms to several months. Finally, the computer includes a UART (universal asynchronous receiver and transmitter) to allow for data transfer to a desktop or portable computer at 300, 1200, 9600, or 76800 baud. If telephone lines are available near testing areas, data from a number of test devices can be retrieved and processed by a single central computer located near the testing site or at some other more remote location.

The computer will collect and process data supplied to it from the optrode sensor. To avoid drift problems associated with absolute measurements, a relative measurement system will be used. At regular time intervals, the computer board will pulse a light-emitting diode (LED) on and off. The LED will generate bursts of light which will be optically coupled to the detector modules. The detector modules will convert the light into a voltage signal which in turn will establish the zero level or baseline to which all other measurements will be relative. During alternate time intervals, the computer board will pulse the LED and also supply pulses of power to the optrode sensor. If hydrocarbon contaminants are present in the area, the sensor will generate light when power is supplied to the sensor. The light produced by the optrode will add to the light generated by the LED, causing the output voltage of the detector modules to be greater during the time interval when the sensor and LED is on as compared to the output voltage produced during the interval when only the LED is on. The difference in voltage levels during these two separate time intervals will indicate the presence and level of concentration of contaminants. This measurement technique will also reduce noise in addition to eliminating the drift associated with absolute measurements. The computer will synchronize the pulsing of the LED and optrode sensor with the measurements made by its analog to digital converter. Voltage signals outside this synchronization period will be rejected as noise, thus increasing the signal-to-noise ratio of the measurements.

To further increase the accuracy of the testing, three separate detector modules will be used in the system. Each of these integrated optical detectors will look at a different portion of the optical spectrum to verify the effect caused by the presence of contaminants. The optical transducer incorporates three separate components—a narrow optical bandpass filter, a silicon photodiode detector, and a low noise current amplifier—into a single miniature package. In addition, the metal case is hermetically sealed for operation in harsh conditions. Incorporating these three elements into a single shielded metal package reduces external noise pickup and reduces the number of optical and electrical connections. The heart of the detector module is a silicon photodiode that converts photons which strike the device into an equivalent electrical current. The photodiode used in this module has a high sensitivity down to 10^{-11} watts of incident optical power, and has an extremely linear response. The useful optical band width of the photodiode extends from 400 to 1100 nm. This is ideal, since the ruthenium complex's ECL emission is from ~500 to 700 nm, with a peak at 607 nm. Current produced by the photodiode is amplified and converted into a voltage by a high gain, high input impedance operational amplifier. The overall response of the detector module is determined by a narrow optical bandpass filter which is placed over the silicon photodiode. The filter has a typical center wavelength accuracy of ± 1.5 nm and 60 dB rejection of signals outside the filter band width.

The complete test data from the optrode sensor will be stored in the computer's memory for later retrieval and analysis. The memory capacity will allow the device to go several months at a time without having to retrieve data.

CONCLUSIONS

To date, this ongoing research project has established the overall feasibility of using electrogenerated chemiluminescence as an analytical technique for detecting trace amounts of aromatic hydrocarbons in water. More specifically, the results obtained from this study have led to a number of conclusions as follows:

- Electrochemiluminescence is a sensitive technique for detecting hydrocarbons in water at part-per-billion concentrations and lower.
- The electrochemiluminescence can be carried out in the solid state as well as in aqueous solution.
- ECL can be used to quantitatively measure the concentrations of trace pollutants.
- The ECL technique can be used to selectively detect petroleum hydrocarbons, such as benzene and toluene, in the presence of hydroxyl-containing species, such as ethylene glycol.
- It will be possible to translate the results obtained in the laboratory to the design of a portable and durable device of reasonably low cost.

FUTURE DIRECTIONS

Research continues on the design and construction of a practical field device. Current efforts seek to determine the scope of the technology's applicability by studying a number of significant factors relating to the use of ECL for pollutants detection in water. The work includes both laboratory and field work and consists of a number of different segments as follows:

- establish the selectivity and ultimate sensitivity of the ECL system to different kinds of compounds, both alone and in mixtures.
- determine the sensitivity of the system to impurities and undesirable quenching mechanisms.
- establish techniques to prevent undesirable quenching mechanisms.
- construct and evaluate a portable device in the laboratory and then perform field tests.
- establish the long-term performance of the engineered device.

A two-year program is currently anticipated, at the end of which a low-cost field device will be available for commercial application.

ACKNOWLEDGMENTS

The financial support for this research, by the Headquarters Air Force Engineering and Sciences Center, Environics Division, Tyndall Air Force Base, Florida, and the National Science Foundation, is gratefully acknowledged.

REFERENCES

1. Peterson, J. L., and G. G. Vurek. *Science* 13:123 (1984).
2. Chudwyk, W. A., et al. *Anal. Chem.* 57:1237 (1985).
3. Wolfbeis, O. S., et al. *Anal. Chem.* 60:2028 (1988).
4. Seitz, W. R. *Anal. Chem.* 56:16A (1984).
5. Kenny, J. E., et al. in *Luminescence Applications*, M. C. Goldberg, Ed. ACS Symposium Series No. 383, The American Chemical Society, Washington, DC, 1989.
6. Kamlet, M. J., et al. *Environ. Sci. & Tech.* 20:690 (1986).
7. Schultz, J. S., et al. *Talanta* 35:145 (1988).
8. King, J. M. H., et al. *Science.* 249:778 (1990).
9. De Luca, M. A., and W. D. McElroy. *Bioluminescence and Chemiluminescence* (New York, NY: Academic Press, 1981).
10. Harvey, E. N. *Bioluminescence* (New York, NY: Academic Press, 1952).
11. Herring, P. J. *Bioluminescence in Action* (New York, NY: Academic Press, 1978).
12. Adam, W., and G. Cilento. *Chemical and Biological Generation of Excited States* (New York, NY: Academic Press, 1982).
13. Cormier, M. J., D. M. Hercules, and J. Lee. *Chemiluminescence and Bioluminescence* (New York, NY: Plenum Press, 1973).
14. McCapra, F. *Proc. Org. Chem.* 8:231 (1973).
15. Bard, A. J., et al. *J. Amer. Chem. Soc.* 103:512 (1981).
16. Bard, A. J., et al. *J. Amer. Chem. Soc.* 95:6582 (1973).
17. Bard, A. J., et al. *J. Amer. Chem. Soc.* 104:6891 (1982).
18. Bard, A. J., et al. *Anal. Chem.* 55:1580 (1983).
19. Buttry, D. A., and F. C. Anson. *J. Amer. Chem. Soc.* 104:4824 (1982).
20. Rubenstein, I., and A. J. Bard. *J. Amer. Chem. Soc.* 103:5007 (1981).
21. Rubenstein, I., and A. J. Bard. *J. Amer. Chem. Soc.* 102:6642 (1980).
22. Krishnan, M., X. Zhang, and A. J. Bard. *J. Amer. Chem. Soc.* 106:7371 (1984).
23. Ege, D. E., P. K. Ghosh, J. R. White, J. F. Equey, and A. J. Bard. *J. Amer. Chem. Soc.* 107:5644 (1985).

CHAPTER 7

Why TCLP, EP-Tox, and the California WET Do Not Derive Data on the Mobility and Transformations of Metals in Soil Systems

James Dragun, John H. Barkach, and **Sharon A. Mason,** The Dragun Corporation, Farmington Hills, Michigan

INTRODUCTION

During the late 1970s, the U.S. Environmental Protection Agency (EPA) developed the Extraction Procedure Toxicity Test Method (EP-Tox). This test method is the predecessor to the Toxicity Characteristic Leaching Procedure (TCLP), and serves as the technical basis for the TCLP. In the early 1980s, the state of California adopted the Waste Extraction Test (WET).

These tests are utilized to classify a waste as being either hazardous or nonhazardous. The classification depends upon the amount of inorganic and/or organic chemicals that can be extracted from the waste by an acidic extractant solution containing an organic ligand.

It is unfortunate, however, that individuals, consulting firms, and other governmental agencies have attempted to utilize these test methods to estimate the leaching potential of metals from soil. They have theorized that the greater the concentration of metal that can be extracted from the soil, the more mobile the metal must be; the more mobile the metal, the more likely that the metal will migrate to groundwater. In other words, they have theorized that the concentration of metals in soil potentially able to migrate to groundwater, such as lead (Pb), zinc (Zn), chromium (Cr), and copper (Cu), is directly proportional to the amount of these metals extracted from soil using the TCLP, EP-Tox, and WET test methods.

The TCLP, EP-Tox, and WET tests can be used only for the classification of hazardous waste; they cannot be utilized to determine the fate (i.e., migration and degradation potential) of any chemical in a soil system. This chapter presents the technical basis for this conclusion.

GENERAL FORMS OF INORGANIC CHEMICALS IN SOIL

The total concentration of any chemical, C_{Total}, in a soil is equal to:

$$C_{Total} = C_{Fixed} + C_{Ads} + C_{Water} \tag{1}$$

where:

C_{Fixed} = concentration of fixed chemical comprising part of the structure of clay and soil minerals, in milligrams (mg) chemical per kilogram (kg) soil.

C_{Ads} = concentration of chemical adsorbed onto the surface of soil minerals and onto organic matter exchange sites, in mg chemical/kg soil.

C_{Water} = concentration of chemical in soil water or groundwater in equilibrium with C_{Ads}, in mg soluble chemical/kg soil.

C_{Fixed} represents the "immobile" fraction of C_{Total}. C_{Ads} and C_{Water} represent the potentially mobile portion of C_{Total}.

In general, when a dissolved element (i.e., C_{Water}) is added to a soil system, it will typically convert to C_{Ads}; then it converts to C_{Fixed}, the relatively fixed or immobile form. Typically, at natural concentrations C_{Fixed} represents at least 90% of the total concentration present in a soil system. This series of reactions is illustrated as follows:

$$C_{Water} \text{ ------> } C_{Ads} \tag{2}$$

$$C_{Ads} \text{ ------> } C_{Fixed} \tag{3}$$

Fixation reactions are those chemical reactions, illustrated by Equation 3, that occur in soil to remove an element from migrating water and transform (i.e., fix) the element in an unavailable or unleachable form.[1,2] The element is immobilized either within the structure of a mineral or at the mineral surface. In other words, fixation attenuates a migrating element.

Adsorption is quantified in the scientific literature, utilizing the parameters from Equations 1, 2, and 3 in two ways. First, adsorption is quantified via the adsorption coefficient, K_{d1}, where:

$$K_{d1} = C_{Ads} / C_{Water} \tag{4}$$

Second, adsorption is quantified via the sorption coefficient, K_{d2}, where:

$$K_{d2} = (C_{Fixed} + C_{Ads}) / C_{Water} \quad (5)$$

Many scientists and engineers, who do not understand the implications of Equation 3, erroneously utilize adsorption coefficients based on Equation 4, which are unrealistically low. The use of Equation 5 typically results in higher adsorption coefficients because it takes into account the reactions illustrated by Equations 2 and 3. This equation is more appropriate for understanding the reactions occurring for metals in soil.

There are three important facts that should be understood concerning the parameters listed in Equation 1.

First, analytical data derived from the chemical analysis of the total metal content of soil (i.e., C_{Total}) relays no information regarding C_{Fixed}, C_{Ads}, and C_{Water} other than the magnitude of their combined concentrations. In other words, if a laboratory report states that a soil contains 125 mg/kg total Pb, this datum cannot reveal if 0.1% of this concentration is potentially mobile (i.e., $C_{Ads} + C_{Water}$) or if 99% is potentially mobile.

At background concentrations, the relative magnitudes of the parameters listed in Equation 1 generally are:

$$C_{Fixed} >> C_{Ads} > C_{Water} \quad (6)$$

The greater part of C_{Total} exists as C_{Fixed} and is immobile. However, this relative ranking may or may not change as C_{Total} increases above the background concentration.

Second, background concentrations represent the total concentration of a chemical present after the soil was formed and weathered. This concentration gives no information on the loading capacity of a soil. The loading capacity can be defined as the maximum amount of chemical that can be added to soil which does not cause water migrating through this soil to contain a harmful concentration of that chemical. In other words, knowing that a soil contains 125 mg/kg total background Pb will not reveal if soil will or will not completely convert an additional loading of 500 mg/kg Pb into an immobile form (C_{Fixed}).

Soil cleanup standards that specify the excavation or treatment of soil containing concentrations of a chemical over a background concentration are usually based on an incorrect premise that the background concentration of a chemical in soil represents a maximum concentration of a chemical which the soil can immobilize. The background concentration only represents the total concentration present after the soil was formed and had undergone some degree of weathering; it gives no indication of the maximum concentration of a chemical which a soil can immobilize; i.e., the loading capacity of the soil.

Third, there is no "universal" analytical method or extractant which is applicable for all forms of chemicals in all soils. The test method employed is dependent upon the individual chemical to be tested, the parameter to be tested (e.g., C_{Fixed} versus C_{Ads}) and the soil type. A number of established, accepted, laboratory methods exist for determining the magnitude of C_{Total}, C_{Fixed}, C_{Ads}, and C_{Water} in soil.[3-7]

Typically for metals, C_{Total} can be measured by wet ashing with a mixture of perchloric, nitric, or sulfuric acids. For metals, C_{Ads} and C_{Water} can be determined using mineral acids (e.g., 0.1 N HCl), organic acids, and chelating agents (e.g., EDTA, DTPA); hot water extractions can be utilized for elements that exist as anions (e.g., B, Mo, Se).

C_{Fixed} can be further fractionated.[8] The analytical method which should be utilized varies with the metal of concern and the form of the metal to be analyzed. For example, the forms of fixed copper (Cu) are expressed in Equation 3:

$$C_{Fixed} = C_{sac} + C_{saom} + C_{sao} + C_{oo} + C_{bo} + C_{ml} \qquad (7)$$

where:

C_{sac} = concentration of Cu specifically adsorbed onto clay.
C_{saom} = concentration of Cu specifically adsorbed onto soil organic matter.
C_{sao} = concentration of Cu specifically adsorbed onto oxides.
C_{oo} = concentration of oxide occluded Cu (i.e., Cu incorporated into the concretions and coatings of oxides).
C_{bo} = concentration of biologically occluded Cu.
C_{ml} = concentration of mineral lattice Cu.

The various extractants and the forms of Cu extracted by each are listed in Table 1.

Table 1. The Forms of Cu Analytically Determined by Sequential Extraction Utilizing Various Extractants

	Extractant				
Form of Copper	$CaCl_2$	Acetic Acid	Potassium Pyrophosphate	Oxalate + UV	Hydrofluoric Acid
C_{Water}	+	+	+	+	+
C_{Ads}	+	+	+	+	+
C_{sac}	–	+	+	+	+
C_{saom}	–	P	P	+	+
C_{sao}	–	P	P	+	+
C_{oo}	–	P	P	+	+
C_{bo}	–	–	–	P	+
C_{ml}	–	–	–	–	+

Source: Reference No. 8.
\+ = Form is extracted by the extractant.
P = Form is partially extracted by the extractant.
– = Form is not extracted by the extractant.

In general, two factors primarily govern fixation reactions in a natural soil-groundwater system: soil Eh and soil pH. In general, for many elements, if Eh is low, the reduced and fixed specie should form; if Eh is high, the oxidized and more mobile specie should form. In general, the solubility of most heavy metals, including As, Cr, and Pb, is inversely related to pH: the amount retained is dependent upon the pH of the soil, with retention increasing with increasing pH.[9]

Because Eh and pH are primary factors governing the fate of metals in soil systems, laboratory tests must control these two factors at levels similar to levels present at the subject site.

What Happens When TCLP and Other Extraction Tests Are Utilized on Soils?

When test methods such as the TCLP, Ep-Tox, and WET tests are utilized to determine the amount of extractable chemical in soil, these tests provide a value, $C_{Extract}$, in which:

$$C_{Extract} = aC_{Fixed} + bC_{Adsorbed} + C_{Water} \tag{8}$$

where:

$C_{Extract}$ = concentration of a chemical extracted from a soil, where $C_{Total} > C_{Extract}$

a,b = fractions

Since a and b are not determined, $C_{Extract}$ provides no information regarding the magnitude of C_{Fixed}, $C_{Adsorbed}$, and C_{Water}; information which is needed to determine the potential mobility and transformation of a chemical in soil.

The technical support documents prepared by the U.S. EPA for these extraction tests do not state that these tests should be utilized to determine the migration and degradation potential of any chemical in a soil system.

The scientific literature reveals that soils react with inorganic and organic chemicals. For example, metals can be immobilized in soils via several soil-chemical reaction pathways.[1,2,10]

- chemisorption
- irreversible penetration of soil-mineral lattice structures
- metal precipitation with subsequent formation of new soil-minerals such as insoluble soil oxides, oxyhydroxides, phosphates, etc.
- metal occlusion by formation of soil oxides, oxyhydroxides, phosphates, etc.

In addition, organic chemicals can be transformed in soils via several reaction pathways[1,11,12] which include:

- microbial degradation
- aqueous oxidation
- aqueous reduction
- aqueous hydrolysis
- surface-catalyzed oxidation
- surface-catalyzed reduction
- surface-catalyzed hydrolysis
- bound residue formation

The TCLP, EP-Tox, and WET tests do not take these reactions into account. In fact, these methods contain no procedures to control soil factors governing reactions of chemicals, such as Eh and pH, at environmentally relevant levels.

Furthermore, when soils are exposed to the extractants utilized by these methods, gross alterations will occur in soil mineralogy, in naturally occurring soil-chemical reactions, and in soil physical/chemical properties. Gross soil alterations result from the interaction of the extractant with soil and will (a) dissolve some soil minerals which serve as the basic "building blocks" of soil, (b) impede crystallization and formation of aluminum hydroxides and other soil minerals while causing structural distortions in newly formed minerals, (c) perturb hydrolytic reactions of aluminum, and (d) desorb, via mass action, metals and organic chemicals from soil adsorption sites which may not normally be desorbed.[13-16]

In simpler terms, the TCLP, EP-Tox and WET tests reverse the natural reactions illustrated in Equations 2 and 3, as illustrated by Equations 9 and 10:

$$C_{Fixed} \text{------>} C_{Ads} \tag{9}$$

$$C_{Ads} \text{------>} C_{Water} \tag{10}$$

In summary, the TCLP, EP-Tox, and WET extractants cause gross alterations in the chemical and mineralogical properties of soil systems. Also, these tests contain no procedures to control soil factors governing reactions of chemicals at environmentally relevant levels. As a result, the data derived from these test methods cannot be extrapolated to actual field conditions and, therefore, cannot be used to assess the mobility and transformations of chemicals in soil.

The Scientific Literature Has No Information Supporting the Use of These Test Methods to Determine the Leaching Potential of Metals in Soils

The authors of this chapter have conducted a detailed review of the scientific literature. The purpose of this review was to determine if any scientist or scientific organization has published any data or information showing that TCLP, EP-Tox, and WET can generate meaningful data concerning the leaching potential of any metal in soil.

The authors searched 55 technical journals published between 1976 and 1991, over 65 conference proceedings and over 50 U.S. EPA reports. No data, no information, and no papers were found showing that these test methods can generate useful data on the leaching potential of metals in soil.

The Scientific Literature Contains Other Test Methods That Can Generate Meaningful Data on the Potential Leaching of Chemicals in Soil

The scientific literature contains at least seven standardized laboratory test methods addressing the leaching potential of chemicals in soils.[17-25] In addition, the scientific literature contains at least 60 methods addressing the chemical analysis of soil, and over 40 methods addressing the biological properties of soil.[5,7,26] Some of these test methods were developed over 100 years ago, and are still scientifically valid and acceptable today.

Because acceptable laboratory test methods now exist, these methods should be utilized to determine the potential migration, degradation, and transformations of metals in soil.

SUMMARY AND CONCLUSIONS

During the late 1970s, the U.S. Environmental Protection Agency (EPA) developed the Extraction Procedure Toxicity Test Method (EP-Tox). The EP-Tox test method is the predecessor to the Toxicity Characteristic Leaching Procedure (TCLP) and serves as the technical basis for the TCLP. In the early 1980s, the state of California adopted the California Assessment Manual Waste Extraction Test (WET).

These tests are utilized to classify how a waste must be managed (i.e., hazardous or nonhazardous). However, during the past few years, individuals, consulting firms, and other governmental agencies have attempted to utilize these test methods to estimate the leaching potential of metals from soil. These groups have theorized that the greater the concentration of metal that can be extracted from the soil, the more mobile the metal must be; the more mobile the metal, the more likely that the metal will migrate to groundwater.

The EP-Tox, TCLP, and WET tests must not be utilized to determine the mobility and degradation potential of any chemical in a soil system. These test methods can be used only for the classification of hazardous waste produced by waste generators for handling and/or disposal.

These test methods are not suitable for the estimation of the leaching potential of metals from soil for five reasons. First, these tests provide no information regarding the magnitude of C_{Fixed}, $C_{Adsorbed}$, and C_{Water}; the parameters which determine the potential migration and transformation of a chemical in soil. Second, the technical support documents provided by the U.S. EPA for these tests show that the Agency does not recommend that these tests be utilized to determine the migration and degradation potential of any chemical in a soil system. Third, the scientific literature contains substantial data showing that these test methods alter soil properties and soil reactions. Fourth, the scientific literature has no information supporting the use of these test methods to determine the leaching potential of metals in soils. Fifth, the scientific literature contains other test methods that can generate meaningful data on the potential leaching of chemicals in soil.

REFERENCES

1. Dragun, J. *The Soil Chemistry of Hazardous Materials.* Hazardous Materials Control Research Institute, Silver Spring, MD, 1988.
2. Marshall, C. E. *The Physical Chemistry and Mineralogy of Soils.* Volume 1—Soil Materials. (New York, NY: John Wiley & Sons, 1964).
3. Baker, D. E., and L. Chesnin. Chemical Monitoring of Soils for Environmental Quality and Animal and Human Health. *Advances in Agronomy* 27:305–374 (1975).
4. Baker, D. E., and M. C. Amacher. The Development and Interpretation of a Diagnostic Soil-Testing Program. Bulletin 826. College of Agriculture, The Pennsylvania State University, University Park, PA, 1981.
5. Hesse, P. R. *A Textbook of Soil Chemical Analysis.* (New York, NY: Chemistry Publications, 1972).
6. Jackson, M. L. *Soil Chemical Analysis—Advanced Course.* 2nd ed., University of Wisconsin, Madison, WI, 1973.
7. Page, A. L., R. H. Miller, and D. R. Keeney, Eds. *Methods of Soil Analysis.* Part 2. Chemical and Microbiological Properties. Second Edition. (Madison, WI: American Society of Agronomy, 1982).
8. McLaren, R. G., and D. V. Crawford. "Studies on Soil Copper. I: The Fractionation of Copper in Soils," *J. Soil Sci.* 24:172–181 (1973).
9. Adriano, D. C. *Trace Elements in the Terrestrial Environment.* (New York, NY: Springer-Verlag, 1986).
10. Lindsay, W. L. *Chemical Equilibria in Soils.* (New York, NY: John Wiley & Sons, 1979).
11. Dragun, J., and C. S. Helling. "Soil- and Clay-Catalyzed Reactions: I. Physicochemical and Structural Relationships of Organic Chemicals Undergoing Free-Radical Oxidation," in *Land Disposal of Hazardous Waste.* Proceedings of the Eighth Annual Research Symposium. EPA-600/9-82-002. 1982.
12. Alexander, M. "Biodegradation of Chemicals of Environmental Concern," *Science* 211:132–138 (1981).
13. Clark, C. J., and M. B. McBride. "Chemisorption of Cu(II) and Co(II) on Allophane and Imogolite," *Clays and Clay Minerals* 32:300–310 (1984).
14. Miller, W. P., D. C. Martens, and L. W. Zelazny. "Effect of Sequence in Extraction of Trace Metals from Soils," *Soil Sci. Soc. Amer. J.* 50:598–601 (1986).

15. Pohlman, A. A., and J. G. McColl. "Kinetics of Metal Dissolution from Forest Soils by Soluble Organic Acids," *J. Environ. Qual.* 15:86–92 (1986).
16. Wang, M. K., J. L. White, and S. L. Hem. "Influence of Acetate, Oxalate, and Citrate Anions on Precipitation of Aluminum Hydroxide," *Clays and Clay Minerals* 31:65–68 (1983).
17. Helling, C. S., and J. Dragun. "Soil Leaching Tests for Toxic Organic Chemicals," in *Test Protocols for Environmental Fate and Movement of Toxicants*. Association of Official Analytical Chemists, Arlington, VA, 1981.
18. U.S. EPA. Office of Pesticides and Toxic Substances. "Soil Thin Layer Chromatography," Test Guideline #CG-1700, in *Chemical Fate Test Guidelines*. EPA 560/6-82-003. Washington, DC: U.S. EPA, 1982a.
19. U.S. EPA. Office of Pesticides and Toxic Substances. "Sediment and Soil Adsorption Isotherm," Test Guideline #CG-1710, in *Chemical Fate Test Guidelines*. EPA 560/6-82-003. Washington, DC: U.S. EPA, 1982b.
20. U.S. EPA. Office of Pesticides and Toxic Substances. "Soil Thin Layer Chromatography," Support Document #CS-1700, in *Chemical Fate Test Guidelines*. EPA 560/6-82-003. Washington, DC: U.S. EPA, 1982c.
21. U.S. EPA. Office of Pesticides and Toxic Substances. "Sediment and Soil Adsorption Isotherm," Support Document #CS-1710, in *Chemical Fate Test Guidelines*. EPA 560/6-82-003. Washington, DC: U.S. EPA, 1982d.
22. U.S. EPA. Office of Toxic Substances. "Soil Thin Layer Chromatography," in Proposed Environmental Standards; and Proposed Good Laboratory Practice Standards for Physical, Chemical, Persistence, and Ecological Effects Testing. *Federal Register* 45(227):77332–77365 (1980a).
23. U.S. EPA. Office of Toxic Substances. "Adsorption," in Toxic Substances Control Act Premanufacture Testing of New Chemical Substances. *Federal Register* 44(53): 16257–16264 (1980b).
24. U.S. EPA. Office of Toxic Substances. "Soil Thin Layer Chromatography," in *Support Document for Test Data Development Standards*. EPA-560/11-80-027. Washington, DC: U.S. EPA, 1980c.
25. U.S. EPA. Office of Toxic Substances. "Adsorption," in *Technical Support Document for Guidance for Premanufacturing Testing: Discussion of Policy Issues, Alternative Approaches, and Test Methods*. Washington, DC: U.S. EPA, 1979.
26. Black, C. A., Ed. *Methods of Soil Analysis*. Part 2. Chemical and Microbiological Properties. American Society of Agronomy, Madison, WI, 1965.

CHAPTER 8

Bioremediation of Petroleum Products in Soil

Christopher J. Englert, Earl J. Kenzie, and **James Dragun** The Dragun Corporation, Farmington Hills, Michigan

INTRODUCTION

Discoveries of petroleum products in soil and groundwater have led to comprehensive nationwide assessments of potential sources. Assessments conducted by regulatory and private-sector groups have identified several potential sources such as underground storage tanks, underground pipes, accidental spills, transportation mishaps, and illegal dumping. Once discovered, remediation of soil and groundwater to levels that are not deleterious to human health and the environment becomes the new focus of attention.

Microbial degradation of petroleum products in soil, either via naturally occurring or facilitated methods, is a process that is successfully used to reduce soil concentrations of the product to acceptable levels. The objectives of this chapter are to present a general discussion of: (1) the composition of the petroleum-degrading microbial population of soil, (2) the soil factors affecting the biodegradation process, (3) the petroleum product chemical structure and its influence on the biodegradation process, and (4) engineered bioremediation systems for treatment of petroleum products in soil.

It is important to note that the principles discussed in this chapter are applicable not only to naturally occurring and facilitated biodegradation that takes place in topsoil, in unsaturated zone subsoil, and in groundwater, but also to a wide variety of engineered systems, such as land treatment facilities, landfills, and groundwater treatment.

THE COMPOSITION OF THE PETROLEUM-DEGRADING MICROBIAL POPULATION

The heterogeneous or heterotrophic microorganisms found in soils include naturally occurring populations that possess the ability to degrade petroleum products.[1-3] This population imparts a large hydrocarbon assimilatory capacity to most soils.

Table 1 lists the genera of hydrocarbon-degrading bacteria and fungi isolated from soil. In decreasing order, *Pseudomonas, Arthrobacter, Alcaligenes, Corynebacterium, Flavobacterium, Achromobacter, Micrococcus, Nocardia,* and *Mycobacterium* appear to be the most consistently isolated hydrocarbon-degrading bacteria from soil. In decreasing order, *Trichoderma, Penicillium, Aspergillus,* and *Mortierella* were the hydrocarbon-degrading fungi to be most often isolated from soil. It is clear that bacteria and fungi are the principal agents of petroleum biodegradation in soil, but the relative contribution of each is not clear.

Spore-forming bacteria generally have a negligible role in biodegradation.[5] Although *Bacillus* strains have been isolated from contaminated soils, this may be due to their persistence in soil and subsequent spore germination during enrichment and isolation procedures. Also, a number of actinomycetes have been shown to have hydrocarbon-degrading abilities; however, these organisms do not seem to compete successfully in contaminated soils.[6] The role of algae and protozoa is poorly documented in the literature and does not appear to be significant.

A unique group of hydrocarbon-degrading bacteria not included in Table 1 is the methanotrophs, which possess a highly specialized C_1 metabolism. The methanotrophs are strict anaerobes and typically metabolize petroleum products at rates one or two orders of magnitude lower than aerobic bacteria. Methanotrophs are ubiquitous in soil and become greatly enriched near natural or anthropogenic seeps of methane-containing natural gas.[4]

Early researchers noted that the number of aerobic bacteria in an agricultural soil increased on application of a crude oil. Although the bacterial numbers increased, specie diversity of aerobic microbes decreased with little effect on the anaerobic microbes. More recent investigations confirm these findings. Microbial numbers and activity are generally enhanced in contaminated soils. Stimulation of microbial activity is positively correlated to increasing amounts of hydrocarbons in soil.[7] Odu[8] reported that soil receiving an application of 39.2% of crude oil possessed the highest number of microorganisms relative to soil receiving less amounts of oil. Pinholt et al.[3] showed that eight months after contamination, oil-degrading bacteria in soil increased tenfold to almost 50% of the total bacterial count. In this case, no pronounced decrease in fungal species diversity occurred, although *Scolecobasidium* and *Mortierella* were selectively enriched, as were, to a lesser degree, *Humicola* and *Verticillium.* Also, Jensen[6] reported that oil-

Table 1. Genera of Hydrocarbon-Degrading Bacteria and Fungi Isolated from Soil

Bacteria	Fungi
Achromobacter	*Acremonium*
Acinetobacter	*Aspergillus*
Alcaligenes	*Aureobasidium*
Arthrobacter	*Beauveria*
Bacillus	*Botrytis*
Brevibacterium	*Candida*
Chromobacterium	*Chrysosporium*
Corynebacterium	*Cladosporium*
Cytophaga	*Cochliobolus*
Erwinia	*Cylindrocarpon*
Flavobacterium	*Debaryomyces*
Micrococcus	*Fusarium*
Mycobacterium	*Geotrichum*
Nocardia	*Gliocladium*
Proteus	*Graphium*
Pseudomonas	*Humicola*
Sarcina	*Monilia*
Serratia	*Mortierella*
Spirillum	*Paecilomyces*
Streptomyces	*Penicillium*
Vibrio	*Phoma*
Xanthomonas	*Rhodotorula*
	Saccharomyces
	Scolecobasidium
	Sporobolomyces
	Sprotrichum
	Spicaria
	Tolypocladium
	Torulopsis
	Trichoderma
	Verticillium

Source: Bossert and Bartha.[4]

treated soils possessed lower bacterial species richness than untreated soils. Populations of *Arthrobacter* and coryneforms such as *Corynebacterium*, *Brevibacterium*, *Mycobacterium*, and *Nocardia* showed strong positive responses to oil contamination.

FACTORS AFFECTING THE BIOREMEDIATION PROCESS

A number of soil factors affect the biodegradation process. The manipulation of these factors is advantageous to neutralize the effects of Liebig's law of the minimum. This law states that the rate of a biologic process such as growth or metabolism is limited by that factor present at its minimum level. The response to an improvement of the factor is not necessarily linear. This law was originally applied only to mineral nutrients, but it is equally applicable to other factors such as temperature and pressure. This section addresses five important soil factors that affect microbial processes.

First, the primary factor limiting microbial growth in soil is the scarcity or absence of a suitable and available source of energy.[9] The great majority of soil microorganisms are heterotrophic and use available organic material for energy. Soil microbiologists have long observed that wherever available energy material is abundant in soil, microbes capable of using that material are usually abundant.[9] In fact, Clark[9] stated that a significant fraction of the soil microbiota must be in a dormant condition a good portion of time because of the inadequacy of the soil's energy supply.

Second, the pH range that is best for soil microorganisms is between six and eight. For most species, the optimum pH within this range is slightly above seven.[9]

Third, soil temperature affects the degradation rate of organic chemicals. Biochemical reactions follow the general rule that the rate of chemical reaction increases as the temperature increases. Microbial activity requires liquid water; as a result, temperature should be above the freezing point of water. Also, many, but not all, microorganisms contain essential enzymes that will be denatured at temperatures of about 50°C; therefore, this temperature represents a reasonable upper limit for microbial activity. Optimum petroleum degradation rates by aerobic bacteria occur at temperatures between 15°C and 30°C.

Fourth, although the scientific literature contains information on how soil moisture content affects microbial population growth in relation to crop production, very little is known about how the soil moisture content affects microbial degradation of organic chemicals. In general, microorganisms need water to support their metabolic processes, and extreme moisture conditions may be unfavorable for the growth of soil microorganisms. However, not all microbial subgroups are equally affected by low soil moisture content, and seldom are individual species eliminated entirely in dry soil.[9]

Fifth, there are at least 11 essential macronutrient and micronutrient elements that must be present in soil in the proper amounts, forms, and ratios to sustain the growth of aerobic bacteria. These include nitrogen, phosphorus, potassium, sodium, sulfur, calcium, magnesium, iron, manganese, zinc, and copper. Several additional micronutrients are necessary to sustain anaerobic bacteria including nickel, cobalt, and sulfur. Therefore, the soil availability and the soil capacity of these macronutrients should be optimized to achieve maximum degradation rates. The availability of nutrients and oxygen has significant effects on petroleum degradation. In particular, nitrogen and phosphorus fertilizers, as well as oxygen, accelerate biodegradation.

Brink[10] gives an excellent discussion on how the concentration of the organic chemical can have a marked effect on its degradation rate. Higher concentrations of some chemicals may lead to faster degradation because of reduced acclimation time and/or a rapid increase in the microbial population. On the other hand, some chemicals that are readily biodegradable at low concentrations will inhibit microbial activity at higher concentration, presumably because of microbial toxicity.

It has been assumed by many scientists that if a chemical were readily biodegradable at a moderate concentration, say a few milligrams per liter, then the same chemical, if present at nanograms per liter to a few micrograms per liter, would also be readily biodegradable. There is uncertainty about this. It may be that when an organic chemical is present at very low concentrations, the low concentration may become a limiting factor in the biodegradation of the organic chemical through, perhaps, a lack of enzyme induction. It may be that this persistence is a consequence of the fact that very low organic chemical concentrations yield a reaction rate so slow that the chemical appears to be nondegrading.

In addition to the above, after being applied to soil, the chemical must first be solubilized before coming in contact with microorganisms. This requirement is controlled by water solubility, soil moisture, soil water movement, adsorption of the compound on soil particles, and slow solution in particulate soil organic matter.

ENZYMES AND BIOREMEDIATION

It is important to note that enzymes are responsible for catalyzing microbial metabolism. Once inside the microorganism's membrane, chemicals will collide with enzymes. Whether or not the chemical transforms as a result of this collision will depend on (1) the chemical binding to the enzyme, and (2) conformation changes at the active site. In most enzymes, chemical-induced conformation changes play a vital role. Reactive chemicals will produce the ideal alignment of catalytic groups, while nonreacting chemicals produce a less favorable alignment. Nonreacting chemicals are not bound at all or fail to produce an alignment that can lead to reaction. If a chemical can induce a suitable conformation change in an enzyme, then enzyme induction occurs and reaction will occur.

THREE GENERAL PATHWAYS OF MICROBIOLOGICAL METABOLISM

Once inside the membrane, chemicals such as petroleum products are catabolized (i.e., degraded) by microorganisms using three general metabolic pathways: aerobic respiration, anaerobic respiration, and fermentation. In aerobic respiration, organic chemicals are oxidized to carbon dioxide and water or other end products using molecular oxygen as the terminal electron acceptor. Oxygen may also be incorporated into products of microbial metabolism through the action of oxidase enzymes. Microorganisms metabolize hydrocarbons by anaerobic respiration in the absence of molecular oxygen using inorganic substrates as terminal electron acceptors. In anaerobic respiration, CO_2 is reduced to methane, sulfate to sulfide, and nitrate to molecular nitrogen or ammonium ion. Hydrocarbon sources are degraded by fermentation using substrate level phosphorylation as the terminal electron acceptor. In other words, the fermentation process occurs independent of oxygen and depends on organic compounds as electron acceptors.

Fermentation results in a wide variety of end products including CO_2, acetate, ethanol, propionate, and butrate.[11]

CHEMICAL STRUCTURE AND BIODEGRADATION

The biodegradation of petroleum products is the modification or decomposition of the product by microbes to produce ultimately microbial cells and carbon dioxide. The petroleum biodegradation process is complex and not fully understood. The potential degradability of petroleum hydrocarbons can be generally estimated based on the structure of the chemicals comprising the product.

The susceptibility of petroleum products to biodegradation varies with the types and size of the components. For example, alkanes of intermediate chain length (C_{10}–C_{24}) are rapidly degraded. However, very long chain alkanes are resistant to biodegradation, and after exceeding a molecular weight of 500 to 600, they cease to serve as carbon sources. Since there are in excess of one hundred components in gasoline or fuel oil, the rate and extent of degradation is not easily predictable. Thus the overall degradability of any petroleum product will depend on the proportion of degradable chemicals of which it is composed. The biodegradation of petroleum products is generally described by the following equation:

$$\text{Petroleum Product} + \text{Bacteria} + O_2 + \text{Nutrients} \dashrightarrow CO_2 + H_2O + \text{Byproducts} + \text{Biomass}$$

Specifically, the biodegradation pathway for alkane petroleum products is:

$$\text{Alkane} \dashrightarrow \text{Alcohol} \dashrightarrow \text{Aldehyde} \dashrightarrow \text{Fatty Acid} \dashrightarrow \text{Acetate via Beta Oxidation} \dashrightarrow CO_2 + H_2O + \text{Biomass}$$

Carbon is distributed in the environment in a variety of chemicals that range from gases (methane and carbon dioxide) to liquids (benzene and toluene) to solids (simple sugars and asphaltic components of crude oil). The biological degradation of these chemicals is a naturally occurring phenomenon.

A number of published studies and reviews discuss in great detail the degradation of petroleum products in relation to chemical structure.[4,10,12,13] In summary, alkanes, alkylaromatics, and aromatics are the predominant chemicals comprising most petroleum products. The n-alkanes, n-alkylaromatic, and aromatic compounds of the C_{10} to C_{22} range are the least toxic and most readily biodegradable. The n-alkanes, alkylaromatic, and aromatic hydrocarbons in the C_5 to C_6 range are biodegradable at low concentrations by some microorganisms, but in most environments they are removed by volatilization rather than by biodegradation. Gaseous n-alkanes (C_1 to C_4) are biodegradable but are used only by a narrow range of specialized hydrocarbon degraders. The n-alkanes, alkylaromatic, and aromatic compounds above C_{22} have extremely low water solubility. Their solid state at physiological temperatures makes microbial transformations of these chemicals slow.

Branching structures typical of asphaltics generally reduce the biodegradation rate, and aromatic compounds are degraded more slowly than alkanes.[14] Branching creates tertiary and quaternary carbon atoms that constitute a hindrance to beta oxidation. Branched alkanes and cycloalkanes of the C_{10} to C_{22} range are less biodegradable than their n-alkane and aromatic analogs. The biodegradation of cycloalkanes requires synergistic cooperation of two or more microbial species. Also, cycloalkanes of C_{10} and below have high membrane toxicity. Highly condensed aromatic and cycloparaffinic systems, with four or more condensed rings, and the partially oxygenated and condensed components of tar, bitumen, and asphalt degrade slowly. Some hydrocarbons and hydrocarbon biodegradation products are highly resistant to ultimate biodegradation; that is, mineralization. Condensed polycyclic aromatics and cycloparaffins, as well as high-molecular-weight alkanes, are mineralized only very slowly.

Petroleum products contain a wide variety of additives such as multipurpose detergents, anti-icing additives, dispersants, flow improvers, pour depressants, oxidation and corrosion inhibitors, octane improvers, antiknock compounds, combustion aids, antistats, biocides, dyes, demisting agents, and gellants. These additives possess a wide array of chemical structures. It is most fortunate that soil microorganisms possess the ability to transform a wide array of organic chemicals (see Table 2). Alexander[15] discussed in great detail the reaction types for chemical transformations, cleavage, and conjugation reactions (see Table 3, Table 4, and Table 5) due to microbial degradation.

Table 2. Organic Molecular Fragments Amenable to Microbial Transformations.

Alcohols	Ketones
Aldehydes	Lactams
Alicyclic aliphatics	Lactones
Aliphatics (saturated)	Nitriles
Aliphatics (unsaturated)	Nitro compounds
Amides	Nitrosamines
Amines	Organoarsenicals
Aromatics (simple substituted)	Organomercurials
Aromatic heterocyclics	Organophosphorus compound
Azides	Organosulfates
Carbamates	Organotins
Carboxylic acids	Oximes
Condensed aromatics	Quaternary ammonium compounds
Dithiocarbamates	Sulfides
Esters	Sulfonic acids
Ethers	Thioamides
Glycosides	Thiol carbomates
Halides	Thiols
Heterocyclics	Ureas
Hydroxamic acids	
Hydroxyl amines	

Table 3. Categories of Microbial Transformations

Category	Reaction[a]
Dehalogenation	$RCH_2Cl \rightarrow RCH_2OH$ $ArCl \rightarrow ArOH$ $ArF \rightarrow ArOH$ $ArCl \rightarrow ArH$ $Ar_2CHCH_2Cl \rightarrow Ar_2C = CH_2$ $Ar_2CHCHCl_2 \rightarrow Ar_2C = CHCl$ $Ar_2CHCCl_3 \rightarrow Ar_2CHCHCl_2$ $Ar_2CHCCl_3 \rightarrow Ar_2C = CCl_2$ $RCCl_3 \rightarrow RCOOH$ $HetCl \rightarrow HetOH$
Deamination	$ArNH_2 \rightarrow ArOH$
Decarboxylation	$ArCOOH \rightarrow ArH$ $Ar_2CHCOOH \rightarrow Ar_2CH_2$ $RCH(CH_3)COOH \rightarrow RCH_2CH_3$ $ArN(R)COOH \rightarrow ArN(R)H$
Methyl oxidation	$RCH_3 \rightarrow RCH_2OH$ and/or $\rightarrow RCHO$ and/or $\rightarrow RCOOH$
Hydroxylation and ketone formation	$ArH \rightarrow ArOH$ $RCH_2R' \rightarrow RCH(OH)R'$ and/or $\rightarrow RC(O)R'$ $R(R')CHR'' \rightarrow R(R')CHOH(R'')$ $R(R')(R'')CCH_3 \rightarrow R(R')(R'')CCH_2OH$
Beta oxidation	$ArO(CH_2)_nCH_2CH_2COOH \rightarrow ArO(CH_2)_nCOOH$
Epoxide formation	/O\ $RCH = CHR' \rightarrow RCH$ --- CHR'
Nitrogen oxidation	$R(R')NR'' \rightarrow R(R')N(O)R''$
Sulfur oxidation = S to = O	$RSR' \rightarrow RS(O)R'$ and/or $\rightarrow RS(O_2)R'$ $(AlkO)_2P(S)R \rightarrow (AlkO)_2P(O)R$ $RC(S)R \rightarrow RC(O)R'$
Sulfoxide reduction	$RS(O)R' \rightarrow RSP'$
Reduction of triple bond	$RC = CH \rightarrow RCH = CH_2$
Reduction of double bond	$Ar_2C = CH_2Ar_2CHCH_3$ $Ar_2C = CHCl \rightarrow Ar_2CHCH_2Cl$
Hydration of double bond	$Ar_2C = CH_2 \rightarrow Ar_2CHCH_2OH$
Nitro metabolism	$RNO_2 \rightarrow ROH$ $RNO_2 \rightarrow RNH_2$
Oxime metabolism	$RCH = NOH \rightarrow RC = N$
Nitrile-amide metabolism	$RC = N \rightarrow RC(O)NH_2$ and/or $\rightarrow RCOOH$

[a]Abbreviations: R = organic fragment, Ar = aromatic, Alk = alkyl, Het = heterocycle.
Source: Alexander.[15]

PREDICTING DEGRADATION RATES

Unlike other physical-chemical reactions and properties of chemicals that may be estimated using mathematical techniques—water solubility, solvent solubility, adsorption, bioconcentration, dissociation constants, rates of hydrolysis, and photolysis—no quantitative, accurate procedure(s) for estimating the rate of biodegradation of chemicals exists at the present time.

Table 4. Microbial Cleavage Reactions

Molecular Fragment	Reaction
Ester	RC(O)OR′ --> RC(O)OH
Ether	ArOR --> ArOH
	$ROCH_2R'$ --> ROH
C -- N bond	R(R′)NR″ --> R(R′)NH and/or --> RNH_2
	$RN(Alk)_2$ --> RNHAlk and/or --> RNH_2
	RNHCH(R′)R″ --> RNH_2
	RNH_2CH_2R' --> RNH
Peptide, carbamate	RNHC(O)R′ --> RNH_2 and/or HOOCR′
	R(R′)NC(O)R″ --> R(R′)NH + HOOCR″
= NOC(O)R	RCH = NOC(O)R --> RCH = NOH
C -- S bond	RSR′ --> ROH and/or ESR′
C -- Hg bond	RHgR′ --> RH and/or Hg
C -- Sn bond	R_3SnOH --> R_2SnO --> $RSnO_2H$
C -- O -- P	$(AlkO)_2P(S^a)R$ --> $AlkO(HO)P(S^a)R$ and/or --> $(HO)_2P(S^a)R$
	$ArOP(S^a)(R)R'$ --> ArOH and/or $HOP(S^a)(R)R'$
P -- S	RSP(O)(R′)OAlk --> HOP(O)(R′)OAlk
Sulfate ester	$RCH_2OS(O_2)OH$ --> RCH_2OH and/or $HOS(O_2)OH$
S -- N	$ArS(O_2)NH_2$ --> $ArS(O_2)OH$
S -- S	RSSR --> RSH

[a]Sulfur or oxygen
Source: Alexander.[15]

Table 5. Categories of Microbial Reactions

Category	Reaction
Methylation	ArOH --> $ArOCH_3$
	$CH_3As(O)(OH)$ --> $(CH_3)_2AsH$ and/or $(CH^2)_3As$
	Hg^{2+} --> CH_3Hg^{3+} --> $(CH_3)_2Hg$
Ether formation	RCH_2R --> R(R)CHOCH(R)R
N-Acylation	$ArNH_2$ --> ArNHC(O)H
	$ArNH_2$ --> $ArNHC(O)CH_3$
	$ArNH_2$ --> $ArNHC(O)CH_2CH_3$
Nitration	ArH --> $ArNO_2$
N-Nitrosation	$(Alk)_2NH + NO_2$ --> $(Alk)_2NNO$
Dimerization	$2ArNH_2$ --> ArN = NAr
	RSH --> RSSR

Source: Alexander.[15]

This state is probably due to the fact that the majority of published research is descriptive and focuses on: (1) identification of the organisms responsible for degradation of specific substances, (2) the metabolic products of such degradation, and (3) classification of metabolic pathways. Quantitative data are scarce and have generally not been compiled. Furthermore, experimental methods for measuring biodegradation rates are not standardized. The variables that control rates are not well understood, as they have not been examined across different classes of chemicals. As a result, the data are not comparable and apply only to a particular set of experimental conditions.

Predictions tend to rely on the expertise of the scientist as he or she applies various rules of thumb based on chemical structure and on findings of published studies having experimental or field conditions similar to the one of concern. Owing to the complexity of the biodegradation process, the development of accurate, quantitative methods to predict the biodegradation rates of organic chemicals will prove to be a formidable challenge to the microbiologist, soil scientist, and environmental engineer for many years.

BIOREMEDIATION TECHNOLOGIES

A number of engineered treatment systems are available for bioremediation of petroleum products in soil. The selection of a bioremediation system must be based on the physical/chemical/biological properties of the product, site constraints, cleanup criteria, and state or local permit requirements. The following will address some of the more common bioremediation alternatives for treatment of petroleum products in soil.

Landfarming

Landfarming is commonly used for treatment of petroleum products and a wide range of liquid and solid waste. Landfarming is an enhanced bioremediation technique in which the petroleum product is spread on soil and biodegraded by soil microbes. Landfarming is effective for the removal of petroleum hydrocarbons, typically reducing total petroleum hydrocarbon (TPH) levels below 1000 mg/kg. Remediation to less than 100 mg/kg may require extended treatment time, addition of microbes, aqueous surfactants, and frequent rototilling or soil rotation.

The steps involved for implementation and treatment of petroleum products or soils by this technique are shown in Figure 1. The first prerequisite for implementation of landfarming is sufficient land area to spread the soil to a depth of 6 to 30 inches. The upper 6 to 12 inches is the treatment zone or zone of incorporation. The soil depth must be adequate to permit rototilling or mechanical aeration of the soil and prevent liner puncture.

Many states require the installation of a geotextile and/or plastic liner beneath the soil to capture leachate. Leachate is moisture which exceeds the holding capacity of the soil and percolates through the soil onto the liner. A 20 to 100 mil high-density polyethylene (HDPE) liner, leachate underdrain piping, and concrete collection sump are used to capture leachate for recirculation or treatment.

An HDPE underdrain system wrapped in filtration cloth and backfilled with a minimum of 12 inches of sand beneath 6 inches of gravel may be installed for collection of leachate if the area will be used for an extended period of time or as required by local regulations. The treatment area should be surrounded by a

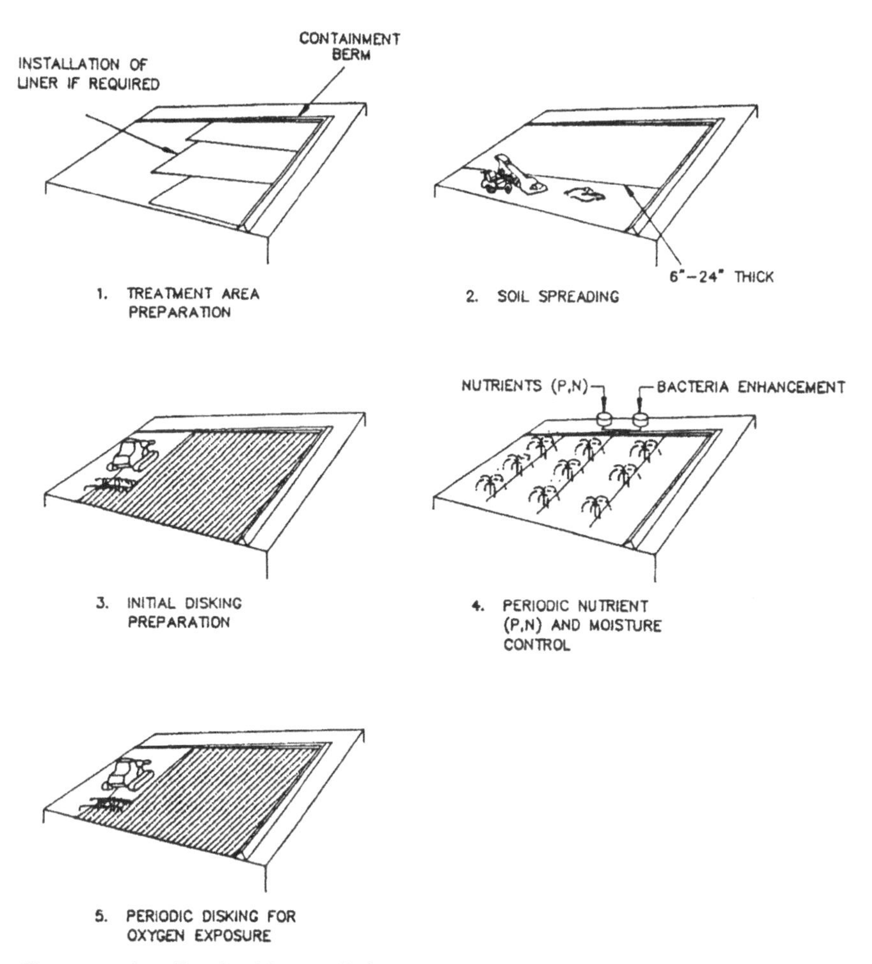

Figure 1. Landfarming bioremediation treatment system.

containment berm to support the liner system and prevent surface runoff from leaving the site as shown in Figure 1.

Landfarmed soil may be covered to minimize release of odors and leachate production. Local air quality regulations may require that the landfarming operation be covered for containment and treatment of fugitive VOC emissions.

Petroleum-laden soil is spread on the liner and rototilled to improve soil texture, increase soil permeability, soil uniformity, and improve transfer efficiency of atmospheric oxygen. A water spraying system may be installed to distribute water conditioned with nutrients and surfactant, to increase soil moisture levels, and maximize microbial growth. The addition of nutrients and other conditioning reagents must be evaluated prior to system design, based on petroleum concentration, indigenous microbial population, contaminant biodegradability, and cleanup criteria.

Landfarming is effective for removal of volatile and some semivolatile petroleum compounds. High molecular weight semivolatile compounds may be refractory and resist biodegradation. During treatment, soils require frequent rototilling and periodic rotation to optimize the rate and extent of biodegradation.

The cost for landfarming is generally low compared with other treatment alternatives. Typical landfarming costs range from $15 to $50 per cubic yard. The treatment time for petroleum-laden soil is from two to six months. Treatment efficiency is high during summer months, with little decrease of petroleum hydrocarbons occurring as soil temperature declines during the fall and winter months. Manpower requirements for landfarming are low. Following construction of the treatment area and spreading of petroleum-laden soils, regular rototilling, occasional soil rotation, and sampling are necessary.

Biopile

A modification of the landfarming process which also utilizes microbial degradation of petroleum products is the biopile. A biopile is generally constructed to include a perforated underdrain used to collect leachate and for withdrawal of air from the pile.

An example of a biopile soil bioremediation system is shown in Figure 2. A vacuum pump is connected to the underdrain to draw air into the biopile and stimulate microbial growth. A biopile may also include a spray or drip irrigation system to optimize soil moisture levels and treatment efficiency.

Local regulations may require the biopile to be covered to capture fugitive VOC air emissions for treatment prior to discharge into the atmosphere. The withdrawn air may be treated using vapor phase-activated carbon, catalytic or thermal oxidation. Biopiles typically produce fewer fugitive emissions compared with landfarming since the net transport of ambient air is into the pile and the withdrawn air is treated.

The treatment time for soil using this alternative is from one to four months. The treatment cost may range from $20 to $70 per cubic yard.

Composting

Composting is a bioremediation process similar to landfarming, but includes the addition of material referred to as a soil amendment to supply energy for microbial growth and degradation of petroleum products. This process is most effective for the removal of organic compounds containing high volatile solids concentrations. The added material or amendment may be hay, grass clippings, leaves, wood chips, straw, sawdust, or manure. The soil amendment is added to increase soil permeability, improve oxygen transfer, improve soil texture, and provide an energy source to rapidly establish a large microbial

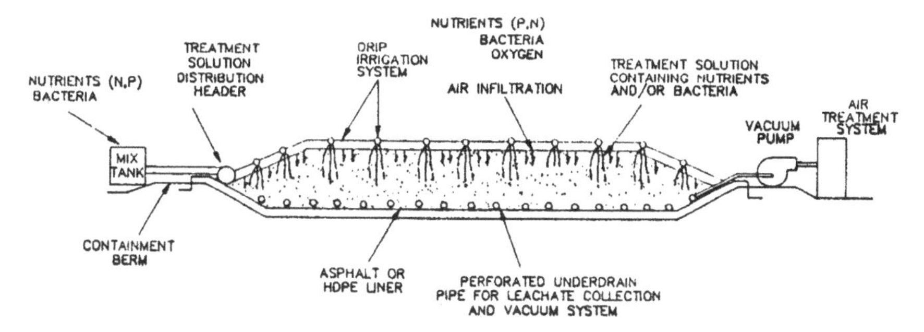

Figure 2. Biopile treatment system.

population. Microbes typically consume both the soil amendment and the petroleum product.

Composting is an aerobic process which through self heating can generate high temperatures within the compost pile, which is necessary for treatment during winter months. Soil temperatures of 50°C are possible within the compost pile under favorable conditions.

A composting process schematic is included in Figure 3 and Figure 4. The amended soil is generally piled in windrows over an HDPE-perforated piping system used to aerate the compost pile. A blower is operated to introduce air necessary for microbial respiration into the compost pile. Low pressure regenerative or centrifugal blowers capable of imparting a vacuum of one-half to ten inches mercury are typically specified. The windrows are rotated periodically to maximize the transfer of oxygen into the pile.

The proper mix of petroleum-laden soil, moisture, nutrients, and soil amendment is critical for establishing the proper balance to satisfy the energy requirements and optimize bioremediation efficiency. The compost pile must be sufficiently porous to allow the passage of air, and have a solids content that is appropriate for the solids handling equipment. A mix that is too wet will reduce porosity, hinder material handling, and increase treatment time.

Bench-scale or pilot-scale testing is often useful to determine the appropriate soil:soil amendment ratio to optimize treatment efficiency. Since there are frequently many treatment variables from site to site, testing is recommended to evaluate the feasibility of bioremediation. Field sampling and testing will determine the extent of biodegradation and potential for compliance with the local cleanup requirements.

The compost process may reduce treatment time for bioremediation of petroleum-laden soil compared with landfarming or biopile techniques. The treatment time for petroleum-laden soils may be from one to four months. The treatment cost may range from $20 to $80 per cubic yard, which is slightly higher than landfarming due to the added cost of the soil amendment, mixing, and handling.

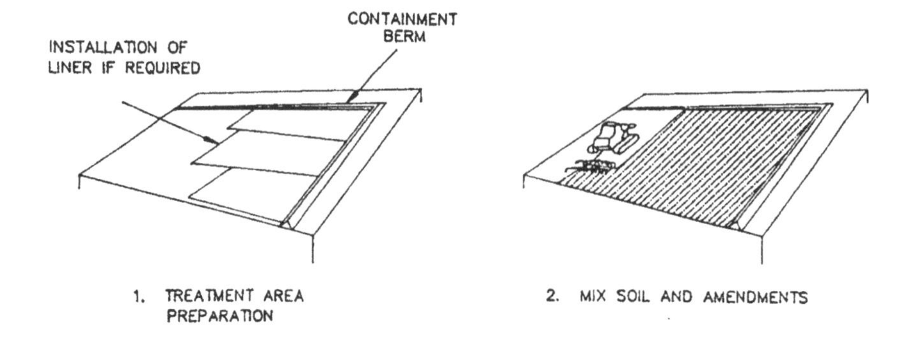

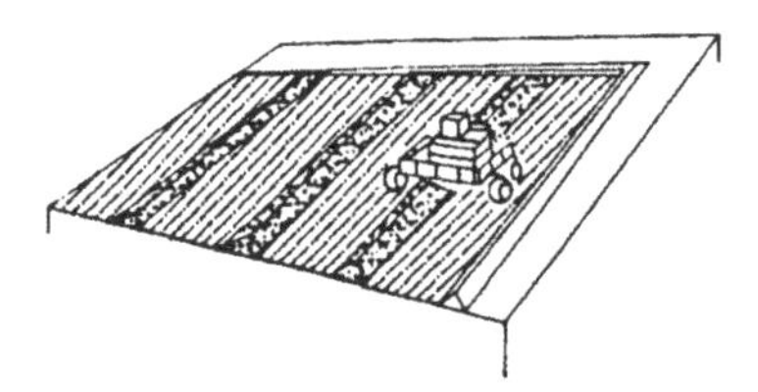

Figure 3. Composting bioremediation system.

In Situ Bioremediation

Use of in situ bioremediation has increased recently for treatment of petroleum-laden soil. The in situ bioremediation process generally allows for ongoing business operations at contaminated sites such as filling stations.

A schematic of an in situ bioremediation system is included in Figure 5. Groundwater withdrawal wells are advanced for collecting affected groundwater which is treated or dosed onto the treatment area to increase soil moisture, nutrient, and oxygen concentrations. The treatment system piping may be buried in the ground or installed at ground elevation. The applicability of this process will depend upon the soil hydraulic permeability, hydrocarbon contaminant, contaminant concentration, soil temperature, soil nutrient level, indigenous microbial population, and cleanup criteria.

This alternative requires aeration of the recirculated or added water to supply oxygen, nutrients, microbes, and other additives necessary for the biodegradation of petroleum products. The most common methods for oxygen addition include: aeration/reinjection of extracted groundwater, soil venting, addition of liquid oxygen, or hydrogen peroxide. Oxygen concentrations up to 500 mg/L are possible

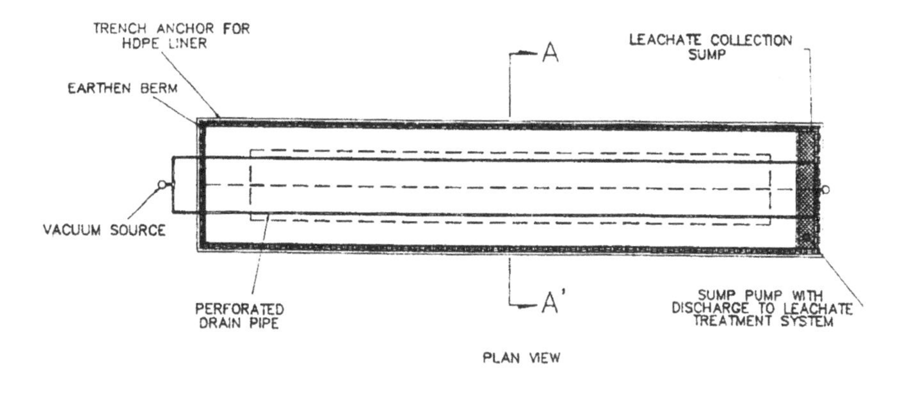

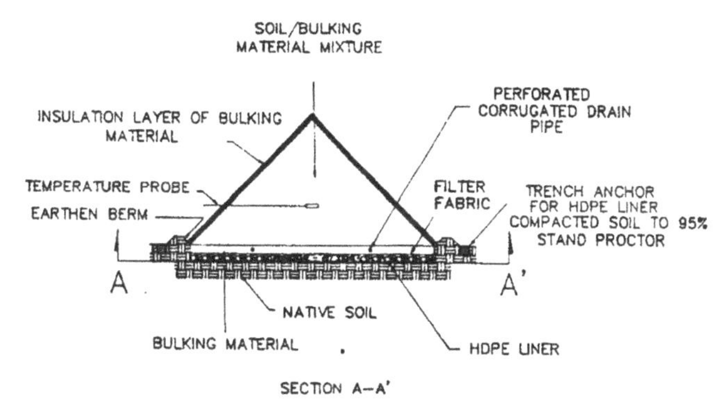

Figure 4. Soil composting system.

using hydrogen peroxide, compared with 12 mg/L using diffused aeration. The high oxygen concentration possible using hydrogen peroxide makes it the preferred alternative.

Limitations of the in situ bioremediation process may include: reduced degradation rate due to soil temperatures, potential for migration of contaminants into groundwater, higher operation and maintenance costs compared with ex situ land treatment alternatives, added treatment time compared with ex situ alternatives, and permit approval in some states.

Bioremediation may pose limitations in terms of the ability of the recovery systems to capture contaminated groundwater and restrictions on reinjection of treated or nutrient enriched waters into the unsaturated soil. Regulations may require the installation of numerous groundwater withdrawal wells, trench drains, or slurry walls to contain and capture affected groundwater for treatment. The limitations must be identified during a site investigation and the engineered bioremediation system must be designed to address and overcome limitations to ensure successful treatment.

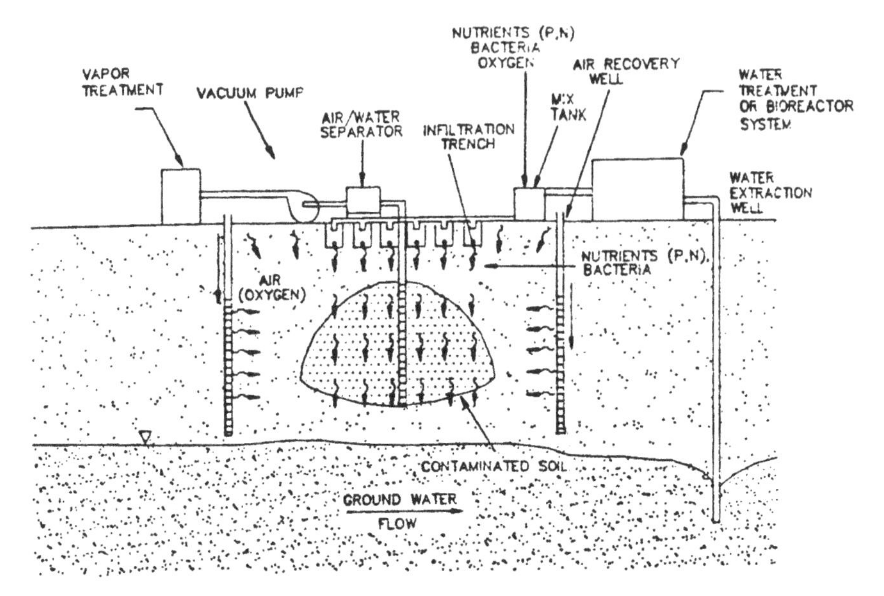

Figure 5. In situ bioremediation system.

The system is constructed using simple construction, irrigation, and well drilling techniques. Material of construction for underdrains, groundwater extraction wells, and injection wells include polyvinyl chloride (PVC), galvanized, and HDPE. The system basis of design should be prepared by an experienced environmental engineer to assure successful operation and completion of the bioremediation project.

In situ bioremediation treatment is a proven technique for the remediation of petroleum products in unsaturated soil. Installation time is dependent upon the system design, groundwater containment strategy, and system size. The remediation period may require from six months to several years, depending upon the extent of contamination and cleanup criteria. The cost of soil treatment using in situ bioremediation is from $30 to $100 per cubic yard.

SUMMARY

Microbial degradation of petroleum products in soil, either via naturally occurring or facilitated methods, is a process that is successfully used to reduce soil concentrations of the product to acceptable levels. The heterogeneous or heterotrophic microorganisms found in soils include naturally occurring populations that possess the ability to degrade petroleum products.

Five important soil factors that affect biodegradation processes include: (1) availability of an energy source, (2) favorable soil pH, (3) soil temperature, (4) availability of soil moisture, and (5) availability of essential macro- and micronutrients.

The potential degradability of petroleum hydrocarbons can be generally estimated based on the structure of the chemicals comprising the product. Branching structures typical of asphaltics generally reduce the biodegradation rate, and aromatic compounds are degraded more slowly than alkanes. Branching creates tertiary and quaternary carbon atoms that constitute a hindrance to beta oxidation.

Enzymes are responsible for catalyzing microbial metabolism. Once inside the membrane, chemicals such as petroleum products are catabolized (i.e., degraded) by microorganisms using three general metabolic pathways: aerobic respiration, anaerobic respiration, and fermentation.

A number of engineered treatment systems are available for bioremediation of petroleum products in soil. The selection of a bioremediation system must be based on the physical/chemical/biological properties of the product, site constraints, cleanup criteria, and state or local permit requirements.

Landfarming is commonly used for treatment of petroleum products and a wide range of liquid and solid waste. Landfarming is an enhanced bioremediation technique in which the petroleum product is spread on soil and biodegraded by soil microbes. Landfarming is effective for the removal of petroleum products typically reducing total petroleum hydrocarbon (TPH) levels below 1000 mg/kg.

Petroleum-laden soil is spread on the liner and rototilled to improve soil texture, increase soil permeability, soil uniformity, and improve transfer efficiency of atmospheric oxygen. A water spraying system may be installed to distribute water conditioned with nutrients and surfactants, to increase soil moisture levels, and maximize microbial growth.

The cost for landfarming is generally low compared with other treatment alternatives. Typical landfarming costs range from $15 to $50 per cubic yard. The treatment time for petroleum-laden soil is from two to six months.

A modification of the landfarming process which also utilizes microbial degradation of petroleum products is the biopile. A biopile is generally constructed to include a perforated underdrain which is used to collect leachate and withdraw air from the pile. A vacuum pump is connected to the underdrain to draw air into the biopile and stimulate microbial growth. A biopile may also include a spray or drip irrigation system to optimize soil moisture levels and treatment efficiency. Biopiles typically produce fewer fugitive emissions compared with landfarming since the net transport of ambient air is into the pile and the withdrawn air is treated.

The treatment time for soil using this alternative is from one to four months. The treatment cost may range from $20 to $70 per cubic yard.

Composting is a bioremediation process similar to landfarming but includes the addition of a soil amendment to supply energy for microbial growth and biodegradation of petroleum products. The proper mix of petroleum-laden soil, moisture, nutrients, and soil amendment is critical for establishing the proper balance to satisfy the energy requirements and optimize bioremediation efficiency.

The treatment time for petroleum-laden soils may be from one to four months. The treatment cost may range from $20 to $80 per cubic yard, which is slightly higher than landfarming due to the cost of the soil amendment, mixing and handling.

Use of in situ bioremediation has increased recently for treatment of petroleum-laden soil. The applicability of this process will depend upon the soil hydraulic permeability, hydrocarbon contaminant, contaminant concentration, soil temperature, soil nutrient level, indigenous microbial population, and cleanup criteria.

This alternative requires aeration of the recirculated or added water to carry oxygen, nutrients, microbes, and other additives necessary for the biodegradation of petroleum products. The most common methods for oxygen addition include: aeration/reinjection of extracted groundwater, soil venting, and addition of liquid oxygen or hydrogen peroxide.

The bioremediation period may require from six months to several years, depending upon the petroleum product, product concentration and cleanup criteria. The cost of soil treatment using in situ bioremediation is from $30 to $100 per cubic yard.

REFERENCES

1. Odu, C. T. I. "The Effect of Nutrient Application and Aeration on Oil Degradation in Soil," *Environ. Pollut.* 15:235–240 (1978).
2. Perry, J. J., and Scheld, H. W. "Oxidation of Hydrocarbons by Microorganisms Isolated from Soil," *Can. J. Microbiol.* 14:403–407 (1968).
3. Pinholt, Y., Struwe, S., and Kjoller, A. "Microbial Changes During Oil Decomposition in Soil," *Holartic Ecol.* 2:195–200 (1979).
4. Bossert, I., and Bartha, R. "The Fate of Petroleum in Soil Ecosystems," in *Petroleum Microbiology,* R. M. Atlas, Ed. (New York, NY: Macmillan Publishing Co., Inc., 1984).
5. Felix, J. A., and Cooney, J. J. "Response of Spores and Vegetative Cells of *Bacillus* spp. in a Hydrocarbon-Water System," *J. Appl. Bacteriol.* 34:411–416 (1971).
6. Jensen, V. "Bacterial Flora of Soil After Application of Oily Waste," *Oikos* 26:152–158 (1975).
7. Dibble, J. T., and Bartha, R. "Effect of Environmental Parameters on the Biodegradation of Oil Sludge," *Appl. Environ. Microbiol.* 37:729–739 (1979).
8. Odu, C. T. I. "Microbiology of Soils Contaminated with Petroleum Hydrocarbons. I. Extent of Contamination and Some Soil and Microbial Properties After Contamination," *J. Inst. Pet.* 58:201–208 (1972).
9. Clark, E. A. "Soil Microbiology," in *The Encyclopedia of Soil Science,* Part 1, E. A. Fairbridge and C. W. Finkl, Jr., Eds., (Stroudsburg, PA: Dowden, Hutchinson, and Ross, 1979).
10. Brink, R. H. "Biodegradation of Organic Chemicals in the Environment," in *Environmental Health Chemistry,* J. D. McKinney, Ed. (Lancaster, PA: Technomic Publishing Company, 1981).
11. Speece, R. E. "Anaerobic Biotechnology for Industrial Wastewater Treatment," *Environmental Science & Technology* 17(9):416–427 (1983).
12. Gibson, D. T. "Microbial Metabolism," in *The Handbook of Environmental Chemistry,* Volume 2, Part A, Reactions and Processes. O. Hutzinger, Ed., (New York, NY: Springer-Verlag, 1980).

13. Ribbons, D. W., and Eaton, R. W. (1981). "Chemical Transformations of Aromatic Hydrocarbons That Support the Growth of Microorganisms," in *Biodegradation and Detoxification of Environmental Pollutants,* A. M. Chakrabarty, Ed. (Boca Raton, FL: CRC Press, 1981).
14. Atlas, R. *Petroleum Microbiology* (New York, NY: Macmillan Publishing Co., Inc., 1984).
15. Alexander, M. "Biodegradation of Chemicals of Environmental Concern," *Science* 211:132–138 (1981).

CHAPTER 9

Mathematical Hydrocarbon Fate Modeling in Soil Systems

Marc Bonazountas and **Despina Kallidromitou,** National Technical University of Athens, Department of Civil Engineering, Athens, Greece

INTRODUCTION

This chapter describes principles, concepts, uses, and limitations of state-of-knowledge mathematical pollutant fate modeling related to hydrocarbon environmental contamination. Mathematical environmental modeling primarily refers to the terrestrial environment and secondarily to other environmental media such as air and water.

The purpose of this chapter is to help readers understand modeling complexities, identify specific modeling packages, and select a documented mathematical model for environmental quality and fate analyses, and assist in environmental decisionmaking, including analyses of control strategies to minimize environmental risk or human exposure from hydrocarbon contaminated sites.

Each section of this chapter is fairly self-contained; it includes an introduction, a discussion of key physical, chemical, and biologic issues related to each modeling category, a discussion of important environmental interactions of organic and inorganic pollutants with the specific environment, an outline of model requirements for input data, and examples of model applications.

The chapter is divided into sections that provide information on legal and regulatory issues related to modeling, including major pathways of pollutants in the environment; sources and emissions of hydrocarbons in the environment;

hydrocarbon fate processes and chemistry, including the mathematics of the various process relations for miscible and immiscible pollutants; general aspects of mathematical modeling; miscible pollutant fate modeling of hydrocarbons and other miscible pollutants, including applications; immiscible mathematical pollutant fate modeling; aquatic equilibrium pollutant fate modeling; and critical parameters affecting pollutant fate modeling of hydrocarbons in soils.

Many terrestrial pollution management programs are being developed within federal agencies, state and local governments, and the private sector. The planning and implementation of these programs should be based on sound technical information and knowledge, including mathematical modeling as a tool for predicting contaminant fate in the environment.

Literature Review

Selected recent studies and literature of pollutant transport and transformation modeling in the soil (vadose) zone are reported in the references section. The traditional differential equation modeling[1] employing convection-dispersion mass transport differential equations[2] has given way to more sophisticated approaches to explain physicochemical and biochemical processes in soil systems via compartmental modeling, stochastic modeling[3] and other techniques.[4–6]

Several journals such as the *Journal of Contaminant Hydrology, Transport in Porous Media, Ground Water, Water Resources Research, Journal of the American Society of Civil Engineers,* and *Contaminated Soils,* report the state-of-knowledge on vadose zone and groundwater modeling. Conferences and workshops such as the biannual Conference on Hydrocarbon Contaminated Soils[7] provide ample information on modeling applications and fate of pollutants from hydrocarbon contaminated sites. Finally, review studies from Bonazountas,[8] Gee et al.,[4] and van Genuchten and Shouse[9] summarize current understanding of vadose and groundwater pollutant transport modeling issues.

Support for this intensive research has been provided by the United States government and agencies as well as international organizations and authorities (e.g., European Commission), given the importance of the vadose zone in agricultural chemical fate migration;[10] hydrocarbon fate from waste sites, including remediation techniques;[11] multimedia and multiple exposure pathway analyses for risk-based decisionmaking processes related to hydrocarbon contaminated sites;[12] design and implementation of states' underground storage tanks assurance funds; and other purposes. The ultimate research effort is to develop *tools* that can *reliably* predict the fate of contaminants in the soil zone and the groundwater in a way that can support environmental decisionmaking when selecting alternative waste sites or conducting environmental auditing and site characterization.

An almost exhaustive detailed literature review of the recent four years on several important issues related to calibration and model validation, mathematics and solution algorithms, pollutant types and characteristics, experimental work, remediation, and other topics has been recently presented by Gee et al.[4] The authors of this publication acknowledge this review. The authors also acknowledge the books edited by Kostecki and Calabrese,[13-15] and Calabrese and Kostecki,[7,16-18] and the books of Dragun[19] and Samiullah,[20] since they provide the most comprehensive information sources on hydrocarbon environmental chemistry and fate, public health effects, and remediation. This information forms the basis of mathematical modeling applications for interested scientists, as it provides chemical property values, soil chemistry reaction rates, toxicological effects parameters, and other migration-, fixation-, adsorption-, biodegradation-, and mobility-parameters important to modeling. Some of this information is also presented in the following sections.

This chapter can be considered as a resource document to aid in the process and to indicate weaknesses that have to be avoided. The following sections, however, only summarize the rapidly expanding body of knowledge on terrestrial hydrocarbon fate modeling, and cannot be considered as providing an in-depth review of available models and practices.

BACKGROUND

Legal and Regulatory Issues of Modeling

Models are applied in a variety of ways to assist in decisionmaking, and will be used to a greater extent in the future. Specific statutes or regulations require the use of models in certain situations (e.g., California LUFT Manual[21]). Additionally, provisions of the National Environmental Policy Act of 1969 (NEPA), as well as judicial decisions constituting NEPA and other environmental statutes, should facilitate the increased use of mathematical models.

There are a number of reasons for the increasingly widespread use of mathematical and computer models in environmental decisions. For example, Congress mandated the use of computer models in the 1977 Amendments to the Clean Air Act. Under the 1977 Amendments, models must be used in connection with the Prevention of Significant Deterioration (PSD) for air quality and for designation of nonattainment areas. Under the PSD program enacted in the 1977 Amendments, a major emitting facility in an area subject to PSD regulations must obtain a permit prior to commencing construction; this must be preceded by an analysis of air quality impacts projected for the area as a result of growth associated with such a facility. The analysis must be based on air quality models specified by regulations promulgated by the Environmental Protection Agency (EPA). The 1977 Amendments also require conferences on air quality modeling every three years, to ensure that the air quality models used in the PSD program reflect the current state-of-the-art in modeling. The same concepts apply under the California actions of selecting or siting or conducting environmental auditing related to petroleum contaminated sites.[21] Mathematical modeling plays a major role in decisionmaking and impact analyses.

The increasing use of quantitative computer models has placed a new burden upon decisionmakers in their review of environmental decisions based on modeling. This burden is a part of the new era in environmental decisionmaking, and reflects the increasing involvement of scientific and technical issues in legal decisions. Problems in judicial review arising from the use of models and other quantitative methodologies in environmental decisionmaking are described by Case.[22]

The problems arising from the use of an environmental model in regulatory practice are twofold. First, it may actually increase the likelihood that a substantially incorrect decision will be reached. This greater probability of error generally can be traced to the inability of environmental decisionmakers to deal with certain aspects of modeling in making such decisions. Second, the use of a model increases the danger that wrong environmental decisions may not be detected and corrected by the reviewing courts.

Additionally, certain institutional aspects of the environmental agencies, the courts, and their mutual relationship further contribute to the difficulties of judicial review in cases relying on environmental models. Such institutional problems include the self-perceived lack of scientific expertise on the part of judges, the lack of judicial access to technical resources to assist in analyzing the issues involved, the limits on the courts' ability to supplement the record, and the traditional deference that the courts give to administrative decisions.

Nevertheless, model development and application will increase in the future, especially in the areas of soil-groundwater pollution and general environmental modeling, where past applications and decisionmaking have reached the level of employment, as in the air environmental analyses. Beyond any legal mandates, increasingly complicated and intractable environmental problems will compel the greater use of quantitative models by environmental decisionmakers. Many experts believe in environmental fate, exposure, and risk modeling, since models contribute to scientific understanding of the environmental quality.[23]

Purpose of Modeling

Modeling fate of chemicals in environmental and soil systems is conducted for three main purposes: (1) assessment of environmental quality; (2) assessment of human exposure from pollutants and risk from pollution; and (3) decisionmaking, including implementation of control strategies for environmental and human protection.

With a soil system as an example, there are four major exposure pathways for contaminants at uncontrolled waste sites: leachate-groundwater, surface water, contaminated soil, and residual waste and air. The environmental setting for an uncontrolled waste site located above the water table is shown in Figure 1[24] and in Figure 2.[25] The potential pathways to human and ecological receptors are depicted in Figure 3. A variation of the above would be a site where the waste was buried below the water table; in this case, the leachate plume and the groundwater are coincident. The exposure pathways are essentially the same in both cases.

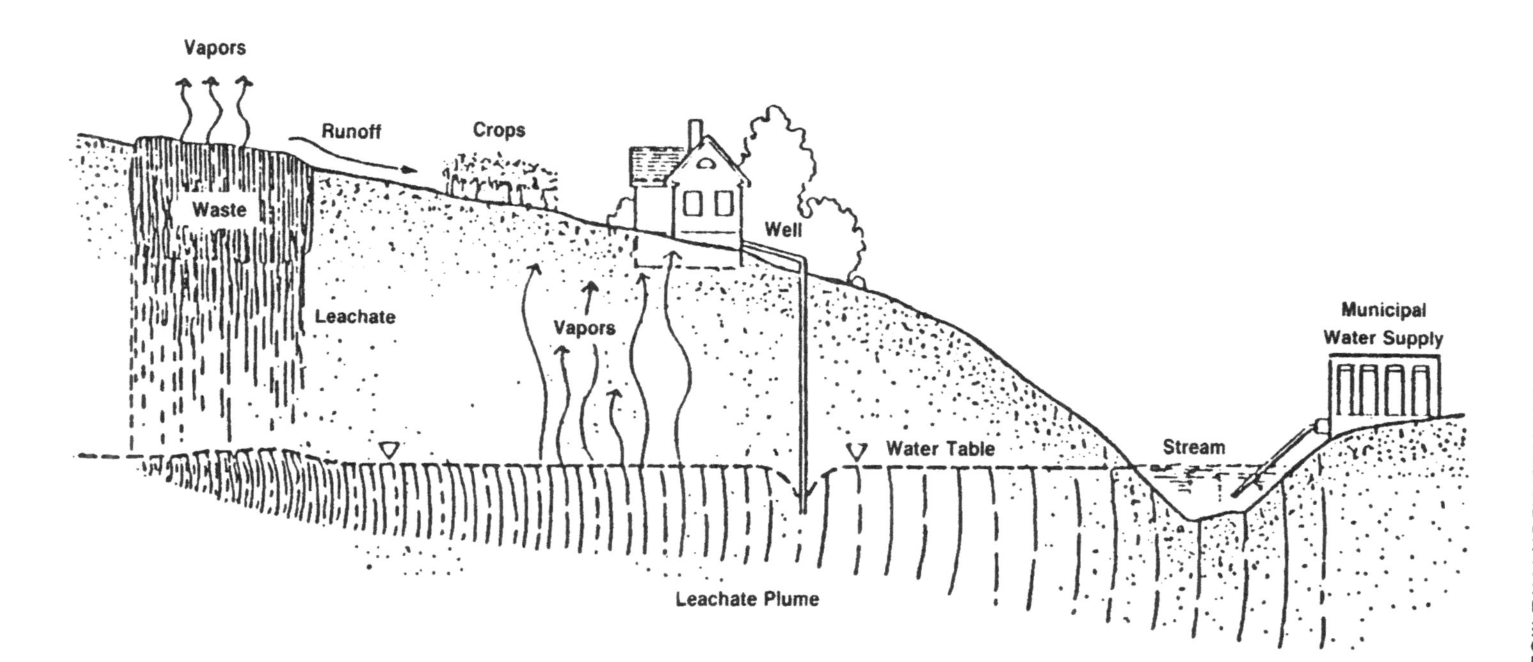

Figure 1. Environmental pathways from a contaminated site (from Ehrenfeld and Bass[24]).

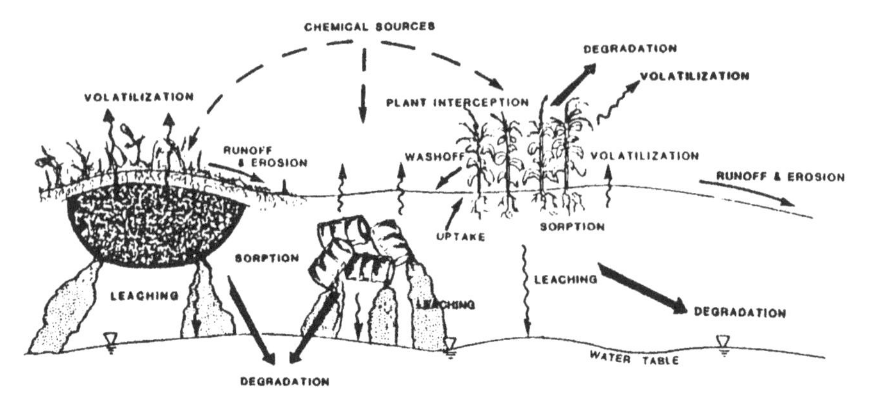

Figure 2. Chemicals in the soil environment (from Donigian and Rao[25]).

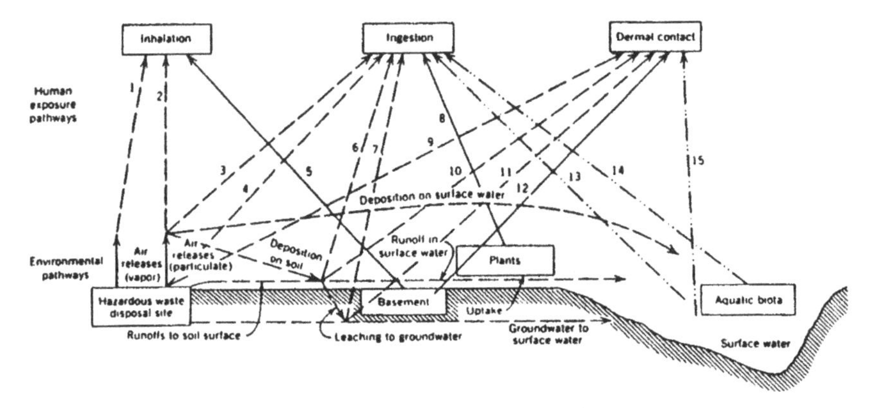

Exposure pathways from hazardous waste site: 1—inhalation of vapors, 2—inhalation of particulate contaminant, 3—swallowing of larger particles, 4—ingestion of soil on site (particularly children), 5—inhalation of vapors from seepage in basements, 6—ingestion of soil off site (particularly children), 7—ingestion of groundwater, 8—ingestion of plants (home gardens), 9—dermal contact with soil on site, 10—dermal contact with soil off site, 11—dermal contact from household use of groundwater (flushing, etc.), 12—dermal contact with seepage in basements, 13—ingestion of surface water, 14—dermal contact with surface water (swimming, etc.), 15—ingestion of aquatic biota.

Figure 3. Exposure pathways from a contaminated site (from Bonazountas[1]).

Remedial actions are designed to reduce exposure of both humans and the environment to acceptable levels either by containing pollutants originating from the site in place, or by removing the hazardous substances from the immediate environment.

Modeling can play an integral role in waste cleanup and other environmental protection studies. A model is a tool which, if applied properly, can greatly assist decisionmakers in effectively dealing with complex issues at uncontrolled polluted sites. Today, five basic modeling categories are important: (1) emission models, to quantify release or pollutant emissions in the environment (e.g., air emissions or leachate from a waste site); (2) fate models, to estimate concentrations of pollutants in the environmental media (e.g., fate of pollutants in the soil and to the groundwater); (3) exposure models, to correlate or convert environmental concentrations to uptake or absorbed doses by humans (e.g., inhalation); (4) risk models, that also account for dose-response models, for the extrapolation of animal carcinogenicity data to humans for the estimation of probable human risks to cancer or other risk types; and (5) cost-effectiveness models (e.g., reduction of human risk) when imposing alternative actions or strategies (e.g., remedial actions at waste sites).

This chapter presents information on fate modeling in soil and groundwater and adjacent environmental media (e.g., air, land, water-bodies) as impacted by hydrocarbons released from waste contaminated sites. Emphasis is placed on the soil-groundwater system or compartment that consists of: (1) the land, (2) the unsaturated soil (vadose) zone, and (3) the saturated (groundwater) zone, as schematically presented in Figure 4. Fate analyses of pollutants in the adjacent media is presented in summary only, since the purpose of this chapter is to support scientists by emphasizing soil pollution and groundwater quality from petroleum and other hydrocarbon contaminations.

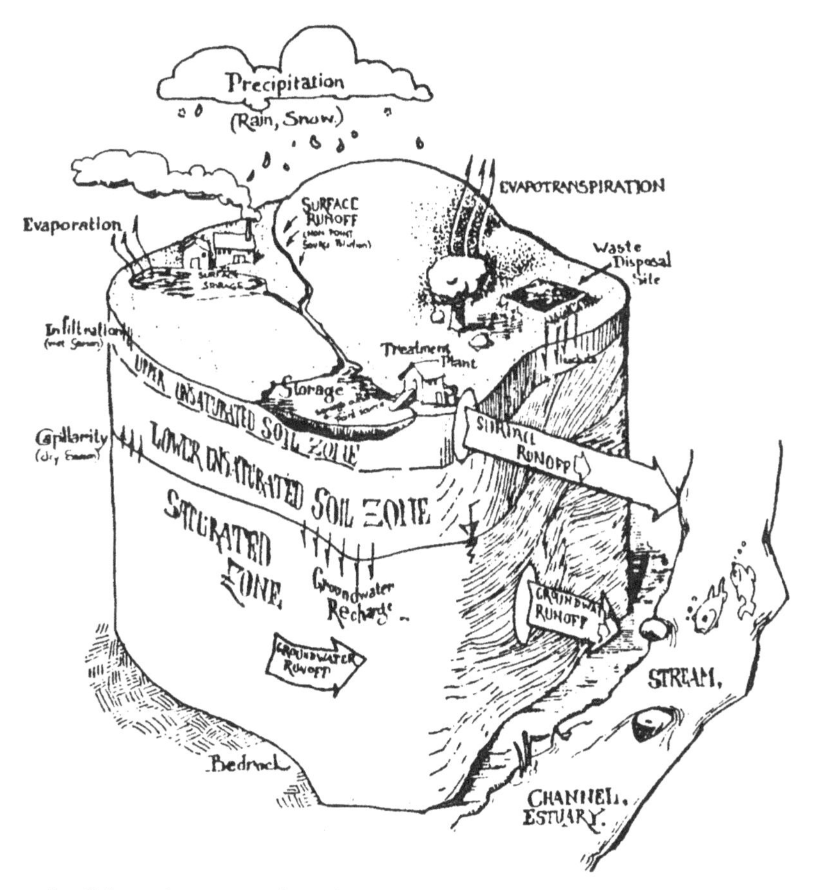

Figure 4. Schematic presentation of the soil compartment (from Bonazountas and Wagner[35]).

Concepts of Environmental Modeling

Information Requirements

Environmental pathway analysis provides the link between quantification of source emissions and assessments of receptor exposure (e.g., human exposure) through estimation of the ambient concentrations of contaminants in the various environmental media. A successful mathematical modeling effort must fairly accurately quantify the relationship between chemical releases into the environment and actual amounts of these chemicals to which the air, soil, and water, as well as humans and other biota, are exposed.

Whether the concern is human health or environmental impact, concentrations of the chemical compounds at user-specified receptors of media of concern must be estimated. This estimation is designated as *mathematical environmental fate modeling*. Mathematical environmental fate modeling generally requires knowledge of: (1) the distribution of the release of the material into the natural environment, (2) the environmental conditions influencing the fate (transport, transformation) of the chemical compounds, (3) the physical and chemical properties of the material, and (4) techniques (models) for analyzing the information gathered, as schematically shown in Figure 5.[26]

Two basic techniques can be employed to investigate environmental pathways: analytic sampling programs and mathematical fate modeling. Sampling programs are costly to design and implement; therefore, computerized mathematical models of environmental processes are frequently used to generate information unavailable by other means, or in order to estimate data (i.e., environmental concentrations) that would otherwise be costly to obtain.

Optimal use of environmental mathematical models necessitates knowledge of the following:

- entry and dynamics of chemicals in the environment
- potential pathways of pollutants
- exposure pathways when confronted with human and biotic exposure
- mathematical modeling concepts
- model application, input data, monitoring, and validation issues.

Entry and Dynamics of Chemicals in the Environment

When identifying pathways and, hence, choosing models, we must consider what becomes of the pollutant as it enters the environmental media. Within any medium, three types of processes, defined as *intramedia* processes, govern the pollutant concentration at each point at any given time:

- *advection:* pollutant mass movement in the medium carrying the pollutant.

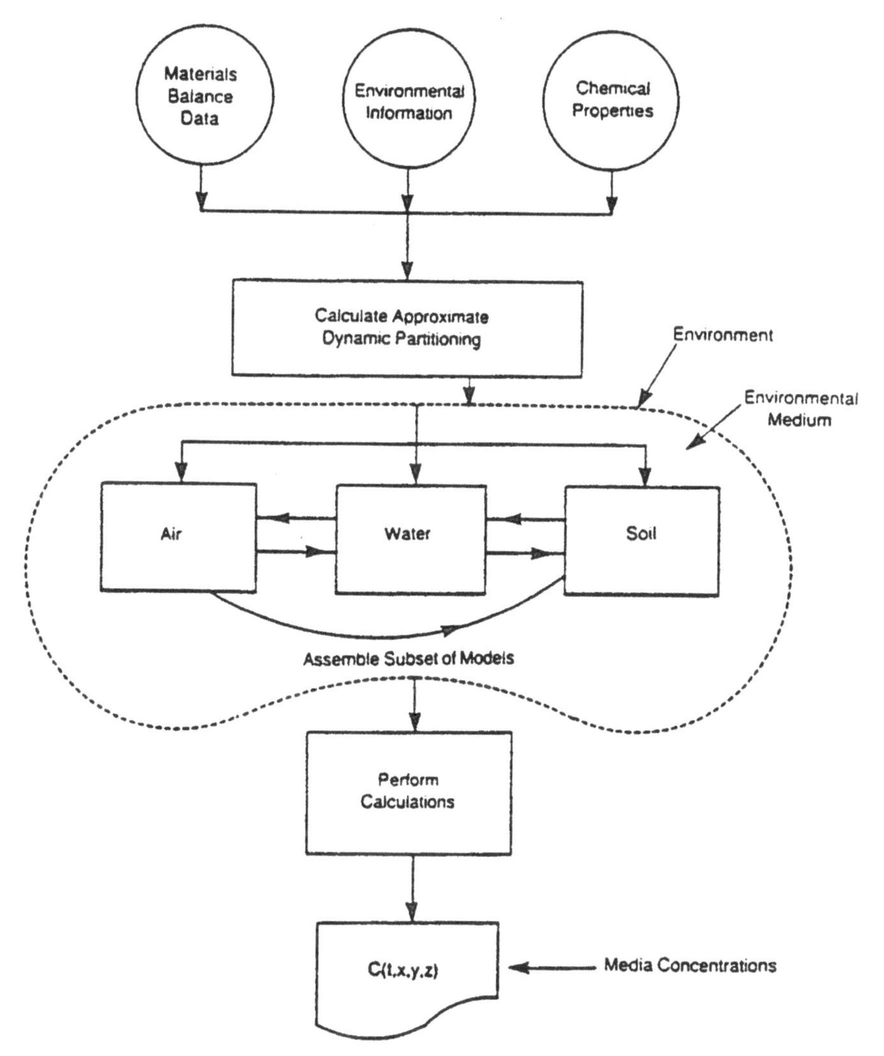

Figure 5. Schematic of environmental modeling.

- *diffusion:* movement or spread of the pollutant, relative to the mass of the medium, as driven by molecular or turbulence-scale dynamics.
- *transformation:* production or consumption of the pollutant, usually driven by the chemical reactions in the medium.

Superimposed on these mechanisms of change operating in the bulk volume of each medium are processes that transfer the pollutant from one medium to another. These processes are called *intermedia transfer* and, as such, are distinguished from transformation processes. (Some conceptual model frameworks lump intermedia transfers with embedded transformation processes.)

Examples of intramedia pollutant transfer are as follows:

- *surface deposition:* rainout (liquid --> solid), washout, fallout and dry deposition
- *evaporation:* codistillation (liquid --> gas) or volatilization
- *adsorption:* liquid (or gas) --> solid and *desorption* [solid --> liquid (or gas)].

In locating and choosing a model, one can simplify fate assessment efforts by delineating (1) the source patterns and (2) the dominant dynamic processes.

Taking the intramedia process first, one can address model criteria by considering the ratio of characteristic time. The advection time is the principal length scale of the domain, L, divided by the average flow speed, u:

$$t_a \simeq L/u$$

Typically, L may be the stream reach distance and u the flow velocity.

The diffusion time is approximated by the random walk hypothesis and is given by:

$$t_d \simeq \Delta^2/2D$$

where Δ is the characteristic transverse direction (e.g., stream depth) and D is the transverse diffusivity, be it turbulent or molecular.

Finally, the transformation time is approximated by:

$$t_t \simeq C/\Delta C_t$$

where ΔC_t is the average rate concentration change due only to transformation (typically a chemical reaction rate), and C is the average concentration of the domain.

If $t_t \ll t_a$ and $t_t \ll t_d$, there is rapid chemical change before any movement occurs. If $t_t \gg t_a$ and $t_d \ll t_a$, there is little chemical change; diffusion spreads the pollutant rapidly, making the mixture homogeneous. If $t_t \simeq t_d \simeq t_a$, all processes act simultaneously.

Taking these cases in order, we see that the first case does not require a model (except possibly a reacting plume in the near field), the second case is approximated by a nonreactive box model, and the third calls for a full reactive diffusion model.

Source geometry, interphase transfer, and time dependencies must be superimposed on the above features. For example, the advection distance would be different for point and area sources. Also, the significance of source location must be considered in light of interphase transfer efficiency. For example, a water discharge containing a pollutant of high volatility and low solubility transfers the problem immediately from one of water modeling to one of air modeling. Implicit in this environmental dynamic considerations are three principles:

- Intramedia processes are largely assessed on the basis of environmental scenarios.
- Intermedia transfers are largely determined by the pollutant's fate properties.
- Chemical transformations can be involved in both of the above.

There are, of course, exceptions to these rules. For example, molecular diffusion is a pollutant fate property but may control an intramedia process. Similarly, rainfall history is an environmental scenario characteristic but may control an intermedia transfer.

An experienced model user can do a simplified systems-level analysis prior to model selection, based on entry characteristics and environmental dynamics of the pollutant. Experienced analysts would advise that it is better to rely on intuition and a few calculations than to construct a formal, logical decision tree for guiding this process. Inexperienced scientists must analyze source characteristics, environmental dynamics, and pollutant fate properties more carefully before processing with model selection and application.

There are no specific rules or procedures for approaching model selection based on entry characteristics and environmental dynamics of pollutants. The only criteria are the model features described in the user's manual, but these are often not objectively reported. Characterization of the sources, the environment, and the fate properties is prerequisite to any such procedure.

Pathways of Pollutants

The two major stages of an environmental modeling pathway analysis are:

I. Examination of background information relevant to the environment
II. The quantitative analysis.

Stage I may apply an initial *scan* effort to identify key factors, such as monitoring data evaluations and geographic setting evaluation, that may indicate which media (air, soil, water) must be modeled, and which pathways and media are likely to be of secondary importance. Principally, Stage I aims at (a) identifying potential (probable) pathways of species, (b) evaluating available data on the specific region or site, (c) identifying what receptors in the area might be affected or are of importance for further consideration, and (d) identifying candidate models for estimating media concentrations.

Stage II, the quantitative analysis, involves (a) collection of monitoring and other site date, (b) prioritization of the important pathways, (c) selection of models to simulate these pathways, (d) compilation of input data for models, (e) performance of the simulations, (f) analysis of model output/results, (g) output validation (using monitoring data) whenever feasible, and (h) design and/or evaluation of control strategies (future actions) for environmental protection which may involve additional model runs.

By carefully considering the fate properties and potential receptors effects during Stage I, the user will gain an understanding of the critical environmental pathways, enabling him to establish priorities for Stage II. For example, for a chemical whose main effect is toxicity to benthic organisms, the pathway leading from the source through the air into the water and to the receptor provides the logic for selecting a model that considers the relevant pathways. Considering relative contributions of multiple release modes, amounts, or dominant fate properties, certain pathways will be preferred to others among a large array of possible combinations. Flow charts or diagrams are useful tools for identifying and establishing the pathway connections between sources and receptors, and thereby defining fate modeling approaches.

Figure 6 is a typical environmental pathway chart designed for a regional pollutant fate modeling application. It shows the three major environmental media (air, soil, and water) and the intermedia and intramedia pathways of pollutants originating from a source (point, area, line, other). In reference to this figure, this chapter later describes models and pathways relevant to the air, surface runoff (land) and water compartments.

Figure 6 shows, for example, that a model user can proceed to (a) separately consider each medium and select the appropriate calculation technique or model that characterizes the behavior in a medium (e.g., air), or (b) consider transfer mechanisms to other media (e.g., water via deposition) from that medium. Experience is a key factor in selecting the appropriate time and space scales for the modeling task. Preliminary half-life estimates for transfer and transformation processes are important to consider. A possible outcome is that chemical transformations or dilution may effectively terminate the pathway within a specific medium.

Aravamudan et al.[27] present five hypothetical examples that illustrate the thinking required to analyze and model the fate of chemicals in multimedia environments. In this chapter, however, we deal with modeling issues and models of only the three single medium environments via key processes (e.g., soil-air via volatilization).

Samiullah[20] reports on the chemical release and environmental pathways of chemicals, including the releases and pathways from nature such as forests that relate volatile hydrocarbons (terpenes) which, during sunny weather conditions, may participate in photochemical reactions exactly as hydrocarbon contaminated sites. He summarizes the work of Klopffer et al.[28] and the research on environmental pathways accomplished at the Umwelbundesamt in Berlin, where environmental media have been subdivided into sectors with likely molecular exchanges arrowed in Figure 7 and described in Table 1.

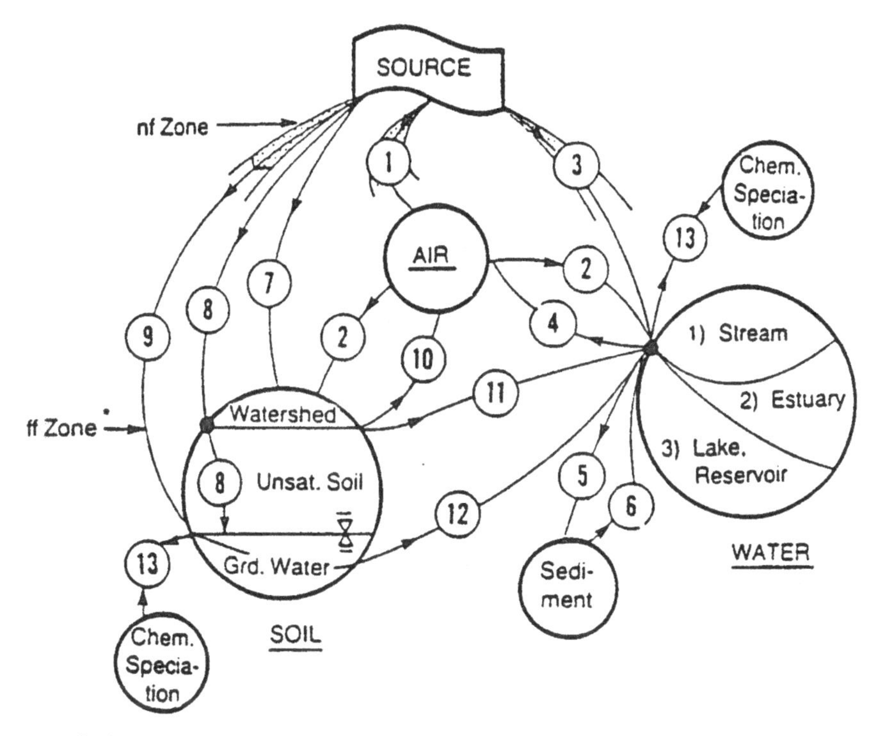

* Paths 1,3,9 have near-field (nf) and far-field (ff) zones

** Degradation, transformation, and out-of/into basin transfer pathways are not shown in this chart.

#1: Source-to-Air
#2: Air-to-Surface (i.e., wet/dry depositions)
#3: Source-to-Water (stream, estuary, lake/reservoir; nf, ff models; overall water quality; pollutant-specific models)
#4: Water-to-Air (e.g., volatilization, codistillation; pollutant-specific models)
#5: Water-to-Sediment (e.g., adsorption, diffusion, pollutant-specific models)
#6: Sediment-to-Water (e.g., desorption, dissolution, dispersion; pollutant-specific models)
#7: Not a real pathway; rather, pollutant input modes to watershed/soil models
#8: Source-to-Soil-to-Groundwater
#9: Source-to-Groundwater (nf, ff models)
#10: Land-to-Air (e.g., volatilization, dust pollutant particle resuspension)
#11: Land-to-Water (e.g., via runoff or sedimentation)
#12: Groundwater-to-Water (e.g., via advection, diffusion)
#13: Aquatic Equilibria (or chemical speciation for water or soil)

Figure 6. Environmental pathways chart (from Bonazountas and Fiksel[26]).

According to Figure 7, for example, an exchange will probably take place between A_1 (gaseous distribution in the troposphere) and A_3 (adsorbed aerosol), or between B_3 (hydrosol) and C_2 (sediment). This figure may be used as the basis of an environmental model for predicting concentrations (C_x) in different phases by solving equations of the *traditional differential type* (see Miscible Pollutant Fate Modeling section for details):

Table 1. Major Pathways of Chemicals in the Environment

	Pathways Between Sectors[a]	Mechanism	pc or env.[b]
	$A_1 \rightarrow A_2$	Turbulent diffusion	env.
	$A_1 \rightarrow A_4$	Dissolution gas → water	pc
	$A_4 \rightarrow A_1$	Volatility from aqueous solution	pc
A	$A_1 \rightarrow A_3$	Adsorption gas → aerosol	pc
	$A_3 \rightarrow A_1$	Volatility from adsorbed state	pc
	$A_3 \rightarrow A_4$	Condensation of water on aerosol	env.
A	$A_1 \rightarrow B_4$	'Dry deposition' (dissolution gas → water)	pc
	$A_1 \rightarrow C_3$	'Dry deposition' (adsorption)	pc
↓	$A_3 \rightarrow C_3 + B_4$	'Dry deposition' (of particles)	env.
B	$A_4 \rightarrow C_3 + B_4$	Rain	env.
+	and		
C	$A_3 \rightarrow A_4 \rightarrow C_3 + B_4$ (direct)	'Raining out'	
	$B_4 \rightleftharpoons B_1$	Turbulent diffusion and mixing	env.
	$B_1 \rightarrow B_5$	Turbulent diffusion	env.
B	$B_1 \rightarrow B_3$	Adsorption	pc
	$B_3 \rightarrow B_1$	Desorption	pc
	$B_3 \rightarrow B_5$	Sedimentation	env.
	$B_1 \rightarrow B_2$	Adsorption	pc
	$B_2 \rightarrow B_1$		
B	$B_4 \rightarrow A_1$	Volatility from aqueous solution	pc
↓	$B_4 \rightarrow A_3$	'White capping'	env.
A	$B_1 \rightarrow C_2$	Adsorption	pc
+			
C	$B_3 \rightarrow C_2$	Sedimentation	env.
	$C_3 \rightarrow C_1$	Adsorption/desorption ('leaching')	pc
C			
	$C_1 \rightarrow {}_3$	Adsorption/desorption	pc
C	$C_1 \rightarrow B_2$	'Leaching'	pc
↓	$C_3 \rightarrow A_1$	Volatility from adsorbed	pc
A		state, aqueous solution	
+		(moist soil) and pure	
B		substance	

Source: Data from Klopffer et al.[28]

[a]For notation of environmental media and sectors, see Figure 3.7, Klopffer et al.[28]; reentry from stratosphere, deep sea, and some other quantitatively less important pathways have been neglected.

[b]pc: Determined by physicochemical properties of the substance and environmental limiting factors.
env.: Determined by environmental (e.g., meteorological) factors only; in some cases the assignment is still ambiguous.

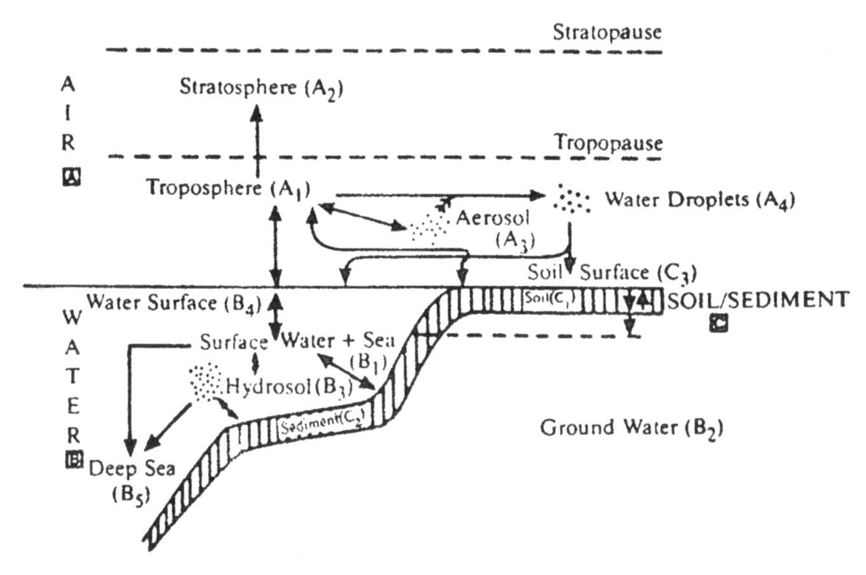

Figure 7. Important pathways for organic chemicals in the environment (from Klopffer et al.[28]).

$$dC/dt = I_x + \Sigma k_i C_i - C_x \Sigma k_j \ (= 0 \text{ at steady state})$$

where I_x is the input rate into the sector x from the *technosphere* (per volume), k_i are rate constants of environmental transfer processes into sector x (first-order process assumed), C_i is the concentration of the substance in the sector donating the substance to x, and k_j are rate constants of environmental processes by which the substance is transferred from x into some other sector. Additional information is provided by Klopffer.[28]

Partitioning Modeling

Some researchers have estimated the fate/partitioning of chemicals in the environment from point, areal, and nonpoint sources with a partitioning model that evaluates the distribution of a chemical between environmental compartments, based on the thermodynamics of the system. The chemical with its environment will tend to reach an equilibrium state between compartments.[29,30]

As Samiullah reports,[20] rather than develop different models for the various media and phenomena occurring in the environment, partition models adopt a more general approach to kinetics of transport, without reference to particular mechanisms such as diffusion, fluid flow, and evaporation. Models consisting of compartmental systems best satisfy these requirements. Such compartments assume that various regions of the ecosystem can be represented by a series of *ideal* volumes in which chemicals move from one volume to the next according to the laws of kinetics. These ideal volumes imply that all property variations are ignored, and perfect mixing is assumed so that the flow has the same properties as the contents of the compartment following transfer and transformation processes of selected type (Figure 8). Such known models are the Klopffer,[31] Neely,[32] Mackay Fugacity,[30] and Wood.[33] They are not studied in this chapter because of their limited application in hydrocarbons fate from from oil-polluted sites.

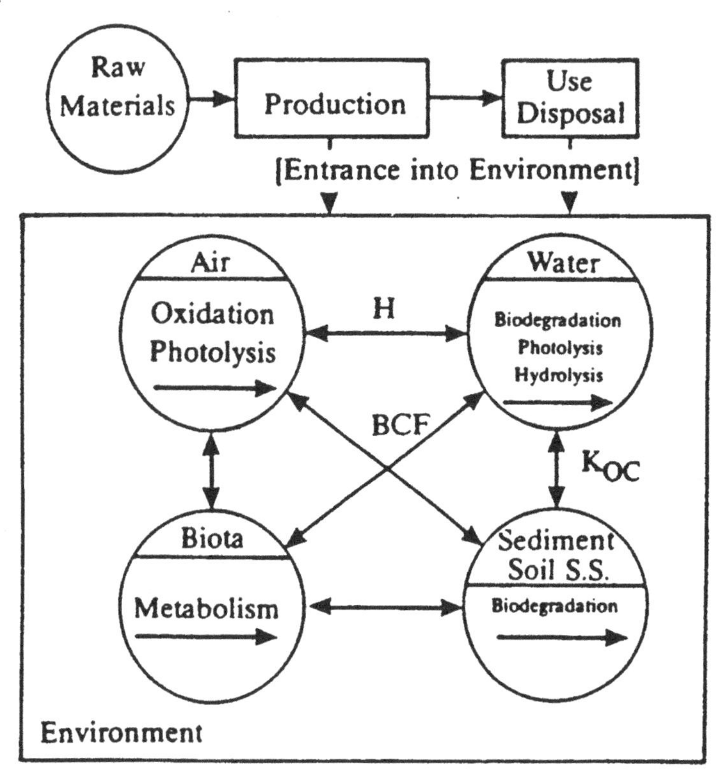

Figure 8. Environmental transfer and transformation processes for chemicals (from Yoshida et al.[28]).

Concepts of Terrestrial Modeling

General

Terrestrial chemicals fate modeling has traditionally been performed for three distinct subcompartments: (1) the land surface (or watershed), (2) the unsaturated soil (or soil) zone, and (3) the saturated (or groundwater) zone of the region (Figure 4). In general, the mathematical simulation is structured around two major processes, the hydrologic cycle and the pollutant cycle, each of which is associated with a number of physicochemical processes. Land surface models also account for a third cycle, sedimentation.

Land surface models describe pollutant fate on land (known as watershed), the unsaturated soil zone of the region and the pollutant contribution to the water body of the area. Unsaturated soil zone models simulate both (1) soil moisture movement and (2) soil-moisture and soil-solid quality conditions of a soil zone profile extending between the ground surface (land) and the groundwater table. Groundwater models describe the fate of pollutants in aquifers, or the saturated zone.[34]

When used properly and with an understanding of their limitations, mathematical models can greatly assist decisionmakers in determining the importance of pollutant pathways in the terrestrial subsurface environment. For this reason, the use of models has grown dramatically over the past decade. However, although the number of terrestrial model types is very large, there are only a few fundamental modeling concepts.

Soil zone modeling is highly complex because of the physical and chemical dynamics of a soil subcompartment—in contrast to those of a water or air subcompartment—that are governed by external (*out-compartmental*) forces such as precipitation, air temperature, and solar radiation. Water and air modeling are generally simpler, because the dynamics of these compartments are governed by *in-compartmental* forces.

Chemical modeling in soil systems provides information in the distribution of elements (e.g., metal species) within a soil matrix consisting of soil-solids, soil-moisture (in the soil zone), or soil-water in groundwater and soil-air (in the soil zone, Figure 9) for the soil zone of a region. The objective is to determine the mass of pollutants in the solid phase, aqueous phase, and/or gaseous phase at a given point and time.

Principal Processes

An evaluation of the fate of organic and inorganic compounds in soil and groundwater requires a detailed consideration of the physical, chemical and biological processes and reactions involved, such as complexation, absorption, precipitation, oxidation-reduction, chemical speciation, and biological reactions to determine the free metal concentration in soil solutions. These processes can affect such characteristics as species solubility, availability for biological uptake, physical transport, and corrosion potential.[36–38]

To describe the complex interactions involved, various kinds of models have been developed. These are, for example, *adsorption* models (which can utilize mathematics involved in specific adsorption, surface complex formation, and ion exchange), *surface complexation* models, *constant-capacitance* models, *cation-exchange* models, and overall *fate modeling* packages that take into account the effects of one or more geochemical processes. One category (*equilibrium* or

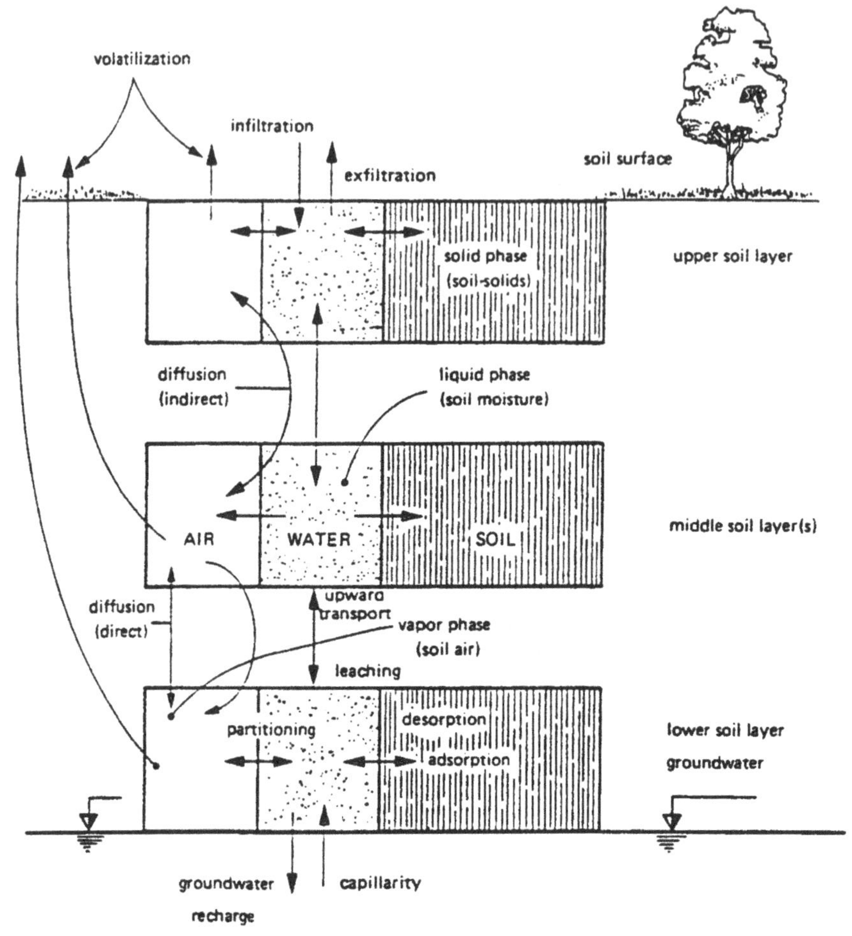

Figure 9. Schematic of phases in the soil matrix.[35]

speciation models) is designed to determine the distribution of inorganic species in the soil water. Chemically-based computer models of soluble trace metal speciation are being employed increasingly in decisionmaking, such as in studies related to sewage and effluents applied to agricultural land.[39]

The transport of particular species in terrestrial systems is of interest to a variety of scientists since measurement or reporting of total concentration of a particular inorganic compound in the soil may be misleading in many environmental management situations. Toxic effects of trace metals, for example, may be affected more by their chemical form than by their total concentration. Therefore, mathematical computer models capable of simulating the distribution of inorganic pollutant species in soil and groundwater systems are valuable tools for analyzing contaminant pathways.[40–44]

Terrestrial Modeling Overview

Terrestrial mathematical fate modeling and models are available for: (1) a dissolved (miscible) organic or inorganic pollutant, (2) an immiscible contaminant (with one or more pollutants), and (3) a metal or an inorganic compound. At this stage of intensive research on terrestrial chemical modeling we can group the prevailing concepts into two major categories: (1) geochemical or, more appropriately, pollutant fate models, and (2) speciation equilibria models. This terminology is not standard, but is employed for convenience and as a means of models classification.

Figure 10 presents the principals of chemical processes related to inorganic pollutant fate modeling, whereas Figure 11 presents the principal controls of free trace metal concentrations in soil solutions as modeled by certain speciation models. The following sections provide information on both pollutant fate and geochemical models in terrestrial systems, with an emphasis in the fate of hydrocarbons.

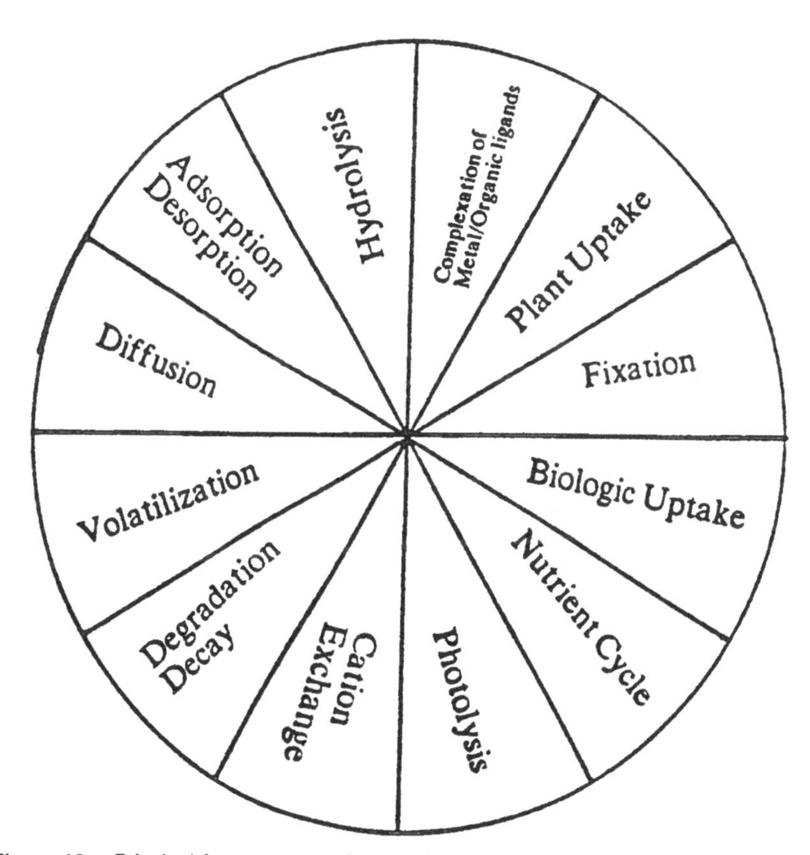

Figure 10. Principal fate processes in organic pollutant fate.

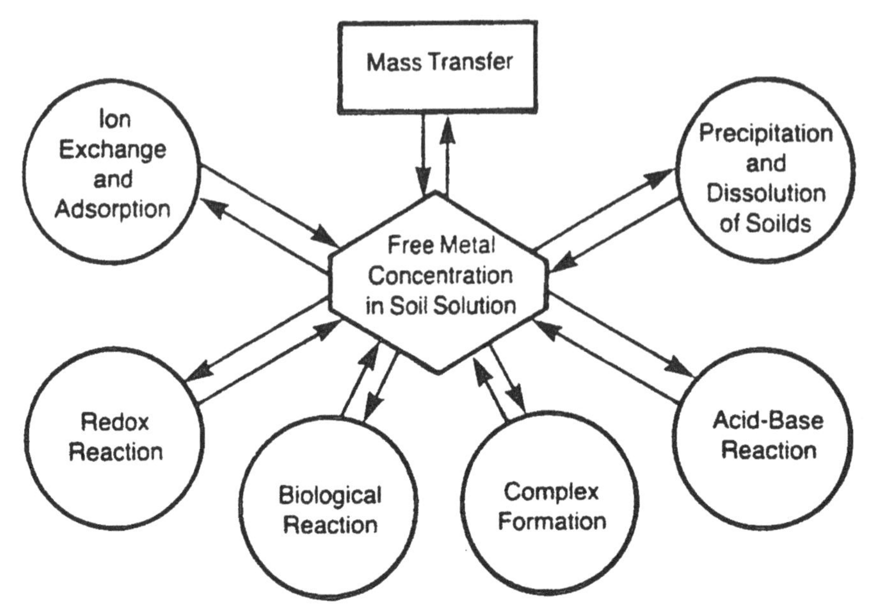

Figure 11. Principal controls of free trace metal concentrations in soil solutions (adapted from Mattigod et al.[45]).

SOURCES AND EMISSIONS

Through numerous human activities, pollutants are released to the soil compartment. The particular particle significantly influences the fate of pollutants in the soil and groundwater zones. Releases include both point source and area loadings. They may be intentional, such as hydrocarbon contaminated sites or landfills and spray irrigation of sewage; unintentional, such as spills and leaks; or indirect, through pesticide drift or surface runoffs.

The point of release may be at the soil surface, or from a source buried deep in the soil. Substances released are in liquid, semiliquid, solid, or particulate form. In some cases a waste material will be pretreated or deactivated prior to disposal, to limit its mobility in soil. The rate of release may be continuous, such as at a municipal landfill; intermittent; on a batch basis such as practiced by some industries; or as a one-off episode such as the uncontrolled disposal of barrels of waste or a spill.

Soil and groundwater contamination are commonly encountered problems at uncontrolled waste sites; they result from the migration of leachate originating from a wide variety of waste management facilities.

Table 2 indicates primary pollutant sources and waste models, and Table 3 indicates the primary and secondary sources and associated pollutants. The primary sources of soil contamination include land disposal of solid waste, sludge, and wastewater; industrial activities; and leakages and spills, primarily of petroleum

Table 2. Sources and Wastes Contaminating Soils

Pollutant Source	Wastewater Impoundments	Solid Waste Disposal Sites	Wastewater Spray Irrigation	Land Application	Injection or Disposal Wells	Septic Tanks and Cesspools	Pits	Infiltration/ Surface Runoff	Leaching from Storage Sites
Industrial									
Wastewater	x		x		x				
Sludge				x					x
Solid waste		x							
Municipal									
Wastewater			x						
Sludge		x		x					
Solid waste		x							x
Household wastewater						x			
Agricultural feedlot								x	
Mining	x	x			x		x	x	
Petroleum exploration					x		x		
Cooling water					x				
Buried tanks and pipelines									x
Agricultural activities								x	x

Table 3. Primary Sources of Soil Contamination and Associated Pollutants[47]

Source	Type of Pollutants
Industrial sources	
Chemical manufacturers	Organic solvents
Petroleum refineries	Chlorinated hydrocarbons
Metal smelters and refineries	Heavy metals
Electroplaters	Cyanide, other toxics
Paint, battery manufacturers	Conventional pollutants
Pharmaceutical manufacturers, paper and related industries	Acids, alkalis, other corrosives; many are highly mobile in soil
Land disposal sites	
Landfills that received sewage sludge, garbage, street refuse, construction and demolition wastes	BOD, inorganic slats, heavy metals, pathogens, refractory organic compounds, plastics; nitrate; metals including iron, copper, manganese suspended solids
Uncontrolled dumping of industrial wastes, hazardous wastes	
Mining wastes	Acidity, dissolved solids, metals, radioactive materials, color, turbidity
	BOD, nutrients, fecal coliforms, chloride, some heavy metals
Agricultural activities	
Agricultural feedlots	
Treatment of crops and/or soil with pesticides and fertilizers; runoff or direct vertical leaching to septic tanks and cesspools	Herbicides, insecticides, fungicides, nitrates, phosphates, potassium, BOD, nutrients, heavy metals, inorganic salts, pathogens, surfactants; organic solvents used in cleaning
Leaks and spills	
Sources include oil and gas wells; buried pipelines and storage tanks; transport vehicles	Petroleum and derivative compounds; any transported chemicals
Atmospheric deposition	Particulates; heavy metals, volatile organic compounds; pesticides; radioactive particles
Highway maintenance activities	
Storage areas and direct application	Primarily salts
Radioactive waste disposal	
Eleven major shallow burial sites exist in U.S.; three known to be leaking	Primarily ^{132}Cs, ^{90}Sr and ^{60}Co
Land disposal of sewage and wastewater	
Spray irrigation of primary and secondary effluents	BOD, heavy metals, inorganic salts, pathogens, nitrates, phosphates, recalcitrant organics
Land application of sewage sludge	
Leakage from sewage oxidation ponds	

products. The solid waste disposal sites include dumps, landfills, sanitary landfills, and secured landfills.

Land disposal sites result in soil contamination through leachate migration. The composition of the substances produced depends principally on the type of wastes present and the decomposition in the landfill (aerobic or anaerobic). The adjacent soil can be contaminated by direct horizontal leaching of surface runoff, vertical leaching, and transfer of gases from decomposition by diffusion and convection. The disposal of domestic and municipal wastewater on land takes place through septic tanks and cesspools; sewage sludge from primary and secondary treatment plants often spreads on agricultural and forested land (land treatment); liquid sewage, either untreated or partially treated, is applied to the land surface by spray irrigation, disposal over sloping land, or disposal through lagooning of sewage sludge.

Landfills are principally disposal sites for municipal refuse and some industrial wastes. Municipal refuse is generally composed of 40% to 50% (by weight) of organic matter, with the remaining consisting of moisture and inorganic matter such as glass, cans, plastic, and pottery. Under aerobic decomposition, carbonic acid that is formed reacts with any metals present and calcareous materials in the rocks and soil, thus increasing the hardness and metal content of the leachate. Decomposition of the organic matter also produces gases, including CO_2 , CH_4, H_2S, H_2, NH_3 and N_2, of which CO_2 and CH_4 are the most significant soil contaminants.

Agricultural practices affect soil quality in many ways, principally by increasing soil nutrients and by rooting chemicals to the groundwater or to the adjacent water bodies via the washload (sediment).

In specific reference to hydrocarbon soil pollution it may be reported that storage of refined petroleum products has given rise to more concern over potential groundwater pollution[46] for the following reasons:

- products are stored in small underground storage containers widely spread across a country
- many of these underground tanks are privately owned, and it is difficult to enforce standards for the storage of petroleum products
- many of the storage tanks are located in populated areas; thus, a leak may rapidly affect other people.

The problems that arise from leaking underground tanks are twofold: (1) groundwater pollution, and (2) potential fire or explosion hazards.

HYDROCARBON FATE PROCESSES AND CHEMISTRY

General

The chemical, physical, and biological properties of a hydrocarbon substance, in conjunction with the environmental characteristics of an area, result in physical, chemical, and biological processes associated with the transport and transformation of the substance in soil and groundwater. These processes govern the fate (ultimate and long term distribution) of the substance in the environment (air, soil, water, biota). Processes and chemistry are described in the following sections, along with representative mathematical estimation methods. A schematic of the physical, chemical, and biological processes influencing hydrocarbon behavior in soil systems is shown in Figure 12. The fate of diesel fuel in soils is described in detail by Dragun et al.[48]

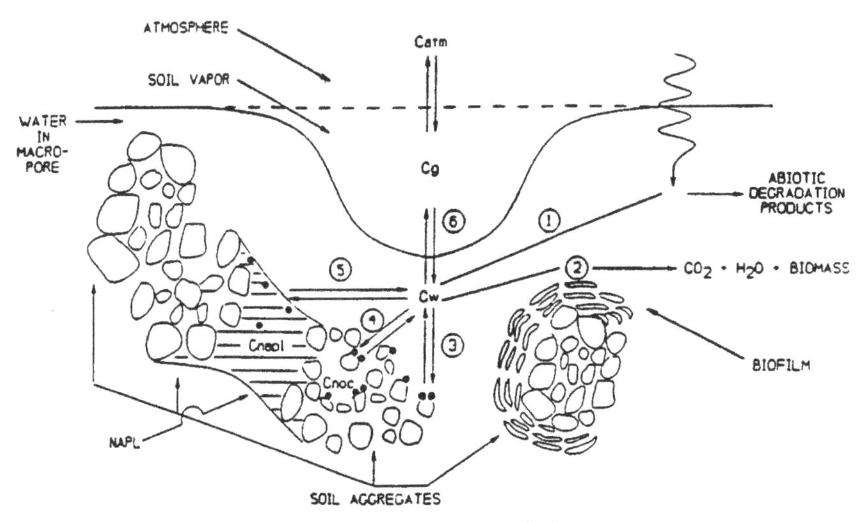

PROCESSES

1. ABIOTIC TRANSFORMATIONS
2. BIOTIC TRANSFORMATIONS
3. MASS TRANSFER FROM AGGREGATE SURFACE INTO THE MACROPORE FLUID
4. MASS TRANSFER FROM AGGREGATE INTERIOR INTO THE MICROPORE FLUID AND DIFFUSION INTO MACROPORE FLUID
5. DIFFUSION OF CHEMICALS THROUGH THE NAPL AND DISSOLUTION FROM NAPL INTO MACROPORE FLUID
6. VOLATILIZATION

DEFINITIONS

NAPL — NONAQUEOUS PHASE LIQUID
C_w — CONCENTRATION OF CHEMICAL IN AQUEOUS PHASE
C_{noc} — CONCENTRATION OF CHEMICAL IN NATIVE ORGANIC CARBON
C_{napl} — CONCENTRATION OF CHEMICAL IN NAPL
C_g — CONCENTRATION OF CHEMICAL IN SOIL VAPOR
C_{atm} — CONCENTRATION OF CHEMICAL IN ATMOSPHERE

Figure 12. Physical, chemical, and biological processes influencing contaminant behavior in soils (from RTI[73]).

Organic contaminants can reach the groundwater zone either dissolved in water or as organic liquid phases that may be immiscible in water. These contaminants travel with the soil moisture and are retarded in their migration by various factors. The subsurface transport of immiscible (nonaqueous phase liquid, NAPL) organic liquids is governed by a set of factors different from those for dissolved contaminants. However, some components of organic liquids can dissolve into the groundwater; therefore, the process of dissolved pollutants has been of primary importance in the past. Metals are subject to different processes, given the presence of organic contaminants.

The rates of each of the environmental important chemical processes are influenced by numerous parameters, but most processes are described mathematically by only one or two variables. For example, the rate of biodegration varies for each chemical with time, microbial population characteristics, temperature, pH, and other reactants. In modeling efforts, however, this rate can be approximated by a first-order rate constant (in units of time).

Soil models tend to be based on first-order kinetic processes; thus, they employ only first-order rate constants with no ability to correct these constants for environmental conditions in the simulated environment which differ from the experimental conditions. This limitation is both for reasons of expediency and due to a lack of the data required for alternative approaches. In evaluating and choosing appropriate soil and groundwater models, the type, flexibility, and suitability of methods used to specify necessary parameters should be considered.

The physical, chemical, and biological behavior of a chemical determines its chemical partitions among the various environmental media, and has a large effect on the environmental fate of a substance. For example, the release into the soil of two different acids (with similar chemical behavior) may result in one chemical mainly volatilizing into the air and the other chemical becoming mainly sorbed to the organic fraction of the soil. The physical behavior of a substance, therefore, can have a large effect on the environmental fate of that substance.

The processes and corresponding physical parameters that are important in determining the behavior and fate of a chemical are different in the case of analysis of dissolved trace-level contaminants, vs. analyses of contaminants from large-scale releases (e.g., spills).

The mechanisms of contamination from various activities and the transport of organic and inorganic contaminants in soil and into the groundwater are described by Bodek et al.,[37] Hall and Quam,[46] Kincaid et al.,[49] Mackay and Roberts,[50] Mercer,[51] Faust,[52] Abriola and Pinder,[53] Fried et al.,[54] Gee et al.,[4] Sudicky and Huyakorn,[55] Wheatcraft and Cushman,[56] Odencrantz et al.,[57] and other researchers.

Nature of Hydrocarbons Fate

Hall and Quam[46] report on the nature of contamination from petroleum products, that one of the most important factors in contamination of groundwater is the extremely low concentration of the product that can give rise to objectional tastes and odors. The specific aspects of contamination may be broadly classified into (a) the formation of surface films and emulsions; and (b) the solubility in water of certain petroleum products.

The problems associated with surface films are minimized due to the ability of aquifers to adsorb or absorb much of the product. However, this phenomenon magnifies the problems associated with the soluble components of the product since hydrocarbons held in this manner are subject to leaching as water passes over them. Surface films may affect the aesthetics and interfere with treatment or industrial processes. They may also be toxic to animal or plant life if they emerge into surface waters.

The water-soluble components of petroleum products which give rise to taste and odor problems are the aromatic and aliphatic hydrocarbons. Phenols and cresols are examples of these compounds, and are known to generate taste and odors at concentrations as low as 0 to 0.01 mg/L. When chlorine is added to drinking water, as in most municipal water supplies, it reacts with the phenols to form chlorophenols which give rise to objectional tastes and odors at concentration as low as 0 to 0.001 mg/L. It is thus apparent that very small quantities of hydrocarbons can give rise to widespread contamination of water resources.

Fate of Organic Hydrocarbons

Major transport and transformation processes for organic dissolved contaminants are advection, dispersion, sorption, and retardation, and chemical and biological transformation. The migration of an immiscible organic liquid phase is governed largely by its density, viscosity, and surface-wetting properties.

Density differences of about 1% are known to influence fluid movement significantly. With few exceptions, the densities of organic liquids differ from that of water by more than 1%. In most cases the difference is more than 10%. The specific gravities of hydrocarbons may be as low as 0 to 0.7%, and halogenated hydrocarbons are almost without exception significantly more dense than water.

Mackay and Roberts[50] designate organic liquids less dense than water as *floaters*, which spread across the water table, and organic liquids more dense than water as *sinkers*, which may plummet through sand and gravel aquifers to the underlying aquitard (relatively impermeable layer) where present.

There is extensive evidence from field studies that low-density organic liquids float on the water table. It is also important to recognize that migration of dense organic liquids is largely uncoupled from the hydraulic gradient that drives advective transport, and that the movement may have a dominant vertical component, even in horizontally flowing aquifers. Such a liquid will sink through the saturated zone as an immiscible phase, displacing the groundwater as it descends (Figure 13).[53] If the contaminant is slightly soluble in water, a plume develops by dissolution of the contaminant liquid retained in the aquifer pores, as well as by dissolution of the pool of contaminant liquid residing on the bottom of the aquifer.

In order to explain the contamination of wells and springs, it often has been found necessary to explain the principles of groundwater flow and the relationship between groundwater and surface waters. In this way, the greater sensitivity of recharge areas can be explained, and the need for early detection of leaks can be stressed. However, for reasons outlined above, this correlation has not always proved to be successful.

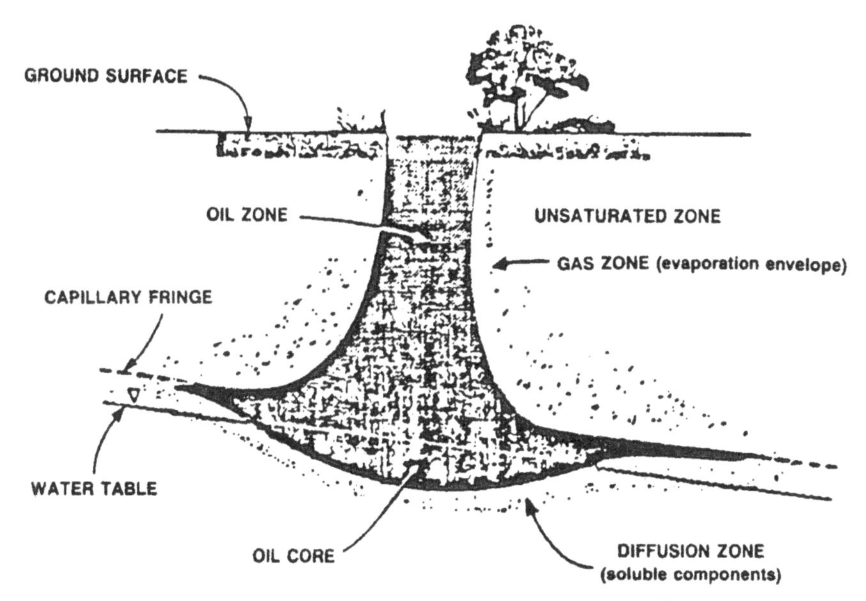

Figure 13. Schematic of petroleum spill (from Abriola and Pinder[53]).

Fate of Inorganic Contaminants

Various paths and interactions of fate are shown in Figure 14,[1] with major mechanisms being the volatilization, solubility, fast aqueous reactions, slow aqueous reactions, speciation, soil interaction, and bioaccumulation.

The interaction of these processes is important to the pollutant fate. For example, chemicals with relatively low vapor pressure, high volatility onto solids, or high solubility are less likely to volatilize and become airborne (fate in the air). In contrast, many chemicals with very low vapor pressures can volatilize at surprisingly high rates because of their low solubilities in water (i.e., high activity coefficients) or low adsorptivity to solids; and thus are likely to volatilize and become airborne.

An important aspect of the inorganics fate is the soil and sediment interaction of the dissolved pollutant phase that includes mechanisms such as: (1) formation of precipitates, (2) adsorption of components onto soil surfaces, (3) modification of speciation by soil constituents (solid and liquid phases), and (4) reactions included by bacteria. All these chemical processes strongly affect environmental fate of pollutants.

The following sections report on individual physical, chemical, and biologic processes as affecting the fate of hydrocarbons in soils and considered in mathematical pollutant fate modeling in soil systems. It is important to notice that mathematical models available are related to:

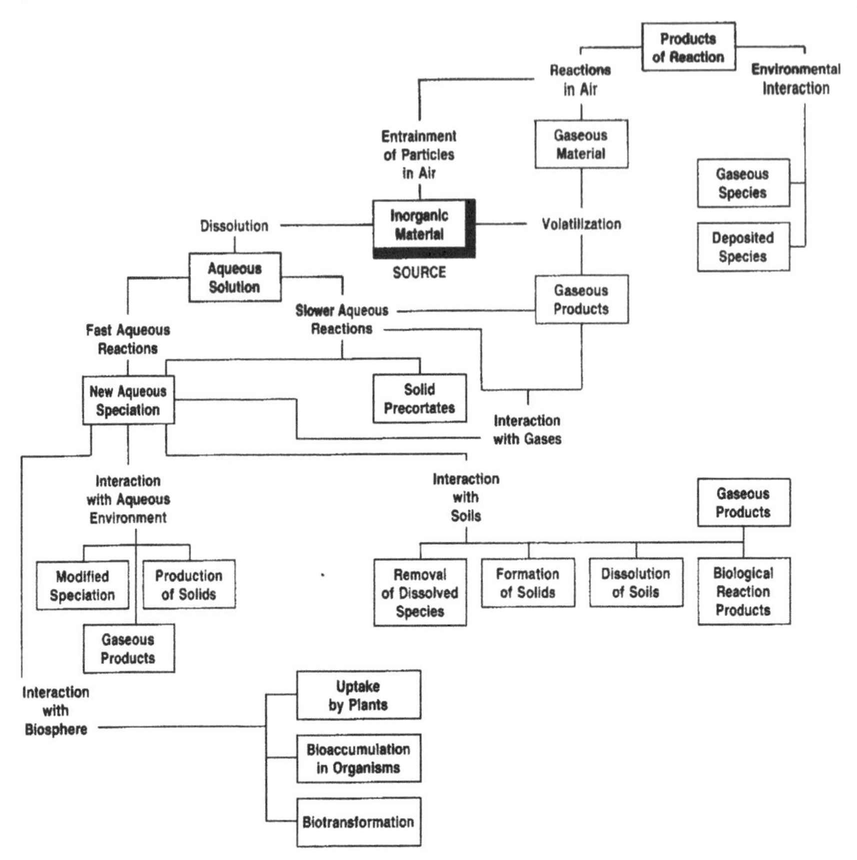

Figure 14. Important inorganic materials fate mechanisms (from Bonazountas in Bodek et al.[37]).

- organic dissolved compounds
- immiscible fluids
- inorganic compounds.

This categorization is not a standard of the literature, but is presented in this chapter to classify the methodologies and processes reported in the literature on mathematical pollutant fate modeling in soil systems, in a unified way.

Miscible Organic Compounds Processes

Physical processes affecting dissolved hydrocarbon organic compound fate in soil systems are mainly: (1) advection, dispersion, diffusion, volatilization, and (2) sorption, ion-cation exchange.

Chemical processes affecting dissolved contaminant migration are mainly: (3) ionization, (4) solubility and hydrolysis, (5) oxidation/reduction, and (6) complexation.

Biological processes affecting dissolved contaminant migration are mainly: (7) bioaccumulation, and (8) biodegration.

Information on the above processes and algorithms related to mathematical modeling is presented below.

Advection, Dispersion, Diffusion, Volatilization

Advection. In sand and gravel aquifers, advection is the dominant factor in the migration of dissolved contaminants. In advective groundwater processes, solutes are transported by the bulk motion of groundwater flowing from regions of the substance where water levels are high to regions where the water levels are low. The magnitude of the driving force is termed hydraulic gradient, and the average linear velocity at which groundwater flows through a granular medium is equal to the product of the gradient and the inherent capability of the medium to transit water.[58] Groundwater velocities in uniform sand and gravel aquifers under natural gradient conditions are typically between 10 and 100 $m/year^{-1}$, with a potential range between 1 and 1000 m/year.

Dispersion. This is dissolved contaminant spread as it moves with the groundwater. Dispersion results from two basic processes, molecular diffusion in solution and mechanical mixing. The process results in an overall net flux of solutes from a zone of high concentration to a zone of lower concentration.

Molecular Diffusion. Molecular diffusion in solution is the process whereby ionic or molecular constituents move under the influence of their kinetic activity in the direction of their concentration gradient. The process of diffusion is often known as self-diffusion, molecular diffusion, or ionic diffusion. The mass of diffusing substance passing through a given cross section per unit of times is proportional to the concentration gradient (Fick's first law).

The rate of diffusion depends on both the nature of the compound and the medium through which it moves. The diffusion coefficient, or diffusivity, was defined for vadose pollution modeling by Tucker and Nelken[59] as:

$$D_{BA} = J_B/\text{grad}[X_B]$$

where:

D_{BA} = diffusion coefficient of compound B in compound A ($cm^2 \cdot s^{-1}$)
J_B = net motal flux of B across a hypothetical plane ($mol \cdot cm^{-2} \cdot s^{-1}$)
grad[X_B] = concentration gradient of B at the hypothetical plane ($mol \cdot cm^{-3} \cdot cm^{-1}$).

The cursive D is used to distinguish this property from the *apparent* diffusion coefficient (D*). Table 4 lists well-known methods for estimating diffusivity into soil water. Additional information is provided by Samiullah.[20]

In saturated groundwater (aquifer), the effective dispersivity is generally much greater than the molecular diffusion coefficient.[59] Scheidigger[60] postulated that the apparent dispersivity is proportional to the flow velocity, and that molecular diffusion may therefore be significant in aquifer systems with very low flow rates. Tucker and Nelken[59] consider that molecular diffusion could be ignored at pure velocities exceeding 0.002 cm/s. Harleman and Rumer[61] cited in Kirda et al.,[62] proposed the following relationship:

$$D_s = f_t D^*_{BW} + \alpha \cdot u^m$$

where D_s is the apparent dispersion coefficient (cm^2/s), f_f = the tortuosity factor (a soil property in the range of 0.01–0.50), u = pore water velocity (cm^2/s), α and m = empirically determined soil constants. BW represents compound B in water. Additional information and relationships are provided by Tucker and Nelken[59] and Samiullah.[20]

Volatilization. Volatilization refers to the process of pollutant transfer from soil to air and is a form of diffusion, the movement of molecules or ions from a region of high concentration to a region of low concentration. Volatilizaton is an extremely important pathway for many organic chemicals, and rates of volatilization from soil vary over a large range. This process is less important for inorganic than for organic chemicals; most ionic substances are usually considered to be nonvolatile.

Many models are available in the literature, and some of these models can be applied only to specific environmental situations and only for chemicals for which they were developed. Obviously all models do not provide the same numerical results when employed to provide answers to a particular problem, so care must be taken in choosing an appropriate unsaturated zone model, or when specifying a volatilization rate. For modeling algorithms and numerical example, the reader is referred to the work of Lyman et al.,[36] Bodek et al.,[37] Bonazountas and Wagner,[35] Jury,[63] and of others listed in the references section of this chapter, and in Table 5.[64] Additional information is reported by Samiullah.[20]

In many cases dispersion may be of minor importance. If dispersion is important, and in the absence of detailed studies to determine the dispersion characteristics of a given field situation, longitudinal and transverse dispersivities must be estimated based on prior field work in similar hydrogeological systems.[50]

Sorption and Ion-Cation Exchange

Sorption. Sorption refers to adsorption and desorption of a chemical onto soil particles. Adsorption is the adhesion of pollutant ions or molecules to the surface or soil solids, causing an increase in the pollutant concentrations on the soil surface over the concentration present in the soil moisture. Adsorption occurs as a result of a variety of processes with a variety of mechanisms, and some processes may

Table 4. Well-Known Methods to Estimate Diffusivity in Soil-Water

Method	Formula	Inputs (Excluding Water Parameters)	Absolute Average Error (%) (Hayduk and Laudie, 1974)
Hayduk and Laudie (1974)[a]	$\partial_{BW} = \dfrac{13 \cdot 26 \times 10^{-5}}{n_W^{1 \cdot 14} V_B^{0 \cdot 589}}$	V_B	5·8 (87 solutes)
Wilke and Chang (1955)	$\partial_{BW} = \dfrac{7 \cdot 4 \times 10^{-8} (\phi_W M_W)^{1/2} T}{n_W V_B^{0 \cdot 6}}$	V_B	8·8 (87 solutes)
Scheibel (1954)	$\partial_{BW} = \dfrac{8 \cdot 2 \times 10^{-8} T}{n_W V_B^{1/3}} \left[1 + \left(\dfrac{3V_W}{V_B} \right)^{2/3} \right]$	V_B	6·7 (87 solutes)
Othmer and Thakar (1953)	$\partial_{BW} = \dfrac{1 \cdot 4 \times 10^{-5}}{n_W^{1 \cdot 1} V_B^{0 \cdot 6}}$	V_B	5·9 (87 solutes)
Reddy and Doraiswamy In: Reid et al. (1977)[b]	$\partial_{BW} = \dfrac{M_W^{1/2} T K'}{n_W (V_W V_B)^{1/3}}$	V_B	20 (96 solutes)
Venezian (1976)	$\partial_{BW} = \dfrac{6 = 10^{-10} T}{n_W (R_M - 0\partial 855)}$ where $R_M = \left(\dfrac{n_D^2 - 1}{n_D^2 + 2} \right) \left(\dfrac{M_B}{\rho_B} \right)^{1/3}$	n_D, ρ_B, M_B	

Source: Tucker and Nelken.[59]

[a]Recommended by Tucker and Nelken.[59]

[b]$K' = 10 \times 10^{-8}$ for $V_W/V_B \leq 1\partial 5$ and $8\partial 5 = 10^{-8}$ for $V_W/V_B > 1\partial 5$.

Table 5. Models to Compute Volatilization

Method	Equations	Difficulty (Calculational)	Information Required: Chemical	Information Required: Environmental
Hartley (1969)	1,2	Low	Saturated vapor concentration Vapor diffusion coefficient in air Latent heat of vaporization Molecular weight of chemical Thermal conductivity of air Gas constant	Humidity Stagnant air layer thickness Temperature
Hamaker (1972b)				
No water loss	3	Low	Vapor diffusion coefficient in soil	Initial concentration
Water loss	4	Low	Vapor pressure of chemical Vapor pressure of water Vapor diffusion coefficient in soil—for chemical and water	Water flux from plot—both liquid and vapor phase
Mayer et al. (1974) (five models)	5	High	Air/soil concentration isotherm coefficient Diffusion coefficient in soil Diffusion coefficient in air	Depth of soil column Air flow velocity Initial concentration Adsorbed concentration Thickness of nonmoving air layer
Jury et al. (1980)	6,7,8	Medium	Gas-phase diffusion coefficient Liquid-phase diffusion coefficient Adsorption parameters Henry's law constant	Total concentration in soil Soil bulk density Adsorbed concentration Volumetric soil water content Chemical concentration in liquid phase Chemical concentration in gas phase Soil air content Water flux from plot
DOW (Swann et al., 1979)	9,10,11	Low	Soil adsorption coefficient Vapor pressure Solubility	

Source: Thomas.[64]

cause an increase of pollutant concentration within the soil solids, not merely on the soil surface.

Adsorption and desorption can drastically retard the migration of pollutants in soils; therefore, knowledge of this process is of importance when dealing with contaminant transport in soil and groundwater. The type of pollutant will determine to what kinds of material the pollutant will sorb. For organic compounds it appears that partitioning between water and the organic carbon content of soil is the most important sorption mechanism.[50]

Adsorption and desorption are usually modeled as one fully reversible process, although hysteresis is sometimes observed. Four types of equations are commonly used to describe sorption-desorption processes: Langmuir, Freundlich, overall, and ion or cation exchange.

The Langmuir isotherm model was developed for single-layer adsorption and is based on the assumption that maximum adsorption corresponds to a saturated monolayer of solute molecules on the adsorbent surface, that the energy of adsorption is constant, and that there is no transmigration of adsorbate on the surface phase. The Langmuir model is described by:

$$ds/dt = k_{sw}\,(s_s - s)\ ;\ s_s = Q^\circ bc/(1+c)$$

The Freundlich sorbtive isotherm is an empirical model expressed by:

$$s = x/m = k \cdot c^{1/n}$$

where:

- ds/dt = temporal variation of adsorbed concentration of compound on soil particles
- s = adsorbed concentration of compound on soil particles
- k_{sw} = Langmuir equilibrium soil-water adsorption kinetic coefficient
- s_s = maximum soil adsorption capacity
- Q° = number of moles (or mass) of solute adsorbed per unit weight of adsorbent (soil) during maximum saturation of soil
- b = adsorption partition coefficient
- t = time
- c = concentration of pollutant in soil moisture
- x = adsorbed pollutant mass on soil
- m = mass of soil
- k = adsorption (partitioning) coefficient
- c = dissolved concentration of pollutant in soil moisture; and
- n = Freundlich equation parameter.

At trace levels, many substances (particularly organics) are simply proportional to concentration, so the Freundlich isotherm is frequently used with $1/n = 1$. For organics, k_{oc} (the adsorption coefficient on organic carbon) is often used instead of k. These coefficients are related by $k = k_{oc}$ (% organic carbon in the solid)/100. Some well-known adsorption isotherm equations are summarized in Table 6.

Ion-Cation Exchange. Ion exchange (an important sorption mechanism for inorganics) is viewed as an exchange with some other ion that initially occupies the adsorption site on the solid. For example, for metals (M_{2+}) in clay the exchanged ion is often calcium:

$$M^{2+} + [\text{clay}]\cdot\text{Ca} <--> \text{Ca}^{2+} + [\text{clay}]\cdot\text{M}$$

Cation exchange can be quite sensitive to other ions present in the environment. The calculation of pollutant mass immobilized by cation exchange is given by:

$$S = EC\cdot MWT/VAL$$

where:

S = maximum mass associated with solid (mass pollutant/mass of soil)
EC = cation exchange capacity (mass equivalent/mass of dry soil)
MWT = molecular (or atomic) weight of pollutant (mass/mole)
VAL = valence of ion (-).

For additional details see Bonazountas and Wagner.[35]

Ionization

Ionization is the process of separation or dissociation of a molecule into particles of opposite electrical charge (ions). The presence and extent of ionization has a large effect on the chemical behavior of a substance. An acid or base that is extensively ionized may have markedly different solubility, sorption, toxicity, and biological characteristics than the corresponding neutral compound. Inorganic and organic acids, bases, and salts may be ionized under environmental conditions. A weak acid HA will ionize to some extent in water according to the reaction.

$$HA + H_2O <--> H_3O^+ + A^-\cdot$$

The acid dissociation constant K_a is defined as the equilibrium constant for this reaction:

$$K_a = [H_3O^+][A^-]/[HA][H_2O].$$

Note that a compound is 50% dissociated when the pH of the water equals the pK_a ($pK_a = -\log K_a$).

Sposito[66] conceded that single-ion activities appear to be of increasing importance in the study of soil-plant relationships and in the surface chemistry of soils.

Table 6. Adsorption Isotherm Equations

Isotherm	Equation[a]	Number of Adjustable Parameters	Asymptotic Properties: Linear at Low c	Asymptotic Properties: Adsorption Max. at High c
Langmuir	$n = KcM/(1 + Kc)$	2	Yes	Yes = M
Freundlich	$n = (Kc)^{\beta}$	2	No	No
Langmuir–Freundlich	$n = (Kc)^{\beta}M/[1 + (Kc)^{\beta}]$	3	No	Yes = M
Redlich–Peterson	$n = KcM/[+ (Kc)^{\beta}]$	3	Yes	No
Toth	$n = KcM/[[1 + (Kc)^{\beta}]1/\beta]$	3	Yes	Yes = M
Multisite Langmuir	$n = \sum_{i=1}^{k} [K_icM_i = (1 + K_ic)]$	2k	Yes	Yes = $\sum M_i$
Sum of Freundlich's	$n = \sum_{i=1}^{k} (K,c)^{\beta i}$	2k	No	No
Dubinin–Radushkevich	$\log n = -\beta[\log^2(Kc)] + \log M$	3	No	No
Modified Dubinin–Radushkevich[b]	$\log n = \begin{cases} \log n^* - \log c^* + \log c & c < c^* \\ -\beta \log^2[Kc/(1 + Kc)] + \log M & c > c^* \end{cases}$	3	Yes	Yes = M

Source: Kinniburgh.[65]

[a] n is the amount adsorbed, c is the equilibrium solution concentration, M is the adsorption maximum, k is an affinity parameter, and β is an empirical parameter that varies with the degree of heterogeneity.

[b] Where (n^*, c^*) is the point where $\partial \log n / 3 \log c = 1$; c^* cannot be calculated explicitly, but is given to a good approximation by $\log c^* \approx (-1/2\beta - \log K$. This estimate can be refined by a modified Newton using $\log c^*_{i+1} = \log c^*_i + (b_i - 1)/2\beta$ where $b_i = [-2\beta/(1 + Kc^*_i)]\log[Kc^*_i/(1 + Kc^*_i)]$ and i is the iteration number.

This was in spite of the fact that the chemical potential of a charged species cannot be defined in a strict thermodynamics sense, and that no general procedure exists through which the activity of a charged species can be related to macroscopic properties without using unmeasurable parameters.[20]

Solubility in Water

Solubility is one of the most significant factors affecting the fate and transport of chemicals in the environment. Chemicals that are highly soluble are quickly distributed by the hydrologic cycle, tend to have relatively low soil adsorption, and bioconcentrate only slightly in aquatic biota. Such compounds are also readily biodegradable by microorganisms.[36]

A variety of solubility estimation methods exist, mostly by the use of regression equations linking the octanol-water partitioning coefficient K_{ow} with solubility.[36,67,68] Partition coefficients may be estimated from structure, and the solubility of hydrocarbons and halocarbons may be estimated by the addition of atomic fragments. Figure 15 illustrates some of the avenues in estimating solubility for organic compounds.[36] A commonly accepted equation is:

$$\log S = -1.12 \cdot \log K_{ow} + 7.30 - 0.15 \cdot t_m \ (\mu mol/L)$$

with

$$r^2 = 0.992, \text{ and } t_m = 25°C$$

where K_{ow} is the octanol-water partition coefficient.

Hydrolysis

Hydrolysis is one of a family of reactions that leads to the transformation of pollutants. Under environmental conditions, hydrolysis occurs mainly with organic compounds. Hydrolysis is a chemical transformation process in which an organic RX reacts with water, forming a new molecule. This process normally involves the formation of a new carbon-oxygen bond and the breaking of the carbon-X bond in the original molecule:

$$R\text{-}X \xrightarrow{H_2O} R - OH + X^- + H^+.$$

Hydrolysis reactions are usually modeled as first-order processes, using rate constants (K_H) in units of $(\text{time})^{-1}$:

$$-d[RX]/dt = K_H[RX]$$

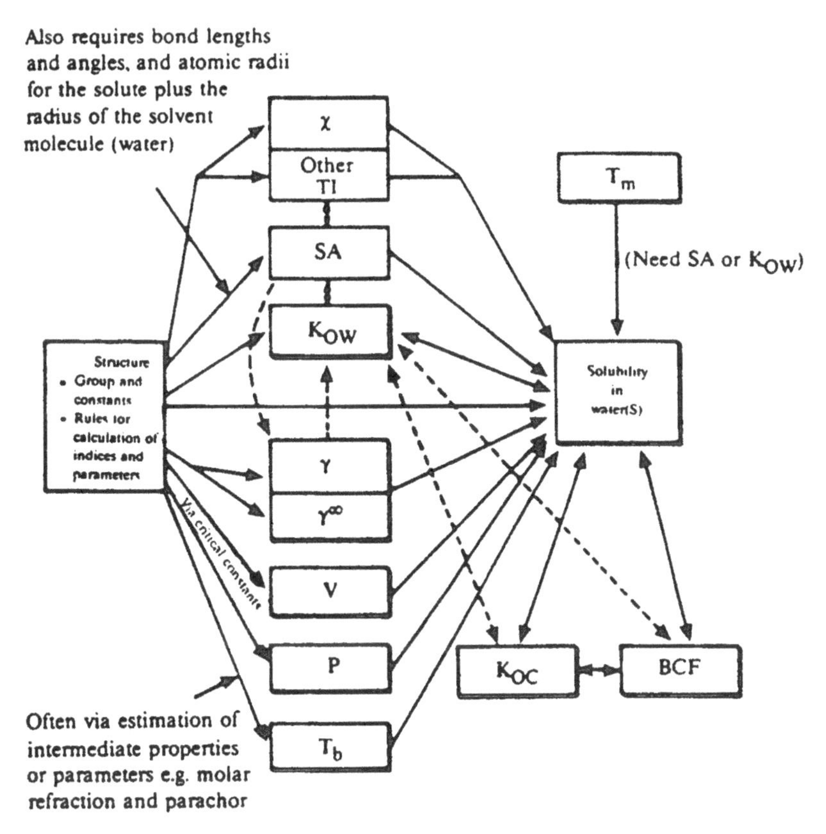

Pathways for estimating the aqueous solubility of organic chemicals. Reproduced from Lyman (1982). BCF, Bioconcentration factor; K_{oc}, Soil-sediment adsorption coefficient; K_{oc}, Octanol-water partition coefficient; P, Parachor; SA, Molecular surface area; T_b, Boiling point; T_m, Melting point; TI, Topological indices; V, Molar volume; γ, Activity coefficient in water; γ^{∞}, Infinite dilution activity coefficient in water; χ, Molecular connectivity parameter, —>, Pathway (unidirectional) leading to S; --->, Other pathways of interest; ⇄, Bidirectional pathways; □, Properties that may be measured in the laboratory (directly or indirectly).

Figure 15. Pathways to estimating solubility (from Lyman et al.[36]).

The first order dependence implies that the hydrolysis of R-X ($t_{1/2} = 0.693/k_T$) is dependent on the R-X concentration.

The rate hydrolysis of various organic chemicals under environmental conditions can range over 14 orders of magnitude, with associated half-lives (time for one-half of the material to disappear) as low as a few seconds to as high as 10^6 years, and is pH-dependent. It should be emphasized that if laboratory rate constant data are used in soil models and not corrected for environmental conditions, as is often the only choice, then model results should be evaluated with skepticism.

Hydrolysis should be distinguished from acid-base reactions, carbonyl hydration (reversible), addition to carbon-carbon bonds (reaction conditions unlikely in the environment) and elimination reactions, which are generally favored by higher temperatures and more strongly basic conditions than are typical of aqueous environments. Elimination may compete with hydrolysis for organic compounds containing good leaving groups such as halides. Hydrolysis comprises a family of reactions involving such diverse compound types as alkyl halides, carboxylic acid esters, carbonates, and nitrites. Table 7 lists some types of organic groups that are generally resistant to hydrolysis and some that are potentially susceptible.[20]

Table 7. Types of Organic Functional Groups

General Resistant to Hydrolysis	Potentially Susceptible to Hydrolysis
Alkanes	Alkyl halides
Alkenes	Amides
Alkynes	Amines
Benzenes/biphenyls	Carbamates
PAHs	Carboxylic acid esters
Heterocyclic PAHs	Epoxides
Halogenated aromatics/PCBs	Nitriles
Dieldrin/aldrin, etc.	Phosphonic acid esters
Aromatic nitro-compounds	Phosphoric acid esters
Aromatic amines	Sulfonic acid esters
Alcohols	Sulfuric acid esters
Phenols	
Glycols	
Ethers	
Aldehydes	
Ketones	
Carboxylic acids	
Sulfuric acids	

Source: Harris.[69]

[a]Multifunctional organic compounds may be hydrolytically reactive if they contain an additional hydrolyzable functional group.

Oxidation-Reduction

For some organic compounds, such as phenols, aromatic amines, and alkyl sulfides, chemical oxidation is an important degradation process under environmental conditions. Most of these reactions depend on reactions with free radicals already in solution and are usually modeled by pseudo-first-order kinetics:

$$-d[X]/dt = K_o'[RO_2^*][X] = K_{ox}[X]$$

where:

X = pollutant
K_o' = second-order oxidation rate constant
RO_2^* = a free radical
K_{ox} = pseudo-first-order oxidation rate constant.

Complexation

Complexation, or chelation, is the process by which metal ions and organic or other nonmetallic molecules (ligands) can combine to form stable metal-ligand complexes. The complex that is found will generally prevent the metal from undergoing other reactions or interactions that would free the metal cation. Complexation may be important in some situations; however, the current level of understanding of the process is not very advanced, and the available information has not been shown to be particularly useful to quantitative modeling.

A variety of models has been used to explain complexation of metal ions by fulvic acid. In the Scatchard method,[70,71] fulvic acid is modeled with distinct classes of site, each with a common ability to complex metal ions. Additional information is provided by Bodek et al.[37]

Bioaccumulation

Bioaccumulation is the process by which terrestrial organisms such as plants and soil invertebrates accumulate and concentrate pollutants from the soil. Bioaccumulation has not been examined in soil modeling, apart from some nutrient cycle (phosphorus, nitrogen) and carbon cycle bioaccumulation attempts. Biotransformation issues in soils are described in details by Valentine.[72]

Biodegradation

Biodegradation refers to the process of transformation of a chemical by biological agents, usually by microorganisms, and it actually refers to the net result of a number of different processes, such as mineralization, detoxication, cometabolism, activation, and change in spectrum. In toxic chemical modeling, biodegradation is usually treated as a first-order degradation process:

$$dc/dt = -K_{DE} \cdot c^n$$

where:

c = dissolved concentration of pollutant soil moisture (μg/mL)
K_{DE} = rate of degradation (day^{-1}), and
n = order or reaction ($n = 1$, first order).

Immiscible Organic Compounds Processes

Density, viscosity, surface wetting, and solubility are the important parameters governing migration of an immiscible organic liquid. According to Mackay and Roberts[50] there is extensive evidence that low-density organic liquids float on the water table. The sinking phenomenon of high-density liquids has been demonstrated in physical model experiments.

The transport of an organic liquid phase also is influenced by its viscosity and its surface-wetting properties compared with those of water. According to Mackay and Roberts and the literature reviewed, halogenated aliphatics tend to spread by capillary action into aquifer media, and to be retained in amounts of about 0.3 to 5% by volume, following the passage of the organic liquid.

An organic liquid of moderately low solubility will contaminate as much as 10,000 times its own volume to its solubility limit. However, organic compounds are rarely found in groundwater at concentrations approaching their solubility limits, even when organic liquid phases are known or suspected to be present.

For the above reasons, it also is difficult to completely remove hydrocarbons from granular aquifers as the capillary attraction between smaller particles may be sufficient to prevent bubbles of, for example, gasoline passing between the grains. Figure 16 shows that a gasoline bubble will continue to move so long as P_{gas}-P_{water} is greater than P_{cap}; otherwise the bubble will be trapped.

Pumping or flushing will not remove all of the contaminant, as water will be defected elsewhere along less resistant paths. The trapped product will then continue to give off water-soluble components. Emphasis has, therefore, been placed upon containment of spills and the early detection of leaks.

Miscible Inorganic Compounds Processes

Important processes and mechanisms of environmental interactions of inorganic materials are volatility, solubility, fast aqueous reactions, slow aqueous reactions, speciation, soil interaction, and bioaccumulation. Processes not already discussed are presented below.

Fast Reactions in Aqueous Media

Reactions of inorganic dissolved species in aqueous media can be classified under the following general categories:

- reaction/dissociation (with solvent molecules)
- substitution reactions (with solvent or dissolved species)
- redox reactions (with dissolved gases or other ions).

Fast reaction can lead to formation of new complex ions (those with different ligands in the coordination sphere and those with a different oxidation state of the

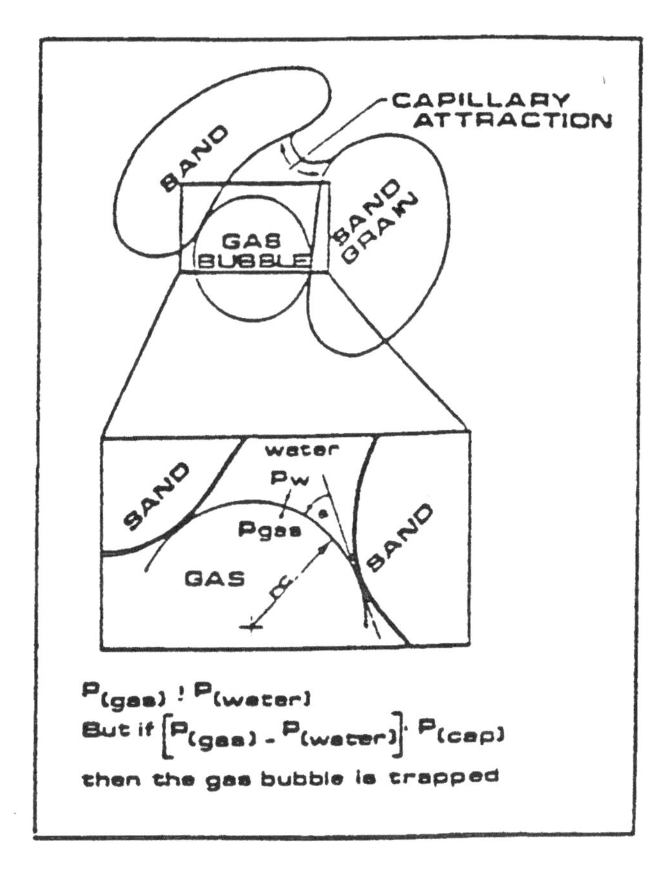

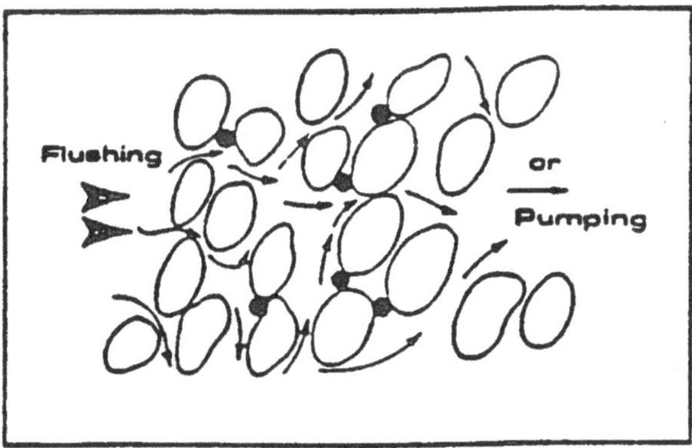

Figure 16. Entrapment of gasoline by capillary forces (from Hall and Quam[46]).

metal centers) and/or other ions. These reactions are important, since the medium that the particular inorganic substance is dissolved in will determine to a great extent the speciation of the particular metal ion. Since the extent of environmental interaction (e.g., soil attenuation, uptake by plants, toxicity) will depend on the chemical form (species) of the metal, knowledge of the behavior with regard to rapid reactions is essential.

In addition, upon mixing of solutions of these metals with the environment (e.g., leachate mixing with groundwater) chemical modification of the species will occur initially via thermodynamically favored rapid reactions and subsequently by the possible slower reactions. Such knowledge is important in the selection and application of aquatic equilibria (or speciation) models described in a later section.

Slow Reactions in Aqueous Media

In the category of slow (slower) reactions in aqueous media, one may consider the following reactions:

- ligand substitution reactions of relatively kinetically inert ions
- electron transfer reactions involving inner sphere mechanisms for relatively kinetically inert ions and some outer sphere reactions
- reactions with dissolved gaseous species/bacterial-catalyzed reduction (e.g., oxidation by dissolved oxygen)
- precipitation of solids by formation of insoluble species via substitution, and/or
- oxidation reactions (e.g., formation of metal hydroxides and metal sulfides).

The above slower reactions impact the speciation and concentration of the inorganic species in the aqueous phase and determine the type and extent of further interactions in the same manner as the fast reactions, although a different time frame will be required to achieve completion of some of these reactions. For the relatively kinetically inert metals (e.g., Cu^{3+}, Fe^{3+}, Cr^{3+} and other such inorganic ions), immediate modifications of the speciation of the ion may *not* occur upon mixing with other aqueous phases (e.g., leachate with groundwater under oxidizing conditions), and the reaction may not proceed prior to the mixture reacting with soil strata.

As such, the time frame for some reactions may indicate that characterizing the speciation in the original aqueous phase (e.g., leachate) will dictate the chemical behavior of the compound with the environment, whereas in other cases, understanding the speciation under the exact conditions of environmental interaction will be required.

Speciation

An evaluation of the fate of trace metals in surface and subsurface waters requires the consideration of speciation, adsorption, and precipitation. These processes can affect metal solubility, toxicity, availability, physical transport, and corrosion potential. As a result of a need to describe the complex interactions involved in these situations, various models have been developed to address a number of specific situations. Steps in speciation calculations are not described in this chapter. Speciation computerized packages (models) are described in a later section.

Soil and Sediment Interactions

Important interactions of dissolved aqueous inorganic species occur with soils. The more important mechanisms of interaction include formation of precipitates, adsorption of components onto soil surfaces, modification of speciation by soil constituents (solid and liquid phase), and reactions induced by bacteria present in soil.

The properties of the surface of the soil particles which will dictate the adsorption behavior can be changed by the acid/base buffer system in contact with the soil. Most of the minerals display amphoteric properties toward solutions; i.e., they behave as weak acid or bases. The buffer capacity of the soil (e.g., related to $CaCO_3$ content) is important with regard to the ability of soil water to fluctuate in pH. Microorganisms can bring important changes in the solubility of soil minerals and other precipitates by altering solution conditions and forming films on surfaces of particles and catalyzing reactions.

Conceptual models for attenuation of species onto soils have been developed which incorporate factors such as electrostatic energy and chemical bonding effects (e.g, van der Waals, dipole effects, and covalent bonding). Langmuir adsorption isotherms have been generated which are consistent with this conceptual model. As previously mentioned, solution chemical equilibria will modify the adsorptive behavior of the soil as well as the species present in solution and its tendency to be adsorbed; therefore, simplified adsorption, attenuation, or speciation models have to be used with caution in inorganic fate modeling.

Bioaccumulation

Much information is available concerning the effects of specific inorganic constituents on a large variety of organisms in specific environments. However, such information has generally been difficult to use in the development of broadly generalized techniques for estimation of effects. Estimation techniques for understanding the impact potential of inorganic compounds on organisms in aquatic or terrestrial environments have not at present been developed.

Terrestrial Plants

If there are exceptions to the general lack of estimation techniques and models relating to inorganic chemical concentrations, biological effects and fate, they are likely to be found in agricultural theory and modeling. The application of soil amendments (fertilizers) as macronutrients (e.g., phosphates, nitrates) or micronutrients (e.g., Mn), or for improving the soil chemical environment to release nutrients (liming), is a relatively well-developed science. At the least, there is an understanding of the optimum ranges of concentrations of agriculturally related chemicals and growth of crop species. This stands in contrast to the paucity of information regarding uptake of organic soil contaminants by plants.

Selected Cases

This section reports selected cases of hydrocarbons fate in soils and related bioremediation practices applied following the thorough understanding of the pollutant migration in the subsurface environment.

Gas Plant Residues in Soils

The fate of hydrocarbons in soils has been studied in a research program to investigate the feasibility of using biological treatment to remediate soils containing residues from manufactured gas plant (mgp) sites, and to develop and verify a rapid assessment protocol for determining the potential for biologically treating specific mgp site soils.[73] This protocol, which features soil characterization, desorption testing, and slurry reactor testing, is cost-competitive with traditional treatability studies and can be completed in approximately half the time. For the first two soils evaluated, total polynuclear hydrocarbon (PAH) concentrations (a critical class of chemicals at mgp sites) were reduced by 95% and 80% to residual plateau concentrations of less than 10 and 5000 mg/kg, respectively. The magnitude of the plateau concentration is not primarily dictated by the availability of nutrients, oxygen, and adequate microbial populations but rather by limitations of mass transfer from the soil-waste matrix to the bulk aqueous phase. Further tests with 13 other mgp soils are in progress, and the results of these will be reported at a later time. The results of sprinkling soil and the fate of naphthalene and phenanthrene are shown in Figure 17.[73]

Fate of Kerosene in Soil Columns

Acher et al.[74] and Yaron et al.[75] conducted laboratory research on fate and soil contamination by a synthetic kerosene soil (Mediterranean red sandy clay). Samples with different moisture contents (0.0, 0.8, 4.0, and 12% w/w) were contaminated by vapor and/or liquid from a mixture containing 5 kerosene components (m-xylene, pseudo-cumene, t-butylbenzene, n-decane, and n-dodecane). The contribution of the different kerosene components to the adsorption, volatilization

A. Naphthalene

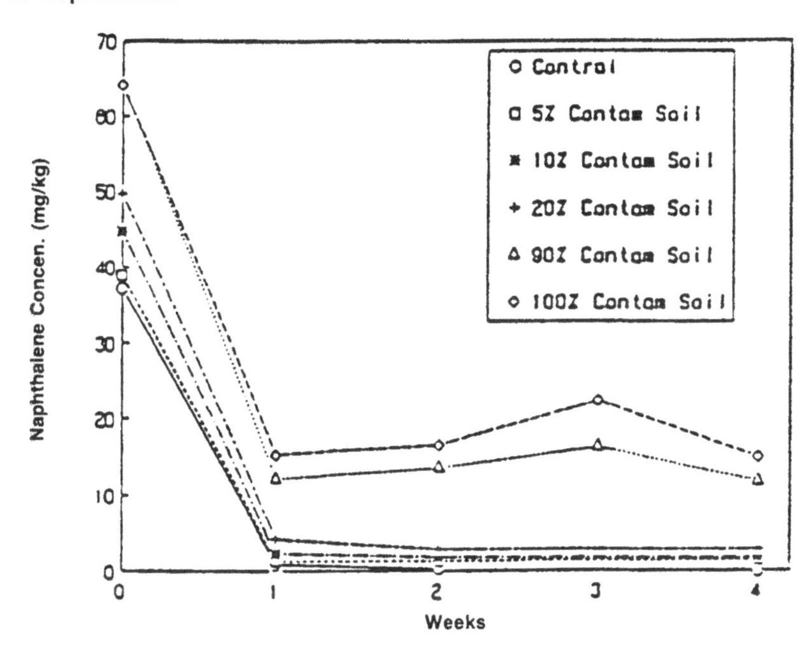

B. Phenanthrene

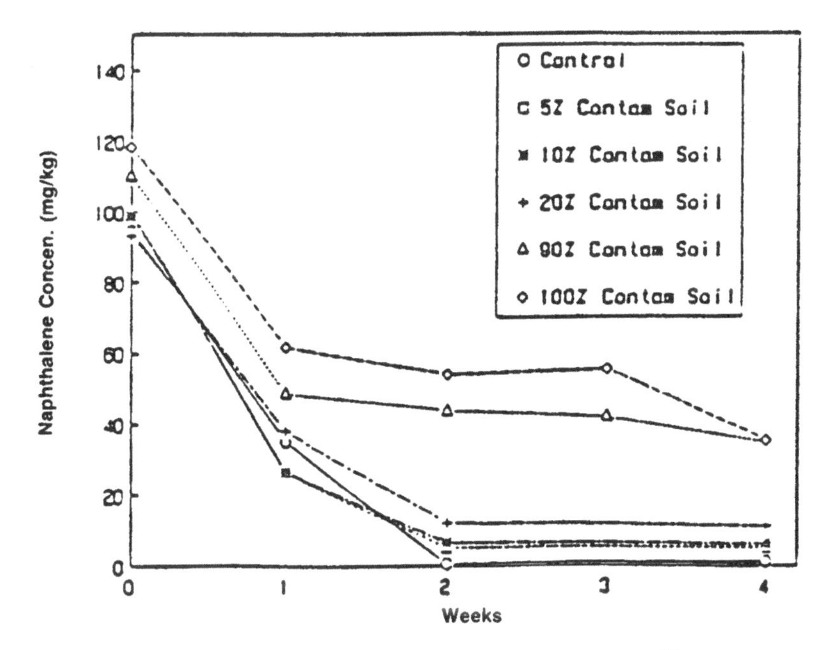

Figure 17. Fate of naphthalene and phenanthrene in soil (from RTI[73]).

and transport processes is described by Acher et al. Vapor adsorption was found to be dependent on the vapor concentration of each component (except for n-decane) and on the soil mositure content. The sorption coefficient of the kerosene components decreased with temperature but showed only a very slight variability between 20° and 34°C in air-dried soil. The volatilization from the soil was high; more than 90% of the aromatic components were desorbed in less than 2 hr. The transport of the kerosene in liquid and vapor phases through the soil column was studied, using amounts of kerosene that were less that 1 mL or more than 10 mL than the retention capacity of the soil column. The increase in the moisture content of the soil increased the rate and the depth of kerosene downward penetration. However, it stopped the vapor movement (at 4%) and the upward liquid movement (at 12%). Among the properties of the kerosene components, volatility seems to be the prime factor which determines kerosene movement once liquid phase movement has ceased. The redistribution of kerosene components in the soil column as affected by soil moisture and time is shown in Figure 18.

MATHEMATICAL MODELING OVERVIEW

General

Mathematical models can greatly assist decisionmakers in determining the importance of pollutant pathways in the environment, as long as they are used properly and with an understanding of their limitation. The use of models has grown dramatically over the past decade, but models are not meant to substitute for good judgment and experience.

Pollutant fate mathematical modeling in soil systems is an area of current intensive research, because of the numerous problems originating at waste sites. The variety of models has dramatically increased during the last decade, but although the variety of models appears to be large, only a few *different* modeling concepts exist, and very few physical or chemical processes are modeled.

Modeling Compartments and Compounds

In general, soil compartment modeling concepts deal mainly with point and nonpoint-source pollution, and therefore, modeling applies to:

1. land (watershed) pollution
2. unsaturated soil zone (vadose or soil), and
3. saturated soil zone (groundwater).

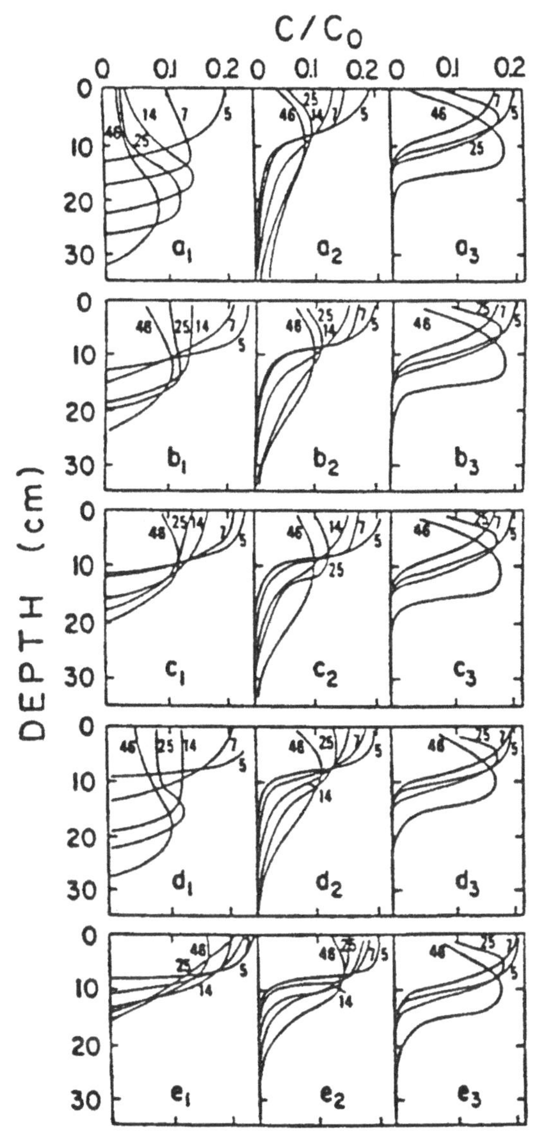

Redistribution of "kerosene" components in soil columns as affected by soil moisture and time. Numbers on curves represent days after "kerosene" application. C = hydrocarbon amount in each soil layer. C_0 = total hydrocarbon amount in the column. (a) = m-xylene; (b) = n-decane; (c) = ps-cumene; (d) = t-butylbenzene and (e) = n-dodecane at (1) 0.0%, (2) 0.8% and (3) 12.0% soil moisture content.

Figure 18. Distribution of kerosene in soil (from Acher et al.[74]).

Modeling is accomplished for:

1. dissolved (or miscible) pollutants (e.g., benzene)
2. immiscible pollutants (e.g., heavy oil)
3. aquatic speciating pollutants.

Modeling for miscible and immiscible pollutants follows comparable patterns of mathematics and approach. Modeling of speciating pollutants enters into chemistry and speciation modeling issues.

Soil compartment modeling is a very complex issue and a major characteristic of a soil subcompartment—as compared to a water or an air subcompartment—is that the temporal physical and the chemical behavior of this subcompartment is governed by *out-compartmental* forces such as precipitation, air temperature and solar radiation, and depth to the groundwater table. This characteristic is also one of the main reasons why soil mathematical modeling can be much more complex than water or air modeling, where the dynamics of the compartment are governed by *in-compartmental* forces. Groundwater modeling can handle a limited number of chemical processes; therefore, a number of aquatic equilibrium models are aimed to fill this chemistry gap.

Soil compartment models (watershed, soil, groundwater) are employed to evaluate pollution originating from various sources, such as hazardous waste or petroleum contaminated sites. Because detailed soil modeling is not always feasible or desirable, a category of ranking models is known in the literature for screening severity of environmental impact originating from waste activities.

There is no scientific reason for a soil model to be an unsaturated soil model only, and not to be an unsaturated (soil) and a saturated soil (groundwater) model. Only mathematical complexity mandates the differentiation, because such a model would have to be, for example, three-dimensional (e.g., Narasimham[76]) and very difficult to operate. Most of the soil models account for vertical flows; groundwater models for horizontal flows.

Model Types

In general, models can be classified into deterministic, which describes the system as a cause/effect relationship, and stochastic, which incorporates the concept of risk, probability, or other measures of uncertainty. Deterministic and stochastic models may be developed from observation, semiempirical and theoretical approaches.

In developing a model, scientists attempt to reach an optimal compromise among the above approaches, given the level of detail justified by both the data availability and the study objectives. Deterministic model formulations can be further classified into simulation models which employ a well-accepted empirical equation that is forced, via calibration coefficients, to describe a system, and analytic models in which the derived equation describes the physics/chemistry of a system.

Modeling Pollutant Releases

Environmental fate models require information on the temporal and spatial distribution of chemical releases. Sources can be described in terms of their dimensionality and releases in terms of temporal distribution. Releases may be continuous and approximately uniform (e.g., hydrocarbon contaminated site), intermittent but predictable (e.g., industrial emissions), cyclic or random (e.g., oil spill). Samiullah[20] categorizes sources and releases as follows:

Dimensionality

Point sources	(smoke stacks)
Line sources	(the aggregated emissions from vehicles moving along a road)
Area sources	(conglomerate domestic fuel emissions)
Volume source	(nebulous concept—perhaps an example could be photochemical smog)

Temporal distribution

Accidents	(instantaneous: kg/event)
Point sources	(kg/s)
Line sources	(kg/s/m^2)

The severity of a substance in a discharge from an industrial source is defined as the ratio of substance concentration either at the source or at some ambient point of interest, to a maximum specified safe concentration level.[77] A source is considered clean unless the severity is expected to exceed unity in more than a given and acceptably small proportion of time; otherwise it is classed as dirty. Classification is based on such parameters as stack emission characteristics and meteorology. Releases may be estimated by a variety of methods, principally:[20]

1. *Measurement*—either by direct sampling or calculation from known application rates.
2. *Materials balance*—can reduce to the simple mass balance where the measured mass of a chemical in products leaving the plant is subtracted from the raw material entering the plant to yield the loss.
3. *Mathematical models*—usually relatively simple.
4. *Ad hoc*—using the most obvious supposition and calculations for a given situation. These include rule-of-thumb bounds on lost fractions of production likely to be economically acceptable.

Solution Algorithms of Models

Without a solution, formulated mathematical systems (models) are of little value. Four solution procedures are mainly followed: the analytical; the numerical (e.g., finite different, finite element); the statistical; and the iterative. Numerical techniques have been standard practice in soil quality modeling. Analytical techniques are usually employed for simplified and idealized situations. Statistical techniques have academic respect, and iterative solutions are developed for specialized cases. Both the simulation and the analytic models can employ numerical solution procedures for their equations. Although the above terminology is not standard in the literature, it has been used here as a means of outlining some of the concepts of modeling.

Generally speaking, a deterministic or stochastic soil quality model consists of two major parts or modules:

- The flow module or moisture module, or hydrologic cycle module, that predicts flow or moisture behavior (i.e., velocity, content in the soil).
- The solute module that predicts pollutant transport, transformation and soil quality.

The above two modules form the soil quality model. The flow module drives the solute module. It is important to note that the moisture module can be absent from the model, and in this case a model user has to input to the solute module information that would have been either produced by a moisture module or would have been obtained from observed data at a site.

Models Categorization

At this stage of intensive research in soil and groundwater quality modeling, it may be reasonable to group the prevailing modeling concepts of the literature into three major categories:

1. Traditional Differential Equation (TDE) modeling
2. Compartmental modeling
3. Stochastic and other types of modeling.

This terminology is not a standard but is employed here for reasons of communication. TDE modeling applies to both the flow or moisture module and the solute module, and a modeling package may consist of one TDE module (e.g., moisture) and a compartmental solute module, or vice versa.

Model Categories of This Chapter

The following sections aim to clarify some key issues related to soil and groundwater models. The following documents provide an overview of this area of science: the publication of Abriola and Pinder,[53] a state-of-the-art publication on multiphase approach in porous media and a series of articles by Mercer and Cohen[78] describing groundwater modeling concepts which are equally applicable to unsaturated soil zone modeling; the monograph of Bachmat et al.[79] listing various models; the work of Bonazountas and Wagner[35] introducing the compartmental soil quality modeling concept and geochemical modeling; and the reference book of Freeze and Cherry[58]; the reference book of NRC[80]; the review article by Gee et al.[4]; the book by Hern and Melancon[10]; the book by Samiullah[20], and other publications listed in the references section.

Reference to the above sources is not meant to exclude other excellent publications presented in reputable scientific journals; rather, it indicates selected basic sources employed to draft the following sections. In the following sections, information is presented on:

- dissolved (miscible) pollutant modeling in soil and groundwater
- immiscible (NAPL) contaminant modeling
- aquatic equilibrium modeling
- existing mathematical modeling packages
- other media (e.g., land, air, water) modeling.

MISCIBLE POLLUTANT FATE MODELING

General

In this section emphasis is placed on the unsaturated soil zone rather than on groundwater modeling. This emphasis is justified by the fact that similar modeling concepts govern both environments, and that the vadose zone is the medium of pollutant transfer to the groundwater.

Unsaturated Soil (Vadose or Soil) Zone Modeling

Soil modeling follows three different mathematical formulation patterns: (1) Traditional Differential Equation (TDE) modeling, (2) compartmental modeling, and (3) stochastic and other types of modeling. Some researchers may categorize models differently as, for example, into numerical or analytic, but this categorization applies more to the techniques employed to solve the formulated model, rather than to the formulation per se. A model has a flow (moisture) module and a quality module.

TDE Modeling

The TDE moisture module (of the model) is formulated from three equations: (1) the water mass balance equation, (2) the water momentum, (3) the Darcy equation, and (4) other equations such as the surface tension of potential energy equation.

The resulting differential equation system describes moisture movement (moisture module) in the soil and is written in a one-dimensional, vertical, unsteady, isotropic formulation as:

$$d[K(\psi)(d\psi/dz+1]/dz = C(\psi)d\psi/dt+S \quad (1)$$

$$\upsilon_z = -K(z,\psi)d\varphi/dz \quad (2)$$

where:

z	= elevation (cm)
ψ	= pressure head, often called soil moisture tension head in the unsaturated zone (cm)
$K(\psi)$	= hydraulic conductivity (cm/min)
$C(\psi = d\theta/d\psi$	= slope of the moisture (θ) versus pressure head (ψ) (cm^{-1}; t = time (min)
S	= water source or sink term (min^{-1})
φ	= $z + \psi$
υ_z	= vertical moisture flow velocity (cm/s).

The moisture module output provides the parameters υ and θ as input to the solute module.

The TDE solute module is formulated with one equation describing the pollutant mass balance of the species in a representative soil volume, dV = dxdydz. The solute module is frequently known as the dispersive-convective differential mass transport equation, in porous media, because of the wide employment of this equation, that may also contain an adsorptive, a decay, and a source or sink term. The one-dimensional formulation of the module is:

$$d(\theta c)/dt = [d(\theta \cdot K_D \cdot dc/dz]/dt - [d(\upsilon c)/dz] - [p \cdot d_s/dt] \pm \Sigma P \quad (3)$$

where:

θ	= soil moisture content
c	= dissolved pollutant concentration in soil moisture
K_D	= apparent diffusion coefficient of compound in soil-air
υ	= Darcy velocity of soil moisture
ρ	= soil density
s	= adsorbed concentration of compound on soil particles
ΣP	= sum of sources or sinks of the pollutant within the soil volume
z	= depth.

Models like the above have been presented by various researchers from the U.S. Geological Survey (USGS) and academia. The above equation has been solved principally:

- numerically over a temporal and spatial discretized domain, via finite difference or finite element mathematical techniques (e.g., Mackay[30]),

- analytically, by seeking exact solutions for simplified environmental conditions (e.g., Enfield et al.,[43]), or
- probabilistically (e.g., Schwartz and Growe[81]).

At this point it is important to note that the flow model (a hydrologic cycle model) can be absent from the overall model. In this case the user has to input to the solute module, the temporal (t) and spatial (x,y,z) resolution of both the flow (i.e., soil moisture) velocity (υ), and the soil moisture content (θ) of the soil matrix. This approach is employed by Enfield et al.[43] and other researchers. If the flow (moisture) module is not absent from the model formulation (e.g., Huff[82]), then users are concerned with input parameters that may be frequently difficult to obtain. The approach to be undertaken depends on site specificity and available monitoring data.

Some principal modeling-specific deficiencies when modeling solute transport via the TDE approach are:

1. Only diffusion, convection, adsorption, and possibly decay can be modeled, whereas processes such as fixation or cation exchange have to be either neglected or represented with the sources and sinks term of the equation because of mathematical complexity.
2. The equation system is applicable mainly to pollutant transport of organics, whereas transport of metals, which can be strongly affected by other processes, cannot be directly modeled.
3. The equation system can predict volatilization only implicitly via boundary diffusion constraints; however, experimental studies have frequently demonstrated an overestimation or underestimation of the theoretical volatilization rate unless a sink or source term is included in the equation.
4. No experimental or well-accepted equation for a process (e.g., volatilization) can be incorporated, since the model has its own predictive mechanism.
5. Pollutant concentrations are estimated only in the soil-moisture and on soil-moisture and on soil-particles, whereas pollutant concentrations in the soil air are omitted.
6. The discretized version of the equation in the case of numerical solutions has a pre-set temporal and spatial discretization grid that results in high operational costs (professional time, computer time) of the model, since input data have to be entered into the model for each node of the grid.

In a modeling evaluation effort, Murarka[83] reports that the currently available coupled or uncoupled models of hydrologic flow and the geochemical interactions are adversely affected by the following factors: difficulties in establishing consistency between the theoretical frameworks, laboratory experiments, and field research; limited basic knowledge about nonequilibrium conditions and phase relations; inadequate existence of geochemical submodels to couple with the hydrologic transport submodels; uncertainties in input data, particularly for dispersion and chemical reaction rate coefficients; and numerical difficulties with model solution techniques.

Compartmental Modeling

Compartmental soil modeling is a concept promoted by the EPA SESOIL model[35] and can apply to both modules. The solute fate module, for example, consists of the application of the law of pollutant mass conservation to a representative user-specified soil element (Figure 6). The mass conservation principle is applied over a specific time step, either to the entire soil matrix or to the subelements of the matrix such as the soil-solids, the soil-moisture, and the soil-air. These phases can be assumed to be in equilibrium at all times; thus, once the concentration in one phase is known, the concentration in the other phases can be considered, whereas phases and subcompartments can be interrelated with transport, transformation, and interactive equations.

Compartmental models may bypass the deficiencies of the TDE modeling because they may handle geochemical issues in a more sophisticated way if required, but this does not imply that compartmental models are better than TDE models. They are simply different. Compartmental models reflect the personal touch of their developers and cannot be formulated under generalized guidelines or concepts.

The moisture module (i.e., driving element) of a compartmental solute model can be either incorporated into the overall model, or can be an independent module as, for example, a TDE module of the literature. At this stage of scientific research a well-developed soil compartment model appears to be SESOIL, Seasonal Soil Compartment Model[35] of the EPA.

SESOIL consists of a dynamic compartment moisture module and a dynamic compartmental solute transport module. The following sections present a demonstration of the basic mathematical equations governing compartmental soil quality modeling. This information has been abstracted from the SESOIL model.

The law of pollutant mass concentration for a representative element can be written over a small time step as

$$\Delta M = M_{in} - M_{out} - M_{trans} \tag{4}$$

The solute (dissolved) concentration of a compound can be related to its soil-air concentration via Henry's law:

$$c_{sa} = c \cdot H/R \cdot (T+273) \tag{5}$$

where

c_{sa} = pollutant concentration in soil-air
c = dissolved pollutant concentration
H = Henry's law constant
R = gas constant, and
T = temperature in °C.

The pollutant concentration of the soil (i.e., solids) can be determined from the sum of the concentrations of the pollutant adsorbed, cation exchanged, and/or otherwise associated with the soil particles, e.g., via adsorption isotherms. One commonly used adsorption isotherm equation is the Freundlich equation:

$$s = K \cdot c^{1/n} \tag{6}$$

where

s = adsorbed concentration of compound
K = partitioning coefficient
c = dissolved concentration of compound, and
n = Freundlich constant.

The total concentration of a chemical in a soil matrix can be calculated from the concentration of pollutant in each phase and the related volume of each phase by:

$$c_o = (n\text{-}\theta) \cdot c_{sa} + (\theta) \cdot c + (\rho_b) \cdot s \tag{7}$$

where

c_o = overall (total) concentration of pollutant in soil matrix
n = soil (total) porosity
θ = soil moisture content
$(n\text{-}\theta)$ = soil-air content or soil-air filled porosity
c_{sa} = pollutant concentration in soil-air
c = pollutant concentration (dissolved) in soil-moisture
P_b = soil bulk density
s = pollutant concentration on soil particles.

The above expressions are input terms to Equation 4, which is then applied for each time step, each subcompartment, and each compartment of the user-specified matrix (Figure 6).

The term M_{in} may reflect input pollution from rain (upper layer), from soil-moisture from an upper layer, and from a lower layer. The term M_{out} reflects pollution exports from the individual compartment, whereas the term M_{trans} reflects all transformation and chemical reactions taking place in the compartment.

All terms can be normalized to the soil-moisture concentration via interconnecting equations such as 5 and 6, which can describe processes such as volatilization, cation exchange, photolysis, degradation, hydrolysis, fixation, biological activity, etc.

The solution of the resulting system of equations is a complicated issue and may require—for computational efficiency and other reasons—development of new numerical solution techniques or algorithms (e.g., SESOIL).

Stochastic, Probabilistic, Other Modeling Concepts

Stochastic or probabilistic techniques can be applied to either the moisture module or the solution of Equation 3 as, for example, the model of Schwartz and Growe,[81] or can lead to new conceptual model developments as, for example, the work of Jury.[63] Stochastic or probabilistic modeling is mainly aimed at describing breakthrough times of overall concentration threshold levels rather than individual processes or concentrations in individual soil compartments.

Coefficients or response functions for these models have to be calibrated to field data, since major processes are studied via a black-box or response function approach and not individually. Other modeling concepts may be related to soil models for solid waste sites and specialized pollutant leachate issues.[84]

Physical, Chemical, Biological Processes Modeled

Modelers should be fully aware of the range of applicability and processes considered by a computerized package. There exists some disagreement among soil modelers as to whether there is a need for increased model sophistication, since almost all soil modeling predictions have to be validated with monitoring data, given the physical, chemical, and biological processes that affect pollutant free in soil systems. Because of the latter consideration, many simplified models may provide excellent results, assuming accurate site-specific calibration is achieved. Nevertheless, model sophistication is reflected in the process modeled, but model selection is mandated by the project needs and data availability.

The important physical processes of a typical soil model are:

1. The hydrologic cycle or moisture cycle of the vadose zone that may encompass the processes of rain infiltration in the soil, exfiltration from the soil to the air, surface runoff, evaporation, moisture behavior, groundwater recharge, and capillary rise from the groundwater. All these processes are interconnected and are frequently referred to as the *hydrologic cycle component* of the model.
2. The *pollutant or solute cycle* that may encompass the processes of advection, diffusion, volatilization, adsorption and desorption chemical degradation or decay, hydrolysis, photolysis, oxidation, cation or anion exchange, complexation, chemical equilibria, nutrient cycles, and others.

3. The *biological cycle* that may encompass processes of biological transformation, plant uptake, bioaccumulation, soil organism transformation and others.

Models in the literature can handle one or more of the above processes and for various pollutants. In general, however, soil models tend to handle the:

1. Hydrologic cycle: temporal resolution of soil moisture, surface runoff and groundwater recharge components, by inputting to the model the net infiltration rate into the soil column.
2. Pollutant and biological cycles: the processes of advection, diffusion, volatilization (diffusion at the soil-air interface), adsorption or desorption (equilibrium) and degradation or decay, which are also the most important chemical processes in the soil zone. All other processes can be lumped together under the source or sink term of Equation 3.

Fortunately—and not unfortunately—no one model exists as yet which simulates all of the physical, chemical, and biological processes associated with pollutant fate in soils. We say fortunately, because such a package would be very data-intensive and difficult to use. Intensive research is required to accomplish the above objective, and the value of the overall product may be questioned by users. A later section presents selected models.

Saturated Soil Zone (Groundwater) Modeling

Saturated soil zone (or groundwater) modeling is formulated almost exclusively via a TDE system consisting of two modules, the flow and the solute module. The two modules are written as:[79]

$$\nabla(\rho k/\mu)(\nabla p - \rho g \nabla Z) - q = d(\varphi\rho)/dt \tag{8}$$

$$\nabla[\rho C(k/\mu)(\nabla p - \rho g \nabla Z)] + \nabla(\rho E)\nabla C = d(\rho\varphi C)dt \tag{9}$$

where:

∇ = Laplace function
C = concentration, mass fraction
E = dispersion coefficient
g = acceleration due to gravity
k = permeability
p = pressure
q = mass rate of production or injection of liquid per unit volume
t = time
Z = elevation above a reference plane
φ = porosity

ρ = density
μ = viscosity.

Mathematical groundwater modeling has been the least problematic in its scientific formulation, but has been the most problematic model category when dealing with applications, since these models have to be calibrated and validated as described later. Actually we have only TDE and some other (e.g., stochastic) formulations.

The proliferation of literature models is mainly due to different model dimensionalities (zero, one, two, three); model features (e.g., with adsorption, without absorption terms); solution procedures employed (e.g., analytic, finite difference, finite element, random walk, stochastic) for Equation systems 7 and 8; sources and sinks described; and the variability the boundary conditions impose.

Some of the principal modeling deficiencies discussed in the previous section (vadose modeling) apply to groundwater models also. In general, (1) there exists no *best* groundwater model, and (2) for site-specific applications, groundwater models have to be calibrated.

The two principal solution methods for Equations 8 and 9 that result in different model categories with substantially different impacts on the level of effort required to run a package are (a) analytical models and (b) numerical models. Employment of the first method results in formulation of expressions applicable to the nodes or the elements of a domain, the number of nodes or elements being user-specified.

In analytical modeling, only averaged data for the entire domain have to be input to the model. The numerical modeling data have to be input for all nodes or elements of the model, a fact that frequently results in high model cost runs, in terms of both professional and computational time. Common numerical solution techniques are the finite difference, the finite element, the method of characteristics, and random walk and their variations. Interested readers are referred to the references for this chapter.

Ranking Modeling

Ranking models are aimed at assessing environmental impacts of waste disposal sites. The first ranking models focused on groundwater contamination from organic pollutants. Later models had a wider scope (e.g., health consideration).

Ranking models rank or rate contaminant migration at different sites, as it is affected by hydrogeologic, soil, waste type, density, and site design parameters. These models are based on questions and answers, and on weighting factors the user has to specify. They are very subjective in their use and have received wide dissemination, because they are easy to use and do not require use of computers. Ranking models do not have a great credibility today when assessing environmental impacts, yet they are available in the literature.

Well-known models are LeGrand,[85] Silka and Swearingen,[86] JRB,[87] MITRE,[88] and Arthur D. Little, Inc.[89] Interested readers should refer to the original publications.

Selected Model Codes

Table 8 lists selected soil and groundwater models and their main features. Table 9 lists limitations and advantages of major model categories. Models listed in Table 8 are documented, operational, and very representative of the various structures, features, and capabilities. For example:

Table 8. Selected Miscible Pollutant Fate Models and Features

	Model Type[a]		Model Formulation[b]					Mathematics[c]		Chemistry Issues[d]					User Concerns[e]			
Model Acronym	Unsaturated Zone	Groundwater	Flow Module	Solute Module	TDE Approach	Compartmental	Statistical, Other	Analytical	Numerical	Organics	Inorganics	Metals	Gaseous Phase	Increased Chemistry	Input Data Req'd.	Calibration	Level of Effort	Application Study
PESTAN	•			•	•			•		•					L	L	L	•
PRZM	•		•	•		•			•	•					M	M	M	•
SCRAM	•		•	•	•				•	•					H	H	H	•
SESOIL	•		•	•		•	•	•	•	•	•	•	•	•	M	M	M	•
AT123D		•		•	•			•		•	•	•			M	L	M	
PATHS		•		•	•			•	•	•	•	•			L	M	M	•
MMT/VVT		•	•	•	•				•	•					H	H	H	•
FEMWASTE		•		•					•	•					H	H	H	•
R. WALK		•		•	•				•	•					H	H	H	•
USGS Models	•	•	•	•	•			•	•	•	•	•	•		H	H	H	•
Other Models																		

[a]The use of complex models (e.g., a numerical soil and groundwater package) that can handle more than one compartment is not always desirable, since generalized packages tend to be cumbersome unless especially designed.

[b]The most representative characteristics are given. The Traditional Differential Equation (TDE) approach applies to the flow and solute module. "Other" includes linear analytic system solutions, for example.

[c]The most representative characteristics are given.

[d]Almost all models can simulate organic, inorganic and metal fate, assuming that a careful calibration via an adsorption coefficient may alter the model output to predict measured/monitored values. However, not all models have by design increased chemistry capabilities (e.g., cation exchange capacity; complexation); therefore, the most representative capabilities are indicated.

[e]L = low, M = medium, H = high input data requirements. In general, numerical models have higher input data requirements and calibration needs and therefore may better represent spatial resolution of a domain. Compartment models provide an optimal compromise. The level of effort is intuitively defined here.

Table 9. Most Important Characteristics of Major Model Categories

Model Category	Advantages	Disadvantages	Comments
TDE type	Clear formulation/capabilities	Rigid model structure; limited capabilities	This has been the traditional computerized modeling approach
Analytical	Easy model use; limited calibration possibilities; limited input data requirements; desk computer use	Rough averaged predictions of pollutant fate; limited application capabilities	To be used as an overall fate (screening) tool
Numerical	Wide range of applications; detailed spatial, temporal resolutions; increased chemistry capabilities	Extensive calibration requirements; input data intensive (nodes, elements, time); require computer use and related skills	Recommended for site-specific applications
Compartmental type	Can be tailored to user's requirements; increased chemistry capabilities; can better meet spatial and temporal domain requirements	Expected user interaction and problem understanding	Today's scientific tendency
Aquatic equilibrium model	Increased chemistry capability	Data intensive; parameters may not be available	Models at a developmental stage
Ranking models	Easy to use with available data	Simplistic approach; output reflects user's intuition	Employed by the EPA, U.S. Army, Air Force, and Navy

1. PESTAN[43] is a dynamic TDE soil solute (only) model, requiring the steady-state moisture behavior components as user input. The model is based on the analytic solution of Equation 3 and is very easy to use, but also has a limited applicability, unless model coefficients (e.g., adsorption rate) can be well estimated from monitoring studies. Moisture module requirements can be obtained by any model of the literature.
2. SCRAM[90] is a TDE dynamic, numerical finite difference soil model, with a TDE flow module and a TDE solute module. It can handle moisture behavior, surface runoff, organic pollutant advection, dispersion, adsorption, and is designed to handle (i.e., no computer code has been developed) volatilization and degradation. This model may not have received great attention by users because of the large number of input data.
3. SESOIL[35] is a dynamic soil compartmental model, with a hydrologic cycle and a pollutant cycle compartmental structure that permits users to tailor the model temporal and spatial resolution to the study objectives. The model estimates the hydrologic cycle components (including moisture behavior) from available NOAA, USDA, and USGS data, and simulates the pollutant cycle by accounting for a number of chemical processes for both inorganic (metal) and organic pollutants.
4. PRZM, Pesticide Root Zone Model[91] developed by the USEPA environmental research laboratory at Athens, Georgia, is a dynamic, compartmental model for use in simulating chemical movement within and below the plant root zone. Time varying transport including advection and dispersion are presented in the program.
5. PATHS[92] is mainly an analytical groundwater model that provides a rough evaluation of the spatial and temporal status of a pollutant fate. The model has its own structure and features, and is a deviation from the TDE, or the compartmental, or stochastic approaches.
6. AT-123D[93] is a series of soil or groundwater analytical submodels, each submodel addressing pollutant transport; in one, two, or three dimensions; for saturated or unsaturated soils; for chemical, radioactive waste heat pollutants; and for different types of releases. The model can provide up to 450 submodel combinations in order to accommodate various conditions analytically.
7. MMT[94] is a one- or two-dimensional solute transport numerical groundwater model, to be driven off-line by a flow transport such as VTT (Variable Thickness Transport). MMT employs the random-walk numerical method and was originally developed for radionuclide transport. The model accounts for advection, sorption, and decay.

The remaining models of Table 8 follow the scientific basic patterns described above, with small variations. All models handle one species at a time, and two soil models (SESOIL, AT-123D) can handle gaseous pollutants also. The U.S. Geological Survey (USGS) has been very active for a number of years in TDE soil and groundwater quality model developments. Models of the USGS are well documented and available in the public domain.

Model Validation Issues

Model selection, application, and validation are issues of major concern in mathematical soil and groundwater dissolved pollutant modeling. For the model selection, issues of importance are: the features (physics, chemistry) of the model; its temporal (steady-state, dynamic) and spatial (e.g., compartmental approach resolution); the model input data requirements; the mathematical techniques employed (finite difference, analytic); monitoring data availability; and cost (professional time, computer time). For the model application, issues of importance are: the availability of realistic input data (e.g., field hydraulic conductivity, adsorption coefficient); and the existence of monitoring data to verify model predictions. Some of these issues are briefly discussed below. Generic steps in the field validation of vadose fate and transport models are extensively described by Hern and Melancon[10] and are documented in Table 10.

Input data have to be compiled and input to the model from site-specific investigations and analyses (e.g., leaching rates of pollutants, soil permeability); national data bases (e.g., climatic data from the NOAA); and other sources (e.g., diffusion rates of pollutants from handbooks). Compilation of input data for site-specific applications have to frequently be obtained from a field monitoring program. Some data categories are pollutant source data, climatic data, geographic data, particulate transport data, and biological data.

Exact knowledge of the physics of the soil system, although essential, is impossible prior to employing any modeling package. Issues of spatial variability of soil properties as related to vadose modeling are reported by Jury.[3] Numerical (e.g., finite difference) TDE soil models, for example, require the net infiltration rainfall rate after each storm, even as an input parameter to their moisture module. The rate can be either a user input or can be generated by another model. The same models require the soil conductivity as a function of the soil moisture content as an input parameter. Its value can be obtained either from field investigations or from laboratory data, or from references, but much uncertainty exists in this area of input data gathering.

Numerical soil models (time, space) provide a general tool for quantitative and qualitative analyses of soil quality, but require time-consuming applications that may result in high study costs. In addition, input data have to be given for each element of the model, and the model has to be run twice, with the number of rainfall events. On the other hand, analytic models obtained from analytic solutions are easier to use, but can simulate only averaged temporal and spatial conditions, which may not always reflect real world situations. Statistical models may provide a compromise between the above two situations.

Model ouput *validation* is essential to any soil modeling effort, although this term has a broad meaning in the literature. For the purpose of this section we can

Table 10. Steps in Field Validation of Soil Fate and Transport Models

Step 1. Identify Model User's Need—The first step in field validation is to obtain a clear understanding of the model user's need, i.e., how will the model be used.

Step 2. Examine the Model—

- **Step 2a.** Detailed examination of the model: The user must precisely define model input data requirements, output predictions, and model assumptions.
- **Step 2b.** Collect Preliminary Data and Performance of Sensitivity Analysis: Preliminary data are required to conduct a sensitivity analysis and determine the most important input variables.

Step 3. Evaluate the Feasibility of Field Validation—Some models cannot be validated in the field, and the validator should consider this possibility.

Step 4. Develop Acceptance Criteria for Validations—The model user must provide criteria against which the model is to be judged.

Step 5. Determine Field Validation Scenario—Many different approaches to field validation are possible. A scenario should be identified and approved by the model user.

Step 6. Plan and Conduct Field Validations Which Should Include the Following Steps—

- **Step 6a.** Select a Site and Compound(s): Consideration of model input requirements, analytical methods, sources of contamination, and site soil characteristics, etc. are among the many factors to consider in selecting a site and compound(s).
- **Step 6b.** Develop a Field Study Design: Development of a detailed field sampling plan for the specific model compound and site.
- **Step 6c.** Conduct Field Study: Implementation of the field plan is not addressed in these guidelines.
- **Step 6d.** Sample Analysis and Quality Assurance: Many analytical procedures are available depending on the chemical and the matrix. Standardized methods should be used together with a sound quality assurance program.
- **Step 6e.** Compare Model Performance with Acceptance Criteria: A comparison must be made between the performance of the model and the user's acceptance criteria using either graphical or statistical techniques.

define validation as *the process which analyzes the validity of final model output;* namely, the validity of the predicted pollutant concentrations or mass in the soil column (or in groundwater), to groundwater and to the air, as compared to available knowledge of measured pollutant concentrations from monitoring data (field sampling).

A disagreement, of course, in absolute levels of concentration (predicted versus measured) does not necessarily indicate that either method of obtaining data (modeling, field sampling) is incorrect or that either data set needs revision. Field sampling approaches and modeling approaches rely on two different perspectives of the same situation.

Important issues in groundwater model validation are the estimation of the aquifer physical properties, the estimation of the pollutant diffusion, and decay coefficient. The aquifer properties are obtained via flow model calibration (i.e., parameter estimation) and by employing various mathematical techniques such as kriging. The other parameters are obtained by comparing model output (i.e., predicted concentrations) to field measurements; a quite difficult task, because clear contaminant plume shapes do not always exist in real life.

Three major input data categories are required for soil and groundwater modeling efforts: climatologic or hydrologic data, soil data, and chemistry data. These data are used as input to models and to validate model output. Climatologic data can be obtained from site-specific investigations or from NOAA or USGS records. Soil data can be obtained from site-specific investigation (e.g., soil hydraulic conductivity) or from USDA data information documents. Chemistry data can be obtained from reference books (e.g., Lyman et al.[36]) or from laboratory analyses (e.g., adsorption coefficient). Data are model-specific and environment-specific.

Samiullah[20] reports on the sensitivity analysis that seeks to identify those issues and key parameters for which the greatest accuracy and precision are needed to reduce uncertainty in model predictions. If the exact mathematical relationship between the model input and output are known, then an analytic approach such as perturbation theory[95] may be applied to sensitivity analysis. If an analytic approach is not possible, then a statistical or nonstatistical numerical approach can be used. Typical is the nonstatistical procedure of varying the value of an input parameter by a given percentage about a nominal value while maintaining all other parameters constant.[68] Values of a sensitivity coefficient may then be calculated expressing the sensitivity of an output parameter to each input parameter. A Spearman's nonparametric partial rank correlation coefficient may be used to remove the effects of all but one input parameter on a given output variable, even though all input parameters are varied simultaneously from one run to another.

Finally, Donigian[96] divided the process of model evaluation into a model construction check, calibration, and post-audit analyses. The model construction check is analogous to model and algorithm examination as defined by most scientists. Calibration is considered to be by Donigian[96] the process of adjusting selected model parameters within an expected range until the difference between model predictions and field data are within selected criteria. Verification provides an independent test of how well the calibrated model is performing. This is achieved by a split-sample testing procedure where model predictions are compared to field observations that were not used in calibration.

Finally, in post-audit, model predictions for a proposed alternative are compared to field observations following implementation of alternatives. Acceptable criteria will reflect both the model capabilities and the assumptions made to present the proposed alternative. Figure 19 shows how errors in soil model input can cause significant discrepancies between observed and predicted data.

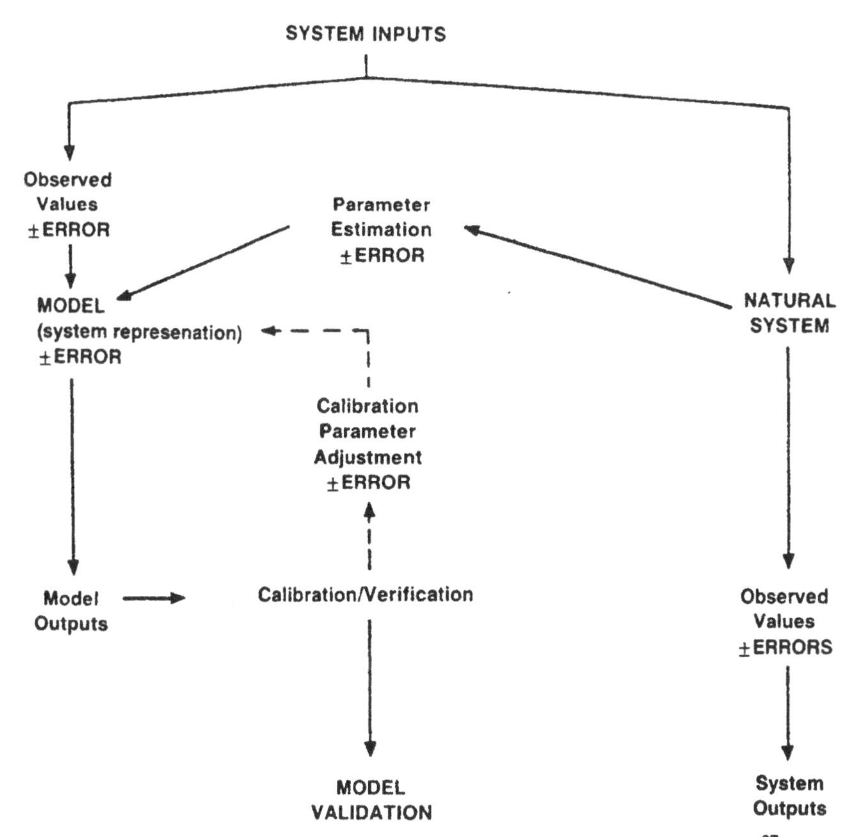

Figure 19. Model and natural system input, output, error (from Dickson et al.[97]).

Modeling Applications

Numerous applications are available in the literature. Some studies are presented in this section. The studies presented are related to leachate mobilization and migration at uncontrolled waste sites: (1) leachate migration to groundwater from land treatment practices, (2) leachate migration to the atmosphere of solvents leaking from barrels buried in the soil zone, (3) leaking underground fuel tanks, (4) biodegradation and biotransformation of hydrocarbons, (5) fate of petroleum in the soil zone, (6) modeling of VOC emissions from land farming practices, (7) probability density functions estimation from solute movement in the soil, (8) mathematical field calibration and validation of modeling simulations, (9) statistical testing of simulations and models performance, and (10) other applications.

Leachate to Groundwater from Land Treatment Practices

Fate of Organics. A site in Montana receives petroleum refinery wastes. Effluents from a number of refinery processes are collected in the wastewater treatment ponds. Sludge from the pond bottom is removed periodically and transported to the land treatment site. At the site the sludge is spread to a depth of several centimeters and is allowed to dry for 2 to 3 weeks. The dried sludge is mixed with the soil to a depth of 15 to 20 cm by a tractor-drawn rototiller. Waste had been applied during several periods starting in 1973. Some applications were to the whole site; others were partial applications to subsections of the site.

The waste composition is reported by the site operators to contain 65% water, 25% oil, 7% solids, and 3% other constituents. Considerable variation in the chemical composition of waste over time is considered likely by plant personnel. A chemical analysis of the waste was performed. The waste was found to contain significant amounts of organic compounds, primarily polycyclic nuclear aromatic hydrocarbons (PAH). The sampling program from which these data were taken did not routinely analyze soil samples for individual organic species. However, a gas chromatography mass spectrometer analysis was performed for one soil sample from the first year. The result of the analysis is presented in Table 11.

The soil in the land treatment area selected for study is a silty clay, 157 cm in depth. Soil permeability is less than 0.15 cm/h (4.1 * 10cm/s), which corresponds to a saturated intrinsic permeability of $4.2 * 10^{-10}$ cm^2.[58] Depth to groundwater is less than 60 m. The general topography in the vicinity of the land treatment

Table 11. Analysis of Soil Core Samples of PAHs

Species	Concentrations[b]: Soil Sample[a] (Year 1, Depth 0–15 cm) (μg/g dry soil)	Waste Sample[c] (μg/mL)
Naphthalene	3.7	76
Acenaphthalene	ND[d]	26
Acenaphthene	2.2	9.8
Fluorene	3.4	24
Phenanthrene	4.9	60
Anthracene	2.4	20
Fluoranthene	3.5	15
Pyrene	2.3	24
Chrysene	10.3	24

[a] The soil core samples were analyzed by GC/mass spectrophotometry with the standard method outlined in EPA-600/7-70-191. An HP fused-silica capillary GC column coated with OV-101 was used for analyses.

[b] Higher molecular weight PAHs (e.g., fluoranthenes, benz pyrenes) were detected but at too low a value (<1 μg/g) to be reliable.

[c] Sample was composite of extractions from all soil samples.

[d] Means not detected (ND).

area is characterized by hills with 0% to 4% and occasionally 15% to 35% slopes. The land treatment area selected for study is nearly flat and level, so little or no runoff occurs. Waste is applied during freeze-free periods in the spring and fall. No other activities take place on the land treatment area during the remainder of the year.

The 40-year average annual rainfall at the site is about 37 cm. The 5-year average (1975–1980) is 41 cm and the 1-year average (July 1979 through June 1980) is 35 cm. Nearly 70% of the average annual rainfall occurs in April through September. May and June are the wettest months, accounting for about 20% of the average annual rainfall, with a secondary peak in September and October. The hottest months are usually July and August, with mean daily maximum temperatures near 31 °C. The coldest months are usually January and February; the long-range daily minimum temperatures are near 11 °C. The freeze-free period averages 129 days a year.

Soil samples were taken from the site in October 1979 and in August 1980. Soil core samples were collected from two depths 0–15 cm below grade and 15–30 cm below grade from both the waste application area and a control area. The control area was as nearly identical as possible to the waste application area except that no waste had been applied. In both the waste application and control areas, samples were collected from 20 equally spaced locations approximately 13.7 cm apart within square-shaped areas of 3000 m^2.

Soil core samples were air-dried prior to analysis. Trace metal analyses were performed on nitric-perchloric acid digests of representative aliquots of the respective soil samples so that reported results do not differentiate between adsorbed and dissolved analyte. Analysis results are expressed as micrograms of analyte per gram of air-dried soil.

Pollutant quantities originating from the site can be input to the model in several ways. If the pollutant is assumed to be present as a concentrated mass, a leaching rate (as a percentage of solubility) can be specified. If the pollutant depends on the rate of precipitation (i.e., acid rain), a rainfall pollutant concentration can be specified. If the pollutant is already mixed into the soil, a total pollutant quantity in each layer can be given. The appropriate type and amount of loading were determined for each disposal site. The loading specified was a function of site history, type of waste, disposal method, etc.

Input data have been compiled from the literature (pollutant data) and from site investigations (climate, soil data) in order to simulate 2 years for each site. At the site, the waste had been applied several times over a 7-year period and had involved considerable variation of both waste composition and application area. The percentage of oil in the soil after the most recent loading (June 1979) was reported by plant personnel to be 7% by weight. Using this weight percentage and waste composition data, an organic pollutant loading rate was calculated. Loadings ranged from 560 $\mu g/cm^2$ (naphthalene) to 150 $\mu g/cm^2$ (anthracene).

For the site, the model SESOIL was run on a monthly basis from October 1978 to September 1980 for two of the most prevalent organic pollutants found at the site: naphthalene and anthracene. For ease of SESOIL application and as a test of the model's behavior, the model was loaded at one time with the total amount of pollutant known to be present in June 1978, although at the site the waste had been applied several times to reach that total. SESOIL-predicted concentrations in soil and soil moisture for a given depth were converted to a total g/g dry soil for the layer.

SESOIL-predicted concentrations and average experimental measured concentrations are presented in Table 12. SESOIL predictions for 1 year later (August 1980) are also given. Again the lower layer results are presented only for completeness, since the experimental lower layer (15–30 cm) and the SESOIL lower layer (20–5000 cm) are not directly comparable. However, the upper layer for SESOIL was specified so as to be comparable with the experimental upper layer, so these upper layer results are expected to agree.

The results for both organics are in modest agreement. The SESOIL results are expected to be higher than those found by the laboratory analysis since no biodegradation was modeled. Biodegradation was expected to occur at this site, but no rate data were available as input to the model.

Fate of Inorganics/Metals. The land treatment site considered is the property of a plastic manufacturing plant. Manufacturing process wastes are treated in a secondary wastewater treatment system centrifuged to yield a sludge whose content is 5% to 10% solids; the resulting sludge is disposed of by land treatment.

In July 1979, 5400 kg/ha of sludge were incorporated into the soil of a clean (i.e., not previously land-cultivated) area of the site. The sludge was injected 12.7 to 20.3 cm below the soil surface, and was subsequently mixed with the soil by ordinary farming methods.

The soil in the land treatment area is silt-loam, with a spatial intrinsic permeability of $7.05 * 10^{-9}$ cm^2 and a surface slope of 3%. Depth of groundwater is reported to be 30–70 cm. The 40-year (1940–1980) average annual rainfall is about 85 cm. The 7-year rainfall (1973–1980) is about 84 cm, and the July 1979–August 1980 rainfall was 79 cm. The average time of rain varies between 0.18 and 0.20/day for the above period. The area receives 84–110 rainstorms per year. The rainy season is 365 days per year. The annual average temperature is 14°C. Almost no surface runoff occurs at the site, due to both the climatic and soil conditions.

Waste application occurred in the spring of 1979. In July 1979, and a year later in August 1980, soil core samples were collected from two depths, 0–15 cm below grade and 15–30 cm below grade, at both the waste application area and the control area. The control area soil was nearly identical to the soil of the waste application area, except that no waste had been applied.

Soil core samples were air-dried prior to analysis. Analyses were performed on nitric-perchloric acid digests of representative aliquots of the respective soil samples, so that reported results represent total metal concentrations and do not differentiate between adsorbed and dissolved analyte.

Chemical data for the model parameters have been obtained from the literature and site-specific investigations. No calibration has been attempted, but only a minor

Table 12. Validation Results for Organics at a Petroleum Waste Site[a]

	All Concentrations in μg/mL (ppm)				
	SESOIL Prediction[b] October 1979	Experimental Values[c,d] October 1979	Difference (%)	SESOIL Prediction August 1980	Predicted % Change
Naphthalene					
Upper soil layer	18.3	3.69	80	11.8	−35
Lower soil layer	0.00312	NC[e]	—	0.0126	+300
Anthracene					
Upper soil layer	5.55	2.36	57	5.39	−2.9
Lower soil layer	0.0000481	NC[e]	—	0.00279	+57

[a]Waste applied to "clean site" during April 1979.
[b]Model depths: upper 0–20 cm; lower 20–5000 cm.
[c]Experimental depths: upper 0–15 cm; lower 15–30 cm.
[d]Experimental concentrations calculated as (concentration in application area)—(concentration in control area).
[e]Lower-depth results not directly comparable (NC).

adjustment of parameters within chemically justifiable limits was undertaken. Predicted concentrations agree reasonably well with those values measured chemically, considering the uncertainty of all parameters affecting pollutant migration in soils. Results are presented in Table 13. A sensitivity analysis has been performed to study soil contamination impacts from sludge application rates, climatologic, soil, and chemistry parameters.

Table 13. Predicted Concentrations of Inorganic Pollutants

Compound	Calculated Amount of Compound Applied (April 1979)	Concentrations (ppm)—July 1979			
		Measured		Predicted	
		0–15 cm	15–30 cm	0–15 cm	15–30 cm
Chromium	2.92 $\mu g/cm^2$	0.80	1.0	0.15	0.005
Copper	3.24 $\mu g/cm^2$	0.20	0.0	0.16	0.0003
Sodium	8.64×10^3 $\mu g/cm^2$	0.89	114.0	35.6	142.0

Migration to the Atmosphere from Buried Solvents

The purpose of this research was to understand via a mathematical modeling effort the long-term potential fate of the leachate of six solvents leaking from buried barrels disposed in soil systems. A barrel was assumed to leak in 1 year. For this investigation, six halogenated organic solvents have been examined:

- Perchloroethylene (tetrachloroethene)
- Methylchloroform (1,1,1-trichloroethene)
- Methylene chloride (dichloromethane)
- Carbon tetrachloride (tetrachloromethane)
- Freon 113
- Trichloroethylene (1,1,2-trichloroethane).

Since it was not the intention to conduct a site-specific study, a number of hypothetical scenarios covering a wide range of climates, soils, and solvents were considered. The methodology developed for the overall assessment is of general use and can be employed for similar and site-specific studies and classes of pollutants.

The actual quantities of pollutant mass removed by each pathway are strongly affected by the climate and soil type. Summaries of the pathways for all six chemicals for a 10-year simulation period, a moderate climate, a silty-loam soil, and three depths of a soil column are presented in Figure 20 and Table 14, and are summarized as follows:

- Of all chemicals studied, Freon is most easily volatilized, whereas methylene chloride is least easily transported to the atmosphere.
- Methylene chloride contributes the most mass to groundwater; Freon 113 contributes the least pollutant mass to groundwater.

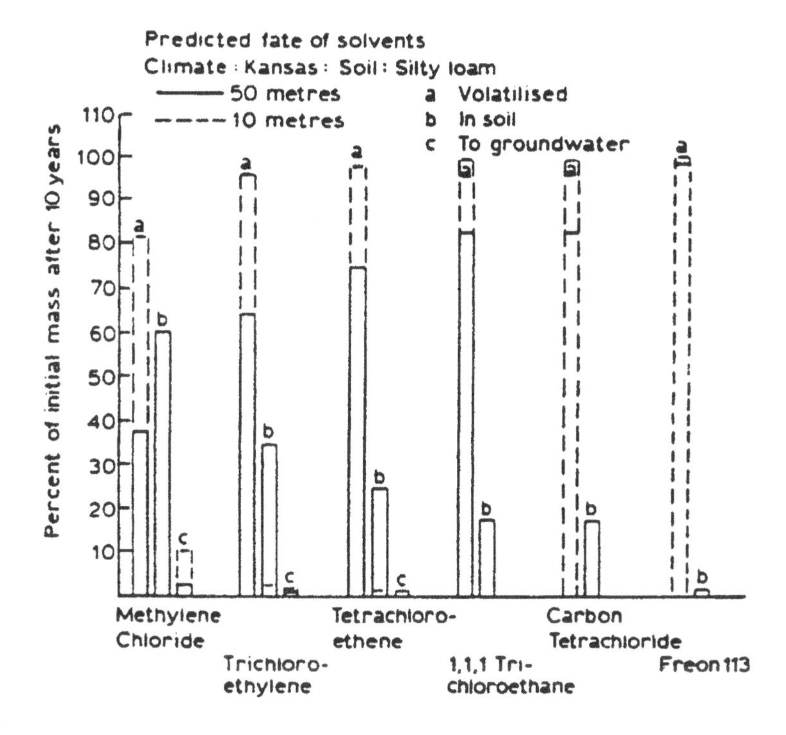

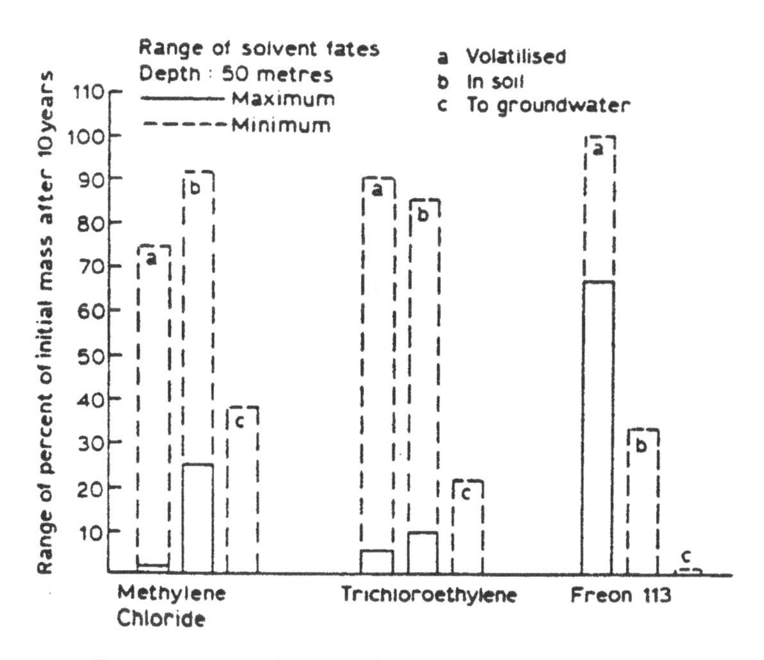

Figure 20. Predicted ranges of solvents fate.

Table 14. Qualitative Fate of Solvents

Chemical	Percentage Volatilized[a]	Percentage Remaining in Soil Column[a]	Percentage Leached to Groundwater[a]
	Depth to Groundwater: 50 m		
Tetrachloroethane	74.9	24.7	0.4
1,1,1-Trichloroethane	82.4	17.3	0.3
Methylene chloride	37.4	60.2	2.5
Carbon tetrachloride	82.4	17.2	0.3
Freon 113	98.5	1.5	00.1
Trichloroethene	64.4	34.7	0.9
	Depth to Groundwater: 20 m		
Tetrachloroethane	88.0	10.6	1.3
1,1,1-Trichloroethane	94.2	4.9	0.9
Methylene chloride	57.0	33.8	9.2
Carbon tetrachloride	94.3	4.9	0.8
Freon 113	99.6	0.3	0.01
Trichloroethene	82.4	14.6	3.0
	Depth to Groundwater: 10 m		
Tetrachloroethene	97.8	1.1	1.1
1,1,1-Trichloroethane	99.3	0.1	0.6
Methylene chloride	81.2	10.3	8.5
Carbon tetrachloride	99.3	0.1	0.6
Freon 113	99.9	0.01	0.01
Trichloroethene	96.0	1.6	2.3

[a]Percentage of mass after 10 years, moderate climate, silty-loam soil. Totals may not add to 100%.

- The other solvents have fates intermediate between Freon 113 and methylene chloride, and are fairly similar to one another. Under moderate conditions, 99% to 64% of their mass volatilized and 0.01 to 3% of their mass reached the groundwater. The remaining mass was captured in the soil column.
- Leaching to groundwater increases for chemicals with low Henry's law constant, low diffusion coefficients and low absorption coefficients. Leaching is generally favored by high rainfall and permeable soils.
- Volatilization is favored for chemicals with high Henry's law constants and high diffusion rates. It is generally enhanced by dry conditions in porous soil. Decreasing soil column depth generally results in increasing volatilization rates up to a certain depth.

Leaking Underground Fuel Tank

An important study on hydrocarbon leaking of underground fuel tanks, including a comparative analysis with a performance analysis on SESOIL and the California Leaking Underground Fuels Manual[21] is presented by Odencrantz et al.[57]

Figure 21 is a schematic illustration of the problem modeled. A thorough sensitivity analysis on soil and other parameters affecting hydrocarbons fate in soil is presented in Table 15.

Table 15. Summary of Sensitivity Analysis in the Form of Time to Predict Peak and Maximum Leachate Concentration

Parameter		Peak Concentration (mg/L)	Time to Peak (months)
Base		0.29	17
Depth to groundwater from source bottom	3 m	0.18	19
	25 m	0.038	8
Climate	Eureka	2.30	3
	Sacramento	1.15	6
	Bakersfield	0.010	32
Biodegradation rate	0 day^{-1}	0.60	18
	0.02 day^{-1}	0.033	6
	0.10 day^{-1}	6.5×10^{-3}	5
Fraction of organic carbon content	0.2%	0.045	32
	2%	1.12×10^{-3}	103
Intrinsic permeability			
1×10^{-8} cm^{-2}		0.53	9
4×10^{-10} cm^{-2}		0.090	31
Disconnectedness index	3.7	0.680	8
	5.0	0.302	7
Source thickness	3 m	0.385	20
	5 m	0.375	32

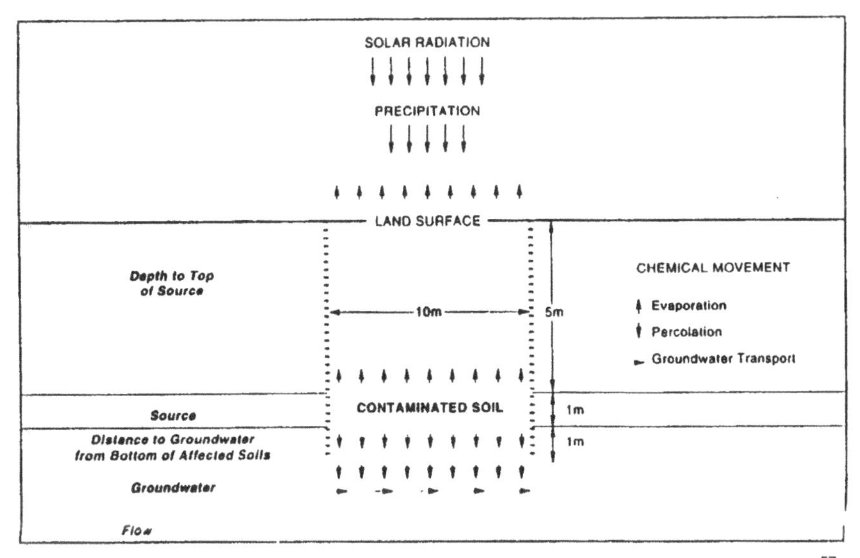

Figure 21. Schematic environment of benzene soil simulations (from Odencrantz et al.[57]).

Figure 22 shows the results of various simulations for benzene (TPH, BTXE) and the impacts of scenario simulations. Major conclusions of the study are:

- The LUFT manual is being applied incorrectly through the State of California, causing large remedial expenditures to address soils affected by TPH which, in the absence of BTXE, may not pose a significant quality threat.
- The methodology developed by LUFT to determine cleanup concentrations of soils can be highly overprotective or underprotective of water quality, depending on the given site conditions. To remediate larger sites in a manner that is both protective of water quality and cost effective, site-specific data collection and modeling assessment should be conducted on a case-by-case basis.
- It is important to consider site-specific variables such as the biodegradation rate and soil organic carbon content, in addition to local climate and depth to the water table, in determining petroleum hydrocarbon cleanup levels.
- SESOIL and AT123D are appropriate models for use in determining acceptable soil cleanup levels protective of water quality.

Biodegradation and Biotransformation of Hydrocarbons

Modeling of biodegradation and biotransformation of hydrocarbons (TCE, PCE, TCE, gasoline, and benzene) in soil and groundwater has been accomplished by Corapcioglou and Hossain[98] by applying two models of the TDE-type. A typical output is shown in Figure 23 for the decay of PCE and the increase of intermediate compounds with time.

The first set of simulations predicts the anerobic biotransformation of dissolved halogenated hydrocarbons such as PCE and TCE through reductive dehalogenation process by indigenous soil biomass. The model assumes abundant quantities of primary substrate, inorganic nutrients, and electron acceptors. The hydrocarbon contaminant is biotransformed as secondary substrates. The kinetics of biotransformation is expressed by Michaelis-Menten equations. Kinetic parameters are obtained by graphical techniques.

The second set of simulations predicts the aerobic biotransformation of hydrocarbon components in an unsaturated-saturated medium. Balance equations for bacterial population and soil oxygen are coupled through a biodegration term. The methodologies developed in the research can be employed to study the feasibility of in situ biological cleaning of aquifers contaminated by hydrocarbons such as TCE, PCE, TCA, gasoline, and kerosene.

Fate of Petroleum in the Soil Zone

Carberry and Lee[99] simulated the fate and transport of petroleum in the unsaturated soil zone under biotic and abiotic conditions. Laboratory scale columns

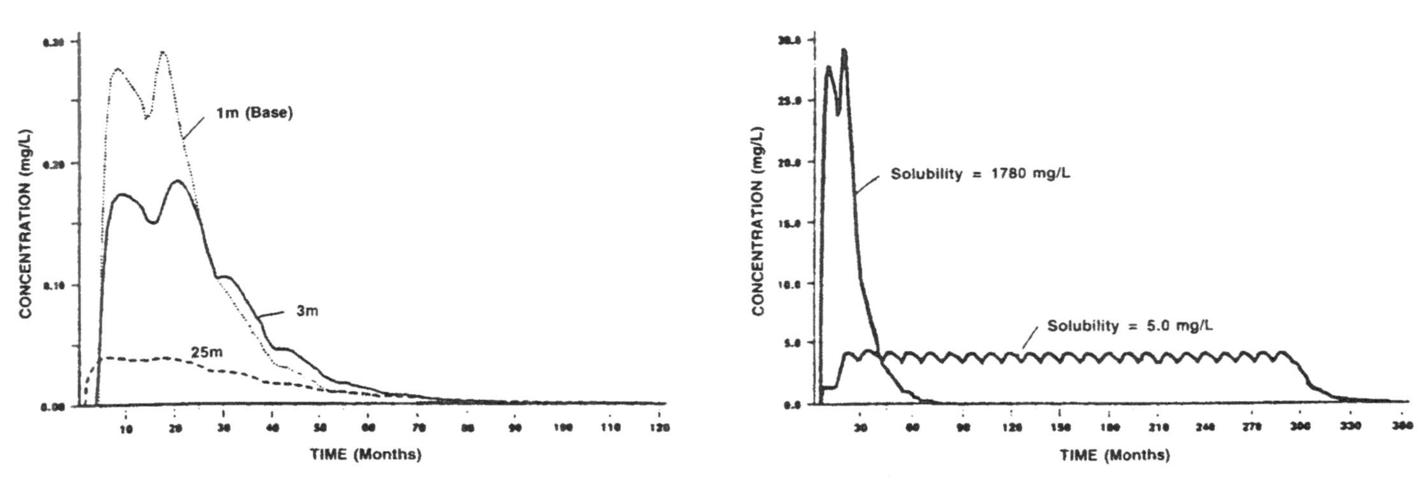

Benzene leachate breakthrough curves at the water table for different values of the distance to the groundwater from the deepest affected soils.

Benzene leachate breakthrough curves at the water table for different values of effective solubility at loading of one-hundred times the base case.

Figure 22. Simulated fate of benzene and correlation analysis (from Odencrantz et al.[57]).

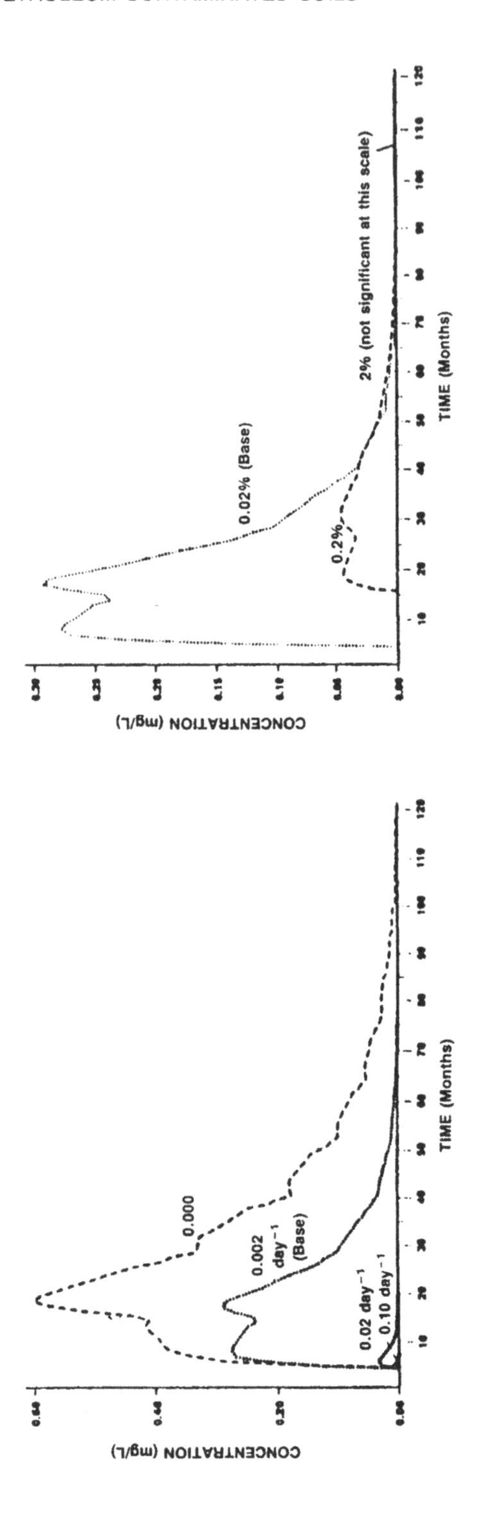

Benzene leachate breakthrough curves at the water table for different values of the biodegradation rate.

Benzene leachate breakthrough curves at the water table for different values of soil organic carbon content.

Figure 22. *Continued*

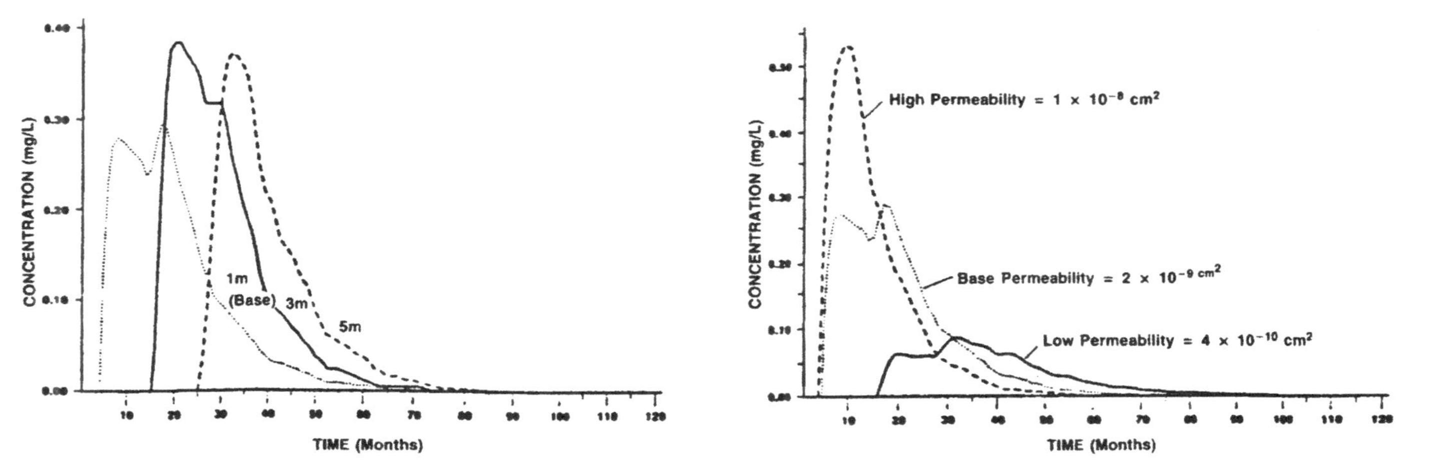

Benzene leachate breakthrough curves at the water table for different values of source thickness.

Benzene leachate breakthrough curves at the water table for different values of intrinsic permeability.

Figure 22. *Continued*

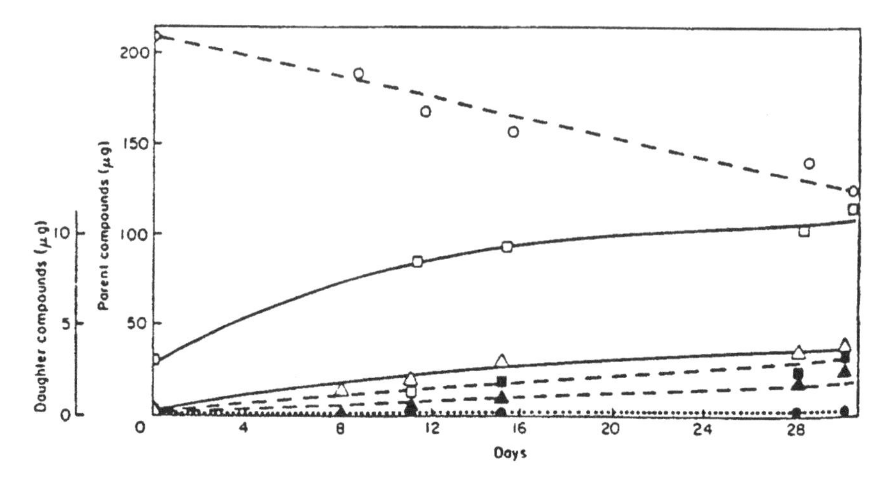

Decay of PCE and increase of intermediate compounds with time (--○--), Tetrachloroethylene (34 days), (—■—), trichloroethylene (43 days); (—△—), *cis* 1,2-dichlorethene (long); (--■--), vinyl chloride (long); (--▲--), 1,1-dichloroethene (53 days); (····●····), *trans*-1,2-dichloroethene (10 ng). Parenthesis denotes half-life.

Figure 23. Simulation of PCE (from Corapcioglu and Hossain[98]).

containing isotropic samples of two-petroleum-contaminated soils were operated to simulate the effects of a petroleum spill, including volatilization vapor from soil. The TDE mathematical approach has been employed. Typical simulations are shown in Figure 24 for dissolved petroleum concentration and in Figure 25 for adsorbed petroleum concentrations.

Modeling VOC Emissions from Landfarming

Cadena et al.[100] evaluated petroleum refinery volatile organic emissions from surface applications of slop soil. Several one-dimensional models have been described and the models of Hartley (JSF) and Thibodeaux (TH) have been applied. These analytic models have the structure of the diffusion dispersion equations presented earlier in this chapter in the section, "Miscible Organic Compound Processes." Pollutant properties used to compute VOC emissions in laboratory conditions are shown in Table 16. Calculated and measured VOC emissions from surface applications of slop soil are shown in Table 17.

Probability Density Function of Solute Movement

Sposito and Jury[101] have set the grounds of stochastic solute modeling in soils in relation to a lifetime probability density function (pdf) for solute movement (e.g., solvents) in the subsurface zone (soil and groundwater). The field-scale measurement and use of these two pdfs in characterizing solute mass transfer and solute concentrations in the subsurface zone were described along with their

Table 16. Pollutant Properties of VOC in Laboratory Conditions Employed for Modeling Purposes

Pollutant	Air Diffusion Concentration D_a (m^2/d)	Henry's Law Constant, K′	Initial Concentration, C_o (g/m^3 of soil)		Water Diffusion Coefficient, D_w (m^2/d)	
			Separator Sludge	Slop Oil	Separator Sludge	Slop Oil
Benzene	0.76	0.000861	122	312	0.00000472	0.00000219
m-Xylene	0.61	0.000266	190	490	0.00000205	0.00000205
o-Xylene	0.61	0.000218	114	196	0.00000438	0.00000205
p-Xylene	0.61	0.000288	88	194	0.00000438	0.00000205
Ethylbenzene	0.67	0.000330	32	94	0.00000438	0.00000205
Naphthalene	0.51	0.000020	120	93	0.00000467	0.00000217

Table 17. Calculated and Measured VOC Emissions from Subsurface Applications of Slop Oil

Pollutant, Medium	Time (h)	VOC Emission Rates ($g/m^2/d$)		
		Measured	JSF	TH
Benzene				
Sand	1	26.006	41.320	54.777
	20	4.397	7.277	12.248
Soil	1	26.957	48.933	65.433
	20	4.553	7.807	14.631
Toluene				
Sand	1	46.051	52.985	71.448
	20	5.935	8.931	15.976
Soil	1	60.652	54.447	73.868
	20	0.993	8.921	16.517
Ethyl benzene				
Sand	1	17.971	8.013	11.083
	20	4.250	1.247	2.478
	1	18.403	8.216	11.459
Soil	20	6.933	1.226	2.563
o-xylene				
Sand	1	8.009	12.223	17.923
	20	1.425	1.745	4.007
Soil	1	8.277	12.573	18.530
	20	2.350	1.706	4.143
m-xylene				
Sand	1	6.929	14.278	20.390
	20	4.544	5.128	11.067
Soil	1	32.918	35.423	51.172
	20	7.983	5.045	11.442
p-xylene				
Sand	1	[a]	[a]	[a]
	20	1.408	2.161	4.559
Soil	1	1.883	14.703	21.081
	20	2.427	2.131	4.713
Naphthalene				
Sand	1	0.179	0.872	2.370
	20	0.101	0.027	0.531
Soil	1	0.048	0.886	2.453
	20	0.101	0.020	0.037

[a]Not calculated.

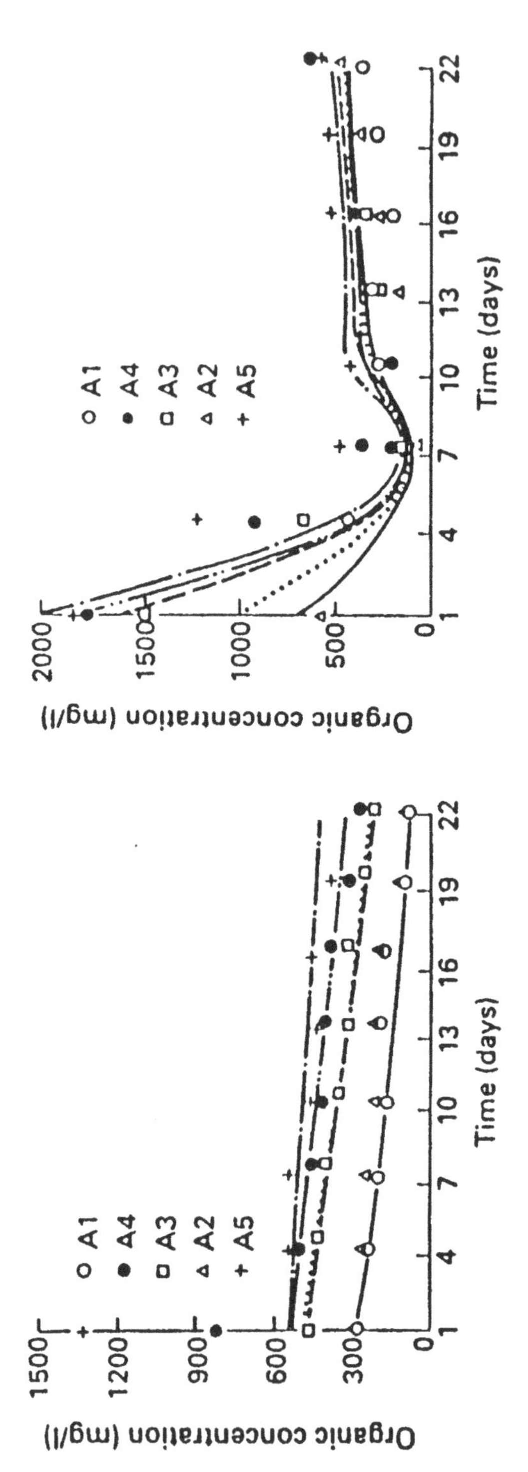

Predicted petroleum liquid phase concentrations (lines) developed by computer model and measured values (points) for fine soil.

Predicted petroleum liquid phase concentrations (lines) developed by computer model and measured values (points) for fine soil during leak experiment.

Figure 24. Fate of petroleum phase in soil.

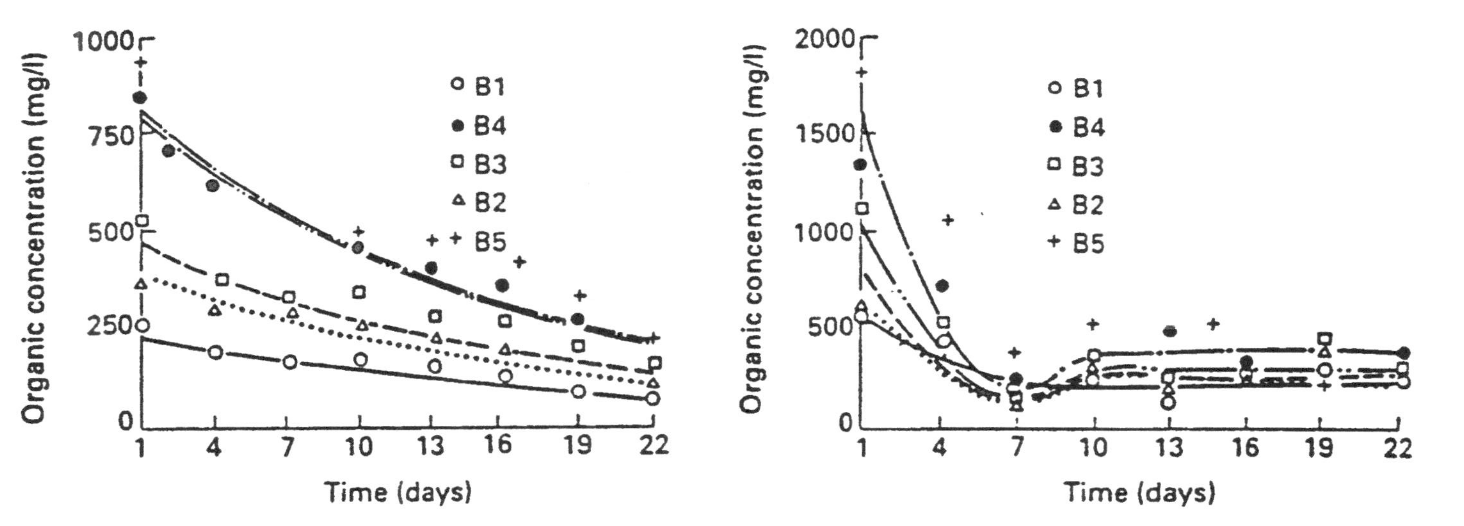

Predicted petroleum liquid phase concentrations (lines) developed by computer model and measured values (points) for fine soil.

Predicted petroleum liquid phase concentrations (lines) developed by computer model and measured values (points) for fine soil during leak experiment.

Figure 24. *Continued*

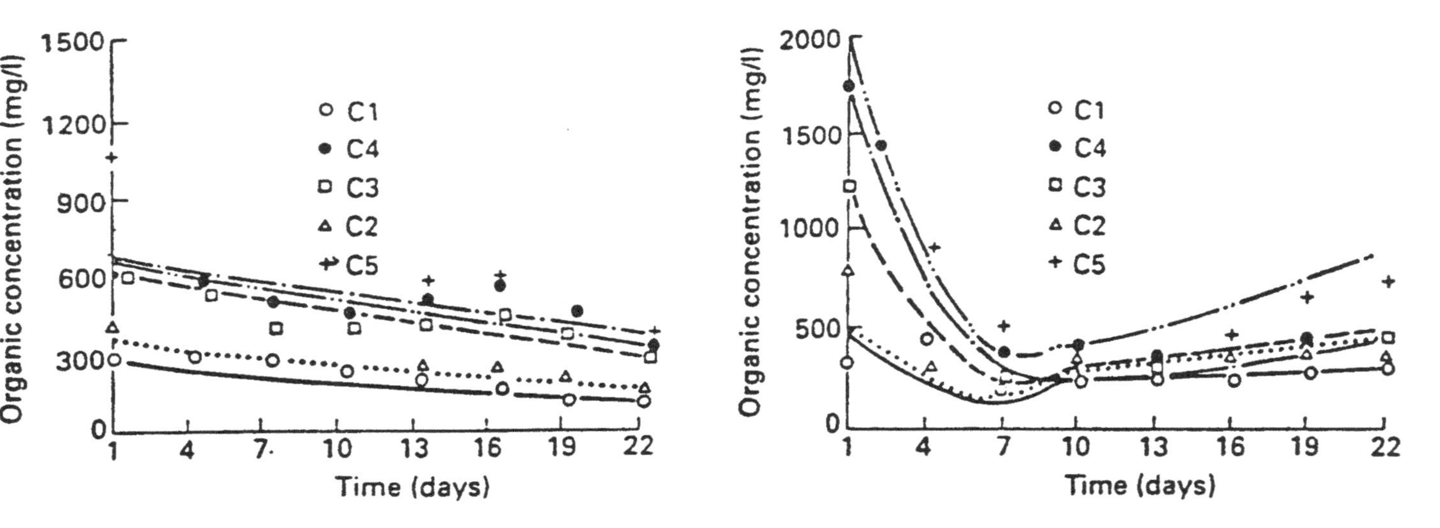

Predicted petroleum liquid phase concentrations (lines) developed by computer model and measured values (points) for coarse soil during spill experiment, abiotic (C) and biotic (D).

Predicted petroleum liquid phase concentrations (lines) developed by computer model and measured values (points) for coarse soil during leak experiment, abiotic (C) and biotic (D).

Figure 25. Fate of petroleum phase in soil (from Carberry and Lee[99]).

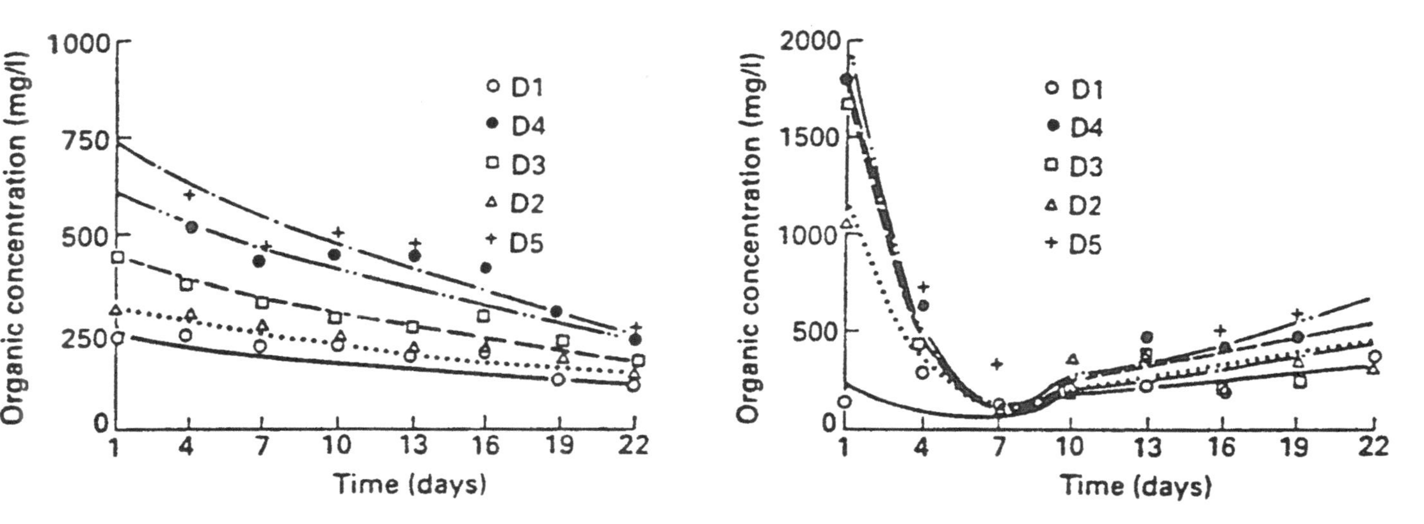

Predicted petroleum liquid phase concentrations (lines) developed by computer model and measured values (points) for coarse soil during spill experiment, abiotic (C) and biotic (D).

Predicted petroleum liquid phase concentrations (lines) developed by computer model and measured values (points) for coarse soil during leak experiment, abiotic (C) and biotic (D).

Figure 25. *Continued*

physical interpretation as lifetime or travel time *spectra*. The relation between time pdf and mechanistic models of subsurface solute transport (e.g., convection-dispersion equation) were illustrated with an application to three-dimensional reactive solute movement in a sand aquifer. Figure 26 illustrates both, the travel time spectra for chloride, carbon tetrachloride and tetrachloroethylene movement in a sand aquifer, and the dimensionless simulated travel time spectra for these three solutes.

Field Calibration and Validation of Models

Jury and Sposito[103] conducted research on field calibrations and validations of solute transport models for the unsaturated soil zone.

Data from two solute transport field experiments were obtained from soil solution samplers and from soil coring and were used to illustrate the parameter estimation problem for two models describing area-averaged vertical transport of mobile solutes. The models, the log-normal transfer function model (TFM) and the convection dispersion equation (CDE), were selected because of their simplicity and because they represent different hypotheses for the governing mechanisms determining longitudinal solute spreading on the field scale.

The two parameters of each model were estimated using three calibration procedures: sum of squares optimization, method of moments, and maximum likelihood. The three calibration procedures yield different estimates for the parameters using a given data set, and different trends in parameter variation at different depths or times in a given experiment. Calculated uncertainty in the parameter estimates was high enough so that no conclusion could be drawn about which model better described the data from the solution sampler experiment, but the TFM provided a better description of the coring experimental data after calibration than the CDE. It was concluded that future experiments on field-scale solute transport through the unsaturated zone will have to monitor movement below 5 m in order to yield a data set suitable for distinguishing between different models. Typical results are shown in Figure 27.

Statistical Tests and Models Performance

A number of statistical techniques can be used to obtain a quantitative measure of the difference between simulated and observed data. Hern and Melancon[10] and Moore et al.[104] discuss different statistical measures and associated aggregate statements useful for evaluating the performance of air quality models and water quality models. The same approach applies to soil modeling.

Basically, two classes of statistical models were selected to generate quantitative measures of performance for the SESOIL tests. The first class is standard linear regression statistics. As Ambrose and Roesch[105] note, regression statistics for observed and simulated data can be calculated by:

$$O^i = a \cdot p^i + b$$

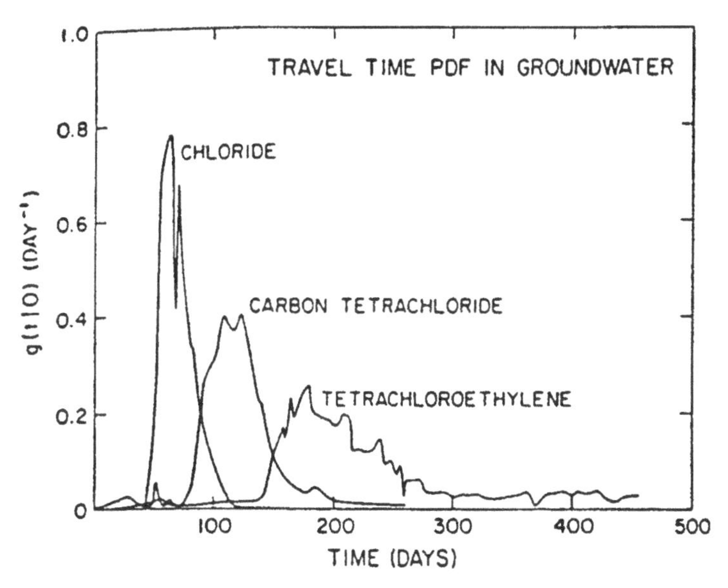

Travel time spectra for chloride, carbon tetrachloride, and tetrachloroethylene movement in a sand aquifer (Gotz and Roberts[102]).

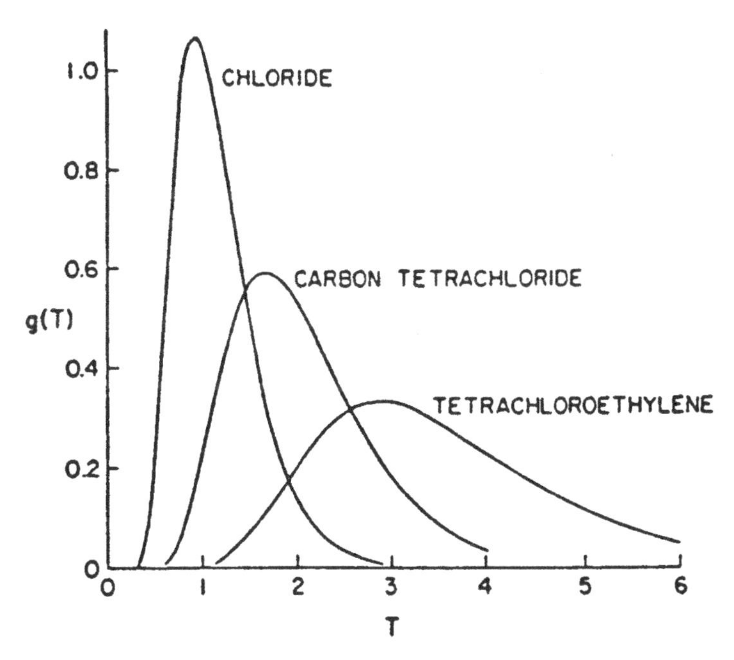

Dimensionless, simulated travel time spectra for the three solutes.

Figure 26. Travel spectra of chloride, carbon tetrachloride and tetrachloroethylene in soil.[102]

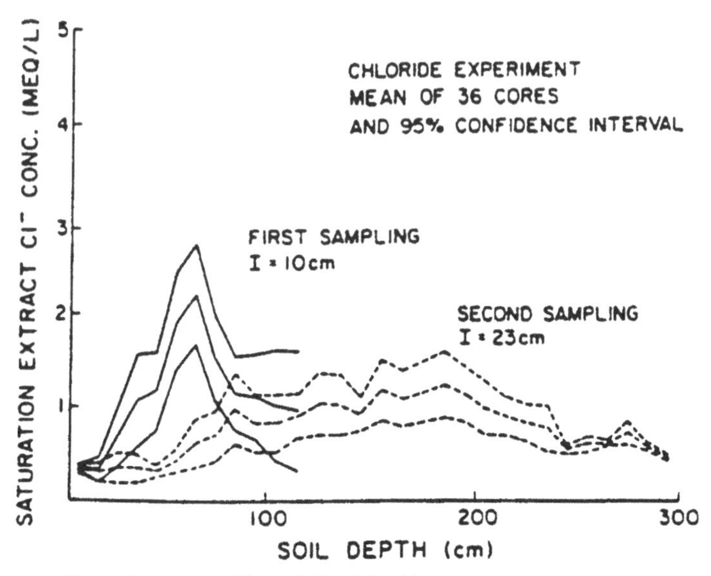

Sample mean (N = 36) chloride saturation extract concentrations and P = 0.05 confidence limits vs. depth. Soil cores taken after 0.1 m (first sampling) and 0.23 m (second sampling) net water applied to the field.

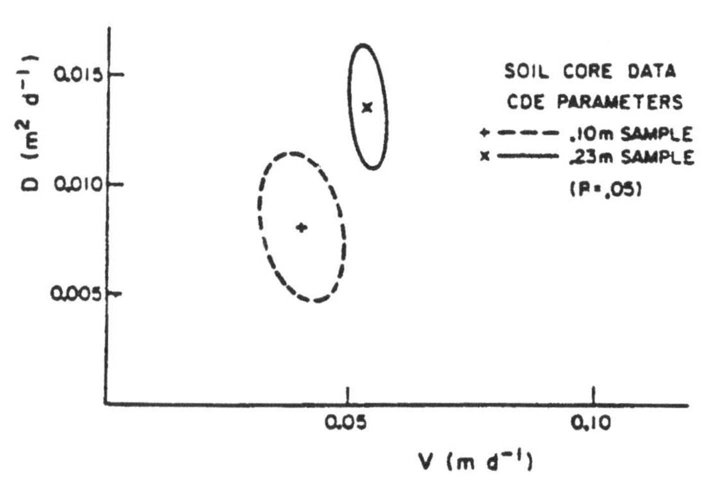

Mean and P = 0.05 confidence interval for CDE parameters, D, V estimated with chloride soil core data and the sum-of-squares method.

Figure 27. Chloride fate modeling and confidence levels (from Jury and Sposito[103]).

where:
O^i = observed data
p^i = simulated data
a = slope
b = intercept

The slope, a, provides a useful measure of the accuracy of model results; values less than 1.0 imply over-prediction by the model, while values less than 1.0 imply under-prediction. The intercept, b, is an indicator of statistical errors. Perfect agreement between the observed data and model predictions occurs when a = 1 and b = 0. The percision of model results can be measured by r, the correlation coefficient. It should be recognized that an implicit assumption made in this analysis is that the predicted values (p^i) are *fixed* and are without error. This assumption arises from the deterministic nature of the models used, where a single value is assigned for each of the model parameters, resulting in a unique predicted value. Since most, if not all, model parameters are spatially-variable, predicted values also have certain error. Thus, computation and interpretation of the linear regression parameters, as shown in the previous equation, may not be strictly valid. Under such conditions, to estimate parameters in linear models Halfom[106] has recently recommended the use of geometric mean functional regression (GMF), which takes into account the errors in both the dependent and the independent variables (in this case, O^i and p^i).

The second class of methods are typically called estimation of error statistics. Ambrose and Roesch state that the average error, E, and its associated relative error, RE, can be used to measure accuracy and systematic errors:

$$E = (1/N) \cdot \Sigma(p^i - O^i)$$

$$RE = E/O$$

where:
O = observed data means
N = number of data point.

The average error gives the absolute amount by which a given quantity is over- or under-predicted, while the relative error gives the percentage of over- or under-prediction.

Other estimation of error statistics that can be calculated include the standard error of estimate, SE, and its coefficient of variation CV, and are also given by Ambrose and Roesch.

A two-step approach was used to estimate and calibrate SESOIL model parameters. Data from the literature, guidance provided in the SESOIL manual, and chemical property parameter estimation methods provided the basis for an initial set of input parameters. SESOIL results based on this parameter set were then compared with observations on depth-averaged soil-water contents, percent aldicarb TTR leached to groundwater, and percent aldicarb TTR remaining in the soil profile over time (Tables 18 and 19). Figure 28 and Table 20 report the PRZM, PISTON, PRESTAN and SESOIL performances.

Table 18. Computed Statistics of SESOIL Confidence

	$O^i = m^{pi} + b$						
Comparison	**m**	**b**	**r^2**	**$\bar{E}$**	**RE**	**SE**	**CV**
SESOIL/observed volumetric soil-water content, percent	0.58	2.02	0.38	0.43	0.07	0.88	0.17
SESOIL/observed aldicarb TTR residues, percent	0.88	−10.22	0.77	16.60	0.47	23.46	0.66

Source: Donigian and Rao.[25]

Table 19. Calibrated SESOIL Cycle and Parameters

Parameter	Initial Value	Calibrated Value
Solubility, mg/L	6000	6000
K_{oc}, cm^3/gm	39	240
Diffusion coefficient in air, cm^2/sec	0.06	0.06
Biodegradation rate, /day	0.0	0.0
Henry's law constant, $M^3atm/mole$	0.33E-08	0.33E-08
Neutral hydrolysis rate, /day	0.0051	0.025

Source: Donigian and Rao.[25]

Table 20. SESOIL Confidence Levels

	$O^i = m^{pi} + b$						
Comparison	**m**	**b**	**r^2**	**E**	**RE**	**SE**	**CV**
SESOIL/observed aldicarb residues, percent remaining	0.78	−1.89	0.76	12.40	0.35	22.32	0.63
PRZM/observed aldicarb residues, percent remaining	0.91	−6.50	0.86	10.80	0.30	17.03	0.48

Source: Donigian and Rao.[25]

Other Applications

Baehr et al.[107] studied and modeled volatile hydrocarbons from the unsaturated soil zone by inducing advective air-phase transport. A mathematical model was developed that allows in situ determinations of air-phase permeability and other parameters affecting hydrocarbon fate.

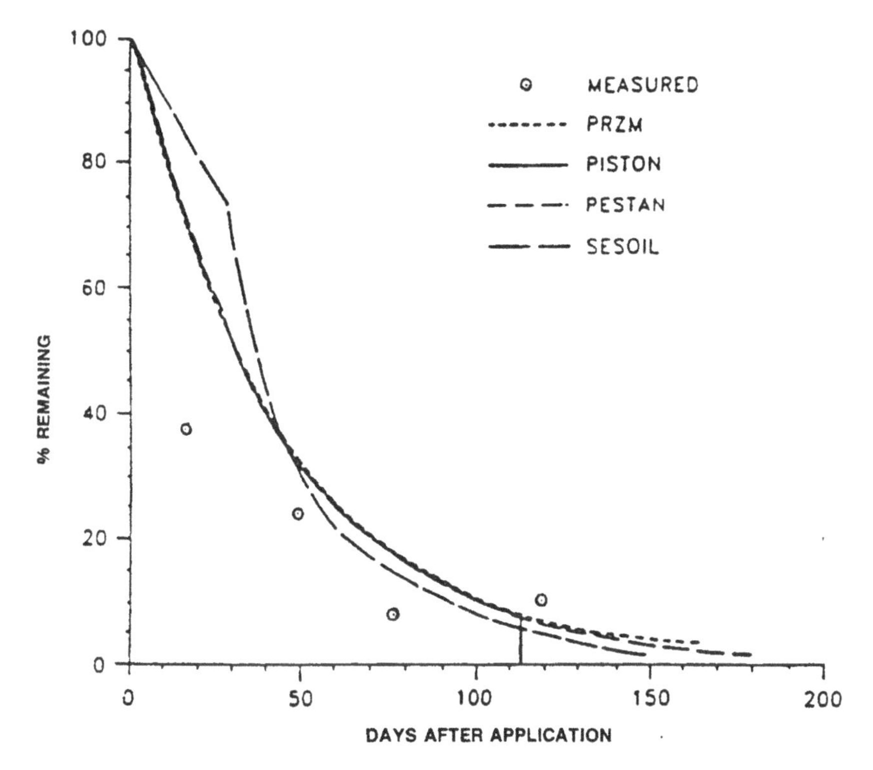

Comparison of PRZM, PISTON, PESTAN, and SESOIL predicted and measured precent aldicarb remaining in the soil column.

Figure 28. Performance of selected soil pollution models (from Ambrose and Roesch[105]).

Grenney et al.[108] developed a mathematical model for the fate of hazardous substances in soils, based on the TDE approach. The model has been applied, tested, and validated for various solvents (e.g., naphthalene) and can be applied for hydrocarbons also.

Gee et al.[4] reviewed the literature on recent studies of flow and transport in the vadose zone, and they provide a lengthy literature on hydrocarbon fate in the vadose and in groundwater.

Sudicky and Huyakorn[55] reviewed the literature for contaminant migration in perfectly known heterogeneous groundwater systems with a focus on contaminant migration and tools that have been developed for quantifying its effects on plume detection, monitoring, and remediation.

Jury et al.[109] modeled (stochastic approach) and evaluated the fate of volatilizing organic chemicals residing below the soil surface. The model identifies those compounds with high potential for loss during a period after incorporation under a certain depth.

Lighty et al.[110] developed a three-dimensional approach for the fundamentals of the three thermal remediation of contaminated soils. Dissolved and gaseous phases were studied.

Shields and Brown[111] studied the applicability of the POSSM model to petroleum product spills. Hillel[112] studied the movement of organic contaminants in soil and Eastcott et al.[113] evaluated modeling aspects of petroleum products in soils. Short[114] studied and modeled movement of contaminants from oily wastes during land treatment, Hern and Melancon[10] report details of vadose zone modeling of organic pollutants, and Dragun et al.[48] provide information on the fate of diesel fuel in soil systems.

The above includes a small number of modeling applications of dissolved hydrocarbons fate in soil and groundwater. Modeling of immiscible hydrocarbons is given in the following section.

IMMISCIBLE POLLUTANT FATE MODELING

General

In an effort to quantify nonaqueous phase liquid (NAPL), or immiscible contaminant migration, from waste sites or oil spills noted with increasing frequency, various mathematical models were developed in the 1980s, and considerable research is continuously directed toward improving our understanding of these immiscible fluid processes in the soil and the groundwater zones of an area. Mercer and Cohen[78] provide a comprehensive review on immiscible pollutant fate modeling. In addition Lenhard and Parker[115] provide a review on measurements and prediction of saturated pressure relationships in three-phase porous media systems.

As reported by Mercer,[51,78] many of the recently developed theoretical concepts and modeling approaches pertaining to this problem had originated in the petroleum industry. However, because of the different physical environment of deep petroleum reservoirs compared to the shallow aquifers, as well as different incentives and areas of concern in the petroleum industry, there is a great need to adopt and extend this work.

According to Abriola and Pinder,[53] Van Dam[116] presented the first detailed analysis of hydrocarbon pollution of groundwater as a two-phase problem. He examined the stages of contaminant infiltration and incorporated a capillary pressure term in his expression for fluid potential. Many researchers followed his work, among them Faust[52] and others reported in the reference section of this chapter. Schwille[117] provides a review on physical, chemical, and biological parameters affecting NAPL migration in porous media with an emphasis on mineral oil products and chlorohydrocarbons.

Modeling Concepts

Four phases separated by distinct interfaces exist in a soil matrix: solid (soil), water, gas, and contaminant. In the real physical system, each phase could possibly be formed by a number of chemical components or species, and mass transfer could occur across phase boundaries. In that respect, a contaminant may be available in four phases: absorbed, dissolved, gaseous, and NAPL. Migration of such a contaminant can be modeled as a three-phase flow process.

Basically there exist two concepts (ways) in modeling three-phase flow:

1. by employing the three-phase flow (air, moisture, NAPL) equations in porous media jointly with a set of relationships
2. by employing the governing equations of a two-phase flow in porous media (e.g., water-NAPL, or air-NAPL, or air-NAPL/water) and by adjusting the coefficients (e.g., relative permeability of fluid) of the equations to reflect the specific problem.

The second approach eliminates the need for an equation governing the third phase (e.g., Faust[52]), but introduces the need for determining the "relative" permeability of the "second" phase (e.g., NAPL-water) to the "first" phase (e.g., air). The model presented by Abriola and Pinder[53] follows the first concept. The model developed by Faust follows the second concept.

Stone[118] proposed a model for estimating three-phase relative permeabilities based on data for two-phase relative permeabilities. In general, two-phase relative permeabilities are determined from laboratory tests on cores; then these results are often fitted to polynomial functions of saturation. Two-phase permeabilities can also be estimated from analytic functions based upon water (or dissolved chemical) characteristics, surface tension, porosity, intrinsic permeability, and other parameters.

As shown from the literature sources of this review, a variety of computer codes exists. Most of the codes consider three phases: air, water, and NAPL, but the necessary input data to drive models do not exist. According to Mercer[51], for example, only one set of relative permeability curves is known in the literature for immiscible contaminants, and this is for the chemical TCE. This type of data does not exist for most solvents and chlorinated hydrocarbons found at spill sites or landfills (see model applications section). In addition to characterizing sites where NAPL exist, in situ water and NAPL saturation need to be measured or determined. Such data do not exist at these sites, or may not be available in the public domain. Therefore, the next major advance in the area of model development may not come until this type of data becomes available.[51]

The following soil and fluid parameters are necessary to predict flow rate and patterns of immiscible fluids in soil matrices:

- chemical and physical properties of the fluids, as, for example, polarity, molecular weight, density, contaminants in fluid (e.g., metal species), viscosity, and solubility in compressed water (water with density of approximately 1.2 g/cm^3).
- chemical and physical properties of soil, as, for example, soil structure, soil texture, soil depths, clay mineralogy, surface and area content, water-holding capacity at 1/3, 1 and 15 bar, effective porosity, and pore-size distribution.

Selected Model Codes

The U.S. Geological Survey in Reston, Virginia, has developed a series of computer codes for NAPL fluids. These codes are available in the public domain and can be easily obtained. Many researchers have developed proprietary codes, information on which is obtained from the publications of their research, as reported below.

Abriola and Pinder[53] presented research on multi-phase modeling of organic contaminants in porous media. The researchers developed a model to describe simultaneous transport of chemical contaminant in three physical forms: as a non-aqueous phase, as a soluble component of an aqueous phase, and as a mobile fraction of a gas phase. The contaminant may be composed of, at most, two distinct components, one of which may be volatile and slightly water-soluble, and another of which is both nonvolatile and insoluble in water.

Equations which describe the system of Abriola and Pinder are derived from basic conservation of mass principles by the application of volume averaging techniques and the incorporation of various constitutive relations and approximations. Effects of matrix and fluid compressibilities, gravity, phase composition, interface mass exchange, capillarity, diffusion, and dispersion are considered. The resulting mathematical model consists of a system of three nonlinear partial differential equations subject to two equilibrium constraints. The three equations describe mass conservation of the water phase, mass conservation of the inert organic species, and mass conservation of the volatile organic species. The equations have five unknowns: two capillary pressures and three mass fractions. To handle the solution of the resultant system, a Newton-Raphson iteration scheme is employed.

In order to apply the finite difference one-dimensional model of Abriola and Pinder to a specific problem, a number of parameters must be evaluated. They include three-phase relative permeabilities, saturation pressure relations, partition coefficients, mixture densities, and viscosities.

Faust[52] has developed a numerical model that describes the simultaneous flow of water and a second immiscible fluid under unsaturated and saturated conditions in porous media. Example applications of the model are performed to demonstrate both proper model function related to solving governing equations and model sensitivity to fluid properties. No field model application or validation is presented, since data such as relative permeabilities and capillary pressures are not available for the type of NAPL and site considered.

Modeling Applications

Hydrocarbons Draining to Aquifer

Faust[52] demonstrates the impact of an undetected leak of a NAPL on a surficial unit considered to be an aquitard. Two base cases were simulated:

- a NAPL more dense than water, where we expect gravity effects to be dominant and create a downward migration
- a NAPL less dense than water, where we expect the contaminant to pool near the water table.

The above two cases and expected conditions are observed in the results of the simulation cited below.

Figures 29 and 30 show contour plots of nonaqueous phase saturation superimposed on a cross section. In one year the dense fluid has long (about 3 months) reached the underlying aquifer that drains the aquitard. A relatively uniform saturation profile has been established below the source. For the nonaqueous fluid

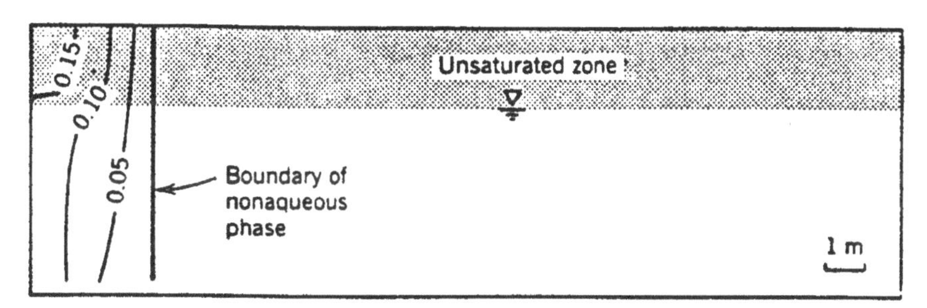

Figure 29. Volumetric saturation of nonaqueous phase with density greater than water: time = 1.16 years (from Faust[52]).

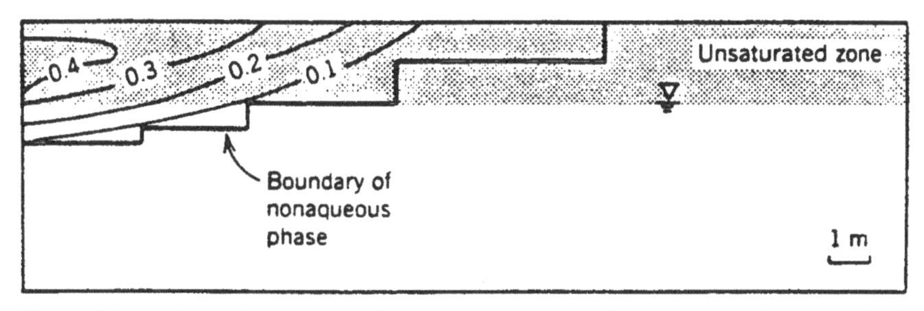

Figure 30. Volumetric saturation of nonaqueous phase with density less than water (950 kg/m^3): time = 1.07 years (from Faust[52]).

that is less dense than water, contaminant saturations are higher (0.44 versus 0.19) and do not extend nearly as far into the saturated zone.

The lighter contaminants have also migrated farther laterally. For this particular example, it is interesting that a large amount of the contaminant remains in the soil zone. The results of Mercer also indicate that the lighter contaminants do not necessarily form a distinct lens above the water table. This example shows that the lens concept may not strictly apply for all possible types of contaminants and hydrogeologic settings.

Figures 31 and 32 illustrate additional migration characteristics for the two cases.

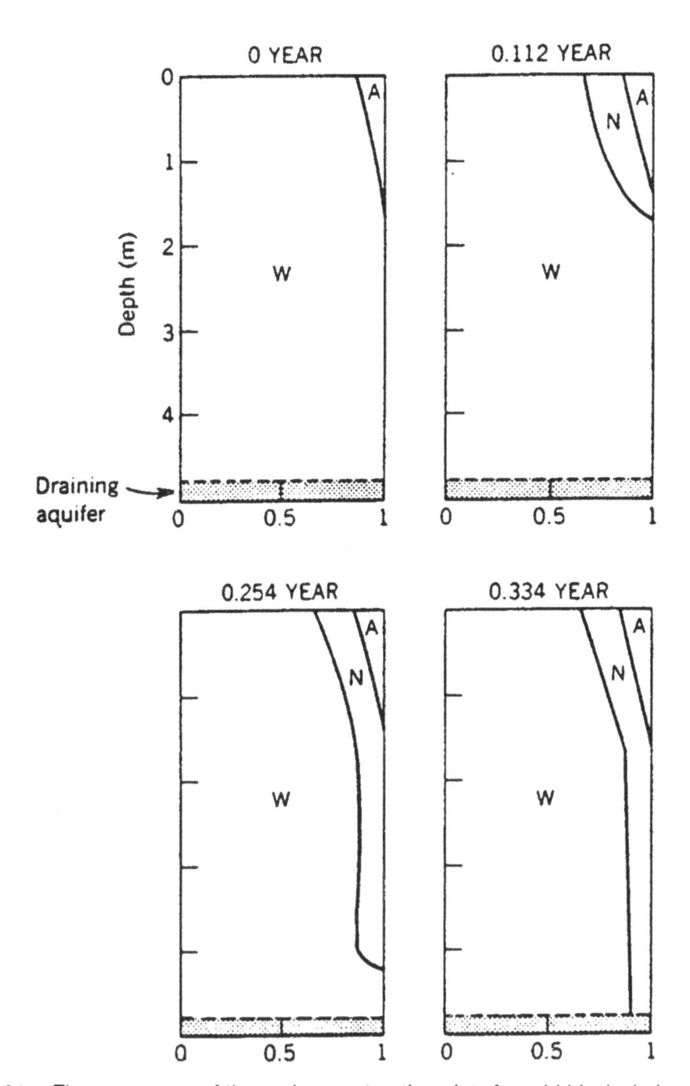

Figure 31. Time sequence of three-phase saturation plots for grid blocks below source; for example, application with NAPL density greater than water (from Faust[52]).

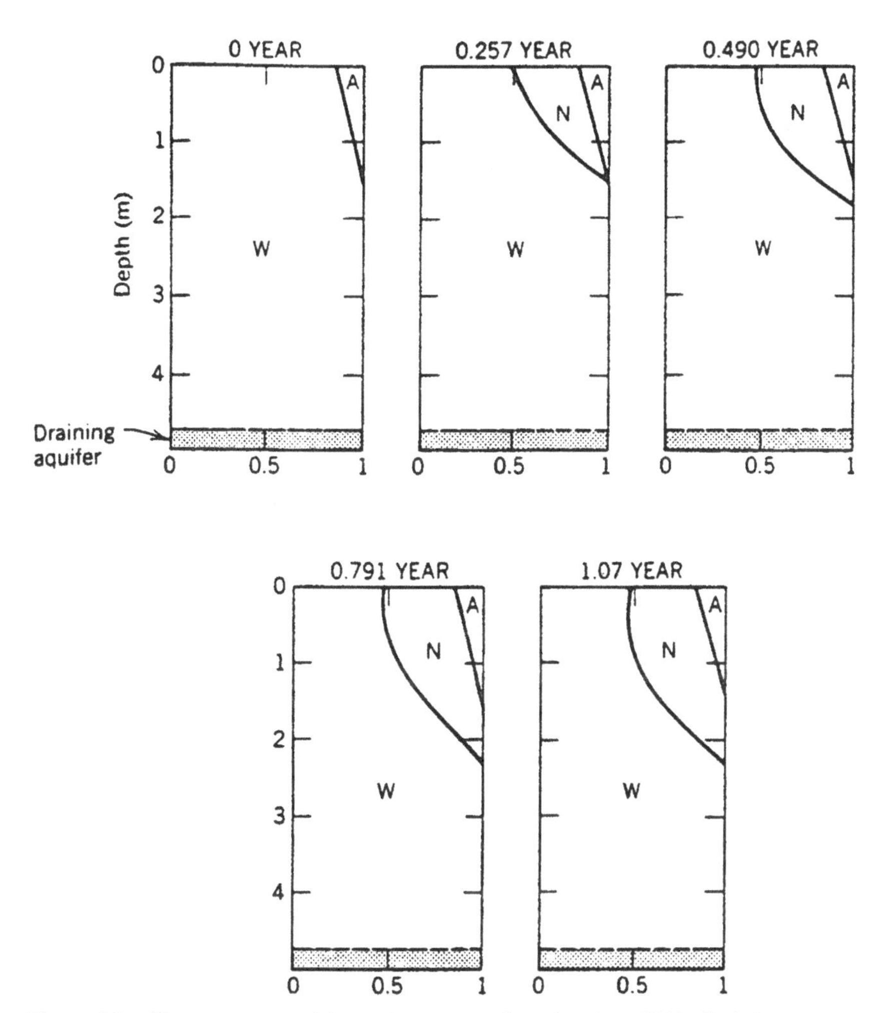

Figure 32. Time sequence of three-phase saturation plots for grid blocks below source: for example, application with NAPL density greater than water (from Faust[52]).

Each figure shows a sequence of three-phase saturations versus depth below the source grid block. For the dense NAPL (Figure 31), a stable saturation profile is established after about three months of leakage. For the higher contaminant, a stable profile is not achieved even after one year. Rather, the two-phase zone continues to expand downward (laterally also, although not shown).

In addition to the two cases above, the sensitivity to density and viscosity was investigated with other simulations by Faust.[52] These results are illustrated in Figure 33. The same data used in the previous cases were used, except density and viscosity of the NAPL were varied. This figure shows depth saturation profiles of the NAPL for different density and viscosity conditions after 0.317 years of leakage. Both increased viscosity and decreased density of the NAPL have a

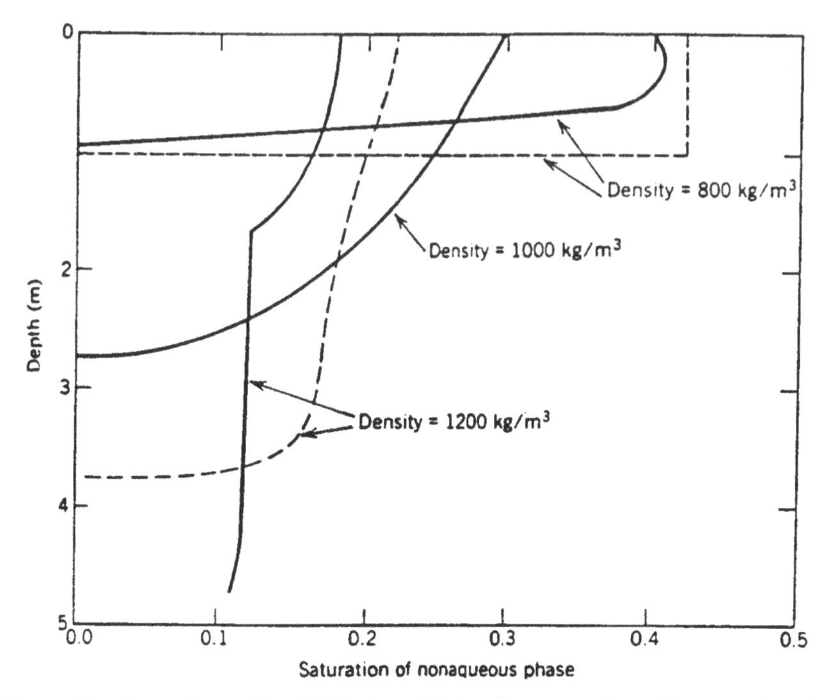

Figure 33. Saturation profile of NAPL for grid below the source: for solid curves, viscosity = 0.001 m/kg·s; for dashed curves, viscosity = 0.004 mg/kg·s (from Faust[52]).

similar effect; i.e., increasing the NAPL saturation near the source. Note that the results show the profile under the source.

Abriola and Pinder[53] similarly demonstrate their model capabilities with the simulation of the fate of one- and two-component organic contaminants in soils. Propagation of the organic liquid front with time is illustrated in their publication (Figure 34) to which the reader is referred.

Volumetric Estimation of Gasoline in Soil

Farr et al.[119] modeled the volume estimation of light nonaqueous phase liquids in porous media. An analytic method is presented, and both the Brooks-Corey and the Van Genuchten equations with parameters derived from laboratory column experiments reported in the literature have been used to relate fluid contents to capillary pressures. Typical simulation results are shown in Figure 35.

Oil Infiltration in Unsaturated Soils

Cary et al.[120] modeled with the TDE approach the infiltration and redistribution of two viscous hydrocarbon oils with viscosities 4.7 and 77 times greater than those in columns of moist silt loam and loamy sand soils. The time required for water and oil to infiltrate two soils and the final volumetric liquid contents

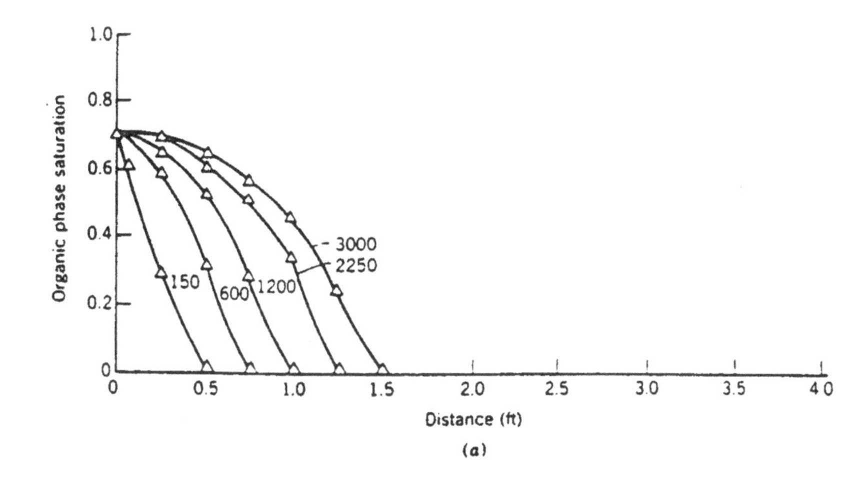

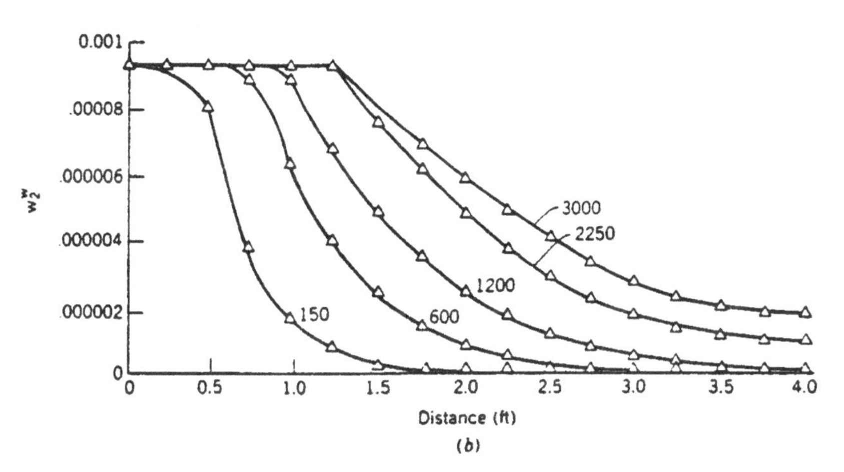

Figure 34. (*a*) Propagation of the organic phase as a function of time elapsed; time for each curve is indicated in seconds. (*b*) Propagation of species 2 in the water phase as a function of time elapsed; time for each curve is indicated in seconds (from Abriola and Pinder[53]).

after 8 hours were measured. The observations were compared to the results predicted by a simple multiphase flow code based on scaling oil flow to unsaturated soil-water flow by using classical theory and relative solutions. A new function for describing the oil permeability was derived and, when used in the code, produced oil content distributions with depth that agreed reasonably well with those observed. The actual oil infiltration time was generally less than predicted, however. The flow of the two organic liquids into moist soils was found to be predictable, based only on classical soil-water relations and the organic liquids' physical properties, under the imposed experimental conditions. The distribution of mineral oil and water in the loamy sand soil 8 hr after adding water to the dry soil and 4 hr after adding mineral oil is shown in Figure 36.

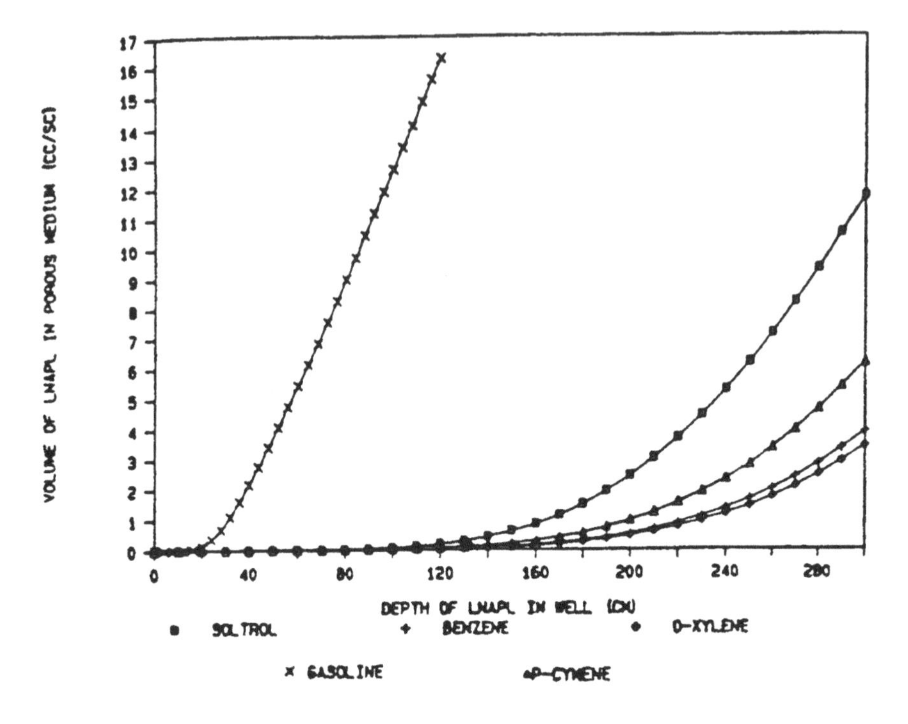

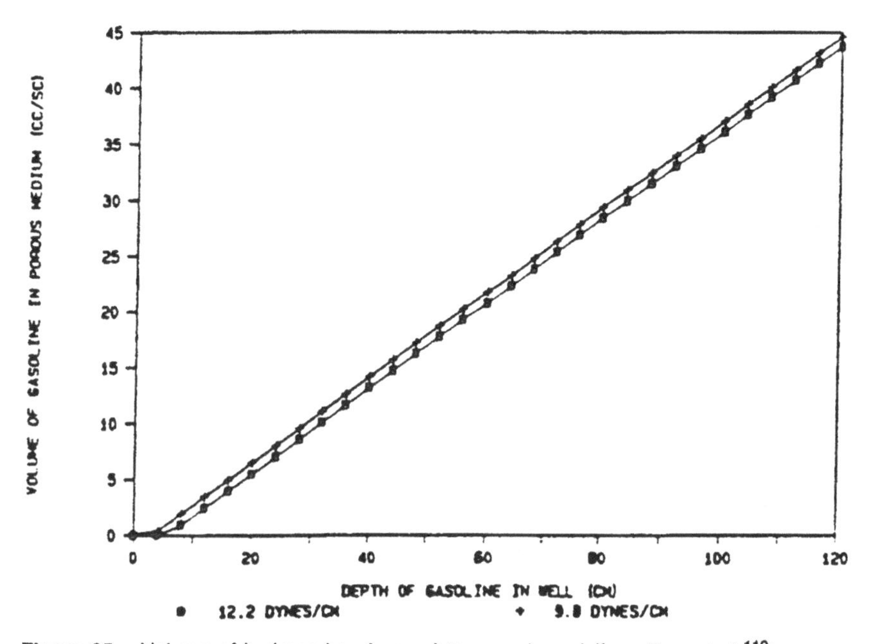

Figure 35. Volume of hydrocarbon in sandstone and sand (from Farr et al.[119]).

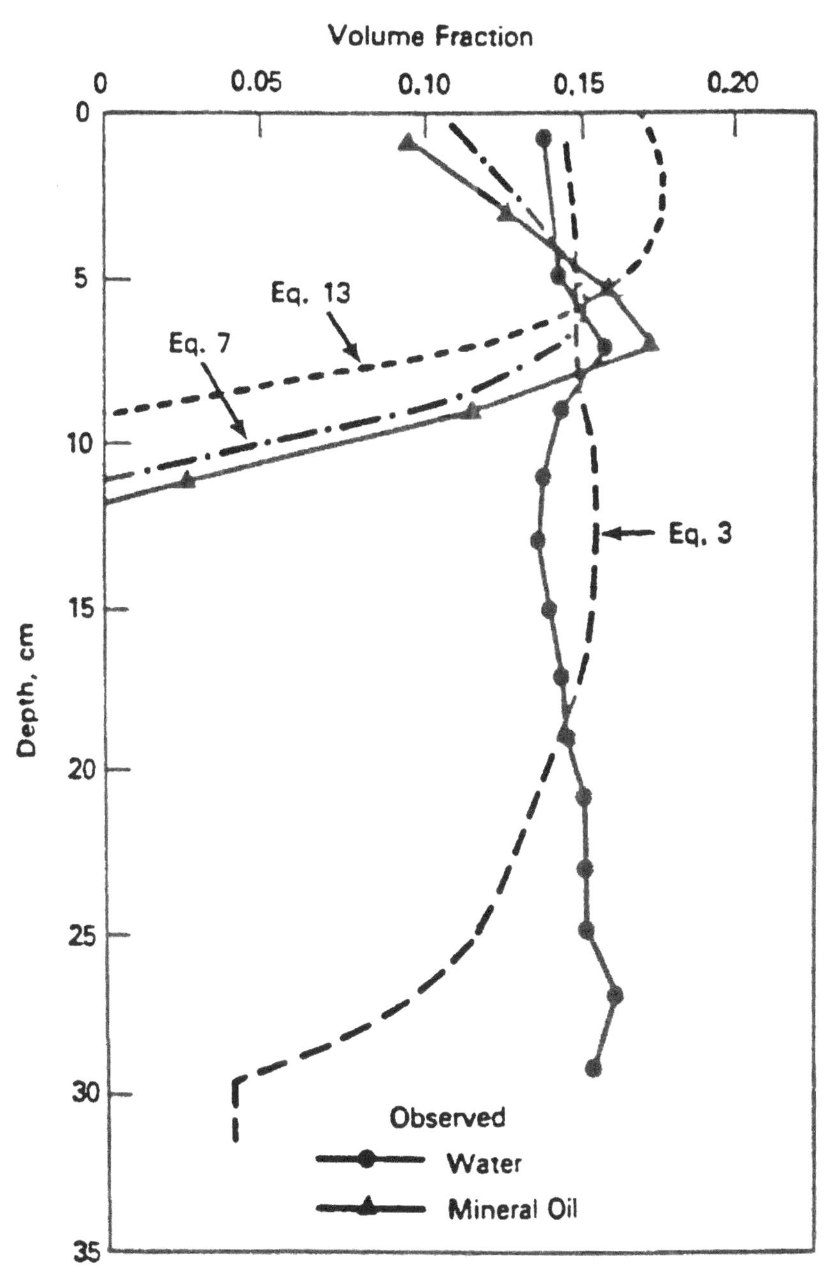

The distribution of mineral oil and water in the loamy sand soil 8 hr after adding water to the dry soil and 4 hr after adding mineral oil. Solid lines connect the observed data; broken curves are predicted results.

Figure 36. Distribution of oil in loamy sand (from Cary et al.[2]).

In Situ Soil Venting

Johnson et al.[121] modeled the cleanup of hydrocarbon contaminated soils by in situ soil venting. The efficiency of any soil venting operation depends significantly on three factors: vapor flow rate, vapor flow path relative to the contaminant distribution, and composition of the contaminant. Simple mathematical models developed were used (TDE approach) as screening tools to help determine if soil venting will be a viable remediation option at any given spill site. The models relate the applied vacuum, soil permeability, and spill composition to the vapor flow rate, velocities, mass removal rates, and residual composition changes with time. In this report the screening models and some sample calculations are presented. The results show the advantages and limitations of venting as a remediation tool, under both ideal and nonideal conditions.

Figure 37 presents predictions for the total mass loss rate Qm(t) and cumulative fraction of the initial spill mass recovered over a period of 400 days. The

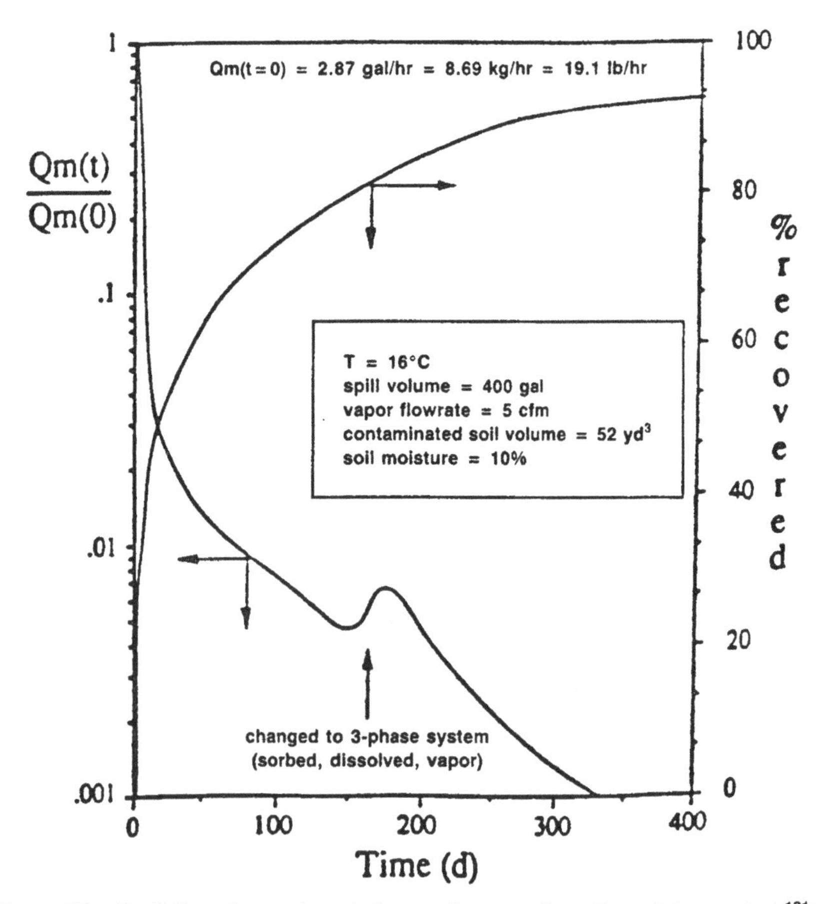

Figure 37. Predicting of mass loss during venting operations (from Johnson et al.[121]).

results show the rate of gasoline removal rapidly decreases over a period of a few days, then continues to decrease less rapidly over the next few months. After a period of 400 days, about 90% of the initial hydrocarbon has been removed.

The change with time of total hydrocarbons (HC) and the benzene, toluene, and xylenes (BTX) concentrations remaining in the soil are presented in Figure 38. BTX and HC concentrations are expressed in terms of dry soil mass, and represent the total mass of BTX and HC in the free liquid, soil moisture, sorbed, and vapor phase. Figure 38 illustrates that venting removes the compounds in the order of their relative volatilities (benzene-toluene-xylenes). Within 200 days the BTX level is reduced to below 1 ppm, while the total hydrocarbon concentrations are reduced from 20,000 ppm to about 1,000 ppm.

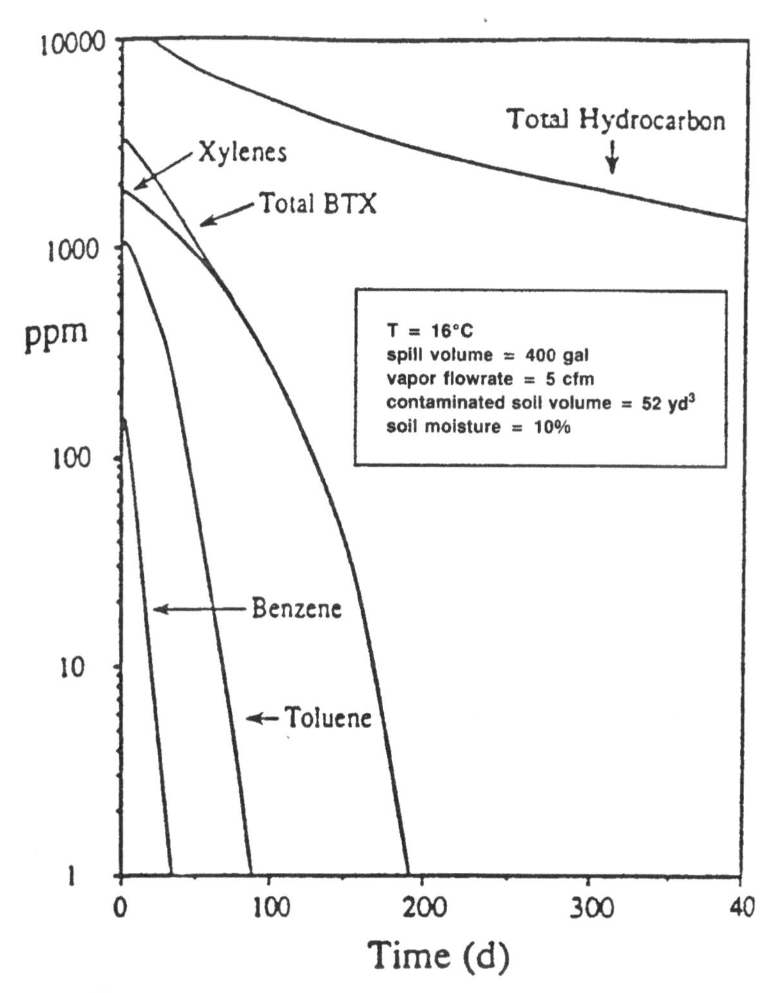

Figure 38. Residual soil concentrations of hydrocarbon following a venting operation (from Johnson et al.[121]).

Free Hydrocarbon Volume in Monitoring Wells

Lenhard and Parker[122] modeled the free hydrocarbon volume from fluid levels in monitoring wells using a scaling procedure that requires knowledge of two-phase air-water saturation-pressure relations and hydrocarbon properties in the environment. Analytic modeling has been employed. Air-water saturation-pressure relations are parametized by either the Brooks-Corey or van Genuchten expression. Parameters of the models are estimated from grain-size distribution for hypothetical soils.

Results reveal that whereas the distance above an oil-water table at which oil saturations become zero may be independent of soil type, estimated light nonaqueous phase liquid (LNAPL) volumes per unit area may differ substantially. Hence, estimates of LNAPL volume cannot be inferred directly from soil LNAPL thickness or well LNAPL thickness data without consideration of effects of soil properties. Furthermore, it is demonstrated that no simple linear conversion scheme can be employed to relate the height of LNAPL in a monitoring well to the LNAPL volume in porous media. Effects of grain-size distribution and well LNAPL thickness on the ratio of actual LNAPL thickness in the aquifer to well LNAPL thickness are shown.

Hydrocarbon Multicomponent Transport in Soil

Kaluarachchi and Parker[123] conducted modeling of multicomponent organic chemical transport in three-fluid-phase porous media. They developed a two-dimensional finite-element model to predict couple transient flow and multicomponent transport of organic chemicals which can partition between nonaqueous phase liquid, water, gas, and solid phases in porous media under the assumption of local chemical solved simultaneously (TDE approach), using an upstream weighted solution method with time-lagged interphase mass-transfer terms and phase densities.

Phase-summed component transport equations were solved serially after computation of the velocity field, also by an upstream weighted finite-element method. Mass-transfer rates were evaluated from individual phase transport equations by back-substitution and corrected for mass-balance errors. A number of hypothetical one- and two-dimensional simulations were performed to evaluate the applicability of the model to predict the transport of slightly soluble and volatile organics in three fluid-phase porous media.

Results indicated that mass-transfer rate and fluid density updating have negligible effects during periods of highly transient nonaqueous liquid phase migration, but become important for long-term simultaneously as cumulative dissolution to the water phase and volatilization to the gas phase account for larger proportions of the total mass. Due to low solubilities of environmentally important organic liquids, the efficiency of organic removal by aqueous-phase dissolution and transport can be very slow. Gas phase diffusion can have a significant influence on the mass transport of organics with large Henry's constants. A typical simulated distribution of TCE concentrations is shown in Figure 39.

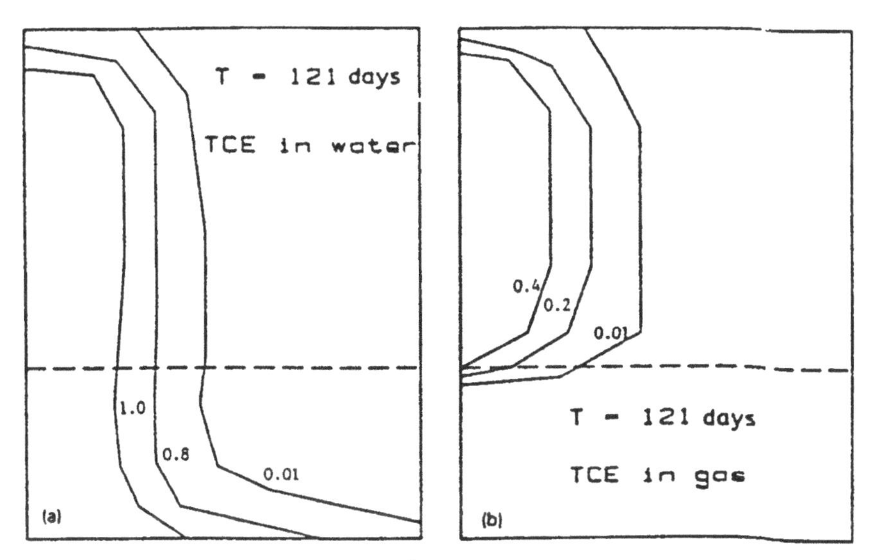

Distribution of TCE concentrations (mg cm^{-3}) at 121 days for example III in: (a) water phase and (b) gas phase.

Figure 39. Distribution of TCE (from Kaluarachchi and Parker[123]).

Oil Entrapment in Rock

Mohanty et al.[124] studied and modeled displacement of oil from an initially oil-filled porous rock. This displacement by water consists of advancement of menisci and rupture of oil connections. In displacements controlled by capillarity, which are typical of oil reservoir floods, these pore-level events are governed by the local geometry, pore topology, and fluid properties, but the pressure field initiates these pore-level events and integrates them with the externally imposed Darcy flow.

This research reports the physics of the pore-level events and their integration on a computational simple model of rock, a square network of pores. The novelty of the approach lies in keeping track of the evolution of the displacement front, and in constructing an approximation of the entire pressure field that carries the information essential for predicting the evolution.

The results give insight into the state of the residual oil saturation and its dependence on pore geometry and the capillary number of displacement. As the capillary number increases, the residual oil saturation decreases and the residual oil blobs tend to be smaller. As the pore size distribution becomes wider, the decrease of residual oil saturation with capillary number becomes smoother.

Hydrocarbons in the Soil Zone

A one-dimensional multiphase mass transport model (TDE, other approach) for the migration of a nonaqueous phase liquid (NAPL) containing sparingly water-soluble organics in the unsaturated soil zone is described. The multiphase NAPL transport (MUNT) model consists of a two-phase immiscible flow model linked to a four-phase chemical transport model. The immiscible flow model incorporates a front-tracking algorithm to determine the front of the invading NAPL as a function of penetration time. The NAPL penetration toward groundwater is shown to be a function of four dimensionless groups: NAPL capillary number, the ratio of the NAPL Reynolds number to the NAPL Froude number, and the ratio of the defending phase to NAPL phase densities and viscosities. Simultaneous predictions for the migration of organic chemicals show that their concentration in the air and aqueous phases at the front can be significant. The prediction of migration of phenanthrene and of benzene are show in Figure 40.

The above includes a limited number of applications available in the literature.

AQUATIC EQUILIBRIUM MODELING

General

An evaluation of the fate of inorganic compounds in soil and groundwater requires more detailed consideration of the chemical and biological processes involved, such as complexation, adsorption, coagulation, oxidation-reduction, chemical speciation, and biological activity, as schematically shown in Figure 41, in the control of free trace metal concentration in soil solutions. The referenced processes can affect, for example, species solubility, availability, physical transport, and corrosion potential.[45]

As a result of a need to describe the complex interactions involved in these situations, various models have been developed to address specific needs. We have, for example, adsorption models, surface complexation models, constant capacitance models, cation-exchange models, and overall fate modeling packages which account for one or more geochemical processes.

As Cederberg et al.[125] report, until recently only a single specific reaction such as ion-exchange or sorption for a small number of reacting solutes has been incorporated into mass transport models. It has also been assumed that the solutes being modeled act independently of the bulk solution composition. Because most contamination sources are actually multicomponent solutions, the need is apparent for models capable of simulating chemical interaction processes.

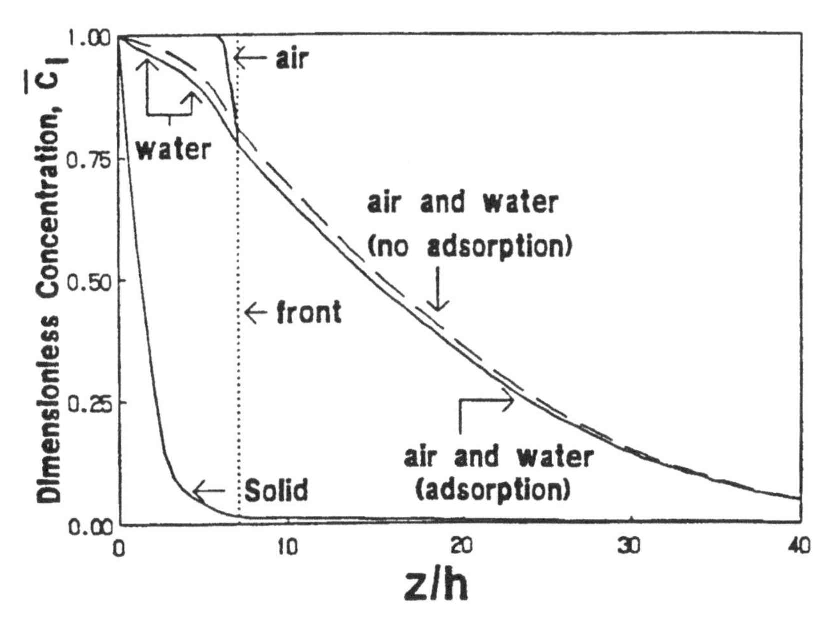

Predictions for the migration of phenanthrene (originating from NAPL and NAPL source) in the air, water, and solid phases of the soil. Two-phase flow of NAPL and air (h = 2 cm; t_d = 5.6 × 10^{10}). Adsorption case (solid curve); no adsorption (dashed curve).

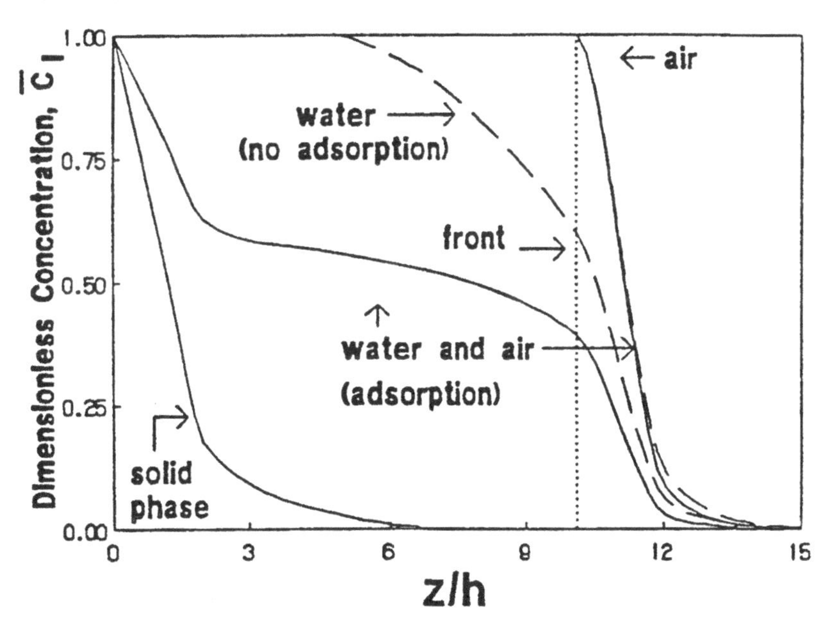

Predictions for the migration of benzene (originating from NAPL and NAPL source) in the air, water, and solid phases of the soil. Two-phase flow of NAPL and air (h = 2 cm; t_d = 5.6 × 10^{10}). Adsorption case (solid curve); no adsorption (dashed curve).

Figure 40. Migration of phenanthrene and benzene NAPLs (from Ryan and Cohen[6]).

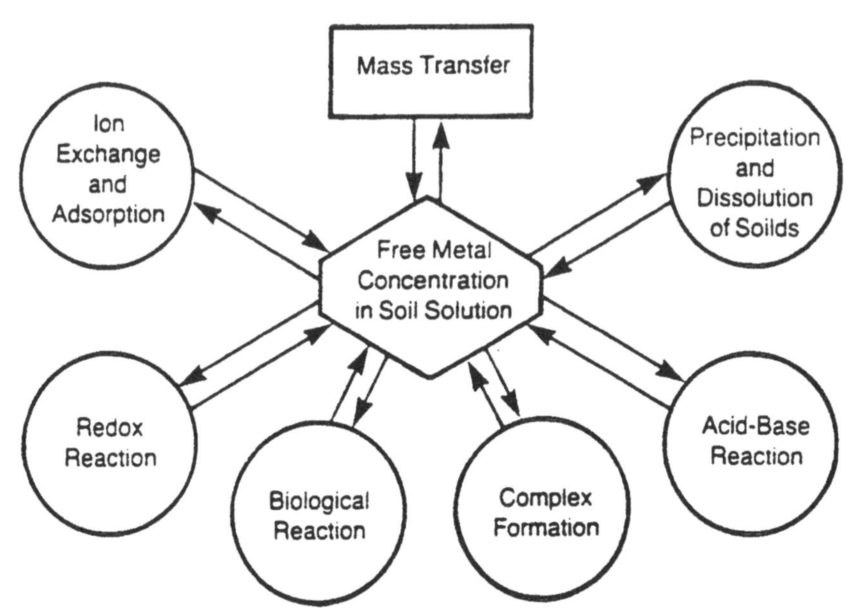

Figure 41. Principal controls on free trace metal concentrations in soil solutions (from Mattigod et al.[45]).

One of these model categories is aiming at providing distribution of inorganic species in the soil water. These models are called "equilibrium" or "speciation" models. Chemically-based computer models of soluble trace metal speciation are being employed increasingly in decisionmaking in studies related to sewage and effluent when applied, for example, to agricultural land (Sposito et al.[126]).

The transport of "species" in terrestrial systems is of interest to a variety of scientists, since measurement or reporting of "total" concentration of a particular inorganic compound in the soil may be very misleading in many environmental managements.[42] Toxic effects of trace metals, for example, may be determined more by their chemical form than by their total concentration (Florence;[127] Allen et al.[128]); therefore, mathematical computer models capable of simulating distribution of inorganic pollutant species in the soil and groundwater systems are valuable tools for contaminant soil/groundwater pathway analyses.

At this state of intensive research in terrestrial chemical modeling it is reasonable to group the prevailing modeling concepts of the literature into two major categories as also earlier reported: the "geochemical" or more appropriately "dissolved pollutant fate models," investigated in soil solutions, and the "speciation equilibria models" for inorganic pollutants. This terminology is not standard, but it is employed here for reasons of communication. Speciation models can simulate the fate of individual dissolved pollutant species and total mass of dissolved pollutants.

The following paragraphs of this section present the status of geochemical speciation equilibria models as applied to inorganic pollutants in soils. It has to be noted that speciation computer codes also account for additional processes, such as redox reaction, adsorption, complexation, and others, since they are related to speciation equilibria. Speciation equilibria models are based on chemical thermodynamic reactions and principles, a research area featuring great progress in recent years. A number of computer codes are available for speciation calculations in soil and groundwater aqueous systems, the most well known of which are presented below.

Excellent state-of-knowledge reviews on chemical equilibria codes (models) of inorganic pollutants in soils are presented by Bodek et al.,[37] Sposito,[129] Cederberg et al.,[125] Kincaid et al.,[49] Miller and Benson,[130] Jennings et al.,[131] Theis et al.,[132] Mattigod and Sposito,[133] Jenne,[134] Nordstrom et al.,[135] and many others. The following paragraphs have benefited from reviews of the above research, and particularly from the work of Sposito. Readers interested in details should refer to the original publications.

Modeling Concepts

Although geochemical speciation modeling was developed relatively recently, more than a dozen comprehensive computer codes are available, as depicted in Figure 42. Kincaid et al.[49] provide an overview of the geochemical code history in which they group models into four major families according to their evolution stage.

The four families of speciation models can be grouped into two categories: (1) "speciation" codes which account for speciation equilibria of inorganic pollutants for a terrestrial water compartment, and (2) "coupled speciation" codes which can simulate both speciation equilibria and solute transport of individual species in the terrestrial (soil, groundwater) environment, both in time and space.

Several investigators have recently addressed the problem of modeling speciation and transport of a multicomponent solution in equilibrium with a soil or groundwater system. Cederberg et al.,[125] Miller and Benson,[130] and Jennings et al.,[131] review the subject and provide detailed literature.

Coupled codes are formulated by two mathematical methods, one described by Miller and Benson[130] and the other by Jennings et al.[131] The former interfaces the computer code for equilibrium distribution of species (e.g., code of Figure 42) with the code for transport, and performs calculations in two steps: (1) species estimation and (2) species transport. This procedure is repeated in time. Jenning's method consists of solving simultaneously a system of equations describing chemical reactions, advective-dispersive transport of reacting contaminants, and interphase mass transfer. The method of solution is irrelevant here.

In summary, equilibrium chemistry models are formulated around two mathematical methods: (1) The first method interfaces the computer code for equilibrium

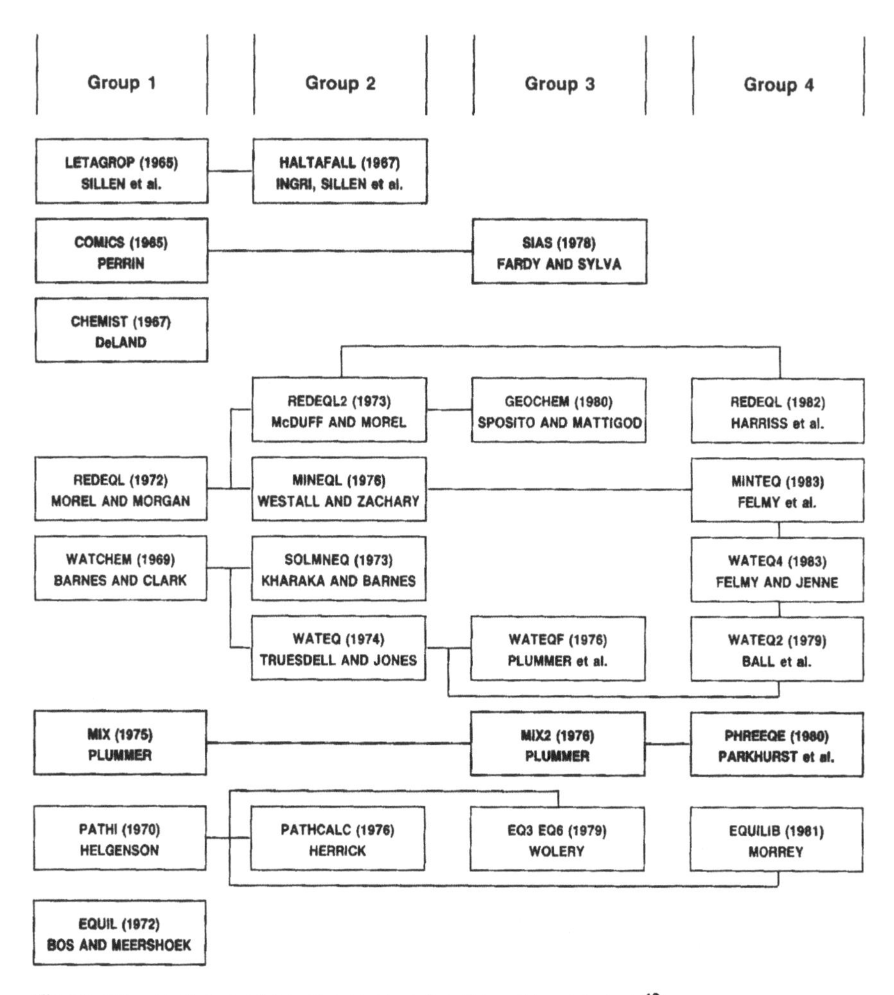

Figure 42. Evolution of geochemical codes (from Kincaid et al.[49]).

distribution of species with the code of transport and performs calculations in two steps; the first is species estimation and the second step is species transport; (2) The second method consists of solving simultaneously a system of equations describing reactions and advective-dispersive transport with interface mass transfer.

The solutions employed do not concern a code user to a great extent.

Coupled speciation codes such as TRANQL (Cederberg et al.[125]), FIESTA (Theis et al.[132]), Kirkner et al.,[136] and CHEMTRN (Miller and Benson[130]) are the subject of current research and are the extension of the speciation codes. Although coupled codes appear to be very powerful tools, they are also very large and require extensive input data. As a result, their use is complex and frequently inefficient (input data requirements and level of effort) for practical solutions to problems.

Coupled codes are not further analyzed in this chapter; rather, emphasis is placed on speciation equilibria (geochemical simple compartment) codes because these models have been more extensively applied and validated, with laboratory and field activity coefficients (input data).

The significant advances made in aquatic chemistry, during the past decade, attest to the availability of an approach to the soil solution that runs parallel with that taken for other natural waters. In that respect there exist many similarities between models developed for soil or groundwater and water environments. For example, the model GEOCHEM[137] has been developed from the water bodies speciation model, REDEQL2, a program created originally by Morel and Morgan and their co-workers at Caltech.[138,139]

Basic geochemical and aquatic equilibria processes modeled by existing computer codes are adsorption/desorption, precipitation/dissolution, reduction/oxidation, hydration and ion-interaction, and aqueous speciation in soil solution. Various mathematical expressions are used to model the above processes.

For the aqueous speciation in soil solution, Sposito[129] reports that "chemical modeling" can be accomplished in two distinct ways: via a "specific interaction" approach or via an "ion-association" approach.

In the former case, the composition of an electrolyte solution is described in terms of the total number of molalities of the stoichiometric components of neutral solutes, while the thermodynamic properties of the solution are expressed with "mean ionic" activity coefficients for neutral solutes introduced into standard chemical thermodynamic equations. In the latter case, the composition of an electrolyte solution is described with the molalities of molecular "species" presumed to exist in solution, whereas the thermodynamic properties of the solution are expressed with "single-species" activity coefficients for the assumed molecular constituents. The ion-association approach appears to be adopted universally by soil chemists, despite its more tenuous relationship with rigorous thermodynamics. For additional details, the reader is referred to the original work of Sposito.

The geochemical databases used are the most important features of a computer code. Databases can provide typical values of kinetic and adsorption rates at typical temperatures (e.g., 25°C). Reasonably good databases are available for GEOCHEM and MINTEQ, among others.

Model selection is a major issue in mathematical geochemical modeling. Factors of importance are study objectives, the features of the model, model input requirements, monitoring data available to validate predictions, model documentation, and cost for an application. It is very difficult to recomend a universal computer code that can handle all of the above factors; therefore, decisionmakers have to first review a number of codes that may meet project requirements, and consequently select the code which best suits their study objectives.

Selected Models/Codes

Geochemical models may be applied for a simplified overall inorganic pollutant fate estimation via an adsorption process (routine), or for many chemical and biological processes that govern the fate of the total pollutant mass from dissolved or partially dissolved (immiscible) fluids. Speciation models estimate the distribution of inorganics in various forms, but certain speciation codes can simulate the fate of individual dissolved pollutant species and the total mass of dissolved pollutants, since they combine fate and speciation codes in one package.

The variety of geochemical (pollutant fate) models has dramatically increased during the last decade. These models can be employed for both organic and inorganic pollutants, since the chemical processes are simulated via their overall equilibrium coefficients (e.g., adsorption coefficients). Speciation computer codes account for additional processes—redox reaction, adsorption, complexation, and others (e.g., James and Rubin[140]). Speciation equilibria models are based on chemical thermodynamic principles. Some of the computer codes available for speciation calculations in soil and groundwater aqueous systems are described by Kirkner et al.[136] and Nelson et al.[141]

Steps in speciation calculations are not of concern in this chapter, which focuses on integrated computerized modeling codes. Excellent state-of-knowledge reviews of chemical equilibria codes (models) of inorganic pollutants in soils are presented by numerous investigators. The following sections have drawn on these reviews. Readers interested in details should refer to the original publications.

Table 21 lists features of known computer codes suitable for geochemical modeling. Of the 12 codes summarized, 6 are very capable in the areas of aqueous speciation, adsorption/desorption, and precipitation/dissolution.[49] In addition, these codes are documented, available in the public domain in the United States, recently updated, and frequently accompanied with a sizable database. Table 22 provides user-related information for the codes presented in Table 21. Many other codes exist in the literature (e.g., BALANCE, Parkhurst et al.[142]), which are not reported here. Two of the best-known geochemical and speciation codes are GEOCHEM and MINTEQ. Coupled geochemical/speciation and transport (codes receiving increased attention) are TRANQL and CHEMTRN. These codes are briefly presented below.

- GEOCHEM is based on the computer program REDEQL2 for calculating equilibrium speciation of chemical elements in a soil solution. The component species are identified as uncomplexed metal cations, the free proton, uncomplexed ligands and the free electron. Single-species activity coefficients are calculated in the program. The model contains critical thermodynamic data for soils, a method for calculating cation exchange, and a correcting method for non-zero ionic strength up to 3 molal. Currently the model stores thermodynamic data for 36 metals and 69 ligands, which form more than 200 soluble complexes and solids. Adsorbed metal species are described in the model, whereas adsorbed ligand species will be considered in the future. GEOCHEM is constantly updated; therefore, interested scientists should contact model developers.

- MINTEQ is similar in structure to MINEQL, which is similar in overall structure to GEOCHEM, since both models originate from REDEQL. MINTEQ is formed from MINEQL and the database of WATEQ.[143] The model includes ion speciation, redox equilibria, calculation of activity coefficients, solubility, adsorption, and mass transfer. The model and its large database are well documented, the latter involving more than 35 metals and 60 ligands.

  As reported in a comparative analysis by Sposito,[129] a principal difference between MINEQL (and consequently MINTEQ) and GEOCHEM is that MINEQL can accept the concentration of any free ionic species, soluble complex, or dissolved gas as input data to be held fixed during a calculation, whereas GEOCHEM can do this only for the activities of H^+, e^-, $CO_2(g)$, and $N_2(g)$. Therefore, if desired, the concentration of Cu^{+2} can be specified as a fixed input datum in MINEQL, a feature providing the code use with a great flexibility in speciation calculations.
- CHEMTRN is a one-dimensional geochemical transport/speciation (coupled) model for solutes in a saturated porous medium. The model includes dispersion/diffusion, advection, ion exchange, formation of complexes and speciation in the aqueous phase, and the dissociation of water. The mass action, transport, and site constraint equations are expressed in differential/algebraic form and are solved simultaneously. This coupled model is at a developmental stage; therefore, no database accompanies the computer code.
- TRANQL is a groundwater mass transport and equilibrium chemistry model for multicomponent systems. The equilibrium interaction chemistry is posed independently of the mass transport equations, which leads to a set of algebraic equations for the chemistry coupled to a set of differential equations for the mass transport. Significant equilibrium chemical reactions such as complexation, ion exchange, competitive adsorption, and dissociation of water may be included in the model. In a recent application, a finite solution is presented first for cadmium, chloride, and bromide transport in a one-dimensional column where complexation and sorption are considered. Second, binary, and ternary ion exchange are modeled and compared to the results of other investigations. According to model developers, results show the model to be a versatile multicomponent transport model, with potential for extension to a wide range of equilibrium reactions.

Table 21. Summary of Geochemical Code Capability, Adaptability and Availability

Criterion	GEOCHEM	REDEQL. UMD	MINTEQ	PHREEQE	EQUILIB	EQ3/EQ6	SIAS	SOLMNEQ	CHEMIST
Code classification	III	III	III	IIIR	II	IIR	I	II	I
Number of elements	36 (94)	3	32	19 (26)	26		10	25	
Aqueous speciation									
Number of ligands/species	69/—	65/—	16/373	8/120	186	20	10	162	(169)
Number of redox species	48 (60)	20	22	6	20	6	(200)	10	Variable
Activity-coefficient correction	DA,HN	DA,SHM	DA	ECH	HN	HN	None	HN	None
Calculation of pH	Yes	Yes	Yes	Yes	Yes	Yes	No	No	No
Method of iteration	N-R	N-R	N-F	N-R,C-F	Pred.	N-R	B-S	B-S	N-R
Adsorption									
Model	J-H	Swiss	Three mdls.	Two mdls.	None	None	None	None	None
Number of species	10	20	No limit	Variable	None	None	None	None	None
Number of surfaces	6	5	3	NA	None	None	None	None	None
Number of species	No limit	None		None	None	None	None	None	None
Precipitation dissolution									
Number of minerals	<500	<500	238	24	200	250	None	158	None
Quantitative mass transfer	Yes	Yes	Yes	Yes	Yes	Yes	None	No	No
Automatic selection of mineral	Yes	Yes	Yes	No	Yes	Yes	None	No	No
Capability for solid solutions	No	No	No	No	None	Yes	None	No	No
Gas generation									
Ammonia	No	No	No	No	Yes	No	No	No	No
Oxygen	Yes	No	Yes	Yes	Yes	Yes	No	No	No
Hydrogen	No	No	No	Yes	Yes	Yes	No	No	No
Carbon dioxide	Yes	Yes	Yes	Yes	Yes	Yes	No	Yes	No

continued

Table 21. *Continued*

Criterion	GEOCHEM	REDEQL. UMD	MINTEQ	PHREEQE	EQUILIB	EQ3/EQ6	SIAS	SOLMNEQ	CHEMIST
Code structure									
Size (32 bit words, K)	434	62	68.1	>64	20.5	363	34	30	25
Modularity	Yes	Yes	Yes	Yes	Yes	Yes	No	No	No
Language	FORTRAN IVG	FORTRAN IV	FORTRAN IV	FORTRAN IVH	FORTRAN IV	FORTRAN 4.6	FORTRAN H	PL 1	FORTRAN
System	IBM 4314 VAX	CDC CYBER 171	UNIVAC 1144 PDP 11/70 VAX	Amdahl DEC VAX	CDC 7600 PDP VAX	CDC 7600 CDC 6600 VAX	IBM/360/65	IBM/360	IBM/360
Other Criteria									
Latest documentation date	1980	1982	1982	1980	1978	1979	1978	1973	1968
Data Base									
Temperature range	No	No	Yes	Yes	Yes	Yes	No	Yes	No
Easily modified	Yes	Yes	Yes	Yes	Yes	Yes	U-S	No	U-S

Source: Kincaid et al.[49]

( ) = maximum, U-S = user supplied, J-H = James-Healy, N-R = Newton-Raphson, B-S = back substitution, Pred. = predictor back substitution, DA = Davies, HN = Helgeson-Nigrini, SHM = Sun-Harriss-Mattigod, ECH = extended Debye-Huckel, I = chemical speciation, II = I + mass transfer by precipitation or dissolution, III = II + adsorption and ion exchange, IIR and IIIR = pseudo kinetics.

Table 22. Advantages and Disadvantages of Chemical Models

Code	Characteristics
GEOCHEM	Available in public domain. Documentation marginal. Recently updated. Models important processes. Includes adsorption. Database probably the largest available. Modularity not yet evaluated.
REDEQL.UMD	Should be available in the public domain soon. Well documented, although not in final form. Recently updated. Models important processes. Database and modularity not yet evaluated.
MINTEQ	Available in the public domain. Well documented. Models important processes. Includes adsorption. Modular construction. Database is the best documented: one of largest available.
PHREEQE	Available in the public domain. Documented. Models precipitation/dissolution. Includes adsorption. Database supplied by user. Modular construction.
EQUILIB	EPRI proprietary code. Well documented. Models precipitation, not adsorption. Database quite extensive, reasonably well documented. Can use other databases. Construction is modular. Method of solution involves unique elements.
EQ3/EQ6	Publicly available. Documentation available for EQ3 only. Updated version of PATHI construction. Contains precipitation, no adsorption but structured for inclusion. Contains capability to model paths of chemical changes.
SOLMNEQ	Publicly available. Documentation old. Does not contain unique characteristics. Database not easily modifiable.
CHEMIST	Publicly available. Old and sketchy documentation. No precipitation/adsorption. No unique modeling characteristics.
CHEMCSMP	Publicly available. Old and very inadequate documentation. May have unique kinetic characteristics but tied to IBM system.
SIAS	Update of an old program. Documentation sketchy. SIAS capabilities covered in better documented, more inclusive codes.

Source: Kincaid et al.[49]

Model Applications

Reports on mathematical modeling on the chemical form of heavy metals in soil solutions are limited.[42,144] This may be due first to the large number of chemical forms in which metals exist and the analytical problems associated with their determination[145] that cannot be easily modeled, and second, to only the recent concerns and efforts on modeling metal species in soil solutions. The use of chemical equilibrium models in decisionmaking is reported by Emmerich et al.,[42] Behel et al.,[146] Matthews,[147] Westall and Hohl,[148] Miller and Benson,[130]

Mattigod and Sposito,[133] Theis et al.,[132] Dowdy and Volk,[150] Kirkner et al.,[136] Jennings and Kirkner,[149] and many others. The above researchers have employed and validated models (e.g., GEOCHEM) with data from field (e.g., Behel et al.[146]) and laboratory [42] analyses.

Example of Model Applications

This section illustrates the use of a model to estimate the chemical speciation of metals in a soil system. The modeling experiment was conducted by the Tennessee Valley Authority and relates to land-treated sewage sludge. It is presented in four sections (Bonazountas, in Bodek et al.[37]).

1. Statement of problem
2. Field monitoring
3. Model application, and
4. Discussion of findings.

Statement of Problem

Anaerobically digested, air-dried sewage sludge was applied (in 1971) in plots of 3.6 by 15.0 meters and was incorporated into the soil to a depth of 15 cm. The application rates were 50, 100, and 200 tons/ha.

Cultivation of the plots began after 12 months with sweet corn (*Zea Mays L. cv.* Silver Queen). At that time, the plots were subdivided into areas of 3.6 by 7.5 meters. Half of each plot received three additional sludge applications—in that year, 12 months later, and 24 months later—at the rate used in the initial application. The other half of each plot received three additional sludge applications—in that year, 12 months later, and 24 months later—at the rate used in the initial application. The other half of each plot received no additional sludge, so that residual effects from the original applications could be monitored. The sludge contained both organic and inorganic pollutants.

Field Monitoring

Before application, the sludge was analyzed for inorganic species and total organic carbon. A field monitoring program was conducted in 1971. The soil of the area is a Sango silt loam soil (Glossic Fragiudult) with a pH of 4.9 and a cation exchange capacity of 7 mol($NH^{+}{}_{4}$)kg^{-1}.

The field monitoring program (collection and analysis of contaminated soils) was repeated in 1977. Soil samples were collected, air-dried, and crushed to less than 2 mm. Water was added to 150-g samples to bring the soil water potential to −0.33 bar, and the samples were then incubated for 7 days at 25°C. Soil solutions were recovered by centrifugation and analyzed for total Cd, Zn, Mn,

Fe, Al, Cu, Ni, Pb, Ca, Mg, Na, K, P, Cl, sulfate, organic carbon, conductivity, and pH. The chemical composition of soil solutions from untreated and sludge-treated soils is shown in Table 23.

Concentrations of Zn, Cd, Mn, Ca, Mg, Na, K, PO_4-P, SO_4-S, and Cl in the soil solution tended to increase with sludge application, particularly in plots that received multiple sludge application. Concentrations of Cu, Ni, Pb, and Fe were generally low (less than 10^{-6}M) at all sludge application rates, but soluble Cu and Ni concentrations in soil treated with sludge at 800 $t \cdot ha^{-1}$ were markedly higher than those in soil receiving lesser amounts of sludge.

Model Application

An attempt was made to duplicate the above conditions and estimate species of chemical in the soil systems via a mathematical model. The GEOCHEM model was used to predict the equilibria of the metals in solution because its database can adequately support simulations of the species presented.

To facilitate both model use and interpretation of the data obtained, GEOCHEM calculations were performed using three sets of conditions: (a) metal-organic complexes were ignored; (b) metal-organic complexes were approximated by a mixture model, and (c) metal-organic complexes were modeled using stability constants obtained experimentally for a sludge fulvic acid (fulvate model). It is almost always necessary to make assumptions of this kind when using a model.

The mixture submodel of GEOCHEM predicted that less than 16% of the total (t) Cd_t, Zn_t, and Mn_t in solution would be complexed with organic and inorganic ligands. Therefore, it was not surprising that exclusion of the mixture model from GEOCHEM has a minor effection the speciation of Cd, Zn, and Mn (Table 24). The results of the overall simulation are shown in Table 25.

The predominant species of Zn in the soil solution was Zn^{+2} (Table 26). Even though sludge applications increased total soluble Zn (Zn_t) from $4*10^{-6}$M to more than $1*10^{-4}$M, 89% to 97% of the Zn_t was present as Zn^{+2}. For most sludge treatments, SO_4 complexes accounted for 5% to 8% of Zn_t. Fulvate complexes (mixture model) with Zn tended to increase with sludge applied, but the amount of Zn_t associated with soluble organic C was always less than 2%. The distribution of Zn species was essentially the same after application of 200 $t \cdot ha^{-1}$ in a single treatment or in four annual 50-t ha^{-1}, increments. Soluble Zn_t increased to a greater extent with the single 200-t ha^{-1} sludge application ($1.7*10^4$M) when compared with four increments of 50 $t \cdot ha^{-1}$ ($1.2*10^4$M).

The activity of Zn^{+2} in the soil solutions was similar to that supported by several Zn solid phases. The activity of Zn^{+2} ranged from approximately 10^{-5} to 10^{-3}M in solutions of pH to 6. Based on stability diagrams, the Zn^{+2} in soil solution could be controlled by one or more of the following solids:[151] (a) $ZnFe_2O_4$ in equilibrium with soil Fe; (b) soil Zn; and (c) Zn_2SiO_4 in equilibrium with amorphous Si. At similar pH values, Zn^{+2} activities obtained from this study were greater than those previously reported for soils incubated with sewage

Table 23. Chemical Composition of Soil Solutions from Untreated and Sludge-Treated Soils

		Sludge applied, t ha^{-1}							
		A[a]				B[a]			
Parameter	Units	0	50	100	200	0	200	400	800
Zn	= 10^{-5}M	0.35	7.94	10.72	16.98	0.40	11.75	16.60	36.31
Cd	= 10^{-8}M	0.98	2.24	3.09	79.4	1.26	3.02	58.9	144.5
Cu	= 10^{-6}M	<1.0	<1.0	<1.0	3.16	<1.0	<1.0	<1.0	6.31
Ni	= 10^{-6}M	<1.0	<1.0	<1.0	<1.0	<1.0	<1.0	<1.0	6.76
Pb	= 10^{-6}M	<1.0	<1.0	<1.0	<1.0	<1.0	<1.0	<1.0	<1.0
Mn	= 10^{-4}M	3.72	3.09	3.89	3.55	3.63	5.37	6.46	7.41
Fe	= 10^{-6}M	<1.0	<1.0	<1.0	<1.0	<1.0	<1.0	<1.0	<1.0
Al	= 10^{-4}M	1.86	1.86	1.86	<1.0	1.86	1.86	<1.0	<1.0
Ca	= 10^{-3}M	1.95	3.63	3.72	4.07	2.19	3.89	4.68	9.12
Mg	= 10^{-4}M	1.20	5.89	4.07	4.37	.95	4.37	4.68	6.31
K	= 10^{-4}M	7.59	19.95	10.23	9.55	8.13	10.0	10.72	11.75
Na	= 10^{-3}M	3.31	3.89	3.89	3.98	2.51	8.32	8.51	9.12
PO_4-P	= 10^{-6}M	2.6	2.9	4.2	15.49	2.57	9.33	26.92	60.26
SO_4-S	= 10^{-3}M	0.26	0.85	0.98	1.15	0.44	0.91	1.66	3.39
Cl^-	= 10^{-1}M	0.60	1.55	1.29	1.23	0.51	0.72	1.07	1.35
Organic C (g m^{-3})		160	130	160	190	170	150	180	250
Conductivity (dS m^{-1})		1.54	2.25	2.14	2.08	1.56	2.03	2.38	3.88
pH		4.7	4.9	5.1	5.6	4.6	5.0	5.2	5.5

Source: Behel et al.[146]

[a]A denotes a single application in 1972; B denotes repeat applications of initial rate in 1972, 1973, and 1974.

Table 24. Effects of Excluding Organic Ligands from Consideration by GEOCHEM on the Percentage of Free Metals in Soil Solution

Sludge Applied[a] ($t\ ha^{-1}$)		Organic Ligands Included[b]	Cd		Zn		Cu		Ni		Mn	
A	B		A	B	A	B	A	B	A	B	A	B
			(Calculated % of Total Present as Free Metal Ion)									
0	0	Fulvate model	67	67	79	80	—[c]	—	—	—	69	73
		Mixture model	92	91	97	96	—	—	—	—	98	97
		None	94	93	98	96	—	—	—	—	98	97
50	200	Fulvate model	69	72	84	84	—	—	—	—	80	81
		Mixture model	85	89	94	94	—	—	—	—	95	95
		None	86	90	95	95	—	—	—	—	96	96
100	400	Fulvate model	66	67	80	81	—	—	—	—	74	78
		Mixture model	84	83	93	91	—	—	—	—	95	84
		None	86	86	94	92	—	—	—	—	95	93
200	800	Fulvate model	64	63	79	78	69	69	—	72	74	76
		Mixture model	82	79	91	87	19	19	—	87	77	80
		None	86	82	93	88	93	88	—	90	94	90

Source: Behel et al.[146]

[a]A denotes 1 application in 1971: B denotes repeat applications of initial rate in 1972, 1973, and 1974.

[b]Fulvate model—calculations based on equilibrium constants for fulvic acid extracted from sewage sludge [48]; Mixture model—mixture model for soluble fulvic acid [31].

[c]Total concentrations below detection limits of analytical methods used.

Table 25. Concentration and Speciation of Mn, Cu, and Ni in Soil Solution as Affected by Rate and Frequency of Sludge Application

Metal	Sludge Applied ($t\ ha^{-1}y^{-1}$)	No. of Applications[a]	Total Concentration ($= 10^{-5}M$)	Free Metal	Metal Complexes SO_4	PO_4	MM[b]
				—% of Total Metal—			
Mn	0	0	37	98	1.8	<0.1	0.3
		0	36	97	2.8	<0.1	0.3
	50	1	31	95	4.2	<0.1	0.3
		4	54	95	4.0	<0.1	0.4
	100	1	39	95	4.9	<0.1	0.5
		4	65	84	6.0	<0.1	9.5
	200	1	36	77	4.5	<0.1	17.9
		4	74	80	8.3	0.2	11.4
Cu	200	1	0.32	19	1.4	<0.1	79.8
	200	4	0.63	19	2.5	<0.1	78.3
Ni	200	4	0.68	87	9.0	0.1	3.3

Source: Behel et al.[146]

[a]Single application in 1971: repeat applications of initial rate in 1972, 1973, and 1974.

[b]MM denotes mixture model for fulvate [31].

Table 26. Concentration and Speciation of Zn and Cd in Soil Solution as Affected by Sludge Rate and Frequency of Application

Sludge Applied ($t\ ha^{-1}y^{-1}$)	No. of Applications[a]	Total Zn ($= 10^{-5}M$)	Free Zn^{+2}	Complexed Zn[b] SO$_4$	Complexed Zn[b] MM[c]	Total Cd ($= 10^{-8}M$)	Free Cd^{+2}	Complexed Cd[b] SO$_4$	Complexed Cd[b] Cl	Complexed Cd[b] MM[c]
			—% of Total—				—% of Total—			
0	0	0.35	97	2.2	0.5	0.98	92	2.7	3.6	1.7
	0	0.40	96	3.5	0.4	1.3	91	4.2	2.9	1.5
50	1	7.9	94	5.3	0.4	2.2	85	6.0	7.5	1.5
	4	11.8	94	5.0	0.7	3.0	89	5.9	3.3	1.9
100	1	10.7	93	6.0	0.9	3.1	84	6.9	6.2	2.4
	4	16.6	91	8.1	1.0	59	83	9.4	4.4	2.8
200	1	17.0	91	6.7	1.7	79	82	7.7	5.6	4.6
	4	36.3	87	11.3	1.3	144	79	13.0	4.6	3.6

Source: Behel et al.[146]

[a]A single application in 1971; repeat applications of initial rate in 1972, 1973, and 1974.

[b]Complexes not given represented <0.1% of total soluble metal.

[c]MM denotes mixture model for fulvate.

sludge.[42] The pH range of the sludge-treated soils was too narrow to evaluate the pH dependence of Zn^{+2}, as predicted by the solubility of known Zn solid phases.

Discussion of Findings

According to Behel et al.,[146] GEOCHEM calculations were close to monitored values and indicated that Zn^{+2} and Cd^{+2} are the predominant species in the soil solution of acid treated with sludges of metal additions. The amount approached the maximum recommended as safe for growth of agronomic crops.

Computer modeling of trace metal equilibria can provide a useful framework for understanding the nature of the multitude of reactions that take place in a soil system. However, a field monitoring program is needed to guide model application with data and calibration parameters.

Model output validation is essential to any soil modeling effort, although this term has a broad meaning in the literature. For the purpose of this section, we can define validation as "the process that analyzes the validity of final model output;" namely, the validity of the predicted pollutant concentrations or mass in the soil column (or in groundwater), as compared to available knowledge of measured pollutant concentrations from monitoring data (field sampling).

Disagreement in the absolute levels of concentration (predicted versus measured) does not necessarily indicate that either method of obtaining data (modeling, field sampling) is incorrect or that either data set should be revised. Laboratory analysis of field samples is difficult and uncertain. Field sampling approaches and modeling approaches produce two different perspectives of the same situation.

Field information, data, camping, and analysis are subject to temporal and spatial uncertainties. Model results are based on assumptions and utilize deterministic approaches; therefore, a disagreement of concentrations is to be expected.

This concludes presentation of a limited number of applications available in the literature.

CRITICAL PARAMETERS OF POLLUTANT MIGRATION

Introduction

A quantitative evaluation of critical parameters affecting dissolved organic and hydrocarbon contaminants fate through soils when conducting mathematical fate modeling is presented by Tucker et al.[152] This information is presented below verbatim from the reported publication.

Degradation, volatilization, advection, and diffusion are processes affecting contaminants in soils. Considering a unit mass of contaminated soil, these processes generally result in a decrease in the concentration of the contaminants from contaminated soils. Degradation processes may result in either a decreased hazard as the contaminants are changed into nontoxic chemicals or an increased hazard as the chemicals are changed into more toxic species.

Volatilization processes, if occurring near the soil surface, will result in a decrease in contamination of the soil-to-soil solution system. Although advection and diffusion (two processes of migration) result in a decrease in contamination of the unit volume as the chemicals migrate, the downgradient soil and water become contaminated. It is possible to determine the fate of the contaminants in the soil by determining the relative rates (half-lives) of the different processes. The shortest half-life is the controlling force. For example, if it takes 50 years for one-half of the contaminant concentration to decrease by migration-related processes, but only 6 months for one-half of the concentration to degrade, then degradation is the controlling process. Conversely, if the contaminants do not degrade or volatilize, then migration via subsurface water is the important process.

If migration is the important process, several critical parameters enhance or retard the rate of migration. Critical parameters include soil properties such as pH, organic carbon content (OC), cation exchange capacity (CEC), and clay content, which influence the retardation of contaminant migration by controlling the adsorption of contaminants to soil constituents. Other soil properties (e.g., porosity, permeability, and water content) influence the rate of water migration and therefore the rate of contaminant migration and are also critical parameters.

Adsorption of organic contaminants generally is found to be strongly correlated with OC. Primarily, the organic contaminants of concern have been pesticides, and the study sites have been farm soils (high in OC). There are, however, soils and sediments that do not contain high amounts of OC (e.g., nonagricultural soils, subsoils, geologic sediments, and aquifers). In these rocks, sediments, and soils, adsorption is influenced by characteristics other than OC. Other important characteristics affecting adsorption include CEC, pH, and clay content.

Land owned by various branches of the Armed Services has been found to be contaminated by organic chemicals used in the manufacture of pyrotechnics, explosive, and propellants (PEP). Usually, the soils do not contain high amounts of OC. The effects of CEC, pH, and clay are more important in controlling adsorption when OC is very low. The objectives of this study described in this paper were to identify the adsorption characteristics of some typical PEP compounds in predominantly low OC soils with a wide range in CEC, pH, and clay content.

Methods

Appropriate county soil surveys (where available) were used to identify the range of soil characteristics typical of Army installations around the country. The characteristics included OC, clay, CEC, and pH. When county soil surveys were not available, state geologic maps were used to identify geologic and geomorphic characteristics to determine possible types of parent materials. Soil surveys from adjacent counties were used to correlate soil types with the appropriate parent material—geomorphic relation found for the installations.

Seventy-one soil series representing 6 of the 10 soil orders (alfisols, aridisols, entisols, mollisols, vertisols, and ultisols) have been identified on the installations. Overall, mollisols and alfisols were predominant.

Values of OC, clay, CEC, and pH were tabulated for soils found on each installation, combined to find the total range of each variable, and separated into low, medium, and high categories. These categories were used as a basis for choosing soil samples to be used in the adsorption study. To minimize the number of possible combinations and to emphasize the effect of each parameter, only low (1) and high (2) categories were used. High values were OC content greater than 2%, clay content greater than 33%, CEC greater than 33 meq/100 g, and pH greater than 7.6. Low values were OC content less than 0.4%, clay content less than 10%, CEC less than 10 meq/100 g, and pH less than 5.2. The resulting number of combinations was 16.

The literature survey revealed an inadequate basis to estimate adsorption of organics to the inorganic fraction of soils. The highest priority was placed on identifying soils for all the combinations involving low OC content. Lower priority was placed on meeting all possible combinations of clay, CEC, and pH at high OC content.

Soil sample acquisition costs precluded sampling from a wide variety of locations. To obtain soil samples that exhibited a wide range in each of the potential controlling parameters, while minimizing soil sample acquisition costs, several horizons (layers) were sampled from a few soil types. Different horizons within a single soil profile typically exhibit characteristic variations in clay content and OC. Therefore, a single soil profile may have high OC and low clay content in one horizon and have low OC and high clay content in another horizon. This plan was designed to take maximum advantage of these characteristic variations in soil horizons.

Soil Collection Methods

Six soils were collected from four states: Florida, Iowa, Arizona, and West Virginia. Two or more samples were collected from four of the soils. The multiple samples were taken from different soil horizons within each soil. Three samples were taken from the Wauchula seres (a spodosol) in Florida, three from the Calcousta series (a mollisol) in Iowa, two from the Grabe series (an alkaline entisol) in Arizona, and two from the Lakin series (an entisol) in West Virginia. Two additional samples were taken from West Virginia, one from the Markland-McGary complex and one from the Chilo series (both alfisols).

Soil Characterization Methods

Soil samples were characterized for parameters, OC, soil reaction (pH), CEC, and particle size distribution (clay). The acid chromate digestion method number

6A1a was used for OC determination. Soil reaction was determined for a soil/water ratio of 1:1. Clay content was determined by a modified pipette method. CEC was calculated from the sum of extractable bases [calcium(Ca), magnesium (Mg), potassium (K), sodium (Na)] and extractable acidity. Extractable bases were replaced by leaching 25 g of soil with 250 mL of 1N NH_4OA_c buffered at pH 7.0. Calcium and Mg were determined by atomic absorption, K and Na by flame emission.

Laboratory Methods

Preliminary experiments were performed to determine optimum contact time required for equilibrium distribution of the explosive compounds between soil and solution, mass balance, and optimum soil mass/solution volume ratios for error minimization. A soil containing approximately 1% OC was used for these tests. If mass balances of 90% of better were achieved, only the initial and equilibrium soil concentrations would be calculated from the differences.

The OC partition coefficient (K_{oc}) for each compound was estimated from the adsorption coefficient calculated by the Henry's law equation:

$$Q_E = K_d C_E \tag{10}$$

where Q_E is the equilibrium soil concentration and C_E is the equilibrium liquid concentration, and

$$K_{oc} = \frac{K_d}{\text{percent OC}} \times 100 \tag{11}$$

for adsorption to the 1% OC soil. The 1% OC soil was a mixture of 22% of a 2.8% organic soil and 78% of an E-horizon spodosol with 0.5% OC, both obtained locally (Gainesville, FL).

The adsorption isotherm tests were performed with the soils listed in Figure 43.

K_d was then estimated for the experimental soils using the same equation and information about the OC. The estimated K_d for a soil-explosive compound combination was used to calculate the appropriate soil mass (M) to solution volume (V) ration using the equation

$$\frac{C_E}{C_O} = \frac{1}{1 + (M/V)K_d} \tag{12}$$

where C_E is the equilibrium solution concentration and C_O is the initial solution concentration. Parameters M and V were varied so that the ratio C_E/C_O when used in the equation

$$R = \left(1 - \frac{C_E}{C_O}\right) 100 \tag{13}$$

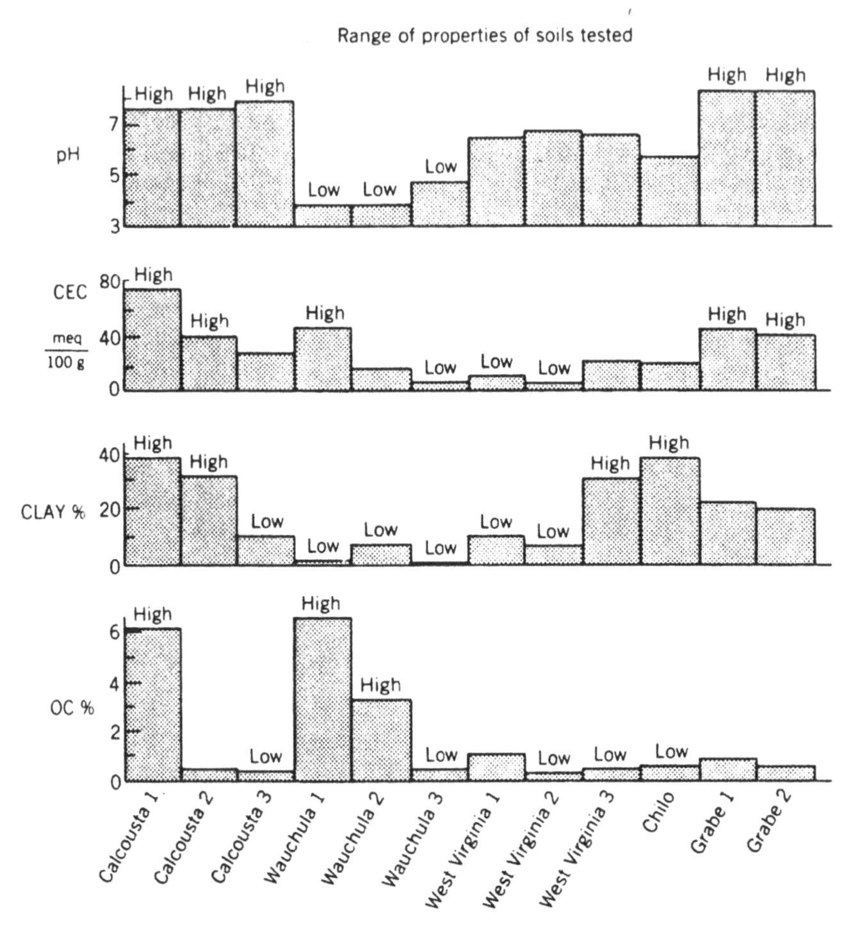

Figure 43. Soil characteristics (prepared for U.S. Army Toxic and Hazardous Materials Agency, Aberdeen Proving Grounds, Maryland).

where R is the percent removal, resulted in values between 20 and 80%. In cases where the soil had a low OC and the explosive compound had a low affinity for the soils, a practical limiting soil/solution ratio of 1 was used.

Data from the batch adsorption experiments consisted of data pairs of equilibrium liquid concentration (C_E) and equilibrium soil concentration (Q_E). A computer program was used to derive the isotherm parameters from the experimental data.

In each equation's fit, the parameters were refined by a standard, unweighted Newton-Raphson algorithm for a least-squares fit. Three adsorption isotherm equations were used to fit the data: Henry's law [Equation (1)], Langmuir

$$Q_E = \frac{(BQ_O)C_E}{1 + BC_E} \tag{14}$$

and Freundlich

$$Q_E = K_F C_E^{1/n} \tag{15}$$

where the variables have the following units:

Q_E = μg/g, equilibrium soil concentration;
C_E = μg/mL, equilibrium solution concentration;
K_d = mL/g, Henry's law constant;
BQ_O = mL/g;
K_F = (μg/g)/(μg/mL); and
n = dimensionless.

Reverse-phase high-performance liquid chromatography (HPLC) retention data on the bonded phase octadecylsilane was used to predict the octanol-water partition coefficients (K_{ow}'s) for trinitrotoluene (TNT) and cyclotrimethylenenitramine (RDX). Log K′ (capacity factor) is linearly correlated with log K_{ow}. The capacity factor was calculated from the equation

$$K' = \frac{t_r - t_0}{t_0} \tag{16}$$

where t_r is the retention time of the compound and t_0 is the retention time of the unretained species.

Aromatic and nitro compounds selected from an extensive tabulation of octanol-water partition coefficients were used to calculate a relation between the log K_{ow} and log K′. The resulting expression was used to calculate log K_{ow} values for TNT and RDX from retention data.

Retention data could not be used to calculated log K_{ow} of nitroguanidine (NGD) since NGD is ionic at all pH values below 10 (pK_a is 10.89). Consequently, log K_{ow} was determined by a flask-shaking method.[153] The partition coefficient was calculated from the equation

$$K = \frac{C_{oct}}{C_{H_2O}} \tag{17}$$

where C_{oct} is the equilibrium concentration in the octanol phase and C_{H_2O} is the equilibrium concentration in the water phase.

Log K_{ow} was determined at three pH values—3, 7, and 9—and in an unbuffered solution to test the dependence of the partition coefficient on pH and to eliminate any possibility of ion pair extraction of the guanidinium ion with the phosphate buffer.

Results

Soil Collection Results

Collected soil samples characteristics spanned a wide range (see Figure 43). The OC content ranged from 6.32%. (Wauchula 1) to 0.10% (West Virginia 2). The clay content ranged from 38.2% (Calcousta 1) to 0.2% (Wauchula 3). CEC ranged from 73.4% (Calcousta 1) to 0.67% (Wauchula 3). The pH ranged from 3.7 (Wauchula 1.2) to 8.8 (Grabe 1.2). Six of the 16 categories sought (a category is one combination of low or high values for the four soil characteristics) could not be filled by any of the collected samples. Three categories (two categories high in OC and clay but low in CEC and one category low in OC and clay and high in CEC) were identified by various soil scientists as not feasible in nature. CEC is directly related to OC and clay contents. A soil with low clay and low OC will generally have low CEC especially if pH is also low. Likewise, a soil with high clay and high OC will generally have a high CEC regardless of pH.

Approximately one-third of the individual extreme characteristics (low or high) could not be met. In fact, soils with extremes in one or two characteristics were easy to find, but soils with the desired combinations of all four characteristics were difficult to find.

Laboratory Results

The equilibrium time experiments indicated that equilibrium was reached in no more than 4 hr of contact time. Concentrations of TNT, RDX, and NGD in water decreased rapidly during the first hour, then leveled off until at least 8 hr. TNT at 4.7 and 0.61 g/mL levels decomposed at longer contact times. A 4-hr contact time was chosen for the rest of the adsorption experiments.

Mass balance totals were 90% or greater except for one test on TNT, which clearly indicated degradation effects after 8 hr. The RDX balances, however, exceeded 100%, probably as a result of an error in the analytic reference material or calibration intercept.

K_{ow} values for RDX and TNT were determined from retention data on HPLC. A linear regression of log K_{ow} versus log K′ for 10 compounds having similar structure yielded the equation

$$\log K_{ow} = 0.66(\log K') + 1.46$$

with r = 0.982. This equation yields log K_{ow} values of 1.1 and 1.9 for RDX and TNT, respectively. Possible errors in these values are estimated to be less than 0.3 log units. The log K_{ow} of NGD was determined in four sets of experiments at pH 3, 7, and 9 unbuffered. NGD is completely ionized at typical pH values for soils. K_{ow} was found to be independent of pH. The experiment with unbuffered solutions produced a slightly different K_{ow}, but the difference was attributed to experimental error. The mean log K_{ow} and standard deviation for all four experiments was −0.915 and 0.04, respectively.

Graphs of K_d versus OC (Figure 44) indicate the relative fit of the data to the curves and the difference in the effect of OC on adsorption of the three chemicals. The regression line fit the NGD data best (R^2 = 0.684) and RDX the poorest (R^2 = 0.293). Although the correlation coefficients are significantly different from zero, OC did not adequately predict the adsorption of the three compounds to soils. The poor fit can be seen readily in the amount of scatter around the three regression lines.

Considering that 7 of the 12 soil samples had OC contents of less than 1%, the poor correlation experienced in these OC regression equations was anticipated.

The stepwise procedure in SAS[155] was run with the measured variables OC, clay content, pH, CEC, and silt content and also with some combinations of measured variables. The combination of variables, NCEC, CEC, APH, and NOC (Table 27) were included to represent interaction terms, since the variables are not entirely independent of each other. CEC, for example, depends on OC, clay, silt, and pH. NCEC, identifying the interation between CEC, OC, clay, and silt, was included to give a term independent of the effects of varying OC and SC contents. The NCEC variable may be interpreted as an indicator of swelling clays such as montmorillonite. CEC tends to decrease with decreasing pH and to increase with increasing pH. The variable CECAPH was included to represent the actual effect of CEC sites at the in situ soil pH.

The third interaction term, NOC, was used to account for the combined effects of OC and clay on adsorption. This term amplifies the effect of clay and diminishes the effect of OC in low OC soils.

Table 27. Regression Equation Variables

OC Clay CEC	$NCEC = \frac{CEC}{(0.01 \times SC + OC)}$	SC = Silt = Clay
pH Silt	$CECAPH = \frac{(CEC \times 0.01)}{10^6(10^{-pH})}$	$NOC = \frac{OC}{Clay}$

Source: ESE.[154]

The first three equations in Table 28 were the statistically best equations (highest R^2). The KRDX and KNGD equations resulted in R^2 values of 0.99 and for

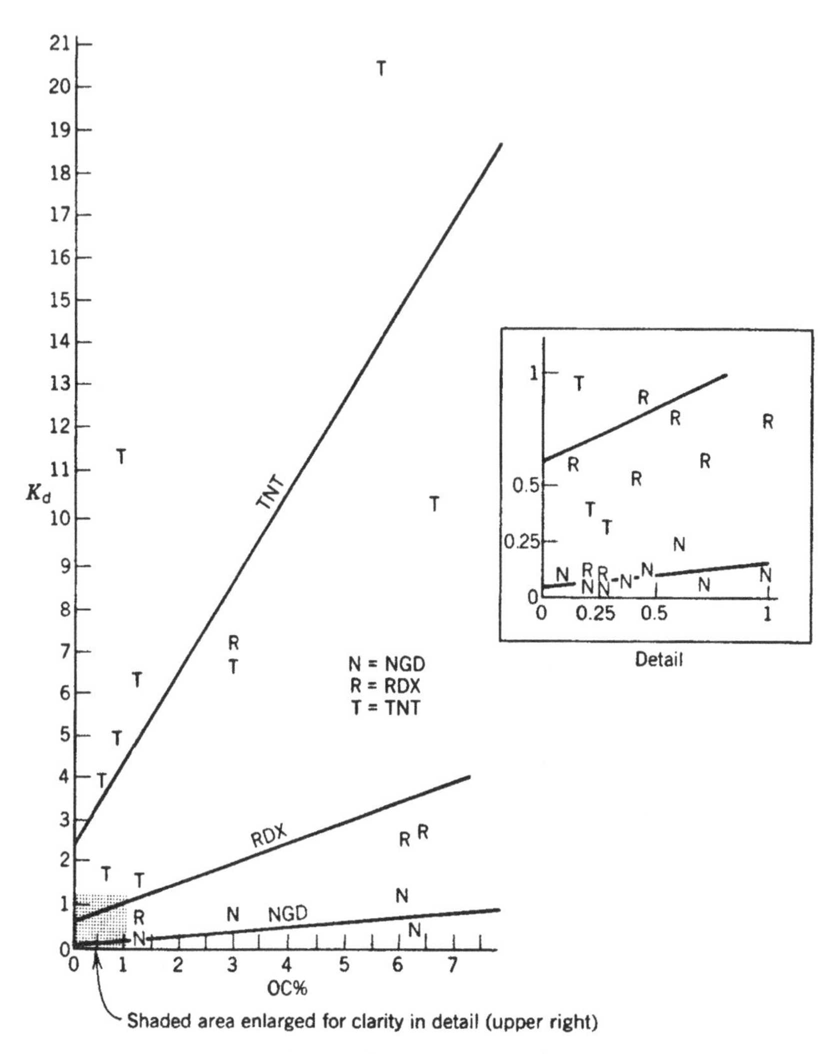

Figure 44. Dependence of PEP contaminant adsorption on soil organic carbon (from Environmental Science and Engineering, 1985).

the KTNT equation, 0.96. Even though these equations exhibited the largest R^2 values they were not considered the best equations. Some of the coefficients did not make sense based on physical chemical reasoning. The coefficients of the silt terms in the KNGD and KRDX equations were negative when a positive coefficient was more scientifically correct. The K values should increase with increasing silt content. The coefficient of CEC in the KRDX equation is also negative, contrary to scientific theory.

Table 28. Regression Equations

Equations Giving the Highest R^2

1. KRDX = 8.5 + 2.24(OC) = 0.058(Clay) − 0.146(CEC) − 1.72(pH) + 0.219(NCEC) + 0.02(CECAPH) − 1.15(NOC) − 0.013(Silt)
   R^2 = 0.99
2. KTNT = −0.606 + 2.22(OC) + 0.097(CEC) + 0.012(CECAPH) − 0.855(NOC)
   R^2 = 0.96
3. KNGD = 0.05 + 0.167(OC) + 0.005(Clay) − 0.09(NOC) − 0.03(Silt)
   R^2 = 0.99

Equations Considered Best

4. KRDX = 12.56 + 1.06(OC) − 2.37(pH) + 0.15(NCEC) + 0.015(CECAPH) − 1.15(NOC)
   R^2 = 0.92
5. KTNT = −0.606 + 2.22(OC) + 0.97(CEC) + 0.012(CECAPH) − 0.855(NOC)
   R^2 = 0.96
6. KNGD = −0.07 + 0.113(OC) + 0.007(Clay)
   R^2 = 0.78

Source: ESE.[154]

The last three equations in Table 27 were considered the best equations, even though they yield slightly smaller R^2 values. The KTNT equation remained the same, however. The NOC term in the KNGD equation was not used because of the small coefficient. The clay term in the KRDX equation was removed because of its small coefficient. The remaining terms represented the most significant and scientifically plausible of all terms considered.

Several insights can be gained by looking at the different steps in the stepwise procedure and also by considering the final output. The first term put into the regression equations is the most significant individual term of all the terms. OC was the first term put into the KRDX and KNGD equations, but CEC was the first term put into the KTNT equation. In the KTNT equation using only CEC, the R^2 was 0.78, higher than the 0.64 for the OC equation.

The coefficient and different terms in the "best" equations indicate the varying effects of the different soil characteristics on the prediction of K. There is a scientific basis for including the measured soil variables and the interaction terms NOC, NCEC, and CECAPH, in the regression equations. The combinations of terms in the best equations, however, may be subject to debate. The stepwise procedure was used because effects of the different combinations of terms could not be anticipated a priori. The stepwise procedure was used to determine the effects statistically, and the physical-chemical significance of the terms was then evaluated. Another test of the validity of the equations considers the effects of error between estimated K_d and measured K_d.

An example of the effect of error in estimating K_d can be seen in different calculated retardation factors for the measured-versus-estimated K_d for RDX in the

Wauchula 3 soil. The measured K_d is 0.07, whereas the calculated K_d is 1.03. Using the equation,

$$R = 1 + \frac{\rho}{n_w} K_d \tag{18}$$

where R is the retardation factor, ρ is the bulk density, and n_d is the volumetric water content.

The retardation factor (R) with K = 0.07 is 1.53, and the retardation factor using K = 1.03 is 8.73, a significant difference. Considering a 0.5-unit difference in retardation as significant, an error in K of greater than 0.07 for a sandy soil or 0.167 for a clayey soil will also be significant.

Discussion and Interpretation

The significance of various processes controlling contaminant migration may be compared by considering the contaminant half-lives in soil that result from the individual processes considered independently. Based on the reasonable assumption that each individual process is first order with respect to contaminant mass per unit surface area, the individual half-lives can be combined to yield the overall half-life:

$$\frac{1}{\gamma ½} = \frac{1}{\gamma ½,\ 1} = \frac{1}{\gamma ½,\ 2} + \ldots + \frac{1}{\gamma ½,\ n} \tag{19}$$

where $\gamma^{½}$, n is the half-life for the nth process considered. The half-life for any individual process is given by

$$\gamma ½ = \frac{0.7 \text{ Mass}}{\text{Loss rate}} \tag{20}$$

where Loss rate is the loss rate attributable to the process and Mass is the initial mass present.

Advecting/Inert Contaminant (e.g., Chloride)

The half-life of an advecting, nonadsorbing contaminant is given by

$$\gamma ½ = \frac{0.7 n_w d}{\rho} \tag{21}$$

where n_w is the water-filled porosity of soil, d is the depth of contaminant penetration, and ρ is the percolation rate of water through soil.

Advecting-Adsorbing Contaminant

The half-life for the advection-adsorption process is

$$\gamma ½ = \frac{0.7 n_w d R}{\rho} \tag{22}$$

where $R = 1 + \left(\frac{\rho}{n_w}\right) K_d/n_w K_d$ is the retardation factor.

Comparing Equations 13 and 14, we see that adsorption is a significant process if the retardation factor R is significantly greater than 1. However, if $K_d > 2n_w/\rho$, then $R > 3$, and adsorption is a critical process. The water-filled porosity of most soils in most climates will average approximately 0.2 in./yr, and soil bulk densities are usually approximately 1.5 g/cm^3. Thus, adsorption will be a critical process if $K_d > 0.3$.

Referring to the results of our adsorption experiments, $K > 0.3$ for TNT in all soils tested. Interpretation of the regression equations suggests that adsorption will be a critical process affecting TNT migration in virtually all soils, as long as CEC > 10 meq/100 g, or OC > 0.25%.

Referring to the results for RDX, $K_d > 0.3$ in all soils tested except Wauchula 3 (OC = 0.18%) and West Virginia 3 (OC = 0.23%). Of all soils tested, only West Virginia 2 (OC = 0.10%) had lower OC than Wauchula 3 and West Virginia 3. It is concluded that adsorption will be a critical process controlling migration of RDX in all soils having OC > 0.25%.

Referring to the results for NGD, $K_d > 0.3$ for Calcousta 1 (OC = 6.04%), Wauchula 1 (OC = 6.32%), and Wauchula 2 (OC = 3.10%). These soils had the highest OC of any soils tested. The regression results also support the conclusion that adsorption of NGD is a critical process only when OC > 3%.

Volatilization

The half-life for volatilization is derived from Equation 12. The total mass of contaminant per unit surface area is

$$\text{Mass} - \rho C_s' d \tag{23}$$

where C_s' is the concentration of contaminant in the bulk soil expressed as g/g dry weight. The loss rate is approximately

$$\text{Loss rate} = \frac{2 D_{ep} C_s'}{d} \tag{24}$$

where D_e is the effective soil pore diffusion coefficient. Based on Jurry et al.[193],

$$D_e = \frac{D_a n_a^{10/3} H + D_w n_w^{10/3}}{n^2(\rho K_d + n_w + n_a H)} \tag{25}$$

where n_a is the air-filled porosity; n is the total porosity ($n = n_a + n_w$); D_a and D_w are the contaminant's molecular diffusion coefficients in air and water, respectively; and H is the contaminant's Henry's law constant expressed in dimensionless units of concentration. Thus, the half-life for volatilization is given by

$$\gamma ½ = \frac{0.35 d^2 n^2(\rho K_d + n_w + n_a H)}{D_a n_a^{13/3} H + D_w n_w^{13/3}} \tag{26}$$

Since $D_a \approx 10^4 D_w$, the volatilization half-life will be small (rapid volatilization) in arid soils ($n_a > n_w$) for contaminants with a large Henry's law constant ($H >> 10^{-4}$). Volatilization is also favored as K_d is small.

The range of long-term average water-filled porosities for a variety of soil textures in a range of climates was calculated by methods presented by Bonazountas and Wagner[156] and are presented in Table 29, $n_a > n_w$ for coarse soils.

Table 29. Range of Soil Moistures as a Function of Soil Texture

Soil Medium	n_w	n_a	Examples
Clay	0.58–0.66	8	
Silty clay	0.42–0.48	14	
Silty clay loam	0.39–0.45	17	Calcousta 1, WVA 3, Chilo 1
Clay loam	0.32–0.38	23	
Loam	0.25–0.32	24	
Silt loam	0.22–0.30	28	Grabe 1, Grabe 2
Silt	0.33–0.41	12	
Sandy clay	0.22–0.27	19	
Sandy clay loam	0.20–0.22	24	Calcousta 2
Sandy loam	0.21–0.26	20	Calcousta 3, WVA 1
Loamy sand	0.05–0.08	26	
Sand	0.05–0.07	29	Wauchula (1, 2, and 3) WVA 2

Source: ESE.[154]

Based on an extensive series of calculations using Equation 18 and comparing the volatilization half-life with the advection-adsorption half-life, it is concluded that volatilization is a critical process for PEP contaminants under the following circumstances.

Contaminant	Volatilization Critical if:
2,4,6-TNT	P < 0.1 in./year, clay < 30%, wilt < 50%
1,3,5-TNB	P < 0.1 in./year in clay loams, loams, silt loams, and all soils having sand > 45%
2,6-DNT	P < 1 in./year, clay < 60%, or P < 10 in./year, sand > 70%
2,4-DNT	P < 1 in./year, silt and clay < 60%.

Volatilization is not of importance for RDX, HMX, and NGD because of their low Henry's law constants.

The equations presented in this section are generally applicable to a wide range of organic contaminants. The application of these equations has been performed for the PEP contaminants in accordance with the scope of this research effort.

Critical Parameters Summary

Generic. The following parameters have been found to be critical on a generic basis regardless of soil-climate-contaminant scenario: (1) soil texture, (2) percolation rate, and (3) depth of contaminant penetration.

The soil texture is simply the particle size distribution; that is, percent sand, silt, and clay. The percolation rate is the average rate of water recharge through soils to the water table. The appropriate averaging time for the percolation rate is the contaminant half-life and is often 1 year or more. The average soil moisture content can be estimated from soil texture and percolation rate using methods presented by Bonazountas and Wagner.[156] They also present values of several parameters that are related to texture and could be considered critical. These are the porosity, effective porosity, field capacity (porosity minus effective porosity), and pore size disconnectedness index. By definition, certain relations between these parameters may be used to calculate one from the other, depending on the data available. Site-specific information would always be preferred over estimates based on texture. A variety of computer models are available to calculate the percolation rate.

In evaluating site-specific contaminant migration issues, it will be necessary to evaluate a variety of other parameters. Depending on the values of these parameters, they may or may not be critical to that situation. These are listed in order of descending relevance:

1. Soil adsorption coefficient.
2. Soil organic carbon context.
3. Soil cation exchange capacity.

4. Soil pH.
5. Depth to water table.
6. Contaminant degradation rate in soil.
7. Contaminant octanol-water partition coefficient.
8. Contaminant Henry's law constant.
9. Contaminant vapor pressure.
10. Contaminant solubility.
11. Contaminant molecular diffusion coefficients in air and water.
12. Contaminant molecular weight and structure.

A variety of relations are available in the literature, many of which have been presented in this report, to estimate one parameter from another. Ultimately, estimates of the adsorption coefficient of the contaminant to the site soils, the contaminant Henry's law constant, degradation rate, and depth to water are required, but the other parameters listed may be used to estimate these, using equations presented here as well as elsewhere.

Site-Specific Critical Parameters. When the adsorption coefficient is very low (0.1 or less), the Henry's law coefficient is low (10^{-4} or less, dimensionless by volume) and the percolation rate exceeds 1 in./year, contaminants move with the water, essentially as inert or conservative constituents.

When the adsorption coefficient exceeds 0.6, adsorption will be a critical process controlling contaminant migration. Considering the contaminants tested experimentally, the following parameters would then be critical:

- TNT and RDX: organic carbon content, CEC, pH
- NGD: organic carbon content

To estimate K_d based on these parameters, one could use the regression relation derived, although considerable caution is advised since their predictive skill has not been tested, or one could compare the properties of site soils with the characteristics of soils tested here and select an appropriate value of K_d based on the experimental results.

When the adsorption coefficient ranges from 0.1 to 0.6, adsorption may or may not be a critical process depending on soil-climate characteristics. Diffusion may be a critical process, especially when the percolation rate is less than 1 in./year and in fine-grained soils. Vapor phase diffusion and the Henry's law constant are critical for TNT and TNB in arid climates ($P < 0.1$ in./year) in soils with less than 30% clay. Vapor phase diffusion of the dinitrotoluenes (DNTs) is important in soils having less than 50% clay where the percolation rate is less than 1 in./year. Vapor phase diffusion may be critical for DNTs at up to 10 in./year percolation in loamy sand and sandy soils.

The importance of vapor phase diffusion is unknown but presumed insignificant for RDX, HMX, and NGD.

MODELING IMPACTS TO OTHER MEDIA

Air Modeling

General

This section describes principles of air quality modeling and summarizes well-documented or Government-approved computerized models as they apply to hydrocarbons fate from waste sites. It is intended to show the complexity and diversity of air modeling.

Pollutants resulting from the use of the dominant energy sources, such as petroleum and petroleum products, often cause a deterioration of the air quality over large or small areas. Most of the pollutants related to petroleum production, processing, and use, have intrinsic toxic or irritative potentials of a rather low order. By contrast, their photochemical reaction-products may affect biological systems even at extremely low concentrations.

Air pollution owes its characteristic intricacy to the large number of sources and their diversity and to the type of pollutant; composition phase; probability of being a precursor of another pollutant; emission rate; status of emission such as heat, volume and momentum rates; height of release; relative proximity to structures; source geometry; motion; and other.

Background

Air quality models analyze the transport and transformation of air pollutants as they migrate from a source and a receptor. These models accept as input: (1) wind flow patterns and turbulence in space and time as functions of topography, urban influences, surface roughness, and weather patterns; (2) resolution of the emissions in space and time; (3) impacts and advection of pollutants from other regions into the region in question; (4) physical-chemical and photochemical reactions of pollutants as they are transported and diffused; (5) surface removal rates and the potential for scavenging of atmospheric pollutants; (6) data from databases sufficient to validated model performance.[157]

The air models produce, as output, estimates of ambient air concentrations and material deposited on surfaces. The air pathway processes that control the fate of these pollutants are transport, diffusion, transformation, and removal. The first of these processes determines where pollutants will be found, and the remaining processes determine their concentration and chemical form. Parameters to be next considered are the numerous atmospheric physical-chemical processes as influenced by topographic factors including relief, roughness, and thermal properties of the surface.

Air pollution pathway processes are significantly influenced by many meteorological and topographical parameters. The transport processes on all scales are governed by wind speed and direction, while diffusion processes are governed by the temperature structure of the atmosphere, wind shear, surface roughness, and terrain features. Physical and chemical transformation processes are influenced by molecular and turbulent diffusion, the presence of water vapor and clouds, and the characteristics of solar radiation. Finally, removal processes are dependent on the nature and extent of precipitating systems (rain or snow), fog layers, turbulence, and surface characteristics.[157]

Pathway Processes

The four significant pathway processes shown in Figure 45 are transport, diffusion, transformation, and removal.

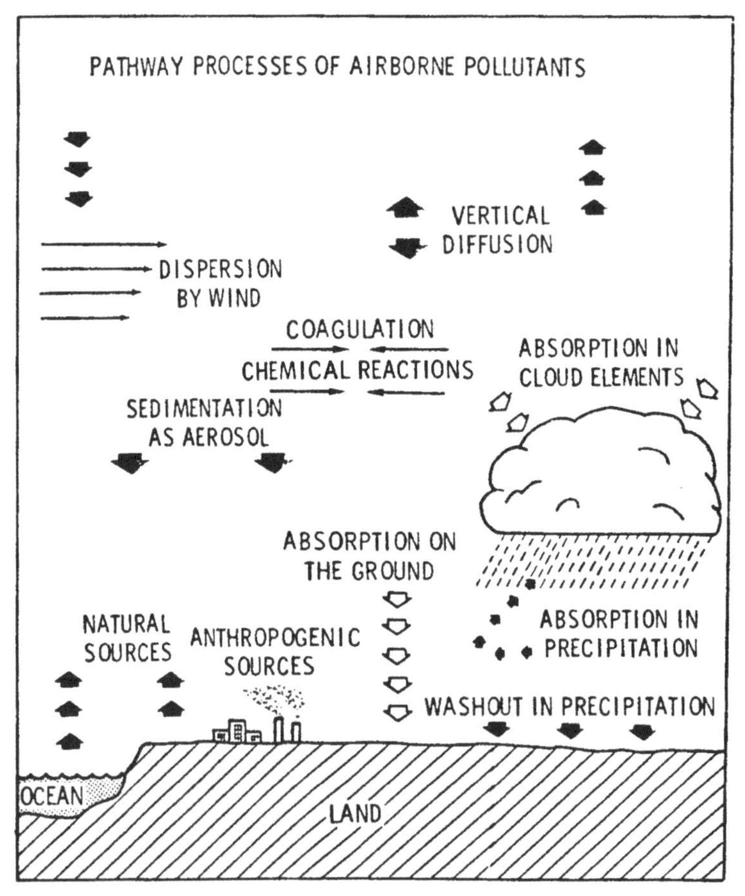

Figure 45. A schematic of the pathway processes of airborne pollutants (from EPA[174]).

Transport is the *movement* of pollutants from a source by ambient winds. Transport determines *where and when* the center of mass of the material will be found. *Diffusion* is the *spread* of pollutants by turbulent mixing. Diffusion always decreases the concentration of the pollutants during transit from the source to the receptor, but the rate of diffusion may vary drastically. *Transformation* is the *change* in physical and chemical characteristics of airborne pollutants. Types of changes involving gaseous materials are reactions producing other gases and gas-to-particle conversions. *Removal* is the reduction of mass of airborne pollutants by deposition. Gases are adsorbed by precipitation elements that ultimately fall to the ground. Gases are also absorbed by vegetation, bodies of water, buildings, and other structural surfaces. Because gases and particles are removed from the air by both precipitation and dry impaction or absorption, the removal processes are classified as *dry* and *wet*. The processes affecting dry deposition can be divided into three regimes as shown in Figure 46.[157] Box 1, in Figure 46 is the atmosphere regime wherein the airborne pollutant is transported by the normal wind and eddy motions. Box 2 is the usually thin layer just above the vegetative canopy or surface elements. Box 3 is the surface receptor interface.

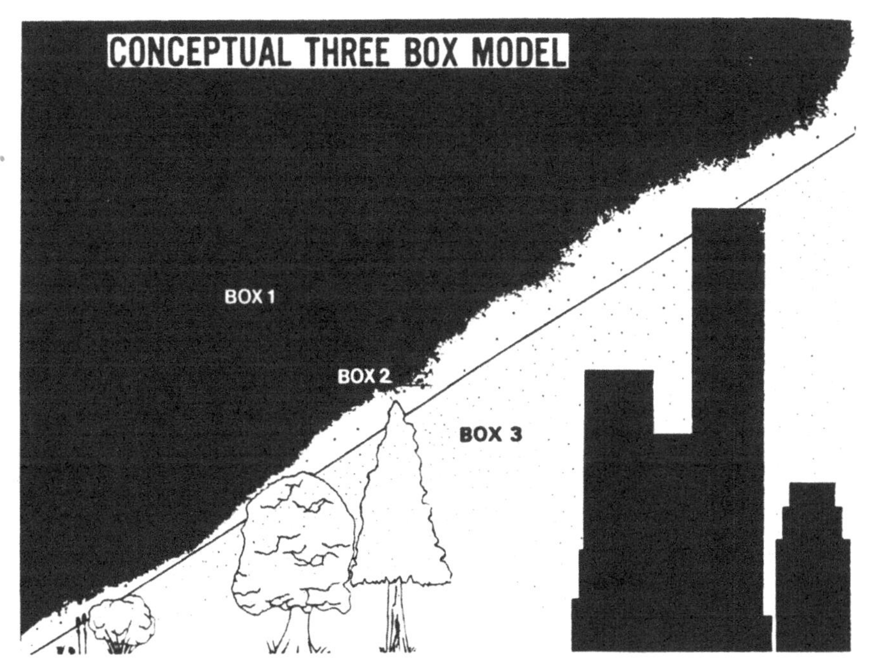

Figure 46. The three box, or regime, concept of dry removal (from Sehmel and Hodgson[190]).

Classification of Models

General. Most models can simulate the consequences of all various source parameters, the occurrence of various weather conditions or changes in receptor characteristics, and air concentrations of pollutants in time and space, as well as the amount of material deposited on the earth's surface by wet and dry deposition processes. The time and space resolution of these quantities depends on the characteristics of the model and the way it is applied. The selection of a model for a given application must consider the following:[158]

- definition of emissions with regard to the space and time in question
- wind flow patterns as a function of location, topography, urban influences, surface roughness and large-scale weather patterns
- physical processes and chemical reactions of pollutants as they are transported and dispersed in the atmosphere
- transport of pollutants from other regions that create background levels for the region in question
- use of databases that are sufficient to calibrate a model and validate it for predictive purposes.

Mathematical models currently used in the air pollution field range from simple empirical models to very complex meteorochemical models. The empirical models are based on a deterministic analysis of air quality, emission, and meteorological data. These models consist of a shallow box with base at the ground top at an assumed upper limit of vertical mixing h(L) and sides positioned to enclose the city as an area source. Also, empirical approaches allow a simultaneous check on data quality through standard statistical tests. Finally, empirical models can usually be formulated and operated at low cost. However, depending on the situation, disadvantages may occur in simulations when using empirical modeling.

Meteorochemical models are derived from the basic physical and chemical principles relating to the processes of transport, diffusion, transformation, and removal. Meteorochemical models are formulated from basic scientific concepts associated with physical and chemical processes occurring in the atmosphere. This formulation provides confidence for various applications and over various ranges of conditions. These models possess computational complexities and require extensive data input and specifications of numerous model parameters.

Empirical and meteorochemical models are usually partitioned according to the model's tendency to emphasize data or physicochemical principles. The distinction between empirical and meteorochemical models is arbitrary. For example, empirical models incorporate varying degrees of physical insight, such as accounting for the transport and the spatial distribution of emissions in the source-receptor relationships. Conversely, meteorochemical models rely on empirically determined parameters such as transformation rates, removal rate constants, and coagulation coefficients. Thus, a family of models exists, ranging from simple rollback models to highly sophisticated interacting aerosol models.

Semi-empirical models, an intermediate class, share the simplicity of empirical models but approach the predictive ability of complex, time-dependent models. In fact, the models commonly categorized as semi-empirical are the most widely used at present. These models contain a mix of empirical and meteorochemical characteristics. Semi-empirical models provide sufficient flexibility to address exposure to inorganic pollutants in the atmosphere at exposure locations from hydrocarbons fate.

Some models that differ with source emission type, geographic extent, and averaging time are included in the list of approved models of the ERA Users Network of Applied Models for Air Pollution (UNAMAP). Some of the important modeling characteristics are time and space scales, the frame of reference, the pollutants and the treatment of multisources, turbulence, topography and model uncertainty.[158]

Using these characteristics as a basis of comparison, the three classes of models are further divided into subclasses. Empirical models are subdivided into rollback and statistical models; meteorochemical into grid, particle and global models; and semi-empirical into Gaussian plumes and puffs, box and multibox, and regional trajectory models. Physical or laboratory models are classed as empirical models since their output is measured data that are usually presented in some statitistical format. Additionally, some of the sophisticated multibox models are classified as meteorochemical rather than semi-empirical.

Table 30 shows the characteristics of model subclasses: rollback, statistical, Gaussian, box and multibox, finite differences, particle, physical, regional and impact. A summary of their characteristics follows.

Rollback Models. Rollback models (Clean Air Act, 1970) are the simplest in air quality modeling. They are based on the assumption that the local concentration of a pollutant above its background level is directly proportional to the emission from a given source.

Although these models have the advantage of being very easy to apply, they are not applicable to reactive pollutants. These models in their simplest form are proportionally relations between emission changes and resultant concentrations, and require a minimum of input data and computations.

Statistical Models. Statistical models have been developed only recently and can establish close relationships between concentration estimates and values actually measured under similar circumstances. These models have low development cost and resource requirements. A statistical model is not applicable beyond the range of conditions included in the data used in its development and optimization. The range of conditions commonly includes variations in meteorological variables, although little variation in spatial distribution of emissions can be

Table 30. Summary Table of Characteristics of Model Subclasses

Model Subclass Mechanisms	Model Class	Geographical Subdivisions	Steady State or Time Dependent	Frame of Reference	Type of Pollutants	Reaction
Rollback	Empirical	Local, regional, national	Steady state	Eulerian	Gases & particles	Nonreactive
Statistical	Empirical	Local, regional	Steady state, time dependent	Eulerian	Gases & particles	Nonreactive, reactive, gas-to-particle
Gaussian plume and puff	Semi-empirical	Local	Steady state, time dependent	Eulerian, Lagrangian	Gases & particles	Nonreactive, reactive
Regional trajectory	Semi-empirical	Regional, national	Time dependent	Lagrangian, mixed, Lagrangian & Eulerian	Gases & particles	Nonreactive, reactive
Box and multibox	Semi-empirical and meterochemical	Local, regional,	Steady state, time dependent	Eulerian, Lagrangian	Gases & particles	Nonreactive, reactive, gas-to-particle
Grid	Meteorochemical	Local, regional	Steady state, time dependent	Eulerian	Gases & particles	Nonreactive, reactive, gas-to-particle
Particle	Meteorochemical	Local, regional	Time dependent	Mixed, Lagrangian & Eulerian	Gases & particles	Nonreactive, reactive, gas-to-particle
Global	Meteorochemical	Global	Time dependent	Eulerian	Gases & particles	Nonreactive, reactive

continued

Table 30. *Continued*

	Treatment of Turbulence	Treatment Plume Additivity	Treatment of Topography	Treatment of Model Uncertainty
Physical	Empirical	Local	Time dependent	Mixed, Eulerian & Lagrangian
Rollback	Well-mixed	Not applicable	Homogeneous to simple terrain	Deterministic
Statistical	Well-mixed	Not applicable	Homogeneous to simple terrain	Stochastic, adaptive
Gaussian plume and puff	Diffusion coefficients	Yes & No	Homogeneous to complex terrain	Deterministic
Regional trajectory	Diffusion coefficients eddy diffusivities	Yes	Non-homogeneous to complex terrain	Deterministic, stochastic
Box and multibox	Well-mixed eddy diffusivities	Yes & No	Homogeneous to simple terrain	Deterministic
Grid	Eddy diffusivities complex formulation	Yes & No	Homogeneous to complex terrain	Deterministic
Particle	Eddy diffusivities	Yes & No	Homogeneous to complex terrain	Deterministic, stochastic
Global	Eddy diffusivities	Yes	Non-homogeneous to complex terrain	Deterministic
Physical	Non applicable	Not applicable	Homogeneous to complex terrain	Deterministic

investigated. Consequently, statistical models are not generally suited for applications involving significant changes in emissions distribution and, as a result, are not transferable to other areas without re-evaluating their specific empirical parameters or coefficients.[157]

Gaussian Plume Models. The most used and most reliable atmospheric models involve the Gaussian plume concept as shown in Figure 47, in which emissions and transport are assumed to be steady-state or pseudo steady-state. Gaussian plume

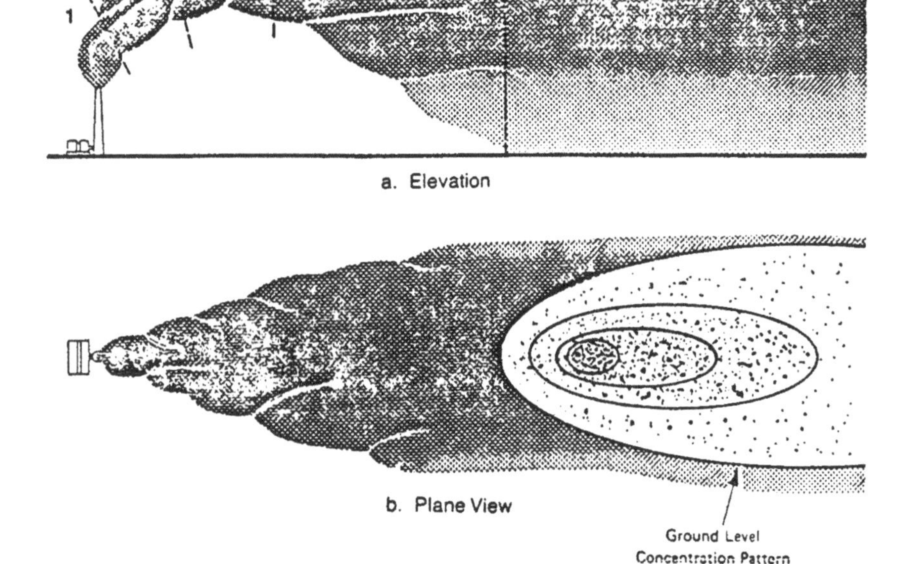

Figure 47. Typical plume behavior and ground-level concentrations in a neutrally stratified atmosphere (from Eschenroeder et al.[159]).

and puff models, which have their mathematical origins in the turbulence and diffusion theory, are the oldest air quality models, except for the simple box models. The Gaussian models have the following properties:

- Concentrations from a continuous steady source vary in direct proportion with the source strength or rate of emission.
- Concentrations vary inversely with the mean wing speed at the source.
- Distance from the source to the receptor and intensity of atmospheric turbulence determine the height and width of a plume at the receptor.

- Concentrations may decrease because of chemical reactions, radioactive decay, and removal.
- Maximum practical time scale is 1–2 hr; maximum space scales, 10–30 km.
- Extra provisions are made to account for calm winds, complex terrain, and low inversion levels.

The most common equation for ground-level concentrations (Figure 47) from an elevated point source is:[157]

$$C(x,y,0;\ H) = A \exp[-(y/\sqrt{2}\sigma_y)^r - (H/\sqrt{2}\sigma_z)^s]$$

The values of the parameters A, S, r are given below:

Bosanquet and Pearson model:

$r = 2$, $s = 1$, $A = Q/(\sqrt{2}\pi\ pq\ x^2U)$, $\sigma_y = qx$, $\sigma_z = \sqrt{2}\ px$

Sutton model

$r = 2$, $s = 2$, $A = 2Q/(C_yC_zx^{2-n}U\pi)$, $\sigma_y = 1/\sqrt{2}\ C_yx^{(2-n)/2}$, $\sigma_z = 1/\sqrt{2}\ C_zx^{(2-n)/2}$

Calder model

$r = 1$, $s = 1$, $A = QU/(2\kappa^2 a\ U_*^2x^2)$, $\sigma_y = (\sqrt{2}\ a_\kappa\ U_*\ x)/U$, $\sigma_z = (\sqrt{2}\ \kappa\ U_* x)/U$

Pasquill model

$r = 2$, $s = 2$, $A = Q/(\pi\sigma_y\sigma_zU)$, σ_y, σ_z = from diffusion data.

where:

$C(x,y,0, H)$ is the ground level (the xy-plane) concentration of pollutant emitted from an elevated point source with effective height $H = h_s + \Delta h$, where: h_s = stack height, Δh = plume rise due to exit velocity and buoyancy.

$U_* = \kappa U(z)\log(z/z_0)$ = friction velocity, where: κ = von Karman's constant, $U(z)$ = mean wind profile, U = mean horizontal wind speed, Z_0 = roughness height

z = characteristic height of the surface layer (vertical height)
q = empirical coefficient
p = numerical coefficient in the eddy diffusivity $K_z = pUz$
a = ratio of lateral velocity fluctuations to the vertical ones

Q = source strength (emission release rate)
σ_y, σ_z = standard deviations of the cloud distribution in the y and z directions, respectively, and C_y, C_z, n = parameters in Sutton's formulation (atmospheric concentration).

A number of different type pollutant sources may have to be analyzed when atmospheric pollutant concentrations and human exposure level are calculated. A model can be selected on the basis of the type of source and receptor to be considered. Generally, the calculation of exposure levels must consider

- point
- area
- line sources.

Box and Multibox Models. The box, column, moving-cell, multibox and integral models are all based on the integral form of the diffusion equation over a two- or three-dimensional volume. This volume may be the region over an urban area, the volume in a deep valley or basin, or a subvolume of one of these regions. The air and pollutants in each box are assumed to be well-mixed, although reaction and removal processes are allowed within the box. Box models assume that pollutants are uniformly mixed throughout a fixed volume (box) of air. The box is taken to extend vertically to the inversion base. Concentrations are then presumed to be proportional to the rates of source emission and inversely proportional to the average residence time and the inversion height.

Multibox models give better resolution in time and space than single boxes. In multibox models, interactions between boxes at the adjacent boundaries are postulated. Box models do not include convergence and divergence of winds; thus, no vertical motion of air. In addition, the concept of an identifiable parcel of air is oversimplified, because such a box cannot exist in a turbulent atmosphere over time scales of interest. Furthermore, large variations can occur in the magnitudes of the flows between cells and, as a result, the mathematical systems may be difficult to solve efficiently.

Finite Difference or Grid Models. Grid models are based on solving the transport and diffusion equations by finite difference approximations. A large body of knowledge exists concerning conventional methods for solving these equations. These models produce urban and regional concentration patterns over an entire grid network. Grid models are generally used to calculate short-term concentrations of reactive pollutants and to calculate concentrations during episodal conditions in urban areas, deep valleys, and mountains. Therefore, the meteorological portion of these complex models is significant. Although grid models are probably the best models to account for such complex phenomena, these models require increased computer time and storage. Other problems with grid models include imprecision in specifying meteorological phenomena and inaccuracies in the chemical components of the model. However, grid models are relatively new and their future appears promising.

Particle Models. Particle models are mixed Lagrangian-Eulerian models, which follow a pollutant passing through an Eulerian grid. In this method, spatial distribution of the pollutant is represented by a large number of Lagrangian particles of constant mass that are advected in a fictitious velocity field consisting of the true velocity field plus a turbulent flux velocity field.[160] The fixed Eulerian grid divides physical space into cells and the particles carry pollution from cell to cell as they are moved by the fictitious velocity field. This field causes the particles to move apart or together, resulting in uneven distributions. In order to satisfactorily simulate spatial distribution of pollution, a large number of particles must be used in each grid cell. The particles are initially placed at random within each cell and the number of particles in a cell depends on the initial concentration specified for that cell. The particle techniques are: Maker-and-Cell (MAC), Particle-in-Cell (PIC), Particle-in-Cell, K-theory (PICK), Atmospheric Diffusion Particle-in-Cell (ADPIC), other particle models.

Global Models. The Global Circulation models (GCM) are large and complex models and require fast, scientific computers with large storage capacities. Using the current GCMs for climate studies presents some computational problems. These models are not relevant to hydrocarbon sites and therefore are not further discussed.

Physical Models. Physical modeling techniques are the wind tunnel, liquid flume, and towing tank, whose greatest asset is the ability to investigate dispersions for configurations too complicated to be economically simulated by mathematical modeling.[157] No examples of hydrocarbon fate modeling via physical modeling are known. Therefore, no further analysis is presented.

Regional Models. Regional models usually pertain to horizontal scales of hundreds of km to 2000 km. The types of simulations being considered for the analysis and assessment of regional problems are the grid, particle, and trajectory models. However, grid models are presently in their early development stages and particle models may be uneconomical. They are not relevant to hydrocarbons' fate from waste sites.

Impacts Models. Knowledge of relationships among such factors as meteorology, transport and diffusion, population exposure, and resulting health effects may be captured in quantitative models. These models describe what is known about the relationship between decision alternatives, state variables, and outcomes. Transformation and removal mechanisms in the atmosphere must be considered for many pollutants to appropriately characterize exposure levels. Depletion mechanisms in the atmosphere include photochemical reactions, wet deposition, dry deposition, and gravitational setting. The causal chain between anthropogenic emissions and consequences is divided into four links, each of which must be quantitatively assessed in order to relate control decisions to consequences. These links are modeled by Bonazountas et al.[158]

Models Review

General. The principal features of some atmospheric dispersion models are outlined in Figure 48. Keddie[162] provided a useful guide to atmospheric dispersal of pollutants and the modeling of air pollution, subject of atmospheric dispersion modeling and long-range transport to air pollutants. Eschenroeder[159] summarized some model intercomparisons, including those between photochemical models, and Doury[163] discussed the limitations to simple mathematical formulae used in atmospheric dispersion studies. Meteorological aspects of global air pollution were reviewed by Munn and Bolin.[164] Ellis[165] has considered acid rain control strategies in the context of some additional long-range transport models.[158]

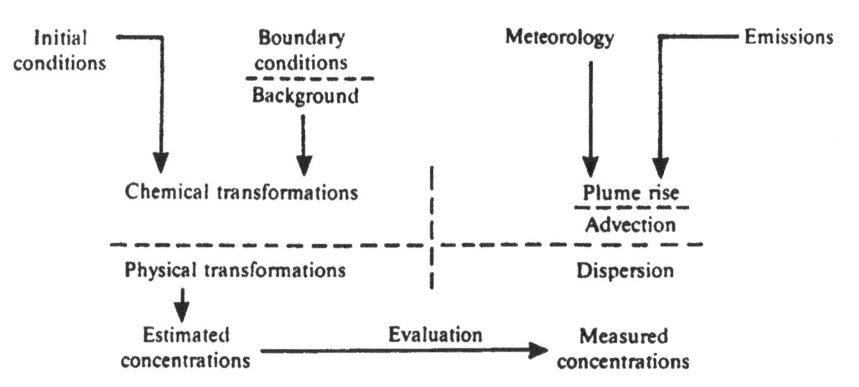

Figure 48. Main elements of atmospheric dispersion models (from Turner[161]).

The models contained in the UNAMAP Series address the wide variety of the problems identified above and have some of the analytical capabilities required to simulate pollutant transport phenomena for assessing human exposure levels. Typical minimum data required for using the models are: wide direction, wind speed, atmospheric stability, atmospheric mixing height data, ambient temperature, precipitation, stack characteristics, sources and receptors, terrain data, and pollutant data.

Each model can be modified to fulfill a specific, limited use. The differences between the models involve such terms as the spreading of plumes as a function of atmospheric stability vertical mixing plume rise source configuration, terrain considerations, acute or chronic exposure. Most of the models are quite different with regard to specific capabilities. The U.S. EPA divides the air quality models into four generic classes:

- Gaussian
- numerical
- statistical or empirical
- physical

Gaussian models are the most widely used, and are recommended for estimating the impact of noncreative pollutants. Numerical models (i.e., grid models or box models) are suggested for urban applications involving reactive pollutants (photochemical smog). Other models can be used for particular applications.

Moreover, the models are categorized by two levels of sophistication:

- Screening techniques (or screening models): they are relatively simple estimation techniques that provide conservative estimates of air quality impacts. They can, in several cases, eliminate from further consideration those sources that clearly do not contribute to ambient concentrations.
- Refined models: they provide a more detailed treatment of physical and chemical processes, require more detailed and precise input data, have higher computational costs, and provide (at least theoretically) a more accurate estimate of the source impact and the effectiveness of different control strategies.

The U.S. EPA also divides the air quality models recommended in its guideline into "preferred" and "alternative." Preferred models are those that the EPA either found to perform better than others in a given category, or chose on the basis of other factors such as fast use, public familiarity, cost or resource requirements, and availability. The preferred models can be used for regulation applications without a formal demonstration of applicability, as long as they are used as indicated by the U.S. EPA.[166]

Alternative models can be used when: (1) a demonstration can be made that the model produces concentration estimates equivalent to the estimates obtained using a preferred model; (2) a statistical performance evaluation has been conducted using measured air quality data and the results of that evaluation indicate the alternative model performs better for the application than a comparable preferred model; (3) there is no preferred model for the specific application but a refined model is needed to satisfy regulatory requirements. Selected Air Quality Models and the characteristics of them are shown on Table 31.[158]

EPA-Preferred Models. The U.S. EPA-preferred air quality models are:[166]

- Buoyant Line and Point Source Dispersion Model (BLP)
- CALINE 3
- Climatological Dispersion Model (CDM 2.0)

Table 31. Features of Selected Air Quality Models

Model Function	APRAC-1	CDM	CRSTER	HIWAY-2	ISC	MPTER
Point source		*	*		*	*
Area source	*	*			*	
Line source	*			*		
Volume source					*	
Short-term	*		*	*	*	*
Long-term		*	*		*	
Plume trapping	*	*	*	*	*	*
Half-life		*			*	*
Gravitational settling					*	*[c]
Aerodynamic effects		*		*	*	
Buoyant plume rise		*	*		*	*
Momentum plume rise					*	*
Stack downwash					*	*
Turbulent deposition						*[c]
Terrain adjustment			*		*	*
Complex terrain						*
Calibration		*				
Urban/rural disp.						
Coefficient[a]	U	U	*	R	*	*
Background calculated	*					
Street canyon flows	*					
Wind speed extrapolation		*	*		*	*
Highway generated						
Turbulence				*	*	
Emission rates function of met. conditions					*	
Emission rate function of time of day	*				*	*
Buoyancy induced						*
Dispersion						
σ_y Function of averaging						
Time and stability						
Inversion penetration						
Factors						
Particle resuspension						
Washout						
No. of sources (max.)[d]	1200	200P 2500A	19	24	100[b]	250
No. of receptors (max.)	625	unlim.	180	50	400[b]	180

continued

Table 31. *Continued*

Model Function	PAL	PTMAX PTDIS PTMTP PTPLU	RAM	TCM	TEM	VALLEY	COMPLEX I,II
Point source	*	*	*	*	*	*	*
Area source	*		*	*	*	*	
Line source	*						
Volume source							
Short-term	*	*	*		*	*	*
Long-term			*	*		*	*
Plume trapping	*	*	*		*	*	
Half-life			*	*	*	*	*
Gravitational settling	*[c]						
Aerodynamic effects	*						
Buoyant plume rise	*	*	*	*	*	*	
Momentum plume rise			*	*	*		
Stack downwash		*	*		*		
Turbulent deposition	*[c]						
Terrain adjustment						*	*
Complex terrain						*	*
Calibration				*			
Urban/rural disp.							
Coefficient[a]	R	R	*	*	*	*	*
Background calculated							
Street canyon flows							
Wind speed extrapolation	*	*	*	*	*	*	*
Highway generated							
Turbulence	*						
Emission rates function of met. conditions							
Emission rate function of time of day			*				*
Buoyancy induced		*					*
Dispersion							
σ_y Function of averaging							
Time and stability					*		
Inversion penetration							
Factors					*		
No. of sources (max.)[d]	180	25 (PTMTP)	250P 100A	Unlimited	300P 50A	50	250
No. of receptors (max.)	30	30 (PTMTP)	150	2500	2500	112	180

Source: Bonazountas and Fiksel.[26]

[a]U = Urban Only, R = Rural Only.

[b]The maximum number of sources and receptors are interrelated: These values are reasonable estimates.

[c]These capabilities are available on a version which is not part of the UNAMAP series of models.

[d]P = Point, A = Area.

- Gaussian-Plume Multiple Source Air Quality Algorithm (RAM)
- Industrial Source Complex Model (ISC)
- Multiple Point Gaussian Dispersion Algorithm/Terrain Adjustment (MPTE)
- Single Source Model (CRSTER)
- Urban Airshed Model (UAM)
- Offshore and Coastal Dispersion Model (OCD).

A detailed presentation of these models is not relevant for this publication. Information can be found in other sources.[158]

EPA Alternative Models. The U.S. EPA list of alternative air quality models is presented below:[160]

- Air Quality Display Model (AQDM)
- Air Resources Regional Pollution Assessment (ARRPA) Model
- APRAC-3
- AVACT II
- COMPTER
- ERT Air Quality Model (ERTAQ)
- ERT Visibility Model
- HIWAY-2
- Integrated Model for Plumes and Atmospheric Chemistry in Complex Terrain (IMPACT)
- LONGZ
- Maryland Power Plant Siting Program Model (PPSP)
- Mesoscale Puff Model (MESOPUFF II)
- Mesoscale Transport Diffusion and Deposition Model for Industrial Sources (MTDDIS)
- Models 3141 and 4141
- MULTIMAX
- Multiple Point Source Diffusion Model (MPSDM)
- Multi-Source Model (SCSTER)
- Pacific Gas and Electric Plume 5 Model
- PLMSTAR Air Quality Simulation Model
- Plume Visibility Model (PLUVUE II)
- Point, Area, Line Source Algorithm (PAL)
- Random Walk Advection and Dispersion Model (RADM)
- Reactive Plume Model (RPM-II)
- Regional Transport Model (RTM-II)
- SHORTZ
- Simple Line-Source Model (GMLINE)
- Texas Climatological Model (TCM)
- Texas Episodic Model (TEM)

Other Models. Many other models are available for air quality applications. The reporting of these models exceeds the purpose of this publication.

Model Application

A typical screening approach for evaluating peak ground-level concentrations resulting from stack emissions would include the use of the model PTPLU. This model of the EPA-UNAMAP series can calculate the maximum ground-level concentrations under a number of meteorological conditions. For a given stack with identified operations parameters, the model will cycle through a number of combinations of wind speeds and atmospheric category and indicate the downwind distance from the stack to where the maximum occurs. The model's output covers the typical range of meteorological conditions, and it indicates the expected short-term (up to one hour) average acute exposure levels that a receptor may encounter.

This screening analysis is useful to quickly estimating peak short-term average levels, and it can be used to determine whether a problem may exist. It generally is geared to calculate a conservative worst case. More accurate levels can be determined by using other models that utilize site-specific meteorological data and can better simulate atmospheric reactivity and atmospheric transport mechanisms such as gravitational settling, deposition and washout.[158]

Water Modeling

General

This section describes principles of water quality modeling and summarizes well-documented computerized models. Emphasis is provided to processes related to hydrocarbon contaminated sites.[167]

Various energy-related activities result in discharges of pollutants to the water environment, via surface run-off or other pathways. The water quality assessment methodology is intended to estimate the impacts on water quality of man-made related activities. Use of observed present conditions seems a reasonable way to establish a baseline. Estimates of impacts on water quality at future times are then based on incremental changes in pollutant inputs to the body of water. These increments can be estimated on the basis of changes in activities in the basin between the present and the future times of interest. Furthermore, since the primary interest is in impacts due to hydrocarbon-related activities, pollutant characteristics from these sources will be emphasized.

Background

Stream or lake reservoir water is principally a mixture of surface run-off (generally low mineral content, high organic content) and groundwater (high mineral content, low organic and dissolved gases content). Its quality depends on the physical, geological, and chemical character of the drainage basin, the permeability of the soils, the climatologic conditions, the percent and type of vegetation, the chemical constituents of precipitation, and the indigenous biota of the water body. Table 32 lists natural processes affecting stream water quality and their related impacts. All of these processes are affected by several factors. Most important are: climatic conditions, geographic conditions, geologic conditions, season of the year, and diurnal variation.

Table 32. Process Affective Natural Water Quality

Process	Principal Water Quality Changes
Precipitation	Dissolved gases (CO_2, N_2, O_x, CO, SO_2), dust particles, smoke particles, bacteria, salt nuclides, dissolved vapors.
Evaporation, evapotranspirtation, volcanic activity forest fires, wind pickup	Water vapor, salt nuclides, vapors from vegetation, organic and inorganic dust particles, bacteria, smoke particles, gases of combustion
Surface runoff	Silt, organic debris, silica, mineral residues of earth materials, soluble and particulate products of biodegradation of organic matter, bacteria, dissolved gases, soil particles
Floodwaters	Silt and other soil materials
Groundwater overflow	Mineralized water, intruded saline water
Groundwater infiltration	Dissolved minerals from surface wash and groundwater outcrop; dissolved gases and compounds from organic life and decay; dissolved gases (CO_2, O_2)
Tidal waters	Increased salinity, dissolved minerals
Biological activity	BOD, nutrients (N,P), other decomposition by-products, nutrient and mineral cycling

Source: Bonazountas and Fiksel.[26]

The total concentration of a pollutant in a body of water is a function of the quantity of the pollutant being discharged from point and nonpoint sources (both natural and anthropogenic), inputs from the stream channel, inputs from sediments, and instream pollutant loads entering the portion of the water body of interest from upstream. If all these inputs are known and if the behavior of the pollutant is understood, then it is in principle possible to predict pollutant concentrations in the water. This is what is attempted in a conventional water quality model for a body of water. However, accounting for all the pollutant inputs in a regional assessment is a very difficult task.[167] Many water constituents are toxic at certain concentrations and interact directly with the biota of the ecosystem, causing death or severe stress and limiting the use of water resources.

Categories of pollutants include: (1) conservative and nonconservative pollutants; (2) conventional and nonconventional pollutants; (3) toxic and nontoxic pollutants.[168]

With growing industrialization and a steadily increasing number of new toxic compounds, there is indeed a great need for development of water quality models that can be used for predicting safety levels and establishing water quality criteria. This modeling effort should take account of: heavy metals, oils and chlorinated hydrocarbons, pesticides, and other organic toxic compounds. Only a few models attempting to describe quantitatively the distribution and effect of toxic substances have been published.

The basis for modeling the distribution and effect of toxic substances in an aquatic ecosystem is knowledge of the process in the ecosystem. However, since the impact of toxic substances on man and aquatic ecosystems is well recognized, it is important to develop such models and use them as environmental management tools.

Pathway Processes

As soon as a chemical is released into the aquatic environment, it is subjected to an array of transport, transfer, and transformation processes. Hydrodynamic transport is described in terms of advective and turbulent forces and occurs simultaneously with transfers to sorbed forms and irreversible transformation process. In time, as a result of dilution, speciation, and transformation, environmental concentration gradients of the chemical are established.

The chemical and physical processes and reactions of a substance are of major importance in water quality modeling as shown in Figure 49. For models of both, (1) individual toxic chemicals, and (2) conventional water quality parameters such as BOD and dissolved oxygen the relevant chemical processes and the environmental effects on the processes must be considered.

Predicting aquatic fate of pollutants involves several steps. The steps described in the remainder of this section include:

- Determination of Fate-Influencing Processes
- Delineation of Environmental Compartments
- Representation of Hydrologic Flow
- Mathematical Representation of Speciation Processes
- Mathematical Representation of Transport and Transformation Processes
- Determination of Pollutant Load and Mode of Entry into the Aquatic Environment.

Prediction of the fate of toxic pollutants requires the user to know which processes act on the toxicant. Figure 50 illustrates the transport and transformation processes which are of potential importance in a lake or other water body.

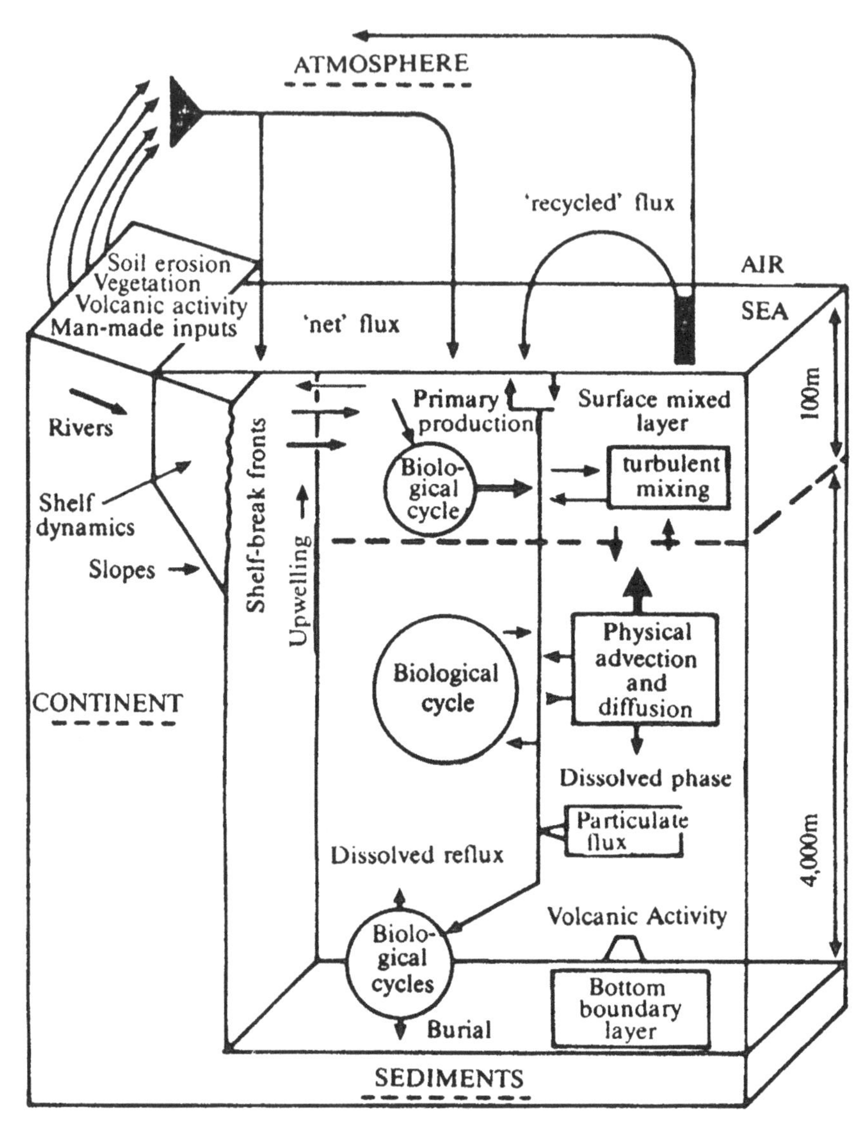

Figure 49. Physical processes: schematic diagram of the major ocean processes for reactive elements (Buart Menard[191]).

Many processes and many sites are involved in delineating the fate of chemicals in the aquatic environment. It is therefore rather obvious that a wide range of factors would be instrumental in affecting both rates of the processes and residence sites in the system. These are: (i) environmental factors (pH, dissolved oxygen, temperature, sunlight, nutrients, mixing, particulate matter), (ii) biota and their metabolic activity, i.e., as modified by (i), and (iii) chemical structure.

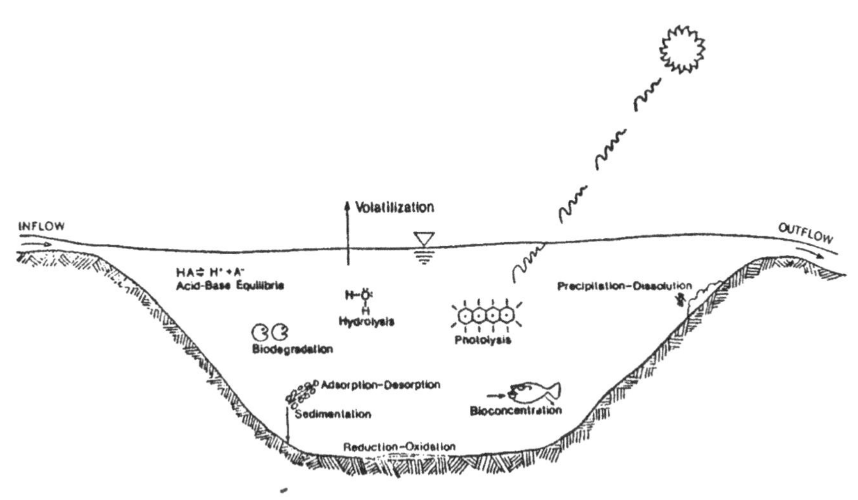

Figure 50. Speciation, transport, and transformation processes in aquatic environment (from Dickson et al.[97]).

The processes that affect the chemical transformation of organic chemicals are depicted in Figure 51 while those that affect storage are shown in Figure 52.

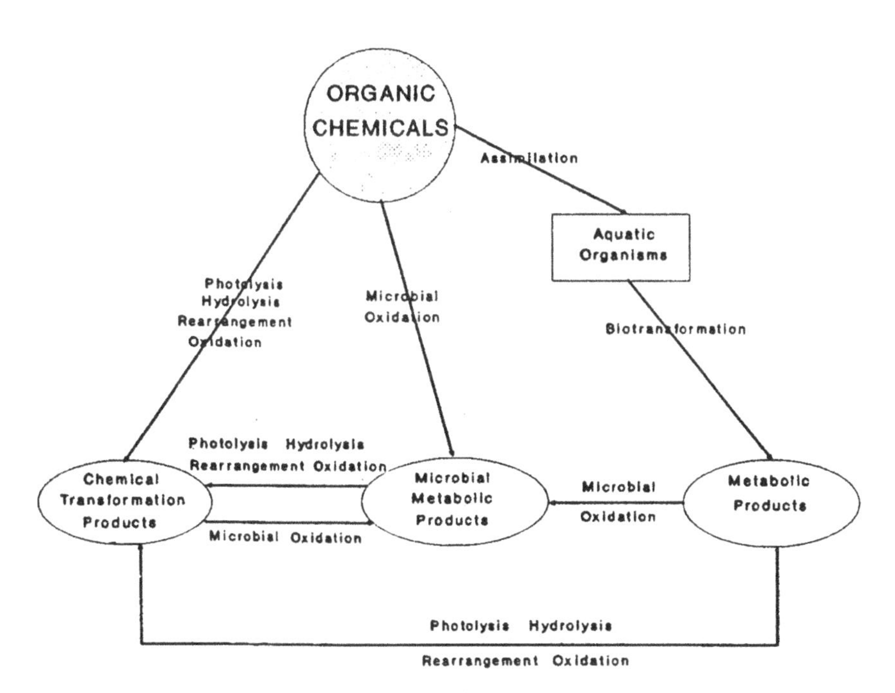

Figure 51. Environmental transformations of organic chemicals.

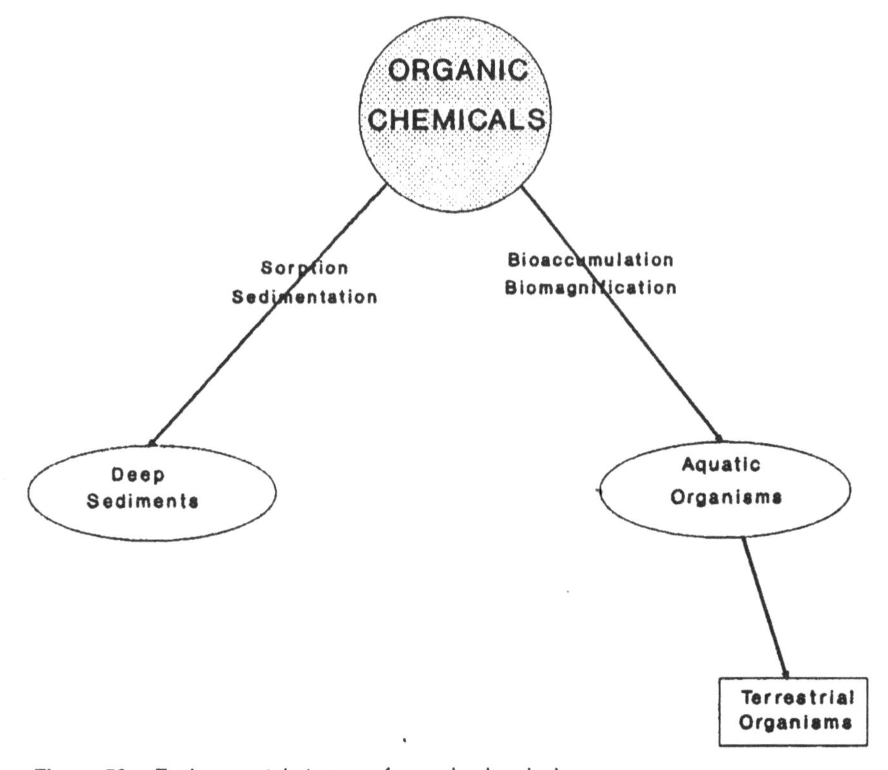

Figure 52. Environmental storage of organic chemicals.

Most chemical processes are described mathematically by one or two parameters that are specific to a particular system. These system-specific parameters must be specified or adjusted to account for all of the environmental factors in the system which can affect the process. For example, the rate of hydrolysis of a toxicant varies for each chemical, and even for one chemical the rate of hydrolysis depends on temperature, pH, and other reactants in the system. However, for a specific system this rate can be described by one rate constant (in units reciprocal time). These parameters may be obtained in three different ways.

The physical behavior of a chemical is important mainly in determining the chemical partitioning among the various environmental media. For example, the reaching into a stream of two different acids (with similar chemical behavior) may result in one mainly volatilizing into the air and the other becoming mainly sorbed on the bottom sediments. The physical behavior of a substance therefore can have a large effect on its environmental fate. The processes and corresponding physical parameters that are important in determining the behavior and fate of a chemical are different in analysis of trace-level contaminants in a water body from analyses of large-scale release (e.g., hydrocarbon spills). The processes of sorption and volatilization are most applicable to trace-level analyses, although they may also play a role in large-scale release analyses. Bulk properties (e.g., viscosity, solubility) are usually only important in simulations involving amounts of contaminants.

Model Selection Criteria

The selection of models for the analysis of exposure to contaminants involves factors addressing a number of issues, not all of which are amenable to expression in specific criteria. Five general steps may be identified in the modeling process:

- Problem Characterization
- Site Characterization
- Model Selection Criteria
- Code Installation
- Model Application.

The five general steps are not the model selection criteria, but rather the overall process by which a problem is identified and a model selected to perform an exposure assessment study. Model Selection Criteria is listed as the third step in this process. The two previous steps, Problem Characterization and Site Characterization, are crucial in the selection of appropriate model. While the steps can be considered sequential in nature, it is important to recognize interactions and feedback mechanisms between them.

Four basic criteria are important for model selection when faced with site-specific water quality studies, such as waste load allocations. They are: (1) the type of processes accounted for in the model; (2) the calculation framework and documentation availability of the model; (3) the availability of input data; and (4) the cost to accomplish the modeling effort, given the study objectives.

Classification of Models

General. In order to isolate those models which may be appropriately applied to a particular water quality/quantity phenomena within a given region of the total basin, several broad categories were defined and used to group models having similar characteristics. These categories are:[26]

- Water Quality Models
- Circulation/Transport Models
- Heated Effluent Dispersion Models
- Toxic Substances Models
- Nonpoint Source Models
- Other Models

These categories delineate the model types. During the past 15 years over a hundred mathematical models have been developed. The major thrust of this modeling research has been in the areas of water quality (primarily euthrophication) and lake circulation or hydrodynamics.

A number of references provide a background on water quality modeling (e.g., Orlob[169]). The U.S. EPA has also recently produced documents addressing methods for predicting concentrations of individual constituents resulting from pollutant loads to aquatic environments (e.g., Delos et al[170]). The U.S. EPA has also funded research to develop mathematical modeling catalogs.[26]

When selecting a water quality model, a user should choose a model which is relevant to the objectives of the simulation (modeling) effort to be undertaken, the type and reliability (precision) of the available monitoring data, and the chemical and physical properties (and reactions) of the substance in the water body.

Water Quality Models. Water quality models can be classified, in general, into the more common deterministic models, which describe the system as cause/effect relationships, and stochastic models, which incorporate the concepts of risk, probability, or other measures of uncertainty. Deterministic model formulations can be further classified into: simulation models that employ well-accepted analytic equations, and differential equation-type models.[26] Water quality models are formulated by employing two or three sets of equations (modules); the hydrologic or flow module, the sedimentation module, and the module comprising a biologic component.

Models can be categorized into zero-, one-, two-, and three-dimensional types.[171] Zero-dimensional water quality models are used to estimate average pollutant concentrations at a particular location as a function of time. One- and two-dimensional models describe the temporal and spatial concentrations of a substance in a water body. Models can be categorized according to their temporal features into steady-state or dynamic, and these features apply to both the flow and the quality aspects of the model. When the model is steady-state in terms of quality (i.e., pollutant load and concentration predictions), as is the EXAMS model,[172] then the user must be careful with its employment, because only long-term concentrations can be predicted.

The numerous models of water quality in streams and estuaries available in the literature cover a wide range of theoretical approaches and solution techniques of their equation systems. These models have been classified into three main concept categories.:

- the traditional numerical differential equation-type (TDE) or hydrodynamic like QUAL-II[173]
- the compartmental models, like EXAMS[172]
- other models (e.g., statistical).

Both model categories can deal with conservative and nonconservative constituent modeling, and can be categorized under the classical water quality or the pollutant specific models. Most common water quality models are shown on Table 33.

Hydrodynamic (TDE) Models. One- and two-dimensional TDE water quality stream models are, in general, far-field models. Roughly speaking, a differential equation or hydromechanic model consists of a system of equations, as shown in Table 34, that are space- and time-dependent and describe the phenomena considered "throughout" the system. In order to solve these equations, the system is desegregated into subsystems of equal properties, and analytic, or numerical

Table 33. Hydrodynamic River Modeling Equations

(1) *Equations of Unsteady Flow* (e.g., one-dimensional)

$$\frac{\partial Q}{\partial x} + \frac{\partial A}{\partial t} - q = 0 \qquad \text{(conservation of mass)}$$

$$\frac{\partial Q}{\partial x} + \frac{\partial}{\partial}\left[\frac{Q^2}{A}\right] = gAS_o - gAS_f - gA\frac{\partial y}{\partial x} + \frac{\tau_x}{\rho}B \text{ (conservation of momentum)}$$

(2) *Equations of Mass Transport* (e.g., one-dimensional)

$$\frac{1}{A}\frac{\partial}{\partial t}(AC) = -\frac{1}{A}\frac{\partial}{\partial x}(QC) + \frac{1}{A}\frac{\partial}{\partial x}(AD\frac{\partial C}{\partial x}) + \text{source} + \text{sink}$$

(3) *Other Equations* (e.g., diffusion, sediment equations)

$$D = \alpha \cdot U* \cong \frac{0.011\ U^2W^2}{U* \cdot y} \cong \beta\frac{Q^2}{U* \cdot r^3}$$

$$C_s = \sum_i \left(\frac{d_i}{y}\right)^{7/6}\left(\frac{t_o}{t_{i_c}} - 1\right) f\left(\frac{U*}{w_i}\right)$$

where:

$$f(U*/w_i) = f(Y/d_i)^{1/6}$$

where:

Q = volume flow rate, A = cross sectional area, q = lateral inflow rate, S_o = channel bottom slope, S_f = friction slope, y = water depth, g = acceleration due to gravity, x = distance along the channel, t = time, t_x = wind stress along the channel, ρ = density of water, B = channel top width, C = concentration of a given substance, D = longitudinal dispersion coefficient, U∗ = bottom shear velocity, α,β = coefficients, r = hydraulic radius of cross section, W = top width of channel, C_s = mean sediment concentration in stream, d_i = incipient particle diameter, t_o = bed shear stress, t_{ci} – critical shear stress for grain size d_i, f() = function, w_i = sediment (d_i) fall velocity

Note: A conservation of mass (continuity) equation can be formulated for the sediment bed load, or suspended load also. The numerical solution of all equations together (water/sediment) becomes extremely complex. This approach, therefore, should be employed with skepticism in particular of estuarine "quality."

Table 34. Water Quality Models

	Environment Simulated[b]								Major Modules[c]				Resolution[d]			Chemistry[a]			Processes[e]						Effort[g]	Reference[f]
	Watershed-River	River (stream)	Lake (reservoir)	Estuary (tidal)	Coast (ocean)	Near/Far Field	Fresh Water	Saline Water	Flow/Hydrology	Sediment	Pollutant	Biotic	Risk	Spatial	Temporal	Mathematical	DO-type	Organics	Inorganics/Metals	Physical	Chemical	Biologic	Speciation	Effect/Food Chain		
CHNTRN		•	•	•	•	N.F	•	•	•	•	•			0.3	D	N.C		•		•	•	•			H	Yeh (1982)
CMRA	•	•				F	•		•	•	•		•	2	D	N.S		•	•	•	•	•			H	Onishi (1980)
CTAP		•	•	•	•	N	•	•		•	•			0.3	S	C		•	•		•				M	Hydro Qual (1981)
DOSAG 1		•				F	•				•			1	S	A	•				•				L	Texas W.B. (1972)
EXAMS		•	•	•	•	N.F	•	•		•	•			0.3	S	C		•		•	•	•			L	Burns et al. (1982)
EXAMS 2		•	•	•	•	N.F	•	•		•	•			0.3	D	C		•		•	•	•			M	Burns et al. (1983)
FETRA		•		•	•	F	•		•	•	•			2	D	N		•		•	•	•			H	Onishi (1981)
HSPF	•	•	•			F	•		•	•	•			1	D	S.N		•		•	•	•			H	Johnson et al. (1980)
MERGE			•		•	N	•	•	•	•	•			1	S	N		•	•	•	•				M	Frick (1981)
MEXAMS		•	•	•	•	N.F	•	•		•	•			0.3	S	C		•	•	•	•	•	•		H	Felmy et al. (1982)
MICHRIV		•				F	•			•	•			1	S	A		•			•				L	DePinto et al.
MIT TWONM		•		•		F	•	•	•		•			2	D	N	•	•		•	•				H	Harleman et al. (1977)
PEST			•				•			•	•			0	D	C		•			•	•			M	Park et al. (1980)
PLUME			•	•	•	N	•	•	•		•			1	S	N		•	•	•					M	Baumgartner et al. (1971)
QUAL-II		•				F	•		•		•			1	S	N	•		•	•	•	•			H	Roesner et al. (1981)
SEDIM		•					•			•				1	S	A				•					L	Bonazountas et al. (1983)

continued

Table 34. *Continued*

	Environment Simulated[b]							Major Modules[c]					Resolution[d]			Chemistry[a]			Processes[e]					Effort[g]		Reference[f]
	Watershed-River	River (stream)	Lake (reservoir)	Estuary (tidal)	Coast (ocean)	Near/Far Field	Fresh Water	Saline Water	Flow/Hydrology	Sediment	Pollutant	Biotic	Risk	Spatial	Temporal	Mathematical	DO-type	Organics	Inorganics/Metals	Physical	Chemical	Biologic	Speciation	Effect/Food Chain		
SERATRA		•	•	•		F	•			•	•			2	D	N		•	•	•	•				H	Onishi et al. (1982)
SLSA		•				F	•			•	•			1	S	A		•			•	•			L	HydroQual (1981)
TODAM		•		•		F	•	•		•	•			1	D	N		•			•	•			M	Onishi et al. (1982)
TOXIC			•				•			•	•	•		0	D	N		•			•	•		•	M	Schnoor (1981)
TOXIWASP		•	•	•		F	•	•		•	•			0.3	D	N		•	•		•	•	•		M	Ambrose et al. (1983)
WASTOX		•	•	•		F	•	•		•	•			0.3	D	N		•	•		•	•	•		M	Connolly (1982)
WQAM		•	•	•		F	•			•	•			0.1	S	A	•	•	•		•	•			L	Mills et al. (1982)
OTHER METHODS																										
—Toxic Sub.			•																						e.g.	Delos et al. (1983)
—Food Chain												•												•	e.g.	O'Connor & Mueller (1981)
DEM				•		F	•		•	•	•			2	D	N	•	•	•	•	•	•			M	Water Resources Engineers Inc. (1974)
ESOO1				•		F	•				•			1	S	A	•			•	•				L	Hydroscience Inc. (1968)
LAKE-3			•			F	•			•	•	•		2	D	N				•	•	•	•	•	M	Thomann, Di Torro, Winfield O'Connor (1975)
DIURNAL		•				F	•				•			1	S	A	•			•	•				M	Hydroscience Inc.
RECEIV-1		•	•	•		F	•				•			2	D	N	•			•	•	•			H	Water Resources Engineers, M&E (1971)
RIBAM		•	•			F	•				•			1	S	A				•	•				M	Raytheon Co. (1974)

continued

Table 34. *Continued*

	Environment Simulated[b]								Major Modules[c]					Resolution[d]			Chemistry[a]			Processes[e]					Effort[g]	Reference[f]
	Watershed-River	River (stream)	Lake (reservoir)	Estuary (tidal)	Coast (ocean)	Near/Far Field	Fresh Water	Saline Water	Flow/Hydrology	Sediment	Pollutant	Biotic	Risk	Spatial	Temporal	Mathematical	DO-type	Organics	Inorganics/Metals	Physical	Chemical	Biologic	Speciation	Effect/Food Chain		
SSM		•	•			F	•				•			1	S	A	•			•	•				L	Hydroscience (1971)
STREAM7B		•				F	•				•			1	S	A	•			•	•				L	Resource Analysis Inc. (1978)
SNSLM		•	•			F	•				•			1	S	A	•								M	Braster Chapta (1978)
FEDBACO3		•	•	•		F	•				•			1	S	N	•			•	•				M	USEPA (1978)
WQRRS		•				F	•				•			1	S	A				•	•	•			L	U.S. Army Corps (1978)
RWQM		•	•			F	•			•	•			1	S	A					•				L	Resources Analysis Inc.
HARO3				•		F	•				•			0.3	S	N	•			•	•				L	Hydroscience Inc. (1974)
SEM		•		•		F	•				•			1	S	A	•			•	•				L	Hydroscience Inc. (1972)

[a]Information obtained from Versan. Inv. (1983) and Bonazountas & Fiksel (1982).[26]

[b]N = near field, F = far field, Water to River Category = multimedia modeling.

[c]Flow/Hydrology module = from the precipitation to the river hydraulics, or dynamic stream flow modeling. Sediment = sedimentation dynamics generated by a module, or as a user input for the quality simulation. Biotic Module = a bioaccumulation, a food chain, or other type of similar models. Risk = a cause effects risk analysis.

[d]0,3 = 0, 1, 2 and 3 dimensional features. In general, compartmental models are either of 0-, or of 0–3 dimensionality. S = steady state, D = dynamic or quasi dynamic (for both the hydro and the pollutant modules). N = numerical solution of differential equations (e.g. finite difference or element), C = compartmental approach, S = simulation modeling, A = analytic solutions of differential equations.

[e]Simulation of metal can be accomplished in a conservative approach; therefore, metal modeling capabilities apply to many models without speciation features (see footnote f).

[f]Biologic processes are referred to chemical/biological degradation of compounds, as contrasted to the effects (biologic species) and food chain biologic processes. Speciation = aquatic chemical equillibria.

[g]H = high, M = medium, L = low.

solution techniques are applied to the entire system, consisting of nodes or elements. Boundary condition equations have to be considered. Many analytic and numerical schemes are available, such as the finite difference method, the implicit multipoint difference method, and the finite element method.

The major advantage of the hydrodynamic models is the detailed and frequently accurate simulation of the hydraulics of the system. However, because of the complicated mathematics involved and the numerical procedures required to solve the equation systems, hydrodynamic models are weak in simulating sediment (e.g., bed quality, bed load, and suspended load) circulations, chemistry, and bioaccumulation effects. The reason for this deficiency is the difficulty to mathematically formulate and efficiently solve in one effort (i.e., with equation system), equation systems governing multiple phenomena such as physical, chemical, and biological.

Furthermore, hydrodynamic models are difficult to be modified by users once the model has been fully developed. This is a disadvantage only for cases where new information has to be incorporated into the model, as, for example, a posterior knowledge of atmospheric contribution or exchange to the stream/estuary system.

In conclusion and generally speaking: hydrodynamic models are strong in physics and weak in chemistry (including sedimentation and biology) of systems. However, if insufficient water-related data are available for a simulation, hydrodynamic modeling might be a good choice. Hydrodynamics models are capable of accepting new information as it becomes available, but only for the purpose of having an improved confidence in the model output, as well as the supplied input data. Most common hydrodynamic models are shown on Table 35.

Compartmental Models. The recent decade tendency is to employ compartmental models, where the water body domain is formed with one or more interactive compartments of any size or shape, a feature not available in the noncompartmental modeling concept. Compartmental models consist principally of water quality and sediment modules. Roughly speaking, a compartmental model consists of a system of differential or algebraic equations that are only time-dependent. That is, the equations are formulated for a specific subsystem (i.e., compartment) or subprocess of the entire physical system. In such a case, an entire system (e.g., river) will be physically represented with multiple interacting compartments and will be mathematically described with interacting equation systems.[158]

Compartment models can be applied in discretized spatial and temporal domains by employing numerical procedures (i.e., subdomains of time steps) comparable to the procedures employed for the hydraulic models. The model spatial domain can be selected to range in size between a few square meters and a number of square kilometers, depending on the desired model output accuracy and the availability of other information (e.g., uniform flora/fauna distribution in the compartment).[174]

In conclusion and generally speaking: compartment models can be stronger in chemistry (including sedimentation and biology) and weaker in the physics of the system. If a sufficient understanding of the stream or estuary hydrodynamics exists for a water quality simulation, water compartment models might be a good choice.

Table 35. Summary Matrix of Surface Water Flow Models

	Model Name													
Characteristics	**HEC-2**	**CHNHYD**	**HEC-6**	**SEDONE**	**HSPF**	**DWOPER**	**DYNHD3**	**EXPLORE-1**	**CAFE**	**WATFLO**	**Leender-tse-2D**	**Leeder-tse-3D**	**DEM**	**HN**
Water bodies														
River	x	x	x	x	x	x	x	x	x	x	x	x		x
Estuary		x		x			x	x	x	x	x	x	x	x
Lake														
Dimensions														
1-dimensional	1L	1L	1L	1L	1L	1L	1L	1L						
2-dimensional							Q2	Q2	2H	2H	2H			
3-dimensional												x		
Time dependence														
Steady-state	x													
Transient		x	x	x	x	x	x	x	x	x	x	x	x	
Solution technique														
Numerical	FD	FD	FD	FD	FD	FD	FD	FD	FD	FD	FD	FD	FD	FD
Availability														
Public domain	x	x	x	x	x	x	x	x	x		x	x	x	
Proprietary										x				x
Support														
Unknown		x		x				x	x		x	x		
HEC	x		x											
U.S. EPA					x		x						x	x
NOAA						x								
Contractual										x				
Other														
Sedimentation			x	x	x								x	
Water quality capabilities					x		x	x			x		x	
Compatibility with specific transport model							x	x	x	x		x		
Autocalibration					x									

LEGEND: 1L = one-dimensional longitudinal; Q2 = quasi-two-dimensional, link-node, horizontal plane; FD = finite difference; FE = finite element.

Other Classifications. Models can be classified according to whether they handle conservative or nonconservative substances. Frequently, they are also classified according to whether they can handle nontoxic or toxic pollutants. Predominant modeling concepts are:[175]

- Nonconservative substance aquatic quality modeling of overall water quality can be performed either on a substance-by-substance approach or by studying such parameters as changes in levels of BOD, DO, algae, ammonia nitrogen, nitrate nitrogen, phosphate, colliform, and pH.
- Conservative substances (e.g., metals) aquatic modeling is performed on a substance-by-substance approach.

Mathematical Modeling

Mathematical deterministic water quality modeling is structured, in general, around three major modules, or sets of equation as shown in Figure 53.

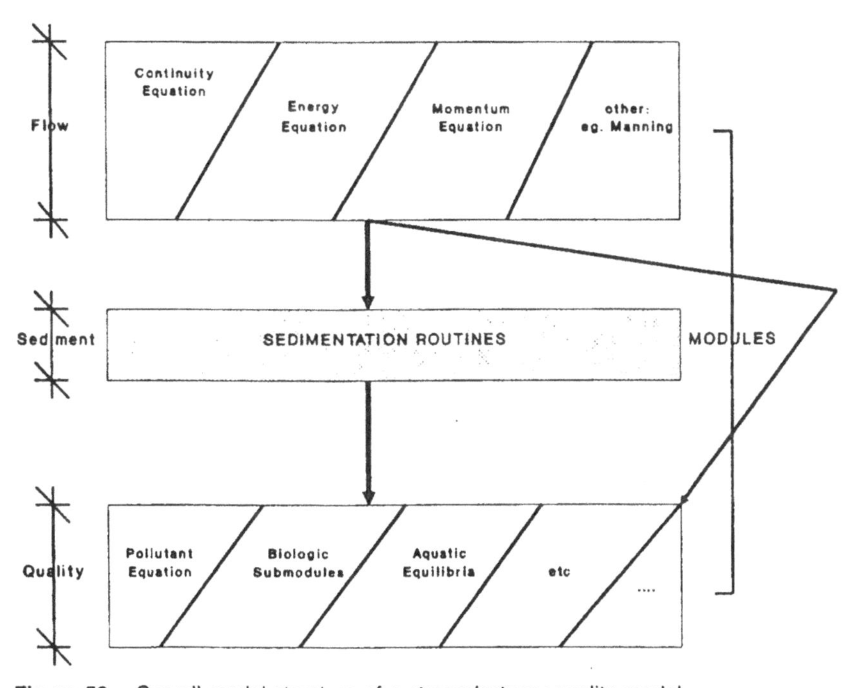

Figure 53. Overall model structure of a stream/estuary quality model.

- A hydrologic or flow module consisting of one (i.e., continuity) or more (e.g., continuity, energy) momentum equation
- A sedimentation module describing sediment bedload and suspended load are being driven by the first (flow) module

- A water quality module consisting of one (i.e., conservation of pollutant mass) or more (i.e., description of various chemical processes) equations being driven by the two previous modules.

Formulated equations for the above three modules are solved either simultaneously or sequentially. In the latter case, the output of the first set becomes an input to the second set, and both, the output of the first and the second set, become inputs to the third set. The sequential solution approach is preferred today because of various mathematical reasons. It is also important to note that either the flow module or the sediment module, or both modules, are frequently absent from most water quality models. These models require that the user input this information using environmental monitoring data rather than module output.[26]

Because the formulation and solution of such equation systems is complicated and difficult to use, various simplifications are made during model formulations. Five important issues in model formulations are:

1. The spatial (i.e., x, y, z) resolution of the model.
2. The temporal (i.e., time) resolution of the model.
3. The mathematical formulation of the model and the computational consideration employed.
4. The process features (i.e., physical chemical, biological) of the model.
5. The kinetic formulation of each feature described.

In reference to the spatial model resolution, models can be categorized into zero-, one-, two-, and three-dimensional frameworks.[171] In reference to the temporal model resolution, water quality models can be categorized according to their temporal resolution features. The following designations are common: (1) Steady-State or Dynamic Models, (2) Long-Term or Short-Term Models.

Model Formulation and Mathematics. Referring back to Figure 51, the most complete approach is to have all three modules (flow, sediment, and pollutants) in a model, in which case the flow module will produce the velocity field of the water body. This flow field is retained for input into both the sediment module and the pollutant module; otherwise, it is a user input to both modules. The equations or equation systems are solved using various mathematical techniques.

All water quality models are based on the conservation of mass and the interaction or behavior of the constituents with the environment. The conservation of mass for the transport of a dissolved contaminate in a water body may be written in differential form as:

$$\partial C/\partial t + \nabla(C\bar{q}) = D\nabla^2 C + \Sigma T + \Sigma r \pm s$$

where:

C = the concentration of dissolved contaminant (M/L^3)
t = time (T)

$\underline{\nabla}$ = the Del operator ($\nabla C = \partial C/\partial x + \partial C/\partial y + \partial C/\partial z$) (1/L)
q = a vector of the x, y, z velocity components (L/T)
D = a dispersion coefficient (L^2T)
ΣT = any phase transfer mechanisms ($M/L^3/T$)
Σr = any reactions ($M/L^3/T$)
s = any source or sink terms ($M/L^3/T$).

The above equation can be written for many interacting processes of the pollutant module and for each process (1, 2, ····, n) of the term Sr of the described equation.

$$dc_1/dt = A(c_1, c_2, \cdots, c_n)$$

$$dc_2/dt = B(c_1, c_2, \cdots, c_n)$$

$$\cdots\cdots\cdots\cdots\cdots\cdots\cdots\cdots$$

$$dc_n/dt = N(c_1, c_2, \cdots, c_n)$$

where A, B, ····, N are algebraic expressions.

It is important to note that for steady-state conditions—in terms of concentration—the above system of differential equations becomes a system of algebraic equation of the type: A = B = ···· = N = O.

As long as the expression A, B, ····, N describes first-order kinetics or processes, simple analytic solutions are feasible; otherwise, solution of equations (SQ-8) with higher order expressions (e.g., c_1^2, c_n^3) becomes a more difficult task.

The formulation of equations for conservative substances applies to non-conservative substances also; however, the reactions of the substance with the environment and/or order substances must be added.

Solution Techniques/Calibrations. Without a solution technique, mathematical models are of little value. Three particular solution procedures are generally followed: the analytical (i.e., exact solution of a differential equation), the numerical (finite difference, finite element solution of differential equations), and the statistical. Analytical techniques are usually employed for simplified and idealized situations (e.g., SLSA-model, DiToro,[176]). Statistical techniques are developed for specialized cases (e.g., CMRA model[177]).

Modelers should be fully aware of the range of applicability and the processes (physical, chemical, biological) considered by a computerized package. The physical processes of a typical water quality model include hydromechanic circulation, sedimentation, pollutant advection, dispersion and sorption water column exchange, heat budget, ice formation considerations, and light attenuation.

The predictive capabilities of an analysis (modeling effort) must be confidently established. This is accomplished through model calibration (or verification), frequently using more than one set of water quality data for the receiving water body. Calibration should also account for single events or critical conditions, where single events may be expressed in terms of variables such as streamflow, temperature, stream sedimentation and water discharges, for which data exist. It is important to note that calibration is usually performed by a comparative analysis of the model output vs. monitoring data. Quality assurance or error analysis (EA) efforts are part of a water quality simulation and calibration effort.

A description of the various categories of water quality models is presented by Orlob;[169] Novotny;[178] Bonazountas and Fiksel;[26] and others.

Use of Toxic Pollutant Models. Selection and use of models which simulate the fate of toxic pollutants water body can range from simple dilution calculations to complex models having multidimensional, dynamic, biologic, or speciation capabilities.[170] Some of the sophisticated models require an extensive knowledge (or input parameters of a compound for: (1) the exchange between bed and water—e.g., particle transport and exchange, diffusion of dissolved material; (2) the governing partitioning process—e.g., metals partitioning; (3) the decay or transformation processes—e.g., biodegradation, photolysis, hydrolysis, volatilization, and bioaccumulation. Assuming that all of the above information is known, sophisticated modeling efforts can be undertaken, if desired. Model selection and use are a function of the data availability and the study objectives.

Modification of Existing Models. Although the models of Table 34 and 35 cover a wide range of applications, model modifications (both theory and code) may be necessary for certain specialized applications. Model modifications may deal with improvement of certain model routines (e.g., incorporation of a volatilization routines into a near-field or mixing zone model), packages, or combining an entire model to another model.

Nonpoint Sources Modeling

General

Much attention has been paid to the problem of point discharges, their effects on receiving waters, and the study and simulation of resulting water quality interactions. Less attention has been paid to the problem of nonpoint source pollution despite the fact that runoff is frequently the most significant determinant of receiving water quality. Runoff can be important, for example, in rural areas where point sources are minor and where intensive farming and other agricultural activities (e.g., feedlot operation) lead to heavy concentrations of pollutants spread over land.

The combined impact of the importance of nonpoint wasteloads and the difficulties inherent in modeling, including their effects, has led to the development of various runoff models such as the Environmental Protection Agency's (EPA) Stormwater Management Model (SWMM), EPA's Nonpoint Source Pollutant

Loading Model (NPS), STORM (Water Resources Engineers/U.S. Army Corps of Engineers), EPA's Agricultural Runoff Management (ARM) Model, EPA's HSPF, and the EPA SESOIL Model.[26]

In general, runoff models are watershed models, designed to estimate nonpoint source pollutant delivery to surface bodies. In order to estimate pollutant delivery, watershed models first determine the hydrologic characteristics of the watershed (runoff soil moisture groundwater recharge by infiltrating precipitation, interflow and snow), then describe its sediment and finally its chemical characteristics.

Watershed models may focus predominantly on surface runoff and erosion processes, but may also simulate subsurface pollutant migration pathways. Nearly all such models simulate erosion processes. They vary widely in both the number of land surface and soil profile chemical phenomena simulated, and in their sophistication in simulating chemical processes. Many operational watershed models incorporate the same, or essentially similar, hydrologic algorithms.

Sources and Emissions

Nonpoint source pollution and its sources are difficult to define and describe. With the exception of the acidification of watersheds by acid precipitation, or other manifestations of the deposition of airborne contaminants, nonpoint sources are associated with a land use. The impact of a given land use (e.g., waste site) will be observed in the watershed where it is encountered, with the exception of subsurface pollutants migrating across surficial drainage boundaries. Among the land uses which have been investigated with respect to their potential to generate nonpoint source pollution are the following:

- agriculture (cropland, pasture, animal feedlots)
- urban (industrial, commercial, residential)
- construction
- silviculture
- mining
- waste disposal.

Heavy metals (especially copper, lead, and zinc) are the most prevalent priority pollutant constituents. Some of the metals are present often enough and in high enough concentrations to be potential threats to beneficial uses.

Sediment is the most widely studied nonpoint source pollutant, not only because of its intrinsic ability to degrade the ecological, aesthetic, and navigational value of surface waters, but also because it can serve as an indicator of associated toxic pollutants, especially pesticides and heavy metals.

Nutrients are generally present in runoff from natural undisturbed land, and are elevated from disturbed land surfaces regardless of the use of fertilizer, but with a few individual site exceptions, concentrations do not appear to be high in comparison with other possible discharges to receiving water bodies. Nutrients associated with fertilizer use contribute to eutrophication in many watersheds.

Coliform bacteria are usually present at high levels in runoff and can be expected to exceed EPA water quality criteria during and immediately after storm events in many surface waters, even those providing high degrees of dilution.

Oxygen-demanding substances (*BOD, COD*) are generally associated with runoff from agricultural and urban watersheds and are present in runoff at concentrations approximating those in secondary treatment plant discharges. If dissolved oxygen problems are present in receiving waters of interest, consideration of runoff controls as well as advanced waste treatment appears to be warranted.

Environmental Factors and Chemistry

This section highlights the important physical, chemical, and other phenomena or processes which interact to form nonpoint source pollution, and which should be considered in watershed modeling.

The major physical phenomena affecting quantity of runoff are: precipitation, infiltration, soil moisture, surface runoff, interflow, groundwater recharge, snowmelt, and the existence of impoundments. The first six phenomena are components of the hydrologic cycle of the soil compartment.

Impoundments and waste sites have topographic features, either natural or man-made, which restrict surface flow, store runoff, and release it at a slower rate than it was received. These included natural depressions, farm ponds, road culverts, etc. As a hydrologic feature they influence the natural lag in the rainfall: runoff relation. More importantly, with respect to nonpoint source pollution, they promote sediment disposition.[35]

The most important process affecting runoff quality is erosion. Erosion results from rainfall and runoff and involves two subprocesses: (1) soil and other particles are removed from the land surface and (2) these particles are transported by overland flow into conveyance systems and water bodies. Since land surface erosion is the principal source of stream sediment, the type of soil, land cover, and hydrologic conditions are major factors in determining the severity and extent of sedimentation problems. Sediment transport by surface runoff is described by the continuity:

$$\partial q_s/\partial x + \rho_s \partial(cy)/\partial t = D_l + D_f$$

where:

q_s = sediment load (mass/width*time)
x = distance
p_s = mass density of sediment particles (mass/volume)
c = concentration of sediment in flow (volume of sediment/volume of flow)
y = flow depth
t = time
D_l = lateral inflow of sediment (mass/area*time)
D_f = detachment or deposition by flow (mass/area*time).

Although sedimentation is storm-event related, its resultant problems are not exclusively quantity problems or water quality problems. Given the broad range of contaminants associated with runoff, and the complexity of soil chemistry, it is not possible to comprehensively discuss the full range of chemical processes occurring in watersheds.

A broad range of environmental contaminants move with soil particles in an adsorbed form. Adsorption is usually characterized by a partitioning coefficient, the ratio of adsorbed contaminant to dissolved contaminant under sediment/water equilibrium. Partitioning coefficients are defined in a variety of ways, but perhaps the most useful is as K_{oc} (organic carbon) which is largely independent of the properties of the soil or sediment. While adsorption is the most important chemical process for many hydrophobic organic compounds and heavy metals, the solubilization of chemicals applied to the land surface, as well as the dissolution of soluble components of the soil itself, is an important process affecting contaminant.

The solubility of many chemicals is affected by the chemistry of the soil water: watersheds severely impacted by acid rain promote the dissolution of many heavy metals from soil; while the solubility of organic chemicals may be affected by the presence of humic acids, and so on. The ARM and CREAMS models include transport of dissolved contaminants in surface and subsurface flow, while the NPS does not. Results of the ARM model applied to the Yazoo River Basin indicate that toxaphene delivery to streams is dominated by dissolved subsurface flow (interflow) rather than surface runoff.[174]

Many organic compounds are degraded by soil microorganisms. The capacity of microorganisms to degrade chemicals depends on the density and type of microbes and the chemical structure. Some chemicals are toxic to microbes in high concentrations. The degradation products are often toxic pollutants in their own right. Nitrogen compounds which contribute to euthrophication, participate in a complex series of microbe-induced reactions which control the cycling of nitrogen in natural and perturbed soil systems. Biodegradation processes are usually modeled as an invariant first-order loss, based on experimental and field studies.

The mobility of most metals is profoundly affected by the pH of soil solutions, which is affected by the chemical composition of the soil as well as anthropogenic influences such as acid rain and agricultural activities. Chemical equilibria models which are not usually a component of a watershed model represent the fundamental approach to simulation of acid-base and oxidation reduction reactions. Watershed models generally incorporate only an a priori estimate of solubility and/or an intuitive representation of the way solubility may change in different soil horizons.

Impacts on Water Quality

The effects of runoff on receiving water quality are highly site-specific. They depend on the type, size, and hydrology of the water body; the runoff quantity and quality characteristics; the designated beneficial use and the concentration levels of the specific pollutants that affect that use. Generally the impacts of the runoff on water quality are rivers, lakes, estuaries, and groundwater.

Mathematical Modeling Approaches

It is technically feasible to rigorously analyze the dynamics of fluid movement through a watershed, applying the theory of unsaturated flow through porous media to describe infiltration (e.g., Freeze and Cherry[58]), the kinematic wave equation for overland flow (e.g., Li[179]), and the theory of saturated flow through porous media to describe groundwater discharge.[58] This approach has been followed to a certain extent by Adams and Kuresu.[180] However, the following factors render such an approach impractical:

- degree of spatial and temporal resolution required for adequate simulation
- inhomogenity of soils and subsurface materials
- resultant input data requirements
- large masses of data to be stored and handled by the computer
- large number of computations required for even limited time periods.

Consequently watershed modelers have adopted a variety of intuitive or phenomenological approaches to the simulation of hydrologic processes. These approaches are based on engineering experience, but some of these approaches involve parameters which are not directly observable. In hydrology, the most suitable models for simulation of contaminant migration require site-specific data (often extensive) for calibration, and there is often considerable uncertainty in their ability to predict what will happen if the watershed is perturbed, as well as the range of hydrologic/climatic conditions to be expected over a projection period.

The nonpoint pollution models fall generally into two categories: the screening (unit loads) planning models and hydrological assessment models.

Screening Models. Screening models are usually simple tools which identify problem areas in a large basin. These models usually rely on assignment of unit loads of pollution to the various lands within the watershed. A unit loading is a simple value or function expressing pollution generation per unit area and unit time for each typical land use. The loads are typically expressed in kilograms per hectare-year. Despite its questionable accuracy, the concept of relating pollution loading to land use categories has found wide application in area-wide pollution abatement efforts and planning.

Loading functions for agricultural areas or waste areas are commonly based on the Universal Soil Loss Equation. Use of the unit load concept presumes that an adequate inventory of land data is available from maps, aerial and terrestrial surveys, remote surveys, and local information. The loading concept is applicable—in most cases—to long term estimates such as average annual loading figures.

Hydrologic Models. Ven Te Chow[181] divided hydrological models into eight categories as shown in Figure 54. Watershed or hydrological models may be broadly categorized by their geometric representation of the watershed (lumped or distributed); temporal scope (discrete storm events or continuous simulation);

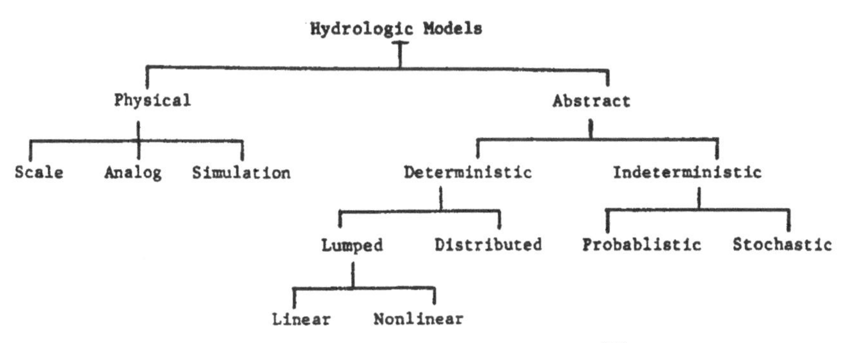

Figure 54. Classification of hydrologic models (from Novotny[178]).

and by their dependence on observable physical inputs or nonobservable parameters (physically based or parametric). The difference between physically based and parametric models is generally one of the degree to which results depend on parameters definable only within the model context. As an example, a concept of a lumped hydrological nonpoint pollution model is shown in Figure 55. The distributed parameter models divide the system into very small finite elements as shown in Figure 56. The principal functional components of the most comprehensive, distributed models are depicted in Figure 57. The most widely used models are in the public domain.

Available Models

Hydrologic Models. The simplest of all hydrologic models for rough estimates of runoff is the Rational Formula (flow rate equals, precipitation, times watershed area, times the runoff coefficient—a fraction between 0 and 1). Another model is the SCS curve number, developed by the USDA Soil Conservation Service[182] to estimate storm runoff from storm rainfall as a function of soil type, vegetation, and antecedent moisture. It has an extensive appreciation in the U.S. The hydrology routines of most operational watershed models are derived from the classical methods of hydrograph separation. One of the earliest and most comprehensive numerical models based on this approach is the Stanford Watershed

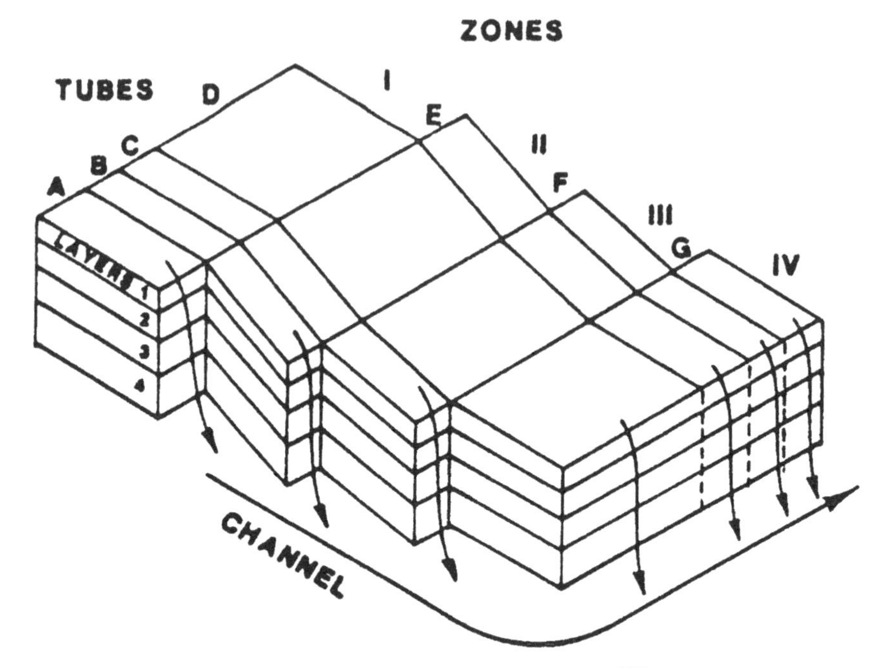

Figure 55. Lumped parameter model concept (from Novotny[178]).

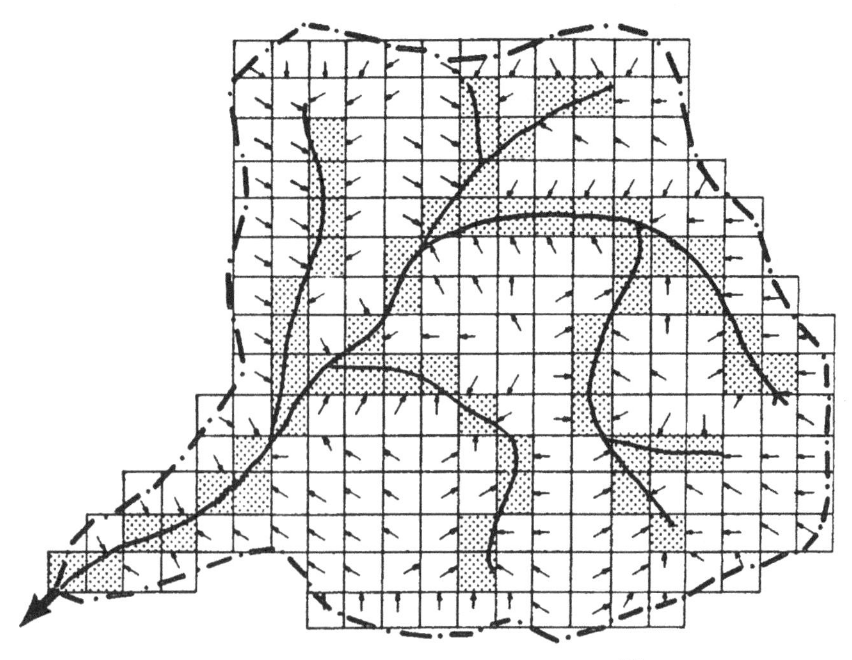

Figure 56. Distributed parameter model concept (from Novotny[178]).

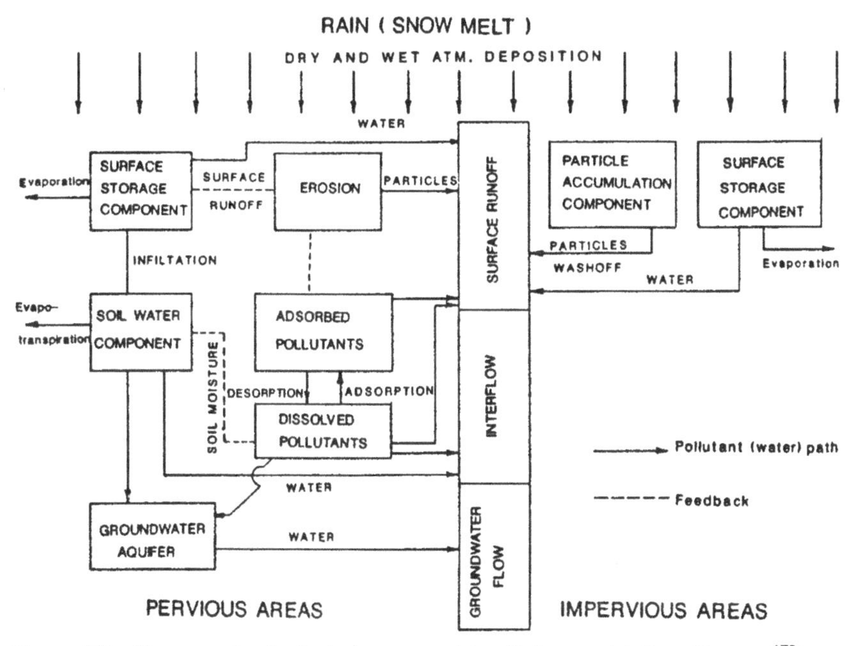

Figure 57. Compounds of a hydrology-nonpoint pollution model (from Novotny[178]).

Model (SWM). This hydrologic model forms the hydrologic underpinning of many nonpoint source of watershed models, including NPS, ARM, and HSPF. A brief description of the most important hydrologic models is given below, while tabular descriptions of the most common NPS models are shown on Table 36.

SWMM: Is designed to simulate nonpoint source runoff, primarily from urban areas. Some processes in the model are represented empirically, while others are represented by finite difference formulations. The SWMM model is an extensive program that models the rainfall/runoff cycle in urban areas in an comprehensive manner. The model is a continuous simulation model. Some methods for calculating infiltration rates are incorporated. The model is capable of simulating multiple catchment and dry-weather accumulation of particulate matter. Flow routing is calculated including storage and back-water effects. Conventional pollutants are simulated, as are arbitrary conservative substances. Erosion is simulated with the USLE. Important features of the SWM are:

- dependence on calibration
- no simulation of channel processes, scale limit of application 25 km^2
- use of Phillip/Horton infiltration equation
- kinematic wave approximattion for overland flow
- lumped parameter approach for water routing

Table 36. Nonpoint Sources Models

Capabilities & Characteristics	MRI	AGRUN	SWMM	STORM	HSPF	NPS	ARM-II	ACTMO	UTM	CMRA	CSU	USDAHL-74	CREAMS
USE													
Agricultural	*	*			*	*	*	*	—	risk	*		*
Watershed	*	*	*	*	*	*	*	*	—	asses.	*	*	*
GEOMETRY													
Lumped	*	*	*	*	*	*	*	*	*	*			*
Distributed											*	*	
TIME SCALE													
Storm event		*	*	*				*			*	*	*
Continuous	*		*	*	*	*	*		*	*			
CHEMICAL PROCESSES													
Adsorption	*		*	*	*		*	*	*	*	*	*	*
Sediment erosion	*	*				*	*	*					*
Nutrient cycle	*	*					*				*	*	*
Conventional pol.	*		*				*						
Conservative pol.	*	*	*				*		*				
Pesticides	*						*						*
INPUTS													
Daily rainfall			*					*					*
Multi yr. precipitation													
Time series	*	*	*	*	*	*	*		*	*			
Hourly rainfall				*									*
Storm precipitation											*	*	
Calibration data	*	*	*	*		*	*	*	*	*			*
OUTPUT													
Nonpoint storm													*
Nonpoint time series	*		*	*	*	*		*	*				
Stream quality													
Time series		*								*	*	*	

- calibrated input parameters available for more than 100 watersheds
- simulation of runoff generated from snow melt.

USDAHL: It demands the specification of many more input parameters than SWM, with the resulting advantage that it relies less upon calibration versus observed hydrographs. USDAHL does not consider snow melt. It is a physically-based, distributed storm event model incorporating many phenomena known or perceived to influence runoff and erosion. The watershed is defined as a sequence of uplands, hill slopes, and bottom lands.

CREAMS: Sediment algorithms, developed by Foster, et al.[183] represent a detailed description of processes known to affect sediment detachment and transport, sacrificing ability to describe temporal variations in the sediment concentration and delivery rate during the storm or spatial variations throughout the watershed (the former capability is available with NPS and ARM while the latter is offered by distributed system models such as CSU and ANSWERS). CREAMS is a physically based model and, as such, does not require extensive calibrations.

UTM: The Universal Transport Model, developed by Oak Ridge National Laboratory, is based on the SWM. Details regarding the erosion/sediment transport algorithms of UTM are not clearly presented in any Oak Ridge UTM reports.[184]

AGRUN: Is a nonpoint source model applicable to agricultural watersheds. Soil loss is estimated using the USLE, infiltration is based on Horton's equation, and streamflow is simulated using a finite-difference procedure. The AGRUN model consists of several compatible programs designed to simulate agricultural runoff, transport, and movement in the receiving stream. The model can simulate multiple catchment and dendritic channel systems. The model is a modification of the SWMM model.

MRI: Nonpoint Source Loading Function. The method is designed to be used for preliminary estimation of nonpoint source loading rates for urban and non-urban areas. The method is empirical and uses the Universal Oil Loss Equation to predict sediment erosion. The loading functions are useful as a "first cut" estimate of nonpoint source loading rates. Only the transport of sediment-attached constituents is considered (e.g., no dissolved contaminants). Various constituents are considered, including pesticides and heavy metals. The method is applicable to a single, small to large, catchment. It can be applied to waste sites related to hydrocarbons fate.

CSU: Li[179] describes the hydrologic component of CSU, Colorado State University Model, a physically based, distributed system, storm event model. The watershed is segmented via uniform grid squares representing overland flow units, and channels (i.e., distributed system). CSU diverges from CREAMS in that it does not lump the whole watershed into a single uniform slope but routes the water from one segment to the next.

STORM: The Storage, Treatment, Overflow, Runoff Model, STORM, is a continuous simulation model that provides an analysis of the quantity and quality of runoff from urban or nonurban watersheds. STORM computes loads and concentrations of six basic water quality parameters and land surface erosion. The purpose of the program is to aid in the sizing of storage and treatment facilities so that the quantity and quality of storm water runoff and land surface erosion may be controlled. It can be applied at hydrocarbon waste sites.

ANSWER: The Areal, Nonpoint Source Watershed Environment Response Simulation, is a model that simulates behavior of watersheds having agriculture as their primary land use. It is a distributed parameter model, primarily event oriented.

Erosion models.

USLE: The Universal Soil Loss Equation (USLE)[185] is the simplest available model for estimating soil loss/erosion from agricultural watershed. It is based on an extensive empirical database (over 10,000 plot-years or runoff data) and was developed to estimate annual soil-loss, not single storm events. It has demonstrated accuracy in estimating monthly soil loss as well.

NPS: The NonPoint Source model[186] uses the SWM as the hydrologic component and is a lumped parameter continuous simulation model. Contaminant is assumed to move with sediment, comprising a predetermined, constant fraction of the sediment. Like SWM, the hydrologic component, the sediment and contaminant parameters must be calibrated to historical data from the watershed.

ARM: The Agricultural Runoff Mode[187] is similar to NPS. Its additional capabilities are: simulation of agricultural practices, incorporation of nutrient and pesticide chemistry (albeit with relatively simple algorithms) and the provision for transport of dissolved contaminant in both overland and subsurface flow. Like NPS, ARM requires extensive calibrations.

ACTMO: The Agricultural Chemical Transport Model[188] is a nonpoint source model based on USDAHL. The erosion submodel is based on the USLE, as modified by Williams.[189] Clay, silt, and sand fractions are eroded, transported, and deposited together, but are subsequently desegregated for consideration of particle size dependent adsorption in the chemical submodel.

Interfaced Runoff/Stream Models. There exists a category of models in which separate descriptions of runoff and channel processes are linked to describe the watershed. Known models are the: UTM (Unified Transport Model), CMRA (Chemical Migration and Risk Assessment Methodology), HSPF (Hydrocomp Simulation Package). The software development and user-oriented features of these models are quite different, but the hydrologic and sediment transport algorithms of each is quite similar.

UTM: The Unified Transport Model[184] is a multimedia model that simulates the movement of a chemical through an inland watershed. The model calculates the concentration of organic and inorganic chemicals in air, water, soil, sediment and biota. It consists of the Atmospheric Transport Model (ATM), the

Wisconsin Hydrologic Transport Model (WHTM), the Terrestrial Ecosystem Hydrology Model (TEHM), and a suite of associated submodels. UTM incorporates plant/water contaminant interaction in greater detail than either CMRA or HSPF, as well as atmospheric dispersion and deposition phenomena (which classifies it as a multimedia model, like ALWAS). UTM has principally been applied to metals.

CRMA: Incorporates[177] ARM chemical processes, plus sophisticated transient flow phenomena in streams, and particle size dependent on sediment transport. CMRA includes a time series analyses routine which can be used in the context of toxic pollutant risk assessment.

HSPF: Is a series of coupled computer codes designed to simulate: (1) water hydrology: (2) land surface runoff; and (3) the fate and transport of pollutants in receiving water bodies. The model is a transient model applicable to rivers and well-mixed (unstratified) reservoirs. HSPF is modular in structure, allowing great flexibility in the consideration of various physical and chemical processes.

Simulation of Chemical Processes

A number of chemical processes are known to affect contaminant distribution in watersheds, and thus nonpoint source pollution migration. The state-of-the-art with respect to incorporation of these chemical phenomena in watershed models is quite limited. Most available models consider adsorption to eroded sediment particles and a total pseudo first-order loss rate, incorporating volatilization and biodegradation as a single lumped process.

- transport of soluble contaminant in surface and/or subsurface water (e.g., CREAMS, ARM, UTM)
- particle size dependent adsorption or portioning coefficients (e.g., ACTMO, CREAMS)
- the effect of adsorption in retarding contaminant infiltration in subsurface flow (e.g., ACTMO, CREAMS, UTM)
- cation exchange (e.g., UTM)
- volatilization (UTM).

Conclusions

The degree to which each of the above described models have been tested is quite variable, and will only be summarized and highlighted here. The hydrological submodels—SCS curve numbers, SWM, CREAMS—have been used successfully in a wide variety of watersheds and may be confidently relied on for prediction if used appropriately. Similarly, the USLE has been extensively tested and shown to be valid for monthly and longer-term average erosion rates. The simulative capabilities of the CREAMS, ARM and CPS are well known from the literature and have been extensively validated.

For evaluating the response of large watersheds to land use, waste sites, or management controls, CMRA, HSPF, and CSU are likely to be most suitable. For evaluating agricultural management practices and their effect on water quality, ARM and CREAMS would be good choices. In the regulatory context ARM has the advantage of producing a continuous output useful in risk assessment and in determining the frequency of exceedance of numerical standards. CREAMS, on the other hand, incorporates a technically more realistic description of erosion processes and pesticide percolation into soils. It is not clear whether this technical sophistication pays off in terms of improved predictive skill. NPS is the logical choice for evaluating urban runoff, and its algorithms may be used in the HSPF context.

ACKNOWLEDGMENTS

Information contained in this chapter has been compiled in recent years under the direct or indirect support of Arthur D. Little colleagues and University of Massachusetts friends, in alphabetical order: Dr. Itamar Bodek, Dr. Aviva Brecher, Ms. Melanie Byrne, Dr. Edward Calabrese, Dr. Alan Eschenroeder, Dr. Paul Kostecki, Dr. Warren Lyman, Dr. Annette Nold, Ms. Anne Novosel, Ms. Joanne Perwak, Ms. Linda Rosen, Dr. Kate Scow, Dr. William Tucker, Dr. Robert Vranka, Dr. Janet Wagner, and many others.

REFERENCES

1. Bonazountas, M. "Mathematical Pollutant Fate Modeling of Petroleum Products in Soil Systems," in *Soils Contaminated by Petroleum,* E. J. Calabrese and P. T. Kostecki, Eds. (New York, NY: John Wiley & Sons, 1988).
2. Cary, J. W., C. S. Simmons, and J. F. McBride. "Permeability of Air and Immiscible Organic Liquids in Porous Media," *Water Res. Bull.* 25(6) (1989).
3. Jury, W. A., D. Russo, G. Streile, and H. E. Abd. "Evaluation of Volatilization by Organic Chemical Residing Below the Soil Surface," *J. Water Resour. Res.* 26:13–20 (1990).
4. Gee, G. W., C. T. Kincaid, R. J. Lenhard, and C. S. Simmons. "Recent Studies of Flow and Transport in the Vadose Zone," Reviews of Geophysical Supplement, pp. 227–239, April 1991. U.S. National Report to International Union of Geodesy and Geophysics, 1987–1990, American Geophysical Union, Washington, DC, Paper No. 91RG00524.
5. Wickramanayake, G. B., N. Gupta, R. E. Hinchee, and B. J. Nielsen. *J. Environ. Eng.* 117(5) (1991).
6. Ryan, P. A., and Y. Cohen. "One-Dimensional Subsurface Transport of a Nonaqueous Phase Liquid Containing Sparingly Water Soluble Organics: A Front-Tracking Model," *Water Resour. Res.* 27:1487–1500 (1991).

7. Calabrese, E. J., and P. T. Kostecki. *Hydrocarbon Contaminated Soils,* Volume I (Chelsea, MI: Lewis Publishers, 1991).
8. Bonazountas, M. "Fate of Hydrocarbons in Soils: Review of Modeling Practices," in *Hydrocarbon Contaminated Soils,* Volume 1 (Chelsea, MI: Lewis Publishers, 1991).
9. Van Genuchten, M. T., and P. J. Shouse. "Solute Transport in Heterogeneous Field Soils," in D. T. Allen, V. Cohen, and I. R. Kaplan, Eds. *Intermedia Pollutant Transport* (New York, NY: Plenum Publishing Corporation, 1989).
10. Hern. S. C., and S. M. Melancon. *Vadose Zone Modeling of Organic Pollutants* (Chelsea, MI: Lewis Publishers, 1986).
11. Blanton, R., and J. Powell. "Onsite Thermal Treatments: A Case Study in Multisource Petroleum Contamination," in E. J. Calabrese and P. T. Kostecki, Eds. *Hydrocarbon Contaminated Soils and Groundwater,* Volume 2 (Chelsea, MI: Lewis Publishers, 1992).
12. Piotrowski, M. "Full-Scale, In Situ Bioremediation at a Superfund Site: A Progress Report," in E. J. Calabrese and P. T. Kostecki, Eds. *Hydrocarbon Contaminated Soils and Groundwater,* Volume 2 (Chelsea, MI: Lewis Publishers, 1992).
13. Kostecki, P. T., and E. J. Calabrese. *Petroleum Contaminated Soils,* Volume 1 (Chelsea, MI: Lewis Publishers, 1989).
14. Kostecki, P. T., and E. J. Calabrese. *Petroleum Contaminated Soils,* Volume 3 (Chelsea, MI: Lewis Publishers, 1990).
15. Kostecki, P. T., and E. J. Calabrese. *Hydrocarbon Contaminated Soils and Groundwater,* Volume 1 (Chelsea, MI: Lewis Publishers, 1991).
16. Calabrese, E. J., and P. T. Kostecki. *Soils Contaminated by Petroleum* (New York, NY: John Wiley & Sons, 1988).
17. Calabrese, E. J., and P. T. Kostecki. *Petroleum Contaminated Soils,* Volume 2 (Chelsea, MI: Lewis Publishers, 1989).
18. Calabrese, E. J., and P. T. Kostecki. *Hydrocarbon Contaminated Soils and Groundwater,* Volume 2 (Chelsea, MI: Lewis Publishers, 1992).
19. Dragun, J. D. *The Soil Chemistry of Hazardous Materials* (Silver Spring, MD: Hazardous Materials Control Research Institute, 1988).
20. Samiullah, Y. *Prediction of the Environmental Fate of Chemicals* (Essex, England: Elsevier Applied Science Ltd, 1990).
21. "Leaking Underground Fuel Tank (LUFT) Field Manual: Guidelines for Site Assessment Cleanup, and Underground Storage Tank Closure," Water Resources Control Board, Leaking Underground Fuel Tank Force, State of California, 1989.
22. Case, C. D. "Problems in Judicial Review Arising from the Use of Computer Models and Other Quantitative Methodologies in Environmental Decision Making," *Environmental Affairs* 10:251 (1983).
23. Swann, R. I., and A. Eschenroeder. "Fate of Chemicals in the Environment: Compartmental and Multimedia Models for Prediction," ACS Symposium Series 225, American Chemical Society, Washington, DC, 1983.
24. Ehrenfeld, J., and J. Bass. "Handbook for Evaluating Remedial Action Technology Plans," U.S. Environmental Protection Agency Report No. EPA-600/2-83-076, August 1983, USEPA, Cincinnati, OH, 1983.

25. Donigian, A. S., and P. S. C. Rao. "Overview of Terrestrial Processes and Modeling," in S. C. Hern and S. M. Melancon, Eds. *Vadose Zone Modeling of Organic Pollutants* (Chelsea, MI: Lewis Publishers, 1986), p. 3.
26. Bonazountas, M., and J. Fiksel. "ENVIRO: Environmental Mathematical Pollutant Fate Modeling Handbook and Catalogue," EPA Contract No. 68-01-5146, draft report, Arthur D. Little, Inc., Cambridge MA, 1982.
27. Aravamudan, K., M. Bonazountas, A. Eschenroeder, et al. "An Environmental Partitioning Model for Risk Assessment of Chemicals: Volume I: Introduction and Summary; Volume II: Narrative Description of the Methodology; Volume III: User's Manual," EPA Research Contract No. 68-01-3857, Task Order 17, 2 Vol., U.S. Environmental Protection Agency, Washington, DC. Final Report ADL-81099-86. Arthur D. Little, Inc., Cambridge, MA, 1982.
28. Klopffer, W., G. Rippen, and R. Frische. "Physico-Chemical Properties as Useful Tools Predicting the Environmental Fate of Organic Chemicals," *Ecotoxicol. Environ. Safety* 6:294–301 (1982).
29. McCall, P. J., D. A. Laskowski, R. L. Swann, and N. J. Dishburger. "Estimation of Environmental Partitioning of Organic Chemicals in Model Ecosystems," in *Residue Ren* 85, F. A. Gunter and J. D. Gunter, Eds., 1983, pp. 231–244.
30. Mackay, D. M. "Finding Fugacity Feasible," *Environ. Sci. Technol.* 13:1218–1223 (1979).
31. Frische, R., G. Esser, W. Klopffer, and W. Schonborn. "Criteria for Assessing the Environmental Behavior of Chemicals: Selection and Preliminary Quantifications," *Ecotoxicol. Environ. Safety* 6:283–293 (1982).
32. Neely, W. B. *Chemicals in the Environment: Distribution, Transport, Fate, Analysis* (New York, NY: Marcel Dekker, 1980).
33. Wood, W. P. "Comparison of Environmental Compartmentalization," OECD Chemicals Group, Working Part of Exposure Analysis (EXPO), Room Doc. 80.21.
34. Anderson, M. P. "Movement of Contaminants in Groundwater: Chemical Processes," in *Groundwater* (Washington, DC: National Academy Press, 1984).
35. Bonazountas, M., and J. Wagner. "Modeling Mobilization and Fate of Leachates Below Uncontrolled Hazardous Waste Sites," in *Proceedings of the 5th National Conference on Management of Uncontrolled Hazardous Waste Sites*, HMCRI, VA, 1984.
36. Lyman, W. J., W. F. Reehl, and D. H. Rosenblatt. *Handbook of Chemical Property Estimation Methods* (New York, NY: McGraw-Hill Book Company, 1982).
37. Bodek, I., W. J. Lyman, W. F. Reehl, and D. H. Rosenbatt, Eds. *Environmental Inorganic Chemistry: Properties, Processes, and Estimation Methods* (New York, NY: Pergamon Press, 1988).
38. Plummer, L. N., B. F. Jones, and A. H. Truesdell. "WATEQF—A Fortran IV Version of WATEQ, a Computer Program for Calculating Equilibrium of Natural Water," *U.S. Geol. Surv. Water Resources Invest.* 76 (1976).
39. Senesi, N., and G. Sposito. "Residual Copper (II) Complexes in Purified Soil and Sewage Sludge Fulvic Acids: Electron Spin Resonance Study," *Soil Sci. Soc. Am. J.* 48:1247–1253 (1984).
40. Bower, C. A., and L. V. Wilcox. "Soluble Salts," in *Methods of Soil Analysis*, C. A. Black, Ed. (Madison, WI: American Society of Agronomy, 1965), pp. 947–948.

41. Cole, C. R., "Information on the MMT/VVT Model To Be Obtained by the Author," Battelle Pacific Northwest Laboratory, Richland, WA, 1980.
42. Emmerich, W. E., L. J. Lund, A. L. Page, and A. C. Change. "Predicted Solution Pharc. Forms of Heavy Metals in Sewage Sludge-Treated Soils," *J. Environ. Qual.* 11(2):182 (1982).
43. Enfield, C. G., R. F. Carsel, S. Z. Cohen, T. Phan, and D. M. Walters. "Approximating Pollutant Transport to Groundwater," USEPA, RSKERL, Ada, OK, unpublished paper, 1980.
44. Uhlman, K., and M. E. Portman. "Groundwater Modeling Without Fear," *Civil Engineer* ASCE, New York, NY, September, 1991.
45. Mattigod, S. V., G. Sposito, and A. L. Page. "Factors Affecting the Solubilities of Trace Metals in Soils," in *Chemistry in the Environment,* ASA Special Publication No. 40, Soil Science Society of America, American Society of Agronomy, Madison, WI, 1981.
46. Hall, P. L., and H. Quam. "Countermeasures to Control Oil Spills in Western Canada," *Groundwater* 14(3):163 (1976).
47. Bonazountas, M. "Mathematical Modeling of Inorganic Contaminants in the Environment: Air, Soil, Water," U.S. Army contract research effort in progress, Arthur D. Little, Inc., Cambridge, MA, 1985.
48. Dragun, J., S. A. Mason, and J. Barkach. "What Do We Really Know About the Fate of Diesel Fuel in Soil Systems?" in *Hydrocarbon Contaminated Soils,* Volume I, E. J. Calabrese and P. T. Kostecki, Eds. (Chelsea, MI: Lewis Publishers, 1991), pp. 149–165.
49. Kincaid, C. T., J. R. Morrey, and J. E. Rogers. *Geochemical Models for Solute Migration,* Volume 1, "Process Description and Computer Code Selection," Report EA-3417, Electric Power Research Institute, Palo Alto, CA, 1984.
50. Mackay, D. M., and P. V. Roberts. "Transport of Organic Contaminants in Groundwater," *Environ. Sci. Technol.* 19(5):384 (1985).
51. Mercer, J. W. "Miscible and Immiscible Transport in Groundwater," *EOS, USGS* 691 (1984).
52. Faust, C. R. "Transport of Immiscible Fluids Within and Below the Unsaturated Zone: A Numerical Model," Geotrans Rep. No. 84-01, Geotrans, Herndon, VA, 1984.
53. Abriola, L. M., and G. F. Pinder. "A Multiphase Approach to Modeling of Porous Media Contamination by Organic Compounds: 1. Equation Development; 2. Numerical Solution," *Water Resources Res.* 21(1):11–18 (1985).
54. Fried, J. J., P. Muntzer, and L. Zilliox. "Ground-Water Pollution by Transfer of Oil Hydrocarbons," *Groundwater* 17(6):586 (1979).
55. Sudicky, E. A., and P. S. Huyakorn. "Contaminant Migration in Perfectly Known Heterogeneous Groundwater Systems," Paper No. 91RF00525. Reviews of Geophysics, Supplement. American Geophysical Union, Washington, DC, 1991.
56. Wheatcraft, S. W., and J. H. Cushman. Paper No. 91RG00527. Reviews of Geophysics, Supplement. American Geophysical Union, Washington, DC, 1991.
57. Odencrantz, J. E., J. M. Farr, and C. E. Robinson. "Transport Model Parameter Sensitivity for Soil Cleanup Level Determinations Using SESOIL and AT123D in the Context of the California Leaking Underground Fuel Tank Field Manual," in *Hydrocarbon Contaminated Soils,* Volume II, P. T. Kostecki, E. J. Calabrese, and M. Bonazountas, Eds. (Chelsea, MI: Lewis Publishers, 1992), pp. 319–342.

58. Freeze, R. A., and J. A. Cherry. *Groundwater* (Englewood Cliffs, NJ: Prentice-Hall, 1979).
59. Tucker, W. A., and Nelken. "Diffusion Coefficients in Air and Water," in *Handbook of Chemical Property Estimation Methods,* W. J. Lyman, W. F. Reehl, and D. H. Rosenblatt, Eds. (New York, NY: McGraw-Hill, 1982), Chapter 17.
60. Scheidigger, A. E. *The Physics of Flow Through Porous Media,* 3rd ed. (Toronto: University of Toronto Press, 1974).
61. Harleman, D. R. F., and R. R. Rummer. "The Dynamics of Salt-Water Intrusion in Porous Media," *Civ. Eng. Dept. No.* 55 (Cambridge, MA: MIT Press, 1962).
62. Kirda, C., D. R. Nielsen, and J. W. Biggar. "Simultaneous Transport of Chlorides and Water During Infiltration," *Soil Sci. Soc. Am. Proc.* 37:339–345 (1973).
63. Jury, W. A. "Simulation of Solute Transport Using a Transfer Function Model," *Water Resources Res.* 18(2):363–368 (1982).
64. Thomas, R. G. "Volatilization from Soil," in *Handbook of Chemical Property Estimation Methods,* W. J. Lyman, W. F. Reehl, and D. H. Rosenblatt, Eds. (New York, NY: McGraw-Hill Book Company, 1982), Chapter 16.
65. Kinniburgh, D. G. "General Purpose Adsorption Isotherm," *Environ. Sci. Technol.* 20:895–904 (1986).
66. Sposito, G. "The Future of an Illusion: Ion Activities in Soil Solutions," *Soil Sci. Soc. Amer. J.* 48:531–536 (1984).
67. Coates, M., I. W. Connell, and D. M. Barron. "Aqueous Solubility and Octanol to Water Partition Coefficients of Aliphatic Hydrocarbons," *Environ. Sci. Technol.* 19:628–632 (1985).
68. Miller, M. M., S. P. Wasik, G. L. Huang, W. Y. Shiu, and D. Mackay. "Relationship Between Water Octanol-Water Partition Coefficient and Aqueous Solubility," *Environ. Sci. Technol.* 19:522–529 (1985).
69. Harris, J. C. "Rate of Aqueous Photolysis," in *Handbook of Chemical Property Estimation Methods,* W. J. Lyman, W. F. Reehl, and D. H. Rosenblatt, Eds. (New York, NY: McGraw-Hill Book Company, 1982), Chapter 8.
70. Scatchard, G. *Ann. N.Y. Acad. Sci.* 51:660–672 (1949).
71. Sposito, G., K. M. Holtzclaw, and C. S. LeVesque-Madore. "Trace Metal Complexation by Fulvic Acid Extracted from Sewage Sludge: I. Determination of Stability Constants and Linear Correlation Analysis," *Soil Sci. Soc. Am. J.* 45:465–468 (1981).
72. Valentine, R. L. "Nonbiological Transformations," in *Vadose Zone Modeling of Organic Pollutants,* S. C. Hern and S. M. Melancon, Eds. (Chelsea, MI: Lewis Publishers, 1986), pp. 223–241.
73. "Biological Treatment of Soils Containing Manufactured Gas Plant Residues," Remediation Technologies, Inc. Pittsburgh, PA, 1990.
74. Acher, A. J., P. Boderie, and B. Yaron. "Soil Pollution by Petroleum Products. I. Multiphase Migration of Kerosene Components in Soil Columns," *J. Contaminant Hydrology* 4:333–345 (1989).
75. Yaron, B., P. Sutherland, T. Galin, and A. J. Acher. "Soil Pollution by Petroleum Products. II. Adsorption-Desorption of 'Kerosene' Vapors on Soils," *J. Contaminant Hydrology* 4:347–358 (1989).
76. Narasimham, T. N. "A Unified Numerical Model for Saturated and Unsaturated Groundwater Flow," Ph.D. thesis, University of California, Berkeley, CA, 1975.

77. Leadbetter, M. R., and W. G. Tucker. "Environmental Assessment of Industrial Discharges Based on Multiplicative Models," *Environ. Sci. Technol.* 15:1355–1360 (1981).
78. Mercer, J. W., and R. M. Cohen. "A Review of Immiscible Fluids in the Surface: Properties, Models, Characterization and Remediation," *J. Contaminant Hydrology* 6:107–163 (1990).
79. Bachmat, Y., J. Bredehoeft, B. Andrews, D. Holz, and S. Sebastian. "Groundwater Management: The Use of Numerical Models," Water Resources Monograph, American Geophysical Union, Washington, DC, 1980.
80. "Groundwater Models: Scientific and Regulatory Applications," National Research Council (Washington, DC: National Academy Press, 1990).
81. Schwartz, F. W., and A. Growe. "A Deterministic Probabilistic Model for Contaminant Transport," U.S. NRC, NUREG/CR-1609, Washington, DC, 1980.
82. Huff, D. D. "TEHM: A Terrestrial Ecosystem Hydrology Model," Oak Ridge National Laboratory, Oak Ridge, TN, 1977.
83. Murarka, L. "Planning Workshop on Solute Migration from Utility Solid Waste," Publ. EQ-2415, Electric Power Research Institute, Palo Alto, CA, 1982.
84. Schultz, D. "Land Disposal of Hazardous Waste," in *Proceedings of the 8th Annual Research Symposium*, F. J. Hutchell, Ed. U.S. Environmental Protection Agency, Cincinnati, OH, 1982.
85. LeGrand, H. E. *A Standard System for Evaluating Waste Disposal Sites*, National Water Well Association, Washington, DC, 1980.
86. Wilka, L. R., and T. L. Swearingen. "A Manual for Evaluating Contamination Potential of Surface Impoundments," U.S. Environmental Protection Agency, Groundwater Protection Branch, EPA 570/9-78-003, 1978.
87. "Methodology for Rating the Hazard Potential of Waste Disposal Sites," JRB Associates, McLean, VA, 1980.
88. "Site Ranking Model for Determining Remedial Action Priorities Among the Uncontrolled Hazardous Substances Facilities," The MITRE Co., McLean, VA, 1981.
89. "Capillary Pressure and Its Effects on Non-Aqueous Phase Liquid Migration and Contamination," report to commercial client, prepared for USEPA, State of New York, Arthur D. Little, Inc., Cambridge, MA, 1982.
90. Adams, R. T., and F. M. Kurisu. "Simulation of Pesticide Movement in Small Agricultural Watersheds," Final Report, Environmental Research Laboratory, U.S. Environmental Protection Agency, Athens, GA, 1976.
91. Carsel, R. F., C. N. Smith, L. A. Mulkey, J. D. Dean, and P. P. Jowise. "User's Manual for the Pesticide Root Zone Model (PRZM): Release 1," Report No. EPA/600/3-84/109, U.S. Environmental Protection Agency, Athens, GA, 1984.
92. Nelson, R. W., and J. A. Schur, "Assessment of Effectiveness of Geologic Oscillation Systems: PATHS Groundwater Hydrologic Model," Battelle Pacific Northwest Laboratory, Richland, WA, 1980.
93. Yeh, G. T., and D. S. Ward. "FEMWASTE: A Finite-Element Model of Waste Transport Through Saturated-Unsaturated Porous Media," Oak Ridge National Laboratory, Environmental Sciences Division, Publ. No. 1462, ORNL-5601, 1981.
94. Foote, H. P. For information contact Battelle Pacific Northwest Laboratories, P.O. Box 999, Richland, WA 00352, 1982.

95. Tomovic, R. *Sensitivity Analysis of Dynamic Systems* (New York, NY: McGraw-Hill Book Company, 1963).
96. Donigian, A. S., Jr. "Model Predictions vs. Field Observations: The Model Validation/Testing Process," in *Fate of Chemicals in the Environment: Compartmental and Multimedia Models for Prediction,* R. L. Swann and A. Eschenroeder. ACS Symposium Series 225, Chapter 8, American Chemical Society, Washington, DC, 1983, pp. 151–171.
97. Dickson, K. L., A. W. Maki, and J. Cairns. *Modeling the Fate of Chemicals in the Aquatic Environment* (Ann Arbor, MI: Ann Arbor Science, 1982), p. 413.
98. Corapcioglu, M. Y., and M. A. Hossain. "Theoretical Modeling of Biodegradation and Biotransformation of Hydrocarbons in Subsurface Environments," *J. Theor. Biol.* 142:503–516 (1990).
99. Carberry, J. B., and S. H. Lee. "Fate and Transport of Petroleum in Unsaturated Soil Zone Under Biotic and Abiotic Conditions," *Wat. Sci. Tech.* 22(6):45–52 (1990).
100. Cadena, F., D. J. Fingleton, and R. W. Peters. "Evaluation of VOC Emissions from Landfarming Operations," *44th Purdue Industrial Waste Conference Proceedings* (Chelsea, MI: Lewis Publishers, 1990).
101. Sposito, G., and W. A. Jury. "The Lifetime Probability Density Function for Solute Movement in the Subsurface Zone," *J. Hydrology* 102:503–518 (1988).
102. Gotz, M. N., and P. V. Roberts. "Interpreting Organic Solute Transport Data from Field Experiments Using Physical Non-Equilibrium Models," *J. Contam. Hydrology* 1:77–93 (1986).
103. Jury, W. A., and Sposito, G. "Field Calibration and Validation of Solute Transport Models for the Unsaturated Zone," *Soil Sci. Soc. Amer. J.* 49 (1985).
104. Moore, G. E., T. E. Stoeckenius, and D. A. Stewart. "A Survey of Statistical Measures of Model Performance and Accuracy for Several Air Quality Models," EPA-450/4-83-001 (Research Triangle Park, NC: U.S. Environmental Protection Agency, 1982).
105. Ambrose, R. B., Jr., and S. E. Roesch. "Dynamic Estuary Model Performance," *J. Environ. Eng. Div.* ASCE, 108 (1982).
106. Halfom, E. "Regression Method in Ecotoxicology: A Better Formulation Using the Geometric Mean Functional Regression," *Environ. Sci. Tech.* 19:747–749 (1985).
107. Baehr, A. L., G. E. Hoag, and M. C. Marley. "Removing Volatile Contaminants from the Unsaturated Zone by Including Advective Air-Phase Transport," *J. Contaminant Hydrology* 4:1–26 (1989).
108. Grenney, J. W., C. L. Caupp, R. C. Sims, and T. E. Short. "A Mathematical Model for the Fate of Hazardous Substances in Soil: Model Description and Experimental Results," *Hazardous Waste and Hazardous Materials* 4(3) (1987).
109. Jury, W. A., D. Russo, G. Streile, and H. E. Abd. "Evaluation of Volatilization Chemicals Residing Below the Soil Surface," Paper No. 89WR01546. Water Resources Research, American Geophysical Union, Washington, DC, 1989.
110. Lighty, J. S., G. D. Silcox, D. W. Persing, V. A. Cundy, and D. G. Linz. "Fundamentals of the Thermal Remediation of Contaminated Soils: Particle and Bed Desorption Models," *J. Environ. Sci. Technol.* 24(5) (1990).
111. Shields, W. J., and S. M. Brown. "Applicability of POSSM to Petroleum Product Spills," in *Petroleum Contaminated Soils,* Volume 1, P. T. Kostecki and E. J. Calabrese, Eds. (Chelsea, MI: Lewis Publishers, 1989).

112. Hillel, D. ''Movement and Retention of Organics in Soil: A Review and a Critique of Modeling,'' in *Petroleum Contaminated Soils,* Volume 1, P. T. Kostecki and E. J. Calabrese, Eds. (Chelsea, MI: Lewis Publishers, 1989).
113. Eastcott, L., W. Y. Shiu, and D. Mackay. ''Modeling Petroleum Products in Soils,'' in *Petroleum Contaminated Soils,* Volume 1, P. T. Kostecki and E. J. Calabrese, Eds. (Chelsea, MI: Lewis Publishers, 1989).
114. Short, T. E. ''Movement of Contaminants from Oil Wastes During Land Treatment,'' in *Soils Contaminated by Petroleum: Environmental and Public Health Effects* E. J. Calabrese and P. T. Kostecki, Eds. (New York, NY: John Wiley & Sons, 1988).
115. Lenhard, R. J., and J. C. Parker. ''Measurement and Prediction of Saturation-Pressure Relationships in Three-Phase Porous Media Systems,'' *J. Contaminant Hydrology* 1:407–424 (1987).
116. Van Dam, J. ''The Migration of Hydrocarbons in a Water Bearing Stratum,'' in *The Joint Problem of Oil and Water Industries,* P. Hepple, Ed. (Amsterdam: Elsevier, 1967), pp. 55–96.
117. Schwille, F. ''Groundwater Pollution in Porous Media by Fluids Immiscible with Water,'' *Sci. Total Environ.* 21:173–185 (1981).
118. Stone, H. L. ''Estimation of Three Phase Relative Permeability and Residual Oil Data,'' *J. Can. Pet. Technol.* 12(4):53–61 (1973).
119. Farr, A. M., R. J. Houghtalen, and D. B. McWhorter. ''Volume Estimation of Light Nonaqueous Liquids Phase in Porous Media,'' *Groundwater* 28(1):February (1990).
120. Cary, J. W., C. S. Simmons, and J. F. McBride. ''Predicting Oil Infiltration and Distribution on Unsaturated Soils,'' *Soil Sci. Soc. Am. J.* 53:335–342 (1989).
121. Johnson, P. C., M. W. Kemblowski, and J. D. Colthart. ''Quantitative Analysis for Cleanup of Hydrocarbon-Contaminated Soils by In Situ Soil Venting,'' *Groundwater* (1990).
122. Lenhard, R. J., and J. C. Parker. ''Estimation of Free Hydrocarbon Volume from Fluid Levels in Monitoring Wells,'' *Groundwater* 28(1) (1990).
123. Kaluarachchi, J. J., and J. C. Parker. ''Modeling Multicomponent Organic Chemical Transport in Three Fluid Phase Porous Media,'' *J. Contaminant Hydrology* 5:349–374 (1990).
124. Mohanty, K. K., H. T. Davis, and L. E. Scriven. ''Physics of Oil Entrapment in Water-Wet Rock,'' SPE Reservoir Engineering, Society of Petroleum Engineering, February 1987.
125. Cederberg, G. A., R. L. Street, and J. O. Leckie. ''A Groundwater Mass Transport and Equilibrium Chemistry Model for Multicomponent Systems,'' *Water Resources Res.* 21(8):1095–1104 (1985).
126. Sposito, G., F. T. Bingham, S. S. Yadav, and C. A. Inouye. ''Trace Metal Complexation by Fulvic Acid Extracted from Sewage Sludge: I. Determination of Chemical Models,'' *Soil Sci. Soc. Am. J.* 46:51–56 (1982).
127. Florence, T. M. ''Trace Metal Species in Fresh Waters,'' *Water Resources Res.* 11:681–687 (1977).
128. Allen, H. E., R. H. Hall, and T. D. Brishin. ''Metal Speciation: Effects on Aquatic Toxicity,'' *Environ. Sci. Technol.* 14:441–443 (1980).
129. Sposito, G., ''Chemical Models of Inorganic Pollutants in Soils,'' *CRC Crit. Rev. Environ. Control* 15(1) (1985).

130. Miller, C. W., and L. V. Benson. "Simulation of Solute Transport in a Chemically Reactive Heterogeneous System: Model Development and Application," *Water Resour. Res.* 19:381–391 (1983).
131. Jennings, A. A., D. J. Kirkner, and T. L. Theis. "Multicomponent Equilibrium Chemistry in Groundwater Quality Models," *Water Resour. Res.* 18:1089–1096 (1982).
132. Theis, D. L., D. J. Kirkner, and A. A. Jennings. "Multi-Solute Subsurface Transport Modeling for Energy and Solid Wastes," Dept. of Civil Engineering, University of Notre Dame, Notre Dame, IN, 1982.
133. Mattigod, S. V., and G. Sposito. "Chemical Modeling of Trace Metal Equilibria in Contaminated Soil Solutions Using the Computer Program GEOCHEM," in *Chemical Modeling in Aqueous Systems,* E. A. Jenne, Ed., Symposium Series No. 93, American Chemical Society, Washington, DC 837-856 (1979).
134. Jenne, E. A. "Chemical Modeling: Goals, Problems, Approaches and Priorities," in *Chemical Modeling in Aqueous Systems,* E. A. Jenne, Ed., Symposium Series No. 93, American Chemical Society, Washington, DC, 1979.
135. Nordstrom, D. K., L. N. Plummer, T. M. L. Wigley, T. T. Wolery, J. W. Ball, E. A. Jenne, R. L. Bassett, D. A. Crerar, T. M. Florence, B. Fritz, M. Hoffman, G. R. Holdren, G. M. Lafon, S. V. Mattigod, R. E. McDuff, F. Morel, M. M. Reddy, G. Sposito, and J. Thrailkill. "A Comparison of Computerized Chemical Models for Equilibrium Calculations in Aqueous Systems," in *Chemical Modeling in Aqueous Systems,* E. A. Jenne, Ed., Symposium Series No. 93, American Chemical Society, Washington, DC, 1979.
136. Kirkner, D. J., H. W. Reeves, and A. Jennings. "Multicomponent Solute Transport with Sorption and Soluble Complexation," *Adv. Water Resources* 7:120 (1984).
137. Sposito, G., and S. V. Mattigod. "GEOCHEM: A Computer Program for the Calculation of Chemical Equilibria in Soil Solutions and Other Natural Water Systems," Department of Soil and Environmental Science, University of California, Riverside, CA, 1980.
138. McDuff, R. E., and F. M. M. Morel. "Description and Use of the Chemical Equilibrium Program REDEQL-2," Technical Rep. EQ-73-02, California Institute of Technology, Pasadena, CA, 1973.
139. McDuff, R. E., and F. M. M. Morell. "Description and Use of the Chemical Equilibrium Program REDEQL-2," Technical Rep. EQ-73-02, W. M. Keck Laboratory, California Institute of Technology, Pasadena, CA, 1974.
140. James, R. V., and J. Rubin. "Applicabilities of Local Equilibrium Assumption to Transport Through Soils of Solutes Affected by Ion Exchange," in *Chemical Modeling in Aqueous Systems,* E. A. Jenne, Ed., Symposium Series No. 93, American Chemical Society, Washington, DC, 1979.
141. Nelson, D. W., D. E. Elrick, and K. K. Tanji, Eds. *Chemical Mobility and Reactivity in Soil Systems,* SSSA Spec. Publ. No. 11, Soil Science Society of America, American Society of Agronomy, Madison, WI, 1983.
142. Parkhurst, D. L., L. N. Plummer, and D. C. Thorstenson. "BALANCE: A Computer Program for Calculating Mass Transfer for Geochemical Reactions in Groundwater," Report USGS/WRI-82-14, U.S. Geological Survey, Reston, VA, 1982.
143. Ball, J. W., E. A. Jenne, and D. K. Nordstrom, "WATEQ2: A Computerized Chemical Model for Trace and Major Element Speciation and Mineral Equilibria in Natural

Waters,'' in *Chemical Modeling in Aqueous Systems,* E. A. Jenne, Ed., Symposium Series No. 93, American Chemical Society, Washington, DC, 1979.

144. Mahier, R. J., F. T. Bingham, G. Sposito, and A. L. Page. ''Cadmium-Enriched Sewage Sludge Application to Acid and Calcareous Soils: Relation Between Treatment, Cadmium in Saturation Extracts, and Cadmium Uptake,'' *J. Environ. Qual.* 9:359–364 (1980).
145. Norvell, W. A. ''Equilibria of Metal Chelates in Soil Solution,'' in *Micronutrients in Agriculture,* J. J. Mortvedt, P. M. Giordano, and W. L. Lindsay, Eds., Soil Science Society of America, Madison, WI, 1972.
146. Behel, D., Jr., D. W. Nelson, and L. E. Sommers. ''Assessment of Heavy Metal Equilibria in Sewage Sludge-Treated Soil,'' *J. Environ Qual.* 12(2):181 (1983).
147. Matthews, P. J. ''Control of Metal Applications Rates from Sewage Sludge Utilization in Agriculture,'' *CRC Crit. Rev. Environ. Control* 14(3) (1984).
148. Westall, J., and H. Hohl. ''A Comparison of Electrostatic Models for the Oxide/Solution Interface,'' *Adv. Colloid Interface Sci.* 12:265–294 (1980).
149. Jennings, A. A., and D. J. Kirkner. ''Instantaneous Equilibrium Approximation Analysis,'' American Society of Civil Engineers, New York, NY, 1984.
150. Dowdy, R. H., and V. V. Volk. ''Movement of Heavy Metals in Soils,'' in *Chemical Mobility and Reactivity in Soil Systems,* D. W. Nelson et al., Eds., SSSA Spec. Publ. No. 11, Soil Science Society of America, American Society of Agronomy, Madison, WI, 1983.
151. Lindsay, W. L., *Chemical Equilibria in Soils* (New York, NY: John Wiley & Sons, 1979).
152. Tucker, W. A., E. V. Dose, G. J. Gensheimer, R. E. Hall, D. N. Koltuniak, C. D. Pollman, and D. H. Powell. ''Evaluation of Critical Parameters Affecting Contaminant Migration Through Soils,'' in *Soils Contaminated by Petroleum: Environmental and Public Health Effects,* E. J. Calabrese and P. T. Kostecki, Eds. (New York, NY: John Wiley & Sons, 1988).
153. ''Test Guideline for Octanol/Water Partition Coefficients,'' Organization for Economic Cooperation and Development, Washington, DC, 1979.
154. ''Mathematical Pollutant Fate Modeling of Petroleum Products in Soil Systems,'' prepared for U.S. Army Toxic and Hazardous Materials Agency, Report AMXTH-TE-TR-85030, 1985.
155. ''SAS User's Guide: Statistics,'' SAS Institute, Inc., Cary, NC, 1982.
156. Bonazountas, M., and J. Wagner. ''SESOIL: A Seasonal Soil Compartment Model,'' Arthur D. Little, Inc., Cambridge, MA, 1981.
157. ''Mathematical Models for Atmospheric Pollutant,'' Appendices A through D, Electric Power Research Institute, Palo Alto, CA, 1979.
158. Bonazountas, M., A. Brecher, and R. Vranks. ''Mathematical Environmental Modeling,'' in *Environmental Inorganic Chemistry,* I. Bodek, W. Lyman, W. Reel, and D. Rosenblatt, Eds. (New York, NY: Pergamon Press, 1988).
159. Eschenroeder, A. Q., G. C. Magil, and C. R. Woodruff. ''Assessing the Health Risks of Airborne Carcinogens,'' Final report EPRI EA-4021, Electric Power Research Institute, Palo Alto, CA, 1985.
160. Zannetti, P. ''Air Pollution Modeling Theories,'' Computational Methods and Available Software, CA, 1990.

161. Turner, D. B. "Workbook of Atmospheric Dispersion Estimates," U.S. Environmental Protection Agency, Office of Air Programs, Research Triangle Park, NC, 1974.
162. Keddie, A. W. C. "Atmospheric Dispersal of Pollutants and the Modeling of Air Pollution," in *Causes, Effects, and Control,* R. M. Harrison, Ed., Royal Soc. Chem. Special Publ. No. 44, pp. 197–207, 1983.
163. Doury, A. "Operational Calculation Aids for Atmospheric Dispersion," *Sci. Tot. Environ.* 25:3–17 (1982).
164. Munn, R. E., and B. Bolin. Review paper. "Global Air Pollution: Meteorological Aspects: A Survey," *Atmos. Environ.* 5:363–402 (1971).
165. Ellis, J. H. "Acid Rain Control Strategies," *Environ. Sci. Technol.* 22:1248–1255 (1988).
166. U.S. Environmental Protection Agency, "Guidelines on Air Quality Models (Revised)," Office of Air Quality Planning and Standards, Research Triangle Park, NC (1986). Report No. EPA/450/2-78-027R.
167. "Conceptual Design of a Regional Water Quality Screen Model," U.S. Department of Energy, DOE/EV/10154-T1, Washington, DC, 1981.
168. Bonazountas, M., K. Scow, J. Green, P. Hughes, and J. Wagner. "Mathematical Modeling of Total Maximum Daily Loads (TMDLs) for Selected Stream Reaches," EPA-68-01-6160/Task 8(3). U.S. Environmental Protection Agency Monitoring and Data Support Division, Washington, DC, 1982.
169. Orlob, G. T., Ed. "Mathematical Modeling of Water Quality: Streams, Lakes, and Reservoirs," International Institute for Applied Systems Analysis (New York, NY: John Wiley & Sons, 1983).
170. Delos, C. G., W. L. Richardson, J. V. DePinto, P. W. Rogers, K. Rygwelski, R. Wathgton, R. B. Ambrose, and J. P. St. John. Technical Guidance Manual for Performing Waste Load Allocations, Book II: Streams and Rivers; Chapter 3: Toxic Substances. U.S. Environmental Protection Agency, Office of Water Regulations and Standards, Water Quality Analysis Branch, Washington, DC, 1983.
171. Zison, S. W., W. B. Mills, D. Deimer, C. W. Chen. "Rates, Constants and Kinetics Formulations in Surface Water Quality Modeling," Final Draft Report. EPA-600/3-78-105. U.S. Environmental Protection Agency, Office of Research and Development, Athens, GA, 1978.
172. Burns, L. A., D. M. Cline, and R. R. Lassiter. "Exposure Analysis Modeling System (EXAMS) User Manual and System Documentation," EPA-600/3-82-023, U.S. Environmental Protection Agency, Athens, GA, 1982.
173. Roesner, L. A., P. R. Giguere, and D. E. Evenson. User's Manual for the Stream Quality Model QUAL-II. EPA-600/9-81-015, U.S. Environmental Protection Agency, Environmental Research Laboratory, Office of Research and Development, Athens, GA, 1981.
174. Swain, W. R., and V. R. Shannon, Eds. "Proceedings of the Second American-Soviet Symposium on the Use of Mathematical Models to Optimize Water Quality Management," EPA-600/9-80-033. U. S. Environmental Protection Agency, Environmental Research Laboratory, Duluth, MN, 1980.
175. International Workshop on the Comparison of the Application of Mathematical Models for the Assessment of Changes in Water Quality in River Basins, Both Surface Water and Groundwater, UNESCO, La Coruna, Spain, 1982.

176. Di Toro, D. M., J. J. Fitzpatrick, and R. V. Thomann. "Water Quality Analysis Simulation Program (WASP) and Model Verification Program (MVP): Program Documentation," EPA/68-01-3872, U.S. Environmental Protection Agency, Environmental Research Laboratory, Duluth, MN, 1981.
177. Onishi, Y., S. M. Brown, A. R. Olsen, M. A. Parkhurst, S. E. Wise, and W. H. Walters. "Methodology for Overland and Instream Migration and Risk Assessment of Pesticides," EPA-600/3-82-024, U.S. Environmental Protection Agency, Environmental Research Laboratory, Athens, GA, 1982.
178. Novotny, V. "A Review of Hydrologic and Water Quality Models for Simulation of Agricultural Pollution," in *Agricultural Nonpoint Source Pollution: Model Selection and Applications,* A. Giorgini and F. Zingales, Eds. (New York, NY: Elsevier Publishing Co., 1986).
179. Li, A. Colorado State University Model. Colorado State University, Water Resources Center, Boulder, CO, 1979.
180. Adams, R. T., and F. M. Kurisu. "Simulation of Pesticide Movement on Small Agricultural Watersheds," EPA-600/3-76-066, U.S. Environmental Protection Agency, Environmental Research Laboratory, Athens, GA, 1976.
181. Chow, V. T. "Hydrologic Modeling," *J. Boston Soc. Civ. Eng.* 60:1–27 (1972).
182. "Hydrology: National Engineering Handbook," U.S. Department of Agriculture, Soil Conservation Service, Engineering Division, Document 210-VI-NEH-4, Amend. 6, Washington, DC, March, 1985.
183. Foster, G. R., L. J. Lane, J. D. Nowlin, J. M. Laflen, and R. A. Young. "A Model to Estimate Sediment from Field-Sized Areaws," in "CREAMS: A Field Scale Model for Chemicals, Runoff, and Erosion from Agricultural Management Systems," W. G. Knisel, Ed., U.S. Department of Agriculture, Conservation Research Report No. 26, USDA-SEA-AR, Tucson, AZ, 1980.
184. Patterson, M. R, T. J. Sworski, A. L. Sjoreen, M. G. Browman, C. C. Coutant, D. M. Hetrick, B. D. Murphy, and R. J. Raridon. "A User's Manual for UTM-TOX, A Unified Transport Model, UTM," Draft Report, Contract No. IAG No. AD-89-F-1-399-0, U.S. Environmental Protection Agency, Office of Toxic Substances, Washington, DC, 1980.
185. Wischeier, W. H., and D. D. Smith. "Predicting Rainfall Erosion Losses," U.S. Department of Agriculture, Agriculture Handbook No. 537, 1978.
186. Donigian, A. S., Jr., and N. H. Crawford. "Modeling Pesticides and Nutrients on Agricultural Lands," EPA-600/2-7-76-043. U.S. Environmental Protection Agency, Environmental Research Laboratory, Athens, GA, 1976.
187. Donigian, A. S., Jr., D. C. Beyerlein, H. H. Davis, Jr., and N. H. Crawford. "Agricultural Runoff Management (ARM) Model, Version II: Refinement and Testing," EPA-600/3-77-098, U.S. Environmental Protection Agency, Environmental Research Laboratory, Athens, GA, 1977.
188. Frere, M. H., C. A. Onstad, and H. N. Holtan. "ACTMO, an Agricultural Chemical Transport Model," U.S. Department of Agriculture, Agricultural Research Service, Headquarters, ARS-H-3, 1975.
189. Williams, J. R. "Sediment Yield Prediction with Universal Equation Using Runoff Energy Factor," in *Present and Prospective Technology for Predicting Sediment Yields and Sources,* U.S. Department of Agriculture, Agricultural Research Service, South Region, ARS-S-40, 1975.

190. Sehmel, G. A., and W. H. Hodgson. "Predicted Dry Deposition Velocities. Atmosphere-Surface Exchange of Pollutants and Gaseous Pollutants," USERDA Symposium Series 38, 1974, pp. 399–422
191. Buat-Menard, P. E. "The Role of Air-Sea Exchange in Geochemical Cycling," NATO-ASI Series 185, (Norwell, MA: D. Reidel Publishing Co., 1986).
192. Yoshida, K. T. Shigeoka, and F. Yamauchi. "Non-Steady State Equilibrium Model for the Preliminary Prediction of the Fate of Chemicals in the Environment, *Ecotox. Environ. Safety* 7:179–190 (1983).
193. Jurry, W. A., R. Grover, W. F. Spencer, and W. J. Farmer. "Modeling Vapor Losses of Soil-Incorporated Triallate," *Soil Sci. Soc. Am. J.* 44:445–450 (1980).

CHAPTER 10

Petroleum Contaminated Soil: Chemistry and Modeling

Sum Chi Lee, Linda Eastcott, Wan Ying Shiu, and **Donald Mackay,** Department of Chemical Engineering and Applied Chemistry, University of Toronto, Canada

INTRODUCTION

In this chapter we review and discuss the issue of modeling, or expressing in quantitative mathematical terms, the fate of individual hydrocarbons present in oil-contaminated soil. This is an important issue to the victims of petroleum contamination, to those who regulate this issue, and to those who may have caused the problem. Subsurface contamination incidents can be very expensive to study, remediate, and compensate. Often there are no obviously satisfactory intervening countermeasures. There can be considerable public inconvenience, economic loss, risk of fire and explosion, distress from odor, and real or perceived adverse health effects. There can be difficult legal problems of compensation, liability, and transfer of ownership of contaminated land.

It is a challenging scientific and engineering issue for several practical reasons:

- The subsurface environment is accessible only with considerable effort and cost; thus, the overall condition of the oil may be difficult to determine. Its future behavior is in even more doubt.
- Although there have been many successful studies of the fate of oil in soils, there is still doubt about the mechanisms and rates of hydrocarbon movement in the oil, air, and water phases present in soils. Nor is the

partitioning or association of oil with soil mineral and organic matter adequately understood. In part, this is because soils are usually heterogeneous.

- Degradation or chemical conversion is known to occur, primarily by microbial oxidation (photolysis being obviously unimportant at depths in the soil) but the rates are not easily predictable because of uncertainties about the numbers and types of microorganisms present and their performance under conditions of limited oxygen and nutrients, and excess carbon. The fate of the degradation products and intermediates is even less understood.
- Crude oils and petroleum products are mixtures of, conservatively, several thousand hydrocarbons with many of indeterminate structure, including substances containing oxygen, sulfur, and nitrogen, and possibly fuel-derived metals such as lead and manganese.

A desirable goal is to establish an ability to make quantitative statements of past and future behavior for field situations in which actual spills have occurred. In only a few cases has it been possible to state, for example, that in a period of one year, 20% of a particular mass of oil spilled was lost by evaporation, 2% by dissolution, and 30% by degradation, and that 3% was altered by photolysis. The underlying justification is that if we understand the condition of the petroleum (i.e., its amount, location, area, depth, and composition) and especially how it has migrated, or will migrate in oil, air, and water phases, we will be in a better position to assess the hazard and devise appropriate remedial measures. In some cases it may be justifiable to take no remedial action, but this should be done for good reasons, i.e., it can be demonstrated that this is the best course of action.

Despite the difficulties listed earlier, it is believed that by a combination of complementary laboratory studies and field studies, it should be possible to develop an improved capability of predicting the fate of petroleum in soils and thus respond to situations of petroleum contamination more effectively and economically.

In summary, the purpose of this chapter is to provide a perceptive on the fate and modeling of soil contamination with petroleum, and to examine the chemistry of hydrocarbons as the determinant of oil behavior. A modeling approach is described and discussed, and some illustrative results are given in the hope that others may follow this approach and thus contribute to an increased understanding of oil behavior in soil.

MECHANISMS OF MIGRATION

We suggest that there are two levels of modeling effort which are required to understand subsurface petroleum migration: (a) the bulk flow or hydrodynamics of oil in which the oil is treated as a simple fluid; and (b) expressions of the fate of individual chemicals present in oil. We focus primarily on the second.

Bulk Flow of Oil

The oil (when introduced from the surface) tends to drain through the unsaturated soil zone, replacing the air and possibly some water. The rate of migration is primarily controlled by the oil's viscosity and the porosity or permeability of the soil. Some secondary influences are the amount of water present and the wetting characteristics of the oil-air-water-soil system.

Most oils are less dense than water, the exceptions being bitumens and chlorinated solvents. The oil thus tends to "float" at the interface between the water-saturated and unsaturated zones and spread horizontally, usually in the direction of groundwater movement. The groundwater level may rise and fall with weather and season, causing the oil to rise and fall in response, but some oil tends to become permanently trapped in the soil and resists being "floated out"; thus, it is possible to have considerable quantities of oil smeared below the groundwater level. The oil may move as a large slug, as smaller slugs, as droplets, or as colloidal or emulsified particles.

It should be noted that denser-than-water fluids will continue to sink through groundwater until they reach an impermeable layer. Other fluids may be appreciably soluble in water; thus, they dissolve in the groundwater and migrate as a solution. To a first approximation, petroleum products are "insoluble" in that the fraction that dissolves is negligible (i.e., a few percent). This is negligible from the point of view of bulk oil behavior. But this small amount is not negligible from the point of view of groundwater contamination. It takes only 1 g of oil to contaminate a cubic meter of water to a level of 1 ppm.

The Fate of Oil: Partitioning into Air and Water

The second and more subtle problem is the estimation and modeling of the behavior of the individual oil components as they partition into air and water phases (evaporate and dissolve) and use these phases as vehicles for migration, often at a very much more rapid rate than occurs to the oil phase. The implications for air and water contamination are obvious.

If the oil is a pure substance, such as benzene, it is straightforward to calculate the concentration in the air and water phases in equilibrium with the hydrocarbon phase. For benzene the saturation air concentration deduced from the gas law and the published vapor pressure is 400 g/m^3, the water concentration (solubility) is 1780 g/m^3, and for the oil phase, the density is some 700,000 g/m^3. Benzene is one of the most volatile and soluble hydrocarbons; thus, more representative concentrations are 1–50 g/m^3 or mg/L in air and water.

If, as is usual, the oil is a mixture, then the solubilities of each component in air and water are dependent not only on the oil composition and temperature but also on the amount of air and water to which the oil has been, or is exposed. As the oil is exposed to increasing volumes of air (or water), it becomes depleted of the more volatile (or soluble) components and it thus appears to become less soluble in air (or water). There is thus no unique solubility for a mixture; it depends on the oil's exposure conditions. A fuel oil may thus contaminate water to the extent of 10 g/m^3 initially (ignoring the presence of oil emulsions), but as it is washed with more water, this concentration drops steadily. But it is not clear if 1 g/m^3 will be reached in 1 day, 1 month, or 1 year. This issue has been discussed by Billington et al.,[1] Shiu et al.,[2] and Mackay et al.[3]

The only method of approaching this problem rigorously is to examine in detail the oil's partitioning chemistry in the hope that if we can determine its composition, and the properties of the components, we can then deduce the bulk properties of solubility and vapor pressure, and how these properties will change with time in the subsurface environment. From these data it is possible to calculate the composition in the air and water near the site and make further assessments of fire and toxicity hazards. We later return to this issue and suggest a method of approach.

Sorption

In the vicinity of a bulk oil phase, the amount of hydrocarbon "adsorbed or absorbed," or to be less precise "sorbed," onto the soil matrix is usually not important because the solid is essentially smothered with oil. Remote from the bulk oil phase where the hydrocarbons are present only in low concentrations in dissolved form in air or water, the phenomenon of sorption becomes very important. The soil has a capacity to sorb hydrocarbons as they migrate through the soil in air or water. Essentially, if the amount sorbed is 10 times the amount dissolved, approximately 90% of the hydrocarbon is immobilized and the average velocity of migration is reduced by a "retardation factor" of about 10. As air and water flow out from a contaminated region, they create a soil zone that is contaminated by sorption. The sorbed concentration builds up until it reaches its saturation level; then any further hydrocarbon passes through to contaminate new downstream material. The migration velocity (and thus area of contamination) of hydrocarbons is therefore dependent on the extent of this sorption phenomenon.

A noteworthy implication is that even if the bulk oil source is removed, the surrounding soil contaminated by sorption will continue to act as a long-term source of contamination to clean water and air.

The evidence indicates that hydrocarbon sorption occurs primarily to organic matter in the soil; that is, humic detritus. If organic matter is present to the extent of 1% or more, it usually dominates sorption. Sorption to wet clays or other mineral matter is relatively less important. At low organic contents, the sorptive capacity of clays and other mineral matter may become important, especially if they are dry. The kinetics of sorption to these porous media may be quite slow. Unfortunately, this capacity is difficult to assess. Organic sorption is easier to predict, the approach being to define an "organic carbon partition coefficient" or ratio of concentration in organic matter to that in water and relate it to the commonly available "octanol-water partition coefficient" or to water solubility.

Methods of estimating organic sorption have been reviewed by Mackay[4] and Lyman et al.,[5] who also give extensive data on hydrocarbon properties. An excellent review is that of Karickhoff.[6] Other sources of information on hydrocarbon properties and partition coefficients are Hansch and Leo,[7] Mackay and Shiu,[8] Verschueren,[9] Zwolinski and Wilhoit,[10] and the handbooks by Howard[11,12] and Mackay et al.[13] Commercial and institutional databases are also available, as are systems for estimating properties when none have been measured (Lyman et al., 1982).

Oil Properties

Central to the issue of understanding the fate of oil in the subsurface environment is its composition and chemistry. Table 1 gives some illustrative properties of selected hydrocarbons at 25°C, the aim being to demonstrate that there is a variation in properties extending over many orders of magnitude. The solubility, vapor pressure, and sorption coefficients of a specific oil reflect the combined properties of the individual constituents, weighted in proportion to the amount present. The chemistry of petroleum is reviewed comprehensively in the text by Speight.[16]

The principal properties of interest are the partitioning properties of vapor pressure (i.e., solubility in air), water solubility, and organic carbon or octanol-water partition coefficients. In general, a hydrocarbon will migrate through air and water if an appreciable fraction of it can partition into these phases. Also of importance are the degradation or reaction properties, especially susceptibility to microbial degradation.

MODELS OF HYDROCARBON FATE

If the spill is of a single substance such as styrene, the bulk properties of the hydrocarbon phase will be constant, and obviously those of pure styrene. For a single component with well-established physical-chemical properties and biodegradability properties, it is relatively easy to assemble and validate a fate model using approaches similar to those which have been used for agricultural chemicals.[17-20] If the oil consists of a few well characterized components, it is still possible to calculate the bulk properties from the composition, and even estimate the fate using slightly more complex models. The difficulty with oils, as Figure 1 illustrates for Norman Wells crude oil, is that they consist of a large number of components, only a few of which have been well characterized, and their individual and collective properties are in considerable doubt. The logical approach, when faced with such a situation, is to break the oil down into a manageable number of components or pseudocomponents. A pseudocomponent may be regarded as a group of hydrocarbons which is categorized for convenience as one compound based upon a set of defined criteria; for example, having similar volatility or water solubility. For toxicological or regulatory reasons, it may be desirable

Table 1. Physical-Chemical Properties of Selected Hydrocarbons at 25°C

Chemical	Molecular[a] Mass g/mol	mp[a] °C	bp[a] °C	Density[a] g/cm³	Molar[b] Volume LeBas cm³/mol	Solubility[c] g/m³	Vapor[c] Pressure Pa	Log K_{ow}[d]
n-Pentane	72.15	−129.70	36.1	0.614	118.0	38.5	68400	3.62
n-Hexane	86.20	−95.00	68.0	0.660	140.6	9.5	20200	4.11
n-Heptane	100.20	−90.60	98.4	0.670	162.8	2.93	6110	4.66
n-Octane	114.20	−56.20	125.7	0.700	185.0	0.66	1880	5.18
n-Nonane	128.30	−54.00	151.0	0.720	207.2	0.122	571	5.65
n-Decane	148.28	−9.60	174.2	0.730	229.4	0.052	175	6.25
n-Dodecane	170.33	−9.60	216.3	0.766	273.8	0.0034	15.7	6.80
n-Tetradecane	198.38	5.86	253.7	0.763	318.2	0.000655	1.27	8.00
Cyclopentane	70.14	−93.90	49.3	0.799	100.0	156	42400	3.00
Cyclohexane	84.20	6.55	80.7	0.779	118.0	55	12700	3.44
Methylcyclohexane	98.19	−126.60	100.9	0.770	140.4	14	6180	2.82
Benzene	78.10	5.53	80.0	0.879	96.0	1780	12700	2.13
Toluene	92.10	−95.00	111.0	0.867	118.0	515	3800	2.69
Ethylbenzene	106.20	−95.00	136.2	0.867	140.4	152	1270	3.13
p-Xylene	106.20	13.20	138.0	0.860	140.4	185	1170	3.15
1,3,5-Trimethylbenzene	120.20	−44.70	164.7	0.865	184.8	48	325	3.58
n-Propylbenzene	120.20	−101.60	159.2	0.862	170.0	55	449	3.69

continued

Table 1. *Continued*

Chemical	Molecular[a] Mass g/mol	mp[a] °C	bp[a] °C	Density[a] g/cm^3	Molar[b] Volume LeBas cm^3/mol	Solubility[c] g/m^3	Vapor[c] Pressure Pa	Log K_{ow}[d]
Naphthalene	128.20	80.20	218.0	1.025	147.6	31.7	10.4	3.35
1-Methylnaphthalene	142.20	−22.00	244.6	1.030	199.7	28.4	7.9	3.87
Biphenyl	154.20	71.00	277.5	0.992	185.0	7.48	1.2	4.03
Fluorene	166.20	116.00	295.0		188.0	1.84	0.008	4.18
Anthracene	178.20	216.20	340.0	1.283	199.0	0.041	0.0008	4.63
Phenanthrene	178.20	101.00	339.0	0.980	199.0	1.29	0.0161	4.57
Fluoranthene	202.30	111.00	383.0		217.0	0.263	0.00121	5.22
Pyrene	202.30	156.00	393.0	1.271	214.0	0.135	0.0006	5.22
Chrysene	228.30	255.00	441.0		251.0	0.002	0.00000385	5.79
Benzo[a]pyrene	252.30	175.00	496.0		263.0	0.0038	0.00000073	6.04
Styrene	104.10	−30.60	145.2		133.0	300	670	3.16
Quinoline	129.20		237.7	1.093	159.5	6110	0.0155	
p-Cresol	108.10	34.8	202.0	1.018	125.6	18000	0.0142	1.94

[a]CRC Handbook of Chemistry and Physics, 64th ed.[14]
[b]Eastcott et al.[15]
[c]Mackay and Shiu.[8]
[d]Hansch and Leo.[7]

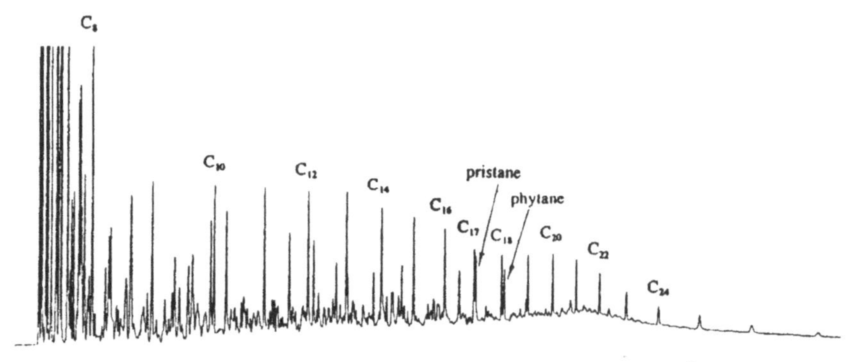

Figure 1. Capillary gas chromatogram of fresh Norman Wells crude oil.

(instead or as well) to focus on only one, or a few, individual hydrocarbons such as benzene or benzo[a]pyrene.

There may thus be three situations to be modeled:

1. A pure, single component spill; e.g., styrene.
2. One or more components in a complex mixture; e.g., benzene in a crude oil spill.
3. The overall fate of all components or pseudocomponents in an oil mixture.

It is believed that in the third situation the best approach is to group all hydrocarbons in the oils into a two-dimensional matrix according to volatility-related parameter and chemical class. In studies of diesel fuel,[21] Norman Wells crude oil,[22] we have suggested a system illustrated by Table 2 in which the hydrocarbon classes are the following five structural groups:

1. Normal alkanes (which are easily identified, have well-defined properties, and are rapidly degraded).
2. Branched alkanes and cycloalkanes (which are difficult to identify and are less degradable).
3. Isoprenoids (which can be regarded as isoalkanes and are very resistant to biodegradation).
4. Aromatics (which are fairly well identified and tend to be much more water soluble than other hydrocarbons).
5. Polars (which are mainly sulfur, oxygen, and nitrogen compounds that are poorly identified and tend to be quite water soluble).

The rows are groups of similar volatility or vapor pressure as indicated by position on the capillary column gas chromatogram, and are labeled as elution groups. The elution groups are designated as C_6, C_7, C_8 up to C_{21}. The class C_8 consists of all compounds that elute after n-heptane (n-C_7) up to and including

Table 2. Composition Matrix for Fresh Norman Wells Crude Oil

Elution Group	Normal Alkanes	Branched Alkanes	Isoprenoids	Aromatics	Polars	Total
C_6	0.98	0.44	0.00	0.00	0.00	1.42
C_7	1.56	0.24	0.00	0.19	0.00	1.99
C_8	2.49	8.10	0.00	0.36	0.00	10.95
C_9	1.85	7.52	0.00	0.21	0.00	9.58
C_{10}	1.84	8.41	0.00	0.24	0.00	10.49
C_{11}	1.36	4.69	0.00	0.38	0.00	6.43
C_{12}	1.37	2.99	0.00	0.50	0.20	5.06
C_{13}	1.96	3.86	0.00	0.52	0.20	6.54
C_{14}	1.46	2.74	0.00	0.68	0.20	5.08
C_{15}	1.33	3.27	0.00	0.76	0.20	5.56
C_{16}	2.23	2.60	0.00	1.16	0.20	6.19
C_{17}	0.84	2.60	0.00	0.70	0.20	4.34
C_{18}	0.82	2.09	0.80	0.83	0.20	4.74
C_{19}	0.77	2.42	0.89	0.96	0.20	5.24
C_{20}	0.65	2.61	0.00	0.90	0.20	4.36
C_{21}	1.11	8.17	0.00	2.55	0.20	12.03
Total	22.62	62.75	1.69	10.94	2.00	100.00

n-octane (n-C_8). The average vapor pressure is thus somewhat larger than that of n-octane because the isoalkanes in this class are mainly C_7 and even C_8, which have higher vapor pressures than the equivalent normal isomer. Benzene, for example, falls between n-C_6 and n-C_7; thus, it falls into the C_7 row and the aromatic column. Pristane falls in the C_{18}, isoprenoid class. Methylpyridine falls in the C_9, polar class. Each pseudocomponent can be assigned average properties of solubility, vapor pressure, and degradability mainly on the basis of the established properties of a hydrocarbon known to be present in the group.

The concept of elution groups can be more clearly illustrated with the gas chromatogram of a synthetic hydrocarbon mixture in Figure 2. The gas chromatogram, which separates the hydrocarbons in approximate order of volatility, has been divided up along the normal alkane peaks which are labeled as C_n, where n is the carbon number of the normal alkane. Methylcyclohexane belongs in the C_8, branched alkane class; toluene in the C_8, aromatic class; and n-octane in the C_8, normal alkane class.

Chemical analyses by gas chromatography (GC), column chromatography, and other techniques enable estimates to be made of the amounts of hydrocarbons in each class. Different petroleum products will correspond to different groupings. A gasoline will contain mostly C_3–C_{10} rows, diesel fuel mainly C_{10}–C_{20}; a crude oil will contain a wide range of components and even higher boiling material. The petroleum refining process essentially represents selection of a pattern of groups within the entire matrix.

If the oil is subjected to evaporation, dissolution in water, and microbial degradation, it is possible to suggest equations describing these process rates for each component or element as a function of its properties. This is obviously best done by computer. One simple approach is to calculate "exposures" to evaporation,

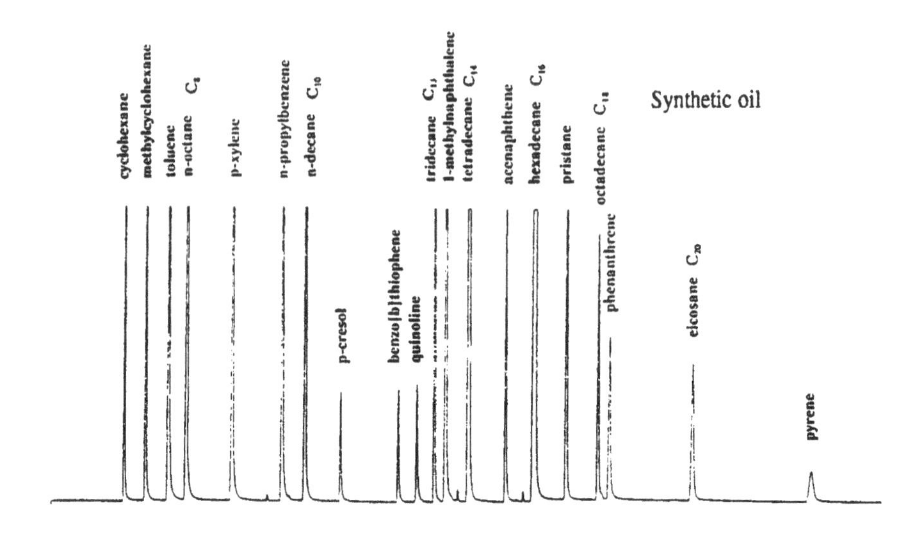

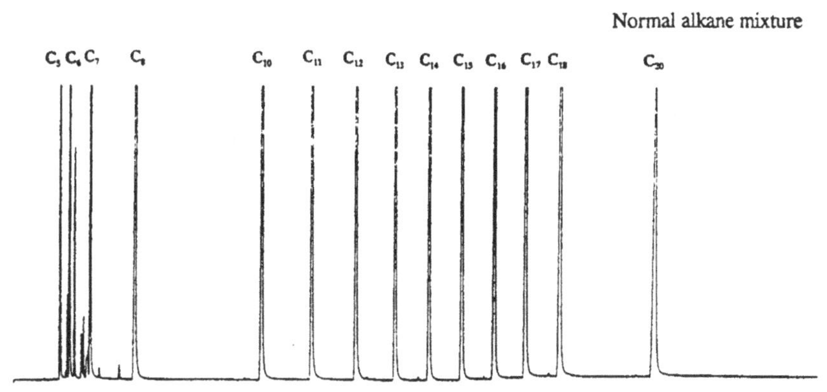

Figure 2. Classification of the hydrocarbons in a synthetic oil mixture according to elution groups.

dissolution in water (i.e., leaching) and degrading reactions, then multiply these "exposures" by a "susceptibility" term to estimate the extent of loss by these three mechanisms, individually and in total. A brief justification follows, the approach being suggested by Eastcott et al.[22]

A Simple Fate Model

If a volume of oil V m^3 contains a volume fraction x of a specific component or pseudocomponent, the partial pressure P (Pa) will be approximately xP^S where P^S is the pure component vapor pressure. Similarly, the water solubility C (mol/m^3) will be approximately xC^S where C^S is the pure component solubility. This is essentially application of a version of Raoult's law. Air in contact with the oil will thus have a concentration of hydrocarbon P/RT mol/m^3 where R is the gas constant 8.314 Pa m^3/mol K and T is absolute temperature K. If the flow rates of air and water from the oil are F_AV m^3/h and F_WVm^3/h, the rates of loss by evaporation and dissolution will be F_AVP/RT or F_AVxP^S/RT, and F_WVC or F_WVxC^S mol/h. F_A and F_W are thus the number of "oil volumes" of air and water which contact the oil each hour with units of reciprocal time. For example, if V is 2 m^3 and F_A is 100 h^{-1}, the air flow is 200 m^3/h. Reactions may also result in the loss of hydrocarbons at a rate Vxk/v mol/h where k is a first-order degradation rate constant h^{-1} and v is the hydrocarbon's molar volume m^3/mol. The amount of hydrocarbon present is Vx/v mol, thus

$$\frac{d\left(\frac{Vx}{v}\right)}{dt} = -\frac{F_AVxP^S}{RT} - F_WVxC^S - \frac{Vxk}{v}$$

Rigorous solution requires that as x changes, its effect on v be included, especially if x is large.

Single Component Spill

In the limit, for a single component spill x is 1.0 and the equation can be integrated directly to give V as a function of time and

$$\frac{dV}{dt} = -\frac{F_AVP^Sv}{RT} - F_WVC^Sv - Vk$$

The groups P^Sv/RT and C^Sv are dimensionless ratios of hydrocarbon concentrations in air to oil and water to oil and can conveniently be designated as H_A and H_W thus

$$\frac{dV}{dt} = -F_AVH_A - F_WVH_W - Vk$$

$$= -V(F_AH_A + F_WH_W + k)$$

Assuming F_A, F_W and k to be constant with time gives V as a function of time and the initial volume V_O.

$$V = V_O\exp(-F_AH_At - F_WH_Wt - kt)$$

The relative importance of each loss process becomes obvious. F_At and F_Wt are the ratio of the total air and water volumes to the oil volume, and can thus be regarded as dimensionless "exposures." For reaction, time is the exposure.

The susceptibility terms are H_A, H_W and k. If estimates are available for F_A, F_W, H_A, H_W, k and t, the volume of hydrocarbon remaining can be estimated.

Multicomponent Spill

For a component in a mixture, the easiest approach is to assume V to be constant and integrate with respect to x, yielding a similar expression which will be only an approximate solution that is similar to that devised for a pure single component

$$x = x_O \exp(-F_A H_A t - F_W H_W t - kt)$$

The half times for evaporation, dissolution and reaction in both single and multicomponent cases are then $0.693/(F_A H_A)$ i.e., τ_A, $0.693/(F_W H_W)$ i.e., τ_W, and $0.693/k$ i.e., τ_R and the overall (shorter) half time τ_O is

$$1/\tau_O = 1/\tau_A + 1/\tau_W + 1/\tau_R$$

The relative rates or proportions are evaporation τ_O/τ_A, dissolution τ_O/τ_W, and reaction τ_O/τ_R. An example of such a calculation is given in Table 3 which shows that for a spill of styrene, the approximate half-lives are

evaporation	1925 hours	i.e., 39%
dissolution	1820 hours	i.e., 41%
reaction	3850 hours	i.e., 20%
total	753 hours (30 days)	i.e., 100%

In this case of relatively rapid exposure to air and water flows, there is a half-life of 30 days which is mainly attributable to losses by evaporation and dissolution.

As is readily apparent by consulting Table 1, the properties of other hydrocarbons vary over a wide range; thus, substitution of these properties can result in markedly different half-lives and in considerable changes in the relative importance of the removal processes.

As also shown in Table 3, it is possible to calculate the approximate concentrations of the hydrocarbons of interest in the receiving media of air and water.

It may be necessary to estimate F_A from mass transfer coefficients, or diffusivity and path lengths and F_W from water flows. These flows may be augmented for remedial purposes during soil venting or washing; thus, F_A and F_W may be "engineered" values.

The primary advantage of this approach is its conceptual simplicity.

DISCUSSION

By sampling oil from spill sites and analyzing the samples to assign amounts to each element, it is possible to examine the fidelity of this model, modify the

Table 3. Sample Calculation with Styrene Spilled in Shallow Soil Subject to Fairly Rapid Water Flow

Physical-Chemical Properties at 25°C

Molecular mass	104.1 g/mol	
Vapor pressure, P^S 670 Pa		$P^S/RT = 0.27$ mol/m^3 (28.1 g/m^3
Water solubility, C^S	2.88 mol/m^3	(300 g/m^3)
Molar volume, v	1.33×10^{-4}m^3/mol	(133 cm^3/mol)

Therefore, the "susceptibility" terms are

$H_A = P^S v/RT = 3.6 \times 10^{-5}$ and $H_W = vC^S = 3.8 \times 10^{-4}$

Pure Styrene

For a spill of pure styrene with an initial volume V_O of 1m^3 the "exposure" terms can be estimated if it can be deduced that the styrene is exposed to an air volumetric flow rate of 10m^3/h, a water volumetric flow rate of 1m^3/h, and degradation at a rate constant k of $1.8 \times 10^{-4}h^{-1}$, F_A is 10 h^{-1} and F_W is 1 h^{-1}. The half times for evaporation, dissolution and degradation are $\tau_A = 0.693/(F_A H_A) = 1925$ h, $\tau_W = 0.693/(F_W H_W) = 1820$ h, and $\tau_R = 0.693/k = 3850$ h, respectively. The overall half time τ_O is about 753 h or 1 month.

After 100 hours of exposure, the remaining volume of styrene is estimated by

$$V = V_O \exp(-F_A H_A t - F_W H_W t - kt)$$

$$V = 1\ m^3 \exp(-10\ h^{-1} \times 3.6 \times 10^{-5} \times 100\ h - 1.0\ h^{-1} \times 3.8 \times 10^{-4} \times 100\ h - 1.8 \times 10^{-4}\ h^{-1} \times 100\ h) = 0.91\ m^3$$

The concentration of styrene in the air and water are the saturation values of 0.27 mol/m^3 (28.1 g/m^3) in air and 2.88 mol/m^3 (300 g/m^3) in water.

Styrene as a Component in a Multicomponent Mixture

If the spill is 1% styrene in a mixture which is primarily less volatile and less soluble, e.g., a diesel fuel, then the same half-lives will apply but the concentrations in the air and water phases will initially be only 1% of saturation and will fall as x decreases.

In this case, after 100 hours x will have fallen from 0.0100 to 0.0091 and the concentration in air will now be

$$0.27 \times 0.01 \times 0.91 \text{ or } 0.0025 \text{ mol/m}^3 \text{ (0.26 g/m}^3\text{)}$$

Note: By assuming F_A and F_W to be constant, the integration is much easier. This implies, however, that as V decreases, the air flow rate VF_A also decreases. If the flow rate is constant (as is likely), then it may be necessary to "segment" the calculation into time periods, increasing F_A to allow for the decrease in V.

equations, and adjust the parameters to reconcile the model and the observed fate. This gives the capability of estimating or predicting future behavior and behavior in other situations. This capability is invaluable for several reasons:

- It enables the changing composition and amount of oil to be estimated as a function of exposure conditions.
- The prevailing exposure conditions or rates can be estimated by analyzing fresh and weathered samples of oil. The inherent ability of a specific subsurface oil to evaporate, dissolve, and degrade can then be measured.

- The amounts of oil that have been removed can be estimated; thus, the amounts that have evaporated (and thus have established concentrations in air), dissolved in water (and thus have established concentrations in surface or groundwater), and permanently degraded can be estimated. This may be valuable in assessing whether or not hazardous conditions existed.
- The capability can be used as an identification or "fingerprinting" technique to determine if a given sample of oil could conceivably have originated from a certain source. For example, a gasoline should readily be distinguishable from a diesel fuel. This should be valuable in litigation, especially when assigning responsibility for spills.
- It provides a rational and systematic method for processing and interpreting potentially large amounts of analytical data obtained in monitoring programs.
- It enables estimates to be made of the fate of specific hydrocarbons of particular toxicological interest. For example, we can focus on the fate of benzene or a specific carcinogen.
- Finally, from a scientific viewpoint it enables authoritative general statements of principle or broad applicability to be made concerning the fate of the oil, instead of merely making case-by-case observations.

A major difficulty is ascertaining, in the contaminated soil volume, the extent of heterogeneity in process rate. We suspect that there is a considerable variation in the rate of evaporation and biodegradation from place to place within the soil. Occasional sampling can thus give a misleading impression of the overall rate of conversion. We refer to this colloquially as the "Danish Blue Cheese" problem. If samples of that cheese were taken of perhaps 1 mm^3 size, it could be deduced that there was either very little, or very intense microbial activity, because the activity is very heterogeneous. Fortunately, that heterogeneity is visible. On the contrary, we have no visible or mental impression of the heterogeneity of oil biodegradation. Regretfully, oil degrading microorganisms are not a visible blue. We believe that it is essential to characterize this effect, both in the laboratory and in the field.

A further problem is the effect of oil concentration. We find that at concentrations in soil below 1 or 0.5% of oil by volume, the degradation rate is fairly independent of oil concentration. However, at higher oil concentrations, the first-order degradation rate must fall and the oil biodegradation half-life must rise. Ultimately, of course, when the oil reaches saturation conditions in the soil, i.e., 30% to 50% oil, biodegradation virtually ceases. This is an important consideration in real spills. The location of the point at which biodegradation starts to be adversely affected by the amount of oil present is not well established. The effect arises because of the toxicity of the large amount of oil to the microbial population, and possibly other factors, such as alteration of the water environment, the depletion of oxygen, and the availability of nutrients relative to the amount of hydrocarbons. It should be possible, we believe, to develop equations expressing biodegradation rate as a function of variables such as hydrocarbon type (i.e., element in the matrix), total concentration of oil, amounts of oxygen and nutrients, and exposure or acclimation time.

CONCLUSIONS

It is suggested that the identification and development of remedial measures and the assessment of hazard, with their legal and regulatory ramifications, will be more effective and economic if they are based on sound understanding of how bulk oils and the specific hydrocarbons present in oils behave in the subsurface environment. This includes understanding the bulk flow properties of the oil as it migrates through the soil interacting with water, and assessing or quantifying the partitioning between air, water, soil, and oil phases and the sorption processes. Often artificial enhancement of these natural processes can be effective remedial measures, examples being venting to encourage evaporation or stimulation of microbial degradation by introducing organisms, nutrients, and oxygen.

The problem of predicting the fate of subsurface petroleum is complex and challenging, but with carefully conducted studies of partitioning, migration, and degradation in both the laboratory and the field, and reconciliation of the findings with improving mathematical models, it should be possible to create and validate an adequate predictive capability. This can form a foundation of scientific knowledge from which effective economic and acceptable remedial measures can be devised within an appropriate regulatory context.

In summary, we believe that a suite of models will emerge describing the fate of oil in soils. Some will treat the migration or flow of the bulk phase under the action of gravity and hydraulic and capillary forces. Others, such as that described here, will treat the gradual decay of specific hydrocarbons over a period of months as they are subject to evaporation, dissolution, and biodegradation. The purpose of these models is not just to describe the science of chemical fate in soil (although this is justification enough), but to contribute to an improved ability to describe, respond to, and mitigate the fate and effects of hydrocarbons in real spill situations.

ACKNOWLEDGMENTS

The authors are indebted to NSERC, the Association of American Railroads, Environment Canada, and Canadian Petroleum Products Institute for their support of work described in this chapter.

REFERENCES

1. Billington, J. W., G. L. Huang, F. Szeto, W. Y. Shiu, and D. Mackay. "Preparation of Aqueous Solutions of Sparingly Soluble Organic Substances. I. Single Component Systems," *Environ. Toxicol. Chem.* 7:117–124 (1988).

2. Shiu, W. Y., A. Maijanen, and D. Mackay. "Preparation of Aqueous Solutions of Sparingly Soluble Organic Substances. II. Multicomponent Systems—Hydrocarbon Mixtures and Petroleum Products," *Environ. Toxicol. Chem.* 7:125–137 (1988).
3. Mackay, D., W. Y. Shiu, A. Maijanen, and S. Feenstra. (1991). "Dissolution of Non-Aqueous Phase Liquids in Groundwater," *J. Contam. Hydrol.* 8:23–42 (1991).
4. Mackay, D. *Multimedia Environmental Models: The Fugacity Approach* (Chelsea, MI: Lewis Publishers, Inc., 1991).
5. Lyman, W. J., W. F. Reeht, and D. H. Rosenblatt. *Handbook of Chemical Property Estimation Methods* (New York, NY: McGraw-Hill Book Company, 1982).
6. Karickhoff, S. W. "Organic Pollutant Sorption in Aquatic Systems," *J. Hydraul. Eng.* (ASCE) 110:707–735 (1984).
7. Hansch, C., and A. Leo. *Substituent Constants for Correlation Analysis in Chemistry and Biology* (New York, NY: John Wiley & Sons, 1979).
8. Mackay, D., and W. Y. Shiu. "A Critical of Henry's Law Constants for Chemicals of Environmental Interest," *J. Phys. Chem. Ref. Data* 10:1175–1199 (1981).
9. Verschueren, K. *Handbook of Environmental Data on Organic Chemicals*, 2nd ed. (New York, NY: Van Nostrand Reinhold. 1983).
10. Zwolinski, B. J. and R. C. Wilhoit. *Handbook of Vapor Pressures and Heats of Vaporization of Hydrocarbons and Related Compounds*, API-44, TRC Publ. No. 101, Texas A&M University, College Station, 1971.
11. Howard, P. H., Ed. *Handbook of Fate and Exposure Data for Organic Chemicals. Vol. I.—Large Production and Priority Pollutants* (Chelsea, MI: Lewis Publishers, Inc., 1989).
12. Howard, P. H., Ed. *Handbook of Fate and Exposure Data for Organic Chemicals. Vol. II.—Solvents* (Chelsea, MI: Lewis Publishers, Inc., 1990).
13. Mackay, D., W. Y. Shiu, and K. C. Ma. *The Illustrated Handbook of Physical-Chemical Properties and Environmental Fate for Organic Chemicals. Vol I.—Monoaromatic Hydrocarbons, Chlorobenzenes and Polychlorinated Biphenyls* (Chelsea, MI: Lewis Publishers, Inc., 1992).
14. Weast, R. C. *CRC Handbook of Chemistry and Physics*, 64th ed. (Boca Raton, FL: CRC Press, 1983–1984).
15. Eastcott, L., W. Y. Shiu, and D. Mackay. "Environmentally Relevant Physical-Chemical Properties of Hydrocarbons: A Review of Data and Development of Simple Correlations," *Oil & Chemical Pollution* 4:191–216 (1988).
16. Speight, J. C. *The Chemistry and Technology of Petroleum* (New York, NY: Marcel Dekker, 1980).
17. Jury, W. A., W. F. Spencer, and W. J. Farmer. "Behavior Assessment Model for Trace Organics in Soil: I. Model Description," *J. Environ. Qual.* 12:558-566 (1983).
18. Jury, W. A., W. J. Farmer, and W. F. Spencer. "Behavior Assessment Model for Trace Organics in Soil: II. Chemical Classification and Parameter Sensitivity," *J. Environ. Qual.* 13:567–572 (1984).
19. Jury, W. A., W. J. Farmer, and W. F. Spencer. "Behavior Assessment Model for Trace Organics in Soil: III. Application of Screening Model," *J. Environ. Qual.* 13:573–579 (1984).
20. Jury, W. A., W. F. Spencer, and W. J. Farmer. "Behavior Assessment Model for Trace Organics in Soil: IV. Review of Experimental Evidence," *J. Environ. Qual.* 13:580–587 (1984).

21. Eastcott, L., W. Y. Shiu, and D. Mackay. "Modeling Petroleum in Soils," in *Petroleum Contaminated Soils,* Vol. I, P. T. Kostecki and E. J. Calabrese, Eds., (Chelsea, MI: Lewis Publishers, Inc., 1989), pp. 63–80.
22. Eastcott, L., W. Y. Shiu, and D. Mackay. Weathering of Crude Oil in Surface Soils. Report No. 90-3 for PACE (Petroleum Association for the Conservation of the Canadian Environment) now Canadian Petroleum Products Institute, 1990.

CHAPTER 11

The Utility of Environmental Fate Models to Regulatory Programs

Michael A. Callahan, Office of Health and Environmental Assessment, U.S. Environmental Protection Agency, Washington, D.C.

The environmental fate of pollutants has been an important issue at least as early as the 1960s, when environmentalists began to raise public consciousness that chemicals may ultimately travel far from their original sources.[1,2] The U.S. Environmental Protection Agency (EPA) itself was founded at least in part to answer the public concern that the environmental fate of chemicals be understood and chemicals be controlled, so that people or the ecosystem would not be unexpectedly exposed to toxic chemicals.

During the 1970s and 1980s, the EPA and other organizations have supported research designed to better understand, and better predict, the fate of chemicals in the environment. Computational tools, usually computer-based programs or models, were developed to help in predictions. Early models predicted the transport or dispersion of the medium itself (i.e., air, water), and used these phenomena as a first approximation of where the pollutant would go. Later, properties of the pollutant, how these properties affected the pollutant's interaction with the environment, and degradation of the pollutant were taken into account. Still, by the late 1970s, most such models were "single-medium" models; that is, they were air models, groundwater models, surface water models, etc., but didn't follow pollutants from one medium to another. In the late 1970s, "multimedia" models began to be developed, either by connecting several single-medium models together, or by designing more sophisticated integrated multimedia models.

Throughout the years of development of environmental fate models, their primary utility in the regulatory process of EPA has remained the same: fate models were, and are, useful tools to help decisionmakers understand *some* of the factors that need to be weighed in a regulatory decision. In order to understand how environmental fate models fit into the regulatory scheme, it is necessary to show the relationships among risk management, risk assessment, exposure assessment, and environmental fate.

RISK ASSESSMENT AND RISK MANAGEMENT

The National Research Council (NRC) in 1983[3] noted that regulatory actions taken by government agencies such as EPA are based on two separate and distinct, albeit related, processes. These are *risk assessment* and *risk management.* According to the NRC, risk assessment is "the use of the factual base to define the health effects of exposure of individuals or populations to hazardous materials and situations." Risk management, on the other hand, is "the process of weighing policy alternatives and selecting the most appropriate regulatory action, integrating the results of risk assessment with engineering data and with social, economic, and political concerns to reach a decision."

The NRC report further describes risk assessment as containing the following four steps: *hazard identification* (the determination of whether a particular chemical is or is not causally linked to particular health effects), *dose-response assessment* (the determination of the relation between the magnitude of exposure and the probability of occurrence of the health effects in question), *exposure assessment* (the determination of the extent of human exposure before or after application of regulatory controls), and *risk characterization* (the description of the nature and often the magnitude of human risk, including attendant uncertainty).

In considering these definitions, some interesting points arise. First, regulatory decisions are not made only on the basis of risk assessment; that is, only on the basis of the factual science of the extent and nature of risk. Rather, regulatory decisions use the scientific analysis (risk assessment) as only one of several inputs needed. "Engineering data," or the practicability of a regulatory control being considered, is a major factor in regulatory decisions, as are "economic concerns," or the cost of implementing the control being considered. Second, exposure assessment is a component of risk assessment, and as such includes a responsibility to determine exposure both before and after regulatory controls. At the time a regulatory decisionmaker is considering alternatives for controlling risks, the "after control" exposures cannot be measured, since they do not yet exist. Exposure assessment must then be used as a predictive tool in estimating potential future exposures. It is this predictive nature of exposure assessment which relies heavily on models, including environmental fate models.

APPROACHES TO EXPOSURE ASSESSMENT

The EPA Guidelines for Exposure Assessment[4] and a recent National Research Council report[5] define exposure as the *contact* with a chemical or physical agent. The magnitude of this contact is determined by measuring or estimating the concentration of an agent available at the visible outer boundaries of the person during some specific time. Once the agent crosses the outer boundary, the amount inside becomes a dose. Exposure assessment is the qualitative or quantitative determination/estimation of the magnitude, frequency, duration, and route of exposure, and often also describes the resultant internal dose.

Figure 1 shows a schematic of environmental exposure, dose, and effects. On the left side of the diagram, there is a source of an environmental pollutant ("environmental situation"), from which chemicals are released into the environment. These chemicals may be degraded or transported, or both, in the environment, leading to various environmental concentrations in environmental media (air, groundwater, etc.). Most environmental fate models simulate the *pathways* chemicals take in the environment and the resultant media concentrations. Note, however, that the pollutant concentrations alone do not constitute exposure. Since exposure is *contact,* there must be an individual or population contacted. The activities that bring the population into contact with the chemical are termed "population habits or "activity patterns." It is at the point where an individual or population contacts the environmental pollutant that exposure occurs.

After the exposure (following the top line in Figure 1 to the right), some of the pollutant may be absorbed, after which there may be changes in levels of the pollutant in body tissues or fluids, or changes in such things as enzyme production. These are measurable changes often referred to as biomarkers. Subsequent health effects may or may not occur.

A regulatory agency such as EPA is especially interested in the point of exposure, since it is at this point that risk reduction makes the biggest impact. Since toxicity is an intrinsic property of a chemical, seldom if ever can the toxic nature of a pollutant be changed. Risk reduction then becomes a matter of reducing exposure, either by reducing the concentrations of the pollutant in the environmental media, changing the form of the pollutant so it is less likely to be absorbed, or by changing population habits so that individuals will no longer come in contact with the pollutant.

The lower line in Figure 1 illustrates three approaches that exposure assessors have used to quantify exposures. First, there is point-of-contact measurement. It is possible to measure the magnitude of contact directly as it happens. The best-known example of this is the radiation dosimeter, a small badge-like device worn in areas where exposure to radiation is possible. The dosimeter effectively measures exposures to radiation while it is taking place, then indicates when a preset level has been exceeded. Another example of point-of-contact measurement of

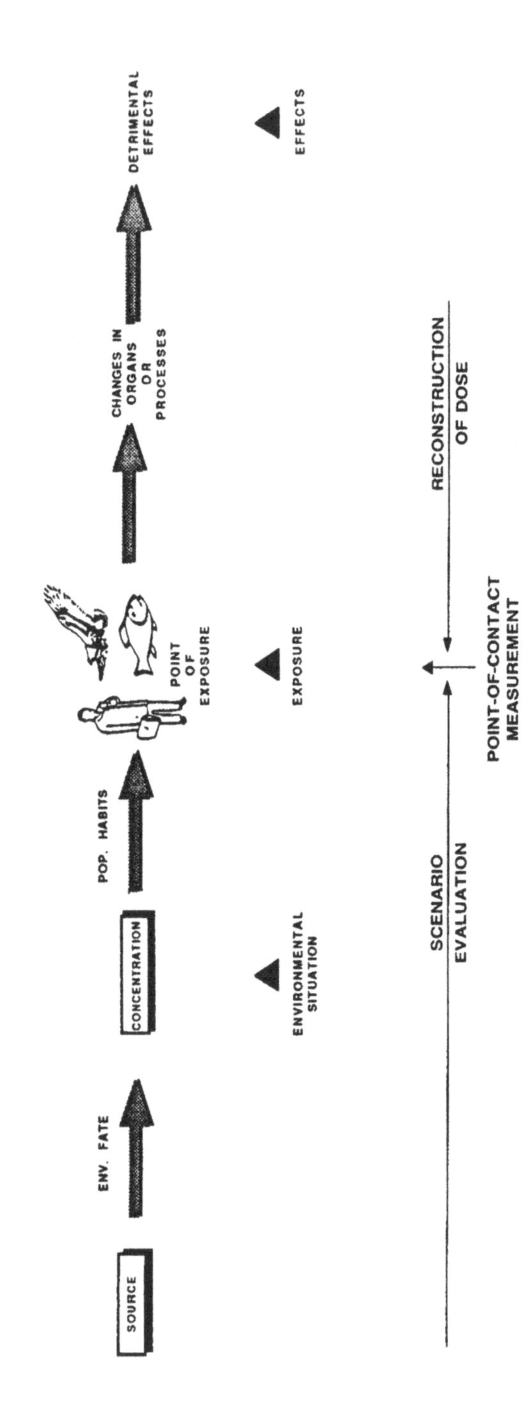

Figure 1. A schematic of environmental exposure, dose, and effects.

pollutant exposure is provided by the Total Exposure Assessment Methodology (TEAM) studies conducted by EPA.[6] In the TEAM studies, a small pump with a collector and absorbent is attached to a person's clothing and measures the exposures to airborne solvents or other pollutants while the exposure takes place. The absorbent cartridges are then analyzed for a variety of chemicals. The key to point-of-contact measurement techniques is that they must be at the interface between the person and the environment and measure the exposure while it is taking place.

A second approach to assessing exposure is by using tissue levels or biomarkers along with pharmacokinetics to reconstruct what absorbed dose must have been at some time in the past. The key to using this approach is an understanding of the relationship between exposure and the observable change within the body. This will many times require some knowledge of pharmacokinetics of specific chemicals, data often lacking at present.

The third approach is scenario evaluation, where one tries to predict, or estimate, exposure based on knowledge of sources, environmental pathways/fate, monitoring results of concentrations in environmental media, modeling results, and combine this with knowledge about population habits. This is an approach widely used for exposure assessment, and often it makes extensive use of environmental fate models.

Each of these approaches has strengths and weaknesses. The point-of-contact measurement approach, for example, can give us the most accurate measurement of what exposure is actually taking place. On the other hand, it is difficult to use this approach for predicting future exposures, since it involves monitoring present activities and exposures. For the same reason, it may be difficult to extrapolate the results from a study of one group in one geographic location to another group elsewhere. The sources of the pollutant are not always clear in this type of study, which is a disadvantage for an agency trying to control sources. Finally, this approach tends to be costly, and methods are not available for studying all chemicals.

The reconstructive approach's strengths are twofold; it can provide a positive indication that pollutants have actually crossed the exchange boundaries after exposure has occurred, and it may provide a good indication of past exposure levels. In order to do this, however, data are needed to link exposure levels to the levels found in the body, and except for a relatively few environmental pollutants, these data are unavailable. This approach does not predict future exposures, and it will not work for all chemicals (e.g., metabolites of other chemicals may cause interferences).

The scenario evaluation approach has a major strength in that it can be used to predict present or future exposures or estimate past exposures. This makes it a powerful tool for risk managers who are evaluating alternatives for possible regulation. Also, it tends to be less costly to use, since it uses whatever data are available. However, it has one glaring weakness: without valid models or (sometimes unavailable) environmental data, or data on population habits, it may give results that bear little resemblance to reality. Many of the models considered for use in scenario evaluation may be difficult to validate. Although an uncertainty assessment is necessary for exposure assessments in general, for detailed assessments using scenario evaluation, an uncertainty analysis is critical.

Tables 1–3 summarize the advantages and disadvantages of each approach relative to the others. The best exposure assessments will normally use a combination of these approaches to reduce uncertainty and add credibility to the assessment.

Table 1. Point-of-Contact Measurement of Exposure

Description:	Direct, real-time measurements of the contact of a chemical or substance with an organism
Examples:	Radiation Dosimeters TEAM study measurements with solvents
Advantage:	Can be best indication of actual exposures in sampled population
Disadvantages:	Sources not always clear Methods not well developed for all chemicals May be costly Results at one location may not apply elsewhere

Table 2. Reconstruction of Dose

Description:	Measurement of chemical or other indications of changes in body tissues, fluids, etc., and relating these measurements back to exposure
Examples:	Biomarkers Calculation of absorbed dose via use of body burden and pharmacokinetics
Advantages:	Can provide direct evidence that chemical has crossed exchange boundaries after exposure Can be a good indication of past exposures
Disadvantages:	Cannot predict future exposures Will not work for all chemicals Sources not always clear Research/data base not well developed

Table 3. Exposure Scenario Evaluation

Description:	Estimation of contact intensity, frequency, duration, and route by estimation of concentration in media and/or estimation of the habits/activities of individuals or populations that bring them into contact with the chemical
Examples:	Estimation of exposure via source estimation, monitoring data, fate models, use of exposure scenarios, etc.
Advantages:	Can be used to estimate past, present, or future exposures Usually less costly Powerful risk management tool to evaluate options
Disadvantages:	Can have limited accuracy, may be misleading May be difficult to validate Relies on data that may not be available

THE ROLE OF ENVIRONMENTAL FATE MODELS IN EXPOSURE ASSESSMENT

As the above discussion indicates, if exposure assessment is to be the determination of exposure, both before and after the application of regulatory controls, it becomes necessary to at least partly use the scenario evaluation approach to exposure assessment. An environmental fate model usually simulates what happens to a pollutant from the time it is released from the source into the environment until it has reached some future point in time where it is distributed as a concentration in one or more media.

For example, EPA's Office of Pollution Prevention and Toxics (formerly Office of Toxic Substances) must evaluate potential risk for new chemicals submitted under the Premanufacturing Notice (PMN) program. Since these chemicals have not yet been manufactured, there are no monitoring data, populations currently exposed, or body burden levels for most of these new chemicals. A predictive assessment, usually involving some form of environmental fate model, is a necessary tool for estimating potential risks from new chemicals.[7-13] Decisionmakers can also use the results from such predictive analyses to estimate the reduction in potential exposure if certain controls are imposed upon the use of the new chemical.

Environmental fate models are often used for existing chemical assessment also, especially where collection of data on ambient concentrations would be prohibitively time-consuming, expensive, or impracticable. EPA's Office of Water uses fate models in Waste Load Allocation studies,[14] where models are useful to calculate the effluent quality required to meet ambient water quality criteria in a receiving stream. EPA's Office of Air and Radiation uses fate models to estimate pollutant dispersion and degradation in evaluation of New Source reviews and State Implementation Plans, to calculate concentrations of pollutants, and to compare these predictions with the National Ambient Air Quality Standards. EPA's Office of Underground Storage Tanks used fate models to help evaluate the environmental benefit of several alternative strategies before proposing a rule for underground storage tank requirements. All EPA's program offices use environmental fate models at one time or another as part of the process of evaluating risk or potential risk.

This does not mean, however, that a fate model *by itself* provides regulatory decisions. A fate model is merely a tool used to assess exposure, either before or after a regulatory control being contemplated. As was discussed above, the risk assessment is only one of the inputs for making regulatory decisions.

SUMMARY

In summary, the role of an environmental fate model in a regulatory agency such as EPA is as a predictive tool useful in estimating both present and potential exposures. Risk management decisions in EPA depend not only on risk assessment, but also on other factors such as cost, practicability, etc. Environmental fate models can be powerful tools in helping assess both current situations and predicting the results of control alternatives being considered, but fate models themselves do not direct decisions. Models are an important part of a much larger process of risk assessment and risk management, and it is this process that the agency decisionmakers use to help them make regulatory decisions.

NOTICE

The opinions expressed in this chapter are the author's and do not necessarily reflect the policy of the U.S. Environmental Protection Agency.

REFERENCES

1. Carson, R. *Silent Spring* (New York: Houghton Mifflin Co., 1962).
2. Graham, F., Jr. *Since Silent Spring* (New York: Houghton Mifflin Co., 1970).
3. "Risk Assessment in the Federal Government: Managing the Process," National Research Council, National Academy Press, Washington, DC, 1983.
4. "Guidelines for Exposure Assessment," *Federal Register* 57FR 22888–22938, May 29, 1992.
5. "Human Exposure Assessment for Airborne Pollutants: Advances and Applications." National Research Council, National Academy Press, Washington, DC, 1990.
6. "The Total Exposure Assessment Methodology (TEAM) Study" (3 vol.), U.S. EPA Report-600/6-87/022a,b,c (1987).
7. "Methods for Assessing Exposure to Chemical Substances, Vol. 1, Introduction," U.S. EPA Report-560/5-85/001 (1985).
8. "Methods for Assessing Exposure to Chemical Substances, Vol. 2, Methods for Assessing Exposure to Chemical Substances in the Ambient Environment," U.S. EPA Report-560/5-85/002 (1985).
9. "Methods for Assessing Exposure to Chemical Substances, Vol. 3, Methods for Assessing Exposure from Disposal of Chemical Substances," U.S. EPA Report-560/5-85/003 (1985).
10. "Methods for Assessing Exposure to Chemical Substances, Vol. 4, Methods for Enumerating and Characterizing Populations Exposed to Chemical Substances," U.S. EPA Report-560/5-85/004 (1985).
11. "Methods for Assessing Exposure to Chemical Substances, Vol. 5, Methods for Assessing Exposure to Chemical Substances in Drinking Water," U.S. EPA Report-560/5-85/005 (1985).
12. "Methods for Assessing Exposure to Chemical Substances, Vol. 6, Methods for Assessing Occupational Exposure to Chemical Substances," U.S. EPA Report-560/5-85/006 (1985).
13. "Methods for Assessing Exposure to Chemical Substances, Vol. 8, Methods for Assessing Environmental Pathways of Food Contamination," U.S. EPA Report-560/5-85/008 (1986).
14. "Technical Support Document for Water-Quality Based Toxics Control," U.S. EPA Report-440/4-85/032 (1985).

CHAPTER 12

An Evaluation of Organic Materials That Interfere with Stabilization/Solidification Processes

M. John Cullinane, Jr., and **R. Mark Bricka,** U.S. Army Engineer Waterways Experiment Station, Vicksburg, Mississippi

INTRODUCTION

Background

The Environmental Protection Agency (EPA) is responsible for evaluating the suitability of hazardous waste and materials for land disposal. Chemical stabilization/solidification (S/S) is one technique that has been proposed as a means for controlling the release of contaminants from landfilled wastes to surface and ground waters. Indeed, S/S of hazardous wastes is recognized in regulations implementing both the Superfund Amendments and Reauthorization Act of 1986 (SARA) and the Hazardous and Solid Waste Act Amendments of 1984 (RCRA).

A variety of S/S technologies have been proposed for treating hazardous wastes. The most commonly applied technologies use portland cement, pozzolan, or portland cement-pozzolan combinations as the primary means of contaminant immobilization.[1,2] A potential problem with using S/S technology involves chemical interferences with the hydration reactions typical of the portland cement and pozzolan processes. Experience in the construction industry has demonstrated that small amounts of some chemicals can significantly affect the setting and strength development characteristics of concrete. Consequently, the concrete industry has developed fairly stringent criteria for the quality of cement, aggregate, water, and additives (accelerators or retarders) that are allowed in concrete.[3]

Of particular concern to S/S technology is the effect of organic compounds on the strength and contaminant immobilization characteristics of the final product. It is well documented that small concentrations of organic compounds,[4,5] sugars,[6] formaldehydes,[7] and various chemical contaminants typically found in hazardous waste affect the setting mechanisms of pozzolanic cements and lime/flyash pozzolans. Roberts[8] and Smith[9] reported on the effects of methanol, xylene, benzene, adipic acid, and an oil and grease mixture on the strength and leaching characteristics of a typical lime/flyash S/S formulation. Smith[9] concludes that there was a good correlation between the effects of organic compounds on lime/flyash pozzolanic systems and the reported effects on the hydration of portland cement. More recently, Chalasani et al.[10] and Walsh et al.,[11] using X-ray diffraction and scanning electron microscopy techniques, reported on the effects of ethylene glycol and p-bromophenol on the microstructure of portland cement hydration products. Ethylene glycol was found to produce significant changes in the microstructure even after a year of curing time.

Purpose and Scope

The purpose of the research described in this chapter is to develop data on the compatibility of organic waste constituents with three binding agents: portland cement, portland cement/flyash, and lime/flyash. Only the results of the unconfined compressive strength (UCS) test are presented at this time. The remainder of the data were to be presented in a comprehensive report scheduled for publication later in 1988.

MATERIALS AND METHODS

The study reported herein was conducted in three phases: (1) preparation of a synthetic wastewater and sludge; (2) addition of a binder and interfering material to the sludge; and (3) UCS testing of cured specimens containing sludge, binder, and interference chemicals.

Synthetic Wastewater and Sludge Production

Initial laboratory tests revealed that a synthetic wastewater containing nitrate salts of cadmium, chromium, and mercury at 600 times the EPA extraction procedure limit, and nickel at 600 times the California limit, could be treated with calcium hydroxide to produce a hydroxide sludge with typical metal concentrations of 86.2, 84.1, 18.8, and 0.137 mg/g (dry weight basis) of nickel, chromium, cadmium, and mercury, respectively.

Typically, the raw sludge contained 8% solids (by weight) and was very fluid. The sludge was dewatered to approximately 30% solids using a rotary drum vacuum filter. A constant moisture content between sludge batches was maintained by adjusting the solids content of the dewatered sludge to 25%, using the supernatant liquid from the sludge production process as a dilution liquid.

Specimen Preparation

The 25% solids content sludge was divided into three 150-gal samples and binder material was added to each at the following ratios.

Binder/Sludge Ratio

Binder	Ratio
Portland cement (Type 1)	0.3:1 cement:sludge
Portland cement (Type 1)/ flyash (Type F)	0.2:1 cement:sludge 0.5:1 flyash:sludge
Lime/flyash (Type C)	0.3:1 lime:sludge 0.5:1 flyash:sludge

After mixing the sludge with the binder, each binder/sludge sample was subdivided into four equal parts. A single organic interfering substance was added to each of the subsamples at ratios of 0.0, 0.02, 0.05, and 0.08 (by weight) interference chemical to binder/sludge material. The subsample to which no interference chemical was added was used as a control specimen. To account for variability between batches, control specimens were prepared each time an interference chemical was processed. This allows for the comparison of UCS results between batches.

Interference/binder/sludge mixtures (I/B/S) were then molded into 2 in. cubes in accordance with ASTM Method C-109-77/86.[12] Because the I/B/S mixture was usually viscous and could not be tamped into the molds, the ASTM method was modified to include vibration of the I/B/S mixture to remove any air pockets that developed during the molding process.

The specimens were cured in the molds at 23°C and 98% relative humidity for a minimum of 24 hr and removed from the molds whenever they developed sufficient strength to be freestanding. After removal from the molds, the specimens were cured under the same conditions for periods of 4, 11, and 28 days. At the end of each curing period, the UCS of the specimens was determined in accordance with ASTM C 109-77/86.[12] A minimum of four replicates were performed for each interference/binder/sludge mixture.

DISCUSSION OF RESULTS

Space limitations do not allow for presentation of all study results, but typical results for selected interference/binder/sludge mixtures are discussed below.

Table 1 presents the results, reported as the percent increase or decrease in 28-day UCS from the control specimen. Figures 1 through 7 provide a graphical representation of selected study results. Figures 1 through 3 present the UCS versus curing time for the phenol-portland cement, oil-portland cement, and grease-portland cement interference/binder/sludge combinations. Each curve shown on these figures represents the strength development curve for one interference material concentration. Figure 4 presents the 28-day UCS versus interference concentration for each of the three binders for the phenol interference. Figure 4 illustrates that increasing the concentrations of the interference chemical does not necessarily affect the different binders to the same degree. Figures 5, 6, and 7 present a graphical representation of the relative effects of interference concentration on the 28-day UCS of each of the three binders.

The data for the 28-day UCS are an indicator of the UCS trends observed at the earlier curing periods. This is clearly illustrated in Figure 1, which is a plot of cure time versus UCS for the phenol/portland cement/sludge material. Although the slope of the curve varies between binder and interference treatments, in most cases the lines of constant interference concentration do not cross.

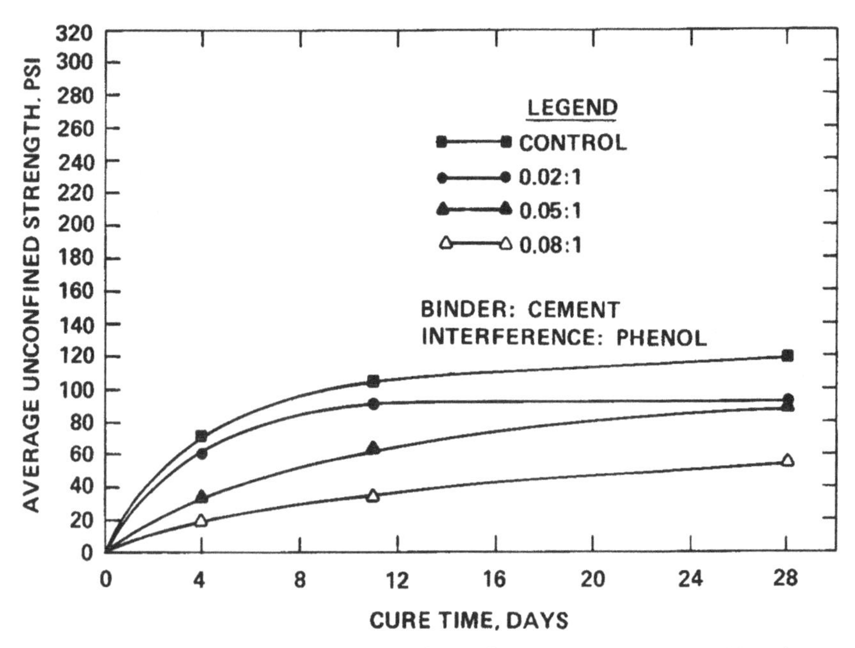

Figure 1. Unconfined compressive strength as a function of curing time and interference material concentration for the phenol-portland cement interference-binder combination.

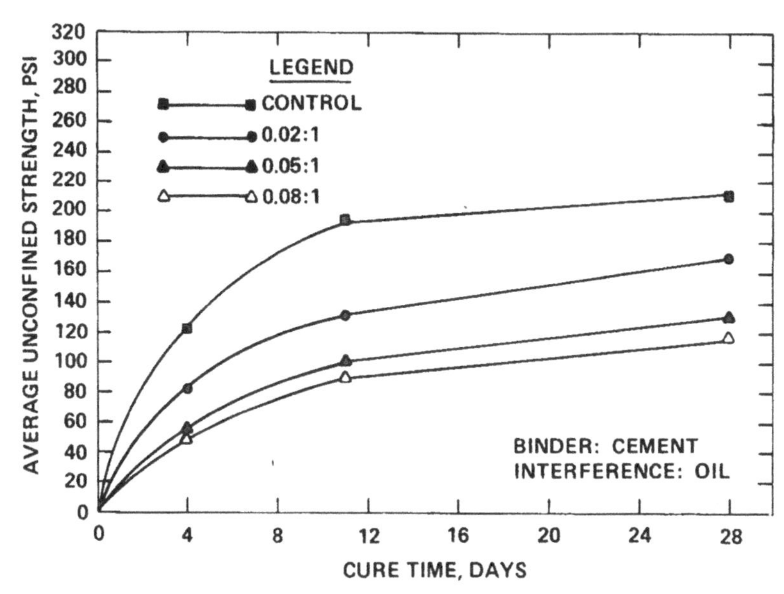

Figure 2. Unconfined compressive strength as a function of curing time and interference material concentration for the oil-portland cement interference-binder combination.

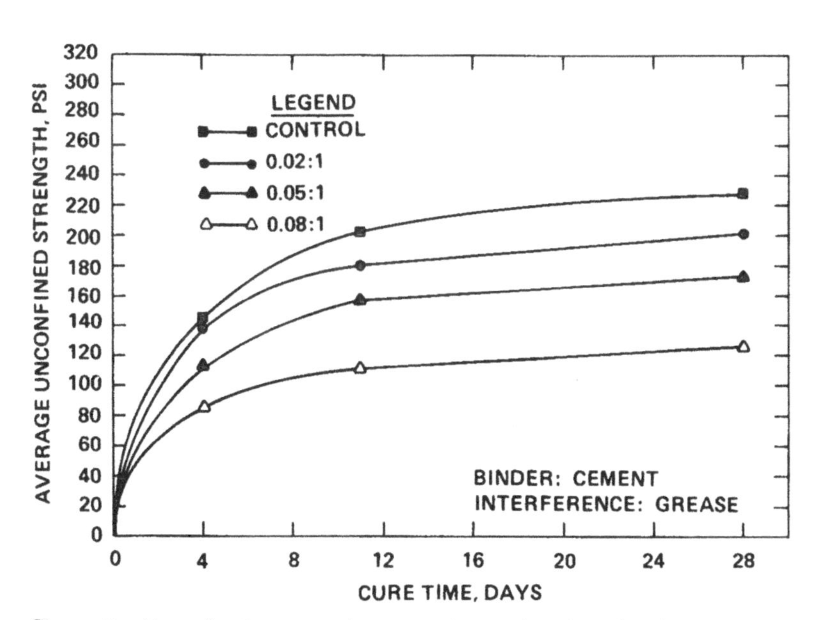

Figure 3. Unconfined compressive strength as a function of curing time and interference material concentration for the grease-portland cement interference-binder combination.

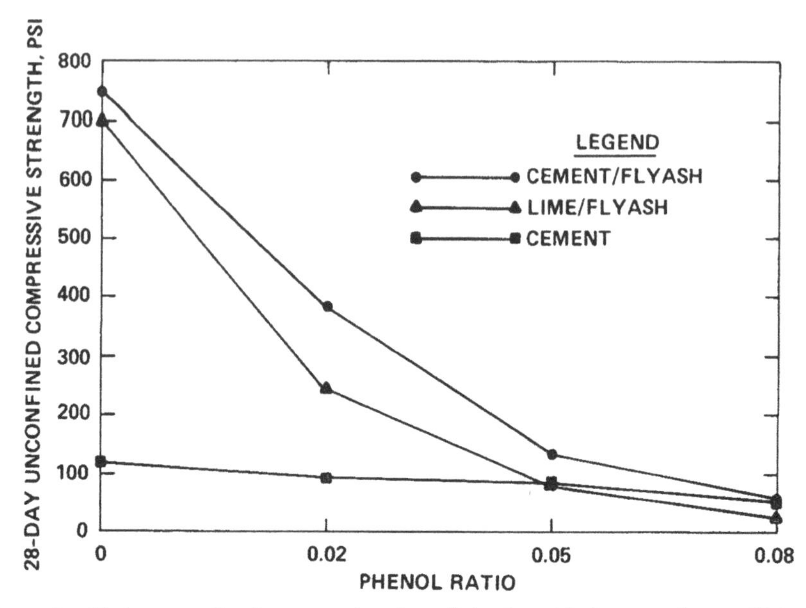

Figure 4. 28-day unconfined compressive strength for three binders as a time and interference material concentration.

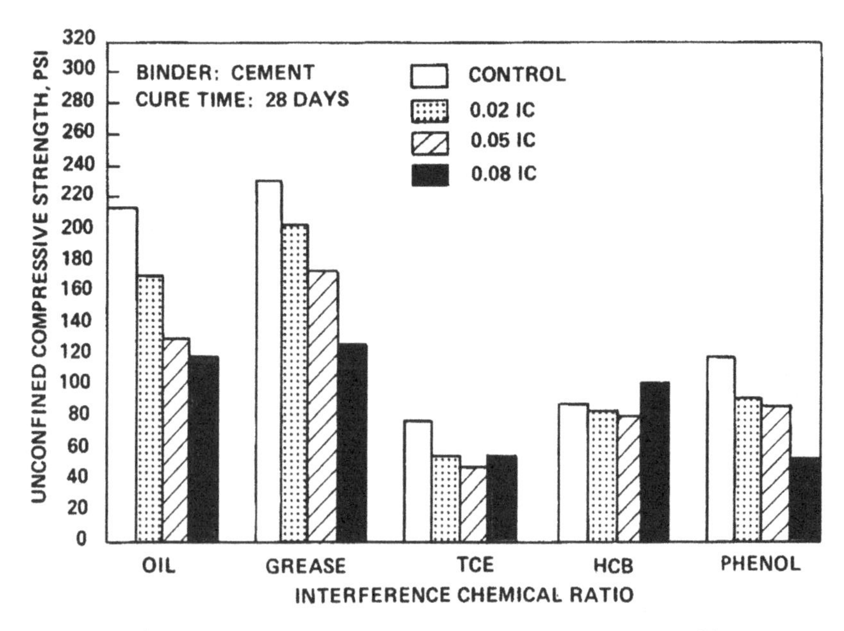

Figure 5. Effect of interference material concentration on the 28-day UCS for the Type I portland cement binder.

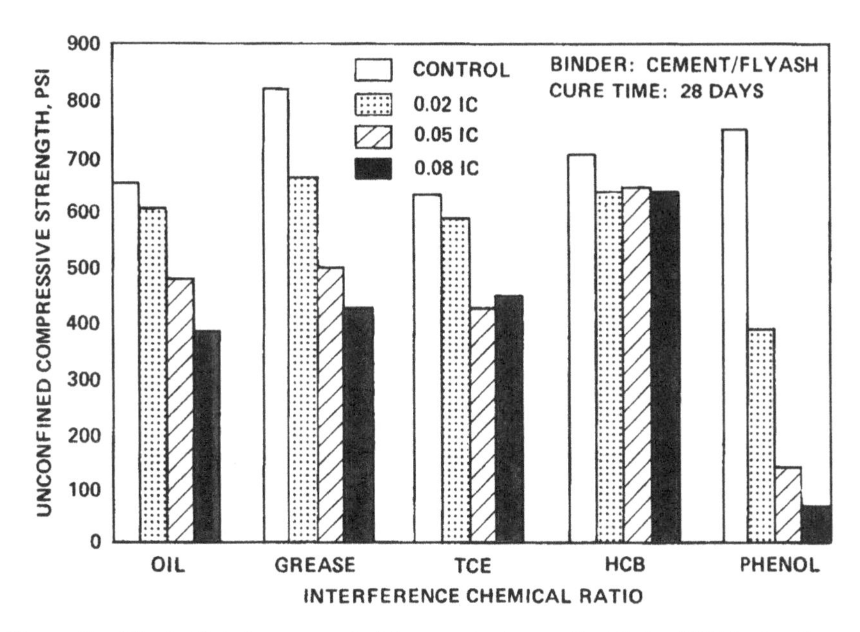

Figure 6. Effect of interference material concentration on the 28-day UCS for the Type I portland cement/flyash binder.

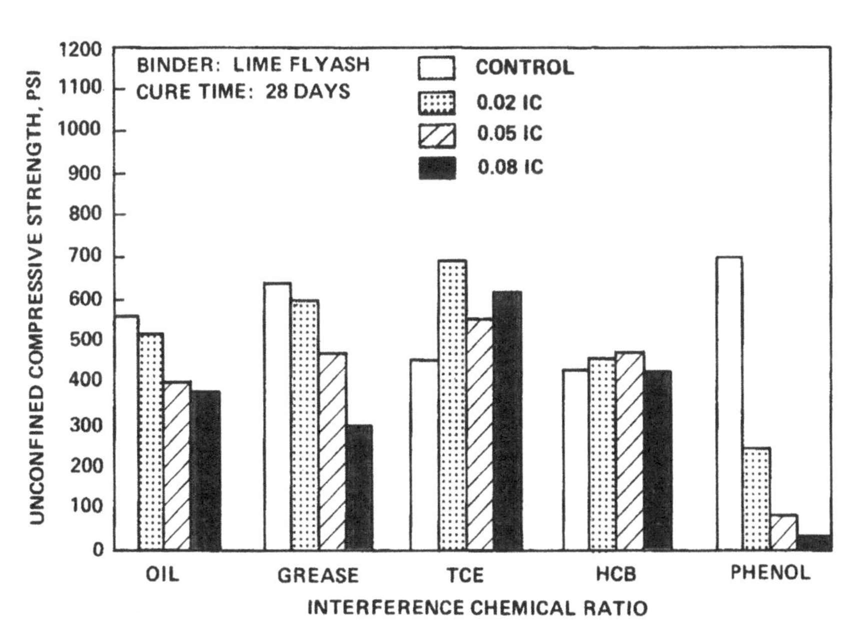

Figure 7. Effect of interference material concentration on the 28-day UCS for the lime/flyash binder.

Portland Cement Binder

The data presented in Table 1 show that the interference effects may be positive or negative, depending on the interfering material and concentration of the interfering material. For portland cement, the UCS generally declined with increasing organic interference concentration. The addition of oil, grease, and phenol resulted in a consistent decrease in the UCS with increasing concentration of interfering material. A 0.08 phenol ratio resulted in a 54% decrease in the UCS of the specimen. The addition of hexachlorobenzene was shown to have only a marginal effect on UCS, with a 15% increase in UCS at the 0.08 ratio. This unexpected result remains unexplained.

Table 1. 28-Day Unconfined Compressive Strength as a Percent of Control Specimen[a]

Interference Chemical	Portland Cement Binder I/B/S Ratio			Cement/Flyash I/B/S Ratio			Lime/Flyash I/B/S Ratio		
	0.02	0.05	0.08	0.02	0.05	0.08	0.02	0.05	0.08
Oil	-20	-38	-44	-8	-28	-42	-7	-27	-32
Grease	-12	-25	-45	-20	-40	-48	-7	-27	-54
Trichloroethylene	-28	-36	-27	-7	-33	-29	+51	+20	+34
Hexachlorobenzene	-5	-6	+15	-10	-9	-10	+6	+9	-1
Phenol	-22	-26	-54	-49	-82	-92	-65	-88	-96

[a]All results reported as percent increase (+) or decrease (−) from the control specimen rounded to nearest whole percent. Average of four replicates.
[b]Interference to binder/sludge ratio.

By comparing the UCS results for the control specimens of different binders, it is evident that the portland cement specimens developed less strength than the lime/flyash or the cement/flyash specimens. Upon closer examination, however, it can also be observed that the total binder/sludge ratio for the portland cement binder is lower than for the other binders.

Portland Cement/Flyash Binder

The portland cement/flyash interference/binder/waste mixture showed a consistent decrease in UCS for all interference materials. In general, the greater the concentration of interfering material, the greater the impact on 28-day UCS. The addition of a 0.08 ratio of phenol resulted in a decrease of over 90% in 28-day UCS. The effect of hexachlorobenzene was not concentration-dependent and resulted in a consistent 9% decrease in 28-day UCS regardless of concentration. The addition of 0.08 ratio of oil resulted in a 42% reduction in UCS.

Lime/Flyash Binder

With the exception of trichloroethlyne and hexachlorobenzene, the addition of organic interfering materials had a consistently negative impact on the 28-day UCS of the lime/flyash interference/binder/sludge mixture. The addition of a 0.08 ratio of phenol resulted in an 80% decrease in 28-day UCS. The addition of oil or grease at a 0.08 ratio resulted in a 32% and 54% decrease, respectively, in 28-day UCS.

The addition of trichloroethylene appeared to result in a gain in 28-day UCS; 51%, 20%, and 34% at the 0.02, 0.05, and 0.08 interference ratios, respectively. The addition of hexachlorobenzene resulted in a slight increase in 28-day UCS for the 0.02 and 0.05 interference ratios, 6% and 9%, respectively. However, a 1% reduction in UCS occurred at the 0.08 interference ratio.

CONCLUSIONS

Several conclusions can be drawn that characterize the effects of the interference materials investigated in this project on the UCS of stabilized/solidified waste materials.

1. The interference chemicals tested had a measurable effect on the setting and strength development properties of the stabilized/solidified waste. The magnitude of the effect depended on the type of binder, the curing time, and the type and concentration of the interfering compound.
2. Stabilized/solidified waste showed decreased UCS development with increasing oil or grease concentrations.
3. Although the waste stabilized with portland cement resulted in lower 28-day strength development, it appears that the concentration of interference material had less effect on the UCS development properties for the portland cement binder than for the portland cement/flyash or lime/flyash binders.
4. Phenol concentrations above 5% resulted in marked decreases in 28-day UCS development for all the binders tested.
5. The chlorinated hydrocarbons evaluated in this study had less effect on the UCS development properties than the other interference materials investigated.

ACKNOWLEDGMENTS

The tests described and the resulting data presented, unless otherwise noted, were obtained from research conducted by the U.S. Army Engineer Waterways

Experiment Station and were sponsored by the U.S. Environmental Protection Agency, Hazardous Waste Engineering Research Laboratory, Cincinnati, Ohio, under Interagency Agreement DW96930146-01. Mr. Carlton Wiles, Hazardous Waste Engineering Research Laboratory, was the EPA project officer. Permission to publish this information was granted by the Chief of Engineers and the U.S. Environmental Protection Agency.

REFERENCES

1. "Guide to the Disposal of Chemically Stabilized and Solidified Waste," U.S. EPA Report-SW-872, Office of Research and Development, Municipal Environmental Research Laboratory, Cincinnati, OH (1980).
2. Cullinane, M. J., L. W. Jones, and P. G. Malone. "Handbook for Stabilization/Solidification of Hazardous Waste," U. S. EPA Report-540/2-86/001, Hazardous Waste Engineering Research Laboratory, Cincinnati, OH (1986).
3. Jones, J. N., M. R. Bricka, T. E. Myers, and D. W. Thompson. "Factors Affecting Stabilization/Solidification of Hazardous Wastes," Proceedings: International Conference on New Frontiers for Hazardous Waste Management, U.S. EPA Report-600/9-85-025, Hazardous Waste Engineering Research Laboratory, Cincinnati, OH (1985).
4. Young, J. F. "A Review of the Mechanisms of Set-Retardation of Cement Pastes Containing Organic Admixtures," *Cement and Concrete Research* 2(4) (1972).
5. Young, J. F., R. L. Berger, and F. V. Lawrence. "Studies on the Hydration of Tricalcium Silicate Pastes. III. Influences of Admixtures on Hydration and Strength Development," *Cement and Concrete Research* 3(6) (1973).
6. Ashworth, R. "Some Investigations Into the Use of Sugar as an Admixture to Concrete," Proceedings of the Institute of Civil Engineering, London, England (1965).
7. Rosskopf, P. A., F. J. Linton, and R. B. Peppler. "Effect of Various Accelerating Chemical Admixtures on Setting and Strength Development of Concrete," *Journal of Testing and Evaluation* 3(4) (1975).
8. Roberts, B. K. "The Effect of Volatile Organics on Strength Development in Lime Stabilized Fly Ash Compositions," Master's Thesis, University of Pennsylvania, Philadelphia, PA (1978).
9. Smith, R. L. "The Effect of Organic Compounds on Pozzolanic Reactions," I. U. Conversion Systems, Report No. 57, Project No. 0145 (1979).
10. Chalasani, D., F. K. Cartledge, H. C. Eaton, M. E. Tittlebaum, and M. B. Walsh. "The Effects of Ethylene Glycol on a Cement-Based Solidification Process," *Hazardous Wastes and Hazardous Materials* 3(2) (1986).
11. Walsh, M. B., H. C. Eaton, M. E. Tittlebaum, F. K. Cartledge, and D. Chalasani. "The Effect of Two Organic Compounds on a Portland Cement-Based Stabilization Matrix," *Hazardous Wastes and Hazardous Materials* 3(1) (1986).
12. *Annual Book of ASTM Standards: Construction, Volume* 04.01, *Cement; Lime; Gypsum,* American Society for Testing and Materials, Philadelphia, PA (1986).

CHAPTER 13

Incorporation of Contaminated Soils into Bituminous Concrete

Karl Eklund, Eklund Associates, Berkley, Massachusetts

Introduction

Since our first paper[1] on the process of incorporating petroleum contaminated soils in asphalt paving mix, there have been several interesting developments in the technology. While a complete discussion of all the variations of the process would both be inappropriate in this context and would involve proprietary technology, this chapter will attempt to give a comprehensive review of the principal approaches.

Asphalt Technology

Asphalt appears naturally in the form of a thick liquid or a solid, but the primary source is as one of the products of the refining of crude oil into useable petroleum products. As commercially available, it is a solid at temperate ambient temperatures and is stored and transported hot.

Various grades and types of asphaltic materials are derived from the refining process, as shown in Figure 1. When liquified by heating and mixed with aggregate, ''hard'' and ''soft'' asphalt makes the ordinary 'asphalt paving mix' or bituminous concrete. This hardens into concrete when it cools to ambient temperatures. ''Cutback'' asphalt is a mixture of asphalt and a volatile petroleum hydrocarbon, to be liquid at ambient temperatures so that it can be used as a

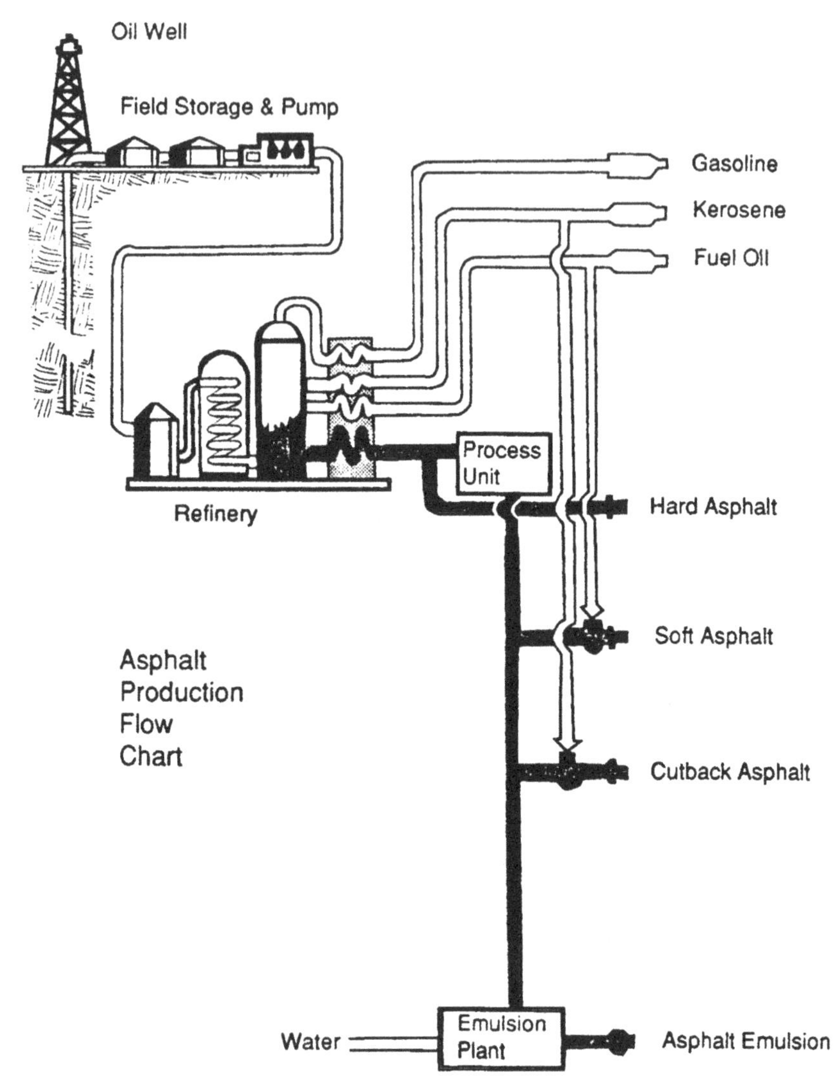

Figure 1. Asphalt production flow chart.

patching material. Because the volatile hydrocarbons escape into the atmosphere, "cutback" asphalt mix is now used only in cold weather.

Asphalt emulsion is a mixture of asphalt, water, and detergents and surface-active agents so as to be a liquid at ambient temperatures. It hardens when the mixture of emulsion and aggregate is compressed sufficiently to allow the asphalt to bind the grains of aggregate.

Paving mixes are characterized by specifying the particle size distribution of the aggregate and the type and quantity of asphalt used.

Bituminous Materials as an Encapsulant for Petroleum Contaminants

Because the asphalts are a product of the same refining process, they are miscible with other petroleum products in any quantity. Petroleum products found as contaminants in water or soil and brought in contact with asphaltic materials will tend to be preferentially adsorbed on the asphalt, as opposed to remaining suspended in water or being adsorbed on wet soil particles. Over a longer time, petroleum adsorbed on an asphalt surface will tend to diffuse into the asphalt asymptotically approaching a bulk absorption. [This explanation is based on development work on a proprietary asphalt emulsion process at American Reclamation Corporation, Southborough, MA. At that time it was not possible to determine the rates of adsorption and absorption.]

In practice, this effect is utilized by replacing the aggregate in an asphalt paving mix production plant with a proportion of petroleum contaminated soil. The variations on this process depend on the type of plant used.

Making Asphalt Paving

Asphalt paving mix processes are broadly divided into two kinds: "hot mix" processes and "cold mix" processes. The "cold mix" plants use either cutback asphalt or asphalt emulsion. Because of the limitations on using cutback asphalt paving mix, it is not considered as a method of incorporation for petroleum contaminated soil.

The basic steps in both the hot and cold processes are shown in Figure 2.

In the hot process the asphalt is in the form of a liquid at 300°–500°F. This is mixed with aggregate in a heated pug mill and either immediately applied or stored in heated silos. Because the mixing process takes place at a temperature significantly above the boiling point of water, any water carried in the aggregate is quickly converted to steam. This would, at best, interfere with the mixing process and, at worst, cause an explosion that could damage the plant. Thus a critical step in the hot process is bringing the aggregate to dryness and to a temperature compatible with the temperature of the asphalt.

Two methods are used to do this. In the older technology of the "Batch Plant" there is a parallel flow rotary kiln used for heating the aggregate. This is shown in Figure 3. Heat is applied by an open flame from a burner at the upper end of the kiln where the aggregate is introduced. As the kiln rotates, the aggregate falls through the flame or hot gases and is moved down to the exit end. There the hot aggregate is transported to the heated pug mill by conveyor. The combustion gases are removed by an exhaust fan to a stack after passing through a fabric filter or 'baghouse.' The main limitations on this process are the air quality requirements which limit the amount of fuel used and the baghouse, which limits the temperature of the exhaust gases that can be handled. Since the exhaust gases are at roughly the temperature of the aggregate batch, plant rotary kilns are typically limited to produce aggregate at 300°F.

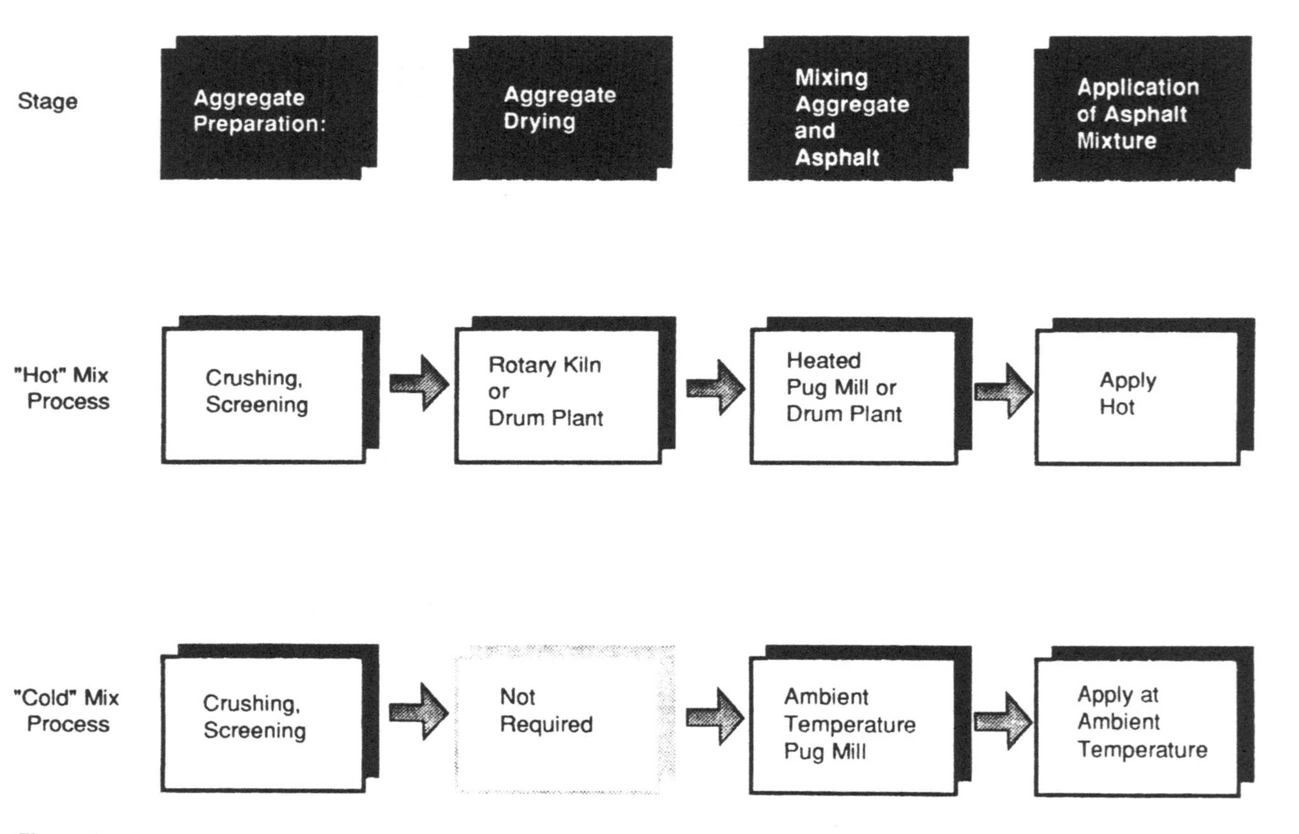

Figure 2. Asphalt paving mix production flow chart.

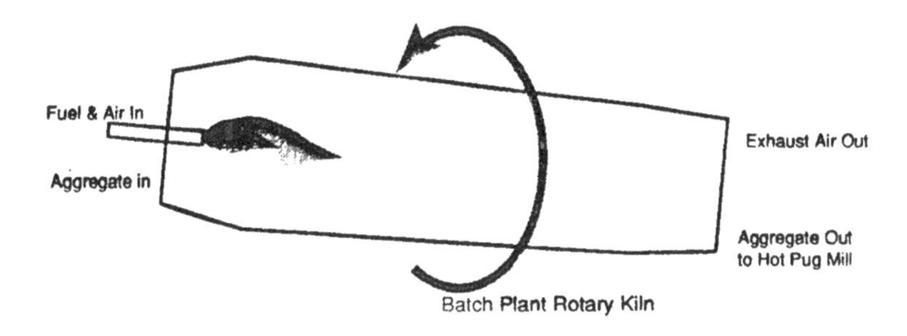

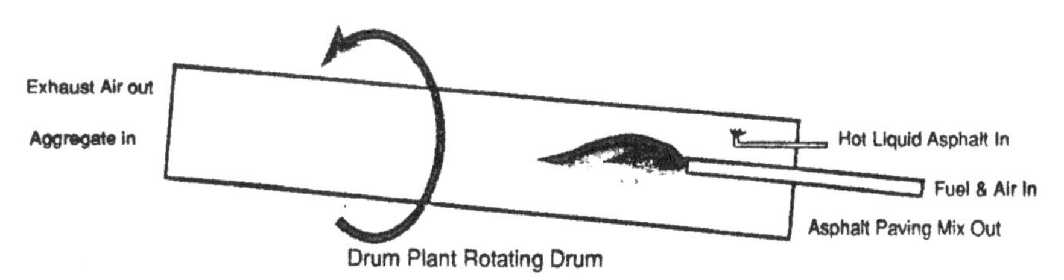

Figure 3. Types of "hot mix" asphalt plants.

The Drum Plant, also shown in Figure 3, represents an improved technology. In the countercurrent version the flame is introduced near the exit of the rotating drum and the asphalt is sprayed on the aggregate beyond the flame. This means that further mixing in a pug mill is not required. Because the exhaust gases exit where the cold aggregate is introduced, the exhaust gases can be considerably lower than the maximum aggregate temperature. This will produce less impact on the air quality control system.

A countercurrent rotary kiln or drum can be used for 'roasting' petroleum contaminated soil to a temperature (say, 600°F) where most of the petroleum contaminants are removed by evaporation. The aggregate produced by this process will have only high-temperature asphalt- or coke-like chars remaining and will be suitable for use as aggregate.

It should be noted that those evaporated hydrocarbons that are not ignited by the open flame in the kiln will escape with the exhaust gases and will not be removed in the baghouse. When petroleum contaminated soils (or recycled waste asphalt paving) are roasted in rotary kilns, some additional air quality control systems will normally be required.

It is possible, however, to recycle crushed and screened waste asphalt paving by mixing it with equal quantities of 600°F aggregate produced in a drum plant. This produces a 300°F mixture that can be introduced into the heated pug mill of a batch plant without excessive generation of petroleum vapors.[2] This same process could be used for petroleum contaminated soil.

The emulsion cold mix process is much simpler and requires considerably less capital equipment. The mixing of the emulsion and aggregate is done at ambient temperatures [down to a temperature that causes freezing of the emulsion] so that there is no necessity to dry the aggregate, which eliminates the need for a kiln or heated pug mill. The emulsion paving mix hardens by compaction rather than cooling, so there is no need for heated storage. Unlike hot mix plants, the equipment for making emulsion mix can easily be made portable.

Environmental Impacts of Incorporating Petroleum Contaminated Soil in Bituminous Paving

Samples of bituminous concrete made using clean aggregate and petroleum contaminated soils have been tested at American Reclamation Corporation and at Applied Environmental Recycling Systems[3] and no significant differences in the leaching of petroleum products from them have been determined, using the leaching procedures of Environmental Protection Agency methods for the E.P. Toxicity and the TCLP tests. Tests using waste oils and oils with other contaminants show that there is a high degree of binding of nonpetroleum contaminants, but this must be evaluated on a case-by-case basis.[3]

The other environmental impacts are to air quality. Both the hot and cold mix processes require some crushing and screening of the soil before processing, so that the aggregate in the process will have a known particle size distribution. If the contaminant is gasoline, there will be a tendency for the gasoline to evaporate during the mechanical agitation of the soil. Attempts have been made to quantify this effect, but the variability in contamination level of typical soils obtained from contaminated sites made the results lack significance. This does not appear to be a problem with less volatile petroleum contaminants.

No other environmental impacts have been noted in the cold mix process.

In the hot mix process, subjecting contaminated soil to heating in a rotary kiln or drum evaporates some of the contaminants in a parallel-flow kiln and a large fraction of the contaminants in a countercurrent drum, where the hot aggregate reaches 600°F. This is generally controlled by limiting the amount of contaminated soil to 10% or less of the total aggregate, or by requiring additional air quality control equipment such as an afterburner.

In hot mix plants that use a substantial fraction of recycled waste asphalt paving (RAP) the contaminated soil can be mixed with the RAP because it will generally produce less of an air quality problem than the RAP.

Comparative Advantages of the Hot and Cold Mix Processes

If a new facility is being designed primarily for processing petroleum contaminated soil, the cold mix process has a significant advantage because of the much lower initial capital costs. Similarly, if a facility is being built for onsite processing at a site contaminated with petroleum, the simplicity and portability of the cold mix process provides an overwhelming advantage.

The disadvantage of the cold mix process is that the paving produced, particularly if the contaminated soil is not primarily sand and gravel, does not generally meet requirements for road building under federal specifications. Because many states and some municipalities follow the federal specifications rather than creating their own, this severely limits the market. Emulsion mix paving has been successfully used for commercial and municipal paving and could be used by highway departments for verges and stormwater control works.[4]

If an existing hot mix facility is being used for the processing of petroleum contaminated soils, the primary advantage is that there is an existing market for hot mix asphalt paving. Thus there is increased assurance to the owner of a contaminated site that his processed soil will be used as paving, rather than ending up as an unmarketed product at a cold mix facility. In addition, the hot mix plant can provide a finish coat of hot mix asphalt over a base of hot or cold mix paving made with contaminated soil that may not have the same appearance as paving made with clean aggregate. [Cold mix paving made with soil with a high iron content can have a dark brown rather than black color. This can affect the marketing potential.]

Regulatory Considerations

Under the federal statute (RCRA) and in many states, petroleum contaminated soil is a solid waste, and incorporation into bituminous paving is considered recycling of a solid waste and recycled as such. In those jurisdictions the primary regulatory concern is air quality requirements.

In some states (e.g., Massachusetts) petroleum contaminated soil is a hazardous waste, and incorporation into bituminous concrete is regulated as recycling of a hazardous waste. This raises the question that a process involving hazardous waste is considered 'treatment' (as opposed to 'recycling') if the effect of the process is to remove the contaminant from the inert matrix so that the *matrix* can be reused.

This is particularly applicable to a facility that roasts petroleum contaminated soil in a rotary kiln or drum in order to reuse the soil as aggregate. It would be possible to interpret the Massachusetts regulations in such a way as to consider that facility a hazardous waste incinerator. The expense of permitting such a facility would make it uneconomic for recycling petroleum contaminated soil. On the other hand, if a substantial fraction of the hydrocarbons evaporated from the soil in the kiln are ignited by the open flame burner heating the kiln, this could be regulated under the provisions for burning waste oil as a fuel. The resolution of this situation in Massachusetts could prove to be a useful precedent if federal regulations are ever changed to make waste oil a hazardous waste.

Conclusions

The recycling of petroleum contaminated soil by incorporation in bituminous concrete has proved to be a viable and economic process in the New England states. Several facilities using both the hot and cold mix processes have operated successfully for over two years without significant environmental problems. There have also been successful examples of onsite processing of soils at major sites. This is currently the preferred process for petroleum contaminated soils in Massachusetts.

Future Developments

Both the hot and cold mix processes are undergoing improvements. There are a number of possible reconfigurations of existing hot mix facilities that can significantly reduce the emissions of unburned hydrocarbons[2] and there are significant improvements to emulsion mix paving by the use of polymeric and other additives to the emulsion.[3] One can expect to see improved versions of both technologies over the next few years.

Acknowledgments

Some of the results described were obtained while the author was with American Reclamation Corporation. This work could not have been done without the advice and consultation of Nathan Wiseblood and the able assistance of Fred Hooper. Information on hot mix processes was obtained from Joseph Hanbury and David Peter of the Simeone Corporation.

References

1. Eklund, K. "Incorporation of Contaminated Soils into Bituminous Concrete," in P. T. Kostecki and E. J. Calabrese, Eds., *Petroleum Contaminated Soils,* Volume I (Chelsea, MI: Lewis Publishers, Inc., 1989), pp. 191–199.
2. Personal communication, David Peter, Simeone Corporation, Stoughton, MA.
3. Personal communication, Glenn Warren, Applied Environmental Recycling Systems, Salem, MA.
4. Personal communication, John Glynn, American Reclamation Corporation, Southborough, MA.

CHAPTER 14

Additive Stabilization of Petroleum Contaminated Soils

Sibel Pamukcu, Department of Civil Engineering, Lehigh University, Bethlehem, Pennsylvania

INTRODUCTION

In 1984 it was estimated that several hundred thousand underground storage tanks, used for storage of petroleum products, are leaking.[1] Amendments to CERCLA (Comprehensive Environmental Response, Compensation and Liability Act) in 1986 show increased recognition of the problem of leaking underground storage tanks (UST). Out of over 3,000,000 USTs in use throughout the country, as many as 500,000 may be leaking petroleum liquids in the ground.

Most petroleum hydrocarbons are considered immiscible with water; therefore, they are primarily transported in the unsaturated or vadose zone in the soil. However, gasoline-range hydrocarbons contain significant quantities of certain compounds which are partially soluble in water. Some of these compounds are carcinogenic and/or Environmental Protection Agency (EPA)-listed hazardous waste components (e.g., benzene, toluene, xylenes). The presence of such compounds in the subsurface environment poses a significant health hazard to the public and environment.

In a number of cases of leakages from underground storage tanks or accidental spillage of petroleum products into the ground, quick remediation of the contaminated ground may be critical. Removal of the soil is a significant part of the overall treatment. However, the high cost associated with excavation and limitations in disposal facilities and receiving landfills often complicate the problem. Physical removal of the contaminated soil from the vadose zone is probably most viable if the contamination is close to the ground surface, and thus would not require very large quantities of soil to be excavated. The technique may also be advantageous over others when remediation is not feasible because of tenacious retention of the contaminant on the soil, and when other physical and chemical parameters of the soil and the contaminant limit effective removal. Once the contaminated soil is removed it has to be disposed of safely as a waste material. This often results in creation of new landfills which ultimately may not serve the purpose of land reclamation and rehabilitation.

There are emerging technologies for in situ cleanup, such as biodegradation or electrokinetic removal.[2-5] However, these are either long-term processes or they are currently at development stages. An economical and fast solution is to stabilize the contaminated soil by mixing it in place with additives and possibly render it suitable for reuse.

The study presented here has been intended to deal with this aspect of the remediation; namely, stabilizing the contaminated soil to render it a useful material, and thus provide an economical and beneficial solution to the problem of cleanup of petroleum contaminated soils. The benefits of the approach are expected to be twofold. First, there is a critical need to reclaim and rehabilitate land in parts of the country where population density and value of land is high. Creating more landfills and waste containment sites (1) uses up available land and (2) threatens the fresh water supply in such areas. The latter will also reduce the utility of land, even those far from the contamination site. Therefore reuse of waste in an environmentally safe and technically sound way is the most attractive solution to waste management problems. The second benefit is that the reclaimed land will be available for use probably with much less effort of site preparation, such as backfilling with borrow materials.

In this study, a form of stabilization/solidification method was applied to fuel oil contaminated soil to bind the oil in a structure formed by the cementing and conditioning action of pozzolanic and earthen materials to produce chemically and physically stable and mechanically handlable new products. If the stabilization of the petroleum contaminated soil is to be successful, the petroleum hydrocarbon should either be dispersed and/or be encapsulated and bound in a monolithic solid of high structural integrity. The end product should be chemically and physically stable and durable, and also easy to handle and work with. In this study, the effort was concentrated on demonstrating the physical integrity of such stabilized materials. Some leachate analysis was also conducted to better assess the differences between several end products with respect to leachability. The results showed that simultaneous effects of the cementing, pozzolanic and sorbent reactions of additives produce a stabilized product with good physical properties and reduced fuel oil content in the leachate relative to that of unstabilized soil.

BACKGROUND

Petroleum Products in Soil

Despite the best efforts of both the petroleum industry and the regulatory community, releases, leakages, and spills of petroleum products occur frequently. It is estimated that in the United States, up to 25% of the underground storage tanks used for storage of petroleum products are leaking.[6] Once a spill or a leakage occurs, the hydrocarbon liquid, under gravity, moves down to the groundwater, partially saturating the soil in its pathway. Upon reaching the groundwater table, the liquid may spread horizontally by migration within the capillary zone. There are three major tasks that need to be performed for remediation and reclamation of the contaminated area: the first is to control the horizontal migration of the contaminant away from the source, the second is the cleanup of the groundwater, and the third is the cleanup of the contaminated zone of soil.

In general, cleanup of the groundwater contaminated with gasoline-range hydrocarbons consists of pumping the water from a well and removal of the floating material. Cleanup of hydrocarbon contaminated soils is usually more complicated. Partially hydrocarbon saturated soil can be a persistent source of contamination of groundwater for decades as water percolates from the surface, or groundwater table fluctuation promotes migration of the soluble compounds.

A petroleum hydrocarbon either degrades or remains unaltered in soil. Degradation comes about by microbial metabolism in which the hydrocarbon may be oxidized to carbon dioxide and water. If the soil is contaminated with gasoline-range hydrocarbon liquids made up of a mixture of volatile hydrocarbons, the liquid state of the hydrocarbon remains in equilibrium with its vapor state. If the soil can be ventilated, more of the liquid state would pass into vapor state, and theoretically the soil can be eventually decontaminated. However, the permeability of the soil and the presence of water are two major limiting factors in accomplishing this type of remediation.

Petroleum products are often attenuated on clay constituent of soil by adhesion and Coulombic interactions.[7] For example, when an advanced front is retained by a clay lense, it is often found that only a minor product layer forms over the groundwater table. Obviously, the degree of oil saturation of a particular soil depends on the physical, chemical, and mineralogical make up of the soil, such as the contact angle between the liquid and the mineral particles. However, in general, as clay content of a soil increases so does its oil retention capacity, and thus the degree of difficulty in remediation.

Applicable Remedial Technologies

The remedial technologies for general soil cleanup can be divided into two categories: in situ technologies (volatilization, biodegradation, electrokinetic application, leaching and chemical reaction, vitrification, passive remediation, isolation/containment); and non-in situ technologies (land treatment, thermal treatment, asphalt incorporation, solidification/stabilization, extraction and treatment, chemical extraction, excavation). For oil contaminated soils, these technologies can be reclassified as follows:

1. physical treatment
2. chemical treatment
3. biological treatment
4. thermal destruction
5. stabilization/solidification

All of these methods have their merits and also limitations in application. The stabilization and solidification technologies work by binding the contaminants within a relatively inert solid matrix which prevents or reduces volatilization and leaching. It also improves the handling and working of the material.[8-11] The resulting matrix may then be used in backfilling or landfilling.

Stabilization of a waste is generally defined as chemical modification of the material to detoxify its waste constituents, which may or may not result in improved physical properties of the material. However, factors such as durability, strength, and resistance to leaching play important roles in predicting long-term performance of the new material. Therefore, improvement of physical properties is essential for long-term integrity of the material, especially if it is being considered for reuse. As defined in several EPA publications,[8,9,12] "stabilization," "solidification," and "fixation" refer to waste treatment which produces the combined effects of: (1) improvement of physical properties; (2) encapsulation of pollutants; (3) reduction of solubility and mobility of the toxic substances. Although each one of the above terms may emphasize one or more of these effects, for all practical purposes the terms have been used interchangeably with little or no error.

There are a number of solidification/stabilization techniques used in the industry for different types of waste materials. Each technique is formulated to be compatible with the specific waste constituents. These techniques can be divided into the following groups:

1. cement-based
2. silicate-based
3. sorbent
4. thermoplastic
5. organic polymer techniques
6. encapsulation techniques
7. vitrification.

Among these, cement- and silicate-based techniques involve well-known pozzolanic or cementation reactions, which result in compounds that act as natural cement. These techniques may offer economy over others because they often utilize other waste products such as fly ash, blast furnace slag, or cement kiln dust as pozzolanic additive.

Silicate-based processes cover a wide range of methods in which the siliceous or pozzolanic material is mixed with other alkaline earths such as lime or gypsum. Although these processes are generally used to stabilize inorganic industrial wastes, they have also been shown to stabilize organic and oily wastes with some degree of success.[13-16] The solidification processes using silicate-based materials generally involve pozzolanic reactions between SiO_2, Al_2O_3, Fe_2O_3, and available calcium in lime. These reactions produce very stable calcium silicates and aluminates which act as natural cement similar to portland cement. In stabilization of organic wastes, both cement- and silicate-based techniques produce a microencapsulating matrix in which the organic component is bound by a combination of chemical reactions and physical isolation. The hydrocarbon component would essentially be fixed or immobilized in such a matrix, which would restrict internal fluid movement also. If the stabilized material is soil-like, a degree of mechanical stabilization, such as compaction, is necessary to ensure low density and formation of a continuous matrix.

The sorbent techniques involve use of certain clay minerals with high specific surfaces to fix and retain hydrocarbon/oil molecules. This retention can take place on outer or inner surfaces of clay minerals. Sodium-montmorillonite, vermiculite, and needle-like shaped attapulgite clays are examples of additives in this technique.[17]

When used with petroleum contaminated soils, the product of these three techniques should be a microencapsulated matrix in which the hydrocarbon/oil components are bound by a combination of physical and chemical isolation.

These components would essentially be fixed or immobilized in this matrix, which would restrict internal fluid movement. With proper mixing and subsequent compaction, a high density homogeneous matrix can be achieved in soils. Chemical reactions may occur between some organics and the inorganic, cementatious additives in soils. Oxidation, along with hydrolysis, is probably the most common reaction for organics in stabilization/solidification systems. Reduction, salt formation and dispersion are also other possible reactions.

In order to assess the integrity of a stabilized or solidified material, a number of physical and mechanical measurements are made. Among these measurements are unconfined compressive strength, permeability, compressibility, density, durability, and leachability. Depending on the intent of use for the final product, either all or a few of these measurements may be conducted. For example, higher unconfined compressive strength may be a good indicator of the improved stability, trafficability, and degree of ease of handling and placement of the material. Lowered permeability may indicate lowered potential for leachability, since low permeability restricts the rate of infiltration of water, reducing the erodibility of a stabilized or solidified compact porous structure. It should be noted that diffusion may also contribute in significant quantity to the leaching out of the contaminants from porous systems.

There is often great variability in these measured properties of solidified or stabilized systems, depending on the type, quantity, and distribution of the waste product(s) and not necessarily on those of the additives.

Geotechnical Properties of Stabilized Product

The main objectives of stabilization/solidification of residual products are: (1) to remove free liquid, thus minimize leachability; (2) to render waste physically stable for handling and placement; (3) to reduce permeability, thus minimize leachability. There are a number of geotechnical parameters used by regulatory agencies to assess performance of stabilized/solidified products. Among these parameters are unconfined compressive strength, permeability, compressibility, durability, and dry density.

Unconfined compressive strength (UCS) is often a good indicator of the integrity of the stabilized material, its trafficability, and degree of ease of handling and placement. In addition, UCS measured over time, during the period of curing, may provide insight to the ongoing cementation and pozzolanic reactions. Furthermore, UCS measurements before and after a mechanical improvement, such as compaction or preloading, may also provide information on the effectiveness of such methods in some stabilization processes.

For purposes of comparison, unconfined compressive strengths of lime stabilized natural soils range from 80 psi (550 kPa) to 1,100 psi (7,584 kPa), depending on the amount of pozzolanic material present, and curing period. Soaked unconfined compressive strength of soil-cement range from 200 psi (1379 kPa) to 1,200 psi (8,273 kPa), depending on soil type and curing period.[18] Freshly prepared solidified material from liquid containing concentrated brine, has been shown to attain 24-hour strength ranging from 7 psi (48 kPa) to 21 psi (145 kPa) with increased lime content.[19] The 24-hour strength of compacted solidified FGD sludge using SFT Terra-Crete process is on the order of 20 psi (138 kPa), where its 28-day strength is about 300 psi (2,068 kPa).[20] These values indicate the variability of the products of various processes involving different waste materials but similar additives for solidification.

Coefficient of hydraulic conductivity or permeability (k) of a solidified/stabilized system may be a good parameter to estimate potential groundwater contamination. The value of $k = 1 \times 10^{-7}$ cm/sec reflects a median value for low permeability clay soil systems. Laboratory studies with pure compacted clays of original water permeability of 10^{-8} cm/sec have shown marked increases in conductivity of these systems when water was displaced by hydrocarbons, such as xylene, gasoline, kerosene, and diesel fuel. Most solidification processes result in material permeability of 10^{-5} to 10^{-6} cm/sec. These numbers can be reduced further if the material is "soil like" or granular, and would undergo volume compression under pressure.

INVESTIGATION

Testing

A testing program was conducted to investigate how much improvement could be accomplished in the physical properties of a petroleum product contaminated kaolinite clay by treatment with additives. A number of additive mixture formulas were tried to obtain optimum results with the measured properties. Georgia kaolinite clay was selected as the soil medium because clays tend to retain hydrocarbons/oils, and their physical and chemical properties appear to be affected more than the other solid constituents of soils. The contaminating petroleum product was selected as No. 6 grade fuel oil (crude oil), since it is only slightly volatile and does not require special precautions to preserve its stability during and after mixing with soil. The additives were portland cement type II, hydrated lime ($Ca(OH)_2$), class-F bituminous fly ash, gypsum (calcium sulfate hemihydrate) and attapulgite clay obtained by crushing palygorskite shale from Florida to passing No. 200 sieve size.

Samples were prepared by mixing the clay with water at the predetermined optimum moisture content of each mixed system before adding the fuel oil. For 10% (by dry weight of clay) fuel oil mixed clay, the optimum moisture content was 27% and the maximum dry density was 13 kN/m^3. The clay was first mixed with the water and then mixed with oil thoroughly. The mixture was left to cure in a closed container for 24 hours to allow for the process of adsorption of oil onto the clay. Since crude oil is only slightly volatile, vaporization of its volatile components during the curing period was regarded as negligible. The contaminated soil was divided into a number of sets. Each set was treated with either single or a combination of additives at different percentages and proportions. The additive amount was varied as 5%, 10%, and 20% by the dry weight of clay. In the combination of reagents, the maximum additive percentage was 25% by dry weight.

The following tests and analyses were conducted on the untreated and treated specimens: unconfined compressive strength (ASTM D2166/D1632), direct tensile strength,[21] Atterberg limits, moisture content, freeze-thaw durability (ASTM D4842), permeability,[22] and oil content analysis of leachate.[23]

Results

Unconfined Compressive Strength (UCS)

The unconfined compressive strength of fuel oil mixed clay is significantly lower than that of the uncontaminated clay, as expected. Figure 14.1 shows the significant

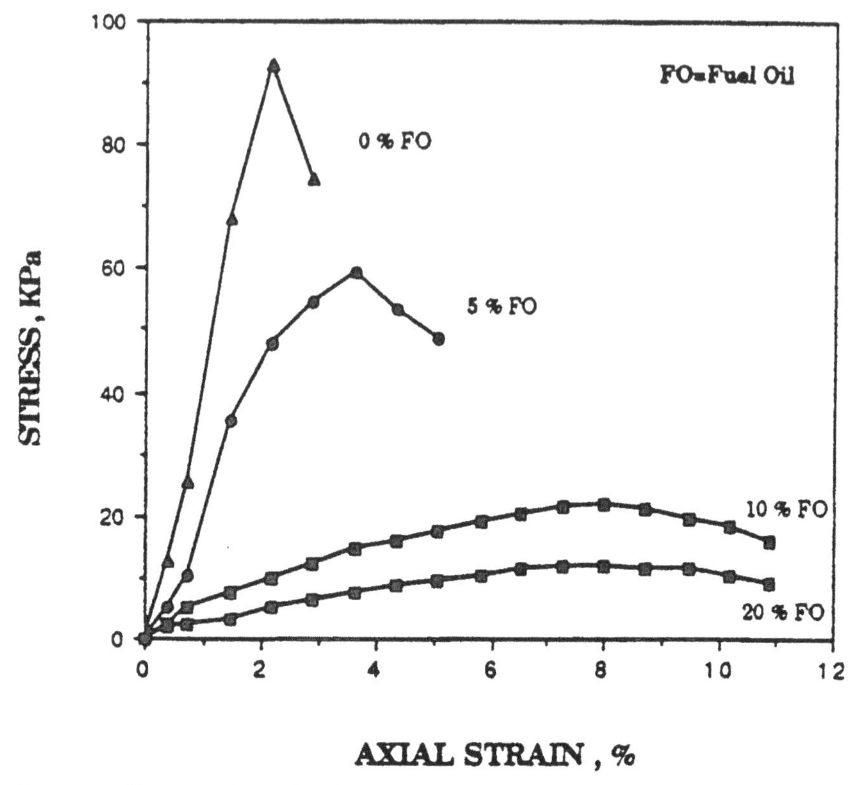

Figure 1. Variation of unconfined compressive strength and stress-strain relation with oil content.

loss of stiffness and strength of the clay with increasing percentage of oil. This has been observed by other investigators also, where the decrease in strength was attributed to increase in the net repulsive forces between clay particles in the presence of neutral nonpolar fluids with low dielectric constants.[24,25]

When the 10% fuel oil contaminated clay is treated with additives, there is a general increase in the strength values, especially for the 7-day constant humidity cured specimens. Table 1 presents the isolated effects of five different agents on the unconfined compressive strength of the contaminated clay. As observed, lime and cement produce marked increase in the strength values which are higher than the strength of the clay without contamination. Figures 2 and 3 illustrate the effects of lime and cement treatment on the unconfined compressive strengths, respectively.

A number of samples treated with combinations of additives were tested also. The best strength values were achieved with the addition of 10% cement, plus 10% lime, plus 5% attapulgite clay, and also 5% cement, plus 10% lime, and 5% fly ash. These results are presented in Table 2.

Table 1. Unconfined Compressive Strength (USC, kPa) Data for Stabilized 10% Fuel Oil Contaminated Kaolinite Clay

Stabilizing Agent	5% Additive Fresh	5% Additive Cured[a]	10% Additive Fresh	10% Additive Cured[a]	20% Additive Fresh	20% Additive Cured[a]
Lime	24.7	50.1	44.8	97.4	109.4	162.9
Cement	49.7	94.5	52.6	112.8	75.2	145.9
Fly ash	16.9	29.6	19.2	36.5	24.7	48.6
Attapulgite	18.4	30.9	37.3	55.1	75.2	92.0
Gypsum	17.0	39.5	32.1	45.3	55.9	88.9

[a]7 days.

The large increase in the strength values with the addition of lime is probably due to the flocculation and agglomeration effect of Ca^{++} ions. Lime initially breaks apart clay clumps, making available more mineral surface for reaction. This dispersion action probably releases the oil from inside the clumps into a more homogeneous mixture of solid, water, and oil phases. Additional lime then produces the gel which sets up while trapping the oil drops inside the solidified structure. This is probably why the addition of lime at a high percentage (20%)

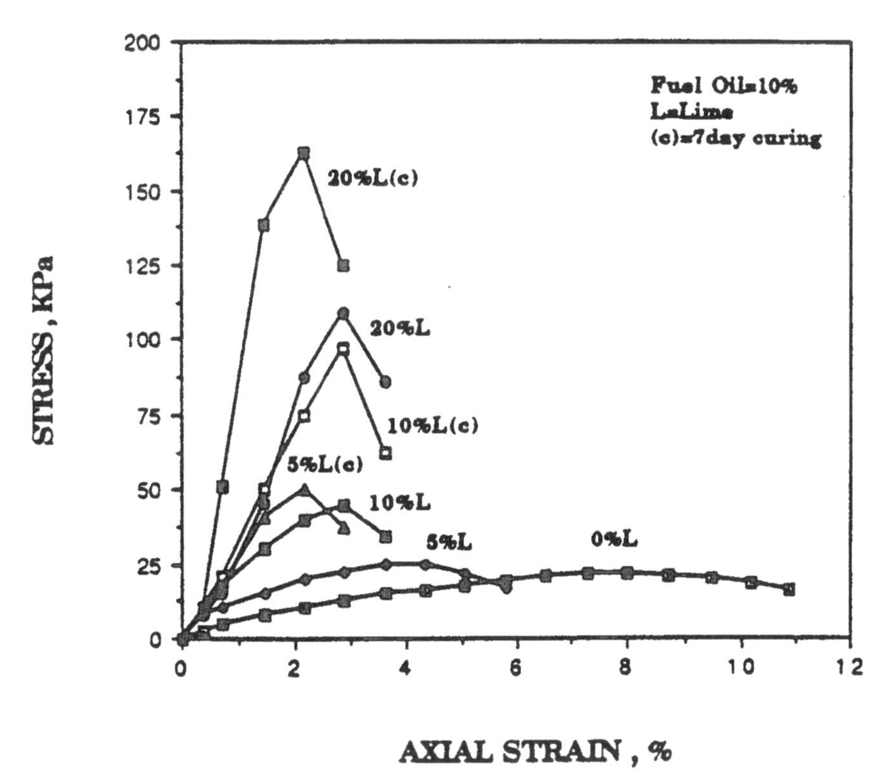

Figure 2. Variation of unconfined compressive strength and stress-strain relation with increased percentage of lime as additive.

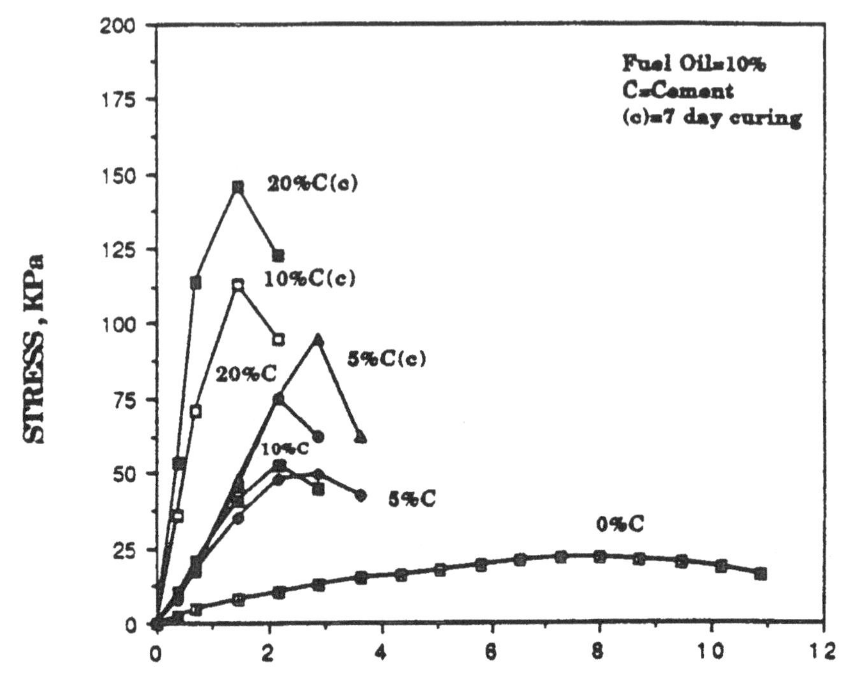

Figure 3. Variation of unconfined compressive strength and stress- strain relation with increased percentage of cement as additive.

resulted in higher strength values than the addition of cement at the same percentage. In this case, the pozzolanic reaction of cement might have been inhibited also, due to the presence of oil which would tend to retain the calcium ions, making it unavailable for the cementation reaction.

Direct Tensile Strength (TS)

The tensile strengths were determined through the "unconfined penetration test" described by Fang and Fernandez.[21] Results of these tests using 10% fuel oil

Table 2. Unconfined Compressive Strength (USC) and Tensile Strength (TS) Data for Stabilized 10% Fuel Oil Contaminated Kaolinite Clay

Stabilizing Agents[a]	UCS (kPa)	TS (kPa)
10% L + 10% FA	52.2	3.5
10% C + 5% FA	42.5	3.0
15% L + 5% FA	83.9	4.0
10% L + 5% C + 5% FA	95.6	3.6
10% C + 10% L + 5% A	115.0	4.7
10% C + 5% G + 5% A	75.3	3.2

[a]G = Gypsum; C = Cement; A = Attapulgite clay; FA = Fly ash; L = Lime.

contaminated clay treated with mixed additives are given in Table 2 also. The tensile strength of 10% oil contaminated clay was measured 1.1 kPa. This value is increased significantly with additive treatment. The best value of tensile strength was again obtained with the combination of cement, lime, and attapulgite clay mixture treatment.

The tensile strength measurements were also used to estimate and compare the variation of cohesion and internal friction angle of the treated materials, as explained in the following section.

Cohesion and Internal Friction Angle

A straight line envelope tangent to the Mohr circles on the right (USC) and the left (TS) of the vertical axis of normal stress versus shear stress representation were utilized to estimate, approximately, the cohesion and internal friction angle variation for the stabilized systems. Figures 4 and 5 illustrate the variations of normalized cohesion and internal friction angle with percent additive, respectively. As observed, additives appear to improve strength by improving the cohesion of the soil only. The internal friction angle appears to be unaffected by the presence of fuel oil. Furthermore, there is little change in the internal friction angle with treatment, except with lime treatment at low percentages.

These results are consistent with explanations given for strength reduction. If the increase in net repulsive forces is responsible for reduced strength, then cohesion component of the shear strength should be affected the most by this phenomenon. The sharp reduction in the internal friction angle at low percent addition of lime is probably due to the pulverization effect of lime, as discussed above. This, in turn, results in reduction of apparent particle size, and also releases oil from inside the clay clumps or 'spherical agglomerations.' The overall effect is then reduction in the internal friction angle. With addition of more lime, clay particles flocculate and agglomerate in a gel-like structure which, subsequently, results in increase in the internal friction angle. Trends of increasing cohesion with addition of cement and lime are similar, whereas fly ash produces no significant change, as expected.

Atterberg Limits

When 10% fuel oil contaminated soil samples were treated with 10% of each of the additives separately, the plasticity index was reduced for all except lime (Table 3). This is attributed to the saturation of kaolinite with a divalent cation (Ca^{++}), which may lead to an edge-to-face flocculation, and therefore higher liquid limit. Each limit test was repeated three times for reliability. It was noticed that the variation of these limits for the untreated clay was high. When the material was treated, the variation reduced significantly, as shown in Figure 6.

The large variation in the measurements of consistency limits of untreated contaminated clay may be due to nonhomogeneous distribution of the nonpolar liquid (fuel oil) in the clay-water system. Slight variations in the distribution of the oil

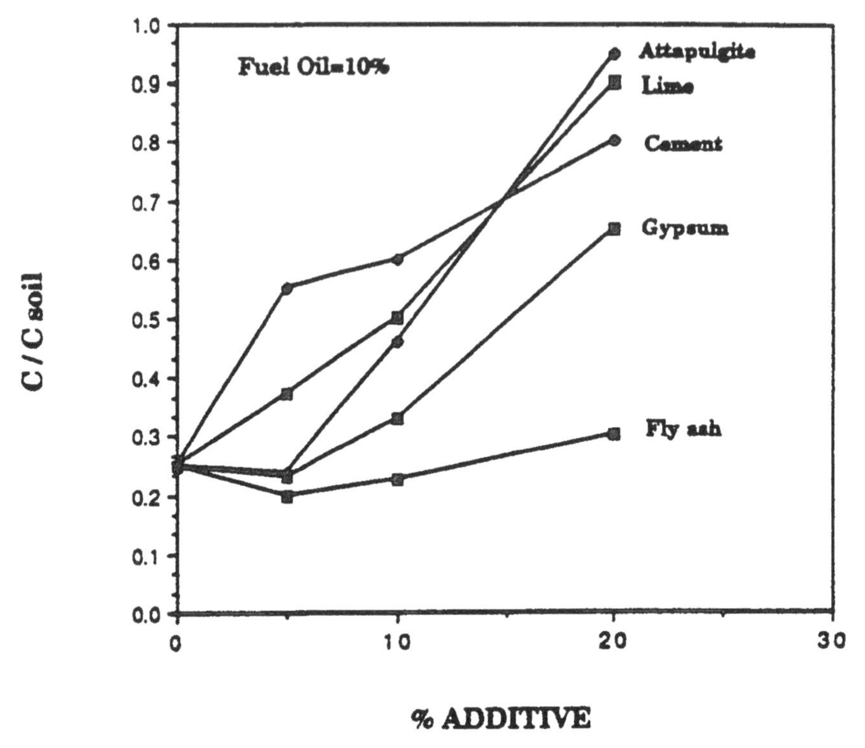

Figure 4. Variation of normalized cohesion of 10% fuel oil contaminated soil with increased percentage of additives.

phase in the soil may result in significant spatial variations of viscosity, density, and moisture content in a given sample mass of clay-water-oil mixture. Therefore, as an example, when running a liquid limit test, the localized oil content at the location where the standard groove (ASTM D4318) is made will likely influence the result for that particular test. However, after stabilization, if the oil becomes bound in a matrix of solidified structure, as expected, the localized

Table 3. Tensile Strength (TS, kPa) Data for Stabilized 10% Fuel Oil Contaminated Kaolinite Clay

	5% Additive		10% Additive		20% Additive	
Stabilizing Agent	Fresh	Cured[a]	Fresh	Cured[a]	Fresh	Cured[a]
Lime	2.7	3.4	2.7	5.2	3.0	9.8
Cement	2.7	5.0	2.7	6.8	4.0	8.7
Fly ash	1.1	2.2	1.1	2.7	1.6	3.3
Attapulgite	1.5	2.5	2.7	3.8	5.0	4.5
Gypsum	1.6	2.5	1.6	3.8	3.5	4.6

[a]7 days.

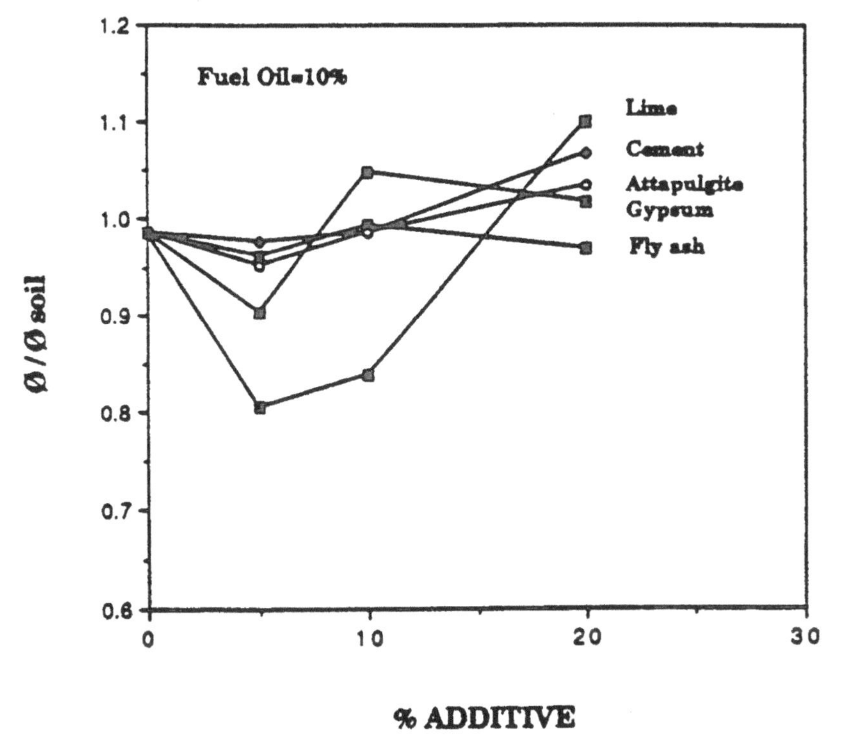

Figure 5. Variation of normalized internal friction angle of 10% fuel oil contaminated soil with increased percentage of additives.

oil should no longer be free to influence the test. This is observed indirectly by the reduced size of bands of variance of liquid limit values obtained for the additive treated clays, as shown in Figure 6.

Dry Density

Addition of fuel oil reduced the dry density of the compacted soil samples, as expected. However, additive treatment improved the density significantly, as shown in Figure 7.

Permeability and Leachate Analysis

Kaolinite clay permeability, measured by a constant head flexible wall permeameter (ASTM D5084), was on the order of 5×10^{-8} cm/sec. When 10% fuel oil was introduced, the permeability increased slightly. However, with treatment the resulting matrices showed up to 10 times decrease in the coefficients of permeability. These results are given in Table 4. It should be noted here that these permeability tests were conducted simultaneously, using seven separate permeameters.

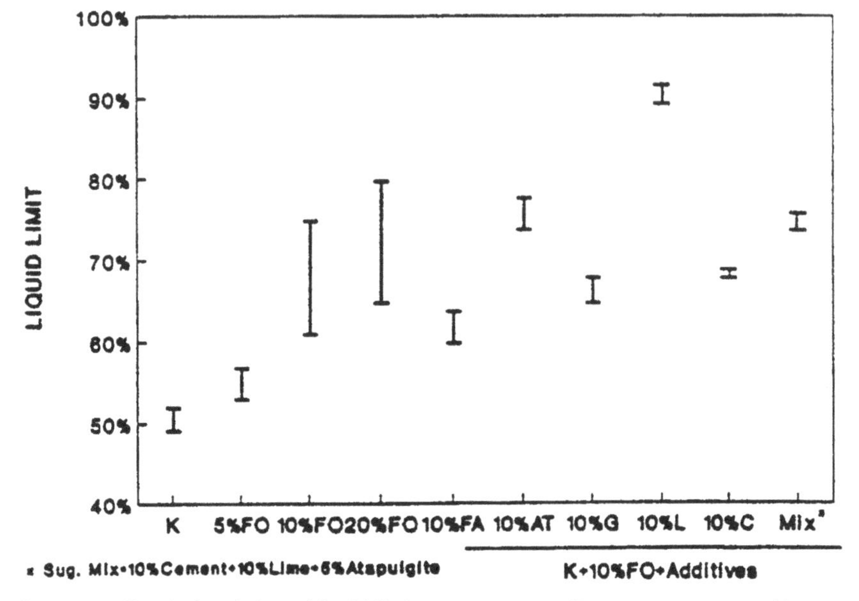

Figure 6. Band of variation of liquid limit measurements (3 measurements each).

The leachates collected from the permeability test of four of these samples were analyzed for oil content. These leachates were collected until after 24 hours of steady-state water permeation through the saturated matrix of each sample. The dissolved or emulsified oil was then extracted from the collected water and quantified, following a standard method for the examination of water and wastewater.[23] The leachate concentrations of oil obtained in this analysis are not necessarily regarded as absolute quantities, but evaluated relative to each other. The assumption in this evaluation is that the initial total oil content is the same for each contaminated mixture specimen tested. This content was set as 10% fuel oil by dry weight of clay when preparing the specimens. However, actual quantitative measurements of content may show slight deviations from this assumption. This may

Table 4. Coefficient of Permeability of Fuel Oil Contaminated and Additive Treated Kaolinite Clay

Specimen[a]	k × 10E-8 (cm/sec)
K	5.2
K + 10% FO	9.3
K + 10% FO + 10% L	1.9
K + 10% FO + 10% C	1.5
K + 10% FO + 10% L + 5% C + 5% FA	0.47
K + 10% FO + 10% C + 5% FA	0.96
K + 10% FO + 10% C + 10% L + 5% A	0.85

[a]A = Attapulgite clay; C = Cement; K = Kaolinite clay; L = Lime; FA = Fly ash; FO = Fuel oil.

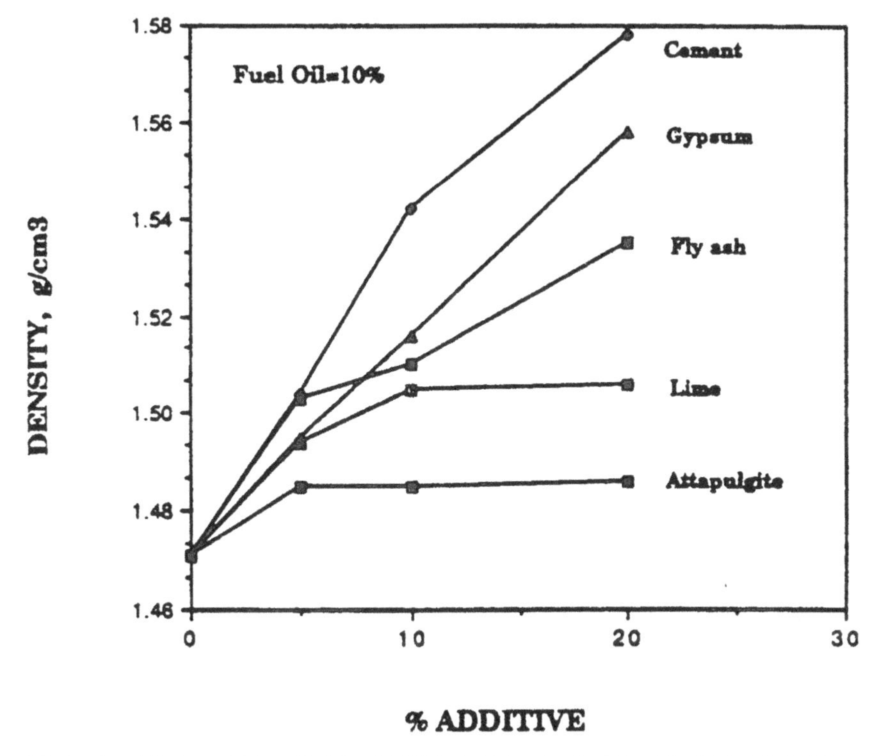

Figure 7. Variation of dry density of 10% fuel oil contaminated soil with increased percentage of additives.

be due to nonuniform distribution of oil in the mixing batch, resulting in poorly replicated specimens or volatilization of some components of the oil during mixing, resulting in a lower initial concentration than expected. Although the reference points may be slightly different for each sample used in the permeability analysis, the variations in leachate concentrations of oil measurements are consistent and significant enough to warrant comparison.

The oil content in the leachate of the untreated soil was 380 mg/L. When the mixture was treated with 10% cement, this number was reduced to 107 mg/L. The leachate from a sample treated with 10% lime showed oil at a concentration of 51 mg/L. The lowest oil content was found in the leachate of soil treated with 10% cement, 10% lime, and 5% attapulgite clay. This concentration was measured as 26 mg/L, which is approximately 93% improvement over the unstabilized soil.

Durability

Finally, the freeze-thaw durability of a few selected samples was measured. These results are given in Table 5. The average cumulative material loss was

Table 5. Freeze-Thaw Durability Test Results

Specimen[a]	Moisture Content (%)	Number of Cycles	Average Cumulative Corrected Relative Mass Loss (%)	Remarks
K + 10% FO	27	1	—	total deterioration
treated with				
10% L	27	1	—	total deterioration
10% L + 5%C + 5%FA	27	1	—	total deterioration
10% C + 5% FA	27	6	8.2	failure along horizontal crack planes
10% C + 10% L + 5% AT	27	5	6.5	failure along horizontal crack planes

[a]K = Kaolinite clay; C = Cement; FO = Fuel oil; L = Lime; FA = Fly ash; AT = Attapulgite clay.

the least for the cement, lime, and attapulgite clay treated soil, which survived four cycles of freeze-thaw prior to failure. It should be noted here that for this and the cement fly ash treated sample, the failure appeared to occur along horizontal planes which were identified as the planes of compaction layers. Poor binding between the compacted layers would cause water entry, which then induces volume change in freeze-thaw and eventual failure. The parts between these planes remained rigid and solidified.

CONCLUSIONS

The additive stabilization of petroleum contaminated soils appears to be a viable method of resource recovery, and may produce economics over other remediation technologies. In all the tests conducted, the treated samples showed significant improvement in their physical properties with respect to those of untreated samples. The following specific conclusions were made:

1. Fuel oil reduced the strength and stiffness of clay soils significantly, even at low percentages of addition.
2. Lime and cement produced marked improvement in the strength, durability, and dry density of the treated samples, in general. Seven-day cured samples of treated clay showed strength nearly double the strength of untreated clay.
3. Plasticity was reduced with additive treatment, except with lime. The large variability in the liquid limit determination was reduced when the contaminated soil was stabilized.

4. Permeability was reduced with additive treatment. The oil content in the leachate of the samples was reduced with stabilization relative to that of unstabilized soil.

In general, the best results in all of the tests conducted were obtained with samples treated with 10% lime, 10% cement, and 5% attapulgite clay. This is probably due to the combined effects of the three different types of stabilization/solidification processes; namely, cementing, pozzolanic, and sorbent reactions represented by each additive. Table 6 summarizes the improvements achieved with this mixture.

Further work is needed to obtain optimum percentages of these additives to create a cost-effective product with the desired physical properties and minimum leachability of oil. Information of long-term leachability under the influence of different pH pore fluids is also needed to fine-tune the recommended formula of additives in stabilizing petroleum contaminated soils.

Table 6. Some Properties of the Suggested Admixture Treated Clay Compared with Untreated Contaminated Clay

	Suggested Mix (10% C + 10% L + 5% AT)	Untreated Soil (K + 10% FO)
UCS (kPa)	115.0	21.8
TS (kPa)	4.7	1.1
LL (%)	75.0	70.9
PL (%)	45.0	33.2
PI (%)	30.0	37.7
Density (g/cm^3)	1.535	1.471
C/C_{soil}	0.65	0.25
ϕ/ϕ_{soil}	1.06	0.98
Freeze-thaw durability (# of cycles to failure)	5	1
Oil content in leachate (mg/L)	25.8	380.0

ACKNOWLEDGMENTS

This work was partially funded by Bethlehem Steel Corporation, Bethlehem, Pennsylvania, and by the Environmental Studies Center of Lehigh University.

REFERENCES

1. Dowd, R. M. "Leaking Underground Storage Tanks," *Environ. Sci. Technol.* 18:(10) (1984).
2. Magazu, D. M., and J. Carbery. "Biodegradation of Petroleum Contaminated Soils,"

Proceedings of the 21st Mid-Atlantic Industrial Waste Conference, (Lancaster, PA: Technomic Publishing Company, 1989), pp. 207-212.

3. Khan, L. I., S. Pamukcu, and I. J. Kugelman. "Electro-Osmosis in the Fine Grained Soil," H. Y. Fang and S. Pamukcu, Eds. *Proceedings of the Second International Symposium on Environmental Geotechnology* (Bethlehem, PA: Envo Publishing Company, 1989), pp. 39-47.
4. Hamed, J., Y. B. Acar, and R. J. Gale. "Pb(11) Removal from Kaolinite Soil by Electrokinetics," *J. Geotechnical Engineering* 117:(2)241-270 (1991).
5. Pamukcu, S., L. I. Khan, and H. Y. Fang. "Zinc Detoxification of Soils by Electro-Osmosis," *Transportation Research Record* 1288:41-46 (1991).
6. Volumetric Tank Testing: An Overview, U.S. Environmental Protection Agency, EPA/625/9-89/009, 1989.
7. Mackenzie, J. M. W. "Interactions Between Oil Drops and Mineral Surfaces," Society of Mining Engineers, AIME, *Transactions* 247:202-208 (September, 1970).
8. Guide to the Disposal of Chemically Stabilized and Solidified Waste. U.S. Army Waterways Experiment Station, SW-872, 1982.
9. Cullinane, M. J., Jr., L. W. Jones, and P. G. Malone. Handbook for Stabilization/Solidification of Hazardous Waste, U.S. Environmental Protection Agency, EPA/540/2-86-001, 1986.
10. Conner, J. R. *Chemical Fixation and Solidification of Hazardous Wastes* (New York, NY: Van Nostrand Reinhold Publishing Company, 1990).
11. Boelsing, F. "Remediation of Toxic Waste Sites: DCR-Technology," Report for Ministry of Economics, Technology and Traffic, Hanover, Germany, 1988, p. 64.
12. Bartos, M. J., and M. R. Palermo. Physical and Engineering Properties of Hazardous Industrial Wastes and Sludges, U.S. Environmental Protection Agency, EPA/600/2-77-139, 1977.
13. Spencer, R. W., R. H. Reifsnyder, and J. C. Falcone. "Applications of Soluble Silicates and Derivative Materials," Proceedings of the Management of Uncontrolled Hazardous Waste Sites, Hazardous Materials Control Research Institute, Silver Spring, MD, 1982, pp. 237-243.
14. Pancoski, S. E., J. C. Evans, M. D. LaGrega, and A. Raymond. "Stabilization of Petrochemical Sludges," Proceedings of the 20th Mid-Atlantic Industrial Waste Conference, Silver Spring, MD, 1988, pp. 299-316.
15. Pamukcu, S., J. B. Lynn, and I. J. Kugelman. "Solidification and Re-Use of Steel Industry Sludge Waste," Proceedings of the 21st Mid-Atlantic Industrial Waste Conference (Lancaster, PA: Technomic Publishing Company, 1989), pp. 3-15.
16. Van Keuren, E., J. Martin, J. Martino, and A. De Falco. "Pilot Field Study of Hydrocarbon Waste Stabilization," Proceedings of the 19th Mid-Atlantic Industrial Waste Conference, Bucknell University, (Lancaster, PA: Technomic Publishing Company, 1987), pp. 330-341.
17. Zarlinski, S. J., and J. C. Evans. "Durability Testing of a Stabilized Petroleum Sludge," Proceedings of the 22nd Mid-Atlantic Industrial Waste Conference, Drexel University, (Lancaster, PA: Technomic Publishing, Company, 1990) pp. 542-556.
18. Pamukcu, S., and H. F. Winterkorn. "Soil Stabilization and Grouting," in *Foundation Engineering Handbook* 2nd ed., H. Y. Fang, Ed., (New York, NY: Van Nostrand Reinhold Company, 1991).
19. Myers, T. E. "A Simple Procedure for Acceptance Testing of Freshly Prepared Solidified Waste," Hazardous and Industrial Solid Waste Testing: Fourth Symposium, ASTM STP 886, 1986, pp. 263-272.

20. Valiga, R. "The SFT Terra-Crete Process," in *Toxic and Hazardous Waste.* Vol. 1, R. B Pojasek, Ed. (Ann Arbor, MI: Ann Arbor Science, 1982).
21. Fang, H. Y., and J. Fernandez. "Determination of Tensile Strength of Soils by Unconfined Penetration Test," *ASTM. STP* 740 (1981), pp. 130–144.
22. Evans, J. C., and H. Y. Fang. "Triaxial Equipment for Permeability Testing with Hazardous and Toxic Permeants," *Geotechnical Testing Journal* 9(3):126–132 (1986).
23. *Standard Methods for the Evaluation of Water and Wastewater.* APHA-AWWA-WPCF, 17th ed., 1989.
24. Moore, C. A. and Mitchell, J. K. "Electro Magnetic Forces and Soil Strength," *Geotechnique.* 24(4)627-640 (1974).
25. Yong, R. N., and M. A. Warith. "Leaching Effect of Organic Solutions on Geotechnical Properties of Three Clay Soils," Proceedings of the 2nd Symposium on Environmental Geotechnology, Vol. 1, (Bethlehem, PA: Envo Publishing Company, 1989), pp. 99–110.

CHAPTER 15

In Situ Vitrification Applications

James E. Hansen and **Craig L. Timmerman,** Geosafe Corporation, Richland, Washington

INTRODUCTION

In Situ Vitrification (ISV) is an innovative, mobile, onsite remediation technology for contaminated solids. ISV has been under development for the U.S. Department of Energy (DOE) since 1980 by Battelle Memorial Institute's Pacific Northwest Laboratories. DOE has licensed the technology to Battelle, who has in turn exclusively sublicensed it to Geosafe Corporation for commercial application purposes.

The ISV technology has been widely published and exhibited. The interested reader is referred to Geosafe for ISV bibliographies and detailed reports on various aspects of the technology. This chapter presents a brief description of ISV, followed by a summary of application and evaluation considerations of interest to regulatory and engineering organizations involved with the comparative evaluation of alternative technologies. The status of ISV technology development and commercialization is also reviewed.

GENERAL PROCESS DESCRIPTION

The ISV process involves in situ electric melting of contaminated solids at very high temperatures, typically in the range of 1,600° to 2,000°C for most soils. Figure 1 illustrates progressive stages of ISV treatment, and Figure 2 presents

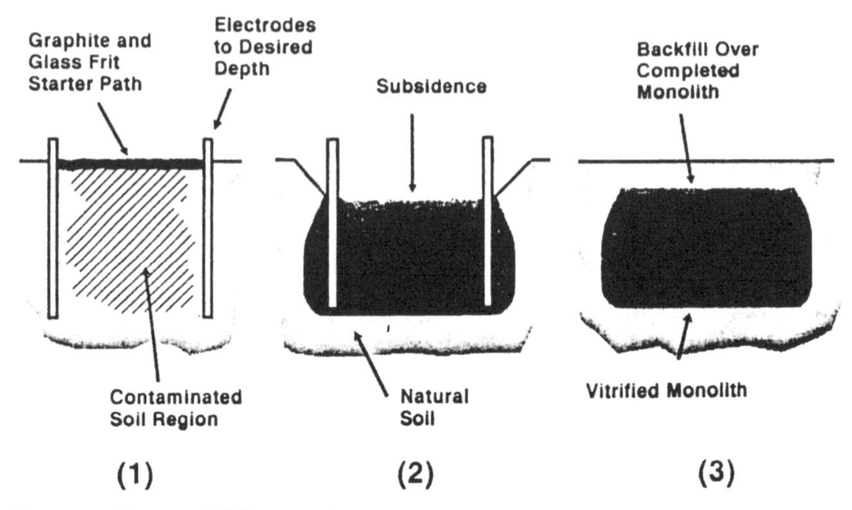

Figure 1. Stages of ISV processing.

typical process conditions. An array of four electrodes is either placed to the desired treatment depth in the volume to be treated prior to treatment (fixed electrodes), or the electrodes are lowered into the treatment volume as the melt progresses (moveable electrodes). A conductive mixture of graphite and glass frit is placed on the surface between the electrodes to serve as an initial conductive

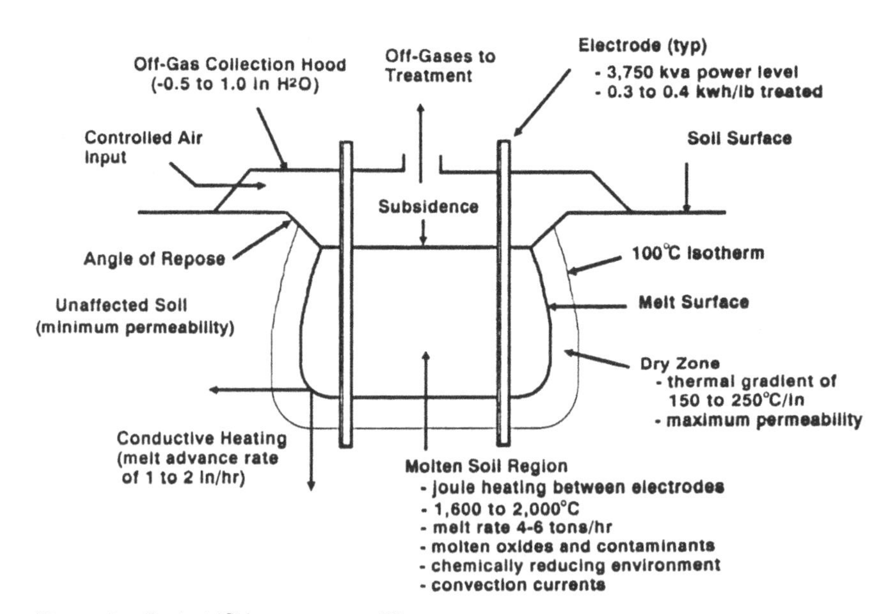

Figure 2. Typical ISV process conditions.

(starter) path. As electric potential is applied between the electrodes, current flows through the starter path, heating it and the adjacent solids to the solid's melting point. Upon melting, typical soils become electrically conductive; thus the molten mass becomes the primary electrical conductor and heat transfer medium allowing the process to continue beyond startup. The molten mass grows downward and outward as long as electric power is applied.

An off-gas collection hood gathers gases that evolve from the treatment zone during processing. Water vapor is usually the predominant evolved gas present in the hood, since most soils contain 15% to 30% moisture above the saturated zone. Secondarily, organic contaminant pyrolysis products and soil decomposition products will evolve to the surface under the collection hood. A large amount of ambient air is allowed to enter the hood, where it supplies oxygen for the combustion of flammable pyrolysis products and for purposes of cooling the hood. The air and other gases are then drawn through an off-gas treatment system to ensure their acceptability for release.

Significant volume reduction (25 to 45 vol% for most soils) occurs as solids particles melt and interstitial void volume is removed. Volume reduction results in a subsidence of the melt surface below the starting grade (see Figure 3). When power is terminated to the melt, it cools to a monolithic, vitrified (glassy with microcrystallinity) residual product which resembles natural obsidian (natural volcanic glass) for most soil applications. Single melts as large as 1,000 tons can be produced by existing large-scale equipment capable of processing 120 tons/day. Adjacent melts fuse together to produce a single impermeable monolithic structure.

Figure 3. Surface of treatment zone showing subsidence volume over 750 ton melt.

Completion of each melt setting involves placement of clean backfill to the desired depth in the subsidence volume.

PROCESS EQUIPMENT

The ISV equipment is mounted on three over-the-road trailers so that it is truly mobile in nature. The equipment is designed for quick interconnection at the site. The ISV equipment system is illustrated in simplified schematic form in Figure 4. Figure 5 presents an aerial view of Geosafe's large-scale commercial system.

The major portion of the equipment system is the off-gas collection and treatment system. A 60-ft diameter off-gas collection hood directs ambient air and evolved gases/vapors from the treatment zone to the off-gas treatment system. This system utilizes quenching, venturi scrubbing, mist elimination, humidity control, filtration, and carbon adsorption unit processes to ensure clean air emissions. The quenching and scrubbing solution is cooled by a self-contained glycol cooling system so that a continuous supply of onsite water is not required.

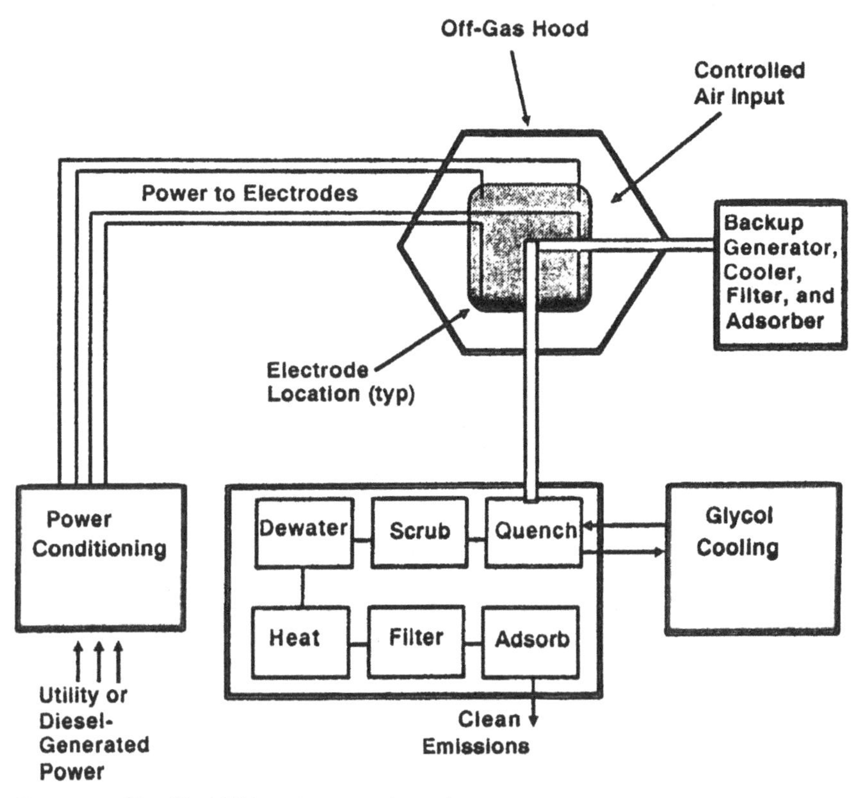

Figure 4. Simplified ISV equipment schematic.

Figure 5. Aerial view of Geosafe large-scale ISV system.

Periodically, contaminants collected in the scrubber solution, filters, and/or carbon beds may be recycled back to a subsequent ISV setting. In this way, only the secondary waste present at the end of the last setting requires further treatment or disposal.

APPLICATION CONSIDERATIONS

Application Types

The process is designed to treat contaminated soil in the ground; however, it may also be applied in a large container. ISV processing is termed "in situ" when the soils are processed where they presently exist, as in a landfill or impoundment. When they are placed in a trench or container for treatment, it is termed "staged" processing. Some applications may involve consolidating contaminated soil by removing and staging some soil on top of existing (in situ) contaminated soil.

Since ISV is a batch or setting type process, its time-operated efficiency increases with depth of processing. The process is most economical when dealing with large quantities (e.g., 300 to 1,000 tons treated/setting of electrodes). Processing depths greater than 10-ft are ideal, but not necessary.

ISV applications may also be categorized relative to the primary location and/or condition of the waste. Such categorization includes: (1) contaminated soil, (2) buried waste, and (3) underground structures. Most ISV development work has focused on contaminated soil applications wherein the contaminated media is primarily soil. The soil has typically become contaminated in such cases through exposure to contaminated liquids. In many cases the contaminated liquid is water that has percolated through impounded or buried waste that may or may not have been removed prior to addressing remediation of the contaminated soil. Contaminated soil applications are relatively straightforward compared to other types of applications, and the ISV technology is considered to be developed and demonstrated for contaminated soil applications.

Buried waste applications address wastes that have been covered by soil such as backfilled impoundments and landfills. Substantial amounts of test work have been performed on a variety of process sludges, ash, and containerized waste. The ISV technology is not considered generically ready for such applications; at this time, a specific test and demonstration plan is necessary for each one.

Buried waste applications involving wastes which were highly heterogeneous at time of burial typically pose a problem of site characterization. It is necessary to know worst-case conditions within the treatment zone to allow appropriate remedial design for the site. Homogeneous wastes, such as some settled lagoon and impoundment sludges and sediments, pose less of a characterization problem. However, the chemical composition of such wastes must be analyzed relative to the soil in the treatment zone to allow prediction and evaluation of melt behavior when the sludge/sediment zones are encountered. In some cases it may be necessary to intermix the soil and waste layers to allow proper treatment. The effect of the wastes on overall residual product chemistry and properties must also be evaluated.

Containerized wastes such as buried drums, crates, and cartons pose additional problems. Whereas the ISV process conditions may be adequate for treating such materials, the site characterization challenge becomes even more severe. The ISV technology is not considered ready for application to such sites at this time except on a test and demonstration basis. ISV is being developed for such applications within the DOE community because of the high cost of alternative technologies. Treatment of such sites may require use of equipment with larger than normal off-gas treatment capacity and/or the use of secondary off-gas containment to protect against unforeseen high gas generation events.

Solid Media

The primary qualification regarding type of soil that may be treated by ISV is whether or not the soil will form and support a melt. ISV test results have indicated that most natural soils may be processed by ISV without modification. Various sludges, sediments, and process tailings have also been successfully tested. For proper application, it is necessary that the soil and/or other solids contain sufficient inorganic material that will remain in the molten state during treatment. It is the molten mass that serves as the electrical conductor during ISV, and the flow of electricity through the melt results in the generation of heat which is then passed into adjacent soil by thermal conduction.

Molten soil must possess sufficient electrical conductivity to allow the process to be performed economically. Electrical conductivity within a soil melt is typically provided by the monovalent alkali earth cations (e.g., sodium, potassium). It is desirable that such cations be present in the 2 to 5 wt% range, which is common for most soils. In the event a soil possesses insufficient molten conductivity, it is possible to obtain the needed conductivity through addition of other materials (e.g., materials that provide Na_2 and/or CaO, such as suitable soil, soda ash, and lime.

The chemical (oxide) composition of the soil is important in determining the quality of residual product produced. Soil is the result of weathering of rocks, and rocks are made up of many minerals (complex metal oxides). Upon melting, minerals decompose to a melt mixture of major oxides, in which silica is predominant for most soils. Silicate melts typically produce a residual product of excellent properties relative to environmental exposure. Other low-silica soils (e.g., limestone/dolomite) have also been treated by ISV to produce a high quality residual product. It is possible to determine the applicability of ISV to various soils by performing and evaluating oxide composition analyses and small-scale melt tests.

Contaminant Disposition

As the high temperature ISV melt moves slowly downward and outward through the contaminated solids, a very steep thermal gradient (150° to 250°C/inch) precedes the melt. At appropriate temperature regimes within this gradient, or within the melt itself, the solids and contaminants undergo change of physical state and decomposition reactions. The possible dispositions of particular contaminants include: (1) chemical and/or thermal destruction, (2) removal from the treatment volume to the off-gas treatment system, and (3) chemical and/or physical incorporation within the residual product. Many site- and application-specific variables affect the disposition of specific contaminants. The primary variables include: (1) contaminant physical and chemical properties, (2) melt chemistry, (3) melt temperature, (4) contaminant dwell time in the treatment zone (in turn dependent on melt viscosity, depth, and other variables), (5) adjacent soil properties, (6) soil moisture content, and (7) extent of overmelting (i.e., amount of soil melted beyond the limit of contamination). Because of the many site-specific variables involved, it is necessary to consider each remediation project individually.

The ISV testing program has indicated that certain classes of contaminants may be expected to undergo basic types of response and ultimate disposition during treatment. Hazardous compounds undergo the phenomenon of pyrolysis (i.e., thermally induced decomposition of compounds into their elements, usually in the absence of oxygen; applicable to organics) and thermal decomposition (applicable to inorganics). For example, chlorinated organics decompose to carbon, hydrogen, and chlorine; nitrates break down into nitrogen and oxygen. In addition to the pyrolysis products, it is also possible that limited quantities of highly volatile materials may evolve from the treatment volume during processing. All materials evolved are captured in a collection hood and are subjected to off-gas treatment processes to ensure all emissions are within regulatory limits.

The solid media itself may also decompose during processing. For example, the inorganic portion of soils, which consists of complex mineral compounds, typically breaks down into major oxide groups such as silica and alumina. Upon cooling of the ISV melt, which is relatively rapid in terms of the time required for minerals to form, a residual product is formed which is glassy (a supercooled liquid of the oxide mixture) and may have varying amounts of crystallinity (from precipitated minerals) present. Such residual product typically has outstanding environmental exposure properties.

Regulatory criteria of interest regarding the residual monolith produced typically relate to: (1) structural, (2) weathering, (3) chemical leaching, and (4) biotoxicity properties. The Environmental Protection Agency (EPA) has performed tests on typical ISV product in these areas.[1] Structural strength tests indicated approximately 10 times the strength of unreinforced concrete, both in tension (ISV values of 4 to 8,000 psi) and compression (ISV values of 30 to 45,000 psi). Freeze/thaw and wet/dry weathering tests indicated the ISV residual was unaffected by repeated exposure. Chemical leaching tests consistently indicated the ISV residual is capable of surpassing the EP Toxicity (EP-Tox) and Toxic Characteristic Leaching Procedure (TCLP) leach tests. The EPA also found the ISV residual to be nontoxic to near-surface life forms.[2]

The above properties of the ISV residual product make it truly unique among remediation alternatives. Because of its unequaled ability to immobilize arsenic, as indicated by TCLP testing, vitrification has been identified as the best demonstrated available technology (BDAT) for arsenic-bearing wastes, as defined in the current Resource Conservation and Recovery Act (RCRA) landban regulations. The ISV residual product is considered to be permanent; that is, capable of withstanding environmental exposure for geologic time periods (e.g., thousands to millions of years).

In typical soil applications, inorganic elements which do not evolve from the melt during processing become part of this residual product through physical and/or chemical incorporation. The reader interested in heavy metals applications considerations is referred to Reference 1.

Tables 1 and 2 present typical results from the ISV development and testing program, indicating performance on various types of contaminants. Table 1 presents organic destruction and removal results, and Table 2 presents heavy metal retention, removal, and leach testing results.

Presence of Water

The presence and movement of water during ISV is a major consideration in evaluating potential applications and in project remedial design. During ISV, the thermal gradient which moves in front of the melt evaporates water within the 100°C isoband that starts less than 1 ft away from the melt. Water vapor moves to the surface through and adjacent to the melt, accomplishing some vapor stripping

Table 1. Typical Organic Destruction/Removal Efficiencies

Contaminant	Concentration (ppb)	Percent Destruction	Percent Removal[a]	Total DRE (%)
PESTICIDES				
4,4 DDD/DDE/DDT	21–240,000	99.9–99.99	>99.9	99.9999
Aldrin	113	>97	>99.9	99.99
Chlordane	535,000	99.95	>99.9	99.9999
Dieldrin	24,000	98–99.9	>99.9	99.99
Heptachlor	61	98.7	>99.9	99.99
VOLATILES				
Fuel Oil	230–110,000	>99	>99.9	99.999
MEK	6,000[b]	>99	>99.9	99.999
Toluene	203,000	99.996	>99.9	99.99999
Trichloroethane	106,000	99.995	>99.9	99.99999
Xylenes	3,533,000	99.998	>99.9	99.99999
SEMIVOLATILES				
PCP	>4,000,000	99.995	>99.9	99.99999
NONVOLATILES				
Glycol	8,000[c]	>98	>99.9	99.99
PCBs	19,400,000	99.9–99.99	>99.9	99.9999
Dioxins	>47,000	99.9–99.99	>99.9	99.9999
Furans	>9,400	99.9–99.99	>99.9	99.9999

[a]Percent removed from off-gas after destruction; percentages are additive for the total DRE.
[b]98% MEK in container, yielding 6,000 ppm in layer of container thickness.
[c]50% Ethylene glycol in container, yielding 8,000 ppm in layer of container thickness.

of other volatiles (e.g., organics) as it moves. The water pathways to the surface are illustrated in Figure 6.

Research studies have indicated that, while a slight steam pressure (1–2 psi) may exist within the dry zone and 100°C isoband, the water flow path is to the surface, as opposed to into the adjacent soil. This occurs because vapor phase permeability within the 100°C isoband is at least several orders of magnitude lower than within the dry zone due to the presence of liquid water within the isoband. Water mass balance experiments have verified that substantially all water present in the treatment zone is removed to the surface during processing.[3]

Since significant energy is required to vaporize water, its presence represents an economic penalty. Therefore, it is economically advantageous for the treatment volume to be as dry as practicable immediately prior to ISV treatment. In a similar manner, consideration should be given to employing means to minimize/eliminate water recharge if ISV is applied in an active water zone. Typical means for limiting recharge include use of barrier walls, well points, or French drains.

Water may also influence processing cost in that its removal from the treatment zone relates directly to processing rate. Since the water vapor flows to the surface through the dry zone and the melt itself, it contributes to the agitation (flow activity) level of the melt. Depending on melt and adjacent soil conditions, processing rate must be controlled to maintain acceptable water vapor generation and removal rates.

Table 2. Typical Inorganic Removal/Retention/Leach Results

Contaminant	Percent Retention	Percent Removal[a]	Total Ret/Rem (%)	Initial Concen.[b]	TCLP[c] Result	Allowable
VOLATILES						
Hg	0	97–>99	>97	5,360	ND	0.2
SEMIVOLATILES						
As	70–85	>99.9	99.98	43,900	0.9	5.0
Cd	67–75	>99.9	99.96	37	0.001	1.0
Co	99–99.9	>99	99.99	17	<0.01	—
Cs	99–99.9	>99	99.99	—	—	—
Pb	90–99	>99.9	99.99	1,550	<0.063	5.0
NONVOLATILES						
Ba	99.9	>99.98	99.9999	185	0.140	100
Cr	99.9	>99.9	99.9999	290	0.020	5.0
Cu	90–99	>99.9	99.99	65,000	3.3	—
Ni	99.9	>99.9	99.9999	47	ND	—
Ra-226	99.9	>99.9	99.9999	6,000 pCi/g	6.4 pCi/g	100[d]
Pu/Th/U	99.99	>99.9	99.99999	—	—	—
Zn	90–99	>99.9	99.99	14,200	<0.05	—

[a]Percent removed from off-gas not retained; thus, percentages are additive for retention and removal.
[b]Concentration in ppm unless otherwise noted.
[c]TCLP values in mg/L unless otherwise noted.
[d]DOE limit for concentration in drinking water.
— = Either not available or not applicable.
ND = Nondetectable.

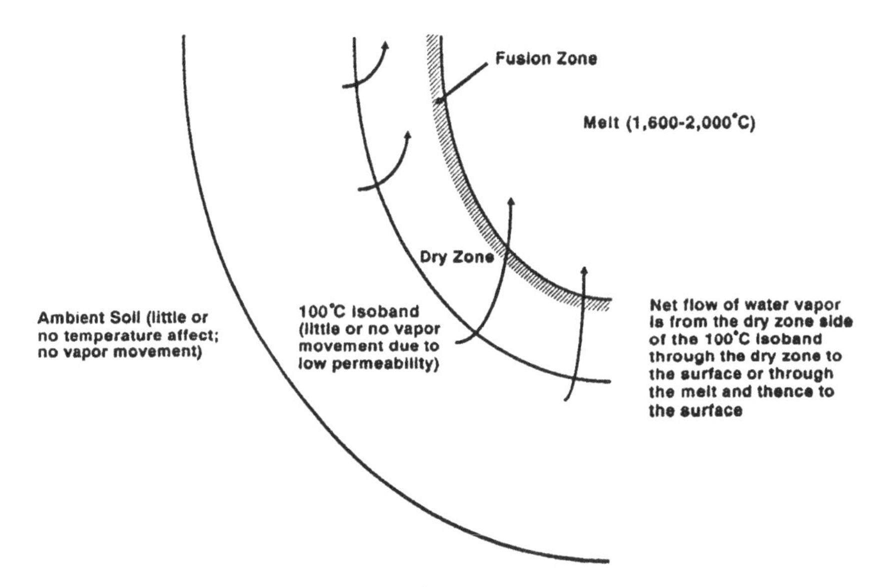

Figure 6. Water vapor pathways to surface.

Effects on Nearby Surroundings

The volume reduction characteristic of ISV processing results in a subsidence volume over the residual product mass (refer back to Figure 3). Such subsidence and adjacent soil sloughing may result in undercutting surface or near-surface structures, unless engineering provision is made to prevent sloughing. It should also be noted that the off-gas collection hood is 60 ft in diameter, compared to a maximum large-scale melt width of 30–35 ft. Thus, a minimum side clearance of 15–20 ft is desirable.

Cooling of the residual monolith may result in some heating of the adjacent soil. During processing, the 100°C isoband extends from less than 1 ft away from the melt to 3–5 ft away (typical case). During cooling, the 100°C isotherm has been observed to extend as far as 10 ft from the melt. This occurrence has the effect of drying out the adjacent soil.

The ISV process has never been observed to induce significant magnetic or electric fields beyond the treatment zone.

Production Rates

The large-scale ISV equipment is capable of a 3.5 MW power level. This corresponds to a maximum soil melting rate in the range of 4–6 ton/hr. This in turn results in a melt advance rate of 1–2 in./hr, depending on soil density. The process is operated 24 hr/day until a melt setting is completed. Downtime for movement of equipment between settings is less than one day.

Other factors, such as allowable water vapor generation and removal rates, may govern processing rates to lower than those possible based on available power level. Allowable production rates can be predicted during remedial design.

Secondary Waste

The ISV off-gas treatment system may collect small quantities of hazardous materials in the quencher/scrubber solution, the HEPA filter, and the activated carbon. These media are either subject to disposal as secondary waste, or they may be placed within a subsequent ISV setting for disposal (of the media) and recycling of the contaminants. In this way the site-wide contaminant destruction/removal/immobilization efficiencies may be maximized and secondary waste minimized. In a similar manner, limited quantities of protective clothing and other site secondary waste may be treated. Demobilization decontamination solution and the quencher/scrubber solution, filters, and activated carbon from the last setting at the site are subject to disposal as secondary waste.

Cost

Typical ISV remediation projects involve the following work elements: (1) site characterization, (2) treatability/pilot testing, (3) remedial design, (4) permitting/compliance analysis and documentation, (5) site preparation, (6) equipment mobilization, (7) onsite vitrification operations, (8) equipment demobilization, (9) site restoration, and (10) delisting and/or long-term monitoring. The cost of these activities is dependent upon specific conditions at the site, and the overall project criteria and objectives. Typical cost ranges are discussed below for those activities that are peculiar to ISV processing.

Treatability/pilot testing is utilized to: (1) demonstrate that the technology is applicable to the specific soil/waste combination at the site, (2) produce contaminant-related performance data necessary to support permitting activities, (3) produce operation-related performance data necessary to support cost estimates and quotations, and (4) produce samples of residual product for use in community relations efforts. Treatability testing involves performance of various physical and chemical tests on actual contaminated materials from the site, followed by engineering-scale ISV melt testing on the materials. The cost of treatability testing for non-PCB, non-dioxin/furan wastes is $25,000 plus the cost of necessary analytical work, which usually falls in the $15,000 to $25,000 range. For PCB and/or dioxin/furan wastes, the base treatability testing cost is $30,000 plus analytical costs in the range of $25,000 to $50,000. These cost ranges do not include data validation costs or other unusual analytical requirements. Treatability testing can usually be completed within 10 to 12 weeks after initiation of the project.

The cost of equipment mobilization and demobilization depends on the transport distance to and from the site. Typical total mobilization/demobilization costs fall in the range of $150,000 to $250,000.

The onsite service cost of ISV processing typically falls in the range of $350 to $450 per ton of material processed. This cost includes all elements of direct and indirect cost, such as labor, materials, energy, equipment amortization, and contractor overhead and profit. The most significant variables affecting this cost include: (1) the cost of electrical power, (2) the amount of water to be removed during processing, (3) depth of processing, and (4) analytical chemistry requirements associated with process control and permit compliance. Given information on these and other pertinent variables, application-specific cost estimates may be developed.

ADVANTAGES AND LIMITATIONS

Advantages

ISV advantages relative to alternative technologies include its capability to: (1) simultaneously process mixed waste types (organic, heavy metal, radioactive), (2) achieve destruction, removal, and immobilization performance beyond regulatory criteria, (3) be performed onsite and in situ, (4) accept significant quantities of rubble and debris in the treatment zone, (5) achieve a significant volume reduction (25% to 45% for most soils), and (6) produce an unequaled residual product with a geologic time life expectancy (thousands to millions of years). ISV also possesses significant differences from other vitrification technologies (e.g., plasma, joule-melter, slagging kiln), many of which are considered advantages.[4]

Limitations

The primary limitations on ISV applications relate to: (1) total organic concentration, (2) water recharge rate, (3) depth of processing, and (4) presence of inclusions. Since organics become gaseous pyrolysis products during ISV, the concentration of organics must be limited in relation to the off-gas collection and treatment equipment capacity. The average allowable concentration for most organics falls in the range of 5 to 10 wt%.

Fully saturated soils may be processed; however, it is economically advantageous to minimize soil moisture content and water recharge rate. These factors influence cost through consumption of energy and impacting of processing rate. Processing rate may be limited by the amount of energy going into water removal, or by operating at less than full power to maintain acceptable water vapor generation and removal rates peculiar to a specific application.

The maximum depth processed by ISV to date is 19–20 ft. Greater depths will be attempted in the continuing ISV development program.

The ISV process is capable of accommodating significant inclusions within the treatment zone (e.g., rocks, roots, drum remnants and other metal scrap, concrete, asphalt, construction debris, etc.); however, the concentration of these must be limited so as to not interfere with proper formation and advancement of the melt. All of the above limitations are subject to consideration during applicability analyses, treatability testing, and project remedial design.

CHARACTERISTICS PERTINENT TO RI/FS EVALUATION CRITERIA

ISV typically fares well in evaluation of short- and long-term effectiveness, permanence, and reduction of toxicity and mobility because of its excellent capability to destroy, remove, and/or immobilize contaminants. It is outstanding in the area of volume reduction in that vitrification is the only known means to achieve significant volume reduction (25% to 45%) in silica-based soils.

ISV is limited in regard to the implementability criterion in that it is commercially available from a single source. The technology is covered by a basic patent which has been licensed by DOE to Battelle, and has been exclusively sublicensed to Geosafe for commercial application. At this time Geosafe owns a single commercial large-scale ISV machine capable of processing 15–25,000 tons of soil per year. The off-gas collection hood portion of that machine is presently undergoing redesign to increase its design capacity.

As noted in the cost discussion above, total ISV project costs are highly site-specific. When total project costs are evaluated, use of ISV may be the most cost-effective alternative for specific sites, particularly when considered in relation to the effectiveness, permanence, and volume reduction criteria.

ISV is considered capable of meeting state and federal Applicable or Relevant and Appropriate Requirements (ARARs) where it is being considered for use. ISV has enjoyed good support by the regulatory community. The onsite and in situ nature of ISV, and the quality of its residual product have resulted in generally excellent acceptance by the public.

DEVELOPMENT AND COMMERCIALIZATION STATUS

U.S. DOE Development Program

The current DOE ISV development program is addressing potential applications at DOE's Hanford, Idaho Falls, Oak Ridge, Rocky Flats, and Savannah River plant sites. The program is addressing contaminated soil, buried waste, and underground tank type applications. Battelle's Pacific Northwest Laboratories provides national coordination of DOE's ISV program. Experimental and demonstration work is being performed by Geosafe, Battelle, ORNL, and EG&G Idaho.

Vapor Retreat Issue

The ISV development and commercialization program has long been plagued by a competitor who claims that vapors generated during ISV treatment do not rise toward the surface as claimed by the ISV technical community, but rather "retreat" into the soil adjacent to and underneath the melt. Geosafe and Battelle have thoroughly investigated these allegations and have found them to be without technical merit and in exact opposition to observed ISV processing performance.[5]

The EPA has participated in the resolution of this issue by requiring treatability test work plans to include the attainment of independently qualified performance data relative to whether or not contaminants move into the adjacent soil during processing. One such recent test on PCB-contaminated soil included a large number of samples (for statistical significance) and an independent data validation effort. The test results confirmed the complete absence of contaminant migration into the adjacent soil.[6]

Moveable Electrode Development

The DOE ISV program has developed a moveable electrode concept for application to very high metals content sites. The need for such electrodes was identified relative to an INEL site containing very high levels of metals in buried waste (20–40 wt%). The original ISV fixed electrode concept, wherein electrodes are placed to full depth in the treatment zone before initiation of processing, was found to be limited in such high metals applications due to shorting between the electrodes caused by pooling of molten metal at the bottom of the melt. Moveable electrodes are lowered into the melt as melt depth increases. In the event of metal pooling in the melt, the moveable electrodes can be maintained at least a minimum distance above the pool, thereby avoiding shorting.

Moveable electrodes avoid the necessity of preplacing electrodes in the treatment zone. They also have the advantage of being made only from graphite, whereas the fixed electrodes are of a combination graphite/molybdenum core construction. At this time all ISV test equipment has been converted to the moveable electrode capability. Battelle successfully demonstrated the use of moveable electrodes at large-scale during an underground tank treatment test in June, 1991.

Geosafe Operational Acceptance Testing Event

Geosafe recently experienced an event during large-scale operational acceptance testing of a new fabric hood design that resulted in significant damage to the hood. The test results indicated that the hood containment fabric was not acceptable for worst-case operating conditions. Geosafe has investigated the cause(s) of the event, and is redesigning the large-scale ISV off-gas collection hood. Geosafe is returning to an all-metal hood design, such as has been the standard during the ISV development program. Commercial ISV field operations are awaiting completion and testing of the new hood.

Cobble Walls for Melt Control

Recent tests employing cobble (2–4 in. diameter rock) barrier walls have shown the concept to be effective in retarding, but not eliminating, melt growth rate in the direction of the cobble wall. The cobble material, which has a high solid density and large void volume between stones, melts slower than soil particles with small void volumes. The objective of these tests has been to explore the use of cobble walls for applications wherein melt shape control is desired for various purposes.

Preferred Remedy Selections

ISV has been selected as a preferred remedy at 10 private, EPA-Superfund, and DOD sites within the U.S. These selections include: (1) Parsons Chemical/ETM Enterprises (EPA-V), (2) Northwest Transformer (EPA-X), (3) Arnold AFB Site 10 (DOD), (4) Rocky Mountain Arsenal M-1 Holding Ponds (DOD), (5) Ionia City Landfill (EPA-V), (6) Crab Orchard National Wildlife Refuge (EPA-V), (7) Anderson Development Company (EPA-V), (8) Crystal Chemical (EPA-VI), (9) Wasatch Chemical (EPA-VIII), and (10) Transformer Service Facility (EPA-X). Remediation contracts currently exist for two of these sites. The others are at various stages of treatability testing, remedial design, or are under negotiation between regulators and responsible parties.

REFERENCES

1. Paxton, J. ''Environmental Protection Agency (EPA), Test Program on Raw, Stabilized and Vitrified Soil, Western Processing Inc.,'' NPDEN-GS-1, Department of the Army, North Pacific Division Materials Laboratory, Corps of Engineers, Troutdale, OR, December 12, 1985.
2. Green, J. C. et al. ''Comparison of Toxicity Results Obtained from Eluates Prepared from Non-Stabilized and Stabilized Waste Site Soils.'' *Proceedings of the 5th National Conference on Hazardous Wastes and Hazardous Materials, April* 19–21, 1988. Las Vegas, NV.
3. Bonner, W. F. and J. L. Buelt. ''In Situ Vitrification: Test Results for a Contaminated Soil Melting Process.'' PNL-SA-16584. Paper presented at the 1989 Incineration Conference, May 1–5, 1989, Knoxville, TN.
4. Hansen, J. E. ''Vitrification Technologies,'' *Immobilization Technology Seminar.* U.S. EPA, Center for Environmental Research Information, Cincinnati, OH; CERI-89-222, 1989.
5. ''Geosafe Corporation Comments on Claims by Larry Penberthy, President of PEI, Inc., Against In Situ Vitrification Technology.'' Geosafe Corporation. November 22, 1990.
6. *Engineering-Scale Test Report for Application of In Situ Vitrification Technology to Soils Contaminated with Polychlorinated Biphenyls at the Northwest Transformer Superfund Site.* GSC 1006, Rev. 1, February 6, 1991. Geosafe Corporation, Kirkland, WA.

CHAPTER 16

Field Studies of In Situ Soil Washing

James H. Nash and **Richard P. Traver,** Chapman, Inc., Atlantic Highlands, New Jersey

INTRODUCTION

Surface and near-surface contamination often serve as the source for groundwater contamination. Percolation of rainwater through spill sites quickly carries soluble and semisoluble contaminants away from the point of origin. Contaminants considered "insoluble" above parts per million nevertheless migrate more slowly. Gross contaminant sources supply pure product that, over many years, flows deeply through unsaturated soils.

Part of the Environmental Protection Agency's (EPA's) Superfund site cleanup research has been directed at washing such contaminated soil with the aid of aqueous surfactant solutions. The research takes two directions. The first is to excavate the soil and mix it in a wash solution. The second research objective concentrates on the application or injection of a surfactant solution into undisturbed soil in situ. A segment of this in situ research is the subject of this project summary.

This demonstration effort grew out of mutual need between the EPA and the U.S. Air Force. From 1982 to 1985, the EPA researched soil washing technology, using surfactants in laboratory studies. Recompacted soils were used in these studies to simulate in situ conditions. Truly undisturbed contaminated soil was not tested up to that time. The U.S. Air Force, as part of its Installation Restoration Program, was seeking processes to clean up 128 fire training pits at Air Force installations. The Air Force selected the Air National Guard base in Camp Douglas, Wisconsin as a candidate site for the EPA to test either excavated or in situ soil washing. The EPA and the Air Force representatives chose in situ washing after further consideration.

THE LABORATORY STUDY

Previous laboratory work identified a 50:50 blend of two commercially available surfactants that work well in removing contaminants from soil. They are Adsee 799 and Hyonic PE-90, sold by Witco Chemical and Diamond Shamrock, respectively. To determine if this same blend would work at the Volk Field fire training pit, contaminated soil samples were collected.

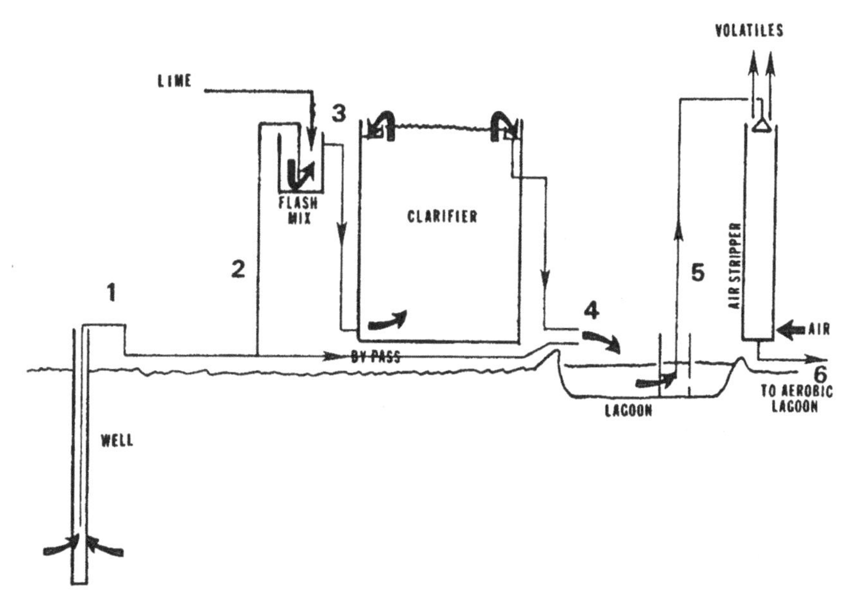

Figure 1. Volk Field pilot treatment for water.

Five physical tests characterized the soil. They were: grain size, TOC, cation exchange capacity (CEC), mineralogy by X-ray diffraction, and permeability. The grain size of the contaminated soil was 98% sand. By X-ray diffraction, alpha-quartz comprised the major portion of the soil with a minor amount of feldspar being present. TOC was as high as 14,900 $\mu g/g$. The cation exchange capacity of 5 mq/100 g was not significant to the contamination levels; however, it did support the X-ray diffraction mineralogic findings. The permeability of the fire pit soil, at 10^{-3} and 10^{-4} cm/sec, was one to two orders of magnitude less than adjacent uncontaminated soil.

Chlorinated hydrocarbons were part of the volatile contamination. Dichloromethane, chloroform, 1, 1, 1-trichloroethane (TCA), and trichloroethylene (TCE) at concentrations up to 3 $\mu g/g$ and total chlorinated solvents up to 3.5 $\mu g/g$ were determined by the VOA procedure. Other hydrocarbons are aliphatic, aromatic, and polar constituents. The level of hydrocarbon contamination is in the hundreds of μg/g based on the laboratory analysis.

Contaminated groundwater from the aquifer below the fire training pit, a significant problem, was also characterized in the lab study. VOA, TOC, and ultraviolet spectroscopy (UV) were used. The investigations determined that the groundwater contains chlorinated and nonchlorinated hydrocarbons in excess of 300 μg/L.

The soil adsorption constant (K) is a measure of a pollutant's tendency to adsorb and stay on soil. A value of 2,000 for PCBs indicates a two-hundredfold greater adsorption (holding power) than benzene at $K = 10$. Benzo(a)pyrene, a toxic substance, and oil have similar values—$K = 30{,}000$–$40{,}000$. Grouping contaminants according to a K value and evaluating removal efficiencies (RE) gives order to an otherwise complex collection of chemical classes. This is a report of the EPAs and the U.S. Air Force's field evaluation of in situ soil washing of compounds having K values between 10^1 and 10^6.

THE FIELD STUDY

The field study was conducted by laying out ten 60 cm × 60 cm × 30 cm pits, dug into the contaminated surface of the fire training area, which served as reservoirs that held various surfactant solutions. Field technicians applied wash solutions into the holes at the maximum rate of 77 L/m^2 per day. The daily dosage was applied in four increments. Since each hole percolated the solutions at different rates, the time interval between doses varied from hole to hole. Testing in three of the pits stopped when the time intervals for the next application approached 10 hours, indicating unacceptable permeabilities being created. Following seven days of washing, the pits received rinses with local, potable well water.

A combination of infrared spectroscopy (IR) and gravimetric determinations of soil extracts was used to evaluate "before and after" contaminant concentrations. To determine contaminant concentration, soil samples were taken after the rinse process, extracted with carbon tetrachloride, and analyzed by IR spectrophotometer for spectral absorbance by the carbon hydrogen bond. The extracts were then air dried and weighed to determine gravimetrically the contaminant concentration (nonvolatile).

The contaminant concentration before soil washing was based on the extracts of soil samples taken adjacent to the test holes. No samples were taken directly from the test holes before washing in order not to bias permeation rates.

Based on both the gravimetric and IR determinations of contaminant concentrations, there was no measurable decrease in contaminants following as many as 14 pore volumes of soil washing in the field tests.

In addition to the soil washing, the field crew conducted a bench-scale groundwater treatment study. From that study a treatment system was assembled and operated which successfully reduced TOC, VOA, and BOD_5 by 50, 99, and 50%, respectively. At these effluent levels, discharge to the local aerobic sewage lagoon was below the Wisconsin Department of Natural Resources' permit limits. A total of 320,000 L of contaminated groundwater was treated at rates of 15,000 to 45,000 L/day.

The bench-scale study investigated the use of lime, alum, ferric sulfate, hydrogen peroxide, polymeric electrolytes, and mineral acids. The application of these chemicals was guided by conventions appropriate to wastewater treatment plants. The resulting water treatment process (shown in Figure 1) was based on the addition of lime at 2 g/L. The lime created a flocculation of iron oxides and organics. The contaminant plume contained up to 52 mg/L iron. Particulate sedimentation in a clarifier, followed by additional residence time in a holding lagoon, reduced the TOC, BOD_5, and VOA to acceptable discharge levels. A final polishing of the volatiles in an air stripper was the final step in the process. Table 1 is a summary of the analytical data.

Table 1. Analytical Tests and Sampling Points Table for the Water Treatment Process

Pt. No.	Description	Tests Performed	Approximate Values, Average or Range
1	Individual well head	volatile organic	10–20 mg/L
		total organic	60–760 mg/L
		chemical oxygen demand	6–500 mg/L
		oil and grease	0.2–46 mg/L
		pH	5.1–6.2
2	Well field effluent	volatile organic	10–20 mg/L
		total organic	250 ± 14% mg/L
		iron	32 mg/L
		pH	6.0 ± 0.2
		chemical oxygen demand	41 mg/L
		flow rate	.25–2 L/sec
3	Flash mixer effluent	total organic (dissolved)	160 mg/L
		suspended solids	350 mg/L
		pH	6.8–9.7
		flow rate	.5–2 L/sec
4	Clarifier effluent	total organic	205 ± 7% mg/L
		suspended solids	13.6–104 mg/L
		pH	7.6
		flow rate	.5–2 L/sec
5	Air stripper feed	volatile organic	3.5–7.0 mg/L
		total organic	151 ± 13% mg/L
		temperature	6–15°C
		flow rate (water)	.95–1.26 L/sec
		oil and grease	3.6 mg/L

continued

Table 1. *Continued*

Pt. No.	Description	Tests Performed	Approximate Values, Average or Range
6	Air stripper effluent	volatile organic	0.3–0.5 μg/L
		total organic	146 mg/L
		flow rate (air)	101 L/sec
		oil and grease	3.6 mg/L
		biochemical oxygen demand	2.5 mg/L
		chemical oxygen demand	180 mg/L
7	Clarifier	suspended solids	4.4 mg/L
8	Clarifier bottom	suspended solids	2331 mg/L
9	Soil	oil and grease	800–16,000 mg/kg

CONCLUSIONS

1. In situ soil washing of the Volk Field fire training pit with aqueous surfactant solutions was not measurably effective. It is likely that this same ineffectiveness would occur at other chronic spill sites that have contaminants with high soil-sorption values ($K > 10^3$).
2. In situ soil washing requires groundwater treatment. Groundwater treatment at this site was very successful with the simple addition of lime. Air stripping effectively removed the volatile organics. Advantages at this site were its remoteness for workable air emission limits that facilitated groundwater treatment operations and a local sewage treatment system owned by the responsible party. TOC levels of the recovered groundwater were reduced to one-half the initial values by precipitation with lime, which allowed for direct discharge to the aerobic treatment lagoons. Obviously, not all waste sites have these favorable conditions.

CHAPTER 17

Land Treatment of Hydrocarbon Contaminated Soils

John Lynch and **Benjamin R. Genes,** Remediation Technologies Inc., West Concord, Massachusetts

Land treatment has been used as a waste treatment and management technology by United States petroleum refineries for more than 25 years. More recently, the technology is being applied as a cost-effective alternative for soil decontamination at Superfund sites and as elements of corrective actions for contaminated soil at Resource Conservation and Recovery Act (RCRA) facilities. This chapter presents information and operational data for a specific Superfund site which utilized land treatment as the final remedial action. The waste material consists of contaminated soils from a wood-treating plant. The constituents of concern are polynuclear hydrocarbons which are also found in petroleum industry waste. Operational data and removal rates are quantified for gross hydrocarbons and specific polynuclear aromatic hydrocarbons.

Land treatment uses, the ability of indigenous bacteria in the soil matrix, and the assimilative capacity of the soil to decompose and contain the applied waste in the surface soil layer (usually the top 15–30 cm or 6–12 in.). The upper soil layer is the *zone of incorporation* (ZOI). Soils in the ZOI, in conjunction with the underlying soils where additional treatment and immobilization of the applied waste constituents occur, is referred to as the *treatment zone.* The treatment zone in the soil may be as much as 1.5 m or 5 ft. Soil conditions below this depth generally are not conducive to oxidation of the applied waste constituents. The transformations, biological oxidations, and immobilization will occur primarily in the zone of incorporation.

BACKGROUND

Wastewaters from a creosote wood-preserving operation had been discharged to a shallow, unlined surface impoundment for disposal since the 1930s. The discharge of wastewater to the disposal pond generated a sludge which is a listed hazardous waste under the RCRA. Due to groundwater contamination of the shallow aquifer at the site by polynuclear aromatic hydrocarbons (PNAs), the state of Minnesota nominated the site for listing on the Superfund National Priorities List in 1982. Since 1982, numerous remedial investigation activities have been undertaken to determine the nature and extent of contamination at the site. Based on the results of these studies and extensive negotiations, the Minnesota Pollution Control Agency (MPCA), the U.S. Environmental Protection Agency (EPA), and the owner of the facility signed a Consent Order in March 1985 specifying actions to be taken at the site.

In general terms, the remedial actions selected by the MPCA and EPA involve a combination of offsite measures to control and mitigate the impacts of contaminated groundwater, and source control measures to treat contaminated soil and sludges. The offsite controls involve a series of gradient control wells to capture contaminated groundwater. The source control measures include onsite biological treatment of the sludges and contaminated soils, and capping of residual contaminants located at depths greater than 5 ft. Costs for onsite treatment and capping were estimated to be $59/ton.

PILOT-SCALE STUDIES

Before the onsite treatment alternative was implemented, bench- and pilot-scale studies were conducted to define operating and design parameters for the full-scale facility. Several performance, operating and design parameters were evaluated in the land treatment studies. These included:

- soil characteristics
- climate
- treatment supplements
- reduction of gross organics and PAH compounds
- toxicity reduction
- effect of initial loading rate
- effect of reapplication.

Three different loading rates were evaluated in the test plot studies: 2%, 5%, and 10% benzene extractable (BE) hydrocarbons. The soils used in the pilot study consisted of a fine sand which was collected from the upper 2 ft of the RCRA impoundment. The soil was contaminated with creosote constituents consisting primarily of PNA compounds. Total PNAs in the soil ranged from 1,000 to 10,000 ppm, and BE hydrocarbons in the contaminated soil ranged from approximately 2% to 10% by weight.

Because the natural soils are fine sands and extremely permeable, it was decided that the full-scale system would include a linear and leachate collection system to prevent possible leachate breakthrough. To simulate the proposed full-scale conditions, the pilot studies consisted of five lined 50-foot square test plots with leachate collection. The studies were designed to maintain soil conditions which promote the degradation of hydrocarbons. These conditions included:

- maintain a pH of 6.0 to 7.0 in the soil treatment zone
- maintain soil carbon-to-nitrogen ratios between 50:1 and 25:1
- maintain soil moisture near field capacity.

Hydrocarbon losses in the test plots were measured using benzene as the extraction solvent. The analysis of BE hydrocarbons provides a general parameter which is well suited to wastes containing high molecular weight aromatics, such as creosote wastes. Reductions of BE hydrocarbons were fairly similar between all the field plots. Average removals for all field plots over four months were approximately 40%, with a corresponding first-order kinetic constant (k) of 0.004/day.

The reduction of PNA constituents was monitored by measuring decreases in 16 PNA compounds. The compounds listed in Table 1 were monitored in the test plots.

Greater than 62% removals of PNAs were achieved in all the test plots and laboratory reactors over a four-month period. PNA removals for each ring class are shown below:

- 2-ring PNA: 80–90%
- 3-ring PNA: 82–93%
- 4+-ring PNA: 21–60%
- Total PNA: 62–80%

Table 2 summarizes first-order rate constants and half-life data for BE hydrocarbons and PNA compounds for the 5% and 10% BE hydrocarbon test plots. With the exception of the 4- and 5-ring PNAs, the table shows that the kinetic values are approximately equal for the 5% and 10% loading rates. In the case of the 4- and 5-ring compounds, the 5% loading rate resulted in higher kinetic rates for these compounds as compared to the 10% loading rate. This difference may have been due to more 2-ring and 3-ring compounds being available to soil bacteria at the 10% loading rate. These compounds may be preferentially degraded by soil bacteria.

Table 1. Compounds Monitored in Test Plots

2 Rings	3 Rings	4, 5 and 6 Rings
Naphthalene	Fluorene	Fluoranthene
Acenaphthylene	Phenanthrene	Pyrene
Acenaphthene	Anthracene	Benzo(a)anthracene
		Chrysene
		Benzo(j)fluoranthene
		Benzo(k)fluoranthene
		Benzo(a)pyrene
Dibenzo(a,h)anthracene		
		Benzo(g,h,i)perylene
Indeno(1,2,3,c,d)pyrene		

Table 2. Comparison of Pilot-Scale Kinetic Data at Two Initial Loading Rates

	First-Order Rate Constant (day)		Half-Life (days)	
	5% Plot	10% Plot	5% Plot	10% Plot
Benzene extractable	0.003	0.003	231	231
2-Ring PAH	0.023	0.023	30	30
3-Ring PAH	0.016	0.016	43	43
4-Ring PAH	0.004	0.001	173	693
Total PNAs	0.009	0.008	77	87

OPERATING AND DESIGN CRITERIA

The pilot-scale studies were successful in developing operating and design criteria for a full-scale system. These criteria are summarized below:

- treatment period can be extended through October
- soil moisture should be maintained near field capacity
- soil pH should be maintained between 6.0 and 7.0
- soil carbon:nitrogen ratios should be maintained between 25:1 and 50:1
- fertilizer applications should be completed in small frequent doses
- initial benzene extractable hydrocarbon contents of 5% to 10% are feasible
- waste reapplication should occur after initial soil concentrations have been effectively degraded
- waste reapplication rates of 2–3 lb of benzene extractables per cubic foot of soil per 3 degradation months can be effectively degraded

The studies suggest that all the loading rates tested are feasible. First-order rate constants were fairly similar between all the test plots, although the intermediate loading rate (5% benzene extractable hydrocarbons) may demonstrate a slightly higher removal of high molecular weight PNA compounds. The higher loading rates, however, showed the greatest mass removals. The selection of an initial loading rate should balance additional land area requirements against time requirements for completing the treatment process. Moderate loading rates (5%) will result in a faster detoxification, whereas higher loading rates will decrease land area requirements.

CONSTRUCTION AND START-UP OF FULL-SCALE SYSTEM

Construction of the full-scale system involved preparation of a treatment area within the confines of the existing RCRA impoundment (Figure 1). The treatment area was constructed on top of the impoundment to avoid permitting a new RCRA facility. If the facility were located outside the impoundment, then a Part B permit would have to be obtained before the treatment facility could be constructed. By locating the treatment area within the confines of the impoundment, the treatment system was considered part of closure of the impoundment. This enabled us to fast-track the cleanup, and avoid the delays associated with permitting a new RCRA unit.

The principal construction activities at the site involved:

- preparation of a lined waste pile for temporary storage of the sludge and contaminated soil
- removal of all standing water in the impoundment
- excavation and segregation of the sludges for subsequent free oil recovery
- excavation of approximately 3–5 ft of "visibly" contaminated soil from the impoundment and subsequent storage in the lined waste pile
- stabilization of the bottom of the impoundment as a base for the treatment area
- construction of the treatment area, including installation of a 100-mil high-density polyethylene (HDPE) liner, a leachate collection system, and 4 ft of clean backfill
- installation of a sump for collection of the stormwater and leachate
- installation of a center pivot irrigation system

As previously discussed, a lined treatment area was constructed because the natural soils at the site are highly permeable. The liner also served as a cap for the residual "nonvisibly" contaminated soils left in place below the liner. Therefore, the treatment area liner serves two functions at the site. The first function is to provide a barrier to leachate from the treatment area. The second is to provide a cap over the residual contaminants that were left in place.

The treatment area was constructed on top of the existing wastewater disposal pond after all visibly contaminated soils were removed. The surface area for treatment is approximately 125,000 ft^2. Containment berms with 3:1 slopes enclose the treatment area and prevent surface runoff from leaving the site.

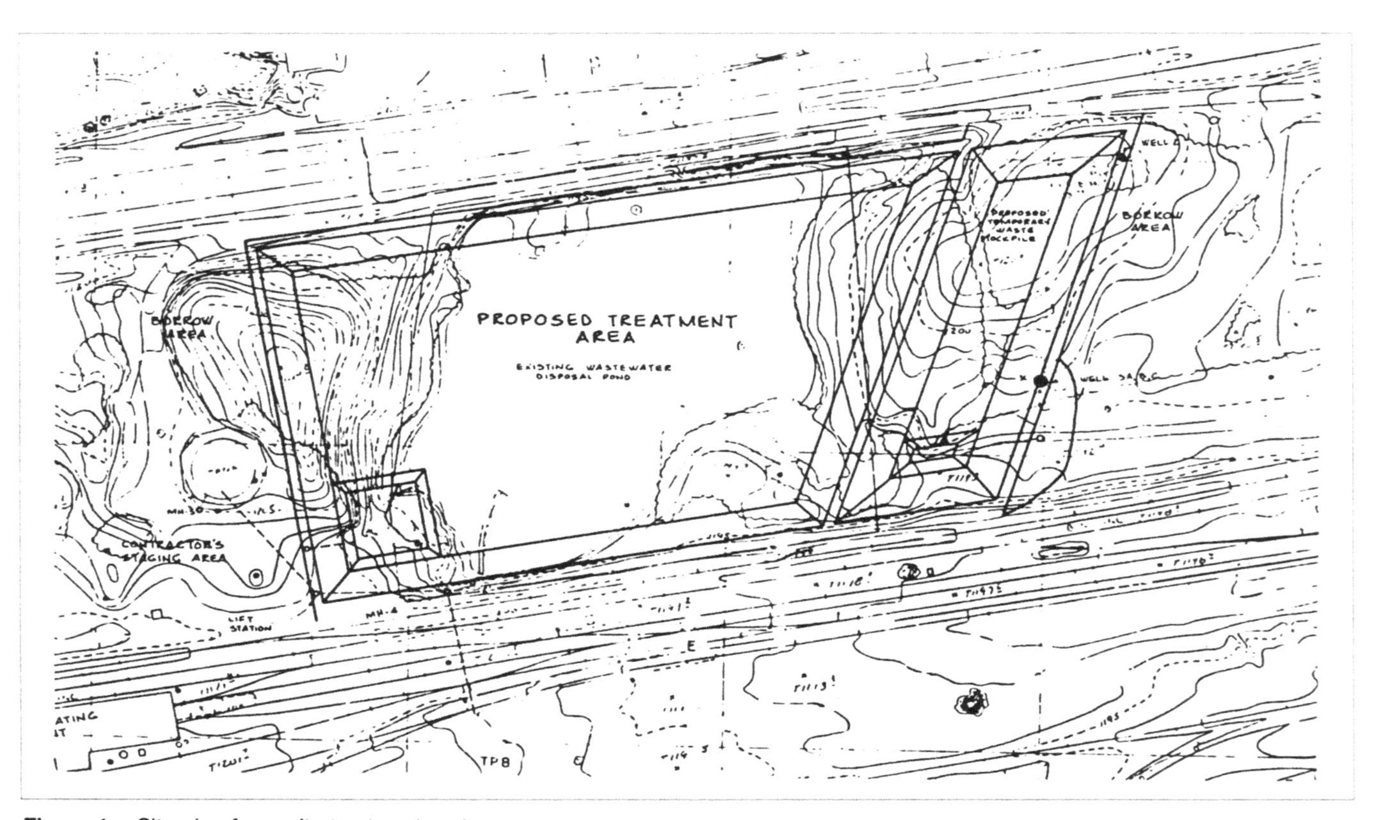

Figure 1. Site plan for onsite treatment system.

The treatment area is lined with a 100-mil HDPE membrane. The base of the liner slopes 0.5% to the south and west. A sump with a 50,000-gallon capacity is located in the southwest corner of the treatment area. A layer of silty sand ballast, 18 inches thick, was placed on top of the treatment area liner. A six-inch gravel layer was placed on top of the ballast. This layer serves as a leachate collection system and as a marking layer for land treatment operations.

The leachate collection system includes two-foot wide leachate collection drains at 100 foot centers. The drains are filled with gravel and perforated pipe to carry leachate from the collection system to the sump. The drains were wrapped in filter fabric to prevent clogging. A two-foot layer of uncontaminated sand was placed above the leachate collection system. This sand serves as an initial mixing layer for the contaminated soils, and is the treatment zone for the full-scale system.

Water in the leachate collection sump is discharged by gravity flow to a manhole, and is automatically pumped via a lift station to a 117,000 gallon storage tank. Water in the storage tank is recycled back to the treatment area via a spray irrigation system. Water in excess of irrigation requirements is discharged to the municipal wastewater treatment plant where it is treated biologically before discharge in compliance with the Publicly Owned Treatment Works (POTWs) National Pollutant Discharge Elimination System (NPDES) permit.

Construction of the waste pile and treatment area was completed in October 1985. In late April 1986, a center pivot irrigation system was installed and 120 tons of manure were spread in the treatment area. Manure loading rates were based on achieving a carbon:nitrogen ratio of 50:1. In addition to nitrogen, the manure provides organic matter which enhances absorption of the hazardous waste constituents.

In May 1986, a three-inch lift of contaminated soil was applied to the treatment area. The target loading rate for startup was a BE hydrocarbon concentration of 5%. The soil was mixed (rototilled) with three inches of native soil to achieve a treatment depth of six inches. This application involved approximately 1200 yd^3 of sludge and contaminated soil. Table 3 summarizes startup data for the full-scale facility. Additional lifts of contaminated soil have been applied in 1988 (see Table 4). To date, a total of 7,500 cubic yards of contaminated soil has been successfully treated at the facility.

The treatment area is irrigated almost daily due to dry weather during the summer months. Irrigation needs are determined from soil tensiometer readings, soil moisture analyses, and precipitation and evaporation records. Typical irrigation rates range from 1/4 inch to 3/8 inch per application. This application rate keeps the soils in the cultivation zone moist without saturating soils in the lower treatment zone. Maintaining soil moisture near field capacity was determined to be a key operating parameter in the pilot-scale studies.

PERFORMANCE OF THE FULL-SCALE FACILITY

Benzene extractable (BE) hydrocarbons and 16 polynuclear aromatic (PNA) compounds are being monitored to evaluate the performance of the facility.

Table 3. Summary of Start-Up Data (5/23/86)

Parameter	Average
Benzene extractables, %	5300
TOC, ppm	29710
TKN, ppm	1367
Ammonia, ppm	2.37
Total phosphorus, ppm	522
Total potassium, ppm	502
pH	7.66
Polynuclear Aromatic Hydrocarbons (PAH), ppm:	
Naphthalene	1148
Acenaphthylene	21
Acenaphthene	1082
Total 2-ring PAH	2251
Fluorene	1885
Phenanthrene	4190
Anthracene	3483
Total 3-ring PAH	9558
Fluoranthene	1575
Pyrene	958
Benzo(a)anthracene and chrysene	837
Total 4-ring PAH	3370
Benzofluoranthenes	368
Benzopyrenes	294
Indeno(1,2,3,c,d)pyrene	111
Dibenzo(a,h)anthracene	100
Benzo(g,h,i)perylene	106
Total 5-ring PAHs	979
Total PAHs	16159

Table 4. Additional Lifts of Contaminated Soil

Years	Application Volumes (cubic yards)
1986	1,200
1987	1,500
1988	1,500
1989	1,500
1990	1,800
	7,500

Figure 2 shows the BE hydrocarbon concentrations measured in the zone of incorporation (ZOI) during the first year of treatment. BE hydrocarbon concentrations

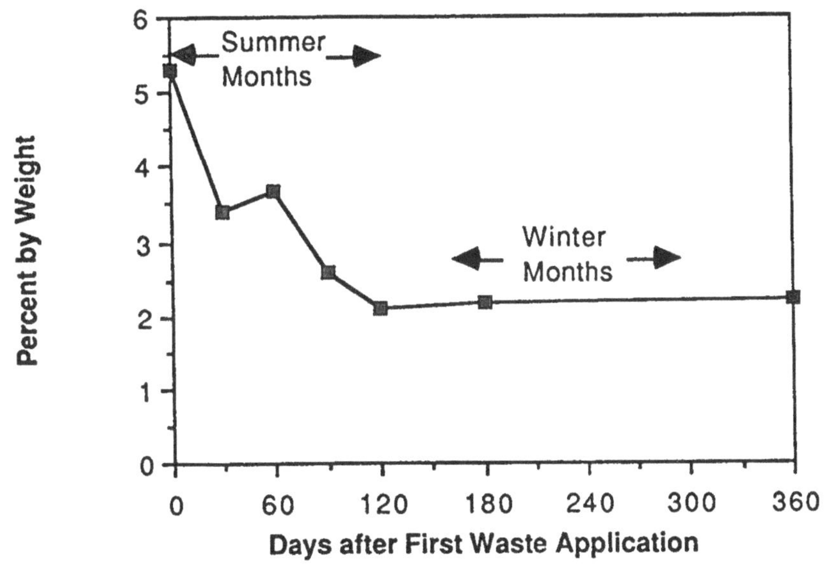

Figure 2. BE hydrocarbons degradation vs time.

decreased approximately 60% over the first year of operation. Most of the decrease occurred during the first 120 days (May through September). Little decrease in BE hydrocarbon concentrations was observed during the fall and winter months. Figure 3 shows the BE concentrations over the five-year life of the land treatment facility.

Figures 4 and 5 show PNA concentrations measured in the treatment facility during the first year of treatment. Figure 3 summarizes data for 2- and 3-ring PNAs. Figure 4 summarizes data for the 4- and 5-ring compounds. Greater than 95% reductions in concentration were obtained for the 2- and 3-ring PNAs. Greater than 70% of the 4- and 5-ring PNA compounds were degraded during the first year of operation. Figure 6 shows the total PAH compound concentrations over the five-year life of the treatment facility.

With the exception of anthracene, all the 2- and 3-ring compounds were degraded below or near detection limits after 90 days of treatment. Greater than 92% of the anthracene present in the waste was degraded during the first 90 days of treatment. Similarly, most of the 4- and 5-ring removals occurred during the first 90 days of treatment. This was expected because the warmest weather occurred during this period.

Table 5 shows average PNA removals measured in the pilot-scale studies and compares them with the full-scale removal efficiencies. Full-scale removal efficiencies were higher than test plot removal efficiencies for every PNA ring class and BE hydrocarbons. However, it must be noted that the full-scale facility operated for 360 days, compared to only 126 days for the test plot units. Table 5 also presents average half-life data for both the test plots and the full-scale unit.

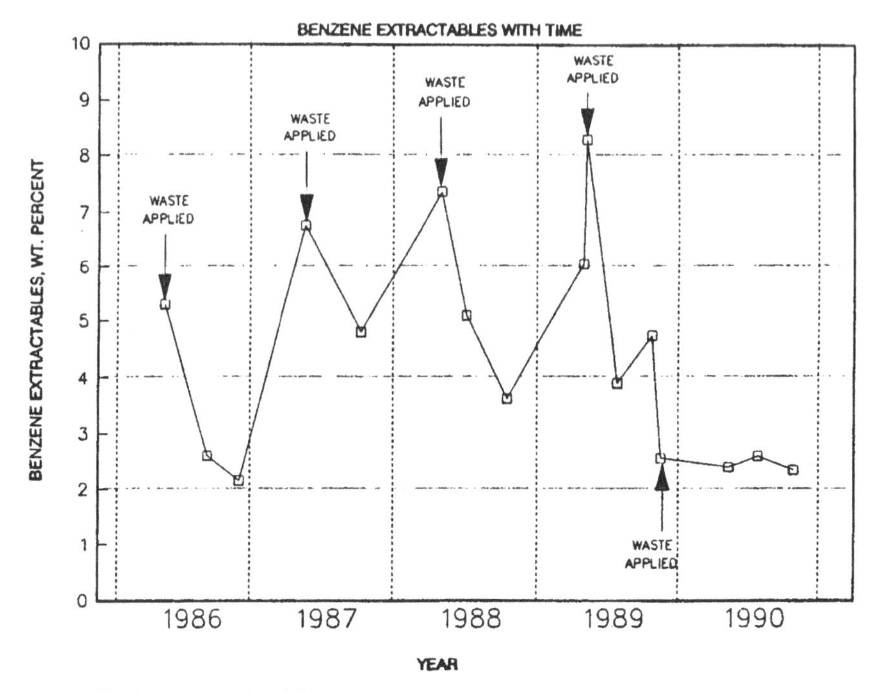

Figure 3. Benzene extractables with time.

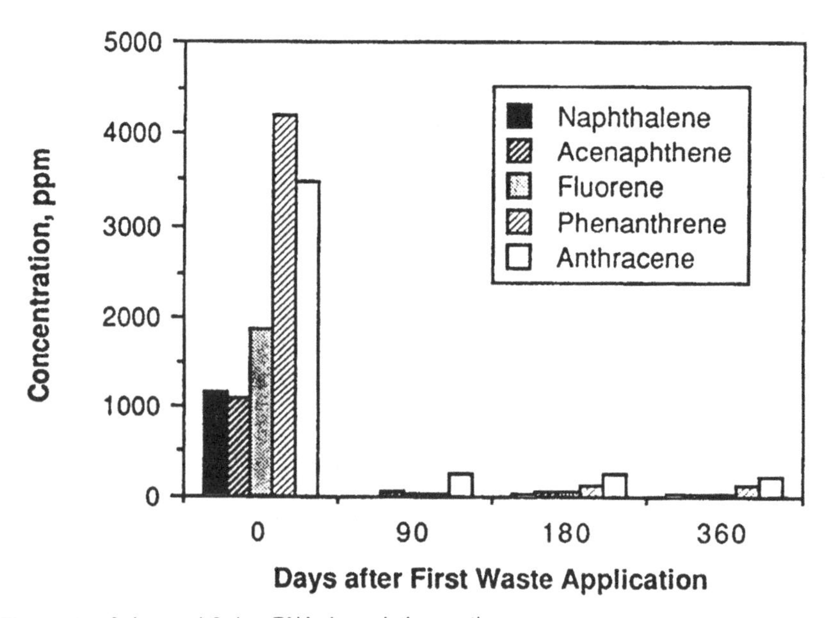

Figure 4. 2-ring and 3-ring PNA degradation vs time.

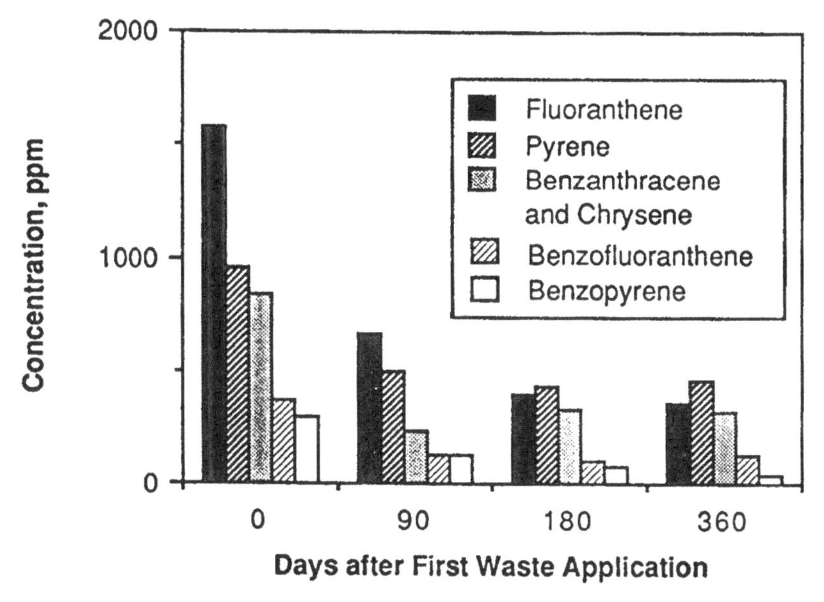

Figure 5. 4-ring and 5-ring PNA degradation vs time.

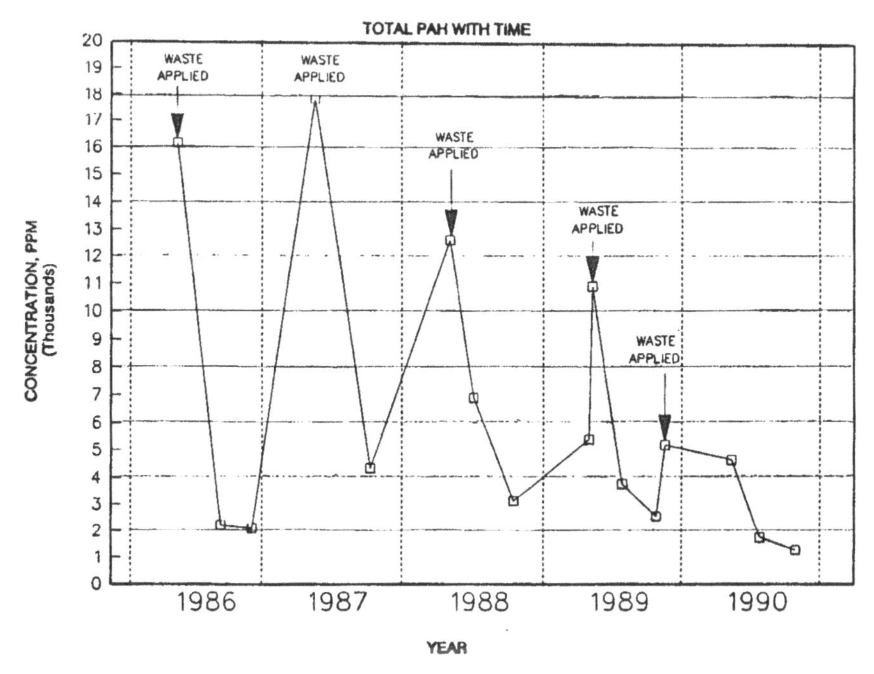

Figure 6. Total PAH with time.

Table 5. Comparison of Full-Scale and Test Plot Removals

Parameter	Avg. Percent Removal Full-Scale[a]	Test Plots[b]	Avg. Half-Life (Days) Full-Scale	Test Plots
2-Ring PAHs	95	93–95	<45	29–33
3-Ring PAHs	95	83–85	45	46–49
4- and 5-Ring PAHs	72	32–60	115	95–226
Total PAHs	90	65–76	65	61–83
BE Hydrocarbons	60	35–56	150	106–202

[a]Removal efficiency calculated after 193 days of treatment.
[b]Removal efficiency calculated after 126 days of treatment.

In summary, the rate and amount of PNA degradation is proportional to the number of rings contained by the PNA compounds (Figure 7). The 2- and 3-ring PNAs degraded most rapidly. The 4- and 5-ring PNAs degraded at slower rates; however, these compounds are strongly adsorbed to soils and are immobilized in the treatment zone of the facility. Table 6 summarizes the removal efficiencies of the treatment system over its five-year operating history. Table 7 summarizes water quality data for the leachate collection system of the facility. Only acenaphthene and fluoranthene were detected in the drain tile water samples. Concentrations for these two compounds were near analytical detection limits.

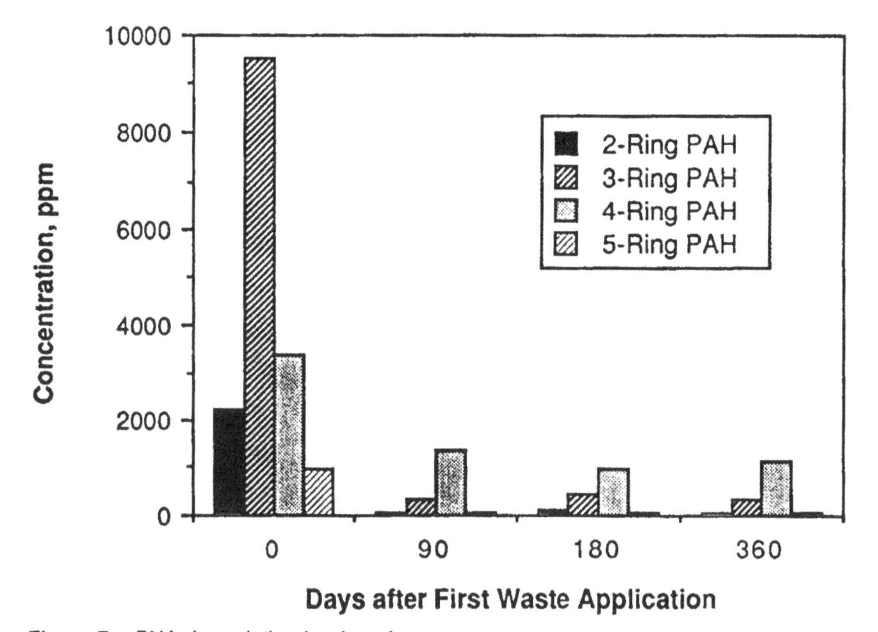

Figure 7. PNA degradation by ring class.

Table 6. Comparison of PAH and Benzene Extractables Removal Efficiencies (Average Percent Removals)

Parameter	Test Plot[a]	1986[b]	1987[c]	Full Scale 1988[d]	1989[e]	1990[f]
2-Ring PAHs	93–95	95	95	97	99.6	89
3-Ring PAHs	83–85	95	91	97	85	81
4+ Ring PAHs	32–60	69	46	41	60	75
Total PAHs	65–76	87	76	75	77	77
BE Hydrocarbons	35–56	59	30	51	43	8.1

[a]Removal efficiencies for the test plots were calculated after 126 days of treatment.
[b]Removal efficiencies for the full-scale facility in 1986 were calculated after 193 days of treatment.
[c]Removal efficiencies for the full-scale facility in 1987 were calculated after 180 days of treatment.
[d]Removal efficiencies for the full-scale facility in 1988 were calculated after 168 days of treatment.
[e]Removal efficiencies for the full-scale facility in 1989 were calculated after 166 days of treatment.
[f]Removal efficiencies for the full-scale facility in 1990 were calculated after 167 days of treatment.
No treatment was assumed to occur between 11/16/89 and 5/10/90.

Table 7. Drain Tile Water Quality

	Concentration, ppb		
Compound	**June 1986**	**August 1986**	**October 1986**
Naphthalene	<1	<1	<1
1-Methylnaphthalene	<1	<1	<1
2-Methylnaphthalene	<1	<1	<1
Acenaphthylene	<1	<1	<1
Acenaphthene	<1	3.7	2.7
Fluorene	<1	<1	<1
Phenanthrene	<1	<1	<1
Anthracene	<1	<1	<1
Fluoranthene	<1	2.1	1.4
Pyrene	<1	<1	<1
Benzo(a)anthracene	<1	<1	<1
Chrysene	<1	<1	<1
Benzofluoranthenes	<5	<1	<1
Benzopyrenes	<5	<1	<1
Indeno(1,2,3,c,d)pyrene	<5	<1	<1
Dibenzo(a,h)anthracene	<5	<1	<1
Benzo(g,h,i)perylene	<5	<1	<1

CONCLUSION

The data developed during this project have shown that onsite treatment of creosote contaminated soils is feasible. Based on the data developed in pilot-scale studies, a conservative design for a full-scale system was developed and constructed. The full-scale unit has matched or surpassed the performance of the pilot-scale unit in degrading creosote organics. The advantages of onsite treatment are that it reduces the source of contaminants at the site in a very cost-effective manner. In addition, it satisfies the developing philosophical approach that EPA has to onsite remedies, and it reduces the liability of the owner/operator due to offsite disposal.

CHAPTER 18

Low Temperature Stripping of Volatile Compounds

Luis A. Velazquez and **John W. Noland,** Roy F. Weston, Inc., West Chester, Pennsylvania

Contamination of soils from past operations involving volatile organic compounds (VOCs), such as from leaking underground fuel storage tanks, has become a major environmental concern. This contamination, if allowed to remain in the soil, can migrate to contaminate underlying groundwater. Roy F. Weston, Inc. (WESTON) has been performing studies on the removal of these contaminants. Low temperature thermal stripping of the VOCs from the soil was found to be an efficient method for removal of these contaminants. Further work in the field of low temperature thermal treatment (LT^3) has resulted in WESTON's design of an innovative low temperature process, LT^3.

TECHNOLOGY DESCRIPTION

WESTON's LT^3 process is a demonstrated technology that provides evaporation of the VOCs, but does not require heating the soil matrix to combustion temperatures. The heart of the technology is the thermal processor, an indirect heat exchanger (hot screws), that is used to dry and heat the contaminated soils up to 450°F, consequently stripping the VOCs from the soil. The indirect heating of the soil minimizes gas flows and reduces the size of the offgas handling equipment. Once the organic contaminants are vaporized, they can either be destroyed through high temperature incineration in an afterburner, or recovered through condensation or adsorption on activated carbon. WESTON has acquired U.S. patent No. 4,738,206 for this technology.

Some of the advantages of the LT^3 process are:

- lower capital and operating costs than incineration
- process is considerably smaller and more mobile than an incinerator with the same throughput
- process can be used for product recovery, which is not possible for incineration
- process results in a processed soil that is suitable for onsite backfill. This eliminates the long-term liability of landfill disposal or onsite containment or storage.
- low temperature minimizes the chance of hazardous heavy metal emissions
- process can be designed to be compact and fully mobile (i.e., fully mounted on three flatbed trailers.)

LT^3 PROCESS DEVELOPMENT

The LT^3 process was developed by WESTON under its current contract with the U.S. Army Toxic and Hazardous Materials Agency (USATHAMA). Bench scale studies of the low temperature thermal stripping concept identified the method as a feasible treatment process. A pilot scale field demonstration of the process using VOC-contaminated soils became the next phase in the development of the technology.[1] In May 1985, USATHAMA contracted WESTON to perform this phase of the work. A site with soil having various VOC contamination (see Table 1) was selected for the study.

WESTON's responsibilities for the project included:

- design of the pilot system
- preparation of test and safety plans

Table 1. Pilot Study, Feed Soil VOCs Concentrations

VOC	Average (ppm)	Maximum (ppm)
Dichloroethylene	83	470
Trichloroethylene	1,673	19,000
Tetrachloroethylene	429	2,500
Xylene[a]	64	380
Other VOCs	14	88
Total VOCs	2,263	22,438

[a]Xylene is not classified as a VOC since its boiling point is approximately 140°C. However, it was included in the study to evaluate the effectiveness of this technology on higher boiling point semivolatile compounds.

- environmental permitting
- equipment selection
- equipment installation and start-up
- performance of the test program
- demobilization and site closure
- preparation of a technical report

Results of this project indicated the following:

- A comparison of the VOCs measured in the feed soil to the VOCs measured in the processed soil and stack gas yielded the following destruction and removal efficiencies:
  —greater than 99.99% removal of VOCs was evidenced in the soils
  —no VOCs were detected in the stack gas, indicating a destruction and removal efficiency (DRE) of 100% for the overall system
- Stack emissions were in compliance with all federal and state regulations (including VOCs, HCI, CO, and particulates).

The demonstrated process represents a significant breakthrough in the treatment of soils contaminated with VOCs. The process represents a unique mix of proven techniques combined in an innovative way to provide an efficient and cost-effective method of treatment. Based on the unqualified success of this demonstration and on the increasing demand for mobile thermal treatment, WESTON, through its affiliate, Weston Services, Inc. (WSI), has begun to provide mobile LT^3 systems. The first of these systems was scheduled to be commissioned in October 1987.

MOBILE LT^3 SYSTEM

The mobile LT^3 system is designed to handle 15,000 lb per hr of contaminated soil based on 20% soil moisture and 1% (i.e., 10,000 ppm) VOCs. The system is comprised of equipment assembled on three flatbed trailers. The feed system trailer is 35 ft long. Each of the other two trailers is 42 ft long. All three trailers are 8 ft wide and 55 in. from the ground to the top of the bed. The total height of the trailers, with the equipment assembled, is under 13.5 ft. Weight and weight distribution is designed so as not to require special road permits. Interconnecting ducts, pipes, conveyors, power and control wires, and equipment that would extend beyond the allowable limits for trailers, such as the afterburner stack, are disconnected and shipped on the trailers during transportation. The process schematic of the WSI mobile LT^3 system is shown in Figure 1. System layout, along with the equipment supplied on each of the trailers, is shown in Figure 2.

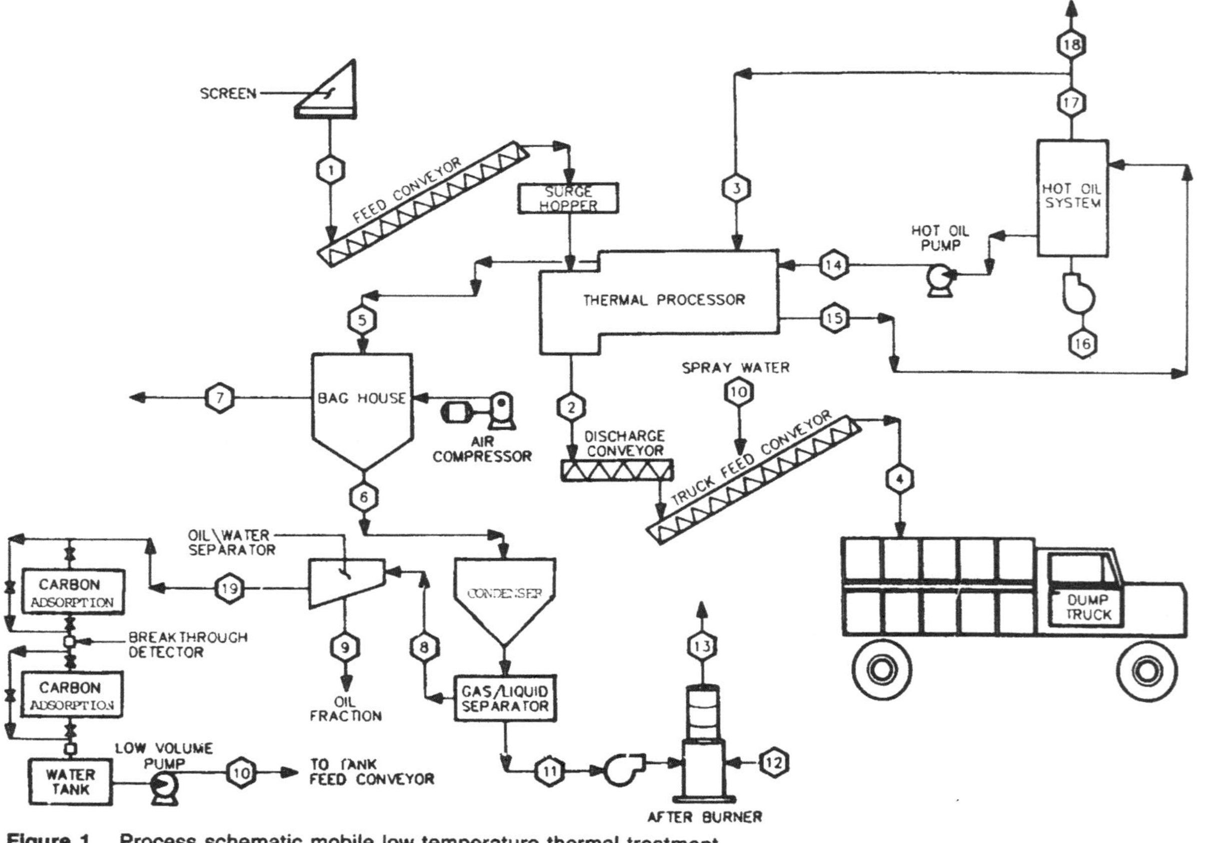

Figure 1. Process schematic mobile low temperature thermal treatment.

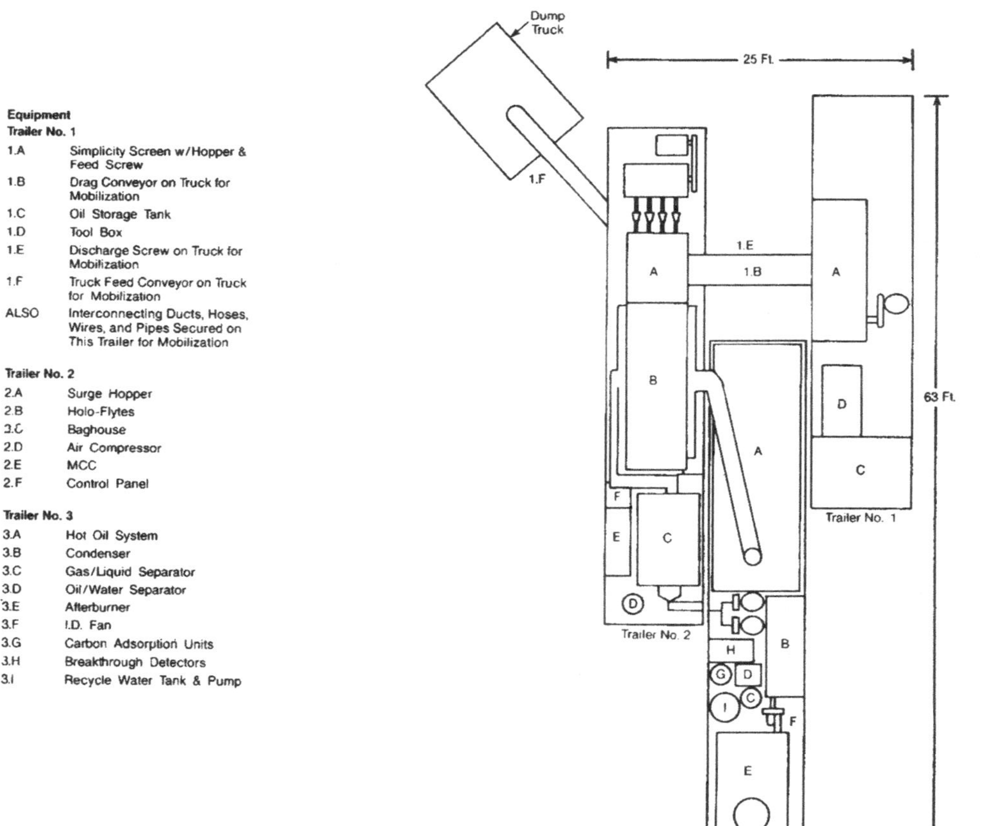

Figure 2. Mobile low temperature thermal treatment.

MOBILE LT^3 PROCESS DESCRIPTION

Excavated soil is fed onto the 4 ft × 10 ft vibrating screen permanently assembled on Trailer No. 1. The soil is screened to a maximum soil topsize of 1 in. to protect the mechanical downstream equipment. This screened soil is conveyed from the hopper, directly under the screen, to a drag flight conveyor. The drag flight conveys the soil up to a surge hopper that completely covers the feed inlet of the top thermal heating unit. This creates a "live bottom" hopper effect to prevent bridging of the soil in the surge hopper. The hopper has a capacity of 100 ft^3 (approximately 45 min of operating capacity). It is designed with the sides hinged at the bottom so they can be folded onto each other for system shipment.

As previously stated, the heart of the LT^3 system is the thermal processor. The thermal processor consists of two indirect heating thermal units (hot screws). These units are arranged in a "piggyback" or series configuration. Each unit is equipped with four screws that are 20 ft long and 18 in. in diameter. Hot oil is circulated through the shafts, flights, and trough of the screw conveyors to indirectly heat the soil from the surge hopper to a maximum of 450°F measured at the outlet of the second unit. Water is sprayed on the treated soil at the truck feed conveyor for cooling and dust control. This moistened treated soil is fed into a dump truck for transport and backfill. The thermal processor is operated under negative pressure to prevent fugitive emissions.

A 6 million BTU/hr hot oil heater is used to provide the required heat for the process. The oil is heated to 640°F and then circulated through the thermal processor. The oil is heated by the combustion offgases from the burner of the heater. A portion of these offgases are directed through the enclosures above the screws to provide an inert atmosphere in the units (i.e., to avoid reaching the lower explosive limit [LEL] of contaminants in air) and to maintain a discharge offgas temperature of 280°F to avoid the formation of condensate in the units.

Processor offgases consisting of water vapors, VOCs, and the used portion of the offgases from the oil heater are passed through a fabric filter (or baghouse) to reduce particulate emissions to less than 0.08 grains per standard dry ft^3 corrected to 12% CO_2. Particulate removed in the baghouse is returned to the surge hopper for treatment. Gases from the baghouse enter a two-stage condenser which knocks out, then cools, the water to a minimum of 140°F.

The gas fraction from the condenser is drawn by the Induced Draft (I.D.) fan and directed into an afterburner. The I.D. fan provides the vacuum required to maintain a negative pressure at the processor. A gas/liquid separator is provided between the gas outlet of the condenser and the I.D. fan in the event there is too much free moisture carryout. The afterburner thermally treats the gases by providing a minimum residence time of 2 sec at 1800°F. The heat of combustion of the VOCs provides some of the heat required for this thermal treatment. Auxiliary fuel (propane) is also burned in the afterburner to maintain a minimum of 1800°F in the afterburner during operation. This is a necessary safety consideration, since the VOC concentration in the feed soil is expected to vary.

The cooled liquid fraction from the condenser is pumped to an oil/water separator to remove any oil present. The oil fraction is collected for offsite treatment and disposal. The water fraction is directed through a two-stage carbon adsorption system which removes any other contaminants in the water. A monitor (i.e., total hydrocarbon analyzer) continuously samples the water from the first carbon adsorption unit to detect breakthrough. In the event of a breakthrough, the units are reversed and the spent unit is replaced. The treated water is collected and sprayed onto the treated soils at the truck feed conveyor for cooling and dust control.

The net result is moist, VOC-decontaminated soil suitable for onsite backfill. This eliminates the long-term liability of landfill disposal or onsite containment or storage.

OPERATIONAL REQUIREMENTS

The following site pad and utilities are required:

- The three trailers require a 26 ft × 63 ft area.
- electrical: 460 V/3 phase/60 Hz/300 amps
- propane: 430 lb/hr
- water: not required
- One senior onsite manager and one technician/operator are required per shift.

LT^3 PROCESS APPLICATIONS IN THE UNDERGROUND STORAGE TANK (UST) MARKETPLACE

There are over two million underground storage tanks in the United States now used to store gasoline (see Figure 3). Literally billions of gallons of engine fuels are stored in these tanks by farms, retail gasoline stations, fleet users, and the military. Refineries and airports generally use above-ground tanks for fuel storage. Some experts estimate that between 75,000 and 100,000 underground storage tanks are now leaking.

On a national basis, a statistical analysis of the underground storage tank population indicated that roughly 5% of the tanks are failing. On a regional basis, the number can be considerably higher. The Maine Department of Environmental Protection estimates that 25% of the state's 10,000 retail gasoline storage tanks have failed. The state of New York estimates that 19% of its 83,000 underground gasoline tanks are leaking.

The environmental disasters associated with leaking underground storage tanks are not limited to contamination resulting from the accidental spillage of extremely toxic materials. A substantial portion of the problem results from gradual leakage during routine storage of motor fuels. For instance:

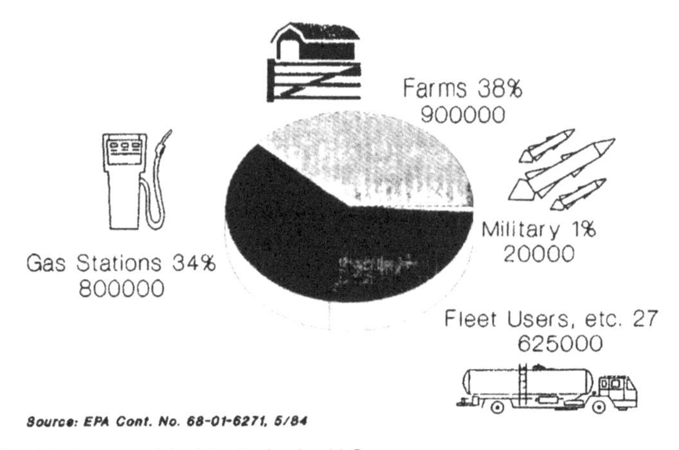

Figure 3. Underground fuel tanks in the U.S.

- Underground storage tanks owned by one petroleum company were found in 1978 to be leaking gasoline at a Long Island service station in East Meadow, New York. An estimated 30,000 gal of gasoline leaked into groundwater supplies, causing odor and safety problems for 27 families living in 25 homes. This company subsequently bought 23 of the homes at 150% of their market value, and settled with the two remaining home owners. With some suits still pending, total costs to date are estimated to be between $5,000,000 and $10,000,000.
- In 1980, a gasoline leak was traced to a service station in the Northglenn suburb of Denver, Colorado. A federal court convicted the petroleum company in June 1981, forcing it to purchase 41 of the homes in the affluent neighborhood at 2.2 times their appraised value. Estimated losses to the petroleum company are approximately $10,000,000.

These are but two examples of environmental disasters where the mobile LT^3 system can be used to remediate contaminated soil problems.

PROCESS DEVELOPMENT AREAS

Research is continuing to determine variables that affect the low temperature stripping of the organics from various soils. Preliminary studies indicate low temperature thermal treatment may be applicable for the removal of relatively high boiling point organics (i.e., up to 600°F). DRE and economic evaluations are being performed using stripping agents (e.g., water, solvent) used not only to enhance the low temperature volatilization of the organic low temperature boilers but also of the high temperature boilers, to determine other applications for the LT^3, and to establish the most efficient, cost-effective methods of operation.

Another research area where the application of the LT^3 is being used is in the recovery of the contaminant. The vaporized contaminants, stripped from the soil from the LT^3 process, are being processed and condensed for recovery.

SUMMARY

Low temperature thermal stripping has proved to be an effective method of treatment for the removal of VOCs from soils. The LT^3 is an innovative technology based on the low temperature thermal stripping principle which is efficient, practical, cost-effective, and versatile.

REFERENCE

1. "Task 11. Pilot Investigation of Low Temperature Thermal Stripping of Volatile Organic Compounds (VOCs) from Soil," Report No. AMXTH-TE-CR-86074, Roy F. Weston, Inc., June 1983.

CHAPTER 19

Barrier Walls to Contain Contaminated Soils

Marco D. Boscardin, GEI Consultants, Inc., Winchester, Massachusetts
David W. Ostendorf, Dept. of Civil Engineering, University of Massachusetts, Amherst, Massachusetts

INTRODUCTION

Cutoff walls and other barrier systems have a long history of use in traditional geotechnical construction for the control of water seepage, flows, and pressures. With the increased awareness of the hazard posed by contaminated soils and groundwater, it was a natural step to employ barrier wall systems to contain and prevent the spread of contaminants. However, the criteria for successful performance of a barrier to contain a contaminant plume are, in general, much stricter than the criteria for a barrier to control water in traditional geotechnical construction. For example, in traditional geotechnical construction, cutoff walls are typically required to reduce water flows or control water pressures so that they are not problems during construction or operation of a facility. As a consequence, small seeps and flows may be acceptable and measures to accommodate small flows and bleed off slight pressures are incorporated in the design. In contrast, a barrier called upon to contain a contaminant plume should ideally permit no passage of contaminants or at least permit passage of contaminants at such a slow rate that the material escaping does not constitute a hazard. The impact of the chemistry of the permeant on the properties, particularly hydraulic properties, of the barrier must also be assessed when the barrier is required to contain contaminants. Most cutoff walls in traditional geotechnical construction must only deal with water with a relatively narrow range of chemistries. On the other hand, barriers used to contain pollutants can encounter a much wider range of pore fluid chemistries which are often much more chemically aggressive than unpolluted water. This chapter examines various cutoff wall systems available from the perspective of construction-related defects and interaction with contaminants.

The impacts of composition and method of construction on performance for barriers based on portland cement, bentonite, asphalt, and chemical binders or grouts are discussed, as are synthetic membrane barriers. Slurry wall, permeation grouting, jet grouting, and vibration beam methods of construction are considered. The effects of organic-based contaminants on material properties and barrier integrity are examined for several cases and a breakthrough model that quantifies the time available for true remediation as a function of chemical compatibility of the contaminant and the barrier is presented.

CUTOFF WALL MATERIALS

Hydraulic barriers have been constructed using a variety of materials. These materials include bentonite, asphalt, portland cement, synthetic membranes and liners, chemical grouts, and composite materials composed of two or more of

Table 1. Conductivity of Barrier Materials with Passive Permeant

Material	K_p (Lab) m/sec × 10^{-10}	K_p (Field) m/sec × 10^{-10}
Bentonite slurry	5	50
Compacted soil bentonite	0.1	10
Cement bentonite	3 (filter cake)	100
Asphalt slurry	0.1	1
Soil cement	0.5	50
Cement grout	0.1 (single well)	1000 (curtain)
Concrete	0.1	—
Plastic concrete	0.1 to 40	—
Butyl rubber	0.001	0.01
Polyvinyl chloride	0.00007	0.01
Chlorinated polyethylene	0.0002	0.01
Ethylene propylene diene monomer	0.002	0.01
Chlorosulfonated polyethylene	0.0004	0.01
High density polyethylene	0.0001	0.01
Silicate grout	47	10000
Acrylamide grout	0.5	10000
Dynagrout T	1	10000
Resorcinol	1	10000

Based on data from References 1–7, 48–50.

the above. In general, these materials exhibit relatively low hydraulic conductivities ($<10^{-9}$ m/s) for laboratory specimens with passive or nonaggressive permeants (see Table 1). In contrast, field test data, again for chemically passive permeants, indicate conductivities 10 to 10,000 times greater than the laboratory values, depending on the material and the method of construction.[3] When an incompatibility exists between the cutoff wall material and the permeant, further increases in conductivity can also occur, as discussed in a later section. General characteristics of each material are briefly described below.

Bentonite

Bentonite slurries, soil-bentonite mixes, and portland cement-bentonite mixes are commonly used in what is termed slurry wall construction to form vertical cutoff walls. An important aspect of the slurry wall construction is the formation of the bentonite filter cake along the sides of the trench. The filter cake forms along the sides of the trench during excavation as long as the level of the slurry is such that flow is directed out of the trench and into the surrounding soil. The filter cake is typically thin, 3×10^{-3} m, but has a low hydraulic conductivity, on the order of 10^{-10} m/s.[7] When sufficient excavation has been performed, the slurry trench is backfilled with either a soil-bentonite mix, a cement-bentonite mix, or a plastic concrete mix. D'Appolonia[8] indicates that the conductivity of soil-bentonite backfill depends on the quantity of bentonite added to the soil and on the grainsize and plasticity characteristics of the soil. For reasonable mixes, hydraulic conductivities of 5×10^{-10} to 5×10^{-9} m/s can be achieved for passive permeants.[8] In contrast, cement-bentonite (CB) mixes typically have conductivities on the order of 10^{-7} to 10^{-8} m/s.[1,5] Hydraulic conductivity of plastic concrete is typically less than 1×10^{-9} m/s.[1,48] Cement replacements such as blast furnace slag and fly ash tend to reduce the permeability of CB and plastic concrete and to reduce their susceptibility to chemical attack.[7,51-54]

Asphalt

The use of asphalt slurries for cutoff walls is relatively new. The slurries are typically proprietary mixtures that include emulsified asphalt, sand, water, and portland cement.[9,10] Laboratory conductivities for asphalt slurries begin at 1×10^{-10} m/s, but rapidly decrease with loss of fluid and curing to less than 1×10^{-12} m/s within a few hours.[5,11] The low conductivity, high cost, and greater resistance to degradation by aggressive permeants makes this an attractive material for use in thin barriers.

Portland Cement

Portland cement-based systems include grouts and conventional concretes, as well as the cement-bentonite mixes described previously. Kosmatka and Panarese[4] indicate passive conductivities in the range of 10^{-12} to 10^{-11} m/s for conventional concretes. However, additives such as polymers and fly ash can reduce the conductivities and increase chemical stability.[12,13,48] Passive hydraulic conductivities for cement grouts are typically 1×10^{-12} m/s or less.[14]

Synthetic Membranes and Liners

Synthetic barrier materials are typically polymer-based and take the form of sheets. There are three basic families of synthetics used in practice; elastomers, thermoplastics, and crystalline thermoplastics (see Table 2).[2] These materials are characterized by very low passive conductivities, $\ll 10^{-12}$ m/s, making their use in thin sheets viable. To form an effective barrier, the sheets must be bonded together to form a continuous unit and installed in such a manner as to prevent tears and punctures from forming and compromising its integrity. In general, the elastomer family is the weakest and most difficult to seam of synthetic barrier materials.[3,15] The thermoplastic family is easier to seam in the field and tends to have somewhat better tear and tensile strength properties than the elastomers.[3] The crystalline thermoplastics, particularly high density polyethylene, tend to resist mechanical tearing and puncturing better and tend to resist aging and exposure related degradation better than the other synthetic barrier materials though seaming may be more complicated.[3,15]

Synthetic barriers are subject to aging and physical and hydraulic property degradation, due to exposure to ultraviolet radiation, ozone, thermal variations, microbial attack, and volatilization.[3] In addition, these materials may be attacked by aggressive permeants and, therefore, compatibility of the barrier with the exposure conditions and the contaminant needs examination. However, the variety of synthetics available means that materials compatible with the various conditions and the various aggressive permeants can be found.

Chemical Grouts

Chemical grouts are liquid solutions designed to harden after injection into a porous medium. The grout, which in theory permeates the voids in the medium, increases the strength and reduces the permeability of the medium as it hardens. Injection of particulate grouts such as cement and bentonite grouts are typically limited in use to rather coarse-grained soils such as gravels and coarse sands with large interparticle voids. Chemical grouts typically have low viscosities and no particulate solids and are therefore able to permeate much finer materials, fine sands to coarse silts, than the particulate grouts.[16] Chemical grouts include silicates, lignosulfites, phenoplasts, aminoplasts, acrylamides, polyacrylamides, acrylates, and polyurethanes.[17] Laboratory conductivities of these materials are typically in the 10^{-10} to 10^{-9} m/s range (see Table 1). Areas of concern with chemical grouts are degree of permeation of the medium, reaction of the grout with the present and future pore fluids, and the potential release of toxic or corrosive materials due to improper mixing or reaction with the soils and its pore fluid.[3,16]

Table 2. Synthetic Membranes

Class	Typical Materials
Elastomers	Butyl rubber (IIR) Chlorinated rubber (CR) Ethylene propylene diene monomer (EPDM)
Thermoplastics	Polyvinyl chloride (PVC) Chlorinated polyethylene (CPE) Chlorosulfonated polyethylene (CSPE) Elasticized polyolefin (ELPO)
Crystalline thermoplastics	Low density polyethylene (LDPE) High density polyethylene (HDPE)

After Folkes.[2]

CONSTRUCTION OF A CUTOFF WALL

The goal during construction of a cutoff wall is to erect a continuous, impermeable barrier (usually but not always vertical) to isolate a contaminated mass of soil from the surrounding, uncontaminated soil and water. This is typically accomplished by extending the barrier to intersect continuous, impermeable (very low permeability) soil or rock layer to effect closure. Equally important, but not considered here, is the construction of an impermeable cap to prevent recharge of the contaminated groundwater and the attendant flows.

Materials with the desired hydraulic and chemical resistance properties can be identified in the laboratory. Unfortunately, many materials do not perform as well in the field as the laboratory studies would indicate (see Table 1), due to the presence of defects and chemical reactions with soils and permeants in the field situation. In this section, the propensity for defects (windows, cracks, or high permeability zones) to occur during and after construction of the various types of barriers is examined. The barrier construction techniques considered include slurry wall methods, vibrating beam methods, injection grouting, and jet grouting.

Slurry Wall Methods

Slurry wall-based methods of creating a vertical barrier are among the most common methods in use. In general, the method consists of excavating a narrow trench, typically 0.6 to 1.5 m thick, down into an aquiclude while using a bentonite-water slurry to keep the trench open and the side walls stable. Slurry trenches in excess of 120 m deep have been constructed,[14] though for contaminant control, slurry trench cutoffs are typically less than 50 m deep.[7] Up to 15 m, backhoes are generally most efficient. Draglines are typically restricted to depths less than 25 m, while clamshells are used for deeper excavation.[8] Once a sufficient length of trench is excavated, the bentonite-water slurry can be replaced by either a soil and bentonite backfill,[8] a plastic, portland-cement concrete,[18] an asphalt slurry,[19] or a synthetic membrane and backfill composite.[20] A variation is the cement-bentonite slurry method,[21] that incorporates portland cement in the slurry which eventually hardens in the trench and does not require backfilling. D'Appolonia,[8] Millet and Perez,[22] Ryan,[23] and Xanthakos[19] describe the construction and specification of slurry walls in more detail.

Evans[1] describes common construction and postconstruction defects encountered in slurry walls. He defines defects as zones of the wall that do not provide the same resistance to groundwater flow and contaminant migration as the good, intact portions of the cutoff. Construction defects include poorly mixed backfill, trapped pockets of slurry, trapped pockets of material that has spalled from the sides of the trench, incomplete keying into the low permeability layer, and loss of the filter cake in portions of the trench wall. Postconstruction defects include cracking due to changes in moisture, temperature, consolidation and stress, as well as increases in conductivity due to chemically aggressive permeants.

A case where a composite, synthetic membrane (high density polyethylene, HDPE) and soil barrier was installed using a modification of the slurry trench technique is described by Druback and Arlotta.[20] The unique feature of this system is the granular material placed within the membrane that lined the top, bottom, and sides of the trench which would permit checking the effectiveness of the interior layer of the liner and allow collection and extraction of any contaminant that passes through the inner layer. Defects in such a system can be associated with separations at seams, and tears and punctures created during installation, backfilling, and postconstruction deformation.[24]

Vibrating Beam Methods

The vibrating beam method of cutoff wall construction, as described by Leonards et al.[5] and Jepsen and Place,[11] consists of driving an injection beam into the ground to a preselected depth, up to 25 m, with a vibratory driver so that the soil is consolidated and displaced during the driving. During extraction of the beam, the void space is pressure grouted with either a cement-bentonite slurry or a slurry based on an asphalt emulsion which hardens in place. The injection beam is a wide flange section with a web on the order of 0.85 m and of sufficient stiffness to minimize bowing under the applied loads. The tip of the beam is fitted with wear plates to protect the beam and injection system, and to create a void of the desired width. Full length grout pipes are attached to the beam to inject the slurry. The beam is also fitted with a fin that acts as a guide to ensure continuity of the wall by running in the grouted void created by the previous penetration. In addition, each beam penetration overlaps the previous penetration by about 10%. In this manner, a thin cutoff wall (on the order of 0.1 m) is constructed. This method is generally used to create vertical walls that intersect an aquiclude for closure. However, Leonards et al.[5] describe a test installation in sandy soil utilizing cement-bentonite walls inclined 45° from the vertical. The use of inclined walls permitted closure without intersecting an aquiclude. The test installation was checked via pumping tests and the effective conductivity for the test cell was 2×10^{-8} m/s, which is approximately the permeability of the cement-bentonite slurry. After the pumping tests, the test section was excavated and inspected. The vertical walls were consistently 0.089 to 0.1 m thick. The inclined walls were thinner, 0.051 to 0.064 m thick, and the intersections of the walls were complete.

Jepsen and Place[11] describe potential defects for cutoffs constructed by the vibrating beam method. They include narrowing, spalling, or collapse of the sides of the wall due to vibrations of the beam as adjacent portions of the wall are constructed, uncertainty associated with intersection of the aquiclude, and control of alignment and verticality of the beam to ensure continuity and to prevent unfilled or ungrouted windows. Jepsen and Place[11] imply that the effective depth for vibrating beam walls is in the range of 14 m. Below 14 m, the potential for windows is greater without special controls on alignment and verticality. However, Leonards et al.[5] and discussions of Jepsen and Place[11] indicate that this method can be used successfully. The presence of numerous boulders or very hard or dense zones is likely to compromise the ability of this method to construct a continuous wall, particularly at depth.

Injection Grouting Methods

Permeation grouting has been used since the early 1800s to control water flow and to increase the strength of the soil.[25] The fluid grout material is typically injected under relatively low pressures to avoid creating fractures, with the aim of filling the void spaces in the soil. After it is in place, the grout will set to bind the soil particles together into a coherent mass and to block water flow through the soil. Grouting to control contaminant transport is especially attractive in several situations including: cases where the barrier must extend to depths greater than those conventionally possible with other techniques, cases where it is desirable to form an impermeable horizon when the natural one is relatively deep, and cases where ground conditions may reduce the practicality of other systems (e.g., many boulders present).[26]

Grouts are usually classified as particulate grouts (e.g., cement and bentonite grouts) and solution grouts (chemical grouts). The particulate grouts are usually restricted to clean, coarse sands and gravels soils,[27] whereas the chemical grouts are capable of penetrating soils with up to 20% coarse silt.[17,27] The grouts are usually injected using parallel, offset rows of holes. The outer rows of holes are grouted first and the inner row(s) is grouted last to ensure complete grouting. The spacing of the holes depends on type of grout, grout viscosity, injection pressure, soil permeability, and rate of grout take.[25]

Defects in grouted barriers usually take the form of zones that have not been permeated with grout and zones where conditions are such that the grout will not set up. In addition, postconstruction defects can develop due to aging of the grout and chemical interaction with the surrounding soil and fluids.[28] The continuity of the grouted barrier is of particular concern. Grout tends to flow along the easiest path available and it is very difficult to achieve the symmetrical, intersecting grout bulbs postulated as illustrated by the cases presented by May et al.[29]

Jet Grouting Methods

Jet grouting is a form of grouting where, rather than permeate the soil voids with minimal disruption of the soil structure, the jetting process is used to mechanically mix the soil and the grout. In general, cement grout is used, but other types of grouts may be employed.[30,31] The method consists of drilling a guide hole to the desired depth, inserting the jetting tool to the bottom of the hole, and then the fluid grout is injected at high pressure as the injector rod is lifted and rotated. The high speed fluid cuts and mixes the cementing agent with the soil to form cylindrical columns 0.7 to 3 m in diameter, depending on the soil type and the specific injection system used, and up to 50 m deep. To form a barrier, a row of intersecting columns must be constructed. Novatecna[30] recommends multiple rows of intersecting columns for depths greater than 15 m, or monitoring and correcting the verticality of the guide holes to ensure continuity. No information was available regarding conductivity of the cement jet grouted soil, but it is anticipated that it would be in the range of portland cement-soil mixes and cement concretes.

Defects in a jet grouted barrier fall into three basic categories: incomplete mixing of the soil and grout, inability of the grout to set up, and gaps or windows in the continuity of the wall. Incomplete mixing can be controlled by jetting pressure, injector rod rotation rate, size and number of injectors, lifting speed, and grout mix and injection rate.[31] Compatibility of the grout mix with the host soil and pore fluid can be checked and altered if need be. The continuity of the wall is the greatest concern. The presence of a guidehole at the location of each cylinder does permit checking the intersection with an aquiclude, and checking and correcting verticality.

The most attractive uses of this technique appear to be to create a floor under a contaminated area in cases where an aquiclude is rather deep and to construct a barrier in areas where space to excavate and mix backfill is very restricted. However, due to the apparent lack of data, caution must be exercised if using jet grouting to create a barrier to contain contaminants.

IMPACT OF AGGRESSIVE CHEMICALS

The above discussion has only considered the hydraulic conductivity of the barriers for passive permeants. However, it is well known that contaminated permeants can react with barrier materials and potentially cause significant increases in conductivity. In this section, the effects of petroleum-based and other organic contaminants on the barrier conductivity are examined briefly.

In general, bentonite-based barriers appear to fare poorly when faced with petroleum products and other organic contaminants.[32,33,52,53] However, proprietary bentonites reported to be more resistant to contaminants are available,[11] and low concentration solutions and certain types of immiscible organic contaminants may not cause as large changes in the bentonite mixes.[1,34,35] Sorptive materials such as fly ash, zeolites, and organically modified clays can also be used to reduce the diffusion rate and to increase the breakthrough time of contaminants through bentonite and cement-based barrier walls.[53,54] Portland cement-based mixes appear to be able to resist oily wastes and sludges.[32,36,37] Of the synthetic membrane materials, high density polyethylene and polyvinyl chloride appear least affected by oily materials.[32,37,38] Asphalts tend to fare poorly in the presence of oily and aromatic hydrocarbon contaminants,[5,37] but Anderson et al.[9] indicate that an asphalt slurry composed of asphalt emulsion, sand, water, and portland cement performed well in tests where xylene, methanol, and creosote oil were used as permeants. Lord et al.[39] indicate resistance of a silicate grout to several organic contaminants and Bodocsi et al.[21] found an acrylamide grout resistant to acetone. However, Malone et al.[28] found that gasoline and oil, as well as other organic permeants, increase the gel times for an acrylate and a silicate grout, thus indicating a potential for compatibility problems.

BREAKTHROUGH MODEL

In a strict sense, any barrier system is temporary. All the barrier systems described above are considered to transmit fluids, though some at very slow rates. The role of a breakthrough analysis is to estimate the time required for a contaminant to breach the barrier. This information can then be used in the planning and timing of more permanent remediation.

Following Ostendorf et al.,[3] the breakthrough model proposed rests on an empirical relationship between cutoff wall permeability, K, and number of contaminated pore volumes, N, discharged through the barrier.

$$K = K_b\left[1 + \left(\frac{K_b}{K_p} - 1\right)(1 - N)^{\alpha}\right]^{-1} \qquad (N<1) \qquad (1a)$$

$$K = K_c\left[1 + \left(\frac{K_c}{K_b} - 1\right)(N)^{-\beta}\right]^{-1} \qquad (N>1) \qquad (1b)$$

As suggested by Figure 1, the empirical conductivity relation simulates the initial decrease of permeability from its passive K_p value to a minimum breakthrough level, K_b, upon the discharge of a single pore volume. The prebreakthrough behavior may be attributed to initial dissolution of the solid matrix by the chemically aggressive permeant with subsequent clogging of the pores. This behavior is modeled using the exponent α. The post breakthrough behavior is marked by a rise to a higher contaminated conductivity, K_c, due to piping and further dissolution of the solid matrix. The post breakthrough behavior is modeled using a second exponent β.

The conductivity determines the specific discharge, v, of permeant under a hydraulic head gradient, S, across the cutoff wall.

$$v = KS \tag{2a}$$

$$v = \frac{\text{discharge}}{\text{gross area}} \tag{2b}$$

The specific discharge may also be cast in terms of the rate of change of pore volume, leading to an integral relation between time, t, and N

$$v = nL\,\frac{dn}{dt} \tag{3a}$$

$$\frac{S}{nL}\int_0^t dt' = \int_0^N K^{-1}\quad dN' \tag{3b}$$

with effective porosity, n, and cutoff wall thickness, L. A straightforward substitution of the empirical conductivity relation (Eq. 1) into Eq. 3b yields the following algebraic relationships between time and pore volume number:

$$t = \frac{nL}{K_bS}\{N + \frac{K_b}{K_p} - 1)[\frac{1 - (1 - N)^{a \times 1}}{\alpha + 1}]\} \qquad (N<1) \tag{4a}$$

$$t_b = \frac{nL}{K_bS}\left(\frac{\alpha + K_b/K_p}{\alpha + 1}\right) \qquad (N = 1) \tag{4b}$$

$$t = t_b + \frac{nL}{K_cS}\left[N - 1 + \frac{K_c}{(K_b - 1)}\left(\frac{1 - N^{1-\beta}}{\beta - 1}\right)\right] \qquad (N>1) \tag{4c}$$

The actual time of breakthrough, t_b, coincides with the discharge of one pore volume through the wall, and is given by Eq. 4b. The expression simplifies considerably for those permeants that do not alter the conductivity of the cutoff wall.

$$t_b = \frac{nL}{K_pS} \qquad \text{(passive only)} \tag{5}$$

Eq. 5 is used to assess the time of breakthrough for essentially pure water or dilute solutions.

Eqs. 1, 2, and 4 may be combined to yield an implicit account of the temporal variations of the specific discharge of the cutoff wall under the action of a

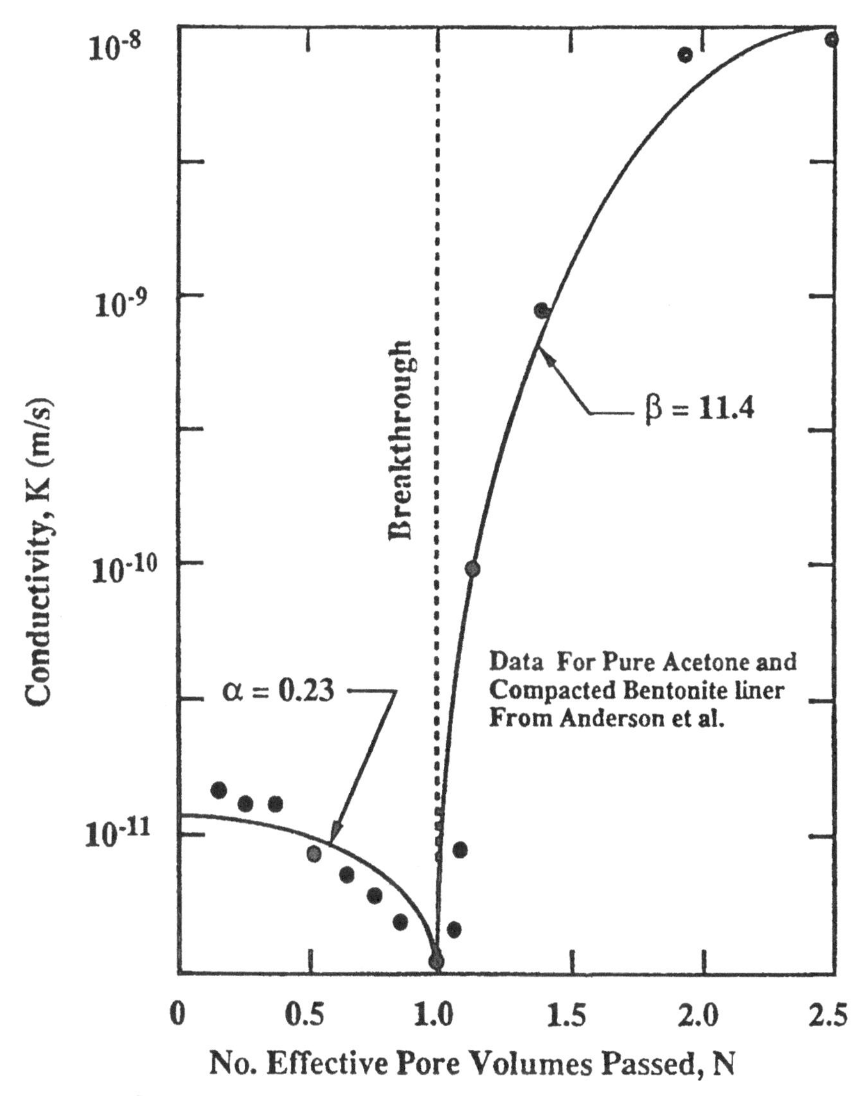

Figure 1. Conductivity vs pore volume, aggressive permeant.

chemically aggressive permeant. Figure 2 shows the expected behavior for the following parameter values:

Hydraulic gradient (S) = 10
Barrier thickness (L) = 1 m
Passive conductivity (K_p) = 1.6×10^{-11}m/s
Breakthrough conductivity (K_b) = 2.6×10^{-12}m/s

$$\text{Contaminated conductivity } (K_c) = 1 \times 10^{-10}\text{m/s}$$
$$\text{Effective porosity } (n) = 0.152$$

In this example, the actual breakthrough time of 2×10^9s (63 years) is followed by another 60-year period of slow discharge until a rapid rise of discharge occurs. Thus, the discharge of relatively significant amounts of contamination occurs abruptly at an effective breakthrough time, t_e, about 120 years after the initial exposure of the barrier to the waste. A working definition of this time follows by setting t equal to t_e when the barrier conductivity (and specific discharge) achieves half of its contaminated value.

$$K = \frac{K_c}{2} \quad (\text{at } t = t_e) \tag{6a}$$

$$t = t_b + \frac{nL}{K_c S}\left[\frac{\left(\frac{K_c}{K_b} - 1\right)^{1/\beta}(\beta = 2) + \frac{K_c}{K_b} - \beta}{\beta - 1}\right] \tag{6b}$$

Eq. 6b is used in the assessment of the time of breakthrough for aggressive contaminants.

Eqs. 5 and 6b suggest that the time of breakthrough varies directly with the effective porosity and wall thickness, and inversely with the conductivity and hydraulic gradient. The porosity effect is most pronounced in channeling, whereby the majority of flow proceeds through a small fraction of the void space, leading to short t_b values. The thickness dependency implies that the synthetic barriers provide no lead time if they fail due to the extreme thinness typically associated with this class of barriers. Larger hydraulic gradients and higher conductivities lead to shorter breakthrough times as well.

TIME OF BREAKTHROUGH ASSESSMENT

Table 3 presents breakthrough times computed for passive permeation through a 1-m-thick cutoff wall under a constant hydraulic gradient of 10, corresponding to a high buildup of mildly contaminated groundwater behind the containment barrier. The alternative materials may be compared on constitutive grounds by contrasting laboratory values for the time of breakthrough. Soil cement, cement grout, lean portland cement concrete, and asphalt slurry (Cases 7, 9, 11, and 6) all offer relatively effective retention of chemically inert wastes in the absence of field imperfections, providing from 15 to 22 years of lead time before the breakthrough of pollution. The bentonite slurry wall (fine backfill), conventional portland cement concrete, acrylamide grout, Dynagrout T, and resorcinol grout alternatives (Cases 1, 12, 14–17) are less effective in passive retention, with breakthrough times of about 1.6 years in magnitude. The bentonite slurry (coarse backfill), cement bentonite, and silicate grout materials (Cases 2–4 and 13) are relatively ineffective, and yield passive breakthrough times ranging from 2 to 5 months.

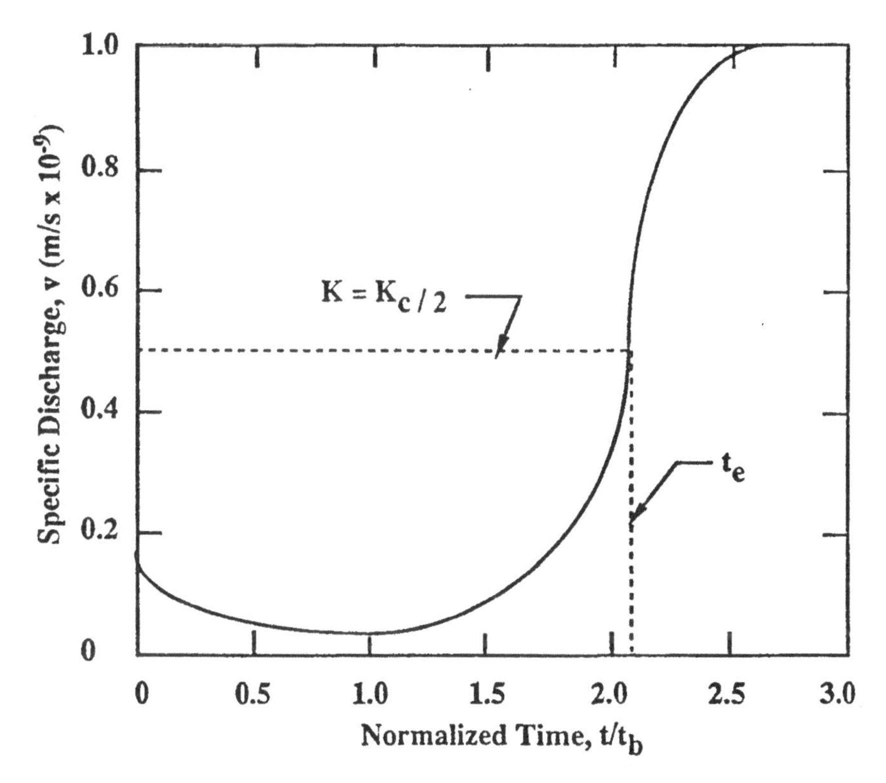

Notes: For Liner Thickness of 1 m and Hydraulic Gradient of 10. Breakthrough Time $t_b = 2.0 \times 10^9$ s. $Kp = 1.6 \times 10^{-11}$ m/s, $K_b = 2.6 \times 10^{-12}$ m/s, $K_c = 10^{-10}$ m/s, n = 0.15.

Figure 2. Specific discharge vs time.

It is important to note the dramatic fall in cutoff wall effectiveness when field improprieties are considered. The chemical grout alternative breakthrough time falls to a scant 1.4 hours (Case 17) due to leaks in the curtain, while the cement grout and soil cement alternatives (Cases 10 and 8) also suffer with t_b values of 14 hours and 2.7 months, respectively. Only the asphalt slurry alternative demonstrates a moderately effective breakthrough time of 1.6 years under field conditions. The decline of field effectiveness underscores the need for detailed specifications of materials and methods, thorough quality assurance procedures, and careful monitoring of cutoff wall performance for leaks and imperfections. Additional field testing of all alternative cutoff wall systems is certainly warranted as well.

Tables 4 and 5 summarize the effects of chemically aggressive permeants on the performance of some of the materials. The varying porosity values cited in Table 4 indicate channeling as a mode of barrier failure, particularly for the

Table 3. Breakthrough Times for Passive Permeants

Case	Barrier (Condition)	K_p m/s × 10^{-10}	n %	t_b s × 10^8	Ref.
1	Bentonite slurry wall (fine backfill)	5	70	1.40	[8]
2	Bentonite slurry wall (coarse backfill)	50	70	0.14	[8]
3	Cement bentonite slurry wall (slag)	10	5	0.05	[18]
4	Cement bentonite slurry wall (no slag)	100	5	0.005	[18]
5	Asphalt slurry (laboratory)	0.1	5	5.00	[9]
6	Asphalt slurry (field)	1	5	0.50	[3]
7	Soil cement (laboratory)	0.5	35	7.00	[2]
8	Soil cement (field)	50	35	0.07	[2]
9	Cement grout (single well, laboratory)	0.1	5	5.00	[40]
10	Cement grout (curtain, field)	1000	5	0.0005	[41]
11	Portland cement concrete (lean mix)	0.1	5	5.00	[42]
12	Portland cement concrete (conventional)	1	5	0.50	[42]
13	Silicate grout (laboratory)	50	5	0.01	[43]
14	Acrylamide grout (laboratory)	0.5	5	1.00	[44]
15	Dynagrout T grout (laboratory)	1	5	0.50	[45]
16	Resorcinol grout (laboratory)	1	5	0.5	[43]
17	Grout (field)	10000	5	0.00005	[46]

Based on passive permeation through a 1-m-thick wall under a hydraulic gradient of 10.

bentonite and asphalt slurry walls (Cases 18, 19, and 22–24). Relatively effective retention is exhibited by the latter barrier in contact with the three organic wastes (effective breakthrough times of 8.6 to 20 years). The bentonite slurry wall resists attack by inorganic acids and bases (Cases 20 and 21), while the soil cement and acrylamide grout alternatives are compatible with separator waste (Case 25) and acetone (Case 33), respectively. Dynagrout T and plastic concrete solid matrices are clogged without piping when subjected to sulfuric acid (Case 31) and scrubber waste (Case 27) for the duration of their testing periods; the longest t_e values (300 and 50 years) are simulated as a consequence. The bentonite slurry wall is relatively ineffective in the retention of pure xylene (Case 18), while the soil cement and silicate grout are degraded by municipal leachate and organics, respectively (Cases 26 and 28). These latter three effective breakthrough times range from 12 days to 5 months. Moderate degrees of effectiveness are exhibited by the remaining cases.

The widely varying response of the cutoff walls, as modeled, to different aggressive contaminants has several implications for the design of effective retention facilities. It is essential that the composition of the waste be known so that a chemically compatible material may be elected. In this regard, particular attention should be paid to immiscible contaminants which will exist at full strength in discrete, density determined locations in the aquifer. The barrier material selected should be tested in a permeameter with its waste to establish data comparable to Figures 1 and 2 with a duration sufficient to observe a rise of conductivity after the breakthrough of contamination. Finally, the aggressive permeant simulations are all based on laboratory data and do not reflect the potentially serious loss of effectiveness inherent in field construction procedures.

Table 4. **Chemically Aggressive Permeant Cases**

Case	Barrier	Permeant (Concentration)	n %	Ref.
18	Bentonite slurry wall	Xylene (pure)	6.8	[33]
19	Bentonite slurry wall	Methanol (pure)	14	[33]
20	Bentonite slurry wall	Hydrochloric acid (pH = 0)	69	[8]
21	Bentonite slurry wall	Sodium hydroxide (pH = 14)	69	[8]
22	Asphalt slurry wall	Methanol (pure)	0.4	[9]
23	Asphalt slurry wall	Xylene (pure)	3.2	[9]
24	Asphalt slurry wall	Creosote oil (pure)	4.5	[9]
25	Soil cement	Separator waste	5.0	[46]
26	Soil cement	Municipal leachate	5.0	[15]
27	Plastic concrete	Scrubber waste	5.0	[18]
28	Silicate grout	Organics	5.0	[39]
29	Dynagrout T grout	Phenol (8%)	5.0	[45]
30	Dynagrout T grout	Potassium ferrocyanide (15%)	5.0	[45]
31	Dynagrout T grout	Sulfuric acid (5%)	5.0	[45]
32	Dynagrout T grout	Perchloroethylene (0.01%)	5.0	[45]
33	Acrylamide grout	Acetone	5.0	[21]

Table 5. **Breakthrough Times for Aggressive Permeants**

Case	K_p m/s × 10^{-10}	K_b m/s × 10^{-10}	K_c m/s × 10^{-10}	α	β	t_b s × 10^8	t_e s × 10^8
18	5.0	5.0	100	—	10.6	0.14	0.15
19	5.0	5.0	280	—	3.73	0.28	0.38
20	5.0	5.0	35	—	5.00	1.38	1.69
21	5.0	5.0	35	—	5.00	1.38	1.69
22	0.051	0.004	0.016	0.28	7.74	2.80	3.87
23	0.13	0.019	0.034	0.36	4.00	6.27	6.27
24	0.29	0.078	0.29	0.18	8.64	2.20	2.71
25	0.5	0.1	0.1	0.23	—	1.75	1.75
26	150	8.0	40	0.23	5.00	0.01	0.02
27	1.8	0.006	0.006	0.23	—	15.84	15.8
28	50	50	8000	—	5.00	0.01	0.01
29	1.0	1.0	100	—	5.00	0.50	0.63
30	1.0	1.0	30	—	5.00	0.50	0.63
31	8.0	0.001	0.001	0.23	—	93.5	93.5
32	1.0	1.0	3000000	—	5.00	0.50	0.50
33	0.5	0.045	0.045	0.23	—	2.89	2.89

For purposes of breakthrough time estimation, representing exponential values (α = 0.23 and β = 5.00) were used in cases where reported data were absent. Based on aggressive permeation through 1-m-thick wall with hydraulic gradient of 10.

SUMMARY AND CONCLUSIONS

A number of barrier materials and methods of construction were examined and their strengths and weaknesses were briefly discussed. Passive hydraulic conductivities were examined and compatibility with organic permeants were noted. In addition, the ability of the various construction methods to create a continuous barrier with few defects was examined. A breakthrough model was also presented to assist in estimates of the time available before contaminants breach the barrier.

Based on the above discussions, several barrier types appear more suited to retain petroleum related contaminants than the others. In general, permeation grouting appears to be a poor choice due to the difficulties in ensuring continuity of the barrier. Slurry wall construction techniques with the appropriate backfill or liner composition should perform satisfactorily. Plastic concrete backfill or a composite of backfill and membrane liner (HDPE or PVC) may be attractive alternatives. With the latter, difficulties with seaming and installation without tears or punctures will require special construction techniques and may hinder its use. Asphalt slurries, particularly those that include portland cement, may also provide satisfactory performance as a slurry trench backfill. However, the high cost of the asphalt slurries has tended to counteract their use in the relatively wide slurry walls. In general, the bentonite-based backfills tend to fare poorly when faced with organic contaminants; however, proprietary mixes of treated bentonite and organically modified clays may provide satisfactory performance under the appropriate conditions. The vibrating beam method of barrier construction creates a much thinner wall than the traditional slurry trench method. Still, use of a low conductivity backfill such as an asphalt slurry may permit satisfactory performance if compatibility problems do not exist. Jet grouting techniques appear to have potential for use in contaminant control, particularly for formation of a horizontal barrier beneath contaminated zones, but more information is needed regarding the continuity of a barrier constructed by this technique.

Crucial factors for success of barriers installed in situ to surround a zone of contamination are compatibility of the materials and the permeant (at the appropriate concentration), compatibility of the construction technique with the barrier materials and the ground conditions encountered, and good specifications and quality assurance controls for construction.

REFERENCES

1. Evans, J. C., "Slurry Trench Cutoff Walls for Waste Containment," *International Symposium on Environmental Geotechnology,* Vol. I, Envo Publishing Company (1986), pp. 303–311.
2. Folkes, D. J., "Fifth Canadian Geotechnical Colloquium: Control of Contaminant Migration by the Use of Liners," *Can. Geotech. J.*, Vol. 19, No 3 (1982), pp. 320–344.
3. Ostendorf, D. W., R. R. Noss, A. B. Miller, and H. S. Phillips, "Hydraulic Containment of Low-Level Radioactive Waste Disposal Sites," Completion Report, UADOE, Environmental Engineering Program, University of Massachusetts (1987).
4. Kosmatka, S. H., and W. C. Panarese, *Design and Control of Concrete Mixtures,* 13th Ed. (Skokie, IL: Portland Cement Association, IL, 1988).

5. Leonards, G. A., J. L. Schemednecht, J. L. Chameau, and S. Diamond. "Thin Slurry Cutoff Walls Installed by the Vibrating Beam Method," *Hydraulic Barriers in Soil and Rock,* ASTM STP 874 (1985) pp. 34–43.
6. Davidson, R. R., and J. Y. Perez. "Properties of Chemically Grouted Sand at Lock and Dam No. 26," *Grouting in Geotechnical Engineering,* ASCE (1982) pp. 433–449.
7. Spooner, P., R. Wetzel, C. Spooner, C. Furman, E. Tokarshi, G. Hunt, V. Hodge, and T. Robinson. *Slurry Trench Construction for Pollution Migration Control* (Park Ridge, NJ: Noyes Publications, 1985).
8. D'Appolonia, D. J. "Soil Bentonite Slurry Trench Cutoffs," *J. Geotech. Eng. Div.*, ASCE, Vol. 106, GT4 (1980), pp. 399–417.
9. Anderson, D. C., K. W. Brown, and J. Green. "Effects of Organic Fluids on the Permeability of Clay Soil Liners," *Proc. Eighth Annual Research Symposium,* USEPA (1982), pp. 179–190.
10. Jogis, H., and R. Bell. "Vibrating Beam Asphaltic Slurry Wall – A Case of Sealing Pond Dikes," National Conference on Hazardous Wastes and Environmental Emergencies, Houston, TX (1984).
11. Jepsen, C. P., and M. Place. "Evaluation of Two Methods for Constructing Vertical Cutoff Walls at Waste Containment Sites," *Hydraulic Barriers in Soil and Rock,* ASTM STP 874 (1985), pp. 45–63.
12. Office of Nuclear Waste Isolation, "Evaluation of Polymer Concrete for Application to Repository Sealing," Report No. ONWI-410, Columbus, OH (1982).
13. Vesperman, K. D., T. B. Edil, and P. M. Berthouex. "Permeability of Flyash and Fly Ash-Sand Mixtures," *Hydraulic Barriers in Soil and Rock,* ASTM STP 874 (1985), pp. 289–298.
14. Littlejohn, G. S. "Design of Cement Based Grouts," *Proc. of Conf. on Grouting in Geotechnical Engineering,* ASCE, (1982), pp. 35–48.
15. Haxo, H. E., R. S. Haxo, N. A. Nelson, P. D. Haxo, R. M. White, S. Dakessian, and M. A. Fong. *Liner Materials for Hazardous and Toxic Wastes and Municipal Solid Waste Leachate* (Park Ridge, NJ: Noyes Publications, 1985).
16. Karol, R. H. *Chemical Grouting* (New York: Dekker, 1983).
17. Karol, R. H., "Chemical Grouts and Their Properties," *Proc. of Conf. on Grouting in Geotechnical Engineering,* ASCE (1982), pp. 359–377.
18. Adaska, W. S., and N. J. Cavalli. "Cement Barriers," *Management of Uncontrolled Hazardous Waste Sites,* HMCRI (1984), pp. 126–130.
19. Xanthakos, P. P. *Slurry Walls* (New York: McGraw-Hill Book Co., 1979), 622 pp.
20. Druback, G. W., and S. V. Arlotta. "Subsurface Pollution Containment Using a Composite System Vertical Cutoff Barrier," *Hydraulic Barriers in Soil and Rock,* ASTM STP 874 (1985), pp. 24–33.
21. Bodocsi, A., I. Minkarah, and B. W. Randolph. "Reactivity of Various Grouts to Hazardous Wastes and Leachates," *Proc. of the Tenth Annual Research Symposium,* USEPA (1984), pp. 43–51.
22. Millet, R. A., and J. Y. Perez. "Current USA Practice: Slurry Wall Specifications," *J. Geotech. Eng. Div.*, ASCE, Vol. 107, GT8 (1981), pp. 1041–1056.
23. Ryan, C. R. "Slurry Cutoff Walls; Application in the Control of Hazardous Wastes,"*Hydraulic Barriers in Soil and Rock,* ASTM STP 874 (1985), pp. 9–23.
24. Nuclear Regulatory Commission. "Trench Design and Construction Techniques for Low-Level Radioactive Waste Disposal," Report No. CR-3144, Washington, D.C. (1983).

25. Herndon, J., and Lenahan, T. *Grouting in Soils, Vol. 1: A State-of-the-Art Report,* FHWA Report No. FHWA-RD-76-26, Washington, D.C. (1976).
26. Malone, P. G., R. J. Larson, J. H. May, and J. A. Boa. "Test Methods for Injectable Barriers," *Hazardous and Industrial Solid Waste Testing: Fourth Symposium,* ASTM SPT 886 (1986), pp. 273–284.
27. Baker, W. H., "Planning and Performing Structural Chemical Grouting," *Proc. of Conf. on Grouting in Geotechnical Engineering,* ASCE (1982), pp. 515–539.
28. Malone, P. G., J. H. May, and R. J. Larson. "Development of Methods for In Situ Hazardous Waste Stabilization by Injection Grouting," *Proc. of the Tenth Annual Research Symposium,* USEPA (1984), pp. 33–42.
29. May, J. H., R. J. Larson, P. G. Malone, and J. A. Boa. "Evaluation of Chemical Grout Injection Techniques for Hazardous Waste Containment," *Proc. of the Eleventh Annual Research Symposium,* USEPA (1985), pp. 8–18.
30. "High Technology in Jet Grouting," Novatecna Company Brochure, 1988.
31. Guatteri, G., J. L. Kauschinger, A. C. Doria, and E. B. Perry. "Advances in the Construction and Design of Jet Grouting Methods in South America," *Proc. Second International Conf. on Case Histories in Geotechnical Engineering,* Vol. II (1988), pp. 1037–1046.
32. Sharma, H. D., and P. Kozicki. "The Use of Synthetic Liner and/or Soil Bentonite for Groundwater Protection," *Proc. Second International Conf. on Case Histories in Geotechnical Engineering,* Vol. II (1988), pp. 1149–1157.
33. Anderson, D. C., W. Crawley, and J. D. Zabcik. "Effects of Various Liquids on Clay Soil-Bentonite Slurry Mixtures," *Hydraulic Barriers in Soil and Rock,* ASTM STP 874 (1985), pp. 93–103.
34. Fernandez, F., and R. M. Quigley. "Hydraulic Conductivity of Natural Clays Permeated with Simple Liquid Hydrocarbons," *Can. Geotech. J.,* Vol. 22, No. 2 (1985), pp. 205–214.
35. Foreman, D. E., and D. E. Daniels. "Permeation of Compacted Clay with Organic Chemicals," *J. Geotech. Eng.,* ASCE, Vol. 112, No. 7 (1986) pp. 669–681.
36. Haxo, H. E., Jr. "Durability of Liner Materials for Hazardous Waste Disposal Facilities," *Proc. 7th Annual Research Symposium on Landfill Disposal: Hazardous Waste,* USEPA (1981) pp. 140–156, as cited by Folkes (1982).
37. Stewart, W. S. *State of the Art Study of Impoundment Techniques,* "USEPA Report No. EPA-600/2-78-196, Cincinnati, OH (1978).
38. Haxo, H. E., N. A. Nelson, and J. A. Miedema. "Solubility Parameters for Predicting Membrane Waste Liquid Compatibility," *Proc. of the Eleventh Annual Research Symposium,* USEPA (1985b), pp. 198–212.
39. Lord, A. E., R. M. Koerner, and E. C. Lindhult. "The Hydraulic Conductivity of Silicate Grouted Sands with Various Chemicals," *Management of Uncontrolled Hazardous Waste Sites,* HMCRI (1983), pp. 175–178.
40. Gulick, C. W., J. A. Boa, and A. D. Buck. "Bell Canyon Test Cement Grout Development Report," Report No. 80-1928, Sandia National Laboratories, Albuquerque, NM (1980).
41. Powell, R. D., and N. R. Morganstern. "The Use and Performance of Seepage Reduction Measures," *Seepage and Leakage from Dams and Impoundments,* ASCE (1985), pp. 158–182.
42. Troxell, G. E., H. E. Davis, and J. W. Kelly. *Composition and Properties of Concrete* (New York: McGraw-Hill, 1968).

43. Spalding, B. P., L. K. Hyder, and I. L. Munro. "Grouting as a Remedial Technique for Problem Shallow Land Burial Trenches of Low-Level Radioactive Solid Wastes," *J. Environ. Qual.*, Vol. 14 (1985), pp. 100–130.
44. Clarke, W. J. "Performance Characteristics of Acrylate Polymer Grout," *Grouting in Geotechnical Engineering,* ASCE (1982), pp. 418–432.
45. Muller-Kerchenbauer, H., W. Friedrich, and H. Hass, "Development of Containment Techniques and Materials Resistant to Groundwater Contaminating Chemicals," *Management of Uncontrolled Hazardous Waste Sites,* HMCRI (1983), pp. 167–174.
46. Davis, K. E., and M. C. Herring. "Laboratory Evaluation of Slurry Wall Materials of Construction to Prevent Contamination of Groundwater from Organic Constituents," *Proc. Seventh National Groundwater Quality Symposium,* NWWA (1984), pp. 491–512.
47. Anderson, D. C., A. Gill, and W. Crawley. "Barrier-Leachate Compatibility: Permeability of Cement/Asphalt Emulsions and Contaminant Resistant Bentonite/Soil Mixtures to Organic Solvents," *Management of Uncontrolled Hazardous Waste Sites,* HMCRI (1984), pp. 126–130.
48. Kahl, T. W., J. L. Kauschinger, E. B. Perry. "Plastic Concrete Cutoff Walls for Earth Dams," Technical Report REMR-GT-15, U.S. Army Corps of Engineers (1991).
49. Bell, L. A. "A Cut Off in Rock and Alluvium at Asprokremmos Dam," *Proceedings of the Conference on Grouting in Geotechnical Engineering,* edited by W. H. Baker, ASCE (1982), pp. 172–186.
50. LaRusso, R. S. "Wanapum Development—Slurry Trench and Grouted Cut-Off," *Grouts and Drilling Muds in Engineering Practice,* (Butterworths, London, 1963), pp. 196–201.
51. Jeffries, S. A. "Bentonite-Cement Slurries for Hydraulic Cut Offs," *Proceedings of the Tenth International Conference on Soil Mechanics, Stockholm,* (A. A. Balkema Rotterdam, 1981), pp. 435–440.
52. Zappi, M. E., R. Shafer, and D. D. Adrian. "Compatibility of Soil-Bentonite Slurry Wall Backfill Mixtures with Contaminated Groundwater," *Superfund '89, Proceedings of the Ninth National Conference,* Hazardous Materials Control Research Institute (1989), pp. 519–525.
53. Mott, H. V. and W. J. Weber. "Solute Migration Control in Soil-Bentonite Containment Barriers," *Superfund '89, Proceedings of the Ninth National Conference,* Hazardous Materials Control Research Institute (1989), pp. 526–533.
54. Evans, J. C., Y. Sambasivam, and S. Zarlinski. "Attenuating Materials in Composite Liners," *Waste Contaminant Systems, Construction, Regulation, and Performance.* Geotechnical Special Publication No. 26, ASCE (1990), pp. 246–263.

CHAPTER 20

In Situ Biological Remediation of Petroleum Hydrocarbons in Unsaturated Soils

Dennis Dineen, Jill P. Slater, Patrick Hicks, and **James Holland,** McLaren Environmental Engineering, Santa Ana, California
L. Denise Clendening, Chevron Oil Field Research Company, La Habra, California

In situ biological remediation of unsaturated soils is a treatment technology that utilizes naturally occurring soil microorganisms to degrade petroleum hydrocarbons to carbon dioxide, water, and humus. Indigenous microorganisms present in the soil are stimulated by providing those elements, usually oxygen and nitrogen, that are limiting the degradation of the petroleum hydrocarbons. Because in situ remediation does not involve excavation, the costs and disruption of excavating soil are eliminated.

Petroleum hydrocarbons with high vapor pressures can be removed from soils very efficiently using in situ vapor extraction. These lighter weight hydrocarbons, which include gasoline, and aromatic additives such as benzene, toluene, and xylene, are extracted as a vacuum is applied to dry wells spaced throughout the contaminated soil. However, petroleum hydrocarbons with lower vapor pressures such as diesel, fuel oil, and crude oil are not easily extractable by this technology.[1]

The in situ biological remediation described in this chapter was developed to remediate soils where excavation would be too expensive or impractical, and where the chemicals are not easily removed by vapor extraction. This in situ bioremediation differs significantly from in situ bioremediation technologies which use "infiltration galleries" to saturate the vadose zone with water containing nutrients and hydrogen peroxide, and then recirculate or discharge the treated water. This in situ bioremediation technology delivers oxygen and nitrogen to the soil in the vapor phase rather than the dissolved phase. Delivery of oxygen and nitrogen in the vapor phase has several distinct advantages over delivery in the dissolved phase:

- Vapor phase delivery maintains unsaturated conditions throughout the affected soil and minimizes the downward migration of chemicals that would occur under saturated conditions.
- Vapor phase delivery provides the soil with atmospheric levels of molecular oxygen (over 20%) compared to the relatively low concentrations of oxygen (less than 800 parts per million) provided by hydrogen peroxide.
- Vapor phase delivery utilizes anhydrous ammonia as a source of reduced nitrogen, thereby reducing the risk of migration of nitrates into the groundwater which could result from excessive application of nitrate fertilizers in water.

Adding oxygen and nitrogen to the soil in situ is accomplished by pulling air through vadose zone wells connected to aboveground pumps and blowers. Oxygen is provided in the air that is pumped from the surface, and reduced nitrogen is provided by injecting low concentrations of anhydrous ammonia into the air stream which passes through the soil.

The design of a successful in situ biological remediation depends on five subsurface parameters:

- Soil microbiology—Petroleum degrading microorganisms must be present throughout the zone where petroleum hydrocarbon concentrations exceed the cleanup standard.
- Soil chemistry—Concentrations of soil nutrients other than oxygen and nitrogen must be adequate to maintain microbiological growth, and no toxic levels of salts or heavy metals can be present.
- Soil physics—Soil air permeability must be adequate to allow movement of oxygen and nitrogen to the affected soil and movement of carbon dioxide away from the affected soil.
- Soil morphology—Soil stratification throughout the affected zone should be well understood to design an effective delivery system.
- Hydrogeology—The depth to groundwater, groundwater flow direction and gradient, the presence or absence of floating product, and petroleum hydrocarbon concentrations in the groundwater should be understood prior to implementation of the in situ bioremediation, to avoid recontamination of cleaned soil from the groundwater.

This chapter describes the results of the bench-scale studies, in field measurements, and the full-scale design of three in situ biological remediation systems in southern California.

SOIL MICROBIOLOGY

The crucial biochemical steps in the breakdown of petroleum hydrocarbons are the oxidation of the straight chained or branched alkanes and the breaking of aromatic rings by oxygenase enzymes.[2] No plants or higher animals are known to have this ability, and relatively few microorganisms possess the enzyme systems necessary to perform these crucial steps.

Surface soils with adequate carbon, oxygen, and nutrients typically contain about ten million to one billion (10^7 to 10^9) microorganisms per gram. Of these, approximately 0.1 to 1.0% are petroleum degraders (10^5 to 10^6). After exposure to petroleum hydrocarbons, the microbial ecology of the soil adjusts so that the number of petroleum degraders increases from 100 to 1,000 times higher (10^6 to 10^8).[3]

Two critical questions for designing an in situ biological remediation in soils in which petroleum hydrocarbons extend to depths up to 50 feet or greater are:

1. Are viable microorganisms present in the soil at the same depths as the petroleum hydrocarbons; and
2. Are these microorganisms capable of degrading petroleum hydrocarbons?

Sites which were candidates for in situ biological remediation were screened to determine whether viable petroleum degrading microorganisms were present throughout the zone of contamination. The following sections describe the methodologies and results of the microbiology screening at a typical site.

Methods and Materials

Samples were collected at 5-foot intervals from the center of the contaminated zone and from an uncontaminated site on the same soil series. Samples were collected in sterile 6-inch brass tubes inside a 2.5-inch diameter split spoon sampler through an 8-inch I.D. hollow stem auger. Upon retrieval of the samples at the surface, soil inside the brass tube was extruded into a sterile bag and stored on ice.

Upon delivery to the laboratory, 10-gram subsamples were extracted and mixed according to standard protocol in a 500-mL flask containing a 0.9% sterile saline solution.[4] One-milliliter aliquots were then taken from the solution and plated on complete agar. Microbial isolates from agar plates used for total cell counts were streaked for purity and further investigated by microscopic analysis for shape and motility. Isolated cultures were then plated on selective media, to allow generic determination of the isolates.

Results and Discussion

At all three sites the total cell counts ranged from 10^2 to 10^5 cells per gram wherever total petroleum hydrocarbon concentrations exceeded 1,000 ppm. Data from uncontaminated soils of the same soil series had cell counts ranging from non-detected to 10^2 cells per gram. There is an apparent correlation between the presence of a carbon source and the presence of aerobic microorganisms at all depths. One notable exception was finding that no microorganisms were present in buried drilling muds, even though the drilling muds had large hydrocarbon concentrations (130,000 ppm). Analyses of these samples for salts and heavy metals did not show any concentration above background levels. The absence of microorganisms in the drilling muds is most likely attributed to the very low oxygen tension in the drilling mud resulting from its low permeability, estimated at 10^{-9} cm/sec. Another possible explanation is the toxicity to microorganisms reported at concentrations above 5% to 10% petroleum hydrocarbons.[5] Subsequent investigation of the isolates indicated that the microorganisms present in the soil were members of the genus *Pseudomonas* and the genus *Arthrobacter*. Members of these genera are the most common petroleum degrading microorganisms.[3]

The results of the initial soil microbiological investigations indicated that indigenous petroleum degrading microorganisms were present throughout the contaminated soil. Additional investigations were then conducted to determine whether these microorganisms could be stimulated to degrade the petroleum hydrocarbons present in the soil.

SOIL PHYSICS

Stimulating indigenous microorganisms to degrade petroleum hydrocarbons depends on the ability to deliver oxygen and reduced nitrogen throughout the contaminated soil. The key parameter controlling vapor movement in the soil is permeability to air (K_{air}). Permeability to air is a function of soil texture (soil particle size distribution), soil moisture, and bulk density. Of these, data on soil texture are most readily available and provide a reasonable indicator of permeability.

Laboratory studies with soil columns have experimentally demonstrated that lighter hydrocarbons are removed rapidly using soil venting techniques. It has also been demonstrated that preferential air paths affect the removal rates.[6] Currently, there is very little predictive capability for evaluating soil venting or air injection systems to determine the field placement of vapor injection or extraction wells. Recently, a few predictive models have been developed to determine the performance of soil venting. However, these models have not been experimentally verified in the field,[7,8] and can only provide guidelines for vapor extraction system design. Therefore, pilot studies were conducted to design the injection and extraction process for the in situ bioremediation systems discussed here.

Predictions of vapor injection and extraction rates were made on the basis of empirical data and on the effective radius of influence of vapor extraction systems.[9] Three 50-foot vadose wells were constructed bisecting the site at a distance of 0, 15, and 45 feet through the contaminated soil. Samples were analyzed for texture, bulk density, moisture content, and permeability. Field tests were then conducted to document the actual permeability in situ.

Methods and Materials

Samples in 6-inch brass tubes taken immediately below the samples analyzed for microbial numbers were subjected to soil physical analyses. In situ permeability tests were conducted by injecting sulfur hexafluoride (SF_6) as a tracer gas into the central well and measuring breakthrough at the 15-foot and 30-foot wells. The tracer gas was injected in the airstream, which was designed to deliver 100 cubic feet of air per minute at a pressure of three to five pounds per square inch. Breakthrough was measured using a specialized gas chromatograph developed for analyzing SF_6 at concentrations as low as 5 parts per trillion (ppt).

Results and Discussion

Vapor phase permeabilities were calculated for two soil strata using data from soil samples which were analyzed for permeability to air in the laboratory using Hazen's equation.[10] Laboratory results predicted an air permeability of 3 cm/sec in the sand stratum and 1 cm/sec in the loamy sand stratum. These permeabilities would result in a transit time of approximately 3 minutes over 15 feet in the sand stratum and 11 minutes in the loamy sand stratum.

Actual breakthrough times of injected air in a second well 15 feet away were measured using SF_6. These data, presented in Figure 1, show breakthrough times of 10 minutes to two hours. The extended breakthrough time of two hours

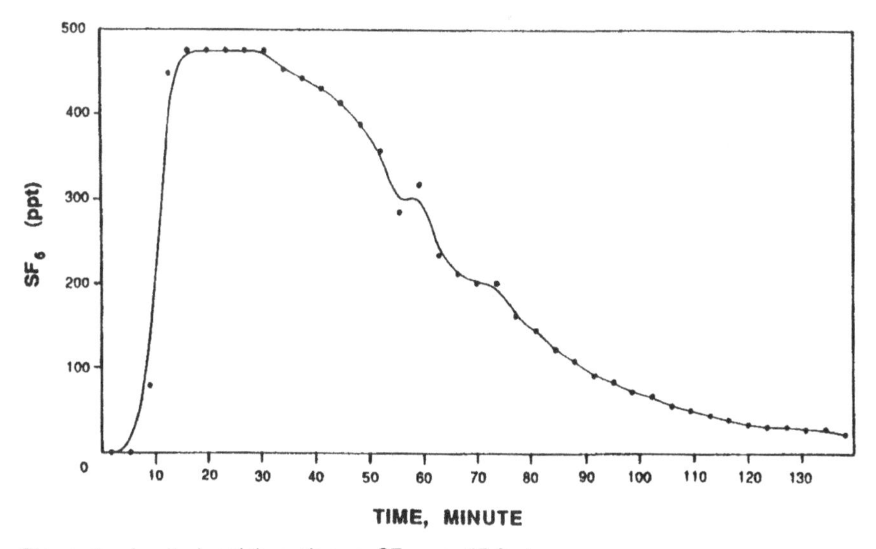

Figure 1. In situ breakthrough over SF_6 over 15 feet.

reflects the multiple pathways taken by air injected into the well. Air moving through the largest pores and/or pores with the least tortuosity travel 15 feet in 10 minutes (0.8 cm/sec), whereas the air moving through the smaller pores and/or the more tortuous pores requires over two hours to travel the same distance (0.06 cm/sec). These data are in general agreement with the empirical data from soil vapor extraction systems.

SOIL CHEMISTRY

In a soil containing petroleum hydrocarbons, biological degradation occurs naturally until the available oxygen and nutrients are consumed. In situ biological remediation stimulates the indigenous petroleum-degrading microorganisms by providing those elements which may be limiting in the soil, notably oxygen, nitrogen, and phosphorus. Once these elements are available in the soil, biological degradation can proceed.

Oxygen is required at a rate of approximately 3.1 pounds of oxygen per pound of hydrocarbon degraded. The maximum amount of oxygen in a well aerated soil is approximately 20%, or 200,000 parts per million (ppm). The maximum amount of oxygen in a saturated soil is approximately 8 ppm. If hydrogen peroxide is used to carry oxygen into the saturated soil, the levels of dissolved oxygen can be increased to 200 to 800 ppm.[11] Since oxygen is usually the most limiting element in contaminated soils, the most efficient system for delivery of oxygen to the soil is in the vapor phase.

Nitrogen is required at a rate of up to one pound of reduced nitrogen per 160 pounds of hydrocarbon degraded.[12] Nitrogen is typically added to the soil as urea or ammonium nitrate, which dissolves in the soil water as ammonium (NH_4^+) and nitrate (NO_3^-). If oxygen is supplied in the vapor phase, reduced nitrogen can also be added in the vapor phase as anhydrous ammonia gas (NH_3). When anhydrous ammonia in the soil air contacts the soil water, the ammonia is dissolved as ammonium ion (NH_4^+). Anhydrous ammonia has been used as a nitrogen source in agriculture for over 40 years. In agricultural applications, ammonia is routinely applied at a rate of 100 to 200 pounds per acre (approximately 100 to 200 ppm) by injecting anhydrous ammonia while disking the soil. Ammonia is toxic to soil microorganisms at concentrations above about 300 ppm.[13]

Phosphorus is generally considered to be the other limiting element in soil bioremediation and is routinely added to above-ground soil bioremediation projects. Phosphorus is very insoluble in most soils, and phosphorus availability decreases below pH 5.5, and above pH 7.0. In southern California, where these demonstrations were conducted, soil pH was 7.5 to 8.5. In this pH range, phosphate availability is decreased even further because of precipitation as calcium and magnesium phosphates. Under these conditions, it is difficult to increase phosphorus availability by adding phosphorus fertilizer.

Other elements are usually not limiting, and there are usually adequate levels in the soil to provide the basic requirements. Bench-scale tests conducted at these sites determined that no elements other than nitrogen and oxygen were limiting bioremediation. Changes in soil microbiology and soil nitrogen levels were measured throughout the treatment.

Methods and Materials

Samples from each of the two major strata were composited separately. Subsamples of approximately 150 grams were transferred from each stratum into 500-mL flasks. One group of 12 replicates from each substratum was a killed control. One additional group of eight replicates from each substratum was treated with 100 ppm anhydrous ammonia and constantly aerated with air, which was bubbled through water to maintain high relative humidity. A third group was similarly aerated, but without the anhydrous ammonia. Four replicates were analyzed on Day 0 for total cell counts. Four replicates from each group were harvested at two weeks and four weeks, and analyzed for total cell counts. In addition, soil was analyzed after four weeks for soil nitrogen to determine the effectiveness of anhydrous ammonia as a nitrogen source.

Results and Discussion

The results of the soil chemical and microbiological studies are shown on Figures 2 and 3. Figure 2 shows the concentrations of all forms of soil nitrogen in the treatment with anhydrous ammonia compared to the treatment with air only. These results show an increase of approximately 50% of all forms of soil nitrogen in the soils treated with anhydrous ammonia. Figure 3 shows the corresponding changes in total cell counts over a four week period. Table 1 summarizes the results of the chemical and microbiological studies.

Data from soil which was treated with 100 ppm anhydrous ammonia in the air stream showed that the use of low concentrations of anhydrous ammonia is an effective mechanism to provide reduced nitrogen to soil microorganisms. Data on changes in microbial populations with treatment show that adding oxygen increased microbial counts by a factor of 10, and that adding both oxygen and reduced nitrogen increased microbial counts by a factor of 100.

Table 1 summarizes the results of the studies to date. Initial concentrations of petroleum hydrocarbons in the soil are 2,000 ppm in the sand, and 6,000 ppm in the loamy sand. Initial cell counts in the sand were 10^5 cells/gram compared to 10^4 cells/gram in the loamy sand. This suggests that the lower petroleum hydrocarbon concentration in the sand is due to the tenfold higher number of petroleum degrading microorganisms. Adding oxygen to the system increased cell counts tenfold in the loamy sand, but not in the sand. This suggests that oxygen was not limiting in the more permeable sand, but was limiting in the deeper, less permeable loamy sand, resulting in a tenfold increase in cell counts.

Adding anhydrous ammonia to the soil increased the cell counts in both the sand and loamy sand. This suggests that nitrogen was limiting at both soil depths,

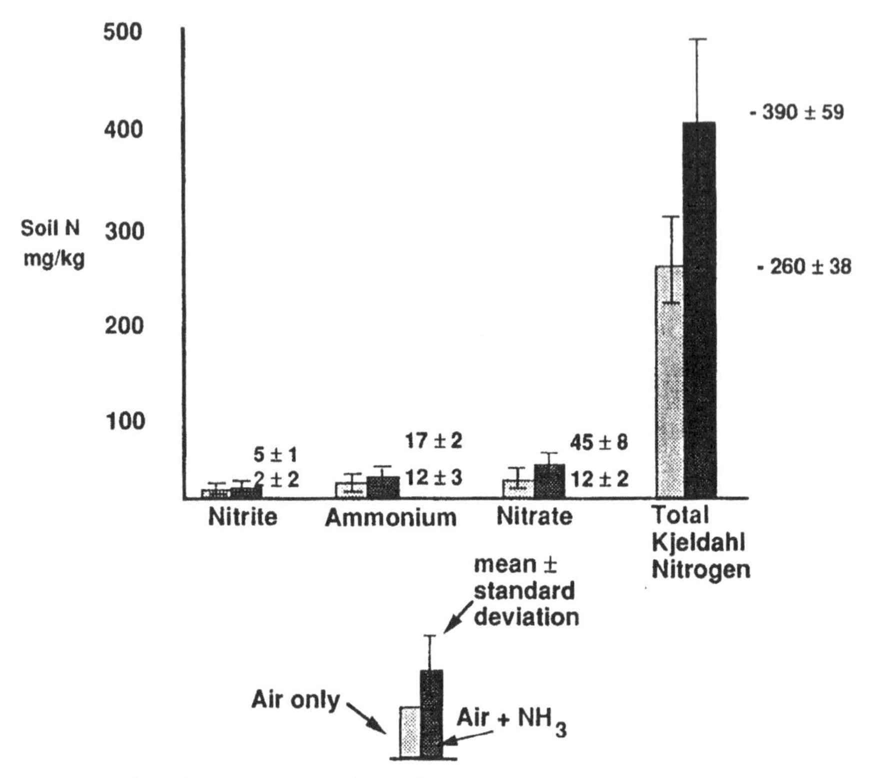

Figure 2. Soil nitrogen concentration with treatment.

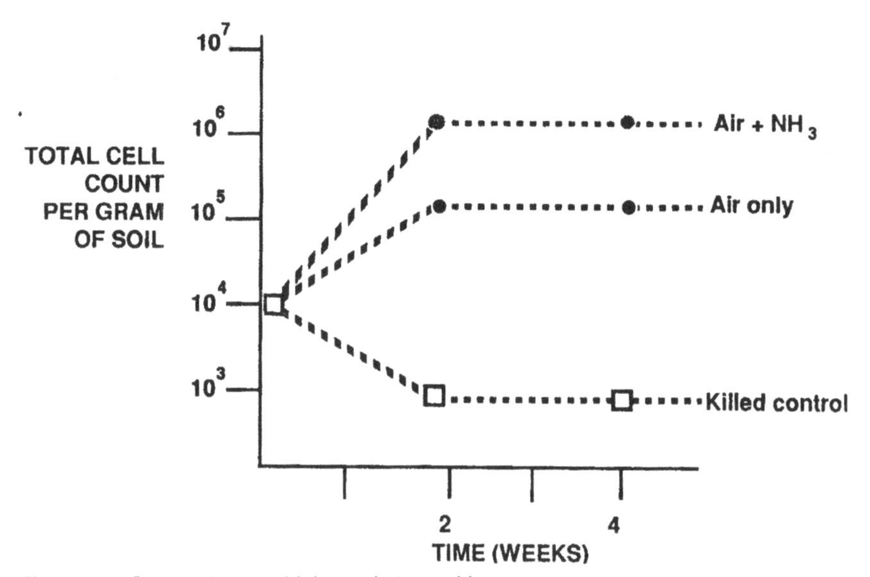

Figure 3. Change in microbial populations with treatment.

Table 1. Summary of Results of In Situ Bioremediation Microbial Studies

Depth	USDA Soil Texture	Soil TPH	Initial Microbial Counts	Microbial Counts After Treatment with Air only	Microbial Counts After Treatment with Air + NH_3
20–35 feet	Sand	2,000 ppm	10^5 cfu/gm	10^5 cfu/gm	10^6 cfu/gm
35–50 feet	Loamy sand	6,000 ppm	10^4 cfu/gm	10^5 cfu/gm	10^6 cfu/gm

cfu/gm = cell forming units/gram soil.

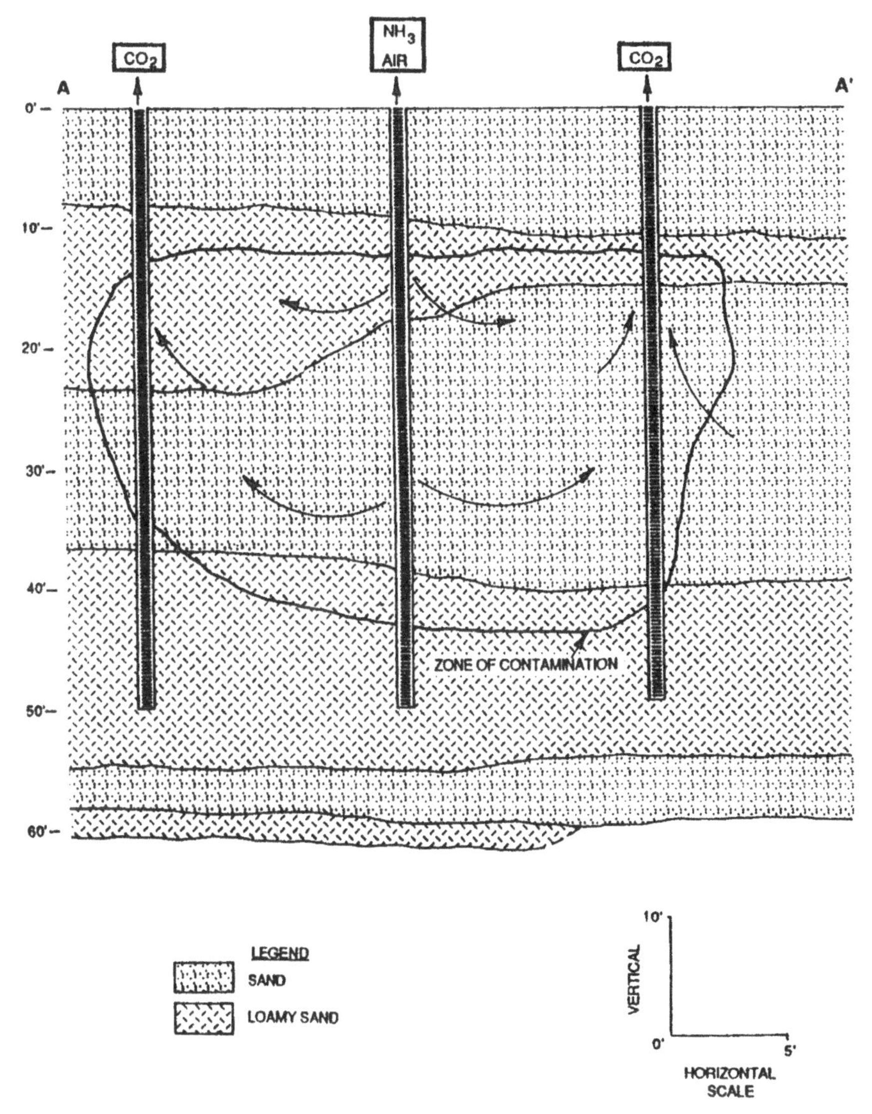

Figure 4. Design of in situ bioremediation using vapor phase application of oxygen and nitrogen.

and that addition of anhydrous ammonia eliminated the nitrogen deficiency, resulting in a tenfold increase in cell counts compared to the well aerated soil.

FULL-SCALE DESIGN AND IMPLEMENTATION

Based on the data from these studies and in situ field measurements, a full-scale bioremediation is being implemented. Air is injected into the soil at a pressure of approximately three pounds per square foot, and anhydrous ammonia is added weekly at 100 parts per million. Additional vapor wells are being installed to cover the entire contaminated area, and to monitor the movement of injected air in the soil.

A cross section of the contaminated soil and schematic of the treatment system are shown in Figure 4.

SUMMARY AND CONCLUSIONS

In situ bioremediation of unsaturated soils involves three very simple, well documented technologies:

- the ability of indigenous microorganisms to degrade petroleum hydrocarbons;
- the use of aboveground pumps and blowers to move vapors through the unsaturated soil; and
- the use of anhydrous ammonia as a source of reduced nitrogen.

Results presented here document that adding anhydrous ammonia to an air stream through the unsaturated soil increases soil oxygen and nitrogen levels, resulting in a hundredfold increase in microbial count. Maintaining viable cell counts at the level of 10^6 to 10^7 is expected to result in a decrease of petroleum hydrocarbons in situ to a cleanup level of 100 ppm.

REFERENCES

1. Hinchee, R. E., D. C. Downey, and E. J. Coleman. "Enhanced Bioreclamation, Soil Venting and Ground-Water Extraction: A Cost-Effectiveness and Feasibility Comparison," *Proceedings of the Conference on Petroleum Hydrocarbons and Organic Chemicals in Ground Water: Prevention, Detection, and Restoration*. National Water Well Association/American Petroleum Institute, 1987.
2. Singer, M. E., and W. R. Finnerty. "Microbial Metabolism of Straight-Chain and Branched Alkanes," in R. M. Atlas, ed., *Petroleum Microbiology* (Macmillan Publishing Co., Inc., 1984).
3. Bossert, I., and R. Bartha. "The Fate of Petroleum in Soil Ecosystems," in R. M. Atlas, ed., *Petroleum Microbiology* (New York: Macmillan Publishing Co., Inc., 1984).

4. Wollum, A. G. "Cultural Methods for Soil Microorganisms," in *Methods of Soil Analysis*. Agronomy Monographs. No. 9, Part 2, 1982.
5. Dibble, J. T., and R. Bartha. "Effect of Environmental Parameters on the Biodegradation of Oil Sludge," *Appl. Environ. Microbiol.* 37:729–739 (1979).
6. Rainwater, K., B. J. Claborn, H. W. Parker, D. Wilkerson, and M. R. Zaman. "Large Scale Laboratory Experiments for Forced Air Volatilization of Hydrocarbon Liquids in Soil," *Proceedings of Petroleum Hydrocarbon and Organic Chemicals in Groundwater: Prevention, Detection, and Restoration*. National Water Well Association/American Petroleum Institute, 1988.
7. Baehr, A. L., and C. E. Hoag. "A Modeling and Experimental Investigation of Induced Venting," *Proceedings of Petroleum Hydrocarbon and Organic Chemicals in Groundwater: Prevention, Detection and Restoration*. National Water Well Association/American Petroleum Institute, 1988.
8. Johnson, P. C., M. W. Kemblowski, and J. D. Colthart. "Practical Screening Models for Soil Venting Applications," *Proceedings of Petroleum Hydrocarbon and Organic Chemicals in Groundwater: Prevention, Detection, and Restoration*. National Water Well Association/American Petroleum Institute, 1988.
9. Krishnayya, A. V., M. J. O'Connor, J. G. Agar, and R. D. King. "Vapour Extraction Systems: Factors Affecting Their Design and Performance," *Proceedings of Petroleum Hydrocarbon and Organic Chemicals in Groundwater: Prevention, Detection and Restoration*. National Water Well Association/American Petroleum Institute, 1988.
10. Burmister, D. M. "The Importance and Practical Use of Relative Density in Soil Mechanics," in *ASTM*, Vol. 48, Philadelphia, PA, 1948.
11. Ward, C. H., J. M. Thomas, S. Fiorenza, H. S. Rifai, P. B. Bedient, J. T. Wilson, and R. L. Raymond. "In Situ Bioremediation of Subsurface Material and Groundwater Contaminated with Aviation Fuel: Traverse City, Michigan," in *Hazardous Waste Treatment; Biosystems for Pollution Control*. Air and Waste Management Association/Environmental Protection Agency Conference, Pittsburgh, PA, 1989.
12. "Manual on Disposal of Refinery Wastes." American Petroleum Institute, Washington, DC, 1980.
13. Tisdale, S. L., and W. L. Nelson. *Soil Fertility and Fertilizers*. (New York: Macmillan Publishing Co., Inc., 1975).

CHAPTER 21

Estimates for Hydrocarbon Vapor Emissions Resulting from Service Station Remediations and Buried Gasoline-Contaminated Soils

Paul C. Johnson, Marvin B. Hertz, and **Dallas L. Byers,** Shell Development, Westhollow Research Center, Houston, TX

I. INTRODUCTION

Soils become contaminated at service stations as the result primarily of leaking underground storage tanks, leaking transport lines, or spills that occur during storage tank filling. Upon detection of a spill, a site investigation is conducted and a remediation plan is formulated. While the specific remediation plan for any site depends on the level of contamination, location of contaminated soil, soil stratigraphy, and other site-specific factors, a typical service station remediation will involve soil excavation, pumping and treating of contaminated groundwater and free-liquid residual gasoline, and in-situ treatment (soil venting or enhanced biodegradation) of the unsaturated zone. Because some hydrocarbon vapors are released to the atmosphere during each stage, it is important to know the range of possible emission levels in order to evaluate the health risk that they may pose to a nearby community.

This chapter is divided into three main sections. In the first section models are developed for computing conservative emissions estimates for each stage of a hypothetical service station cleanup, which consists of tank excavation and replacement, a pump-and-treat operation that removes contaminated groundwater and free-liquid residual gasoline, and in-situ soil venting of the remaining contaminated soil. A model that estimates the emissions for the case in which gasoline-contaminated soils are left in place is also presented. In the second main section these models are used to estimate the benzene emissions associated with this hypothetical remediation, and with leaving the soils in-place (no treatment). The vapor fluxes are used as air dispersion model inputs in the third main section, and ambient air concentrations are calculated for a nearby community. For comparison, the ambient air hydrocarbon concentrations due to hydrocarbon vapor emissions from undisturbed underground gasoline-contaminated soils are also computed.

II. MODEL DEVELOPMENT

II.1 Vapor Equilibrium Models

An integral part of any vapor transport model is the calculation of vapor concentrations at the source based on measured residual soil contamination levels, contaminant composition, soil properties (organic carbon content, soil moisture content), and environmental factors (temperature). Two main approaches are used in vapor transport models, but rarely is their use justified by the authors. Before presenting vapor emissions models, therefore, it is useful to briefly review the various methods for calculating vapor concentrations and justify the approach used in this work.

The influence of soil type, moisture content, chemical type, temperature, and residual soil contaminant levels has been the focus of studies by Chiou and Shoup,[1] Spencer,[2] Poe et al.,[3] and Valsaraj and Thibodeaux.[4] In each study the effect of the parameters listed above on the equilibrium vapor concentration above a soil matrix was studied for a single component. Briefly, changes in the moisture content significantly influence the vapor concentration when the soils are "dry"; that is, the moisture content is less than that required to provide a complete monolayer coverage of water molecules on the soil particle surfaces. This corresponds roughly to the "wilting point" of a soil, and for sandy soil types is in the 0.02 to 0.05 g-H_2O/g-soil moisture content range. It has been observed that the sorptive capacity of soil is greatly increased when the soil is dry.[1,2] When the contaminant concentration is low enough that free adsorptive sites are available on the soil ($\approx < 100$ mg-contaminant/kg-soil), the adsorbed contaminant/vapor equilibrium can be modeled by a modified Brauner-Emmett-Teller (BET) equation.[5] If the moisture content is great enough that there is more than a monolayer of water molecules adhering to the soil surface, then the vapor equilibrium appears to be governed by the partitioning between four phases: vapor, dissolved in the soil moisture, sorbed to the soil particles, and free-residual (when concentrations are great enough).[2] More often than not, the moisture content of soils buried more than a foot below ground surface will be greater than the wilting point, so we will focus on modeling the partitioning of contaminants in this moisture content regime.

As stated above, components in the residual contaminant partition between vapor, adsorbed, soluble (dissolved in soil moisture), and free-liquid (or solid) residual phases. Mathematically, this can be described for any component i:

$$\frac{M_i}{M_{soil}} = y_i\left[\frac{\alpha_i P_i^v \epsilon_A}{RT\rho_{soil}} + \alpha_i\frac{M^{HC}}{M_{soil}} + \frac{M^{H_2O}}{M_{soil}} + \frac{k_i}{M_{w,H_2O}}\right] \tag{1}$$

where:

M_i = total moles of i in soil matrix
y_i = mole fraction of i in soil moisture phase
α_i = activity coefficient for i in water
k_i = sorption coefficient for i [(g-i/g-soil)/(g-i/g-H_2O)]
P_i^v = pure component vapor pressure of i [atm]
ϵ_A = vapor-filled void fraction in soil matrix [cm^3-air/cm^3-soil]
ρ_{soil} = soil matrix density [g/cm^3]
R = gas constant (=82.1 cm^3-atm/mole-K)
T = absolute temperature [k]
M^{HC} = total moles of free-liquid residual contaminant
M^{H_2O} = total moles in soil moisture phase
M_{soil} = mass of soil matrix [g]
M_{W,H_2O} = molecular weight of water [18 g/mole]

The first term on the right-hand side of Equation 1 represents the number of moles of i in the vapor phase, the second represents the number of moles of i in the free-liquid residual phase, the third term is the number of moles of i dissolved in the soil moisture, and the last term is the number of moles of i sorbed to the soil particles. In writing Equation 1 we assume equilibrium between an ideal gas vapor phase, an ideal mixture free-liquid hydrocarbon phase, and a nonideal soil moisture phase. When contaminant levels are great enough that a free-liquid (or solid) residual phase is present, then Equation 1 must be solved iteratively, subject to the condition that $\Sigma\alpha_i y_i = 1$.

Once Equation 1 is solved, the vapor concentration in equilibrium with the contaminant/soil matrix, $C_{i,v}^{eq}$, [mass-i/volume-vapor] is obtained from:

$$C_{i,v}^{eq} = \frac{\alpha_i y_i M_{w,i} P_i^v}{RT} \tag{2}$$

where $M_{w,i}$ denotes the molecular weight of component i. In the limits of low and high residual contaminant soil concentrations Equation 1 reduces to forms that do not require iterative solutions. In the low concentration limit (i.e., no free-liquid or solid precipitate phase present), Equation 2 becomes:

$$C_{i,v}^{eq} = \frac{H\, C_{i,soil}}{[(H\epsilon_A/\rho_{soil}) + \theta_M + k_i]} \tag{3}$$

where:

H = Henry's law constant $(= \alpha_i P_i^v M_{w,H_2O}/RT)$[(g-i/cm^3-vapor)/(g-i/g-H_2O)]
$C_{i,soil}$ = residual contamination level of i [g-i/g-soil]
θ_M = soil moisture content [g-H_2O/g-soil]

In the high residual contaminant concentration limit Equation 2 becomes:

$$C_{i,v}^{eq} = \frac{x_i P_i^v M_{w,i}}{RT} \tag{4}$$

where x_i is the mole fraction of component i in the free-liquid residual phase. For mixtures composed of compounds with similar molecular weights, x_i is roughly equal to the mass fraction of compound i.

Equations 3 and 4 are the two most commonly incorporated in vapor transport models. Note that Equation 3 predicts vapor concentrations that are proportional to the residual soil concentration of each species and are independent of the relative concentrations of each chemical species in the contaminant, while the vapor concentrations predicted by Equation 4 are independent of residual soil concentration levels and depend only on the relative concentrations of species. Due to its mathematical characteristics, rather than any model validation, transient transport models[6,7] most often incorporate Equation 3. Steady-state landfill emission models[8] often utilize Equation 4. It is important to recognize that these models are only valid for specific limiting conditions, and generalization to other concentration ranges can produce very misleading results. For example, Equation 3 predicts that vapor concentrations always increase with increasing residual contaminant levels, but realistically the equilibrium vapor concentration of any compound cannot exceed its saturated vapor concentration $(= P_i^v M_{w,i}/RT)$.

Figure 1 compares vapor concentrations predicted by Equations 1, 3, and 4 for the regular gasoline defined by Table 1. The required chemical parameters (vapor pressures, octanol-water partition coefficients, water solubility values) can be found in Johnson et al.[9] Example model parameters for a sandy soil are:

f_{oc} = organic carbon fraction = 0.002
θ_M = soil moisture content = .05 g-H_2O/g-soil
ϵ_T = total void fraction = 0.35 cm^3-vapor/cm^3-soil
ρ_{soil} = soil bulk density = 1.60 g/cm^3

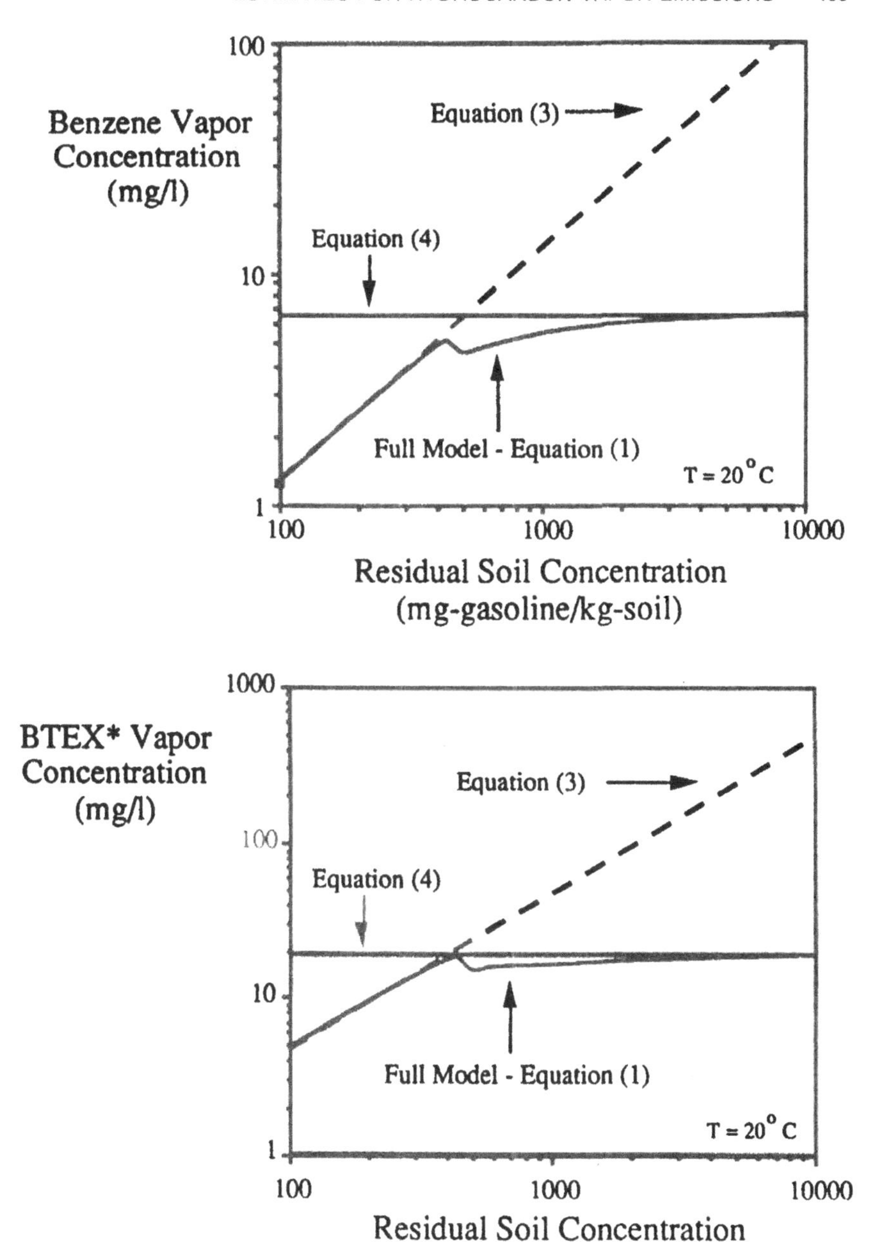

*denotes the sum of benzene, toluene, ethylbenzene, and xylenes vapor concentrations

Figure 1. Comparison of vapor concentration prediction models.

Table 1. Composition of a Regular Gasoline

Compound Name	Mw (g)	Weight Fraction	Mole Fraction	P_i^v(20°C) (atm)
propane	44.1	0.0005	0.0001	8.50
isobutane	58.1	0.0085	0.0137	2.93
n-butane	58.1	0.0259	0.0415	2.11
trans-2-butene	56.1	0.0019	0.0032	1.97
cis-2-butene	56.1	0.0018	0.0030	1.79
3-methyl-1-butene	70.1	0.0010	0.0013	0.96
isopentane	72.2	0.0916	0.1181	0.78
1-pentene	70.1	0.0032	0.0042	0.70
2-methyl-1-butene	70.1	0.0068	0.0090	0.67
2-methyl-1,3-butadiene	68.1	0.0068	0.0092	0.65
n-pentane	72.2	0.0628	0.0810	0.57
trans-2-pentene	70.1	0.0138	0.0184	0.53
2-methyl-2-butene	70.1	0.0129	0.0171	0.51
3-methyl-1,2-butadiene	68.1	0.0003	0.0004	0.46
cyclopentane	70.1	0.0185	0.0245	0.35
2,3-dimethylbutane	86.2	0.0111	0.0120	0.26
2-methylpentane	86.2	0.0515	0.0556	0.21
3-methylpentane	86.2	0.0314	0.0340	0.20
n-hexane	86.2	0.0411	0.0444	0.16
methylcyclopentane	84.2	0.0214	0.0237	0.15
2,2-dimethylpentane	100.2	0.0077	0.0071	0.11
benzene	78.1	0.0172	0.0205	0.10
cyclohexane	84.2	0.0059	0.0065	0.10
2,3-dimethylpentane	100.2	0.0063	0.0058	0.072
3-methylhexane	100.2	0.0099	0.0092	0.064
3-ethylpentane	100.2	0.0168	0.0156	0.060
n-heptane	100.2	0.0356	0.0331	0.046
methylcyclohexane	98.2	0.0055	0.0052	0.048
2,2-dimethylhexane	114.2	0.0046	0.0038	0.035
toluene	92.1	0.0899	0.0908	0.029
2-methylheptane	114.2	0.0028	0.0023	0.021
3-methylheptane	114.2	0.0062	0.0051	0.020
n-octane	114.2	0.0647	0.0528	0.014
2,4,4-trimethylhexane	128.3	0.0015	0.0011	0.013
2,2-dimethylheptane	128.3	0.0003	0.0002	0.011
ethylbenzene	106.2	0.0205	0.0180	0.0092
p-xylene	106.2	0.0153	0.0134	0.0086
o-xylene	106.2	0.0221	0.0194	0.0066
n-nonane	128.3	0.0155	0.0112	0.0042
3,3,5-trimethylheptane	142.3	0.0033	0.0022	0.0037
n-propylbenzene	120.2	0.0346	0.0268	0.0033
1,3,5-trimethylbenzene	120.2	0.0201	0.0156	0.0024
1,2,4-trimethylbenzene	120.2	0.0061	0.0047	0.0019
n-decane	142.3	0.0343	0.0224	0.0036
methylpropylbenzene	134.2	0.0210	0.0146	0.0010
dimethylethylbenzene	134.2	0.0173	0.0120	0.00070
n-undecane	156.3	0.0078	0.0046	0.00060

continued

Table 1. ***Continued***

Compound Name	Mw (g)	Weight Fraction	Mole Fraction	P_i^v(20 : C) (atm)
1,2,4,5-tetramethylbenzene	134.2	0.0511	0.0354	0.00046
1,2,3,4-tetramethylbenzene	134.2	0.0053	0.0037	0.00033
1,2,4-trimethyl-5-ethylbenzene	148.2	0.0191	0.0120	0.00029
n-dodecane	170.3	0.0050	0.0027	0.00040
naphthalene	128.2	0.0041	0.0030	0.00014
methylnaphthalene	142.2	0.0061	0.0040	0.00005
Total		0.996	0.999	

For these values, Figure 1 indicates that Equation 3 is applicable below a residual soil contamination level of about 500 mg-gasoline/kg-soil. Above this residual concentration level, however, Equation 3 predicts increasing vapor concentrations with increasing residual levels, while the complete model predicts that vapor concentrations become independent of the residual concentration level. This limiting behavior is predicted by Equation 4. One must be very careful when using transport models based on a "three-phase model," such as Equation 3, because they will overpredict vapor concentrations and emission rates for many situations. Usually there are no internal checking procedures in these models to ensure that unrealistic vapor concentrations are not being predicted.

Throughout this chapter, Equation 4 is used to predict equilibrium vapor concentrations because it is most applicable for the residual concentration levels encountered at typical service station spill sites.

II.2 Vapor Flux Models

In the following analysis, models are presented for computing conservative estimates for emissions resulting from processes associated with typical service station remediations. Specifically, we will consider emissions associated with tank excavation, in-situ soil venting, and the pumping of groundwater and free-liquid hydrocarbons.

II.2a Emissions Associated with Tank or Soil Excavation

During the excavation of leaky storage tanks, tank backfill material (usually pea gravel) is removed from the tank area and placed in a pile. In some states the pile of contaminated soil can remain uncovered as long as the excavation does not cease for more than an hour (a "one-hour working pile"). Figure 2 illustrates this operation. Both the excavated soil pile and the empty tank pit are potential sources of hydrocarbon emissions. In addition, during the excavation process fine particulate material may be released to the atmosphere. In this study transport of particulate materials is neglected.

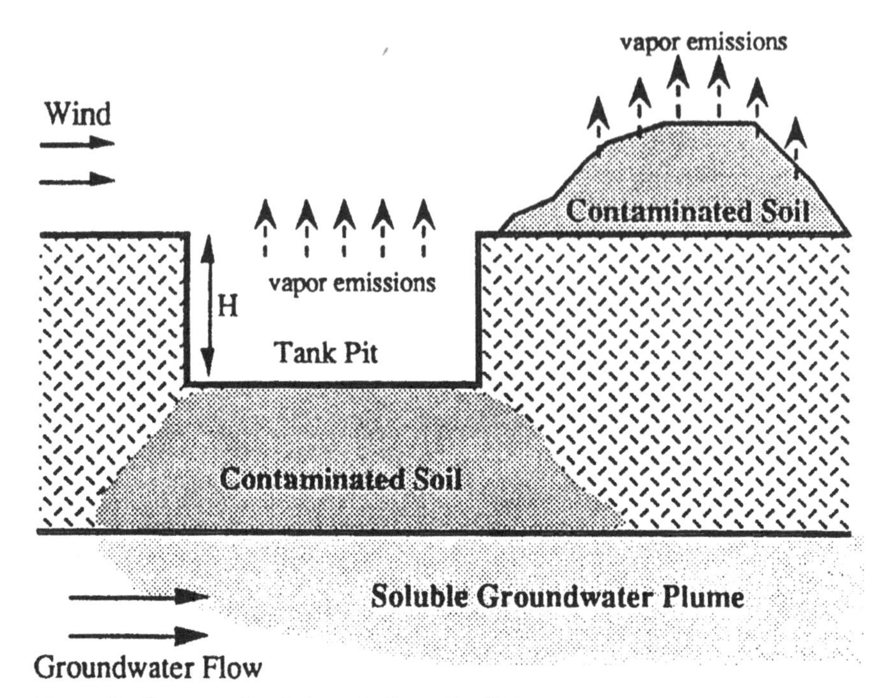

Figure 2. Sources of emissions during excavation.

We can estimate emissions from an excavated pit by assuming that vapor transport is limited by diffusion upwards from the pit bottom through a quasi-stagnant layer of air. At ground surface the vapors are swept away by the wind. For this situation, the vapor flux in the pit can be obtained as a solution to the diffusion equation:

$$\frac{\partial C_{i,v}}{\partial t} = D_i^o \frac{\partial^2 C_{i,v}}{\partial z^2} \tag{5}$$

subject to the following boundary and initial conditions:

$$\begin{aligned} C_{i,v} &= 0; & t &= 0 \\ C_{i,v} &= 0 & z &= H \\ C_{i,v} &= C_{i,v}^{eq} & z &= 0 \end{aligned} \tag{6}$$

where:

$C_{i,v}$ = vapor phase concentration of species i [g/cm^3]
t = time [s]
z = distance above the pit bottom [cm]
H = depth of excavation [cm]
D_i^o = vapor phase molecular diffusion coefficient in air [cm^2/s]

and $C_{i,v}^{eq}$ is the vapor concentration of species i in equilibrium with the contaminant/soil matrix, and as discussed is given by Equation 4. Experimentally measured D_i^o values are available in the literature for many compounds of interest; they can also be predicted reliably from kinetic theory formulas.[10] The solution to Equation 5 is:

$$C_{i,v}(t,z) = C_{i,v}^{eq}\left\langle\left(1 - \frac{z}{H}\right) - \sum_{n=1}^{n=\infty} \frac{2}{n\pi} \exp\left(-\frac{n^2\pi^2 D_i^o t}{H^2}\right) \sin(n\pi z/H)\right\rangle \quad (7)$$

and the corresponding vapor flux, $\mathfrak{F}_{pit}$, from the pit is:

$$\mathfrak{F}_{pit} = -D_i^o \frac{\partial C_{i,v}}{\partial z}\bigg|_{z=0} = \frac{D_i^o C_{i,v}^{eq}}{H}\left\langle 1 + \sum_{n=1}^{n=\infty} 2(-1)^n \exp\left(-\frac{n^2\pi^2 D_i^o t}{H^2}\right)\right\rangle \quad (8)$$

The maximum vapor flux, $\mathfrak{F}_{pit,max}$, occurs at steady state ($t \to \infty$):

$$\mathfrak{F}_{pit,max} = \frac{D_i^o C_{i,v}^{eq}}{H} \quad (9)$$

Note that the assumptions used in this model (nondiminishing source, zero concentration boundary condition) lead to conservative emissions estimates. Emissions during excavation may be higher than those predicted by Equation 9 due to the atmospheric mixing within the pit induced by digging machinery. For most cases, however, Equation 9 should provide a good estimate of the average emissions during the excavation and tank replacement stage.

II.2b Emissions Associated with Exposed Piles of Contaminated Soil

Estimating emissions from an excavated pile of contaminated soil requires a more complex analysis. When fresh contaminated soil is placed on the pile, volatilization occurs rapidly from the soil layers on the outside of the pile. As layers of soil near the surface dry out, the vapor emission rate decreases because vapors must diffuse through a longer path to reach the atmosphere. For this process:

$$\frac{\partial \epsilon_A C_{i,v}}{\partial t} + \frac{\partial \rho_b C_{i,s}}{\partial t} = \frac{\partial}{\partial y} D_{i,v}^{eff} \frac{\partial C_{i,v}}{\partial y} \quad (10)$$

where:

ϵ_A = air-filled void fraction in soil [cm³-air/cm³-soil]
ρ_b = soil bulk density [g/cm³]
t = time [s]
$D_{i,v}^{eff}$ = effective porous media vapor diffusion coefficient for species i [cm²/s]

$C_{i,s}$ = residual concentration of species i in soil [g-i/g-soil]
y = depth into contaminated soil pile [cm]
$C_{i,v}$ = concentration of species i in vapor phase [g/cm^3-vapor]

The air-filled void fraction ϵ_A, is a function of the soil moisture content θ_M, the total residual level of hydrocarbons in soil $C_{T,s}(=\Sigma C_{i,s}$ [g-i/g-soil]), and the total void fraction ϵ_T:

$$\epsilon_A = \epsilon_T - \frac{\theta_M \rho_b}{\rho_{H_2O}} - \frac{C_{T,s}\rho_b}{\rho_{HC}} \tag{11}$$

where ρ_{H_2O} and ρ_{HC} denote the liquid densities of water and the hydrocarbon mixture. The effective porous media vapor diffusion coefficient $D_{i,v}^{eff}$ is generally calculated by the Millington-Quirk expression:[11,12]

$$D_{i,v}^{eff} = D_i^o \frac{\epsilon_A^{3.33}}{\epsilon_T^2} \tag{12}$$

Again, D_i^o denotes the molecular vapor diffusion coefficient for species i in air.

Without simplifying assumptions, Equation 10 must be solved numerically because $C_{i,v}$ is dependent on composition, not just $C_{i,s}$. In addition ϵ_A and $D_{i,v}^{eff}$ will change with time due to the drying process. Fortunately, for our purposes (emissions from gasoline-contaminated soils) we are typically interested in estimating emissions of benzene, which happens to be one of the more volatile compounds in gasoline (see Table 1). As a result, benzene volatilizes at a much greater rate than the majority of gasoline components. Based on this observation, we can model this situation as the volatilization of a volatile compound from a relatively nonvolatile mixture. Therefore, we assume that:

ϵ_A = constant
$D_{i,v}^{eff}$ = constant
$C_{T,s}$ = constant
$C_{i,v} = (C_{i,s} M_{w,T}/C_{T,s} M_{w,i}) M_{w,i} P_i^v/RT$ (Equation 4)

where $M_{w,T}$ and $M_{w,i}$ denote the molecular weights of the hydrocarbon mixture and component i, respectively. We also adopt the following initial and boundary conditions:

$$\begin{aligned} C_{T,s} &= C_{T,s}^o & t &= 0 \\ (C_{i,s}/C_{T,s}) &= x_i^o & t &= 0 \\ C_{i,s} &= 0 & y &= 0 \\ C_{i,s} &= C_{i,s}^o & y &= \infty \end{aligned} \tag{13}$$

The solution to Equation 10, subject to the assumptions discussed above, yields the following expression for the flux, $\mathfrak{F}_{soil}$, of volatile species i:

$$\mathfrak{F}_{soil} = D_{i,v}^{eff} C_{i,v}^{eq} \frac{1}{\sqrt{\pi\alpha t}} \tag{14}$$

where:

$$\alpha = \frac{D_{i,v}^{eff}}{\epsilon_A + \dfrac{\rho_b RT(C_{T,s}/M_{w,T})}{P_i^v}} \tag{15}$$

As expected, Equation 15 predicts that the emission rate decreases with time. The average flux, $\mathfrak{F}_{soil,avg}$, betweent $t = 0$ and $t = \tau$ is:

$$\mathfrak{F}_{soil,avg} = D_{i,v}^{eff} C_{i,v}^{eq} \frac{2}{\sqrt{\pi\alpha\tau}} \tag{16}$$

Equation 16 is expected to provide good emissions estimates for volatile compounds, until significant depletion of the less volatile gasoline components occur. In §III.1c emission rates measured during laboratory experiments are compared with predictions from Equation 16.

II.2c Emissions from a Soil Venting Operation

Figure 3 depicts a typical soil venting operation. Vapors are removed from the soil at a volumetric flowrate Q_{vent}, and then are treated by a vapor treatment unit, which may consist of a vapor incinerator, catalytic oxidizer, carbon bed, or diffuser stack. Of these four options, the greatest emission rate of any compound i occurs when the vapors are untreated and discharged through a diffuser stack at a rate $\mathfrak{F}_{i,untreated}$ equal to:

$$\mathcal{E}_{i,untreated} = Q_{vent} C_{i,vent} \tag{17}$$

where $C_{i,vent}$ denotes the vapor concentration of species i in the extraction well. The greatest vapor concentration that can be obtained at any time during venting is the equilibrium concentration, $C_{i,v}^{eq}$, defined by Equation 4. When the vapors are treated by a process with a destruction/removal efficiency η, then the emission rate $\mathfrak{F}_{i,treated}$, will be reduced to:

$$\mathcal{E}_{i,treated} = (1 - \eta) Q_{vent} C_{i,vent} \tag{18}$$

Typical gasoline-range hydrocarbon destruction efficiencies for incinerators and catalytic oxidizers are >0.95.

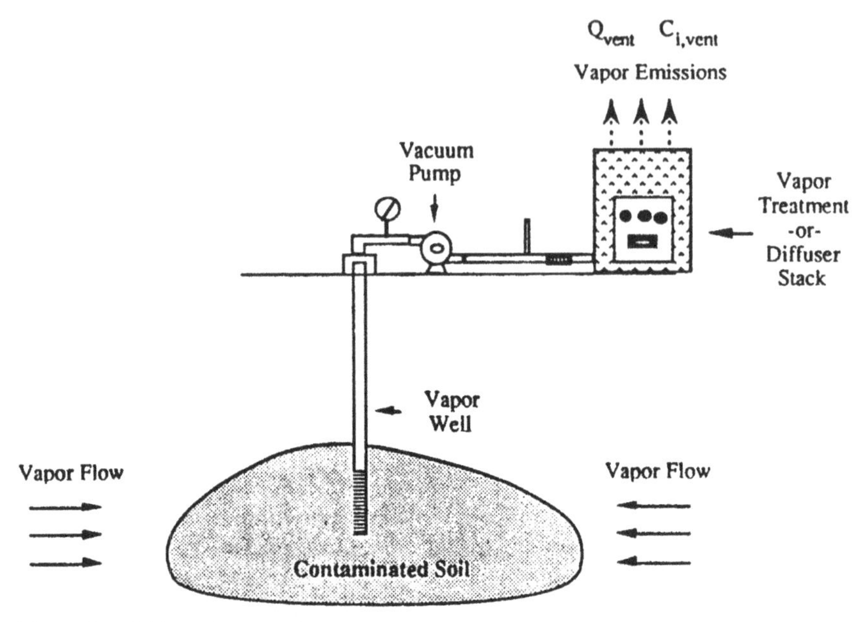

Figure 3. Soil venting operation.

As mentioned above, the most conservative emissions estimates for venting operations are obtained by using Equation 17 for the emission rate and Equation 4 for the vapor concentrations. While this approach might provide good estimates of the emission rate at the start of venting, vapor concentrations decrease with time during venting due to changes in the composition of the residual and due to mass-transfer resistances.[9] To account for this behavior, we calculate a time-averaged emission rate, $\mathcal{E}_{i,avg}$, and average vapor concentration, $C_{i,avg}$, based on the time period for remediation, τ_{vent}, and the mass of compound i removed during this period, m_i:

$$\begin{aligned} \mathcal{E}_{i,avg} &= m_i/\tau_{vent} \\ C_{i,avg} &= m_i/Q_{vent}\tau \end{aligned} \tag{19}$$

Typical venting vapor flowrates are $10 < Q_{vent} < 200$ ft^3/min. For gasoline spill remediations, typical total hydrocarbon vapor concentrations can be as great as 300 mg/L at start-up, but then usually decrease to <50 mg/L.

II.2d Emissions from Groundwater Pump and Treat Operations and Free-Liquid Product Recovery

A pumping operation is pictured in Figure 4, where groundwater and free-liquid product are being removed. If submersible electric pumps are used, then emissions will be minimal. Often, however, injector pumps are used. These work

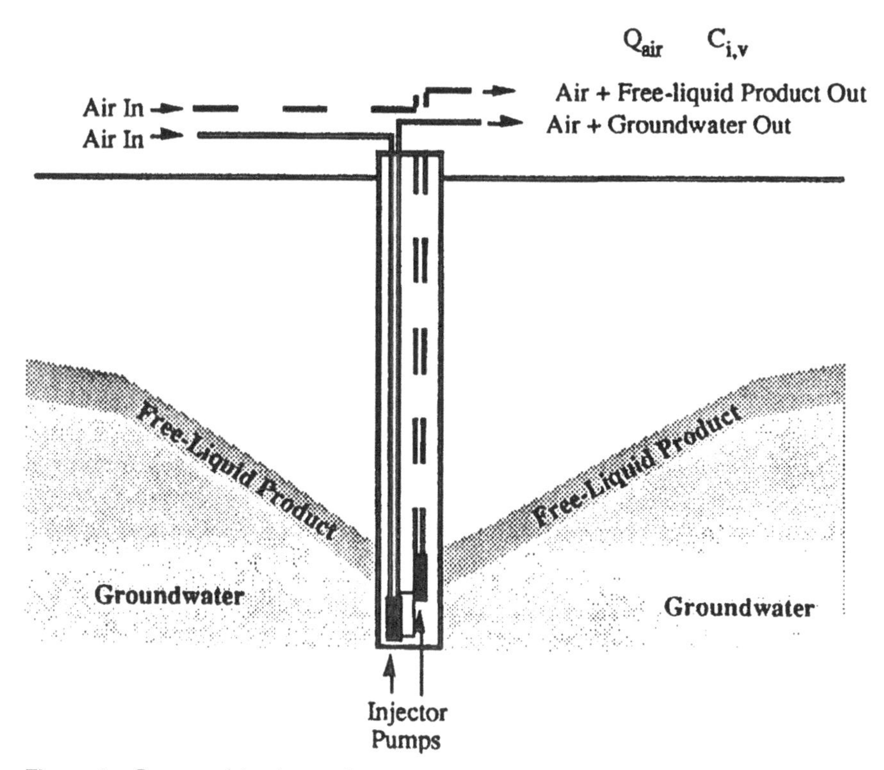

Figure 4. Pump and treat operation.

by displacing fluid with air, and as the fluid is driven to a collection and treatment system aboveground, the air is allowed to escape untreated to the atmosphere. The air flowrate (at 1 atm), Q_{air}, required to obtain a given fluid pumping rate, Q_f, is:

$$Q_{air} = \left[1 + \frac{H[m]}{10.3\ m}\right] Q_f \text{ (pumping groundwater)}$$

$$Q_{air} = \left[1 + \frac{H[m]}{12.9\ m}\right] Q_f \text{ (pumping free-liquid hydrocarbon)} \tag{20}$$

where H is the depth to groundwater expressed in m. The maximum vapor emissions, $\mathcal{E}_{i,pump}$, from the pumps will then be:

$$\mathcal{E}_{i,pump} = Q_{air}\ C_{i,v} \tag{21}$$

where $C_{i,v}$ again denotes the hydrocarbon vapor concentrations in the air. These conservative emission estimates are obtained by assuming that vapor and liquid phases are in equilibrium. In this case $C_{i,v}$ is given by Equation 4 for free-liquid product pumping, and Equation 2 for contaminated groundwater pumping. Soluble levels of hydrocarbons in groundwater are typically an order of magnitude less than saturation, so more realistic levels (such as 1 ppm benzene) might also be used.

II.2e Emissions Associated with Aboveground Water Treatment

Aboveground water treatment systems usually consist of carbon beds, air strippers, or aerobic biotreaters. If a carbon bed is used, emissions will be insignificant because the effluent water must be cleaned to a discharge limit (often $\approx$ 1 ppb) and all contaminant is transferred from the groundwater to the carbon bed. Biotreaters utilize air spargers to maintain high dissolved oxygen levels, and hence they produce emissions during operation. Those emission levels are difficult to estimate, but at worst, they could be only as great as the emissions from an air stripper, which will be modeled in the following analysis. Figure 5 presents a basic air stripping operation. Groundwater with a soluble hydrocarbon level, $C_{i,soluble}$, enters the unit at a flowrate, Q_f, and is then contacted with clean air

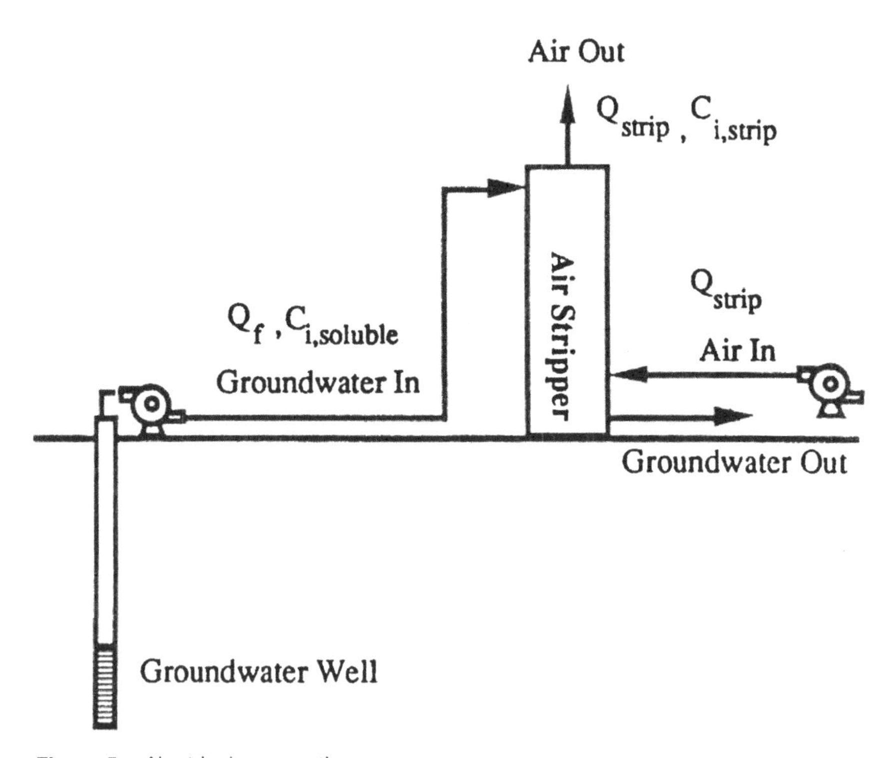

Figure 5. Air stripping operation.

entering at a flowrate, Q_{strip}. Maximum emissions rates occur when the air-stripping efficiency is 100%, and there is no vapor treatment system connected to the air stripper. In this case the emissions rate, $\mathcal{E}_{i,strip}$, and exit vapor concentration, $C_{i,strip}$, are:

$$\begin{aligned} \mathcal{E}_{i,strip} &= Q_f \, C_{i,soluble} \\ C_{i,strip} &= Q_f \, C_{i,soluble}/Q_{strip} \end{aligned} \tag{22}$$

II.2f Emissions from Buried Gasoline-Contaminated Soils

For comparison, we estimate the emissions emanating from gasoline-contaminated soils that are left in place. For the case pictured in Figure 6, where there is a nondiminishing vapor source located at distance H below ground

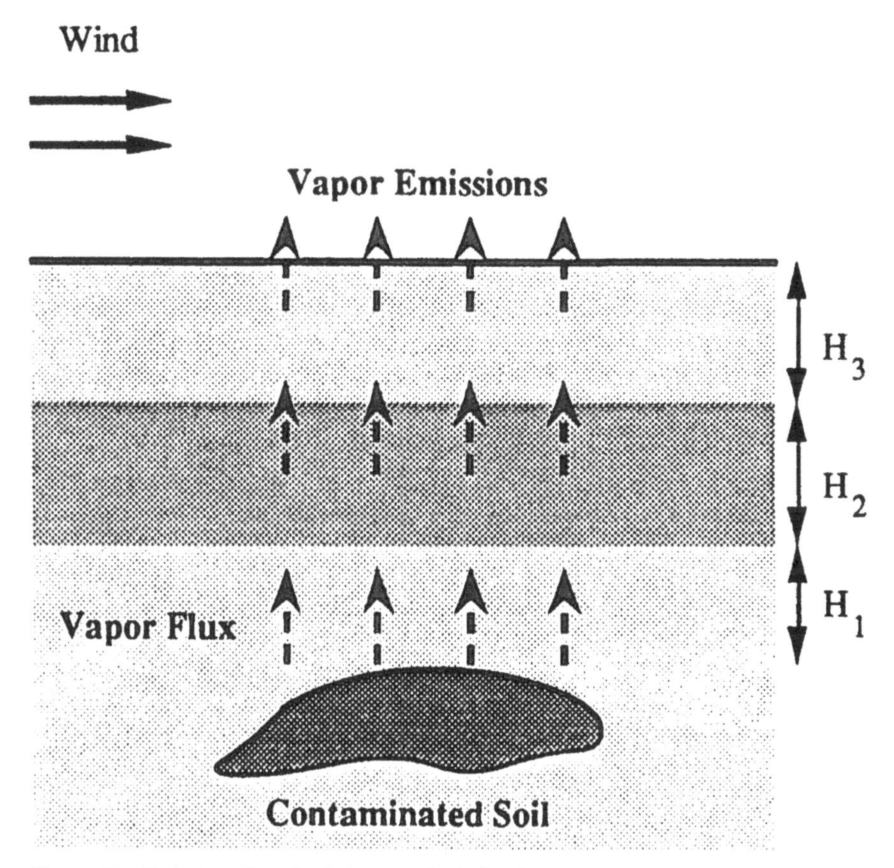

Figure 6. Emissions from buried contaminated soils.

surface and no biodegradation of the vapors (conservative assumptions), the steady-state one-dimensional solution to the governing diffusion equation yields the following expression for the vapor flux, $\mathfrak{F}_{i,undisturbed}$:

$$\mathfrak{F}_{i,undisturbed} = C_{i,v}^{eq} \frac{1}{\sum_n \frac{H_n}{(D_{i,v}^{eff})_n}} \quad (23)$$

where $C_{i,v}^{eq}$ is given by Equation 4, and H_n and $(D_{i,v}^{eff})_n$ are the thickness and effective porous media vapor diffusion coefficient (see Equation 12) for each distinct soil layer above the source. Equation 23 is a generalized form of an equation that is often used for estimating diffusive vapor fluxes from landfills.[8,13] In writing Equation 23 we assume that the vapor concentration is zero at ground level (this yields the maximum emissions rate), and any contribution from diffusion through the soil moisture is insignificant. The latter is a valid assumption for volatile hydrocarbons, but is not necessarily applicable for very nonvolatile compounds.[7]

III. EMISSION RATE CALCULATIONS

To illustrate the use of the models presented in §II.2, benzene vapor emissions estimates are calculated below for a hypothetical service station cleanup consisting of a tank excavation and replacement, soil venting, free-liquid product pumping, and a groundwater pump and treat operation. It will be assumed that gasoline containing 1 mole % (≈1% by weight) benzene is the source of soil and groundwater contamination. "Fresh" gasolines usually contain between 0.5% to 3% by weight benzene.. Table 2 summarizes the results presented in this section; these are inputs to the air dispersion models discussed in §IV.1.

Table 2. Summary of Benzene Vapor Emissions Predictions[a]

Source	Flux (g/cm²-d)	Emission Rate (g/d)	Vapor Conc. at Source (g/cm³)
Empty Tank Pit[b]			
—maximum steady-state	7.7×10^{-5}	58	NC
—three-day average	2.9×10^{-5}	22	NC
Soil Pile[c]			
—1 hr average for 8 hrs/d	3.0×10^{-3}	370 (24 h average)	NC
Soil Venting[d]			
—untreated vapors, constant source	NC	6532	3.2×10^{-6}
—treated vapors, constant source ($\eta = 0.95$)	NC	327	1.6×10^{-7}
—untreated vapors, 500 gal spill	NC	84	4.2×10^{-8}
—treated vapors, 500 gal spill ($\eta = 0.95$)	NC	4.7	2.1×10^{-9}

continued

Table 2. ***Continued***

Source	Flux (g/cm²-d)	Emission Rate (g/d)	Vapor Conc. at Source (g/cm³)
Pumping[e]			
—gasoline-saturated groundwater	NC	560	3.2×10^{-6}
—groundwater with 1 ppm benzene	NC	32	1.8×10^{-7}
—free-liquid gasoline	NC	0.26	3.2×10^{-6}
Groundwater Treatment[f]			
—carbon beds with 1 ppb discharge	NC	0.1	NC
—air-stripper with gas.-sat. water	NC	1965	2.4×10^{-7}
—air-stripper with 1 ppm benzene water	NC	109	1.3×10^{-8}
Leaving Soils In Place[g]	5.9×10^{-6}	2.19	NC

NC = not calculated, or not applicable.
[a]These values correspond to specific conditions (see below).
[b]T = 20°C, 1 mole % benzene, 3 m deep excavation, 6.1 m × 12.2 m area.
[c]T = 20°C, 1 mole % benzene, 10000 mg/kg TPH initial residual gasoline, 6.1 m × 6.1 m area.
[d]T = 20°C, 1 mole % benzene, 1400 1/min (50 SCFM) vapor extraction rate.
[e]T = 20°C, 1 mole % benzene, 18 mg/l saturated conc. in groundwater, 76 l/min (20 gpm) groundwater pumping rate, 0.038 l/min (0.01 gpm) free-liquid product pumping rate.
[f]T = 20°C, 1 mole % benzene, 18 mg/l saturated conc. in groundwater, 76 l/min (20 gpm) groundwater pumping rate, 5660 l/min (200 SCFM) air flowrate.
[g]T = 20°C, 1 mole % benzene, 5 m depth to contamination, no surface seal, 6.1 m × 6.1 m area.

II.1a Emissions Associated with Excavated Pits

Emissions from the empty pit are estimated by assuming vapors diffuse upward from the pit bottom through a stagnant layer of air and are swept away by the wind at ground level. Equations 4 and 9 were used to generate the benzene emissions estimates that appear in Figure 7 for a range of excavation depths and benzene mole fractions. For the air dispersion analyses appearing in §IV.2 the emissions rate corresponds to the parameter values:

x_i	= mole fraction of benzene in the hydrocarbon spill	= 0.01
T	= absolute temperature	= 293 K (20°C)
P_i^o	= vapor pressure of pure benzene	= 0.10 atm
$M_{w,i}$	= molecular weight of benzene	= 78 g/mole
R	= Universal Gas Constant	= 82.1 cm³-atm/mole-°K
D_i^o	= vapor phase diffusion coefficient in air [cm²/d]	= 7270 cm²/d
H	= depth of excavation [cm]	= 305 cm (10 ft)

The result is:

$$\mathcal{F}_{pit,max} = 7.7 \times 10^{-5} \text{ g/cm}^2\text{-d}$$

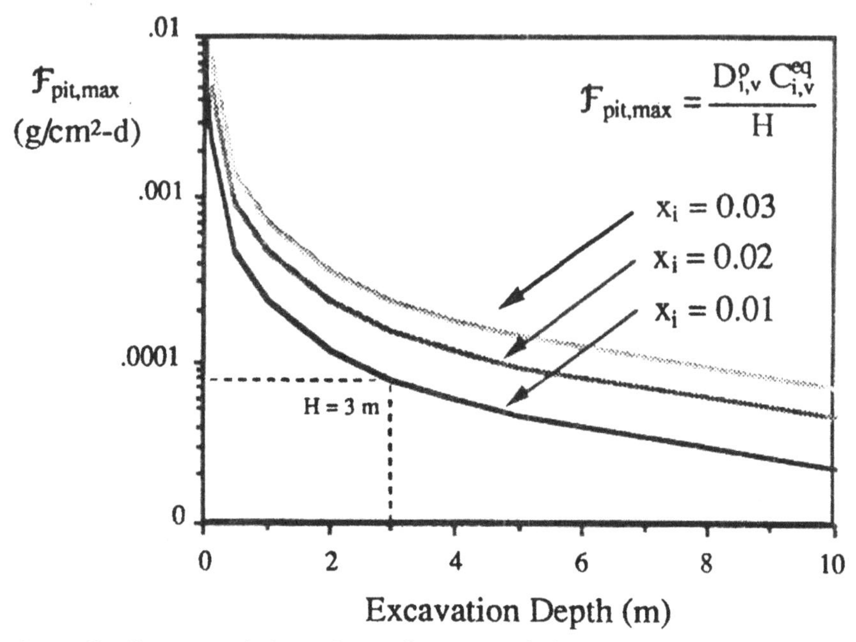

Figure 7. Benzene emission estimates for excavated pit.

For a 6.1 m by 12.2 m (20 ft by 40 ft) pit area, this corresponds to a maximum benzene emission rate $\mathcal{E}_{pit,max}$:

$$\mathcal{E}_{pit,max} = 58 \text{ g/d}$$

A less conservative, but more realistic estimate can be obtained from Equation 8 which models the transient period before a steady-state flux is established. The three-day average flux $<\mathcal{F}_{pit}>_{3\text{-day}}$, and average emission rate $<\mathcal{E}_{pit}>_{3\text{-day}}$ are:

$$<\mathcal{F}_{pit}>_{3\text{-day}} = 2.9 \times 10^{-5} \text{ g/cm}^2\text{-d} \qquad <\mathcal{E}_{pit}>_{3\text{-day}} = 22 \text{ g/d}$$

III.1b Emissions from Exposed Piles of Contaminated Soil

Figure 8 presents emission rates predicted by Equations 16, 15, 11, and 5 for the following parameter values:

ϵ_T	= total void fraction in soil	= 0.38
ρ_b	= soil bulk density	= 1.6 g/cm^3
θ_M	= soil moisture content	= 0.05 g-water/g-soil
ρ_{HC}	= hydrocarbon liquid bulk density	= 0.80 g/cm^3
$M_{w,T}$	= equivalent molecular weight of gasoline	= 100 g/mole

ρ_{H_2O}	= bulk density of water	= 1.00 g/cm^3
$C_{T,s}$	= total hydrocarbon concentration in soil	= 0.01 g-gasoline/g-soil
x_i	= benzene mole fraction in gasoline	= 0.01

and the values given above in §III.1a for P_i°, $M_{w,i}$, R, T, and D_i°. The one-hour average flux $<\mathfrak{F}_{soil}>_{1\text{-hour}}$ for these values is:

$$<\mathfrak{F}_{soil}>_{1\text{-hour}} = 3.0 \times 10^{-3} \text{ g/cm}^2\text{-d}$$

For a pile with a 37 m^2 (a 20 ft × 20 ft pile) surface area, that is only uncovered for the duration of an 8 h shift each day, the daily average emission rate $<\mathcal{E}_{soil}>$ is:

$$<\mathcal{E}_{soil}> = 370 \text{ g/d}$$

This is calculated by multiplying the one-hour average by eight.

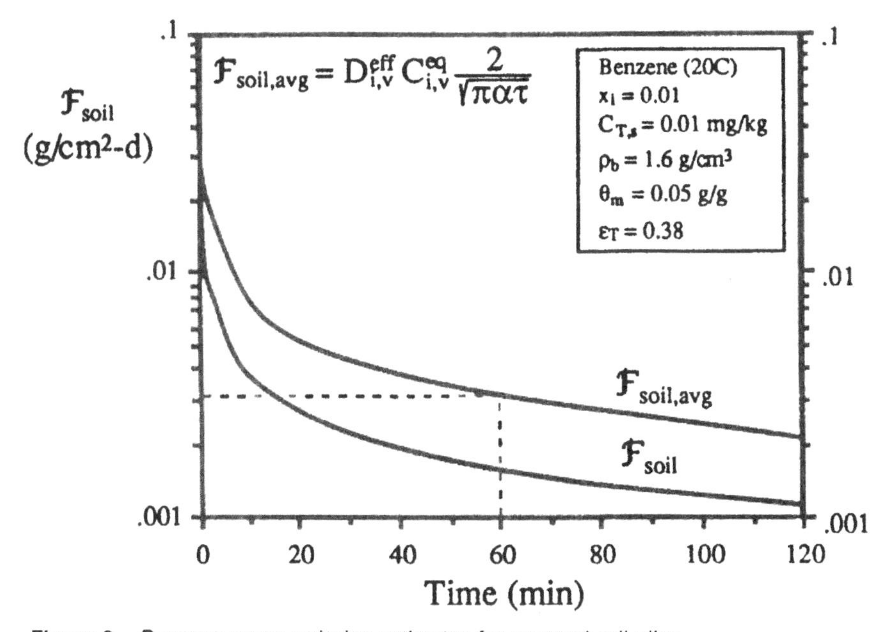

Figure 8. Benzene vapor emission estimates for exposed soil piles.

III.1c Soil Pile Emissions Experiments

Soil pile emissions experiments were conducted with the apparatus pictured in Figure 9. To a coarse (1.2 mm diameter) sand was added enough water and

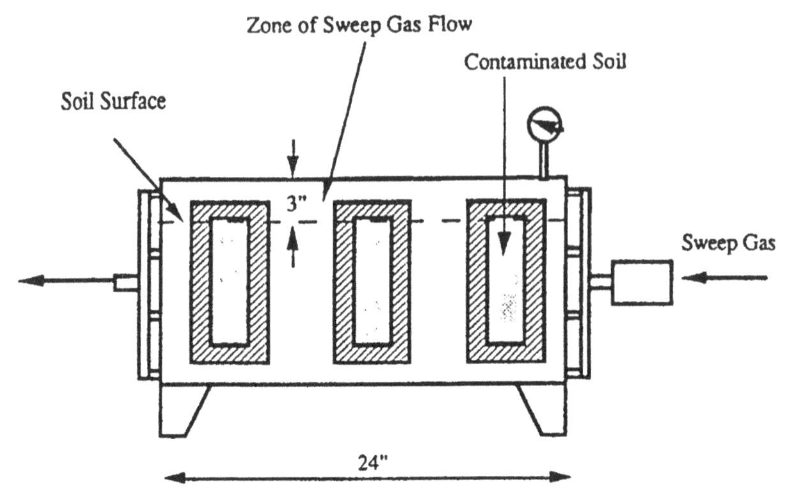

Figure 9. Gasoline emissions experiment apparatus.

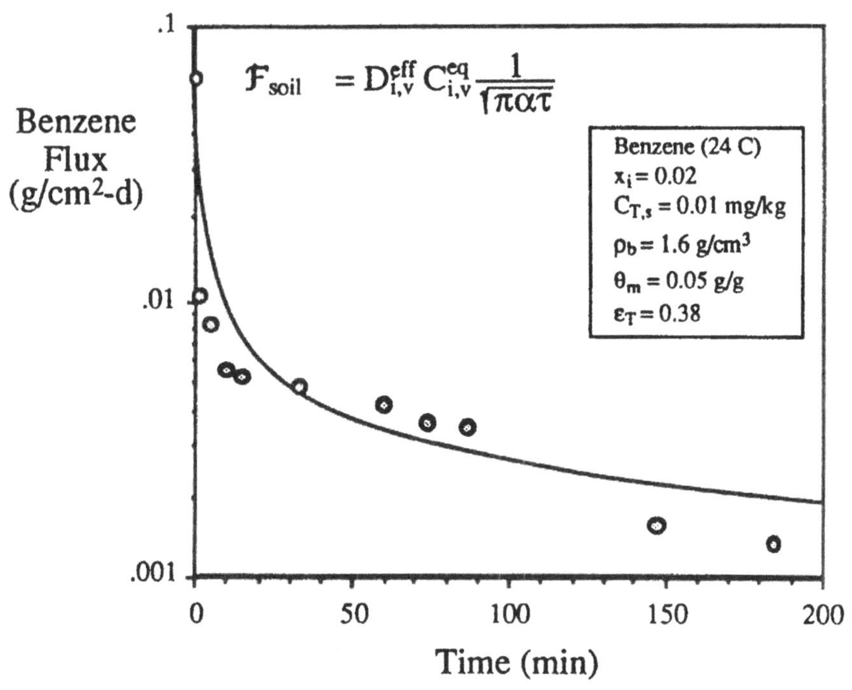

Figure 10. Comparison of measured and predicted benzene fluxes from gasoline contaminated soil.

gasoline to create a 5% moisture content, 10,000 mg-gasoline/kg-soil gasoline-contaminated soil. This soil was placed in the tank, a N_2 sweep gas flow started, and effluent benzene vapor concentrations were monitored with time. The vapor concentration data, sweep gas flowrate, and tank dimensions were used to calculate the vapor flux rates presented in Figure 10. The sweep gas velocity was 3.4 cm/s (0.08 mph), which was determined experimentally to be great enough to ensure that the flux was not dependent on the sweep gas flowrate. An approximate analysis of the gasoline is contained in Table 1, where the initial mole fraction of benzene is 0.0205. The sweep gas temperature was 24°C.

The measured benzene vapor emissions from this experiment are compared with the predictions from Equation 14 in Figure 10. For these conditions, the model reasonably predicts the observed emissions, at least to the accuracy that is desired for emissions estimation purposes.

III.1d Emissions from Soil Venting Operation

Figure 11 presents emission rates for a range of vapor flowrates (Q_{vent}) and vapor concentrations ($C_{i,vent}$). In §IV.2 we use the emission rates corresponding to Q_{vent} = 1400 L/min (50 ft³/min), and the maximum benzene vapor concentration for soils contaminated with gasoline containing 1 mole percent benzene (3.24 mg/L). The worst case is the situation in which vapors are discharged untreated, and there is a nondiminishing benzene source. Equations 17 and 4 predict that the emission rate $\mathcal{E}_{untreated}$, and vapor concentration in the exit gas $C_{v,untreated}$, will be:

$$\mathcal{E}_{untreated} = 6532 \text{ g/d}$$

$$C_{v,untreated} = 3.2 \times 10^{-6} \text{ g/cm}^3$$

If the vapors are treated by a vapor treatment unit with a destruction efficiency $\eta = 0.95$, then the benzene emission rate $\mathcal{E}_{treated}$, and vapor concentration in the exit gas $C_{v,treated}$ are:

$$\mathcal{E}_{treated} = 327 \text{ g/d}$$

$$C_{v,treated} = 1.6 \times 10^{-7} \text{ g/cm}^3$$

It is typically observed that vacuum-well hydrocarbon vapor concentrations decrease during venting, so the values presented above should be regarded as worst-case estimates. To calculate the worst-case average vapor emissions during a soil venting project, we use Equation 19. Suppose that venting removed 100% of the benzene from a 500 gal gasoline spill, which contained 1% by weight benzene, over a six-month period. Then the average emission rate $\mathcal{E}_{avg}$, for the venting operation is calculated with Equation 19 to be:

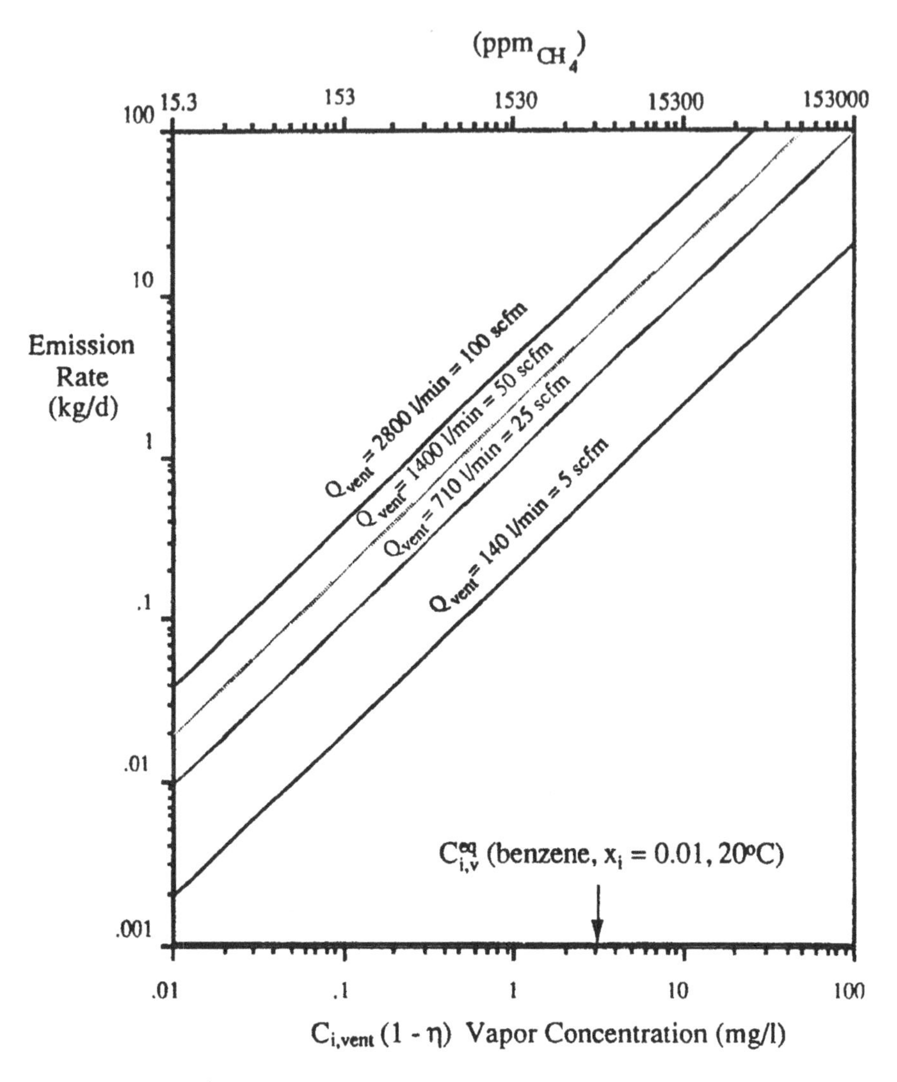

Figure 11. Soil venting emissions estimates.

$$\mathcal{E}_{avg} = 84 \text{ g/d (if untreated vapors are emitted)}$$

$$\mathcal{E}_{avg} = 4.7 \text{ g/d (for a vapor treatment efficiency } \eta = 0.95)$$

Note that these numbers are several orders of magnitude less than the conservative estimates predicted by Equation 17.

III.1e Emissions from Groundwater Pump & Treat Operations and Free-Liquid Product Recovery

Equations 3, 4, and 21 predict conservative emission rate estimates for groundwater and free-liquid product pumping systems that use injector (air displacement) pumps. Calculations were performed for three different cases: (a) pumping gasoline saturated groundwater with a benzene concentration of 18 ppm (equilibrium value for groundwater in contact with gasoline containing 0.01 mole fraction benzene), (b) pumping gasoline-contaminated groundwater with a more realistic benzene level of 1 ppm, and (c) pumping free-liquid product gasoline containing 1 mole percent benzene. At service stations, groundwater pumping rates are generally on the order of 76 L/min (20 gal/min), and average free-liquid product pumping rates might be 0.01 gal/min (860 gal over a two-month period). The results for these pumping rates, and a 6 m (20 ft) depth to the water table are:

(*a*) *gasoline-saturated groundwater,* $Q_f = 20$ *gpm*

$$\mathcal{E}_{\text{pump}} = 560 \text{ g/d} \qquad C_v = 3.2 \times 10^{-6} \text{ g/cm}^3$$

(*b*) *1 ppm benzene in groundwater,* $Q_f = 20$ *gpm*

$$\mathcal{E}_{\text{pump}} = 32 \text{ g/d} \qquad C_v = 1.8 \times 10^{-7} \text{ g/cm}^3$$

(*c*) *free-liquid gasoline pumping,* $Q_f = 0.01$ *gpm*

$$\mathcal{E}_{\text{pump}} = 0.26 \text{ g/d} \qquad C_v = 3.2 \times 10^{-6} \text{ g/cm}^3$$

III.1f Emissions Due to Aboveground Water Treatment

Aboveground water treatment systems may be composed of carbon beds, biotreaters, or air-strippers. Of these options, a carbon-bed system will produce the lowest vapor emissions, because discharge water must meet a minimum water quality standard, which is often ≈ 1 ppb benzene. Even if all the benzene volatilizes from the water as it leaves the carbon beds, the emission rate for a 76 L/min (20 gal/min) operation with effluent water containing 1 ppb benzene calculated from Equation 22 is:

$$\mathcal{E}_{\text{carbon bed}} = 0.1 \text{ g/d} \qquad C_{v,\text{max}} = 1.8 \times 10^{-10} \text{ g/cm}^3$$

The maximum benzene vapor emission rate for the aboveground treatment systems comes from an air-stripper (with no subsequent vapor treatment). Equation 22 predicts the emission rates for a 100% efficient air-stripper. Emissions predictions for a wide range of flowrates and concentrations appear in Figure 12a. The relationship between air flowrate, water flowrate, and the effluent vapor

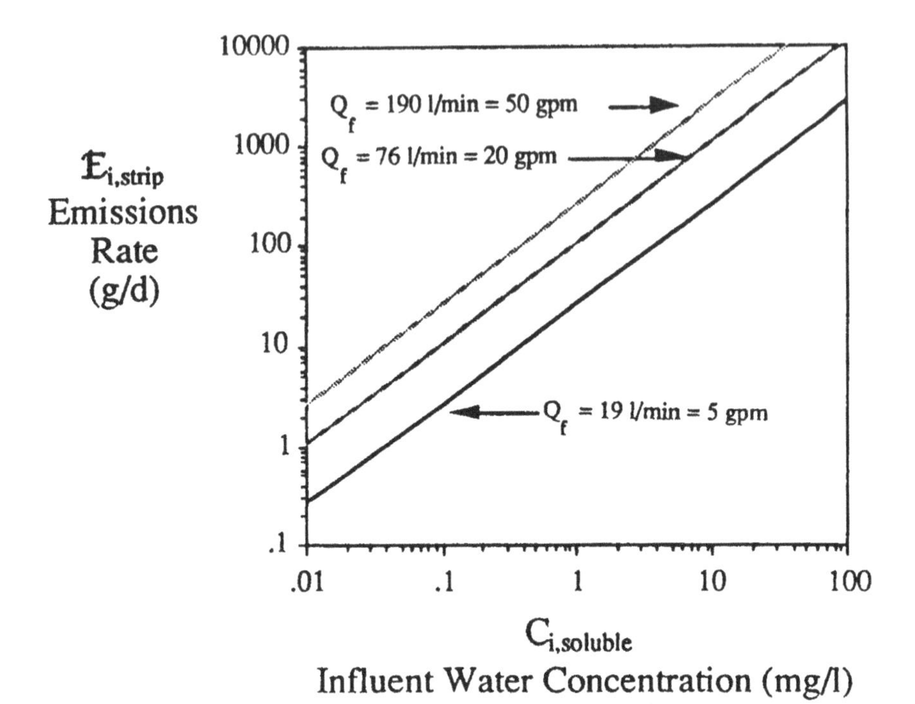

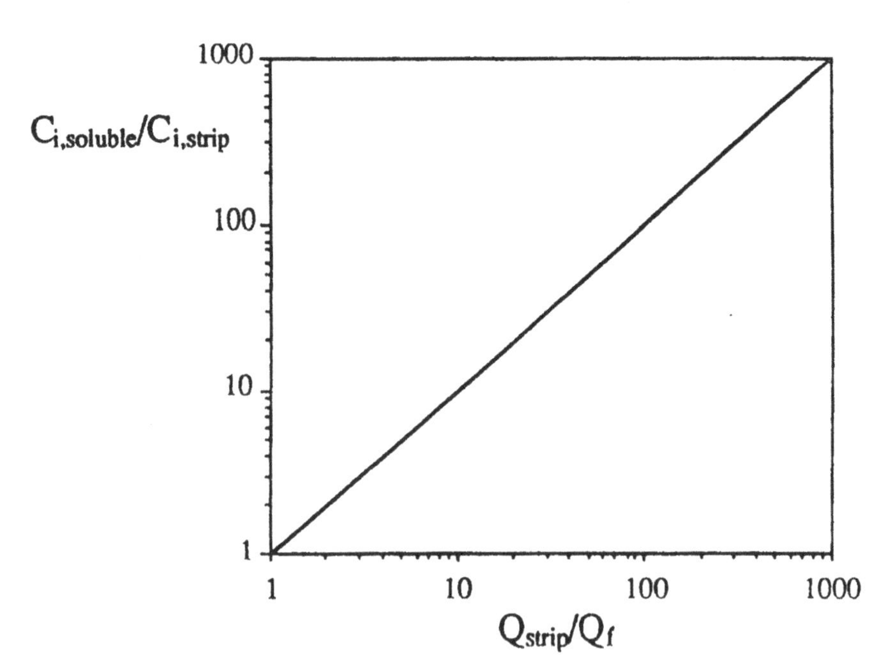

Figure 12. Air stripper emissions.

concentrations are shown in Figure 12b. The emission rates from a 76 L/min (20 gal/min) treatment of gasoline-saturated groundwater (18 ppm benzene), and groundwater containing a more realistic 1 ppm benzene, are predicted by Equation 22 to be:

$$\mathcal{E}_{\text{sat gw}} = 1965 \text{ g/d} \qquad C_v = 2.41 \times 10^{-7} \text{ g/cm}^3$$

$$\mathcal{E}_{\text{1 ppm gw}} = 109 \text{ g/d} \qquad C_v = 1.33 \times 10^{-8} \text{ g/cm}^3$$

It has been assumed that the air flowrate into the stripper is 5660 L/min (200 ft^3/min, $Q_{strip}/Q_f = 75$).

III.1g Emissions from Buried Gasoline-Contaminated Soils

Figure 13 presents benzene emission estimates for gasoline-contaminated soils left in-place as predicted by Equation 23 for the following conditions:

$\epsilon_T = 0.38$
$\theta_M = 0.05$
$D_{i,v}^o = 7270 \text{ cm}^2/\text{d}$
$T = 20°C$

These conservative estimates are for the situation in which relatively homogeneous soils lie above the gasoline-contaminated soil, and there is no low permeability surface cover (i.e., paving, clay liner).

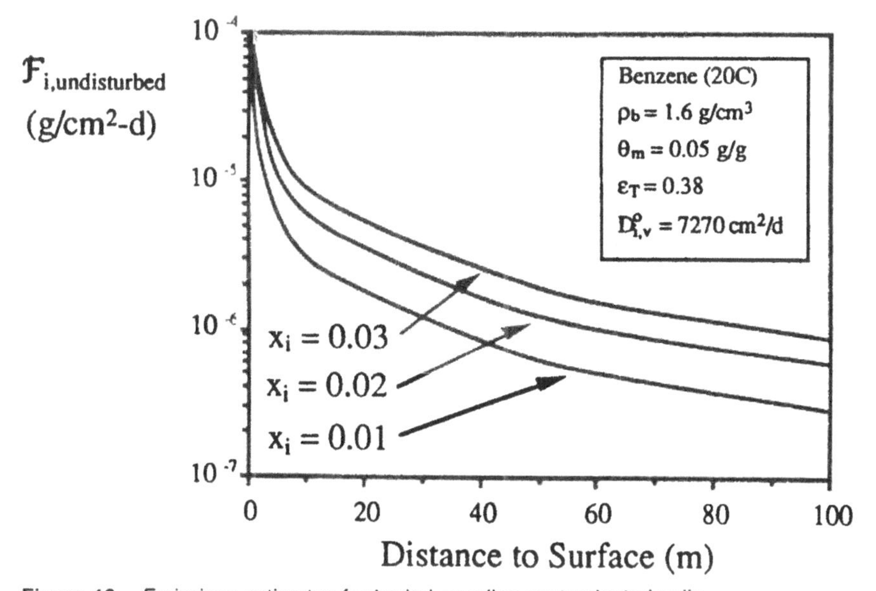

Figure 13. Emissions estimates for buried gasoline-contaminated soils.

IV. AIR DISPERSION MODELING

IV.1 Model Description

The EPA-recommended Industrial Source Complex (ISC) Model, which can be used to predict ambient concentrations from both stack and area sources for receptors located at least 100 m from a source, was used to perform the air dispersion modeling. For this analysis, a polar-coordinate grid was generated with receptors located on concentric rings at radii of 100 m, 300 m, 500 m, 700 m, and 1000 m from each source. Thirty-six receptors, at 10 degree intervals, were located on each ring, making a total of 180 receptors. The ISC model employs standard gaussian dispersion algorithms to calculate the hourly concentration at each receptor. If an entire year of meteorological data is input to the model, then a full year of hourly concentrations can be calculated for each receptor. These hourly values can then be averaged to give an annual average for each receptor, or they can be sorted and ranked to give maximum 1-, 8-, and 24-h averages. For this modeling, meteorological data from Long Beach, California for the year 1964 was used. Other required meteorological parameters include wind speed and direction, ambient temperature, mixing height, and stability class. For area sources, the size and height of the release as well as the emission flux are required input parameters. For stack releases, the stack height and diameter and exit gas velocity and temperature are needed. Modeling options selected for this

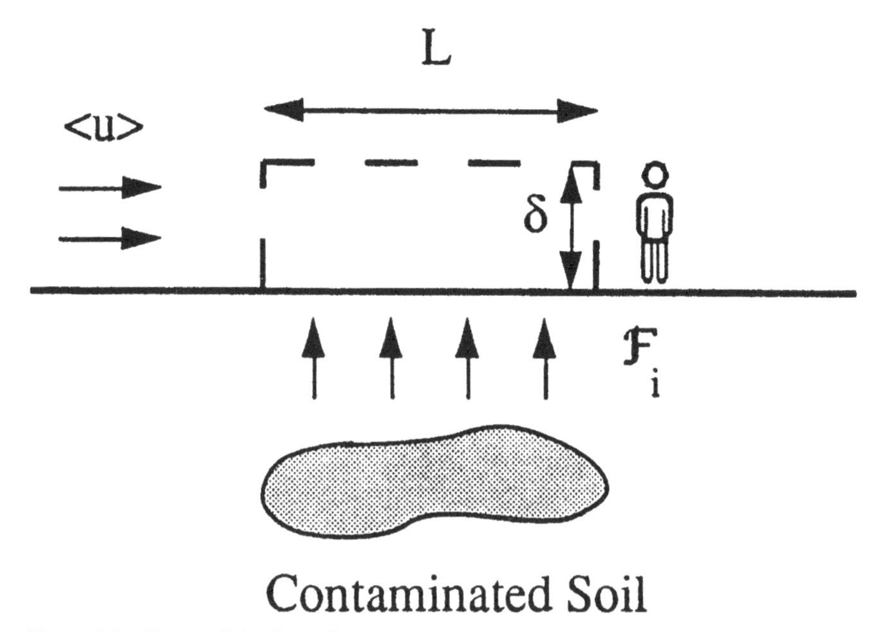

Figure 14. Box model schematic.

study included gradual plume rise, buoyancy-induced dispersion, and stack downwash. The source was assumed to be located in an urban environment.

The "box model" illustrated in Figure 14 was used to predict conservative ambient concentration estimates for receptors located on the downwind edge of the area sources considered in this chapter. It was assumed that vapors within the box were "well-mixed," and the height of the box, δ, was set to be equal to a human breathing zone height − 2 m. Given the wind speed $<u>$, breathing zone height δ, length of emission source parallel to the wind direction L, and emissions flux $\mathfrak{F}_i$, then the ambient concentration $C_{i,amb}$, is:

$$C_{i,amb} = \frac{\mathfrak{F}_i L}{\delta <u>} \tag{24}$$

IV.2 ISC Model Predictions

ISC model predictions are summarized in Table 3. Each scenario modeled is described briefly below.

Table 3. Air Dispersion Modeling Results

		Benzene Vapor Concentrations (μg/m³) Radial Distance					
Scenario	Duration	100 m	300 m	500 m	700 m	1000 m	Box Model
1. excavation[a]		0.39	0.07	0.04	0.03	0.02	24
(70 y average)	3 d	<0.0001	<0.0001	<0.0001	<0.0001	<0.0001	<0.01
2. excavation[a]		0.15	0.03	0.02	0.01	0.01	9.2
(70 y average)	3 d	<0.0001	<0.0001	<0.0001	<0.0001	<0.0001	<0.01
3. soil pile[b]		7.5	1.3	0.78	0.57	0.41	480
(70 y average)	3×8 h	<0.001	<0.0001	<0.0001	<0.0001	<0.0001	<0.01
4. venting, untreated vapors[c]		6.5	0.9	0.36	0.2	0.11	NA
(70 y average)	180 d	0.05	<0.01	<0.01	<0.01	<0.001	NA
5. venting, untreated vapors[c]		0.08	0.01	<0.01	<0.01	<0.01	NA
(70 y average)	180 d	<0.0001	<0.0001	<0.0001	<0.0001	<0.0001	NA
6. venting, treated vapors[c]		0.12	0.04	0.02	0.01	<0.01	NA
(70 y average)	180 d	<0.001	<0.001	<0.001	<0.0001	<0.0001	NA
7. venting, treated vapors[c]		<0.01	<0.001	<0.001	<0.001	<0.001	NA
(70 y average)	180 d	<0.0001	<0.00001	<0.00001	<0.00001	<0.00001	NA
8. free-liquid pumping[c]		<0.001	<0.001	<0.001	<0.001	<0.001	NA
(70 y average)	60 d	<0.00001	<0.00001	<0.00001	<0.00001	<0.00001	NA
9. gasoline-saturated GW pump[c]		0.64	0.08	0.03	0.02	<0.01	NA
(70 y average)	5 y	0.07	<0.01	<0.01	<0.01	<0.001	NA
10. 1 ppm benzene GW pump[c]		0.04	<0.01	<0.01	<0.001	<0.001	NA
(70 y average)	5 y	<0.01	<0.001	<0.001	<0.0001	<0.0001	NA
11. air-stripper gas-sat. GW[c]		1.2	0.25	0.10	0.06	0.03	NA
(70 y average)	5 y	0.09	0.02	<0.01	<0.01	<0.01	NA
12. air-stripper 1 ppm benzene GW[c]		0.07	0.01	<0.01	<0.01	<0.01	NA
(70 y average)	5 y	<0.01	<0.001	<0.001	<0.001	<0.001	NA
13. buried soils left in-place		0.044	<0.01	<0.01	<0.01	<0.01	0.94
(70 y average)	70 y	0.044	<0.01	<0.01	<0.01	<0.01	0.94

[a]peak 24 h concentration.
[b]peak 8 h concentration.
[c]annual average concentration.
[d]nondiminishing source located 5 m below uncovered ground surface.

IV2.a Emissions Associated with Excavated Pits

Scenario 1: The maximum estimated emission rate from the empty pit was estimated to be 58 g/d from a 6.1 m by 12.2 m pit area at a temperature of 20°C (see §III.1a). Since these emissions are hypothesized to last for three days, the peak 24-h concentration was assumed to approximate the highest exposure that might result during the three days. A ground level emission height was assumed.

Scenario 2: A more realistic average emission rate estimate from the pit was calculated to be 22 g/d (see §III.1a). As for Scenario 1, the pit was assumed to be 6.1 m by 12.2 m, the temperature to be 20°C, and the emission height to be ground level. As in Scenario 1, the 24-h peak exposure was determined.

IV.2b Emissions from Exposed Piles of Gasoline-Contaminated Soils

Scenario 3: The estimated one-hour average flux rate at ground level from the 6.1 m by 6.1 m excavated soil pile was calculated to be 3.0×10^{-3} g/cm^2-d in §III.1b, for a temperature of 20°C. It was assumed that the pile would remain exposed for an 8 h shift, and then be covered at night. Therefore, an 8-h peak concentration was calculated to represent the exposure from the pile.

IV.2c Emissions from Soil Venting Operations

In a typical soil venting operation, vapors are removed from the soil by a vacuum pump and then treated by a vapor treatment unit before being released to the atmosphere. In areas with strict emissions regulations, vapor incinerators and catalytic oxidizers are often used. In some areas, diffuser stacks are still allowed. In §III.1d four emission rates were estimated for the venting operations: (a) a worst-case situation in which it was assumed that there was a nondiminishing gasoline source and the vapors were released without any treatment; (b) a more realistic case in which it was assumed that 100% of the benzene from a 1900 L (500 gal) gasoline spill was removed over a six-month period, again with no vapor treatment; (c) emissions from case (a), except that they are treated by a unit with a vapor destruction efficiency $\eta = 0.95$; and (d) same as case (b), except the vapors are treated by a unit with a vapor destruction efficiency $\eta = 0.95$. It should be noted that vapor incinerator and catalytic oxidizer units are quite capable of achieving destruction efficiencies $<95\%$, so the 95% value used in these calculations should be regarded as a worst-case (very conservative) estimate.

All of these scenarios were modeled assuming that the vapor was released from a 4.6 m (15 ft) stack with a 0.6 m (2 ft) diameter. The total vapor flowrate was assumed to be 1400 L/min (50 SCFM) for all cases.

Scenario 4: A worst-case untreated emission rate of 6532 g/d, and 20°C release temperature were assumed.

Scenario 5: The more realistic six-month average untreated emission rate of 84 g/d, and 20°C release temperature.

Scenario 6: A worst-case treated (95% treatment efficiency) emission rate of 327 g/d, and a release temperature of 370°C.

Scenario 7: A more realistic, treated (95% treatment efficiency) benzene emission rate of 4.7 g/d, and a release temperature of 370°C.

IV.2d Emissions from Groundwater Pump & Treat Operations and Free-Liquid Product Recovery

Groundwater and free-liquid product recovery systems utilize various pumping hardware, including submergible air-injected pumps. These pumps work by displacing fluid with air, and while the fluid is collected aboveground, the air is generally emitted untreated. Some volatilization may occur during the period that air and fluid are in contact, so the released air will be a source of emissions. The worst case (maximum emission rate) occurs when the air and fluid phases equilibrate during pumping. Scenario 8 represents the case in which the discharged air is in equilibrium with free-liquid product containing 1 mole % benzene (at 20°C), whereas Scenarios 9 and 10 describe the cases of pumping gasoline-saturated groundwater (18 ppm benzene for 1 mole % benzene in gasoline) and groundwater containing a more realistic 1 ppm benzene, respectively. All scenarios assume that the vapor is released from a 0.64-cm diameter stack at a height of 5 cm above ground. Since pump and treat operations are expected to last for a few years, the annual average was used to estimate exposure levels. In summary, the following scenarios were analyzed:

Scenario 8: An average benzene emission rate of 0.26 g/d (see §III.1e) for the pumping of free-liquid gasoline at a rate of 0.038 L/min (860 gal over a two-month period) and at 20°C temperature.

Scenario 9: Pumping gasoline-saturated groundwater containing 18 ppm benzene at a rate of 76 L/min (20 gpm), which results in a maximum emission rate of 560 g/d at 20°C (see §III.1e).

Scenario 10: Pumping groundwater containing 1 ppm benzene at a rate of 76 L/min (20 gpm), which results in a maximum emission rate of 32 g/d at 20°C (see §III.1e).

IV.2e Emissions Due to Aboveground Water Treatment

The maximum vapor emissions for aboveground water treatment systems occur in air-stripping systems. In §III.1f benzene emissions for air-strippers were calculated and presented in Figure 12. For the air dispersion modeling, the following situations were selected:

Scenario 11: A benzene emission rate of 1965 g/d for an air-stripper treating gasoline-saturated groundwater (18 ppm benzene for 1 mole % benzene in gasoline) at a throughput rate of 76 L/min (20 gpm). The air flowrate through the stripper is 5600 L/min (200 SCFM).

Scenario 12: A benzene emission rate of 109 g/d for an air-stripper treating groundwater containing 1 ppm benzene at a throughput rate of 76 L/min (20 gpm). The air flowrate through the stripper is 5660 L/min (200 SCFM).

IV.2f Emissions from Buried Gasoline-Contaminated Soils

In §III.1g benzene emission estimates for gasoline-contaminated soils left in-place were calculated and presented in Figure 13. For the air dispersion modeling, the following situation was selected:

Scenario 13: The case of a nondiminishing gasoline source containing 1 mole % benzene at 20°C located 5 m below ground surface. The soil is homogeneous and there is no low permeability surface cover (i.e., paving, clay liner).

IV.3 "Box Model" Predictions

As mentioned above, the box model was used to calculate ambient vapor concentrations close to the source. Maximum vapor concentrations will be encountered on the downwind edge of an area source, and for the purpose of generating very conservative exposures, we will assume that a receptor is located at this downwind edge. For the sample calculations a 2.2 m/s (5 mph) wind speed and a 2 m box height were used. The length, L, of the source in the windward direction was taken to be the greatest "edge" length of the source. For example, L = 12.2 m for the excavated pit, and L = 6.1 m for the contaminated soil pile. Values were not computed for the stack sources (air-stripper, soil venting), because we feel that this model is not appropriate for such sources. The "box model" predictions are listed in Table 3 for comparison with the results from the ISC model. As expected, the box model predictions are always greater than the ISC model predictions listed in Table 3.

V. EXPOSURE CALCULATIONS AND CONCLUSIONS

The vapor concentrations listed in Table 3 can be combined with a breathing rate, body weight, duration of exposure, and potency factor to predict the additional human health risk due to each of the sources described in this chapter. Given the current debate over the proper choice of poetency factors, breathing rates, and demographics, however, calculated risks are not presented in this chapter. For comparison though, the IRIS[14] potency factor for benzene (0.029 kg-d/mg), which is based on a 70 y exposure for a 70 kg person, yields a $>1.0 \times 10^{-6}$ risk for average lifetime benzene vapor concentrations $>0.12\ \mu g/m^3$, for a 20 m^3/d breathing rate. In order to compare this value with the predicted concentrations presented in Table 3, one must consider that the exposure duration during any of these processes is much less than 70 y. For example, the maximum duration of exposure from soil pile emissions is only 0.066 y (3 × 8 h). In Table

3, therefore, equivalent 70-y average vapor concentrations are also presented for each scenario. Based on these values, the only 70-y average vapor concentration $>0.12\ \mu g/m^3$ is due to the emissions from contaminated soils left in place ($0.94\ \mu g/m^3$). It is important to note, however, that these predictions correspond to the most conservative assumptions (uncovered, permeable soils, nondiminishing source), and a specific depth-to-contamination. It should be clear that for other soil types, surface covers, and contamination depths, the predicted aboveground vapor concentrations will be low enough so as to not pose a significant health risk. For this reason, future cleanup criteria should be influenced by site-specific environmental risk assessments, which evaluate all significant migration pathways and the potential impact on people and the environment.

In summary, models have been presented for estimating the hydrocarbon vapor emissions for typical service station remediation operations. A range of example scenarios was presented in the text, emission rates were calculated, and air-dispersion models were used to calculate the corresponding vapor concentrations in nearby communities. It is important to remember that these models incorporate conservative assumptions and are intended for use as screening tools to determine whether the emission rates or local vapor concentrations are potentially "large" or "small" as compared to values specified in regulatory requirements.

REFERENCES

1. Chiou, G. T., and T. D. Shoup. "Soil Sorption of Organic Vapors and Effects of Humidity on Sorptive Mechanism and Capacity," *Environ. Sci. Technol.*, 19, 1196–1200, 1985.
2. Spencer, W. F. "Distribution of Pesticides Between Soil, Water, and Air, Pesticides in the Soil: Ecology, Degradation, and Movement," Michigan State University, East Lansing, Michigan, 1970.
3. Poe, S. H., K. T. Valsaraj, L. J. Thibodeaux, and C. Springer. "Equilibrium Vapor Phase Adsorption of Volatile Organic Chemicals on Dry Soils," *J. Hazard. Mat.*, 19, 17–32, 1988.
4. Valsaraj, K. T., and L. J. Thibodeaux. "Equilibrium Adsorption of Chemical Vapors on Surface Soils, Landfills, and Landfarms—A Review," *J. Hazard. Mat.*, 19, 79–99, 1988.
5. Brauner, S., P. H. Emmett, and E. Teller. *J. Am. Chem. Soc.*, 60, 309, 1938.
6. Jury, W. A., W. F. Spencer, and W. J. Farmer. "Behavior Assessment Model for Trace Organics in Soil: I. Model Description," *J. Environ. Qual.*, 12, 558–564, 1983.
7. Jury, W. A., W. F. Spencer, and W. J. Farmer. "Behavior Assessment Model for Trace Organics in Soil: II. Application of Screening Model," *J. Environ. Qual.*, 13, 573–579, 1984.
8. Shen, T. "Estimating Hazardous Air Emissions from Disposal Sites," *Pollution Engineering*, 13(8), 31–34, 1981.
9. Johnson, P. C., M. W. Kemblowski, and J. D. Colthart. "Practical Screening Models for Soil Venting Applications," in *Proceedings of the NWWA/API Conference on Petroleum Hydrocarbons and Organic Chemicals in Groundwater: Prevention, Detection, and Restoration*, November 9–11, Houston, TX, 1988.

10. Bird, R. B., W. E. Stewart, and E. N. Lightfoot. *Transport Phenomena*, (New York, NY: John Wiley & Sons, 1960), pp. 511–512.
11. Millington, R. J., and J. M. Quirk. "Permeability of Porous Solids," *Trans. Faraday Soc.*, 57:1200–1207, 1961.
12. Bruell, C. J., and G. E. Hoag. "The Diffusion of Gasoline-Range Hydrocarbon Vapors in Porous Media, Experimental Methodologies," in *Proceedings of the NWWA/API Conference on Petroleum Hydrocarbons and Organic Chemicals in Groundwater: Prevention, Detection, and Restoration*, November 12–14, Houston, TX, 1986.
13. "Superfund Exposure Manual," U.S. Environmental Protection Agency, Office of Remedial Response, Washington, D.C., EPA/540/1-88/001, 1988, p. 16.
14. IRIS (Integrated Risk Information System), Office of Health and Environmental Assessment, Environmental Protection Agency, "Benzene," last revised 3/1/88.

CHAPTER 22

Letting the Sleeping Dog Lie: A Case Study in the No-Action Remediation Alternative for Petroleum Contaminated Soils

Evan C. Henry and **Michael E. F. Hansen,** Environmental Services Unit, Bank of America, Orange, California

INTRODUCTION

In selected circumstances where geological and hydrological conditions are favorable, a "no-action remediation" may be considered the appropriate alternative to the cleanup of petroleum contaminated soils. This case study presents the rationale for leaving some petroleum contaminated soil in place and untreated, based on the understanding of site conditions and potential impacts as drawn from environmental investigations performed at the site.

The subject site is located in the greater Los Angeles Basin within the flood plain of the Santa Ana River. As such, the site is underlain primarily by coarser-grained river-deposited alluvium. Despite the proximity to the river at a distance of approximately one-half mile, depth-to-groundwater is greater than 100 feet due to the combination of natural conditions and long-term groundwater withdrawal from the area.

The site is located within a municipality which adopted its own regulations for underground storage tanks prior to the passage of the legislation by the state of California. Therefore, the local agency has primacy to regulate underground storage tanks as long as the intent of the state legislation is served by those local regulations. Within the local municipality, as an adjunct to existing authority over the fire safety aspects of underground storage tank operation, the municipal fire prevention department was given the task of administering the underground tank regulation. The local fire department, although not governed by the state or county health departments, has relied heavily on technical support from state and county agencies for assistance in evaluation of the environmental considerations.

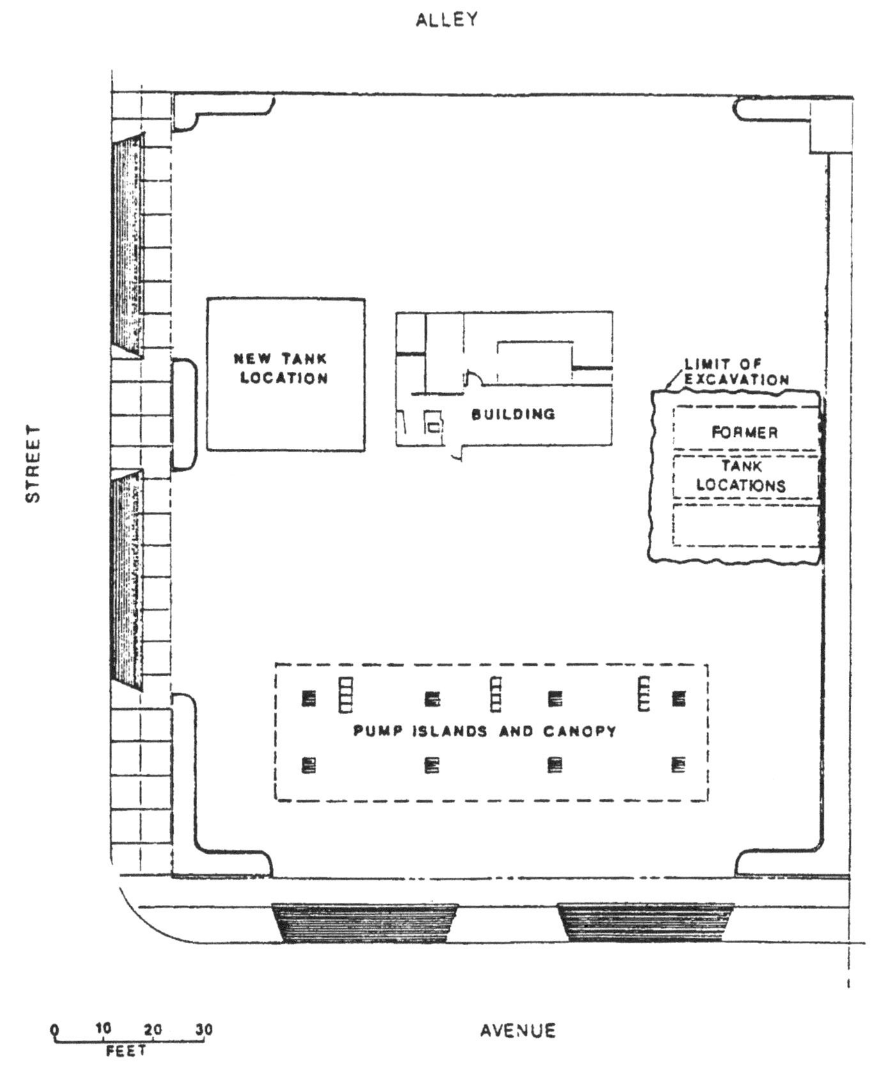

Figure 1. Site plot plan.

DISCOVERY

Three 8000-gallon steel tanks were excavated and removed from the location shown in Figure 1, with the intention that they would be replaced with three dual wall tanks within the same excavation. During removal it was observed that the tank backfill soils exhibited a strong gasoline odor, particularly in the vicinity of the tank turbine and fill ports at one end of the tanks. Soils were removed from around the tanks and stockpiled adjacent to the excavation on a paved part of the station site. Based on visual observations the fuel hydrocarbons in the soils appeared to be from over-spillage during filling at unsealed fill boxes or minor piping leakage at the western end of the tanks (see Figure 1). Six soil samples were obtained according to local regulatory requirements from beneath the former location of the three tanks. The locations of the soil samples are illustrated in Figure 2. Analysis of these soil samples indicated that hydrocarbons were present below the western ends of the tanks at a sampling depth of 14 feet. Hydrocarbon concentrations ranged from nondetected at a detection limit of 1 to 7200 mg/kg, and were identified as gasoline in character. Based on a regulatory action limit of 100 mg/kg for fuel hydrocarbons in the soils, it was evident that the presence of hydrocarbons below the tanks had to be addressed and potentially remediated.

INITIAL REMEDIATION ACTIONS

From field observations during the tank removal process the extent of hydrocarbons in the soils appeared to be limited. Although the depth to which hydrocarbons extended could not be determined at the time, the lateral extent appeared to be confined to an area of approximately 15 by 30 feet. In keeping with a policy of active remediation at the time of tank removal, the decision was made to excavate all affected soils to the maximum practical depth. The excavation was extended to a depth of 20 feet in the western end of the excavation. No additional excavation was performed in the eastern end, where hydrocarbons had not been detected in the soils. The depth of 20 feet was approximately 8 feet below the bottom of the original excavation left from the tank removal. Approximately 500 cubic yards of soil were removed from the excavation and stockpiled on the site. These soils were subsequently aerated in batches to reduce the concentrations of volatile organics to acceptable concentrations for disposal at a local landfill as a nonhazardous material.

Additional soil samples were obtained at the practical limit of excavation, which upon analysis indicated that fuel hydrocarbons were still present at concentrations of 1000 to 2000 mg/kg. As these concentrations exceeded the agency standard of 100 mg/kg, the local regulatory agency required further characterization. With the understanding that it would be practically infeasible to remove additional

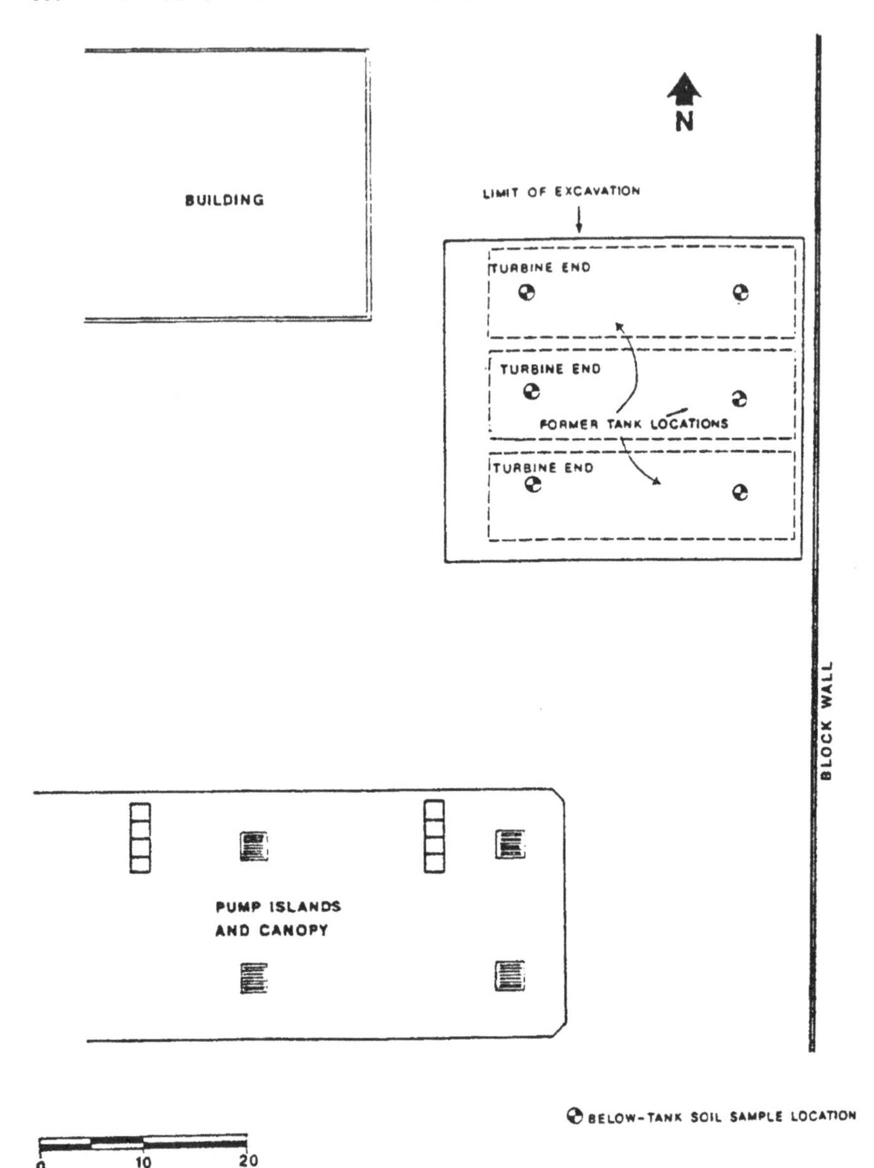

Figure 2. Excavation soil sample locations.

soils without endangering the adjacent building and canopy structure, the active remediation by excavation of soils was halted. As this particular station is one of the highest sales volume facilities in the company retail network, there was a strong emphasis on returning the station to active service as soon as possible. To expedite the process, the new tanks were installed in an excavation in an unaffected area on the opposite end of the site (Figure 1). Because the former tank excavation presented a potential safety hazard and to assist in access for subsequent investigation, the former tank excavation was backfilled with crushed rock. At the time of backfilling, four vertical standpipes of 4-inch diameter polyvinylchloride (PVC) well screen were installed within the crushed rock backfill to serve as vapor sampling ports and potential vapor extraction wells for use in the future, if necessary.

SUBSURFACE CHARACTERIZATION

To characterize the subsurface geologic conditions and the distribution of fuel hydrocarbons in the subsurface, four vertical borings were drilled in the area of the former tank excavation. One boring was located within the excavation and was drilled through the crushed rock backfill. Three other borings were drilled immediately adjacent to the tank excavations through natural materials. The locations of the four borings are shown in Figure 3. The borings were drilled with hollow stem auger drilling equipment, with soil samples obtained with a 2-inch inner diameter split-tube drive sampler outfitted with brass soil sample retainers.

Based upon onsite drilling observations, natural soils consisted of a variety of materials typical of river-deposited alluvium including silty sand, sandy silt, sand and gravel. In general, silty sand or sandy silt was encountered from the ground surface to a depth of 25 feet; coarse sand and gravel were encountered to an approximate depth of 65 feet. The sand and gravel were underlain by sand at an approximate depth of 65 feet. A schematic subsurface cross-section is shown in Figure 4. No groundwater was encountered at the maximum drilled depth of 65 feet.

Analysis for hydrocarbon presence for the three borings drilled just outside the excavation indicated that the hydrocarbon concentrations for the 20 and 50 foot samples were below the detectable limits. This indicated that there had been no significant lateral spreading of hydrocarbons with depth to areas adjacent to the excavated area. Hydrocarbons at a concentration of 3000 mg/kg were detected in the soil sample retrieved at the 20-foot level (just below the level of crushed rock in the excavation) from the one boring drilled within the tank excavation. In contrast, the samples obtained at 55 and 65 feet showed concentrations to be below detectable limits. The presence of gravel with cobbles prohibited acquisition of soil samples suitable for laboratory analysis between depths of 20 and 50 feet in all of the borings.

Based on the results of the soil sampling, it appeared that the presence of hydrocarbons was confined to a relatively shallow zone of limited areal extent, as shown in the cross-section in Figure 4. Although sampling limitations prohibited the determination of the exact bottom of the affected soils, the sampling indicated that the bottom of the hydrocarbons did not extend below a depth of 55 feet. Based on field screening of soil samples with a portable photoionization detector at the time of drilling, it appeared that the base of the hydrocarbons was located at a depth of approximately 25 feet. At this level there was a stratigraphic contact between overlying sandy silt and underlying sand and gravel.

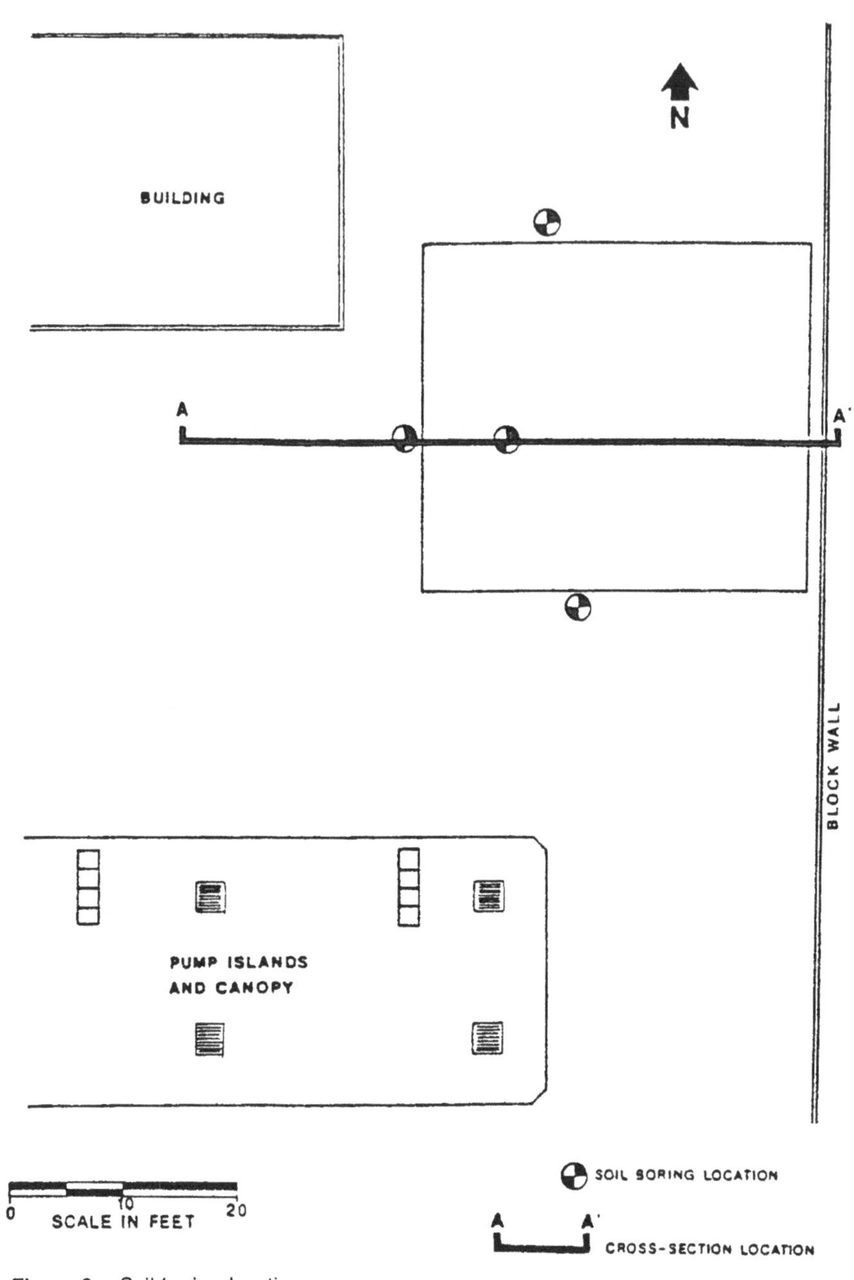

Figure 3. Soil boring locations.

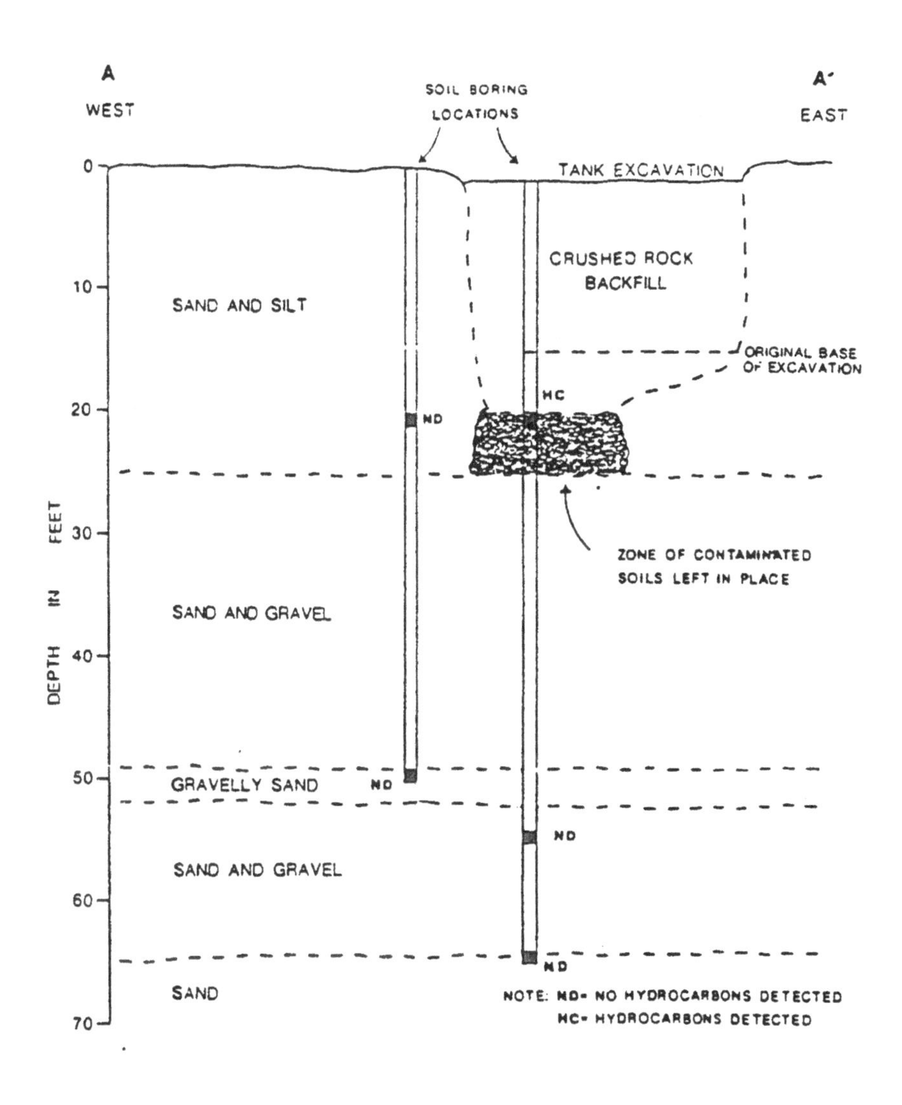

Figure 4. Schematic cross-section (A-A′) of tank excavation.

A report of investigation was prepared based on these findings, which acknowledged that hydrocarbon-contaminated soils were present in the subsurface but that the extent of these hydrocarbons was limited. The conclusions presented based on the results of the investigations included:

1. Hydrocarbons have apparently not spread laterally beyond the immediate area of the tank excavation, as indicated in the data retrieved from the three borings located adjacent to the tank excavation area.
2. Hydrocarbons were not detected at depths immediately below the tank excavation, thereby indicating downward vertical migration hydrocarbon was of limited extent.
3. Based on available regional groundwater data, groundwater under water table conditions appeared to be at least 100 feet deeper than the deepest detected presence of hydrocarbons.

Recommendations developed from the site investigation included:

1. No further onsite investigation was recommended, and
2. As the extent of hydrocarbon in the subsurface was apparently limited laterally and vertically, further remedial action beyond the initial removal of soil during tank excavation (which had already been accomplished), was not recommended.

AGENCY RESPONSE

Following the lead of the county health department standards, the local regulatory agency rejected the conclusions and recommendations of the investigation. The agency responded with its hard-line policy that any and all soils with hydrocarbon concentrations above the set limits of 100 mg/kg would require remediation. Such remediation would have to be undertaken (either started or potentially completed) prior to obtaining a permit to operate the station. From a timing perspective, when this decision by the agency was received, the new tank installation was almost complete. Therefore, waiting for remedial actions to be undertaken prior to receiving the permit to operate would have severely impacted the reopening date of the station. The reasoning for the "hard-line" promoted by the regulatory agency was that this was their only tool for ensuring that remediation took place. Once a permit to operate was approved, the agency had to undertake legal action to close the station if adequate remediation of the soil contamination was not subsequently undertaken.

Despite the agency attitude toward requiring remediation, based on the practicalities of the site, it was felt that a no-action alternative for remediation was the most prudent. In discussion with the agency personnel, it was negotiated that the station could be reopened with a written commitment from the company to address the fate and transport of the remaining hydrocarbons in the soils in a timely manner. Remediation needs would be determined based on the results on the fate and transport study. The intent of the fate and transport study was to estimate the disposition of the hydrocarbons over time if left in place.

FATE AND TRANSPORT ASSESSMENT

The purpose of the fate and transport investigation was:

1. To acquire research and technical data to address the environmental fate and potential for migration of residual hydrocarbons within shallow soils at the tank excavation site; and
2. To determine the potential for adverse impact on groundwater quality in the area.

The scope of work for the investigation involved two aspects: research of the technical aspects of hydrocarbon presence and migration in unsaturated soils, and assessment of site-specific characteristics relative to the current technical understanding of hydrocarbons fate and transport in the subsurface. Research included a literature search to develop an understanding of hydrocarbon characteristics and potential for migration in soils.

DISCUSSION OF FATE AND TRANSPORT OF HYDROCARBON IN THE SUBSURFACE

It is inferred that the hydrocarbon presence detected in the subsurface during the tank removal process was primarily the result of spillage at the unsealed tank fills during transfer of product from tank trucks to the underground storage tanks. This inference is the result of observation of hydrocarbon staining of tank backfill materials at the fill end of the excavation and the lack of hydrocarbon in the subsurface at the other end of the tanks.

From spillage near the ground surface, hydrocarbon product is inferred to have moved vertically downward within the tank backfill around and ultimately to beneath the underground storage tanks. Sufficient hydrocarbon was present to have moved from the backfill materials into the underlying natural soils.

The downward migration of liquid hydrocarbon product typically takes place where sufficient hydrocarbon product is present to allow for liquid flow. As product moves downward through the soil, a small amount attaches itself to the soil particles contacted and remains behind the main body of product. Where the spill is small relative to the surface area available for contact in the zone of migration, the body of product is exhausted on the way down until the degree to which it saturates the soil reaches a relatively low point called the "immobile" or "residual" saturation. At this point the product essentially stops moving.[1]

The volume of soil required to immobilize a given amount of product depends on two primary factors: (1) the porosity of the soil; and (2) the nature of the hydrocarbon as reflected in its characteristic "maximum residual saturation." At or below its maximum residual saturation, the product will not move as a liquid in the soil. The residual saturation for various hydrocarbon products has been empirically derived and is estimated as follows: gasoline—10%; diesel and light fuel oil—15%; and lube and heavy fuel oil—20%.[1]

With the above residual saturation percentages, theoretical hydrocarbon concentrations (in units of milligrams per kilogram, mg/kg) were calculated for various soil porosities. The hydrocarbon concentrations were calculated using the following formulae:

$$\text{(1) Hydrocarbon Concentration} = \frac{\text{Weight of Hydrocarbon}}{\text{Weight of Hydrocarbon plus Weight of Soil}}$$

(2) Weight of Hydrocarbon = Unit Weight of Water (62.4 lb/ft)
× Specific Gravity of Gasoline (0.80)
× Residual Saturation (0.10)
× Porosity (varies from 0.20 to 0.60)

(3) Weight of Soil = 146 lb/ft

The theoretical concentrations are listed in Table 1. A unit weight of 146 pounds per cubic foot for sandy and silty soils[2] and a specific gravity of 0.80 for gasoline were used in the calculations.

From Table 1, the possible range of residual hydrocarbon concentration in the soils is from approximately 3,400 to 60,864 mg/kg. The concentration of hydrocarbons in the soil at a minimum would have to exceed the low value of this range for movement of the hydrocarbon as a liquid. The maximum concentration of hydrocarbons detected in the soils still remaining in place was 3,000 mg/kg. Based on comparison of the concentration of product in the soil to the possible range of residual saturation, the hydrocarbon in the soil was not anticipated to be mobile as a liquid.

From the initial investigation, no hydrocarbons were detected in samples outside the tank excavation area nor at depth. Samples were not obtained directly

Table 1. Theoretical Residual Hydrocarbon Concentration[a,b]

Soil Porosity (percent)	RESIDUAL SATURATION (percent of total)					
	5	10	15	20	25	30
20	3,400	6,800	10,200	13,600	17,000	20,400
30	5,100	10,200	15,300	20,400	25,400	30,500
40	6,800	13,500	20,300	27,000	33,800	40,500
50	8,512	17,024	25,536	34,304	42,560	51,072
60	10,144	20,288	30,432	40,576	50,720	60,864

[a]All results are reported in milligrams per kilogram.
[b]For estimated specific gravity of gasoline equal to 0.80 and for estimated unit weight of silty sand of 146 lbs/cubic foot (Lambe and Whitman[2]).

beneath the excavation between 20 and 55 feet, due to the presence of coarse gravel in the formation which precluded the acquisition of samples with the drive sampler. However, no hydrocarbons were evident in the relatively finer grained soils at depths of 55 and 65 feet. The data from these samples indicated that hydrocarbons had not migrated to this depth. Field observations during drilling also supported the conclusion that hydrocarbons had not migrated below a depth of 25 feet. Therefore, the bottom of the immobile residual hydrocarbons was projected to be greater than 75 feet above the projected depth to groundwater of 100 feet.

The locations of the borings and the sample depths are considered sufficient to have detected whether any lateral spreading had occurred, or if alternate pathways for migration were present. There were no soil layers of lower permeability which could possibly have altered the potential for downward migration of hydrocarbons. In one documented case, the mass transfer of hydrocarbons through sand/clay interfaces was not found to produce appreciable lateral spreading.[3] This further supports the conclusion that the extent of the presence of residual hydrocarbons in the subsurface at the site was limited and localized.

Although in the liquid phase hydrocarbons are immiscible with water, some hydrocarbons and other associated organic compounds are slightly soluble in water. As a result, it is possible for some hydrocarbon constituents to be carried downward if percolating waters come into contact with the hydrocarbon. At the station, the soils beneath the tank excavation were described variably as dry, slightly moist, or moist. No samples were described as wet, which would be taken as indicative of being saturated enough to allow for water flow to the maximum drilled depth of 65 feet. Since the tank excavation was paved, infiltration of surface water into the affected subsurface area has been eliminated. Therefore, it is not expected that any water will enter the soils to serve as a vehicle for further downward migration of dissolved organic compounds. Hydrocarbons also have the potential for migration in the soils in the vapor state. The amount of vapor generation from liquid hydrocarbon is a function of the vapor pressures of the individual organic compounds, and the total pressure of the atmosphere (in this case the soil atmosphere). The "lighter" organics will volatilize faster, thereby decreasing the overall concentration by increasing the relative proportion of "heavier" compounds in the residual product. This process is frequently referred to as "weathering."

The relative percent of certain organic compounds found in gasoline are listed in Table 2. Table 2 also lists the percentage of selected compounds relative to the total hydrocarbon concentration detected in the soil vapor from the crushed rock excavation backfill. The percentage of certain organic compounds was significantly less in the soil vapor from the crushed rock excavation fill than in the product itself. In addition, the percentage of benzene was significantly less than toluene and xylene in the vapor samples. The vapor pressure of benzene is much higher than that of toluene and xylene, and therefore it will volatilize more rapidly, resulting in a lower residual percentage of this compound being present over time. In addition, the three compounds as a group are generally lighter than many of

Table 2. Relative Concentrations of Hazardous Compounds in Gasoline and Percentages Found in Soil Vapors

Compound	Relative Concentration in Product (% by Weight)	Relative Concentration in the Soil Vapor from Crushed Rock Backfill[a] (% by Weight)
Benzene	0.81 to 1.35	0.013
Toluene	5.92 to 12.3	0.11
o-xylene	1.94 to 2.05	NA[b]
m-xylene	3.83 to 3.87	NA
p-xylene	1.54 to 1.57	NA
Total Xylene	7.31 to 7.49	0.36
Cyclohexane	0.17 to 0.36	NA

Source: Reference No. 4
[a]Data reported is from vapor wells installed in the tank excavation area.
[b]NA = Not Available.

the polyaromatic hydrocarbons found in gasoline. As a result, the percentage of these compounds relative to the total amount of residual hydrocarbon was expected to diminish over time. The hydrocarbon product appears to have been in the soil for a relatively long time, at least long enough to allow the escape of a significant percentage of the volatile aromatic compounds present in the soils in the vapor state.

From field observations using a portable photoionization detector, no significant hydrocarbon vapors were detected in soil samples taken from borings drilled adjacent to the tank excavation. However, by qualitative analogy, a reading of "no detection" on the photoionization detector was taken to indicate a relatively low, if not undetectable, level of hydrocarbons in the soil vapor of the natural soils outside the excavation. In contrast, the organic content of the soil vapor in the newly emplaced crushed rock backfill was relatively high.

Transport of volatile organic compounds through the soil is a function of the concentration of the gas at the source, and a diffusion coefficient. The diffusion coefficient in soil is a function of the soil characteristics which affect the tortuosity of the path which the organic molecules must follow. The soil characteristics affecting the tortuosity include the air volume of the soil and the porosity of the soil. The contrast in the inferred soil vapor concentration of the surrounding natural soils and the crushed rock backfill was probably the result of differences in the diffusion coefficient of the two soil materials. The natural materials at the 20-foot level were characteristically a silty sand with up to 50% silt. The tank backfill was crushed rock with particle size characteristic of fine gravel. No specific estimates were developed, but it was inferred that the crushed rock would be more transmissive of vapors than the finer-grained natural soils as indicated by the levels of vapors seen in the samples from the crushed rock backfill area.

The hydrocarbons which remain in the soils are expected to biodegrade over time. It is well known that many organic compounds, including hydrocarbons found in gasoline, are broken down by naturally occurring bacteria in the soils.

The hazardous compounds such as benzene and toluene are also broken down aerobically.[5] Under aerobic biodegradation, the nonhazardous end products of organic breakdown are water and carbon dioxide. Since hydrocarbons occur within the unsaturated zone well above the water table, they will be subject to long-term aerobic conditions and are expected to diminish in concentration over time, due to biodegradation.

CURRENT STATUS

The preceding discussion was presented in a report of the fate and transport investigation. The report was submitted to the local regulatory agency for their review, comment, and action.

A period of over four years has transpired between the submittal of the report of the environmental fate and transport to the regulatory agency and the current case study report. During that time, the retail service station has been in full operation with the new tank and piping system in compliance with the current local and state regulations. In addition, the station was sold to another retail gas station operator with no known impacts on the sale or continued operation of the station. No specific further actions have been requested by the local regulatory agency in response to the report submittal.

It is our opinion that the above-outlined actions represent satisfactory environmental compliance and are an appropriate remedial alternative. The environmental conditions were quantified through scientific investigation. Based on the understanding of the potential fate and transport of the residual subsurface hydrocarbons, it was concluded that no detrimental impact would result from leaving the affected soils in place. As such, in this case the "no-action alternative" was the appropriate response.

REFERENCES

1. Mull, R., "Migration of Oil Products in the Subsoil with Regard to Groundwater Pollution by Oil," in Jenkins, Ed. *Advan. Water Poll. Res., Proc. Int. Cong.* (Oxford: Pergamon Press, Inc., 1971).
2. Lambe, T. W., and R. V. Whitman, *Soil Mechanics* (New York: John Wiley & Sons, Inc., New York, 1969).
3. Kuhlmeier, P. D. D., and G. L. Sunderland, "Distribution of Volatile Aromatics in Deep Unsaturated Sediments," in *Proceedings of Conference on Characterization and Monitoring of the Vadose (Unsaturated) Zone,* National Water Well Association, November, 1985, pp. 198–214.
4. Marley, M. C., and G. E. Hoag, "Induced Soil Venting for Recovery/Restoration of Gasoline Hydrocarbons in the Vadose Zone," in *Proceedings Petroleum, Detection and Restoration, National Water Well Association,* November, 1984.
5. Patterson, J. W., and P. F. Kodukala, "Biodegradation of Hazardous Organic Pollutants," *Chem. Eng. Progress,* 77:4, April 1981, pp. 48–55.

CHAPTER 23

A Proposed Approach to Regulating Contaminated Soil: Identify Safe Concentrations for Seven of the Most Frequently Encountered Exposure Scenarios

Dennis J. Paustenbach and **Renee Kalmes,** ChemRisk Division, McLaren/Hart Environmental Engineering, Alameda, California
James D. Jernigan, Amoco Corporation, Chicago, Illinois
Rená Bass and **Paul Scott,** ChemRisk, McLaren/Hart Environmental Engineering, Springfield, Missouri

INTRODUCTION

There are only a few federal EPA cleanup guidelines for chemicals in soil, including PCBs,[1] lead,[2] and 2,3,7,8-tetrachlorodibenzo-p-dioxin (TCDD).[3] Although these cleanup guidelines are intended to protect human health, they are by their very nature non-site-specific. With over 2,000 different potential soil contaminants at hundreds of sites throughout the United States, both regulators and the regulated community face an almost overwhelming task of deciding "how clean is clean."

Many state regulatory agencies have recently published guidelines for cleanup levels of some soil contaminants,[4–8] particularly petroleum contaminated soils. The stated rationale for many of the state guidelines is to prevent contamination of groundwater and/or to protect human health, although in many instances the scientific underpinnings for some state criteria are elusive. For example, the so-called "Type B" cleanup criterion for benzene in soils in the state of Michigan was derived[9] simply by multiplying the groundwater criterion (1 ppb) by a factor of 20. In many states, informal or formal guidelines specify that remediation is required when soil concentrations of total petroleum hydrocarbons (TPH) exceed 100 ppm, while others require cleanup to background or to the limits of the analytical detection limits.[4-6]

The problems associated with the multitude of guidelines and criteria that are currently being developed or proposed in the United States are that: (1) the basis or rationale for some of the cleanup levels is not clearly presented, (2) approaches to setting cleanup levels vary between states, (3) cleanup objectives often vary among agencies within the same state, and (4) the cleanup levels are usually set in a haphazard way, rather than in a systematic and scientifically defensible manner.[6] The cost and inefficiencies of this situation are not trivial for the regulated community and society as a whole. For example, even a three-fold difference in cleanup levels (which is often considered of negligible regulatory significance) can result in over a 10-fold difference in the cleanup costs. Even in the case of relatively nonvolatile, nonmobile chemicals such as TCDD, the difference in costs to remediate soils to 1 ppb, 10 ppb, or 50 ppb can be dramatic (Figure 1). Thus, if a generally accepted and scientifically defensible methodology could be developed to more accurately identify cleanup levels for contaminated soils, remediation monies could be efficiently and properly allocated.

One goal of the Council for the Health and Environmental Safety of Soils (CHESS) is to establish a generic method for deriving soil cleanup levels.[10] CHESS represents an ambitious effort to optimize the scientific process, and has brought together as many as 30 of the nation's experts on contaminated soil. Since the group's inception in 1986, these experts have exhaustively evaluated all known techniques for setting soil cleanup levels and have written a critique of those approaches.[11] The group is currently working with the American Petroleum Institute to develop a computer model which will account for all the factors which should go into setting site-specific and chemical-specific criteria. This model will probably be completed by mid-1992. The approach discussed here is not in conflict with the goals of CHESS, but rather, it represents a way for agencies to use the methodologies they are suggesting to establish regulatory guidelines.

The objective of this chapter is to propose a method which should allow state and other agencies to quickly propose cleanup levels for sites having contaminated soils. The heart of the approach is to set seven different soil cleanup guidelines for each chemical which address the most typically encountered exposure scenarios. The rationale and approach for setting each guideline are presented. Two examples of how the approach can be applied are included. Each type of guideline applies to a particular land use scenario or satisfies a need to protect a particular environmental receptor (Table 1). In short, based on our nearly 20 years of experience, we believe that seven guidelines, rather than one or two, may be necessary for states to properly regulate contaminated soils. Several of the types of soil cleanup guidelines (Table 1) address exposure scenarios which are of significant concern to the public and to regulatory agencies. For example,

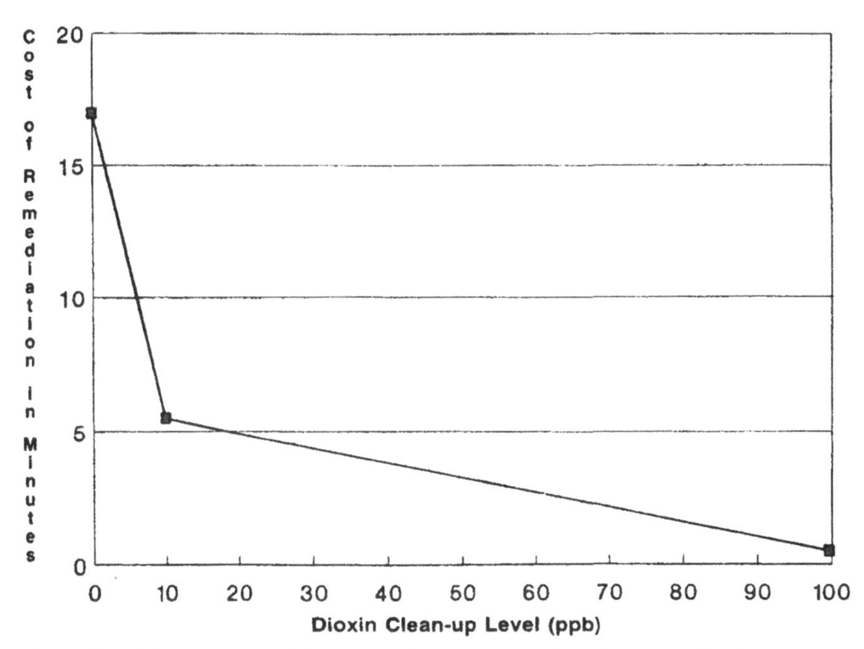

Figure 1. Difference in costs to remediate soils from TCDD to various levels.

residential, recreational, agricultural, and industrial exposure scenarios are settings which often must be evaluated to satisfy regulatory agencies.[12,13] However, other issues and scenarios should also be considered. For example, contaminated soil can affect groundwater quality. This has been an important environmental issue for nearly two decades, and is very dependent on site-specific considerations such as the geology of the site and depth to groundwater, rather than classic exposure pathways. Other reasons for cleaning soil are to control the runoff hazard and to protect grazing wildlife; scenarios which we attempt to address. During the past four decades, these have become important issues, and will continue to be so in the coming years.[14]

The guidelines presented here were not intended to be site-specific nor were they intended to be rigorous evaluations of the possible health hazards. Instead, the objective was to identify concentrations for seven common exposure scenarios for each contaminant at which, when encountered, no further evaluation would

Table 1. The Seven Most Common and Important "Types" of Exposure Scenarios Which Require Cleanup Guidelines

• Residential	• Protection of wildlife
• Industrial	• Protection of groundwater
• Recreational	• Runoff hazard
• Agricultural	

need to be conducted. We believe that this approach can help agencies respond more efficiently, since they could rapidly dismiss those sites which pose a negligible hazard. Using this approach, when contaminant concentrations exceed the screening-level, or safe concentration, a site-specific assessment would be conducted. Although it may appear cumbersome to set seven soil cleanup levels for each chemical, this is surely better than the prospect of having 30–50 separate state guidelines which attempt to address only two or three different scenarios.

DESCRIPTION AND RATIONALE FOR SEVEN SOIL CLEANUP CRITERIA

Residential Scenario (Nonvolatiles)

When setting a soil cleanup guideline for residential areas, incidental soil ingestion by children is nearly always the most important factor for the nonvolatile chemicals and, frequently, for the volatiles. This has been shown in several assessments of contaminated soils.[12-17] Particularly with the nonvolatile chemicals, where contamination may be limited to the top 6 to 12 inches of soil, incidental soil and dust ingestion will often represent from 60% to 85% of the risk.[18] Dermal contact with contaminated soil is usually the second most important source of uptake in residential scenarios, followed by the ingestion of garden vegetables grown in contaminated soil. Inhalation of suspended particulates rarely contributes more than 1% to 2% of the total uptake and risk.[12,13,19-22]

Residential Scenario (Volatiles)

For the volatile chemicals, inhalation due to vaporization from soil and the ingestion of soil will constitute the primary sources of exposure (risk), particularly if the contamination is recent. Occasionally, even when the contamination is below the ground surface (as is often the case with leaking underground storage tanks), inhalation can pose an acute health hazard or an explosion hazard, especially in basements located near heavily contaminated soil. In these settings, the second most important route of human exposure is usually dermal contact. Due to their relatively short environmental half-lives, volatile compounds rarely enter the food chain. In general, the primary reason for remediating sites with relatively high quantities of volatile chemicals is the threat of contamination of shallow groundwater aquifers or the threat to aquatic species, rather than direct human exposure.

Industrial or Occupational Scenario

The most significant routes of exposure at industrial sites are normally dermal contact with soil and incidental soil ingestion by adults.[20,22] Dermal contact tends to be the predominant pathway because adults do not intentionally ingest soil; however, due to hand-to-mouth contact, they may ingest 0–10 mg/day or more.[16,94] Because many industrial sites are contaminated with volatile organic chemicals, which tend to have high mobility in soils, protection of groundwater is usually the driving force behind establishing cleanup levels. Thus far, it has become commonplace for most agencies to set two cleanup guidelines for soil; one for residential and the other for industrial/occupational settings.[1,8,12]

Recreational Scenario

Sources of soil contamination at recreational areas can be numerous. In many instances, recreational areas, such as parks, baseball fields, playgrounds, etc., are former industrial properties which have been abandoned and deeded to the city. In other cases, recreational areas were former municipal or industrial landfills which have been capped and considered safe. In other instances, the use of sludge from municipal waste treatment facilities, paper and pulp processing, and other processes has been used as soil amendment material on recreational lands because it contains nutrients required for plant growth such as nitrogen, phosphorus, calcium, and minor trace elements.[19,23] Sometimes, the use of pesticides and herbicides on recreational lands to control poison ivy, poison oak, etc., has been a source of long-term contamination. Like the residential scenario, health risks at recreational sites are driven primarily by incidental soil ingestion and dermal contact with soil by children.

In areas where there has been a widespread application of sludge as fill or soil amendment (fertilizer), runoff may be the biggest potential environmental and human health concern for recreational lands. The potential problem is that runoff of persistent contaminants on topsoil via suspended particulates can be transported to rivers, lakes, streams, and harbors and can pose a threat to both wildlife and people.[17,24,25]

Agricultural Scenario

Although it might be anticipated that the ingestion of vegetables or cereals grown in contaminated soil would be the primary concern in an agricultural exposure scenario, usually the most significant route of human exposure involves uptake of soil by grazing animals.[26,27] This is considered an indirect route of uptake of the contaminant. To the surprise of many persons, although animals will ingest potentially contaminated crops and vegetation, the primary problem is that grazing cattle ingest an average of 0.9 kilograms of soil per day.[27,28] The ingestion of soil can represent 80% of the cow's uptake of persistent chemicals (such as dioxins and furans), since weathering processes usually will remove the majority of the particles which deposit onto crops. For the nonvolatile chemicals, the human hazard can be greater than expected because persistent, lipophilic chemicals accumulate in beef tissue and fat, resulting in a potentially significant dose to humans who eat the meat and drink the milk. The uptake of volatile chemicals by grazing animals via soil usually poses no significant hazard at low concentrations.

Runoff of contaminated soils into streams, creeks, and lakes is the second most important concern when evaluating an agricultural scenario. This is not surprising, given the millions of acres treated with a myriad of chemical products and the proximity of most farmland to streams, lakes, and other waterways. For example, the U.S. Department of Agriculture (USDA) estimated that from 1968 to 1977, 11 million acres of land were treated annually with mirex bait,[29] yet it is a relatively small volume pesticide compared to others. Thousands of tons of many chemicals, especially herbicides, are applied annually to agricultural soils. As before, compounds that are lipophilic and environmentally persistent (e.g., stable) pose the most significant hazard because of their propensity to bind strongly to soil, which becomes sediment and then accumulates in the tissues of aquatic organisms.[17]

Protection of Wildlife Scenario

Concern over the potential risks to wildlife has become a major issue at contaminated sites in recent years and is likely to become even more so in the next decade.[30,31] Uptake of contaminated soil by fish, mollusks, game birds, song birds, deer, and endangered species tend to be the most important considerations in setting cleanup levels based on protection of wildlife.[32] Issues such as the migratory patterns of birds, changes in biological diversity, increased susceptibility to predators, etc., tend to make these assessments very complex. Adverse effects on the predator food chain can also be significant,[31] and therefore should also be evaluated.

Protection of Groundwater

The primary concern at sites contaminated with small to moderate molecular-weight chemicals is the contamination of groundwater. Clearly, the amount of contaminant in the soil, the water solubility of the contaminant, the potability of water from the aquifer, and the toxicity of the chemical dictate the level of concern.[17,33] During the 1980s, the threat of contamination of groundwater by soil was the primary reason for remediating most nonresidential sites which had high levels of chemicals. Recent evidence suggests that it may not be technologically possible to remediate groundwater aquifers which have become contaminated and, therefore, this may not be as important a rationale in the 1990s.[34]

Runoff Hazard

The possibility that persistent chemicals can be present for decades in soils and subsequently be transported through runoff to streams will drive the cleanup at a large number of sites in the 1990s. The recent emphasis on PCBs, dioxins, furans, metals, and PAHs in sediments is evidence that this will be one of the major issues in the coming years. As noted previously, the assessment of agricultural lands will frequently be driven by the runoff hazard.[35]

KEY EXPOSURE PARAMETERS

To identify the most accurate soil cleanup guideline or standard, the most appropriate exposure factors need to be selected. This has proved to be a significant task since high quality information on exposure is often not available; on the other hand, the range of likely or reasonable values can often be identified.[36] The EPA has suggested numerous exposure factors in a number of guidance documents,[2,7,37] but these are often quite conservative. These conservative factors, if combined together, can produce unrealistic estimates of exposure.[17] A more reasonable alternative is the use of best-estimate values and to account for the range of plausible values using Monte Carlo methodologies.[16,38,39,94] This approach should accurately predict the range of likely values for the human uptake of chemicals with more confidence than approaches based on point estimates.[16,39,117]

Typically, when developing health-based cleanup levels, one or two key parameters tend to "drive" the risk assessment at sites having contaminated soil.[16] That is, one or two exposure parameters tend to account for a vast majority (75% to 90%) of the risk (Table 2). The following section discusses the seven different scenarios which we believe deserve specific guidance. The key parameters, or "drivers," that dictate the soil cleanup levels are identified and discussed.

Soil Ingestion Rate

Foremost among the exposure factors is the rate of soil ingestion, particularly among children. Incidental soil ingestion occurs at all ages as a result of hand-to-mouth activities. However, it is widely accepted that children approximately 2 to 6 years old are the primary age group that consumes a potentially significant amount of soil.[14,16,42,43] Several literature surveys have been conducted which attempt to characterize what is known about soil ingestion rates among children. The majority of published estimates of soil ingestion rates for adults and children are presented in Table 3.

The most thorough and rigorous studies to date are those by Calabrese and co-workers[43] and van Wijnen and colleagues.[62] In the Calabrese study, seven different tracer elements were quantitatively evaluated in the stools of 65 school children, aged 1 to 4 years old. Although this study originally reported that the three most reliable tracers were Al, Si, and Y, as validated by a supplemental adult study,[58] a more recent analysis by the same group[59,60] indicated that Zr and Ti are probably the only reliable tracers, and therefore the bulk of the published work probably overestimates the true value. A value of 20–50 mg/day probably represents a conservative, yet realistic, soil ingestion rate for children.

In the van Wijnen study,[62] two tracer elements (Al and Ti) were quantitatively measured in the stools of children ages 1 to 5. For children 1 to 2 years of age, the mean ingestion rate ranged from 0 to 90 mg/day. For children aged 3 to 5, the mean ingestion rate ranged from 0 to 60 mg/day.

The validity of soil ingestion estimates for children derived from soil tracer methodologies has recently been questioned.[59,60] By calculating the soil recovery

Table 2. Factors That Typically "Drive" Soil Cleanup Levels for the Most Common Exposure Scenarios

Residential
- Soil/dust ingestion
- Dermal uptake
- Garden vegetables ingestion
- Dermatitis hazard

Recreational
- Soil ingestion
- Dermal contact
- Runoff

Industrial
- Soil ingestion
- Dermal uptake
- Inhalation of fugitive dust
- Groundwater/runoff

Agricultural
- Uptake via crops
- Uptake by grazing animals
- Runoff (aquatic)
- Groundwater hazard
- Fugitive dust

Wildlife
- Uptake by birds
- Adverse effects on predator food chain
- Effects on development and reproduction

Table 3. The Range of Estimates for Soil Ingestion by Humans for Various Age Groups Which Have Been Published Over the Past 20 Years (mg/day)

	Age Group			
Researcher	**Infant[a]**	**Toddlers[b]**	**Children[c]**	**Adults[d]**
Baltrop[44]		100		
Lepow et al.[45]		100		
Day[46]		100		
Duggan and Williams[47]		50		
NRC[28]		40		
Kimbrough et al.[12]	0–1000	1000–10,000	100–1000	100
Hawley[48]	0	90	21	57
Binder et al.[49]		180		
Bryce-Smith et al.[50]		33		
Schaum[51]		100–5000		
USEPA[52]		100		
Clausing et al.[53]		56		
LaGoy[54]		100		
Paustenbach[18]		25–50		0–10
Lipsky[55]		240		
Sedman[56]		50–200		
USEPA[37e]		200	100	100
Calabrese et al.[43,57–59]		9–40	1–10	0–31
Davis et al.[61]		25–80	25–80	
Van Wijnen et al.[62]		0–90		
Sheehan et al.[13]	0	33	10	10
Calabrese and Stanek[116]	0	pica child		

[a]Infants have usually been defined as children aged 0–2 years and who receive significant adult supervision.
[b]Toddlers have usually been defined as children aged 2–4 or 2–6 years.
[c]Children are usually defined as 4–12 years of age.
[d]Adults are often considered 13–75 years of age.
[e]The current USEPA position of soil ingestion.

variances and soil detection limits for each tracer, it was shown that virtually all recent studies[43,49,61,62] provide little convincing evidence upon which to derive a precise estimate of soil ingestion. Of the three or four major studies, only the soil ingestion rates (16 and 55 mg/day) reported by Calabrese et al.[58] for two out of seven tracers were within acceptable statistical norms.[59] Thus, a conservative estimate of an average soil uptake of about 40–50 mg/day appears to be credible. While this soil ingestion rate is less than the 200 mg/day rate currently recommended by the EPA for the typical child,[15,37] it should be noted that the EPA value predates the work of Calabrese et al. and van Wijnen et al. The recent paper by Calabrese and Stanek[116] suggests that some children can ingest as much as 2,000 to 10,000 mg/day of dust and/or dirt. Prior estimates have suggested that 1 in 300 to 1 in 1,000 children are afflicted with geophasia or pica.

The majority of soil ingested by adults is thought to occur via hand-to-mouth contact (such as during smoking) and soil on produce.[18] As such, adults are expected to ingest quantities of soil which are perhaps 2- to 10-fold less than children.[14,39] Although there are inadequate data to characterize the distribution of soil ingested by adults, a value of 10 mg of soil per day is expected to be conservative.

Dermal Contact Rate

As mentioned previously, the second major driver for most soil cleanup levels is dermal contact. While there are several factors to consider, one of the more important ones is the dermal contact rate or soil loading factor. The soil contact rate is an estimate of the amount of soil that adheres to a given area of human skin over a specific period of time. As presented in Table 4, several studies have estimated or evaluated the amount of soil or dust that is likely to be in contact with skin.

Lepow et al. found that an average of 0.5 mg soil/cm^2 skin could be removed from the hands of children aged 2 to 6 years old.[45,65] Soil samples were collected from 10 children's hands by pressing a preweighed self-adhesive label on the hand surface. Roels et al. determined the amount of lead on the hands of 11-year-old rural school children by pouring dilute nitric acid over the palm and fingers of the dominant hand.[63] Although the amount of soil on the hand was not determined by the authors, others have estimated this value by dividing the amount of lead on the hand by the concentration of lead in the soil. Using this method, the South Coast Air Quality Management District (SCAQMD) estimated about 120 mg soil on the hand.[68] Assuming that the surface area of one hand is 342 cm^2, and that about 60% of the hand was sampled in the SCAQMD study, they estimated a soil loading rate of about 0.6 mg/cm^2.

Schaum[51] suggested using a range of 0.5 to 1.5 mg/cm^2 (applicable to all the areas of skin) for adults and children, based on the results of the Lepow et al.[65] and Roels et al.[63] studies, respectively. The estimate of the upper end of the range of Roels et al. is higher than the SCAQMD figure, due primarily to differences in the estimates of the hand surface area sampled in the study.

Table 4. Summary of Soil Loading Rates on the Palms of Hands of Human Test Subjects

Soil Loading Rate (mg/cm^2)	Type of Soil	Age and Sex	Reference Source
1.45	Commercial potting soil	Male adult	USEPA[2]
2.77	Kaolin clay	Male adult	USEPA[52]
1.48	Playground dirt and dust	Males 10 to 13 years old, rural areas	Modified from Roels et al.[63]
1.77	Playground dirt and dust	Males 10 to 13 years old, urban areas	Modified from Roels et al.[63]
0.6	Unsieved	Adult male	Driver et al.[64]
0.5	Not available	Children, −4 years old	Lepow et al[65]
0.33	House dust	Children, 1–3 years old	Gallacher et al.[66]
0.11	House dust	Adult woman	Gallacher et al.[66]
0.2	Not available	Adults	Que Hee et al.[67]

Que Hee et al.[67] examined the retention of soil on the hands of a small adult. Based on this study and several assumptions regarding soil particle size and adherence to the skin, the California Department of Health Services (CDHS) estimated that an average of 31.2 mg soil adhered to the palm of a small adult.[69] Using 168 cm^2 as the surface area of the palm of a small adult, CDHS estimated a soil contact rate of 0.2 mg soil/cm^2 skin.[69] Driver et al.[64] surveyed various soils of different organic content for their ability to adhere to human hands (adult male). They determined that the average amount of unsieved soil retained on human hands was 0.6 mg/cm^2.

The data from each of these studies can be expected to overestimate the amount of soil that will adhere to the skin surface of other body parts (e.g., forearms, neck, and head) because it is often just the palms that are in direct contact with soil surfaces. This assumption has been generally confirmed in studies of agricultural workers.[70] In addition, these soil loading rates are almost certain to overestimate the actual average exposure because: (1) not everyone works in community gardens; (2) many persons wear gloves when working intimately with dirt; (3) gardeners work directly with soil primarily during only planting and weeding; (4) most people do not garden each day; and (5) the number of days of precipitation during the gardening season further diminishes the frequency of exposure.[18] Thus, a reasonable estimate for soil contact rate appears to be 0.5 mg soil/cm^2 skin.

Inhalation of Suspended Particulates

Contrary to what most citizens believe, inhalation of suspended particulates will rarely, if ever, constitute a route of entry of sufficient magnitude to dictate cleaning contaminated soils.[12,13,19,20] To calculate the inhalation load, one has to estimate the suspension of soil/dust into the air. Usually, only a fraction of the total suspended particulates (TSP) is assumed to be due to contaminated soil. For example, in the dioxin risk assessment conducted by the Centers for Disease Control (CDC) on Times Beach, an average air concentration of TSP of 0.14 mg/m^3 was assumed,[12] and most of it was thought to be from the contaminated site. When actual field data from Missouri and elsewhere were considered, TSP measurements were often much less than this value. For example, the EPA data showed[12,71] that TSP for a rural area was about 0.070 mg/m^3. For other geographical locations, concentrations of 0.15 mg/m^3 for urban environments and 0.10 mg/m^3 for rural environments have been observed.[14] These estimates appear appropriate for establishing soil cleanup levels in the absence of site-specific data.

Fraction of Soil from the Contaminated Source

Typically, only a fraction of soil ingested by a person in any one day will be from a contaminated source. In addition, for any particular site, usually only a fraction of the surface soil is contaminated. In setting de minimis or screening-level soil guidelines, choosing one fraction over another can be problematic since it is site-specific. However, it is reasonable to assume that the fraction of ingested, inhaled, or contacted soil that originates from the contaminated source is that fraction of a day spent at the source itself, whether it is a residence, a workplace, or a recreational park. Data from national time-use studies performed at the University of Michigan and analyzed by the U.S. EPA are useful for assessing exposure at impacted residential areas. These data indicate that adults spend approximately 40% of their waking hours at home,[37] while children and youths are expected to spend approximately 47% of their waking hours at the residence.[37] Workers can conservatively be expected to be exposed to soils at the workplace for 8 hours per 16 awake hours per day (50%).

Exposure Duration

People generally spend only a fraction of their lifetime in any one location. U.S. census data indicates that the average residence time is 9 years in any one house.[15] That same report states that the 90th percentile of the population remains in one residence for 30 years.[15] We suggest that this value be used to establish screening-level soil cleanup criteria at residential sites. For an industrial scenario, a 25-year exposure duration is considered an upper-bound value[7] and would appear to be adequately conservative for purposes of risk assessment.

Bioavailability

Environmental contaminants are able to cross biological barriers with varying degrees of efficiency. When chemicals are bound to soil, the efficiency of uptake decreases even more. The bioavailability, i.e., the percentage of a chemical in soil that is absorbed by humans, is governed primarily by (1) the physico-chemical properties of the contaminant, (2) the environmental matrix in which it is present, and (3) the nature of the biological membrane. Chemicals in soil are usually absorbed to a lesser degree than the chemicals in pure form.[18,72,73] Although this chapter addresses screening-level soil cleanup guidelines, due to the simplicity of the concept there is no reason to ignore chemical-specific bioavailability in setting those guidelines if the appropriate data are available.

In the absence of chemical-specific data, an approach similar to that used by the state of Michigan could be adopted.[74] In that state, oral bioavailability is assumed to be 100% for volatile organic compounds and 10% for semivolatiles and metals bound to soils. In the absence of chemical-specific data, dermal bioavailability is assumed to be 10% for volatile compounds and 1% for semivolatiles and metals.[74] Although this is a fairly gross approach to the problem, it seems adequate until more data are available.

Environmental Fate and Transport

Although not an exposure factor per se, understanding a chemical's environmental fate and transport is essential to properly characterize the hazards of contaminated soil. The first step in understanding the environmental hazard is to examine the physical and chemical properties of that compound.[14] In general, chemicals introduced into the environment may adsorb to soils, dissolve into bodies of water, leach from soil into migrating water, volatilize from soil and water into the atmosphere, or be taken up from soil by vegetation.[33,75] These are known as "transport" processes. A chemical may also undergo photo- or microbial degradation to other products. These are called "fate" processes.

Environmental fate and transport are dictated largely by a compound's physico-chemical properties.[33,76-80] These characteristics govern the ability of a chemical to move from one matrix to another. In short, transport and distribution of all these compounds in the environment is determined to a large extent by the following physico-chemical properties:

- water solubility
- organic carbon coefficient (K_{OC})
- vapor pressure
- Henry's law constant
- octanol/water coefficient (K_{OW})

The physico-chemical properties of concern that govern transport in soil are water solubility and organic carbon coefficient (K_{OC}). The K_{OC} is a measure of relative sorption potential and indicates the tendency of an organic chemical to be adsorbed to soil or sediment. It provides an important measure of a chemical's mobility in soil because it is largely independent of soil properties.[81]

Typically, K_{OC} values can range from one to ten million, with higher values indicative of greater sorption (binding) potential. A low K_{OC} indicates a greater potential for leaching from soil into groundwater, followed by rapid transport through an aquifer. In contrast, a higher K_{OC} suggests a relatively lower potential for leaching into groundwater over a period of time.

K_{OC} values representative of chemicals that are immobile to very mobile in soil[82,83] are as shown in Table 5.

Table 5. Representative K_{oc} Values

Mobility	K_{oc} Coefficient Range (Dimensionless)
Immobile	$K_{oc} > 2,000$
Low	$K_{oc} > 500$ and $< 2,000$
Intermediate	$K_{oc} > 150$ and < 500
Mobile	$K_{oc} > 50$ and < 150
Very Mobile	$K_{oc} < 50$

Vapor pressure and Henry's law constant are two measures of chemical volatility that are important in estimating releases to ambient air. Vapor pressure is an important determinant of the rate of vaporization from contaminated soil, but other factors including temperature, wind speed, degree of adsorption to soil, water solubility and soil conditions are also important.

Henry's law constant combines vapor pressure, water solubility, and molecular weight to characterize a chemical's ability to evaporate from aqueous media, including moist soils.[81] Typical values for Henry's constant (H_a) for compounds ranging from low to high volatility are as shown in Table 6.

The n-octanol-water partition coefficient (K_{OW}) describes the tendency of a nonionized organic chemical to accumulate in lipid tissue and to sorb onto soil particles or onto the surface of organisms or other particulate matter coated with organic material.[16] Although a powerful tool for understanding organic chemicals, it is not as good a predictor for inorganic chemicals, for metal organic complexes, or for dissociating and ionic organic compounds.[84] No published scale for interpretation of K_{OW} data is currently available. However, high K_{OW} values reflect lower water solubilities and higher bioconcentration factors.

ACCEPTABLE LEVELS OF RISK

The level of risk considered acceptable will ultimately define the soil cleanup level. In setting generic cleanup concentrations, the levels of acceptable risk are usually dictated, in part, by whether the risk is voluntary or involuntary. Traditionally, the residential scenario, and most others, are considered involuntary settings while industrial sites can be considered voluntary.

Table 6. Typical Henry's Constant Values

Volatility	Range of Henry's Constant (H_a) (atm-cu.m/mol)
Low	$H_a < 3 \times 10^{-7}$
Intermediate	$H_a > 3 \times 10^{-7}$ and $< 10^{-5}$
Moderate	$H_a > 1 \times 10^{-5}$ and $< 10^{-3}$
High	$H_a > 1 \times 10^{-3}$

Noncarcinogens

For noncarcinogens, the EPA and others have usually used the Reference Dose (RfD) for a specific chemical as the basis for establishing its soil cleanup levels.[2] For the purpose of setting de minimis screening-level soil guidelines, perhaps regulatory agencies should allow one-tenth the RfD to be the maximum allowable dose to be absorbed due to contaminated soil. Given the conservatism of many of the uncertainty factors used to establish RfDs, this value should adequately protect human health. The basis for setting acceptable contaminant levels for soils to protect the environment or wildlife involves the evaluation of several endpoints. For example, to protect wildlife against the adverse effects of a noncarcinogen, perhaps 1/100th the animal no-observed-adverse-effect-level (NOAEL) would be appropriate. Similarly, 1/100th the aquatic NOAEL could be used as a basis for establishing soil cleanup concentrations when protection of aquatic species is the primary concern. An alternative (or complementary) level of acceptable risk could involve setting the soil level so that the steady-state chemical concentration in fish would ensure that no more than one-tenth the human RfD for that could be due to eating fish near the contaminated site. The theoretical human receptor could be represented by the 80th percentile fish-eater. This level would most likely account for the possibility of additivity of chemicals, as well as other factors.

The contamination of sediment poses an additional environmental concern. One approach could involve ensuring that the steady-state concentration of the chemical in a bottom-feeding fish (e.g., catfish) not exceed two times the background uptake of the chemical or result in a human dose not exceeding 1/10th the Acceptable Daily Intake (ADI). For groundwater protection, the goal might be to prevent the contaminant from reaching groundwater, with perhaps 50% of the Maximum Contaminant Level (MCL) as the key endpoint.

Carcinogens

Establishing acceptable levels of risk for carcinogens has been a topic of considerable debate within the risk assessment community. Inasmuch as soil cleanup levels are often "driven" by the presence of carcinogens, it is important to understand the regulatory history of "acceptable risk." It is a common misperception within risk assessment that all occupational and environmental regulations have, as their goal, a theoretical maximum cancer risk of 1 in 100,000 (1×10^{-5}) or 1 in 1,000,000 (1×10^{-6}). When a risk falls above this level, the public and the media quite often misinterpret this as a serious threat to public health. In a 1987 journal article, the former commissioner of the Food and Drug Administration, Dr. Frank Young, discussed this misunderstanding.[85]

> The risk level of one in one million is often misunderstood by the public and the media. It is not an actual risk; i.e., we do not expect one out of every million people to get cancer if they drink decaffeinated coffee. Rather, it is a mathematical risk based on scientific assumptions used in risk assessment. The FDA uses a conservative estimate to ensure that the risk is not understated. We interpret animal test results conservatively and we are extremely careful when we extrapolate risks to humans. When the FDA uses the risk level of one in one million, it is confident that the risk to humans is virtually nonexistent.

In short, the cancer risk levels assumed by some policy makers to represent trigger levels for regulatory action, actually represent levels of risk that are so small as to be of negligible concern.

Recent reviews[87] indicate that the theoretical cancer risks associated with currently enforced environmental regulations are in the vicinity of 1 in 100,000, not 1 in 1,000,000. In a retrospective review of the use of cancer risk estimates in 132 federal decisions, Travis et al.[86] examined the level of cancer risk that triggered regulatory action. The authors considered three measures of risk: individual risk (an upper-limit estimate of the probability that the most highly exposed individual in a population will develop cancer as a result of a lifetime exposure), size of the population exposed, and population risk (an upper-limit estimate of the number of additional cases of cancer in the exposed population). Travis et al.[87] found that for exposures resulting in a small-population risk, the level of risk above which agencies almost always acted to reduce risk was approximately 4×10^{-3}. For large-population risks (the entire U.S. population) agencies typically acted on risks of about 3×10^{-4}. For effects on small populations, regulatory action was never taken for individual risk levels below 1×10^{-4}. For large-population effects, the de minimis risk level dropped to 1×10^{-6}. Consequently, the level of acceptable individual risk is usually dictated by the size of the exposed population.

Recently, final revisions to the National Contingency Plan[88] have set the acceptable risk range between 10^{-4} and 10^{-6} at hazardous waste sites regulated under CERCLA. In the recently promulgated Hazardous Waste Management System Toxicity Characteristics Revisions (55 FR 11798–11863), the U.S. EPA has stated that:

> For drinking water contaminants, EPA sets a reference risk range for carcinogens at 10^{-4} to 10^{-6} excess individual cancer risk from lifetime exposure. Most regulatory actions in a variety of EPA programs have

generally targeted this range using conservative models which are not likely to underestimate the risk.

Interestingly, the U.S. EPA has selected and promulgated a single risk level of 1 in 100,000 (1×10^{-5}) in the Hazardous Waste Management System Toxicity Characteristics (TC) Revisions (55 FR 11798–11863). In their justification, the U.S. EPA cited the following rationale:

> The chosen risk level of 10^{-5} is at the midpoint of the reference risk range for carcinogens (10^{-4} to 10^{-6}) targeted in setting MCLs. This risk level also lies within the reference risk range (10^{-4} to 10^{-6}) generally used to evaluate CERCLA actions. Furthermore, by setting the risk level at 10^{-5} for TC carcinogens, EPA believes that this is the highest risk level that is likely to be experienced, and most if not all risks will be below this level due to the generally conservative nature of the exposure scenario and the underlying health criteria. For these reasons, the Agency regards a 10^{-5} risk level for Group A, B, and C carcinogens as adequate to delineate, under the Toxicity Characteristics, wastes that clearly pose a hazard when mismanaged.

Few state regulatory agencies have adopted a one in one million (1×10^{-6}) risk criterion in making environmental and occupational decisions. The states of Virginia, Maryland, Minnesota, Ohio, and Wisconsin have employed, or propose to use, the 1 in 100,000 (1×10^{-5}) level of risk in their risk management decisions (personal communications with state agencies, 1990). The state of Maine Department of Human Services (DHS) uses a lifetime risk of 1 in 100,000 as a reference for nonthreshold (carcinogenic) effects in its risk management decisions regarding exposures to environmental contaminants.[89] Similarly, a lifetime incremental cancer risk of 1 in 100,000 is used by the Commonwealth of Massachusetts as a cancer risk limit for exposures to substances in more than one medium at hazardous waste disposal sites.[90] This risk limit represents the total cancer risk at the site associated with exposure to multiple chemicals in all contaminated media. The state of California has also established[91] a level of risk of 1 in 100,000 for use in determining levels of chemicals and exposures that pose no significant risks of cancer under the Safe Drinking Water and Toxic Enforcement Act of 1986 (Proposition 65). Workplace air standards developed by the Occupational Safety and Health Administration (OSHA) typically reflect theoretical cancer risks of about 1 in 1,000 (1×10^{-3}) or greater.[92]

ESTABLISHING SAFE SOIL CLEANUP GUIDELINES: TWO EXAMPLES

Our conceptual approach to setting screening-level soil guidelines can be illustrated by examining two chemicals which have been of significant concern when found in soil: 2,3,7,8-tetrachlorodibenzo-p-dioxin (TCDD) and the volatiles benzene, toluene, and xylene (BTX), which are indicator chemicals for gasoline.

TCDD

Environmental Fate and Distribution

The chemico-physical properties of TCDD (Table 7) indicate that this chemical is virtually water insoluble, has a relatively high lipid solubility, has a high soil-binding constant (K_{OC}), almost no vapor pressure, a low Henry's law constant, and a fairly substantial bioconcentration factor (BCF). Thus, TCDD is relatively nonvolatile and essentially immobile in soil. Moreover, one would expect to find TCDD in sediment rather than in the water column, and its bioaccumulation potential in the food chain would be very high.

Table 7. Chemical and Physical Properties of TCDD

Chemical	Physical Property
Molecular weight (g/mol)	322
Water solubility (mg/L)	2.0×10^{-4}
Lipid solubility (log K_{ow})	6.72
Soil adsorption constant(K_{oc})(mL/g)	3,300,000
Vapor pressure (mm Hg)	1.7×10^{-6}
Theoretical bioconcentration factor (BCF)	5,000–50,000

Residential, Industrial, and Recreational

The method for deriving health-based soil cleanup levels for the residential, industrial, and recreational situations is very similar. In many ways, calculating these levels is a risk assessment in reverse. First, potential exposure routes are identified and exposure parameters are identified to quantitatively describe exposure potential. By examining the physico-chemical properties of the chemicals of interest, one can determine which exposure routes will be significant (e.g., ingestion and dermal contact could be significant for TCDD, whereas inhalation of vapors would not). Second, the relative contribution of each of the routes of exposure to the total dose can be determined by assigning a hypothetical chemical concentration in soil (e.g., 1 ppm) and apportioning exposure according to standard exposure equations (see below). By doing this, the relative contribution of each exposure pathway can be used to calculate the amount of total dose that can safely be received via each route of exposure.[32]

The following formulae were used to calculate the lifetime average daily dose (LADD) for humans exposed to soil via ingestion (Equation 1), dermal contact (Equation 2), inhalation of vapors (Equation 3), and inhalation of suspended particulates (Equation 4). Inhalation Equation 3 is used for evaluating BTX inhalation exposure and Equation 4 is used to evaluate TCDD inhalation exposure

$$LADD = HSC \times ED \times EF \times IR \times CF \times FC/BW \times LF \quad (1)$$

$$LADD = HSC \times ED \times EF \times CF \times FC \times [TO + (TI \times IDF)] \times SA \times M/BW \times LF \quad (2)$$

$$LADD = CA \times ED \times EF \times BR \times CF/BW \times LF \quad (3)$$

$$LADD = HSC \times ED \times EF \times BR \times TSP \times CF \times FC \times PDa \times [TO + (TI \times IDF)]/BW \times LF \quad (4)$$

Where:

LADD = Lifetime Average Daily Dose
ED = Exposure Duration (years)
EF = Exposure Frequency (fraction of a year)
IR = Ingestion Rate (mg/day)
CF = Conversion Factor (kg/mg)
FC = Fraction of Soil Originating from Source
TO = Percent of Time Spent Outdoors
TI = Percent of Time Spent Indoors
IDF = Indoor Dust Factor
SA = Skin Surface Area (cm^2)
M = Mass of Soil Adhering to Skin (mg/cm^2)
BR = Breathing Rate (m^3/day)
CA = Modeled air concentration (mg/m^3) based on 1 mg/kg soil concentration
TSP = Concentration of Total Suspended Particulates (mg/m^3)
PDa = Fraction Deposited to Lung Tissue
BW = Body Weight (kg)
LF = Average Lifetime (years)
HSC = Hypothetical Soil Concentration (1 mg/kg)

The parameters to be used in these calculations for the residential, occupational and recreational scenarios are presented in Tables 8 through 11.

Finally, by substituting a dose that represents an acceptable level of risk (see above) in the standard exposure equations and solving the equation for soil concentration, a chemical concentration in soil that should result in no adverse health effects can be derived. For TCDD, the relative contribution of the oral, dermal, and inhalation (particulates) pathways to the total dose under an industrial scenario was 16%, 81%, and 2%, respectively. The conceptual approach is discussed in significant detail in the EPA manual.[32] Multiplying these proportions by the Risk Specific Dose (RsD) at 10^{-5} risk (1.0×10^{-10} mg/kg-day), based on the recent re-evaluation of the Kociba et al. bioassay (1978) of TCDD[23,93] for the following results in apportioned RsDs as follows:

$$\text{Oral: } 1.7 \times 10^{-10} \text{ mg/kg-day}$$
$$\text{Dermal: } 8.4 \times 10^{-10} \text{ mg/kg-day}$$
$$\text{Inhalation: } 2.3 \times 10^{-11} \text{ mg/kg-day}$$

These values can then be substituted for the Lifetime Average Daily Dose (LADD) in the above equations and the soil concentration, Cs, can be calculated as follows:

$$Cs_{oral} = (\text{Apportioned RsD} \times BW \times LF)/(IR \times CF \times B \times FC \times EF \times ED) \quad (5)$$

$$Cs_{dermal} = (\text{Apportioned RsD} \times BW \times LF)/(M \times SA \times [TO + (TI \times IDF)] \times FC \times CF \times B \times EF \times ED) \quad (6)$$

$$Cs_{inhal} = (\text{Apportioned RsD} \times BW \times LF)/(BR \times TSP \times FC \times PDa \times CF \times B \times EF \times ED\ [TO + (TI \times IDF)] \quad (7)$$

Thus, using this approach, the screening-level soil cleanup guideline for TCDD at an industrial site would be approximately 50 ppb. This is smaller than the value of 130 ppb recently proposed by Paustenbach et al.[94] and such differences are to be expected since the procedures suggested here are intended to provide fairly conservative (rather than precise) risk estimates. Soil cleanup guidelines for TCDD for the residential and recreational scenarios are presented in Table 12. It is important to note that ingestion of homegrown vegetables was not included in the residential cleanup levels because several studies suggest that uptake and translocation of TCDD from soils to plants is negligible, and does not present an appreciable health risk to humans.[95-98]

Table 8. Exposure Parameters Used in This Evaluation to Estimate the Human Uptake of Chemicals Due to Ingestion of Contaminated Soil (Used in Equation 1)

	Scenario		
Parameter	Residential	Occupational	Recreational
Exposure duration (ED)	30 years	25 years	30 years
Exposure frequency (EF) (fraction of year)	0.962 days/days	0.343 days/days	0.011 days/days
Average lifetime (LF)	70 years	70 years	70 years
Body weight (BW)	70 kg	70 kg	70 kg
Ingestion rate (IR)	12 mg/day	10 mg/day	10 mg/day
Conversion factor (CF)	1×10^{-6} kg/mg	1×10^{-6} kg/mg	1×10^{-6} kg/mg
Fraction of soil originating from source (FC)	0.47	0.5	1

Table 9. Exposure Parameters Used in This Evaluation to Estimate the Human Uptake Via the Skin Following Contact with Contaminated Soil (Used in Equation 2)

Parameter	Scenario		
	Residential	Occupational	Recreational
Exposure duration (ED)	30 years	25 years	30 years
Exposure frequency (EF) (fraction of year)	0.962 days/days	0.343 days/days	0.011 days/days
Average lifetime (LF)	70 years	70 years	70 years
Body weight (BW)	70 kg	70 kg	70 kg
Conversion factor (CF)	1×10^{-6} kg/mg	1×10^{-6} kg/mg	1×10^{-6} kg/mg
Percent of time spent outdoors (TO)	19%	100%	100%
Percent of time indoors (TI)	81%	0%	0%
Indoor dust factor (IDF)	0.75	0.75	0.75
Skin surface area (SA)	1980 cm^2	1980 cm^2	1980 cm^2
Mass of soil adhering to skin (M)	0.5 mg/cm^2	0.5 mg/cm^2	0.5 mg/cm^2
Fraction of soil originating from source (FC)	0.5	1	0.5

Table 10. Exposure Parameters Used in This Evaluation to Estimate the Uptake by Humans of Contaminated Soil in Suspended Particulates Via Inhalation (Equation 4)

Parameter	Scenario		
	Residential	Occupational	Recreational
Exposure duration (ED)	30 years	25 years	30 years
Exposure frequency (EF) (fraction of year)	0.962 days/days	0.343 days/days	0.011 days/days
Average lifetime (LF)	70 years	70 years	70 years
Body weight (BW)	70 kg	70 kg	70 kg
Breathing rate (BR)	20 m^3/day	11 m^3/day	20 m^3/day
Concentration of total suspended particulates (TSP)	0.25 mg/m^3	0.1 mg/m^3	0.1 mg/m^3
Conversion factor (CF)	1×10^{-6} kg/mg	1×10^{-6} kg/mg	1×10^{-6} kg/mg
Fraction of soil originating from source (FC)	0.47	0.5	1
Fraction deposited to lung tissue (PDa)	0.25	0.25	0.25
Percent of time spent outdoors (TO)	19%	100%	100%
Percent of time indoors (TI)	81%	0%	0%
Indoor dust factor (IDF)	0.75	0.75	0.75

Table 11. Exposure Parameters Used in This Evaluation to Estimate the Uptake by Humans Due to the Inhalation of Vapors from Contaminated Soil (Used in Equation 3)

Parameter	Scenario		
	Residential	Occupational	Recreational
Exposure duration (ED)	30 years	25 years	30 years
Exposure frequency (EF) (fraction of year)	0.962 days/days	0.343 days/days	0.011 days/days
Average lifetime (LF)	70 years	70 years	70 years
Body weight (BW)	70 kg	70 kg	70 kg
Breathing rate (BR)	20 m^3/day	11 m^3/day	20 m^3/day
Conversion factor (CF)	1×10^{-6} kg/mg	1×10^{-6} kg/mg	1×10^{-6} kg/mg

Table 12. Suggested Cleanup Levels for TCDD in Soil Based on the Methodology Proposed

Site	Cleanup Levels
Residential	25 ppb
Industrial	50 ppb
Recreational	200–1000 ppb
Agricultural	400–10,000 ppb
Groundwater	10,000 or greater
Runoff	200 ppb[a]
Wildlife	100 ppt

[a]Based on a 200 acre site.

Agricultural

As mentioned above, agricultural cleanup levels tend to be driven by the ingestion of soil by grazing animals. The potential uptake of TCDD by grazing cattle and dairy cows has been examined extensively elsewhere.[27] Their examination suggested that soil cleanup levels for TCDD under an agricultural scenario could range from 400 to greater than 5,000 ppb, depending on the availability for forage as an animal feed source, the extent cattle are pastured, and the length of time the animals are held in feed lots prior to slaughter. For example, lactating dairy cows are rarely pastured, and some form of supplemental feeding is almost always employed.[27] The fattening period for nonlactating cows may be as long as 150 days, during which time animals can gain as much as 60% to 70% in body weight; such lipophilic chemicals as TCDD will therefore be diluted in the expanding body fat pool. In some settings, such soil concentrations may not pose a hazard to the animal, or humans who eat them, but the runoff hazard might be too great to be acceptable.

Runoff

Runoff of TCDD-contaminated soil into nearby streams or lakes can sometimes become the controlling factor for establishing soil cleanup levels. The potential for surface runoff is important because relatively small amounts of TCDD-contaminated soil may eventually contaminate nearby streams and produce potentially excessive levels of TCDD in fish.

Due to TCDD's low water solubility, low vapor pressure, and strong tendency to adsorb to organic material, it does not readily dissolve in runoff waters. Consequently, it is not easily transported across the soil surface in a dissolved phase.[99] However, since TCDD is bound tightly to the soil matrix, surface transport from TCDD-contaminated areas may be possible through erosion of soil-bound TCDD.

One of the most common methods used to estimate the rate of surface runoff is the Universal Soil Loss Equation (USLE)[100] and sediment delivery ratios (SDR).[2] Although originally designed to estimate runoff of soils from agricultural lands, modifications and adaptations of this equation have been devised to accommodate a wide variety of chemical contamination scenarios.[15,101] The application of the SDR was intended to correct for the overestimation of soil loss produced by the USLE.[31,102] It was developed from empirical data that quantified sediment movement in streams. However, the equations did not differentiate between sediments contributed by bank or gully erosion and actual soil movement from the land surface to the stream.[102] The results of this equation, presented below, can be used to calculate the amount of soil-bound TCDD that might run off to a neighboring stream and accumulate in sediments:

$$SL = Rf \times Ef \times SLf \times Cf \times SPf \tag{8}$$

Where:

SL = Site soil loss rate (kg/acre/yr)
Rf = Rainfall/runoff factor $(\text{years})^{-1}$
Ef = Erodibility factor (kg/acre)
SLf = Slope-Length factor
Cf = Cover/management factor
SPf = Support practice factor

Parameters used to estimate soil loading to a stream neighboring a hypothetical site are based on U.S. EPA guidance[99] and are summarized in Table 13. The cumulative concentration of TCDD in sediment is then,

$$CSDC \times \sum_{i=1}^{n} \text{ of } SDCi$$

$$SDCi = (SC \times SL \times SA \times SSD)/(WA \times WA \times WSD) \tag{9}$$

Where:

CSDC = Cumulative Sediment Concentration (μg/kg)
SDCi = Sediment Concentration in year i (μg/kg)
SC = Soil Concentration (μg/kg)
SL = Site Soil Loss Rate (kg/acre-year)
SA = Site Area (acres)
SSD = Site Sediment Delivery Ratio
WA = Watershed Area (acres)
WL = Watershed Soil Loss Rate (kg/acre-year)
WSD = Watershed Sediment Delivery Ratio

Parameters used to estimate cumulative sediment concentration are presented in Table 13. Uptake of TCDD by fish was calculated using the following equation and the bioaccumulation index methodology described by Goeden and Smith[103] and Cook.[104]

$$FC = CSDC \times Lf \times BI \times 1/OCf \quad (10)$$

Where:

FC = Fish Concentration (μg/kg)
CSDC = Cumulative Sediment Concentration (μg/kg)
Lf = Lipids Factor (0.03)
BI = Bioavailability Index (0.15)
OCf = Organic Carbon Factor (0.1)

The catfish was selected in this analysis as a representative bottom-feeding fish that is commercially and recreationally desirable by fishermen and consumers.

Table 13. Parameters Used to Estimate Soil-Loading Due to Runoff to a Stream Neighboring a Hypothetical Residential or Industrial Site

Site	Parameters
Rainfall runoff factor	375 $years^{-1}$
Erodibility factor	390 kg/acre
Slope-length factor	1.5
Cover/management factor	0.011
Support practice factor	1
Rate of soil loss from site	2413 kg/acre/yr
Area of site	200 acres
Sediment delivery ratio for site	0.77 (dimensionless)
Area of watershed	9000 acres
Rate of soil loss from watershed	2413 kg/acre/yr
Sediment delivery ratio for watershed	56.8

The potential for exposure to TCDD-contaminated soil through the consumption of fish from a neighboring stream was calculated using the following equation.

$$LADD = FC \times CR \times CKf \times EDy/BW \times LF \quad (11)$$

Where:

LADD = Lifetime Average Daily Dose (μg/kg-day)
FC = Fish Concentration (μg/kg)
CR = Fish Consumption Rate (kg/day)
CKf = Cooking Loss Factor
BW = Body Weight (kg)
EDy = Exposure Duration (days when fish are contaminated)
LF = Lifetime (days)

Parameters used to estimate exposure through ingestion of catfish caught at a stream neighboring a hypothetical residential or industrial site are summarized in Table 14.

Using the assumptions and equations described above, together with the RsD at 10^{-5} risk, an acceptable or de minimis soil concentration of TCDD for areas where runoff is a concern would be approximately 200 ppb.

Groundwater, Wildlife, and "De Minimis"

To set soil guidelines for TCDD to protect groundwater is essentially unnecessary. The chemico-physical properties of TCDD show that it is essentially immobile in soil and virtually not water-soluble, so groundwater is unlikely to be threatened by TCDD. The only exception might be in those situations where the TCDD is present (in solution) in solvents in soil. In such cases, the TCDD has been shown to move through soil in the solvent (e.g., co-elution).

Because of the diversity of wildlife populations, it is difficult to determine a non-species-specific soil cleanup level for TCDD. The health risks to birds and deer exposed to TCDD in soil was recently evaluated for sludges contaminated with TCDD which were applied to soil.[23,93] Primary food sources (e.g., insects and earthworms) for the receptor species need to be identified in these assessments and bioconcentration factors of TCDD for those food sources and absorption coefficients estimated. As shown in the papers by Keenan et al.,[23] the calculations can be quite complex and are outside the scope of this chapter. However, the conclusions of other assessments is that soil concentrations of approximately 100 ppt to 10 ppb have been shown conservative and health-protective to protect most species.

In summary, the seven screening-level soil cleanup guidelines for TCDD can range from 0.01 ppb to about 10,000 ppb (Table 12). Clearly, there will be instances where, for example, recreational or industrial sites will also have a runoff

Table 14. Exposure Parameters Used to Estimate Uptake of TCDD by Adults Due to the Ingestion of Fish Caught at a Stream Neighboring a Hypothetical Residential or Industrial Site

Parameters	Value
Fish consumption rate	1.48 g/day
Proportion of TCDD remaining after cooking	0.5
Exposure duration	58 years
Body weight	68.5 kg

problem. In those cases, it may be appropriate to default to the more conservative cleanup guideline.

BENZENE, TOLUENE, XYLENE

Unlike TCDD, the chemico-physical properties of benzene, toluene, and xylene (BTX) (Table 15) indicate that they are relatively mobile in soil and have a relatively high potential for volatilizing into ambient air. Soil binding is much less tenacious for these three chemicals, as is their potential for bioconcentration

Table 15. Select Chemical and Physical Properties of Benzene, Toluene, and Xylene (BTX) Often Needed to Evaluate the Environmental and Human Health Risks

Chemical	B	T	X
Water solubility (mg/L)	1000	515	160
Lipid solubility	2.0	2.8	3.0
Soil adsorption constant (K_{oc})	70	2.5	—
Vapor pressure (mm Hg)	98	28	6.7
Theoretical BCF	24	50	2.2
Soil half-life (days)	20	5	5
Odor threshold	3	2.5	1.0
Flash point (°C)	−11	10	29

or biomagnification.[14] Therefore, the pertinent exposure pathways for these chemicals include potential exposures via soil ingestion, dermal exposure to soil, and inhalation of vapors. As volatiles are not likely to remain adhered to particulates during air transport, inhalation of particulates is not a viable exposure pathway and is therefore not evaluated.

For all three chemicals, a health-conservative assumption of the amount absorbed through the skin of 100% is assumed. Based on the total risk via all routes combined, health-based cleanup levels are back-calculated from the estimated risk levels. Exposure and risks are estimated in a similar fashion to that employed for TCDD, based on a hypothetical soil concentration of 1 mg/kg. However, since toluene and xylene are noncarcinogenic, the daily doses of these chemicals are not averaged over a lifetime to determine LADDs, but rather they are calculated for the duration of exposure as Average Daily Doses (ADDs). The exposure estimates are combined with dose-response information to provide quantitative estimates of carcinogenic and noncarcinogenic health risk via each route individual and via all routes combined.

Potential exposures for benzene via soil ingestion dermal contact, and inhalation are calculated in a similar fashion as TCDD, as shown in Equations 1, 2, and 3. Potential exposure for toluene and via ingestion (Equation 12), dermal contact (Equation 13), and inhalation (Equation 14) are estimated as follows:

$$ADD = HSC \times EF \times IR \times CF \times FC/BW \quad (12)$$

$$ADD = HSC \times EF \times CF \times FC \times [TO + (TI \times IDF)] \times SA \times M/BW \quad (13)$$

$$ADD = HSC \times CA \times EF \times BR \times CF/BW \quad (14)$$

Where:

ADD = Average Daily Dose (mg/kg-day)
EF = Exposure Frequency (fraction of a year)
IR = Ingestion Rate (mg/day)
CF = Conversion Factor (kg/mg)
FC = Fraction of Soil Originating from Source
TO = Percent of Time Spent Outdoors
TI = Percent of Time Spent Indoors
IDF = Indoor Dust Factor
SA = Skin Surface Area (cm^2)
CA = Modeled Air Concentration (mg/m^3)
M = Mass of Soil Adhering to Skin (mg/cm^2)
BW = Body Weight (kg)
HSC = Hypothetical Soil Concentration (1 mg/kg)

The parameter values for the soil ingestion dermal contact, and inhalation equations are shown in Tables 9, 10, and 11, respectively.

The CA term in Equations 3 and 14 is modeled from the hypothetical soil concentration based on two steps. Firstly, an emissions model is used to estimate the emission rate of the chemicals from the soil based on various properties of the soil and chemicals. The model chosen for this analysis is the Behavior Assessment Model (BAM). BAM accounts for all three transport mechanisms in soil (vaporization, capillary action, liquid-phase diffusion). BAM is often considered appropriate for these situations because:

- The vapor emission rates predicted by the model are based on several conservative loss pathways, such as transport of a chemical subject to volatilization at the soil surface and leaching in the soil column via evapotranspiration

- The model always conserves mass
- The model takes into account the time-varying depletion of the contamination in soil, since only a finite amount of chemical is initially present.

The BAM can incorporate a microbial or chemical decay rate specific to the contaminants of concern in estimating vapor flux from soil. This factor is conservatively set equal to zero (i.e., no decay) since contaminant-specific decay rates are generally not available or there is a wide range of reported values in the scientific literature. In addition, all soil properties are assumed to be uniform and constant throughout the soil column. Chemical-specific and site-specific soil parameters used in this analysis are presented in Tables 16 and 17, respectively.

In order to predict average flux rates with BAM, the following variables need to be estimated:

- Ct = total soil solute concentration (mg/cm^3)
- Ve = the effective solute adjective velocity (cm/sec)
- De = the effective diffusion coefficient (cm^2/sec)
- He = the effective stagnant air boundary coefficient (cm/sec).

$$Ct = (S)(Rho)(1 + Grav)/1000 \text{ g/kg} \tag{15}$$

The effective seepage velocity, Ve, can be computed using the following equation:

$$Ve - Jw/R1 \tag{16}$$

Where:

$$R1 = (Rho)(Kd) + Pw + (Pa)(Kh)$$

The effective diffusion coefficient, De, is estimated by:

$$De = Dg/Rg + D1/R1 + (\alpha)(|Ve|) \tag{17}$$

Where:

$$Dg = (Dg_{air})(Pa^{10/3}/Pt^2)$$
$$Rg = Pa + [(Pw + (Pw + (Rho)(Kd))/Kh]$$
$$D1 = (D1_{water})(P1^{10/3}/Pt^2)$$
$$R1 = (Rho)(Kd) + Pw + (Pa)(Kh)$$

The effective transport coefficient, He, of gas across the stagnant air boundary layer is defined by Jury et al.[105] by the expression:

$$He = Dg_{air}/[(Rg)(d)]$$

Where:

$Rg = Pa + [(Pw + (Rho)(Kd))/Kh]$
$d = (D_{wv(air)})(Rho_{wv(sat)})(1 - RH)/[(2)(E)(Rho_{water})]$

The average depth of contamination was assumed to occur at the surface down to a depth of 305 cm for all chemicals of concern. The soil is assumed to be initially uncontaminated below 305 cm.

The average depth of contamination, time duration, Ct, De, He, and Ve for each chemical of interest are then entered into BAM to obtain soil flux rates. The flux rates predicted by the BAM for BTX are as follows:

- Benzene = 2.76×10^{-10} mg/cm^2-sec
- Toluene = 2.55×10^{-10} mg/cm^2-sec
- Xylene = 2.60×10^{-10} mg/cm^2-sec

The second step is to develop a "Box Model" so as to estimate an air concentration from the emission flux rate calculated above. A box model typically uses the concept of a theoretical enclosed space of a box over the area of interest. The model assumes the emission of compounds into the box with their dilution based on a calm wind speed. Airborne concentrations for this enclosed space are then calculated and used as onsite air contaminant concentrations. The box model fails to fully take into account the various processes of dispersion, and may lead to the prediction of relatively high exposure concentrations. The box model is based on a mass balance expression that assumes that the soil-gas flux is instantly mixed with the air flowing across a specified area. Mathematically, this is expressed as the mass rate entering the "exposure box," divided by the volumetric rate of air that flows through the exposure box:

$$CA = \frac{Js \times A_{emit}}{V_{wind} \times A_{wind}} \tag{18}$$

Where:

CA = Air concentrations in mg/m^3
Js = Emission rate (mg/m^2-sec) [milligram per square meter second (mg/m^2-sec)]
A_{emit} = Emitting area (m^2)
V_{wind} = Wind velocity (m/s)
A_{wind} = Area of wind (m^2)

Table 16. Physical and Chemical Properties for Benzene, Toluene, and Xylene (BTX)

Chemical	Organic Soil Carbon-Liquid Partition Coefficient (K_{oc}) (mL/g)[a]	Henry's Law Constant (H) (atm-m^3/mol)[a]	Dimensionless Henry's Constant (K_h)[b] (cm^3liq/cm^3gas)	Solid-Liquid Partition Coefficient (K_d)[c] (mL/g)	Air Diffusion Coefficient (Dg_{air})[d] (cm^3/cm-sec)	Water Diffusion Coefficient (Dl_{water}) (cm^3/cm-sec)
Benzene	83	5.59×10^{-3}	0.23	0.83	0.087	7.47×10^{-6}
Toluene	300	6.37×10^{-3}	0.26	3	0.078	7.47×10^{-6}
Xylene	240	7.04×10^{-3}	0.29	2.4	0.072	7.47×10^{-6}

[a]USEPA.[52]
[b]$K_h = H/[(Ru)(T)]$ where Ru = 8.21E-05 atm-m^3/mole-K and T = 293K.
[c]$K_d = (K_{oc})(f_{oc})$ where $f_{oc} = 0.01$.
[d]Sheehan.[13]

Table 17. Soil Specific Parameters Used in BAM Model for Estimating Vapor Loss from Soils

Parameter	Description	Value	Source/Rationale
Rho	Dry soil bulk density	1.88 g/cm^3	Gravel
Pt	Total soil porosity	0.31 cm^3/cm^3	Pt = Pw + Pa
Pw	Volumetric water content	0.155 cm^3/cm^3	Assumed[a]
Pa	Volumetric air content	0.155 cm^3/cm^3	Assumed[a]
foc	Organic soil carbon	0.01	Assumed
Rho_{water}	Density of water	1 g/cm^3 liq	—
$Rho_{wv(sat)}$	Density of saturated water vapor	1.73E-05 g/cm$^3_{gas}$	—
Grav	Gravimetric water content	8.24E-02	Grav = (Pw)(Rho_{water})/Rho
Jw	Water evaporation rate	1.08 × 10^{-6} cm/sec	Assumed
E	Net evaporation rate	– Jw	
Dwv_{air}	Diffusion coefficient of water vapor through open air	0.23 cm$^3_{gas}$/cm-sec	—
RH	Relative humidity of stagnant zone	0.5	Assumed
Ru	Universal gas constant	8.21E-05 atm-m^3/mole-K	—
T	Temperature	20°C	Assumed
t	Time-span for average flux calculation	70 years	Assumed
Z	Depth of contamination	10 feet	Assumed

[a]50% air content and 50% water content.

For this assessment, assumptions were made regarding the length, width, and height of the exposed area above the soil contamination. The emitting area (A_{emit}) was assumed to be 1.02E + 05 m^2, the wind velocity (V_{wind}) 1 mph, and the area of wind, 548 m^2. The resulting air concentrations can then be used in Equations 3 and 14 to estimate LADD and ADD values for the inhalation of vapors.

Safe cleanup levels for BTX are derived from the exposure estimates in two different ways. Since benzene is considered to be a human carcinogen, carcinogenic risks are calculated based on the Slope Factor (SF) of 2.9E − 02 (mg/kgday)$^{-1}$ published in the IRIS database. This value is considered appropriate for the inhalation, dermal and ingestion exposure routes. Toluene and xylene are noncarcinogenic; therefore, safe cleanup levels are calculated in the form of a

Hazard Quotient (HQ), for a specific exposure route, and a Hazard Index (HI) for all routes combined. A HQ represents the ratio of the estimated exposure of a chemical to its RfD and a HI is the sum of all route-specific HQs. Toluene has published RfDs of 2E01 mg/kg-day via ingestion and 5.7E − 01 mg/kg-day via inhalation. Xylene has RfDs of 2E + 00 for ingestion and 8.6E − 02 for inhalation.

In order to estimate health-based cleanup levels (HBCLs) for these chemicals, a simple ratio approach is used to relate total risk or HI to soil concentration. This is permitted, since all of the above equations used in the exposure and risk calculations are linear. The following relationship is therefore true for each chemical under each exposure scenario:

$$\text{Risk or HI} = \text{CS} \times \text{K} \tag{19}$$

Where:

Risk or HI = Cancer risk or Hazard Index for all routes combined (unitless)
CS = Chemical concentration in soil (mg/kg)
K = Constant $(\text{mg/kg})^{-1}$.

Since plausible cancer risks and HIs are known for each chemical under each scenario at a soil concentration of 1 mg/kg, rearrangement of the above equation gives values of K for each chemical under each scenario. By solving the equation for C_{soil} at a risk level of 1 in 100,000 or an HI of 1, these values of K can be used to generate HBCLs. HBCLs for each of the chemicals under each scenario are presented in Table 18.

The soil concentrations presented in Table 18 assume that two other factors do not come into play: odor threshold and/or significant degradation of groundwater. The odor threshold for BTX has been studied by several investigators; the weight-of-evidence suggests the BTX is detectable at approximately 30 ppm in air to the unacclimated nose. Acceptable soil concentrations for BTX developed for the protection of groundwater must be developed on a site-specific basis. Depending on the depth to groundwater and potential use of the aquifer, protection of groundwater may be the driving concern for establishing soil cleanup level. This is especially true for benzene.

Soil cleanup levels for protection of wildlife, like protection of groundwater, will be very site-specific for BTX because of the variety of species that could potentially be involved. The setting of an agricultural and runoff guideline is much less difficult with BTX, compared to TCDD, because (1) the potential for plant uptake is low, (2) the volatility and short half-life of BTX in soil usually will preclude significant soil runoff. Consequently, concentrations in excess of a few hundred ppm will not usually present a significant problem (except, perhaps, phytotoxicity).

Table 18. Health Based Soil Cleanup Levels for BTX for Seven Typical Exposure Scenarios

Area	Benzene	Toluene	Xylene
Residential	2.5 ppm	2,000 ppm	300 ppm
Industrial	14 ppm	10,000 ppm	1,400 ppm
Recreational	250 ppm	170,000 ppm	25,000 ppm
Agricultural	400 ppm	2,000 ppm	1,000 ppm
Groundwater	site-specific	site-specific	site-specific
Runoff	site-specific	site-specific	site-specific
Wildlife	site-specific	site-specific	site-specific

COMPARISON TO CURRENT STATE GUIDELINES

The soil cleanup guidelines presented here are higher than many that have been adopted in many states. Although there are no formal or official (i.e., regulatory-sanctioned) health-based cleanup levels for TCDD in soil, cleanup levels of 1 ppb for residential areas and 20 ppb for industrial areas were adopted at several sites throughout the United States during the 1980s.[94,106] These informal cleanup levels are primarily a result of a paper published in 1984 by the U.S. Centers for Disease Control (CDC)[12] and a subsequent counter-paper by Paustenbach et al.[20] The CDC paper, which discussed an approach to setting an acceptable soil concentration of TCDD in residential settings, used a number of conservative exposure assumptions in the calculations. The paper concluded that "one part per billion of 2,3,7,8-TCDD in soil is a reasonable level at which to begin consideration of an action to limit human exposure to contaminated soil."[12] This recommendation formed the basis for the cleanup criteria of 1 ppb TCDD which has been applied for the cleanup of residential sites.

During the past eight years, several authors have identified health-based cleanup levels for TCDD in soil using risk assessment methods.[12,19,20,27,107–111] The most frequently cited studies are the risk assessments for Times Beach, Missouri, by the Centers for Disease Control (CDC),[12] the U.S. Environmental Protection Agency,[100] and Paustenbach et al.[20] In an effort to address public health concerns in residential settings like Times Beach, Kimbrough et al.[12] concluded that 1 part per billion (ppb) of 2,3,7,8-TCDD in soil was a reasonable level at which to begin consideration of action to limit human exposure. Several limitations and uncertainties have been identified in the Times Beach risk assessments.[20,112–114] Notable among these was the failure to consider site-specific conditions and the application of overly conservative exposure estimates for direct exposure pathways such as soil ingestion and dermal contact.[20,94,103] Other risk assessments, such as that by Paustenbach et al.,[20] concluded that TCDD soil levels as high as 10 ppb and 1,000 ppb might be acceptable for most residential and industrial sites, respectively, as long as surface soil runoff was negligible.

In a recent paper by Paustenbach et al.,[94] allowable levels of TCDD in residential and industrial soils were determined using Monte Carlo techniques and the most recent data for estimating human uptake of TCDD through dermal contact, soil ingestion, inhalation of fugitive dust, and the consumption of contaminated fish. For a residential site, concentrations of TCDD in soil as high as 19 ppb did not exceed a lifetime incremental cancer risk of 1 in 100,000 for the typically exposed individual. Using a Monte Carlo analysis of the key exposure parameters used in the analysis, TCDD soil concentration for the 75th and 95th percentile person were 12 ppb and 7.5 ppb (10^{-5} risk), respectively. At an industrial site where exposures to workers were limited to dermal contact, soil ingestion, and inhalation of fugitive dust and consumption of fish by an offsite receptor was considered, TCDD concentrations in soil could range between 131 and 579 ppb (10^{-5} risk), depending on the amount of time a worker spends outdoors in direct contact with soil. The acceptable range of TCDD concentrations in industrial soils was not significantly reduced when the consumption of fish from a neighboring waterway by offsite receptors was considered. Since the results of the rather complex Monte Carlo analysis yielded values similar to those calculated using the screening technique, we are confident that the method is appropriate for nonvolatile chemicals.

The differences among regulatory cleanup levels for BTX in soils in the United States is much larger[4-6] than those for TCDD; primarily because BTX contamination is more frequently encountered. A review of state cleanup levels for hydrocarbon contaminated soils has recently been published[6] in which regulatory agency representatives were asked to respond to a series of questions regarding policy on cleanup of hydrocarbon contaminated soil. The survey revealed that total petroleum hydrocarbon (TPH) is the most common guidance level for cleanup, ranging from 10 to 10,000 ppm. A majority of the states also evaluate BTX or some other measurement of volatiles. Table 19 summarizes the range of cleanup values used by different regulatory agencies for petroleum contaminated soils. Due to the broad range, one can only state that the cleanup levels we recommend in this chapter are consistent with nearly all published guidelines. Although the rationale for many of these state guidelines is seldom made clear, it appears that protection of groundwater is the primary concern. The two key advantages of the methodology proposed in this chapter is that the rationale for the soil guideline is clearly defined and a quantitative approach for determining acceptability is clearly presented.

With the exception of New Jersey and Michigan, few states consider different exposure scenarios when setting soil guidelines. The New Jersey Department of Environmental Protection and Energy has recently issued a set of draft guidelines[115] that attempts to use a logical and scientifically based approach to setting soil cleanup levels for both residential and industrial settings, and includes draft standards for the cleanup of the interior surfaces of buildings.[115]

Table 19. Range of Regulatory Cleanup Levels for Petroleum Contaminated Soils[a]

Chemical	PPM
TPH	Background–500
Benzene	0.025–130
Toluene	0.3–200
Xylene	1–50
Total BTEX	1–100
TPH diesel	100–10,000
TPH gasoline	10–1,000
Ethylbenzene	1–68
EDC	0.8

[a]Source: Bell et al.[6]

DISCUSSION

Our recommendation to set seven different soil guidelines for each chemical contaminant is one based on more than 15 years of experience at evaluating how local, state, and federal agencies have struggled with the "how clean is clean?" issue. The process of setting soil cleanup standards has clearly been an arduous one for the United States because, after nearly 20 years, less than five nationally recognized soil cleanup levels are available.

The intent of our approach is to establish a rapid and efficient method for identifying soil contaminant concentrations (or quantities) below which it is very unlikely that health hazard could be present. Having identified acceptable or safe concentrations for the seven most routinely encountered exposure scenarios, then agencies or companies could efficiently dismiss those settings which are not worthy of further evaluation and certainly don't deserve a thorough quantitative risk assessment. The approach is not unlike that currently used by industrial hygienists who rely upon the American Conference of Governmental Industrial Hygienists (ACGIH) Threshold Limit Values (TLV) to determine whether the airborne concentration of a chemical in the workplace is likely to pose a health hazard to the typical employee. Like the TLVs, some degree of professional judgment would need to be used when relying upon these seven guidelines to make the final determination that the site or incident poses no significant risk, but the benefits of developing such a set of criteria would surely outweigh the problems which are inherent in the current process where numerous agencies develop guidelines in an inconsistent manner, each with a different objective.

This approach could be the next logical step in the program which has been initiated by CHESS. It is anticipated that they will have developed a software package which will rapidly calculate site-specific soil cleanup levels based on the best environmental fate and transport models, human exposure parameters, and toxicology data. This model, which would clearly represent the state-of-the-art in exposure assessment, could serve as an invaluable tool for calculating the seven guidelines advocated here. Having used the model to calculate soil cleanup guidelines for perhaps as many as 100 chemicals for the seven exposure scenarios, a booklet much like the one published by the ACGIH (which contains the TLVs) could be published and distributed. Like the TLVs, the guidelines could be revised, updated, or expanded annually based on new information or the efforts of the group. As experience is gained, modifications to the formulae upon which the guidelines were developed could be made.

Although the computer program being derived by CHESS would allow state or federal agencies to conduct their own site-specific risk analyses in a relatively rapid and efficient manner, we believe that a booklet of values is still needed. We believe that many of these professionals will simply not have the time to learn how to use the software package, or the time to use it in their daily practice. It is also possible that as the CHESS software package is updated annually, not everyone will be able to stay abreast of the changes in the software or formulae. Further, the written guidance provided in the TLV-style booklet, which could also be a part of a soils booklet, has proved to be invaluable guidance to those who rely upon the guidelines to make routine and rapid decisions. It seems to us that for the initial screening evaluation, most health professionals would welcome the simplicity and efficiency of being able to look up in a table the chemical and particular exposure scenario of interest to them, so that they could reach an opinion as to whether they had a small, large, or negligible problem. We are confident that if a booklet of values were developed and approved each year by CHESS, or a similar group of acknowledged professionals, this would provide an important mechanism for bringing order to what is rapidly becoming a morass of conflicting guidelines, rules, regulations, and recommendations.

REFERENCES

1. James, R. C., Nye, A. C., Millner, G. C., and Roberts, S. M. "Risks from Exposure to Polychlorinated Biphenyls (PCBs) in Soil: An Evaluation of the TSCA Soil Cleanup Guidelines for PCBs," *Regul. Toxicol. Pharmacol.* (1992, in press).
2. Interim Guidance for Cleanup of Lead in Soils. U.S. Environmental Protection Agency, Office of Emergency and Remedial Response. OSWER Directive 9355.4-02. Washington, DC, September 7, 1989.
3. Johnson, B. R. Letter to David Wagoner, Director, Waste Management Division, EPA Region VII, 1987.
4. Bell, C. E., P. T. Kostecki, and E. J. Calabrese. "State of Research and Regulatory Approach of State Agencies for Cleanup of Petroleum Contaminated Soils," in P. T. Kostecki and E. J. Calabrese, Eds., *Petroleum Contaminated Soils,* Vol. 2 (Chelsea, MI: Lewis Publishers, Inc., 1989), pp. 73–94.
5. Bell, C. E., P. T. Kostecki, and E. J. Calabrese. "An Update on a National Survey of State Regulatory Policy: Cleanup Standards," in P. T. Kostecki and E. J. Calabrese, Eds., *Petroleum Contaminated Soils,* Vol. 3 (Chelsea, MI: Lewis Publishers, Inc., 1990), pp. 49–72.
6. Bell, C. E., P. T. Kostecki, and E. J. Calabrese. "Review of State Cleanup Levels for Hydrocarbon Contamination," in P. T. Kostecki and E. J. Calabrese, Eds.,

Hydrocarbon Contaminated Soils and Groundwater (Chelsea, MI: Lewis Publishers, Inc., 1991), pp. 77–89.

7. Human Health Evaluation Manual, Supplemental Guidance: Standard Default Exposure Factors. U.S. Environmental Protection Agency, Office of Soil Waste and Emergency Response, Washington, DC (1991).
8. "Soil Cleanup Guidelines," New Jersey Department of Environmental Protection, Office of Science and Technology. Trenton, NJ (1991).
9. "Selected Type B Cleanup Criteria," Michigan Department of Natural Resources, Waste Management Division, Lansing, MI, 1990.
10. Kostecki, P. T., and E. J. Calabrese. *Hydrocarbon Contaminated Soils and Groundwater,* Vol. 1. (Chelsea, MI: Lewis Publishers, Inc., 1991).
11. Kostecki, P. T., and E. J. Calabrese. "Council for Health and Environmental Safety of Soils," in P. T. Kostecki and E. J. Calabrese, Eds., *Petroleum Contaminated Soils,* Vol. 2 (Chelsea, MI: Lewis Publishers, Inc., 1989), pp. 485–495.
12. Kimbrough, R., H. Falk, P. Stehr, and G. Fries. "Health Implications of 2,3,7,8-Tetrachlorodibenzo-p-dioxin (TCDD) Contamination of Residential Soil," *Risk Anal.* 5:289–302 (1984).
13. Sheehan, P. J., D. M. Meyer, M. M. Sauer, and D. J. Paustenbach. "Assessment of the Human Health Risks Posed by Exposure to Chromium-Contaminated Soils," *J. Toxicol. Environ. Health* 32:161–201 (1991).
14. Paustenbach, D. J. "A Methodology for Evaluating the Environmental and Public Health Risks of Contaminated Soil," in P. T. Kostecki and E. J. Calabrese, Eds., *Petroleum Contaminated Soils,* Vol. 1 (Chelsea, MI: Lewis Publishers, Inc., 1989), pp. 225–261.
15. Superfund Exposure Assessment Manual. U.S. Environmental Protection Agency, Office of Remedial Response. EPA/540/1-88/001. Washington DC, April, 1988.
16. Paustenbach, D. J. "Important Recent Advances in the Practice of Health Risk Assessment: Implications for the 1990s," *Regul. Toxicol. Pharmacol.* 10:204–243 (1989).
17. Paustenbach, D. J. "A Comprehensive Methodology for Assessing the Risks to Humans and Wildlife Posed by Contaminated Soils: A Case Study Involving Dioxin," in *The Risk Assessment of Environmental and Human Health Hazards: A Textbook of Case Studies,* D. J. Paustenbach, Ed. (New York, NY: John Wiley & Sons, 1989), pp. 296–330.
18. Paustenbach, D. J. "Assessing the Potential Environment and Human Health Risks of Contaminated Soil," *Comments Toxicol.* 1:185–220 (1987).
19. Eschenroeder, A., R. J. Jaeger, J. J. Ospital, and C. P. Doyle. "Health Risk Assessment of Human Exposures to Soil Amended with Sewage Sludge Contaminated with Polychlorinated Dibenzodioxins and Dibenzofurans," *Vet. Hum. Toxicol.* 28:435–442 (1986).
20. Paustenbach, D. J., H. P. Shu, and F. J. Murray. "A Critical Analysis of Risk Assessments of TCDD Contaminated Soil," *Regul. Toxicol. Pharmacol.* 6:284–307 (1986).
21. Motto, H. L., R. H. Daines, D. M. Chilko, and C. K. Motto. "Lead in Soils and Plants: Its Relationship to Traffic Volumes and Proximity to Highways," *Environ. Sci. Technol.* 4:231–237 (1970).
22. Paustenbach, D. J., T. T. Sarlos, B. L. Finley, D. A. Jeffrey, and M. J. Ungs. "The Potential Inhalation Hazard Posed by Dioxin-Contaminated Soil," *J.A.W.M.A.* 41(10): 1334–1340, October, 1991.

23. Keenan, R. E., M. M. Sauer, F. H. Lawrence, E. R. Rand, and D. W. Crawford. "Examination of Potential Risks from Exposure to Dioxin in Sludge Used to Reclaim Abandoned Strip Mines," in D. J. Paustenbach, Ed., *The Risk Assessment of Environmental Hazards: A Textbook of Case Studies* (New York: John Wiley & Sons, 1989), pp. 935–998.
24. Rand, G. M., and S. R. Petrocelli. *Fundamentals of Aquatic Toxicology* (New York, NY: McGraw-Hill, 1985).
25. Paustenbach, D. J. "Health Risk Assessments: Opportunities and Pitfalls," *Columbia J. Environ. Law* 14(2):379–410 (1989).
26. Fries, G. F. "Potential Polychlorinated Biphenyl Residues in Animal Products from Application of Contaminated Sewage Sludge to Land," *J. Environ. Qual.* 11:14–20 (1982).
27. Fries, G. F., and D. J. Paustenbach. "Evaluation of Potential Transmission of 2,3,7,8-Tetrachlorodibenzo-p-Dioxin-Contaminated Incinerator Emissions to Humans via Foods," *J. Toxicol. Environ. Health* 29:1–43 (1990).
28. "Lead in the Human Environment," National Research Council, Washington, DC (1980).
29. Hayes, W. J., and E. R. Laws, Jr. *Handbook of Pesticide Toxicology* (New York: Academic Press, Inc., 1991).
30. Charters, D. "Current Uses and Abuses of Environmental Effect and Fate Data," *Proceedings of the Annual Meeting of ASTM* (Atlantic City: American Society of Testing and Materials, 1991).
31. Assessment of Risks from Exposure of Humans, Terrestrial and Avian Wildlife, and Aquatic Life to Dioxins and Furans from Disposal and Use of Sludge from Bleached Kraft and Sulfite Pulp and Paper Mills. U.S. Environmental Protection Agency, Office of Solid Waste and Emergency Response, Washington, DC, 1990.
32. Exposure Factors Handbook. U.S. Environmental Protection Agency, Exposure Assessment Group. Office of Health and Environmental Assessment. Washington, DC, 1989.
33. Conway, R. A. *Environmental Risk Analysis of Chemicals* (New York: Van Nostrand-Reinhold, 1982).
34. "Throwing Good Money at Bad Water Yields Scant Improvements," *Wall Street Journal*, May 15, 1991, p. 1.
35. Bopp, R. F., M. L. Gross, H. J. Tong, S. J. Monson, B. L. Geck, F. C. Moser. "A Major Incident of Dioxin Contamination: Sediments of New Jersey Estuaries," *Environ. Sci. Technol.* 25:951–956 (1991).
36. Harris, M., K. Conner, V. Lau, and D. J. Paustenbach. "The Range of Reasonable Exposure Factors for Use in Risk Assessment," *Risk Anal.* (in review, 1992).
37. Risk Assessment Guidance for Superfund. Volume 1. Human Health Evaluation Manual (Part A). U.S. Environmental Protection Agency, Office of Emergency and Remedial Response. Washington, DC, 1989.
38. Finkel, A. M. *Confronting Uncertainty in Risk Management: A Guide for Decision-Makers* (Washington, DC: Center for Risk Management, Resources for the Future, 1990).
39. Paustenbach, D. J., D. M. Meyer, P. J. Sheehan, and V. Lau. "An Assessment and Quantitative Uncertainty Analysis of the Health Risks to Workers Exposed to Chromium Contaminated Soils," *Toxicol. Ind. Health* 7(3):159–196 (1991).
40. Paustenbach, D. J., M. Layard, R. Wenning, and R. E. Keenan. "Risk Assessment of 2,3,7,8-TCDD Using A Biologically Based Cancer Model: A Reevaluation of

the Kociba et al. Bioassay Using 1978 and 1990 Histopathology Criteria," *J. Toxicol. Environ. Health,* 34:11-26 (1991).

41. Harris, M., and D. J. Paustenbach. "An Assessment of a Pentachlorophenol-Contaminated Site," *Toxicol. Environ. Health* (in review) (1992).
42. Cooper, M. *Pica* (Springfield, IL: Thomas, 1957), pp. 65–66.
43. Calabrese, E. J., R. Barnes, E. J. Stanek, H. Pastides, C. E. Gilbert, P. Veneman, X. Wang, A. Lasztity, and P. T. Kostecki. "How Much Soil do Young Children Ingest: An Epidemiologic Study," *Reg. Toxicol. Pharm.* 10:123–137 (1989).
44. Barltrop, D. "Sources and Significance of Environmental Lead in Children," Proceedings Int. Environ. Health Aspects of Lead, Commission of European Communities. Center for Information and Documentation, Luxembourg (1973).
45. Lepow, M. L., L. Bruckman, R. A. Rubino, S. Markowitz, M. Gillette, and J. Kapish. "Role of Airborne Lead in Increased Body Burden of Lead in Hartford Children," *Environ. Health Persp.* 99–102 (1974).
46. Day, J. P., J. E. Fergusson, and T. M. Chee. "Solubility and Potential Toxicity of Lead in Urban Street Dust," *Bull. Environ. Contam. Toxicol.* 23:497–502 (1975).
47. Duggan, M. J., and S. Willams. "Lead-in-Dust in City Streets," *Sci. Tot. Environ.* 791–797 (1977).
48. Hawley, J. K. "Assessment of Health Risk from Exposure to Contaminated Soil," *Risk Anal.* 5(4):289-302 (1985).
49. Binder, S., Sokal, D. and Maughan, D. "Estimating Soil Ingestion: The Use of Tracer Elements in Estimating the Amount of Soil Ingested by Young Children," *Arch. Environ. Health* 41(6):341-345 (1986).
50. Bryce-Smith, D. "Lead Absorption in Children," *Phys. Bull.* 25:178–181 (1974).
51. Schaum, J. Risk Analysis of TCDD Contaminated Soil. U.S. Environmental Protection Agency, Office of Health and Environmental Assessment, Office of Research and Development, Washington, DC (1986).
52. Superfund Public Health Manual. U.S. Environmental Protection Agency, Office of Emergency and Remedial Response, Washington, DC. EPA/540/1-86/060. October, 1986.
53. Clausing, P., B. Brunekreef and J. H. van Wijnen. "A Method for Estimating Soil Ingestion by Children," *Int. Arch. Occup. Environ. Health* 59:73–82 (1987).
54. LaGoy, P. "Estimated Soil Ingestion Rates for Use in Risk Assessment." *Risk Anal.* 7:355–359 (1987).
55. Lipsky, D. "Assessment of Potential Health Hazards Associated with PCDD and PCDF Emissions from a Municipal Waste Combustor," in D. J. Paustenbach, Ed. *The Risk Assessment of Environmental and Human Health Hazards: A Textbook of Case Studies* (New York, NY: John Wiley & Sons, 1988), pp. 631-686.
56. Sedman, R. M. "The Development of Applied Action Levels for Soil Contact: A Scenario for the Exposure of Humans to Soil in a Residential Setting," *Environ. Health Perspect.* 79:291–313 (1988).
57. Calabrese, E. J., P. T. Kostecki and C. E. Gilbert. "How Much Soil do Children Eat? An Emerging Consideration for Environmental Health and Risk Assessment," *Comments Toxicol.* 1(3):229–241 (1987).
58. Calabrese, E. J., E. J. Stanek, C. E. Gilbert, and R. M. Barnes. "Preliminary Adult Soil Ingestion Estimates: Results of a Pilot Study," *Reg. Tox. Pharm.* 12:88–95 (1990a).
59. Calabrese, E. J. and E. J. Stanek. "A Guide to Interpreting Soil Ingestion Studies. I. Development of a Model to Estimate the Soil Ingestion Detection Level of Soil

Ingestion Studies," *Reg. Toxicol. Pharm.* 13:263–277 (1991).
60. Calabrese, E. J., and E. J. Stanek. "A Guide to Interpreting Soil Ingestion Studies. II. Qualitative and Quantitative Evidence of Soil Ingestion," *Reg. Toxicol. Pharm.* 13:278–292 (1991).
61. Davis, S., P. Waller, R. Buschbom, J. Ballou, and P. White. "Quantitative Estimates of Soil Ingestion in Normal Children Between the Ages of 2 and 7 Years: Population-Based Estimates Using Aluminum, Silicon, and Titanium as Soil Tracer Elements," *Arch. of Environ. Health* 45(2):112–122 (1990).
62. Van Wijnen, J. H., P. Clausing, and B. B. Brunekreef. "Estimated Soil Ingestion by Children," *Environ. Res.* 51:147–157 (1990).
63. Roels, H. A., J. P. Buchet, R. R. Lauwerys, P. Braux, F. Claeys-Thoreau, A. Lafontaine, and G. Verduyn. "Exposure to Lead by the Oral and Pulmonary Routes of Children Living in the Vicinity of a Primary Lead Smelter," *Environ. Res.* 22:81–94 (1980).
64. Driver, J. H., J. J. Konz, and G. K. Whitmyre. "Soil Adherence to Human Skin," *Bull. Environ. Contam. Toxicol.* 43:814–820 (1989).
65. Lepow, M. L., L. Bruckman, M. Gillette, S. Markowitz, R. Robino, and J. Kapish. "Investigations into Sources of Lead in the Environment of Urban Children," *Environ. Res.* 10:415–426 (1975).
66. Gallacher, J. E., P. C. Elwood, K. M. Phillips, B. E. Davies, and D. T. Jones. "Relationship Between Pica and Blood Lead in Areas of Differing Lead Exposure," *Arch. Dis. Childhood* 59:40–44 (1984).
67. Que Hee, S. S., B. Peace, C. S. Scott, J. R. Boyle, R. L. Bornschein, and P. B. Hammond. "Evolution of Efficient Methods to Sample Lead Sources, Such As House Dust and Hand Dust, in the Homes of Children," *Environ. Res.* 38:77–95 (1985).
68. Multi-Pathway Health Risk Assessment Input Parameters Guidance Document. South Coast Air Quality Management District. Prepared by Clement Associates, Inc., Washington, DC, June, 1988.
69. The Development of Applied Action Levels for Soil Contact. California Department of Health Services (CDHS). Final Draft, 1987 (cited in SCAQMD, 1988).
70. Knaak, S. B., I. Yutaka, and K. T. Maddy. "The Worker Hazard Posed by Re-Entry into Pesticide-Treated Foliage: Development of Safe Reentry Times, with Emphasis on Chlorthiophos and Carbosulfan," in D. J. Paustenbach, Ed., *The Risk Assessment of Environmental Hazards* (New York: John Wiley & Sons, 1989), pp. 797–844.
71. Trijonis, J., J. Eldon, J. Gins, and G. Berglund. Analysis of the St. Louis RAMS Ambient Particulate Data, EPA Report 450/4-80-006a. Produced by Technology Service Corporation under EPA Contract 68-02-2931 for the Office of Air, Noise, and Radiation of the U.S. Environmental Protection Agency, Washington, DC (1980).
72. Goon, D., N. S. Hatoum, J. D. Jernigan, S. L. Schmitt, and P. J. Garvin. "Pharmacokinetics and Oral Bioavailability of Soil-Adsorbed Benzo[a]pyrene (BaP) in Rats," *Toxicologist* 10:218 (1990).
73. Goon, D., N. S. Hatoum, M. J. Klan, J. D. Jernigan, and R. G. Farmer. "Oral Bioavailability of 'Aged' Soil-Adsorbed Benzo[a]pyrene in Rats," *Toxicologist* 11:345 (1991).
74. State of Michigan Draft Risk Assessment Guidelines. Committee on Risk Assessment. Michigan Council on Environmental Quality, Lansing, MI, May 17, 1990.
75. Haque, R., Ed. *Dynamics, Exposure and Hazard Assessment of Toxic Chemicals* (Ann Arbor, MI: Ann Arbor Science, 1980).

76. Thibodeaux, L. J. *Chemodynamics. Environmental Movement of Chemicals in Air, Water, and Soil* (New York: Wiley, 1979).
77. Beck, L. W., A. W. Maki, N. R. Artman, and E. R. Wilson. "Outline and Criteria for Evaluating the Safety of New Chemicals," *Regul. Toxicol. Pharmacol.* 1:19-58 (1981).
78. Mackay, D., and S. Paterson. "Calculating Fugacity," *Environ. Sci. Technol.* 15:1006-1014 (1982).
79. Bergmann, H. L., R. A. Kimmerle, and A. W. Maki. *Environmental Hazard Assessment of Effluents* (New York, NY: Pergamon Press, Inc., 1986).
80. Woltering, D., and W. Bishop. "Assessing the Environmental Risks of Detergent Chemicals," in D. J. Paustenbach, Ed., *The Risk Assessment of Environmental Hazards* (New York, NY: John Wiley & Sons, 1989), pp. 345-390.
81. Lyman, W. J., W. F. Reehl, and D. H. Rosenblatt. "Handbook of Chemical Property Estimation Methods," *Environmental Behavior of Organic Compounds* (New York: McGraw-Hill, 1982.)
82. Dragun, J. *The Soil Chemistry of Hazardous Materials* (Silver Spring, MD: Hazardous Materials Control Research Institute, 1988).
83. Dragun, J. "The Fate of Hazardous Materials in Soil (What Every Geologist and Hydrologist Should Know): Part 2," *Hazardous Materials Control* 1(3):41-65 (1988).
84. Environmental Assessment Technical Handbook. U.S. Food and Drug Administration, Center for Food Safety and Applied Nutrition and the Center for Veterinary Medicine, Washington, DC, 1984.
85. Young, F. A. "Risk Assessment: The Convergence of Science and Law," *Regul. Toxicol. Pharmacol.* 7:179-184 (1987).
86. Travis, C. C., and H. A. Hattemer-Frey. "Human Exposure to 2,3,7,8-TCDD," *Chemosphere* 16(10-12):2331-2342 (1987).
87. Travis, C. C., S. A. Richter, E. A. Crouch, R. Wilson, and E. Wilson. "Cancer Risk Management. A Review of 132 Federal Regulatory Decisions," *Environ. Sci. Technol.* 21(5):415-420 (1987).
88. National Oil and Hazardous Substances Pollution Contingency Plan, 40 CFR Part 300. Environmental Protection Agency. Washington, DC (1990).
89. Policy for Identifying and Assessing the Health Risks of Toxic Substances. Maine Department of Human Services, Environmental Toxicology Program, Division of Disease Control, Bureau of Health, February 1988.
90. Draft Interim Guidance for Disposal Site Risk Characterization—In Support of the Massachusetts Contingency Plan. Massachusetts Department of Environmental Quality Engineering, Office of Research and Standards, October 3, 1988.
91. Safe Drinking Water and Toxic Enforcement Act of 1986 (Proposition 65). California Health and Welfare Agency, Office of the Secretary, Sacramento, CA, 1986.
92. Rodricks, J. V., S. M. Brett, and G. C. Wrenn. "Significant Risk Decisions in Federal Regulatory Agencies," *Regul. Toxicol. Pharmacol.* 7:307-320 (1987).
93. Keenan, R. E., J. W. Knight, E. R. Rand, and M. M. Sauer. "Assessing Potential Risks to Wildlife and Sportsmen from Exposure to Dioxin in Pulp and Paper Mill Sludge Spread on Managed Woodlands," *Chemosphere* 20:1763-1769 (1990).
94. Paustenbach, D. J., R. J. Wenning, V. Lau, N. W. Harrington, D. K. Rennix, and A. H. Parsons. "Recent Developments on the Hazards Posed by 2,3,7,8-Tetrachlordibenzo-p-dioxin in Soil: Implications for Setting Risk-Based Cleanup Levels at Residential and Industrial Sites," *J. Toxicol. Environ. Health* 36:103-148 (1992).

95. Paustenbach, D. J. "The Current Practice of Health Risk Assessment: Potential Impact on Standards for Toxic Air Contaminants," *J.A.W.M.A.* 40:1620–1630 (1990).
96. Wipf, H. K., E. Homberger, N. Neuner, U. B. Ranalder, W. Vetter, and J. P. Vuilleumier. "TCDD Levels in Soil and Plant Samples from the Seveso Area," in *Chlorinated Dioxins and Related Compounds*, O. Hutzinger et al., Eds. (New York, NY: Pergamon Press, Inc., 1982), pp. 115–126.
97. Jensen, D. J., M. E. Getzendaner, R. A. Hummel, and J. Turley. "Residue Studies for (2,4,5-tri-chlorophenoxy)acetic Acid and 2,3,7,8-Tetrachlorodibenzo-p-dioxins in Grass and Rice," *J. Agric. Food Chem.* 31:118–122 (1983).
98. Stevens, J. B., and E. N. Gerbec. "Dioxin in the Agricultural Food Chain," *Risk Anal.* 8:329–335 (1988).
99. Reducing Risk: Setting Priorities and Strategies for Environmental Protection. U.S. Environmental Protection Agency, Science Advisory Board, Washington, DC, 1990.
100. Risk Analysis of TCDD Contaminated Soil. U.S. Environmental Protection Agency, Office of Solid Waste and Emergency Response, Washington, DC, 1984.
101. Dioxin Transport from Contaminated Site to Exposure Locations: A Methodology for Calculation of Conversion Factors. U.S. Environmental Protection Agency, Office of Research and Development, 1985.
102. Knighton, M. D., and R. M. Solomon. "Applications and Research in Sediment Delivery and Routing Models in the USDA Forest Service," in S. S. Y. Want, Ed., *Sediment Transport Modeling* (New York: American Society of Civil Engineers, 1989), pp. 344–350.
103. Goeden, H. E., and A. H. Smith. "Estimation of Human Exposure from Fish Contaminated with Dioxins and Furans Emitted by a Resource-Recovery Facility," *Risk Anal.* 9(3):377–383 (1989).
104. Cook, P. M., A. R. Batterman, B. C., Butterworth, K. B. Lodge and S. W. Kohlbry. Laboratory Study of TCDD Bioaccumulation by Lake Trout for Lake Ontario Sediments, Food Chain and Water. Draft Copy, Chapter 6. Environmental Research Laboratory, U.S. Environmental Protection Agency, Duluth, MN, 1990.
105. Jury, W. A., W.F. Spencer, and W.J. Farmer. "Behavior Assessment Model for Trace Organics in Soil: I. Model Description," *J. Environ. Qual.* 12(4):558–564 (1983).
106. Gough, M. "Human Exposures from Dioxin in Soil—A Meeting Report," *J. Toxicol. Environ. Health* 32:205–245 (1991).
107. Kociba, R. J., D. Keyes, J. Beyer, R. M. Carreon, C. E. Wade, D. A. Dittenber, R. P. Kalnins, L. E. Frauson, C. N. Park, S. D. Barnard, R. A. Hummel, and C. G. Huminston. "Results of a Two-Year Chronic Toxicity and Oncogenicity Study of 2,3,7,8-Tetrachlorodibenzo-p-dioxin in Rats," *Toxicol. Appl. Pharmacol.* 46:279–303 (1978).
108. Nauman, C. H., and J. L. Schaum. "Human Exposure Estimation for 2,3,7,8-TCDD," *Chemosphere* 16:1851–1856 (1987).
109. Birmingham, B., A. Gillman, D. Grant, J. Salminen, M. Boddington, B. Thorpe, I. Wile, P. Toft, and V. Armstrong. "PCDD/PCDF Multimedia Exposure Analysis for the Canadian Population: Detailed Exposure Estimation," *Chemosphere* 19 (1–6): 637–642.
110. Heida, H., M. van den Berg, and K. Olie. "Risk Assessment and Selection of Remedial Alternative, the Volgemeerpoider Case Study," *Chemosphere* 19:615–622 (1989).

111. Di Domenico, A. "Guidelines for the Definition of Environmental Action Alert Thresholds for Polychlorodibenzodioxins and Polychlorodibenzofurans," *Reg. Toxicol. Pharm.* 11:118–123 (1990).
112. Houk, V. "Uncertainties in Dioxin Risk Assessment," *Chemosphere* 15:1875–1881 (1985).
113. Kimbrough, R. D. "Estimation of Amount of Soil Ingested, Inhaled, or Available for Dermal Contact," *Comments Toxicol.* 1:177–184 (1987).
114. Fishbein, L. "Health-Risk Estimates for 2,3,7,8-Tetrachlorodibenzo-dioxin: An Overview," *Toxicol. Ind. Health* 3:91–134 (1987).
115. Preliminary Draft Cleanup Standards, New Jersey Department of Environmental Protection and Energy, Division of Hazardous Waste Management, Trenton, NJ (1991).
116. Calabrese, E. J., and E. S. Stanek. "Distinguishing Outdoor Soil Ingestion from Outdoor Dust Ingestion in a Soil Pica Child," *Regul. Toxicol. Pharm.* 15: 83–85 (1992).
117. Copeland, T. L., A. Holbron, M. Harris, and D. J. Paustenbach. "A Risk Assessment of a Former Wood Treatment Site," *J. Toxicol. Env. Health* (in review), 1992.
118. Keenan, R. E., R. J. Wenning, A. H. Parsons, and D. J. Paustenbach. "A Reevaluation of the Tumor Histopathology of Kociba et al. (1978) Using 1990 Criteria: Implications for the Risk Assessment of 2,3,7,8-TCDD Using the Linearized Multistage Model," *J. Toxicol. Environ. Health,* 34: 279–296 (1991).

CHAPTER 24

Review of Present Risk Assessment Models for Petroleum Contaminated Soils

Paul T. Kostecki, Edward J. Calabrese, and **Holly M. Horton,** Northeast Regional Environmental and Public Health Center, University of Massachusetts, Amherst

Seven risk assessment methodologies were evaluated as a preliminary means of developing a risk assessment methodology that can be applied to petroleum contaminated soils. By determining the strengths and weaknesses of these methodologies, a stronger risk assessment process can be developed.

A risk assessment methodology should include variables which clearly represent the hazard level, exposure level, and the level of risk at a particular site. Criteria used to assess the strengths and weaknesses of the hazard, exposure, and risk analyses are discussed in the following section.

CRITERIA USED TO ASSESS RISK ASSESSMENT METHODOLOGIES

Hazard Analysis

The toxicities of contaminants at a site are determined during the hazard analysis. Animal studies and human studies, when available, are used to derive toxicity values based on dose-response relationships and statistical models. The manner in which toxicity values are derived should be compared among the different methodologies. If the toxicity values are based on already existing standards, incorporation of safety factors or modifiers may be necessary.

Chemicals at a site can have diverse short-term and long-term effects. In order to account for these varying effects over time, noncarcinogenic acute and chronic toxicity should be determined as well as carcinogenic chronic effects. In determination of acute toxicity values for children, body weight factors may be used to modify toxicity values originally derived for adult toxicity.

Pharmacokinetic factors, when available, can account for differences in absorption and metabolism due to different exposure routes (i.e., water ingestion versus air inhalation of a chemical). Without inclusion of pharmacokinetic factors, the same toxicity value used for different exposure routes may not adequately represent the hazard level. For example, toxicity due to ingestion of a chemical may be grossly different from toxicity due to inhalation of the same chemical.

Since many sites are contaminated with a multitude of chemicals, a procedure for evaluation of mixtures of chemicals at a site is important. A standard procedure in a risk assessment methodology would aid in identification of the most toxic chemicals, which can then represent the overall toxicity of a mixture. In this manner, fewer chemicals need to be individually evaluated, thereby streamlining the risk assessment process.

Exposure Analysis

The exposure analysis evaluates the means by which humans encounter chemicals originating from a contaminated site. Exposure routes may include air and dust inhalation, water ingestion, soil ingestion, dermal absorption, and ingestion of crops, livestock, or fish which have been exposed to a contaminant from the site. Not all of these exposure routes will be important for a particular site; however, a useful risk assessment methodology would have the capacity to include any of these factors when relevant for a specific site.

Environmental fate analyses may include variables such as air transport, surface water or groundwater transport, and bioconcentration of chemicals in plants, fish, and livestock. This type of analysis is important because it can indicate not only present exposure routes but future exposure routes as the chemicals travel through the environment. Within environmental fate analyses, chemical interactions are also important and may include factors such as solubility, vapor pressure, half-life, and soil adsorption characteristics. Site-specific characteristics such as geology, climate, soil type, and location of aquifers, surface waters, and runoff areas have a direct influence on the environmental fate of chemicals at a site. A specific environmental fate model may be used in this step which will dictate the types of measurements and analyses necessary.

Risk Analysis

In the risk analysis, the human exposure level for a chemical and the toxicity resulting from such exposure is compared to a critical toxicity value for the chemical. The critical toxicity value generally represents an acceptable exposure level for the chemical. If the critical toxicity value is exceeded, remedial actions are recommended.

The risk analysis should include additivity of the hazard level for chemicals which have the same toxic effect (i.e., erythrocyte hemolysis, nerve damage, liver damage, etc.). When two or more chemicals exert the same effect, the level of risk should be reduced so that a safe level is determined for all of the chemicals. Similarly, the total carcinogenic risk for all carcinogens of concern at a site should be considered so that the total cancer risk is acceptable. In cases where synergistic effects are known, this information can also be incorporated into a risk analysis.

Humans may be exposed to one chemical from multiple sources due to air, dust or water transport; and fish, crop, or livestock uptake. Exposure via different media should be considered in order to evaluate the cumulative toxicity from all media.

All of the criteria used to evaluate the seven risk assessment methodologies are listed in Table 1. These criteria attempt to evaluate the most important variables involved in the hazard, exposure, and risk analyses.

Seven methodologies were reviewed and include: the California Site Mitigation Decision Tree,[1] Rosenblatt et al.,[2] EPA Superfund,[3] Ford and Gurba,[4] New Jersey Method as proposed by Stokman and Dime,[5] the State of Washington

Table 1. Criteria Used to Evaluate Risk Assessment Methodologies

HAZARD ANALYSIS
1. Derivation of toxicity values
2. Noncarcinogenic acute toxicity
3. Noncarcinogenic chronic toxicity
4. Carcinogenic chronic toxicity
5. Assessment of mixtures
6. Body weight factor
7. Pharmacokinetic factors
EXPOSURE ANALYSIS
1. Air inhalation
2. Dust inhalation
3. Water ingestion
4. Soil ingestion
5. Dermal absorption
6. Crop uptake
7. Livestock uptake
8. Fish uptake
9. Environmental fate
10. Half-life factor
11. Site-specific factors
RISK ANALYSIS
1. Additivity of the hazard level for chemicals having a common toxic effect
2. Additivity of the hazard level for multimedia exposure
3. Synergistic effects

Cleanup Policy,[6] and California's Leaking Underground Fuel Tank (LUFT) Field Manual.[7] Each methodology is described and critically evaluated. A summary table follows each methodology (Tables 2–7) along with a final tabular summary (Table 9), directly comparing each of the seven methodologies.

CALIFORNIA SITE MITIGATION DECISION TREE

Summary

The California Site Mitigation Tree[1] consists of four major steps: (1) identification of toxic substances, (2) determination of Applied Action Levels (AALs), (3) environmental fate modeling, and (4) risk analysis.

The identification of toxic substances involves review of chemical information concerning substances at the site. The available data are evaluated for quality and adequacy. The chemicals of interest are classified as carcinogens or noncarcinogens.

AALs are defined as media-specific levels of a substance which, if exceeded, present a significant risk. The AALs are derived from No Observed Adverse Effect Levels (NOAELs) for noncarcinogens and from risk factors for carcinogens. Standard body weight, standard intake, pharmacokinetic, and uncertainty factors are included in the derivation of the AALs.

Mathematical models are utilized to determine the environmental fate of a chemical of interest. Soil adsorption, bioconcentration, and emission rates are evaluated in this step. Exposure levels determined from the environmental fate analysis are used in the risk analysis.

The risk analysis involves comparison of exposure levels with AALs for specific chemicals and media. Multimedia exposures and additive toxic effects are taken into account. If exposure levels are found to exceed the AAL for any chemicals, a significant risk is deemed to exist.

Strengths and Weaknesses

The first step in the California risk assessment process involves identification and separation of the toxic substances from the nontoxic substances at the site. Pertinent toxicity information is reviewed for the chemicals of interest. Guidelines are offered for evaluating the quality and adequacy of data, and the chemicals are classified as carcinogens or noncarcinogens.

Problems with the toxicity analysis include: (1) no rating system is offered for ranking chemicals having a variety of toxic effects and physical/chemical properties, (2) it is unclear how the review of toxicity information is integrated into the risk assessment methodology, and (3) no process is offered for narrowing down chemicals from a complex mixture of substances.

AALs are determined in the second step of the risk assessment. AALs are media-specific levels of a substance which, if exceeded, presents a significant risk. AALs are specific for chemical and media, and can be applied to carcinogens or noncarcinogens. AALs for noncarcinogens are derived from NOAELs, while AALs for carcinogens are derived from risk factors for specific chemicals. The NOAEL or risk factor is used to determine a Maximum Exposure Level (MEL). MELs are modified by standard intake, pharmacokinetic, and uncertainty factors to determine the AAL:

Epidemiological studies which determine a NOAEL are considered to be the best basis for development of an AAL.

$$\text{MEL (mg/day)} = \frac{\text{NOAEL (mg/kg/day)}}{\text{Uncertainty Factor (10)}} \times \text{adult body weight (70 kg)}$$

The uncertainty factor of 10 is designed to protect the more sensitive members of the population. AALs are then developed for specific media using the MEL:

$$\text{AAL}_{\text{water}}\text{(mg/L)} = \frac{\text{MEL (mg/day)}}{\text{Average daily intake of water (2 L)}} \times \text{Pharmacokinetic Factors (PF)}$$

$$\text{AAL}_{\text{air}}\text{(mg/m}^3\text{)} = \frac{\text{MEL (mg/day)}}{\text{Average daily intake of water (20 m}^3\text{/day)}} \times \text{Pharmacokinetic Factors (PF)}$$

Pharmacokinetic factors (PF) may include differences in absorption, distribution, metabolism, and excretion due to different routes of exposure. AALs are not determined for exposures via soil ingestion or dermal absorption.

If epidemiological studies are unavailable for the chemical of interest, NOAELs from long-term animal bioassays can be used. An uncertainty factor of 100 is used to account for uncertainties in animal to human extrapolation:

$$\text{NOAEL}_{\text{human}}\text{(mg/kg/day)} = \frac{\text{NOAEL}_{\text{animal}}\text{(mg/kg/day)}}{100} \times \text{PF}$$

AALs may be derived from Threshold Limit Values (TLVs), which are designed to protect the worker from toxic agents in occupational settings. TLVs are time-weighted average exposures based on an eight-hour workday and five-day work week. MELs derived from TLVs only utilize TLVs based on chronic effects, and the TLV is extrapolated to a 24-hour day and seven-day week. An uncertainty factor of 10 to 100 is applied:

$$\text{NOAEL (mg/day)} = \text{TLV (mg/m}^3\text{)} \times \frac{\text{8 hours}}{\text{24 hours}} \times \frac{\text{5 days}}{\text{7 days}} \times \frac{\text{47 years}}{\text{70 years}} \times \frac{\text{20m}}{\text{day}}$$

$$\text{MEL (mg/day)} = \frac{\text{NOAEL (mg/day)}}{\text{Uncertainty Factor (10 to 100)}}$$

AALs for air and water exposure are then derived from the TLV for air exposure. The validity of this type of extrapolation should be questioned. Extrapolations from air to water exposures and from occupational to lifetime exposures may introduce significant errors. TLVs account for intermediate exposures, while the AALs must account for lifetime, continuous exposure.

In cases where no chronic toxicity data is available, NOAELs can be estimated by extrapolation from subchronic toxicity data. In subchronic studies, animals are exposed to a toxic agent for a fraction of their lifespan (usually 90 days). An uncertainty factor of 10 is used to account for uncertainties in extrapolation from subchronic to chronic data:

$$\text{NOAEL}_{\text{chronic}}(\text{mg/kg/day}) = \frac{\text{NOAEL}_{\text{subchronic}}(\text{mg/kg/day})}{\text{Uncertainty Factor (10)}}$$

$$\text{NOAEL}_{\text{human}}(\text{mg/kg/day}) = \frac{\text{NOAEL}_{\text{animal}}(\text{mg/kg/day})}{\text{Uncertainty Factor (100)}} \times \text{PF}$$

NOAELs may also be developed from acute toxicity studies which establish an LD_{50} value. In this case, an uncertainty factor of 1000 is used.

$$\text{NOAEL}_{\text{chronic}}(\text{mg/kg/day}) = \frac{\text{Oral LD}_{50}(\text{mg/kg})}{1000}$$

In case of little or no toxicity data, AALs can be determined using structure activity analyses. This type of analysis involves a regression analysis which compares a specific predictor variable with toxicity. Examples of predictor variables are octanol/water partition coefficients, molecular weight, solubility, density, boiling point, vapor pressure, etc.

Problems with this type of analysis include the fact that the predictor variable must be carefully chosen and be closely related to the toxic properties of the specific chemical. If the predictor variable does not adequately represent the toxicity of the agent, significant errors may result.

The development of AALs for carcinogens is based upon epidemiological studies or long-term animal studies which delineate a level of risk:

$$\text{MEL (mg/day)} = \frac{\text{Risk Factor (mg/day)}}{\text{Level of Risk } (10^{-6} \text{ or less})}$$

It is unclear how the formula for determination of the MEL establishes a safe level of exposure. The term 'risk factor' is defined as the probability of additional cancers which could result from the exposure of a population to a substance of concern.

The environmental fate of a chemical is analyzed in detail using various mathematical models. Environmental fate is evaluated for the factors of soil adsorption, bioconcentration, and emission rates. The exposure levels of the specific chemicals derived from environmental fate modeling are then used in the risk analysis.

The risk analysis compares the AAL for a chemical in a specific media with concentration levels of the chemical at the site. Three types of comparisons are utilized in the risk analysis.

The first comparison compares the level of exposure in one media to the AAL for the same media:

$$\frac{\text{concentration in one medium}}{\text{AAL in one medium}}$$

The second comparison evaluates cumulative exposure to one chemical over different media.

$$\sum_{\text{medium}=1}^{m} \frac{\text{concentration}_m}{\text{AAL}_m}$$

The third comparison evaluates cumulative exposure to different chemicals which exert the same toxic effect over different media.

$$\sum_{\text{substance}=1}^{s} \sum_{\text{medium}=1}^{m} \frac{\text{concentration}_{m,s}}{\text{AAL}_{m,s}}$$

If any of the three ratios exceeds 1, risk management should be initiated.

The risk analysis is strong in its consideration of multiple exposure pathways and cumulative effects for one toxic endpoint. However, only air and water exposure pathways are assessed. In addition, if NOAELs are unavailable for certain chemicals it may be difficult to assess the risk posed by such chemicals using this risk assessment methodology. Alternatives for such a situation include use of TLVs or structure-activity analyses, which may introduce significant errors. In addition, the California risk assessment method does not utilize already established relevant/applicable ambient standards which could speed up the risk analysis.

Conclusions

The California Site Mitigation Tree is useful in its evaluation of multiple exposure pathways (air and water), additive toxic effects, and environmental fate modeling. Toxic effects are evaluated for both carcinogens and noncarcinogens, while the environmental fate modeling considers soil adsorption, bioconcentration, and emission rates.

Weaknesses of this methodology include the fact that exposures via soil ingestion or dermal absorption are not evaluated. In addition, there are no guidelines for assessment of complex mixtures of chemicals, such as petroleum products.

Since every chemical in a complex mixture cannot be evaluated, a means of determining the most hazardous substances is necessary.

If NOAELs are unavailable from scientific studies they must be derived from TLVs or structure-activity analyses, which, again, can introduce serious errors. The use of relevant, ambient standards, when available, is not suggested in the California methodology as a means of streamlining the risk assessment process. Therefore, the California risk assessment plant tends to be time-consuming and not applicable to exposures resulting from soil ingestion or dermal absorption (Table 2).

Table 2. Summary of the California Site Mitigation Decision Tree Methodology with Respect to the Evaluation Criteria

HAZARD ANALYSIS	
Toxicity value	NOAEL[a]
	—epidemiologic
	—chronic
	—subchronic
	TLV
	LD_{50}
	Structure activity analysis
Noncarcinogenic acute toxicity	Yes
Noncarcinogenic chronic toxicity	Yes
Carcinogenic chronic toxicity	Yes
Assessment of mixtures	No
Body weight factor	Yes
Pharmacokinetic factor	Yes
EXPOSURE ANALYSIS	
Air inhalation	Yes
Dust inhalation	Yes
Water ingestion	Yes
Soil ingestion	No
Dermal absorption	No
Crop uptake	Yes
Livestock uptake	Yes
Fish uptake	Yes
Environmental fate	Yes
	—soil adsorption
	—bioconcentration
	—air transport
	—water transport
Half-life factor	No
Site-specific factors	Yes
RISK ANALYSIS	
Additivity of toxic effect	Yes
Multi-media exposure	Yes
Synergistic effects	No

[a]See text for description of toxicity values.

AN ENVIRONMENTAL FATE MODEL LEADING TO PRELIMINARY POLLUTANT LIMIT VALUES FOR HUMAN HEALTH EFFECTS[2]

Summary

Preliminary Pollutant Limit Values (PPLVs) are acceptable concentrations of a pollutant at a site. PPLVs are derived from Single Pathway PPLVs (SPPPLVs) for a specific contaminant. SPPPLVs represent linear exposure pathways from the site to the human receptor (e.g., soil → water → human or soil → plant → human, etc.). SPPPLVs are based upon the acceptable daily dose of the pollutant (D_T), partition coefficients, standard intake factors, and a body weight factor.
The hazard posed by a specific contaminant is represented by D_T which is based on already established standards, Allowable Daily Intake (ADI), MCL, or TLV, or derived from the no effect level (NEL) or LD_{50} from animal studies. The exposure level is represented by partition coefficients which may include octanol-water partition coefficient (K_{ow}), soil organic carbon adsorption coefficient (K_{oc}), water solubility, saturation vapor density, and others. All SPPPLVs which represent critical exposure pathways are factored into the calculation of PPLVs to determine the cleanup level for a specific contaminant.

Strengths and Weaknesses

The PPLV is a temporary, nonregulatory value that is based on information available in the literature. The PPLV model is flexible and can be modified to account for site-specific conditions. It represents an acceptable cleanup level for a contaminant, and the derivation of this value involves the following major steps.

- Pollutants and pathways are identified.
- An acceptable daily dose of pollutant, D_T, and partition coefficients are determined.
- SPPPLVs are calculated for all potential exposure pathways.
- Critical pathways for each pollutant are selected.
- The PPLV is derived by normalization of SPPPLVs.

In the initial step, pollutants and pathways are identified; however, further guidelines on this step are not given. In addition, guidelines are not offered on how to evaluate complex mixtures of chemicals.

The acceptable daily dose of toxicant, D_T, may be obtained from six sources. Listed in order of preference, these values are: (1) ADI; (2) MCL for drinking water; (3) TLV; (4) Lifetime No Effect Level (NEL_L); (5) Ninety day No Effect Level (NEL_{90}); and (6) Acute Toxicity (LD_{50}).

The TLV must be converted from 5 days/week exposure to 7 days/week exposure by dividing by (7/5 = 1.4). A safety factor of 100 is used to protect sensitive members of the population who would normally not work. The breathing rate, RB, for a 70-kg person doing light work, 12.1 m 3/8 hours, is multiplied times the TLV. Therefore, the TLV is converted as follows:

$$D_T = (TLL \times RB)/1.4 \times \text{Body Weight} \times 100$$

The No Effect Level from a lifetime animal study, NEL_L, is adjusted by a safety factor of 100 to account for interspecies differences. The No Effect Level from a 90-day subchronic study, NEL_{90}, requires a safety factor of 1000 due to the shorter period of exposure and interspecies differences. The LD_{50} from an acute toxicity animal study is used when none of the other values are available. A conversion factor is multiplied times the LD_{50} to yield a safe limit for continuous body concentration.

The D_T value based on one of the latter six factors (ADI, MCL, TLV, NEL_L, NEL_{90} or LD_{50}) represents the acceptable daily intake of a noncarcinogenic threshold pollutant. It is unclear how D_T should be determined for a carcinogenic nonthreshold pollutant. In addition, D_T values for ingestion and inhalation exposure are assumed to be the same. This assumption may introduce significant error for some pollutants.

Each exposure pathway involves a linear compartment model through which the pollutant passes (e.g., soil → air → human). The PPLV model assumes that between two adjacent media, the pollutant is partitioned in a constant manner and all compartments are assumed to be at equilibrium.

One assumption of the PPLV model is that the pollutant is conserved in all compartments. Half-life or biodegradation factors are not considered and, therefore, overestimates of risk may be predicted for chemicals with short half-life factors.

Partition coefficients may not be available for some pollutants or may have to be estimated on scanty data. The value of a partition coefficient may also vary with local site conditions such as soil type, climate, etc. If no data are available to estimate the partition coefficient, the method suggests that K = 1.

The PPLV methodology does not specifically address exposure via soil ingestion or dermal absorption. However, new information or different exposure variables can be incorporated into the standard equation.

The derivation of SPPPLV is based on the acceptable daily dose of pollutant (D_T), partition coefficients for a specific exposure pathway, standard intake factors, and a body weight factor. If one or more intermedia transfers occurs in an exposure pathway (e.g., soil → plant → human), the SPPPLV has the following form:

$$C_i = \frac{BW \times D_T}{(Km_1m_2 \times Km_2m_3 \ldots Km_xm_y) \times (f \times DFI)}$$

C_i = acceptable pollutant concentration in one exposure pathway (mg/kg)
BW = body weight factor (kg)
D_T = acceptable daily dose of pollutant (mg/kg/day)
Km_xm_y = partition coefficients for exposure pathway

f × DFI = standard intake factor (kg/day) (fraction of total diet represented by food of a given type × daily food intake)

If the pollutant reaches the human receptor without an intermediate compartment (e.g., water → human), the SPPPLV has the following form.

$$C_F = \frac{BW \times D_T}{W_i}$$

W_i = Standard intake factor (e.g., water intake)

Partition coefficients are not needed in direct transfer of pollutant from site to human, as in the case of pollutants originating in drinking water.

Once the critical exposure pathways are identified, the appropriate SPPPLVs are used to determine the PPLV.

$$C_F = \frac{1}{\sum_{i=1}^{n} \frac{1}{C_i}}$$

C_F = final acceptable pollutant concentration for all exposures
C_i = acceptable pollutant concentration in one exposure pathway

The example is given that if C_i was 10 ppm by water ingestion, 5 ppm by fish ingestion, and 20 ppm by crop ingestion, the PPLV would be: $C_F = 1 - (1/5 + 1/10 + 1/20) = 2.86$ ppm. The C_F represents the cleanup level for the specific pollutant, based on all critical exposure pathways.

Conclusions

The PPLV method is easy to use, and evaluates multiroute exposures and intermedia transfers of pollutants. The use of partition coefficients simplifies the mathematical modeling necessary in other methodologies for exposure pathway analysis. The PPLV model is made to be flexible in order to incorporate new information as it becomes available.

An acceptable daily dose of pollutant, D_T, is based on ADI, MCL, TLV, NEL, or LD_{50} values for threshold noncarcinogenic pollutants. The derivation of D_T for carcinogenic pollutants is not detailed. In addition, D_T values are the same for ingestion and inhalation exposures, despite the fact that the hazard level of a pollutant may vary with route of exposure.

The PPLV methodology does not address the problem of how to classify and assess the risks presented by a large, diverse mixture of chemicals at one site. The possibility of additive toxic or carcinogenic effects is also not evaluated. If numerous chemicals at one site exert a cumulative systemic effect, the hazard could be much greater than that represented by the PPLV.

The advantages of use of partition coefficients to determine exposure levels is that this simplifies the need for modeling of release rates of chemicals from a site. Major assumptions which should be noted are (1) all compartments (media) are assumed to be at equilibrium, and (2) no decomposition of the chemical occurs. The first assumption may be violated as a result of local site conditions. At a particular site, the standard partition coefficients may not represent the actual transfer ratio between two media, due to such factors as soil type, weather patterns, etc.

The second assumption could present problems if the chemical of interest has a short half-life. A chemical with a short half-life would present less of a hazard than a chemical with a longer half-life. Therefore, the PPLV could overestimate or underestimate risk.

Partition coefficients may be difficult to locate for some chemicals, and these values may have to be based on a single literature value. Since the environmental fate of the pollutants is based solely on this value, the PPLV cleanup level is very sensitive to errors in determination of partition coefficients.

Exposure via soil ingestion and dermal absorption is not specifically addressed; however, as previously noted, these factors could be incorporated into the standard formula.

The PPLV methodology is site-specific, and evaluates the potential for a pollutant to proceed from its point of origin through specific pathways to a target receptor. The calculations are easy to use; however, some error may be introduced when major assumptions of the model are violated. The authors emphasize, however, that the model is flexible and new data should be incorporated to account for site-specific conditions (Table 3).

SUPERFUND HEALTH ASSESSMENT MANUAL

Summary

EPA's *Superfund Health Assessment Manual*[3] proceeds from a relatively simple qualitative assessment of available information to a more detailed quantitative risk assessment. The public health evaluation involves three major components:

- Baseline site evaluation
- Public health assessment of the no-action alternative
- Development of design goals and estimation of risk for remedial alternatives

These three components are outlined in the following summary.

Baseline Site Evaluation

The initial step involves determination of the extent of contamination at a site and may be qualitative (Level 1) or quantitative (Level 2). A Level 2 assessment

Table 3. Summary of the U.S. Army Methodology with Respect to the Evaluation Criteria

HAZARD ANALYSIS	
Toxicity value	ADI[a]
	MCL
	TLV
	NEL_L
	NEL_{90}
	LD_{50}
Noncarcinogenic acute toxicity	Yes
Noncarcinogenic chronic toxicity	Yes
Carcinogenic chronic toxicity	No
Assessment of mixtures	No
Body weight factor	Yes
Pharmacokinetic factor	No
EXPOSURE ANALYSIS	
Air inhalation	Yes
Dust inhalation	Yes
Water ingestion	Yes
Soil ingestion	No
Dermal absorption	No
Crop uptake	Yes
Livestock uptake	Yes
Fish uptake	Yes
Environmental fate	Yes
	—partition coefficients
	—bioconcentration
	—air transport
	—water transport
Half-life factor	No
Site-specific factors	Yes
RISK ANALYSIS	
Additivity of toxic effect	No
Multi-media exposure	Yes
Synergistic effects	No

[a]See text for description of toxicity values.

is required at sites in which migration of contaminants is possible, exposure has occurred or is imminent, and/or a large population resides near the site.

Public Health Assessment of the No-Action Alternative

After the baseline site evaluation has indicated whether a Level 1 or Level 2 assessment will be conducted, the regulatory agency must next assess the no-action alternative.

Level 1 assessment of no-action alternative. The Level 1 assessment involves reviewing available information on types and amounts of chemicals, toxic effects, proximity of human populations, likelihood of chemical release and migration, and potential for human exposure. A comprehensive exposure pathway analysis is required which evaluates the following four criteria:

- source and mechanism of chemical release
- environmental transport medium
- human exposure point
- feasible human exposure route

If a complete exposure pathway is found to exist (all four elements are present) then a more detailed Level 2 assessment should be conducted. Both short term and long term exposure pathways should be considered. A Level 1 assessment is conducted when no migration of contaminants is probable over the short term or long term. Remedial actions after a Level 1 assessment generally involve source control measures based on relevant/applicable standards.

Level 2 assessment of no-action alternative. The Level 2 assessment of the no-action alternative involves five analytical steps:

- selection of indicator chemicals
- assessment of exposure pathways
- estimation of human intakes
- toxicity assessment of indicator chemicals
- characterization of human health risks

The procedures required for these five steps are outlined in the following sections.

1. Selection of indicator chemicals: When a site is contaminated with more than 10 chemicals, a subset of indicator chemicals is selected based upon toxicity, mobility, persistence, and concentration. The indicator chemicals are ranked using the following formula:

$$\text{Indicator score}_i = \text{Concentration}_i \times \text{Toxicity constant}_i$$

Toxicity constants for noncarcinogens are derived from the minimum effective dose (MED) for chronic effects, severity of effects rating, standard factors for body weight, and oral or inhalation intake. Toxicity constants for potential carcinogens are based on the dose at which a 10% incremental carcinogenic response is observed (ED_{10}) and standard intake and body weight factors. Intake factors for soil toxicity constants are based on the assumption of 100 mg/day of soil consumed from ages 2 to 6, averaged over a 70-year life span.

Potential carcinogens and noncarcinogens are scored and selected independently due to the fact that indicator scores for these two groups are not on comparable scales. Indicator scores (IS) are summed for each chemical in different media (air, water, soil) to obtain one IS value for each chemical. The IS values are then ranked, and the 10 to 15 top-scoring chemicals within the two groups become the preliminary list of indicator chemicals.

Other chemical properties which are considered prior to selecting the final set of indicator chemicals are water solubility, vapor pressure, Henry's law constant, organic carbon partition coefficient, and half-life in various media. The final selection should consider the latter properties in the context of the particular exposure pathways at the site.

2. Assessment of exposure pathways: A combination of site monitoring data and environmental modeling results is used to estimate exposure pathways. Because site monitoring data alone will not reveal future conditions, environmental fate modeling is recommended by EPA. Simple environmental fate equations are provided, and more sophisticated computer models are referenced in the manual. EPA suggests that simple models using conservative assumptions be used for Superfund sites.

   Exposure pathways must be identified based on the four elements listed in Level 1 assessment (release source, environmental transport media, exposure point, and exposure route). If one of the latter characteristics is missing, the exposure pathway is incomplete.

   EPA suggests that chemical releases be quantified in terms of release rates to predict environmental fate. Air, soil, surface water, and groundwater release modeling are described in the manual. Environmental fate and transport modeling is then conducted based on release rates.

   The projected concentrations of the indicator chemicals at the exposure points are compared to relevant/applicable standards. If all of the indicator chemicals have relevant/applicable standards, then a full quantitative risk characterization is not required. EPA established a hierarchy of standard values to be applied to Superfund sites:

   - relevant/applicable standards for appropriate exposure pathway (EPA considers drinking water maximum contaminant levels (MCLs), state water quality standards, and ambient air quality standards to be the only relevant/applicable standards)
   - toxicity values based on EPA's Health Effects Assessment (HEA)
   - other standards or criteria such as water quality criteria

   If standard values exist for all of the indicator chemicals, the Level 2 assessment of the no-action alternative is complete. The ambient standards can then be used in the development of target concentrations for remedial alternatives. If ambient standards do not exist for all of the indicator chemicals, exposure point concentrations are used to calculate chemical intakes for which risk is estimated.

3. Estimation of human intakes: Subchronic and chronic daily chemical intakes (SDI and CDI) are based on the previously determined short-term and long-term concentrations of the indicator chemicals at the exposure points (step 2). Intake is defined as the amount of contaminant taken into the body per unit body weight per unit time (mg/kg/day). Intakes are calculated separately for each chemical

in each medium (air, groundwater, and surface water). Groundwater and surface water intake are summed for each chemical to yield the total oral intake. The following formulas are used to calculate intakes:

$$\text{Subchronic daily intake (SDI)} = \begin{array}{c}\text{Short-term}\\ \text{concentration}\end{array} \times \begin{array}{c}\text{Standard}\\ \text{intake factor}\end{array}$$

$$\text{Chronic daily intake (CDI)} = \begin{array}{c}\text{Long-term}\\ \text{concentration}\end{array} \times \begin{array}{c}\text{Standard}\\ \text{intake factor}\end{array}$$

Standard Intake Factors Used by EPA

	Adult	*Child*
Avg. body weight	70 kg	10 kg
Amt. of water/day	2 L	1 L
Amt. of air/day	20 m^3	5 m^3
Amt. of fish/day	6.5 g	—

EPA assumes 100% absorption of the contaminant after intake, due to uncertainties in present data. In addition, formulas for less common exposure routes (soil ingestion, dermal absorption) are not included in the manual. If such an exposure route is important at a Superfund site, EPA suggests that the regulatory agency contact EPA headquarters for guidance on a case-by-case basis.

4. Toxicity assessment of indicator chemicals: For noncarcinogenic indicator chemicals, EPA has determined the acceptable intake for subchronic exposure (AIS) and the acceptable intake for chronic exposure (AIC) for specific media. These values are listed in the appendix of the manual. The AIS and AIC values are derived from data obtained by animal studies and human epidemiological studies, when available. EPA states that these values are designed to protect sensitive populations, but this is not further explained.

   For carcinogenic indicator chemicals, carcinogenic potency factors (CPF) have been determined by EPA, and represent lifetime cancer risk per mg/kg/day. The CPF is the estimated upper 95% confidence limit of the carcinogenic potency of a chemical.

   The latter critical toxicity values (AIS, AIC, and CPF) are to be used in the risk characterization of the indicator chemicals (step 5). If critical toxicity values are not available for all of the indicator chemicals, EPA suggests that its headquarters be contacted for guidance.

5. Characterization of human health risks: The risk characterization involves a comparison between the estimated daily intakes (SDI or CDI) and the critical toxicity values (AIS, AIC, or CPF) for the indicator chemicals.

   For a set of noncarcinogenic indicator chemicals, a hazard index is derived from the summation of the individual risks from each noncarcinogen:

$$\text{Subchronic Hazard Index} = \sum_{i=1}^{n} \frac{\text{subchronic daily intake}_i}{\text{acceptable intake, subchronic}_i}$$

$$\text{Chronic Hazard Index} = \sum_{i=1}^{n} \frac{\text{chronic daily intake}_i}{\text{acceptable intake, chronic}}$$

The hazard indices for inhalation and oral exposure for all noncarcinogenic indicator chemicals are summed to assess the effects of multiple exposure pathways. The hazard index takes into account that multiple subthreshold exposures could result in an overall adverse effect. EPA emphasizes that the hazard index is not a predictor of incidence or severity; rather, it is an indicator of acceptable or unacceptable exposure levels.

EPA recognizes that the application of a hazard index to a mixture of chemicals having varying effects could overestimate risk. Therefore, if the hazard index exceeds 1, the compounds should be segregated by critical effect (described in the manual), and separate hazard indices derived for each critical effect (i.e., neurotoxic, hemotoxic, etc.). If any of the individual hazard indices exceed 1, remedial action will be necessary.

The risk characterization for carcinogenic indicator chemicals is determined using the following formula:

$$\text{Total Carcinogenic Risk} = (\text{chronic daily intake}_i \times \text{carcinogenic potency factor}_i)$$

The carcinogenic risks from various exposure routes (oral, inhalation) are also summed to yield a total carcinogenic risk. The total carcinogenic risk for the site, based on the indicator chemicals, is used to design remedial options based on a target carcinogenic risk.

The Level 2 assessment of the no-action alternative is complete at this point. If remedial action is deemed necessary, the public health evaluation proceeds to the next section (see Figure 1).

Development of Design Goals and Estimation of Risk for Remedial Alternatives

The final state of the health assessment process involves the following steps:

1. Indicator chemicals are reviewed in light of the specific remedial actions to be undertaken. Some indicator chemicals may be treated more easily with certain cleanup methods.
2. Exposure pathways are reviewed for the remedial alternative to be used, because new exposure pathways could be created by the remedial option.
3. Target concentrations are determined at human exposure points.
4. Target release rates are estimated.
5. Chronic risk from noncarcinogens is assessed.
6. Potential short-term health effects of remedial alternatives are assessed.
7. Effects of remedial alternative failure are assessed.

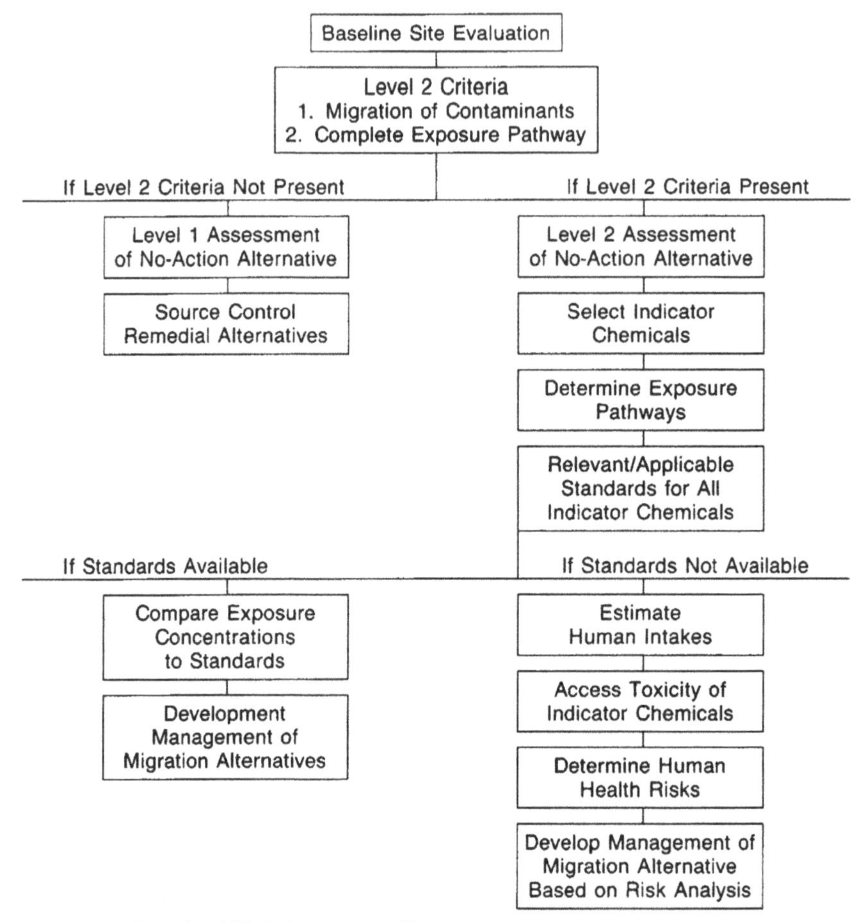

Figure 1. Superfund Risk Assessment Process.

The most important steps for risk assessment are steps 3 and 5. The goal of step 3 is to determine a target concentration range for each indicator chemical. The target concentrations are calculated using relevant/applicable standards or a target risk range. If all of the indicator chemicals have ambient standards, these are used as the basis for the target concentration range. If the indicator chemicals do not have applicable standards, target concentrations are calculated using toxicity and intake values.

Potential carcinogens are evaluated first because their target concentrations are usually lower than those for noncarcinogens. The target total carcinogenic risk at Superfund sites must fall between 10^{-4} to 10^{-7}. Various methods may be used to apportion carcinogenic exposure among multiple carcinogens and multiple exposure routes. One method suggested by EPA is to divide the target carcinogenic risk level (10^{-4} to 10^{-7}) by the number of carcinogenic indicator chemicals. This will yield an individual target carcinogenic risk for each carcinogen. The individual carcinogenic risk is then divided by the carcinogenic potency factor for that chemical to determine the target chronic daily intake level. The target daily intake is then divided by standard intake factors to yield the target long-term concentration. These formulas are outlined below:

$$\frac{\text{Total Target Risk } (10^{-4}\text{–}10^{-7})}{\text{Number of Potential Carcinogens}} = \text{Target risk for each chemical}$$

$$\frac{\text{Individual Target Risk}}{\text{Carcinogenic Potency Factor}} = \text{Target chronic daily intake}$$

$$\frac{\text{Target Chronic Daily Intake}}{\text{Human Intake Factor}} = \text{Target concentration}$$

The target concentrations for each carcinogenic indicator chemical are used to design the remedial alternatives for the site.

A second approach suggested by EPA is to let one or two extremely potent carcinogens drive the design process. One or two indicator chemicals may be so potent that exposure to them must be extremely low (e.g., dioxin). In this case, the total carcinogenic risk will fall within the target range for the other carcinogens as a result of remedial actions designed to reduce the most potent carcinogen.

Once remedial alternatives have been assessed for reduction of carcinogenic risk, these remedial alternatives should be checked to ensure that noncarcinogenic risk is reduced as well. Only chronic risk is considered, because long-term remediation is the goal of this process. The hazard index is determined for each critical effect as described in the Level 2 assessment of risk characterization:

$$\text{Hazard Index} = \frac{\text{chronic daily intake}_i}{\text{acceptable intake chronic}_i}$$

The hazard index should be less than 1 but if it exceeds 1 for any endpoint, the remedial design should be altered to reduce this risk.

Strengths and Weaknesses

EPA's Superfund Health Assessment Manual provides a comprehensive step-by-step procedure for assessing the health risks posed by contaminated sites. Since many hazardous waste sites contain a multitude of chemicals, EPA has offered guidelines for selection of the most toxic indicator chemicals which are evaluated in a risk analysis. The number of indicator chemicals can be as high as desired (EPA suggests 5–10), and should include both carcinogens and noncarcinogens, if present at the site.

Chemicals at the site are ranked according to concentration and toxicity values provided by EPA. If no toxicity value has been determined by EPA for a specific chemical, it can be derived using data from animal studies when available (MED or ED_{10}). The top 10–15 scoring chemicals in each class (carcinogens and non-carcinogens) are then further evaluated based on half-life, water solubility, vapor pressure, Henry's law constant, and organic partition coefficients. These factors are considered in the context of the particular exposure pathway. Therefore, the final indicator chemicals reflect the characteristics of the chemicals in their environment and various toxic effects.

EPA's methodology for selection of indicator chemicals is useful for sites contaminated with a mixture of chemicals, such as petroleum products. Toxicity constants used in selection of indicator chemicals are evaluated for soil ingestion exposure. EPA assumed that 100 mg of soil is ingested per day for children ages 2 to 6, and this value was averaged over a 70-year lifespan. This value could be altered if new information suggests more or less soil ingestion for children and/or adults.

EPA offers guidelines for a comprehensive, yet possibly time-consuming, exposure pathway analysis, including modeling of environmental fate, and transport and release rates for subchronic and chronic exposures. The amount of time needed to complete this portion of the risk assessment may be significant, and could slow down implementation of remedial response measures.

Formulas are provided for estimation of environmental fate and transport and release rates. Computer models which can be used in this step are referenced in the manual. The level of sophistication achieved by an individual regulatory agency may depend on equipment available and familiarity with this type of analysis. It is important that the assumptions of the environmental fate model be evaluated for relevance to the specific site under investigation

For soils contaminated with a complex mixture of substances, analysis of environmental fate and release rates would appear to be very important. Certain chemicals may be released slowly over a lifetime and travel through various exposure pathways (air, water, or soil). For this reason, the site should be evaluated for risks posed over a human lifetime or longer. EPA suggests that simple models with conservative assumptions are sufficient for Superfund sites.

The exposure point concentrations of the indicator chemicals are compared to relevant/applicable standards when available. If standards are not available, guidelines for risk assessment are provided using toxicity data, exposure data, and standard assumptions for intake and body weight. Values for subchronic and chronic daily intake (SDI and CDI) are estimated and multiroute exposures (water ingestion and inhalation) are taken into account.

One problem with the latter procedure is that formulas for less common exposure pathways such as soil ingestion or dermal absorption are not included. This is unfortunate, considering that soil ingestion was taken into account by EPA in the selection of indicator chemicals. EPA stated that there was a desire to include soil toxicity constants in step 1 so that chemicals in the soil could be considered when selecting indicator chemicals. Unfortunately, this was not followed up in the determination of daily chemical intakes to be used in the risk characterization. EPA does emphasize that soil ingestion can be an important exposure pathway for children playing near contaminated sites, yet does not further assess this risk in the manual.

EPA assumes 100% absorption of all chemicals after intake, due to uncertainties in present bioavailability data. If further information on the percent of a chemical absorbed becomes available, EPA suggests that this factor be taken into account. Overestimates of risk can be derived if absorption is less than 100% and, therefore, should be carefully evaluated for specific chemicals and exposure pathways (i.e., some indicator chemicals may bind tightly to soil particles and be less likely to be absorbed in the intestine).

The acceptable daily intakes for subchronic (AIS) and chronic (AIC) exposures have been determined by EPA for many chemicals. It is unclear how EPA derived these values, except that they are based on data from animal studies concerning intake and toxic effect. Carcinogenic potency factors (CPF) have been determined by EPA and represent the lifetime cancer risk per mg/kg/day.

If AIS, AIC, or CPF values are not available for some indicator chemicals at a particular site, the risk analysis cannot be completed. For example, AIS, AIC, or CPF values are not available for benz-(a)-anthracene, cadmium, dichlorobenzene, heptane, hexane, isobutane, isopentene, 1-pentene, and xylene. All of these compounds are major constituents in petroleum products.

The risk characterization involves comparison of intake estimates (SDI or CDI) with acceptable exposure levels (AIS or AIC) or risk based on CPFs for indicator chemicals. Soil ingestion and dermal absorption exposure pathways are not considered in this step.

Additivity of carcinogenic risk from multiple carcinogens is taken into account, as well as additivity of chronic noncarcinogenic effects for a specific end point. The total carcinogenic risk from all routes of exposure must fall within a target range of 10^{-4} to 10^{-7}. It should be noted, however, that the target range for carcinogenic risk involves only the selected indicator chemicals which have CPF values. Other carcinogens cannot be adequately evaluated and may pose a significant health risk.

Conclusion

EPA's *Superfund Health Assessment Manual* provides a useful quantitative risk assessment procedure for sites contaminated with a mixture of chemicals. Major problems with this methodology include lack of consideration of exposures due to soil ingestion or dermal absorption, and values used in the risk analysis (AIS, AIC, and CPF) may not be available for many constituents in petroleum products.

The EPA methodology is useful in its careful selection process for indicator chemicals including relevant factors such as half-life, water solubility and vapor pressure. Carcinogenic and noncarcinogenic chemicals are ranked separately which allows for risk assessment of both carcinogenic effects and noncarcinogenic subchronic or chronic effects. Toxicity constants are provided for soil ingestion exposure in the selection of indicator chemicals.

A comprehensive exposure pathway analysis is detailed by EPA including environmental fate, transport and release rates. Such an analysis, although time-consuming, may be valid for sites in which long-term effects are probable.

The total carcinogenic risk posed by all of the carcinogenic indicator chemicals must fall within a target range of 10^{-4} to 10^{-7}. However, chemicals not having carcinogenic potency factors cannot be adequately assessed. In addition, noncarcinogens cannot be properly evaluated if acceptable intake factors are not available.

The strength of the EPA Superfund manual lies in its long-term exposure pathway analysis, indicator chemical selection process, and risk assessment formulas for carcinogens and noncarcinogens. These portions of the methodology could be utilized in a risk assessment process which does take into account less common exposure pathways such as soil ingestion and dermal absorption. In this manner, the health risk assessment of soils contaminated with petroleum products could be adequately evaluated (Table 4).

HEALTH RISK ASSESSMENTS FOR CONTAMINATED SOILS

Summary

Two types of risk analyses were presented in "Health Risk Assessments for Contaminated Soils."[4] The risk assessment formula for acute toxicity utilized the allowable daily intake (ADI) to represent the toxicity of noncarcinogenic contaminants. The ADI was modified by body weight and daily soil intake to determine the soil criteria for a specific contaminant.

The risk analysis formula for chronic toxicity involved dividing an acceptable cancer risk by a unit carcinogenic risk (UCR) established by the U.S. EPA Carcinogen Assessment Group. The factor was then modified by lifetime average soil intake and half-life of the contaminant to yield a soil criteria for a specific carcinogen.

Strengths and Weaknesses

The risk assessment methodology for contaminated soil developed by Ford and Gurba[4] evaluates both acute and chronic toxicity. Since children are believed to ingest the highest levels of soil over a brief time period (ages 2–5) compared to older age groups, they may be a high-risk group for exposure to contaminated soils. For this reason, Ford and Gurba developed a risk analysis formula to evaluate acute toxicity of soil contaminants for children, along with a chronic toxicity formula for carcinogens.

The acute toxicity formula is based on the allowable daily intake (ADI), originally developed to represent the adult intake of a food additive at which no lifetime health effects would occur. Ford and Gurba modified the ADI by body weight and daily soil ingestion for children to yield the following formula:

Table 4. Summary of the EPA Superfund Methodology with Respect to the Evaluation Criteria

HAZARD ANALYSIS	
Toxicity value	AID[a] AIC CPF
Noncarcinogenic acute toxicity	Yes
Noncarcinogenic chronic toxicity	Yes
Carcinogenic chronic toxicity	Yes
Assessment of mixtures	Yes
Body weight factor	Yes
Pharmacokinetic factor	No
EXPOSURE ANALYSIS	
Air inhalation	Yes
Dust inhalation	Yes
Water ingestion	Yes
Soil ingestion	No
Dermal absorption	No
Crop uptake	No
Livestock uptake	No
Fish uptake	Yes
Environmental fate	Yes —air transport —water transport
Half-life factor	Yes
Site-specific factors	Yes
RISK ANALYSIS	
Additivity of toxic effect	Yes
Multi-media exposure	Yes
Synergistic effects	No

[a]See text for description of toxicity values.

$$SC = ADI \times \frac{1000}{SI} \times BW$$

SC = soil criteria (mg/kg)
ADI = allowable daily intake (mg/day)
1000 = conversion factor (g/kg)
SI = soil ingestion (g/day)
BW = body weight adjustment (10 kg/70 kg)

The ADI value represents the toxicity of the contaminant, the SI value represents the exposure, and the BW value adjusts the ADI for children.

Since many of the pollutants in contaminated soils may be carcinogens, a different risk analysis was developed for lifetime, chronic toxicity. The term "lifetime allowable daily intake" (LADI) was derived by dividing an acceptable cancer risk (1×10^{-6}) by a unit carcinogenic risk (UCR). UCRs are established by the U.S. EPA's Carcinogen Assessment Group and are expressed as an excess cancer risk from a lifetime of ingestion of 1 mg/kg/day of a carcinogen. The LADI was modified by lifetime average soil intake and half-life of the contaminant to yield the following formula:

$$SC = LADI \times \frac{1000}{LASI} \times \frac{\frac{t}{2}}{70}$$

LADI = lifetime allowable daily intake (mg/kg/day)

$$= \frac{\text{Risk } (1 \times 10^{-6})}{\text{Unit Carcinogenic Risk (UCR)}}$$

1000 = conversion factor (g/kg)
LASI = lifetime average soil intake (g/kg/day)
t/2/70 = half-life correction factor

The toxicity of a specific carcinogen is represented by the LADI, the exposure is represented by LASI, and a half-life correction factor adjusts for contaminants which may biodegrade significantly over a lifetime.

One limitation of the latter two methodologies is that the values ADI and UCR may not be available for many of the contaminants in petroleum products. In addition, both analyses do not evaluate site conditions, such as geologic or geographic factors, type of soil, and seasonal conditions. Contaminants in soil may react with soil particles by binding more or less strongly and may migrate to drinking water as a result of geologic, geographic, and seasonal influences. The environmental fate of soil contaminants is determined by site conditions which are not taken into account in Ford and Gurba's methodologies.

Soil ingestion is considered to be the most important route of exposure by Ford and Gurba, while other routes of exposure are ignored. One weakness of these risk analyses, therefore, is that the health risk due to exposure via contaminated drinking water, dust inhalation, or dermal absorption is not evaluated.

The effectiveness of the two risk analyses will be reflected by the careful choice of individual constituents which are evaluated in the formulas. For a mixture of pollutants such as petroleum products, the constituents analyzed must represent the overall toxicity of the mixture. If this is not the case, the soil criteria may not establish a safe level of cleanup. A thorough review of the toxicity of constituents in petroleum products should be conducted prior to use of the latter methodologies.

Conclusions

The health risk assessments for contaminated soils developed by Ford and Gurba have included the important factors of acute and chronic toxicity, daily and lifetime soil ingestion, and half-life adjustments for contaminants which biodegrade significantly during a lifetime. Toxicity of the contaminants was evaluated by the terms ADI or UCR, which may not be available for many of the constituents of petroleum products. Both risk analysis formulas failed to evaluate the health risks resulting from exposure via contaminated water, dust inhalation, or dermal absorption. Site conditions, including geologic or seasonal factors, were ignored despite the fact that these variables can strongly influence the environmental fate of soil contaminants. Finally, if the soil contaminant consists of a mixture of chemicals, as is the case for petroleum products, the constituents evaluated in the risk assessment formula must represent the overall toxicity of the mixture.

Table 5. Summary of the Ford and Gurba Methodology with Respect to the Evaluation Criteria

HAZARD ANALYSIS	
Toxicity value	ADI[a]
	CPF
Noncarcinogenic acute toxicity	Yes
Noncarcinogenic chronic toxicity	Yes
Carcinogenic chronic toxicity	Yes
Assessment of mixtures	Yes
Body weight factor	Yes
Pharmacokinetic factor	No
EXPOSURE ANALYSIS	
Air inhalation	No
Dust inhalation	No
Water ingestion	No
Soil ingestion	Yes
Dermal absorption	No
Crop uptake	No
Livestock uptake	No
Fish uptake	No
Environmental fate	No
Half-life factor	Yes
Site-specific factors	No
RISK ANALYSIS	
Additivity of toxic effect	No
Multi-media exposure	No
Synergistic effects	No

[a]See text for description of toxicity values.

SOIL CLEANUP CRITERIA FOR SELECTED PETROLEUM PRODUCTS

Summary

The New Jersey Department of Environmental Protection's "Soil Cleanup Criteria for Selected Petroleum Products"[5] describes a soil cleanup methodology based on a few individual constituents of petroleum products which pose the greatest threat to public health. The most hazardous constituents were identified to be the carcinogenic polycyclic aromatic hydrocarbons (CaPAHs) and benzene. Acceptable soil contaminant levels (ASCL) were determined based on lifetime soil ingestion, a 1×10^{-6} cancer risk, and carcinogenic potency factors for individual constituents. The ASCL was compared to residual soil levels of CaPAHs and benzene resulting after soil cleanup to 100 ppm total petroleum hydrocarbons, as reported in the literature. For residual soil levels yielding a greater than 1×10^{-6} cancer risk, a lower soil cleanup level was proposed. With the exception of used motor oils over 10,000 km, CaPAHs and benzene levels were below the concentration which would exceed a 1×10^{-6} cancer risk after cleanup to 100 ppm total petroleum hydrocarbons.

Strengths and Weaknesses

The New Jersey Department of Environmental Protection's "Soil Cleanup Criteria for Selected Petroleum Products" examined increased cancer risk as a result of lifetime ingestion exposure to soils contaminated with petroleum products. The risk assessment and soil cleanup objectives were presented only for individual chemical constituents of petroleum products which have the highest toxicity, have the ability to migrate, and/or are present in significant amounts. Petroleum product constituents were reviewed for chemical/physical properties, health effects, environmental effects, and carcinogenic properties. The authors determined that the carcinogenic polycyclic aromatic hydrocarbons (CaPAHs) and benzene were the major constituents of concern.

As noted by the authors, however, the concentrations of CaPAHs and benzene vary depending on the type of crude oil and the fractionation process used. Benzo-(a)-pyrene (BaP) exists in small quantity relative to other CaPAHs, although used petroleum products are often enriched in BaP and other CaPAHs. Information on benzene levels in petroleum products is limited, with the highest levels found in gasoline.

Once the primary chemical constituents were identified, these were utilized in the risk assessment methodology. The authors based their methodology on reports in the literature which examined residual soil levels of CaPAHs and benzene after soil cleanup to 100 ppm total petroleum hydrocarbons. The residual soil levels were compared to acceptable soil contaminant levels (ASCL) based on a 1×10^{-6} cancer risk, lifetime soil ingestion, and carcinogenic potency factors established for various compounds by the U.S. EPA Cancer Assessment Group (CAG).

$$\text{ASCL} = \frac{A}{C} \times \frac{1000}{L}$$

A = acceptable cancer risk $-$ 1×10^{-6} (one in a million)
C = carcinogenic potency factor (U.S. EPA CAG)
for BaP = 11.53 $(\text{mg/kg/day})^{-1}$
for benzene = 0.0052 $(\text{mg/kg/day})^{-1}$
1000 = conversion factor (g/kg)
L = lifetime average daily soil intake

While it is impossible to assess the risk posed by every constituent in petroleum products, the validity of limiting the risk assessment to CaPAHs and benzene should be examined. Care should be taken to adequately and thoroughly assess the toxicity characteristics of the major constituents in petroleum products as a means of identifying the most hazardous elements. The authors of the present methodology did not go into great detail on their rationale for choosing CaPAHs and benzene as the constituents of concern. It is important to explain why the latter two constituents are more hazardous than the many other chemicals in petroleum products.

The carcinogenic potency factor for BaP was used to represent all of the CaPAHs in petroleum products, due to the fact that carcinogenic potency factors do not exist for other CaPAHs. Therefore, the ASCL is based only on BaP, and assumes that all of the other CaPAHs are as toxic as BaP. No justification of this assumption was presented by the authors, except that it was consistent with the U.S. EPA's approach to estimating cancer risk from exposure to mixtures of CaPAHs. It is possible that other CaPAHs may be more toxic or have synergistic effects in the mixture.

The ASCL is based on lifetime soil ingestion; yet the authors contended that it is also based on inhalation of dust. It is unclear how inhalation is taken into account by the model, and this is not explained in the methodology. The method also does not consider the serious threat to public health resulting from contamination of groundwater or surface water via migration of toxic constituents.

The soil cleanup criteria does not take into account biodegradation, volatilization, or half-life of the individual constituents. Considering the fact that benzene is highly volatile and may have a relatively short half-life in soil, this factor should be accounted for in the methodology. Other toxic constituents may also biodegrade in significant amounts over time.

Conclusions

The risk assessment method developed by Stokman and Dime (New Jersey Department of Environmental Protection) focused on increased cancer risk as a result of lifetime ingestion of soil contaminated with petroleum products. Exposure via contaminated drinking water due to migration of toxic constituents was not assessed and is one limitation of this methodology. In addition, the soil cleanup criteria did not take into account biodegradation, volatilization, or half-life of the individual constituents. Half-life can be very important for a constituent such as benzene, which is very volatile.

CaPAHs and benzene were chosen as the major constituents of concern in petroleum products. Although this type of risk assessment simplifies the problem of dealing with numerous compounds in petroleum products, the individual constituents should be chosen carefully and clearly be representative of the overall mixture. In this case, it was not clearly proved that CaPAHs and benzene would be representative of the total mixture, with regard to lifetime toxicity. However, the individual regulatory agency may choose its own representative constituents based on a thorough and complete hazard analysis of petroleum product constituents.

If residual soil levels after cleanup to 100 ppm total petroleum hydrocarbons are below the ASCL for the individual constituents, the risk assessment is complete. However, the effect of residual soil levels of the remaining petroleum compounds will depend to some extent on the type of soil and on geologic, geographic, and seasonal conditions. If environmental factors at the site are not taken into account, and/or the individual constituents were not chosen carefully enough, residual soil levels of other constituents not initially focused on could have long-term toxic effects. In summation, the effectiveness of this methodology depends on how representative the individual constituents used in the ASCL are for the overall mixture and environmental conditions at the site (Table 6).

Table 6. Summary of the New Jersey Methodology with Respect to the Evaluation Criteria

HAZARD ANALYSIS	
Toxicity value	CPF[a]
Noncarcinogenic acute toxicity	No
Noncarcinogenic chronic toxicity	No
Carcinogenic chronic toxicity	Yes
Assessment of mixtures	Yes
Body weight factor	No
Pharmacokinetic factor	No
EXPOSURE ANALYSIS	
Air inhalation	No
Dust inhalation	No
Water ingestion	No
Soil ingestion	Yes
Dermal absorption	No
Crop uptake	No
Livestock uptake	No
Fish uptake	No
Environmental fate	No
Half-life factor	No
Site-specific factors	No
RISK ANALYSIS	
Additivity of toxic effect	No
Multi-media exposure	No
Synergistic effects	No

[a]See text for description of toxicity values.

FINAL CLEANUP POLICY

Summary

The Washington Department of Ecology's Final Cleanup Policy[6] involves three levels of cleanup for a contaminated site: (1) Initial Cleanup Levels consist of total cleanup or partial cleanup to eliminate an "imminent" public health threat.

Total Cleanup is achievable when site characteristics include well-defined contamination boundaries, concentrated substances, and/or limited extent of contamination. If only partial cleanup is implemented, it is followed by (2) Standard/Background Cleanup Levels, which offer guidelines for soil, water, and air cleanup. The levels for soil and water are based on multiples of the appropriate drinking water quality standard, water quality background, or soil background. The cleanup levels for air are based on OSHA/WISHA limits for air quality, ambient air quality, or air background levels. The feasibility of the latter types of cleanup is evaluated in a Preliminary Technical Assessment based on site-specific characteristics. If Standard/Background Cleanup Levels are unachievable, (3) Protection Cleanup Levels are implemented, based on multiples of the appropriate water quality standard, water quality background, soil background, or site-specific information, followed by predictive modeling. For contaminated soils with a threat to air, "Dangerous Waste Limits" are used to establish cleanup levels. Follow-up includes long-term monitoring to verify that no threat to public health remains (Figure 2).

Strengths and Weaknesses

The Washington Department of Ecology's Final Cleanup Policy presents a framework to determine the cleanup levels for a contaminated site which is a threat to public health. The cleanup policy is divided into three levels: (1) Initial Cleanup, (2) Standard/Background Cleanup, and (3) Protection Cleanup.

Initial Cleanup Levels of surface water, groundwater, soil, and air are to be implemented when the contaminated site is an "imminent threat to public health or difficulty of cleanup increases significantly with time." The remedial options within the Initial Cleanup are total cleanup or partial cleanup. Total cleanup is only for situations having well-defined contamination boundaries, concentrated substances, or limited extent of contamination. Except for determination of contamination boundaries, no guidelines are offered for determination of concentrated substances or limited extent of contamination. Once it is determined whether the latter three characteristics exist, no further guidelines are given for total cleanup. Therefore, determination of total cleanup is vague, and no concrete guidelines are offered for risk assessment of the contaminants.

Partial cleanup is the next option if total cleanup is not undertaken to "eliminate imminent public health and environmental hazards by only removing those portions of the known contamination that represent an immediate hazard or that significantly increase the difficulty of eventual cleanup." Neither "imminent threat" nor "immediate hazard" are further defined, which results in very subjective determinations for this cleanup level. No further guidelines are discussed, so it is unclear what exactly a partial cleanup involves.

The second level of the cleanup policy is Standard/Background Cleanup Levels, which are implemented when total cleanup is unachievable or contaminants pose a threat to public health over the long term. The Standard/Background Cleanup Levels for soil, groundwater, and surface water are based on 10× the appropriate drinking water standard, and 10× the water quality background or soil

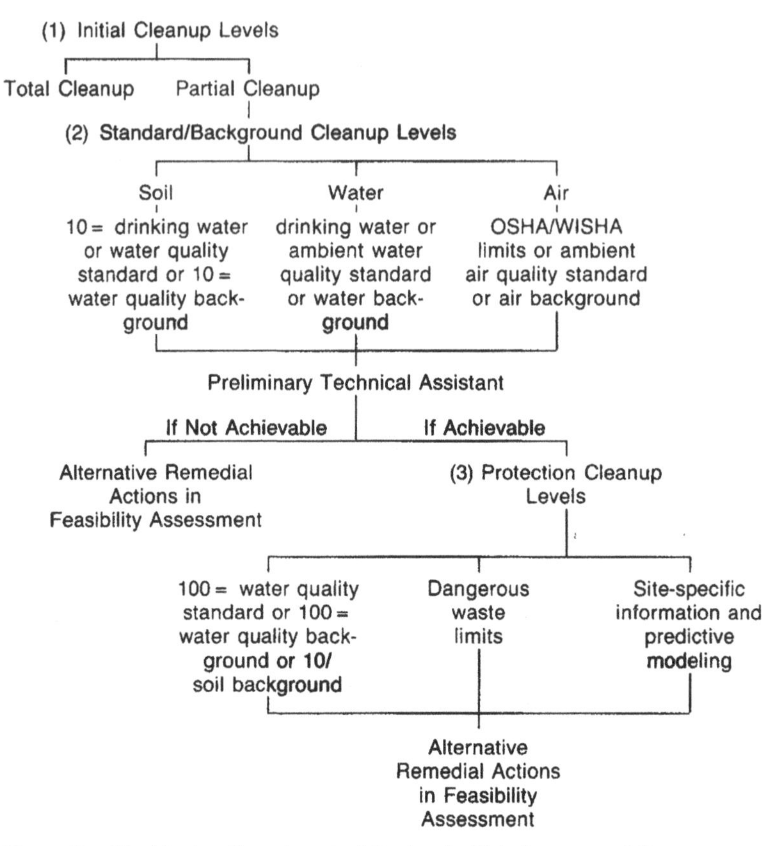

Figure 2. Washington Department of Ecology's Risk Assessment Process.

background. No justification is offered for choosing 10× the water quality standard and, therefore, this level appears to be arbitrary.

Soil background levels are used if no water standards exist. This assumes that background levels are already known, which may not be the case for numerous contaminants. A standard procedure for measuring soil background is not given and, therefore, any measurement undertaken after contamination will be variable, depending on the area used for background level measurement and type of equipment available.

The Standard/Background Cleanup Level for air is based on OSHA/WISHA limits for air quality/ambient air quality or background levels. The latter criteria also assume that these standards exist for specific contaminants, or that background levels are known.

The technical feasibility of the Standard/Background Cleanup Levels is evaluated in a Preliminary Technical Assessment, which considers presence of sole source aquifers, barriers to contaminant migration, sorptive properties of soil, contaminant mobility, and depth to groundwater. Unfortunately, these determinations are not elaborated upon for degree of importance and type of measurements, and are not integrated into a risk assessment methodology. The Preliminary Technical Assessment as described in the cleanup policy is neither complete nor quantitative.

The cleanup policy states, "If the Standard/Background Level is achievable . . . [based on the Technical Assessment] . . . it is used to evaluate the alternative remedial actions in the Feasibility Assessment." A Feasibility Assessment is not described in the cleanup policy and, therefore, alternative remedial actions are not offered.

If Standard/Background Levels are not achievable, the third level in the cleanup policy, Protection Cleanup, is implemented. Protection Cleanup Levels are defined using one of the following: (1) specified multiples of the appropriate water quality standard or background ($100\times$ water quality standard, $100\times$ water quality background, or $10\times$ soil background); (2) Dangerous Waste Limits; or (3) site-specific information on contaminant migration characteristics, leaching tests, or biologic tests and predictive models.

As previously mentioned, the use of multiples of water quality standards or background is not justified and, therefore, is arbitrary. The term "Dangerous Waste Limits" is not clearly defined, and site-specific tests are vague. Which specific tests should be used and how the results should be weighed is unclear. Predictive modeling is suggested, yet no particular models are mentioned or discussed in terms of applicability to risk assessment. The Protection Cleanup Levels are then used to evaluate the "alternative remedial actions in the Feasibility Assessment," which are not further defined in the cleanup policy.

Conclusions

Overall, the Washington Department of Ecology's Final Cleanup Policy offers a useful outline for a step-by-step general assessment of a contaminated site. The cleanup policy considers the three exposure pathways of soil contamination, water contamination (surface and groundwater), and air contamination resulting from hazardous waste at one site. The above three pathways are important when considering spills of petroleum products in soil. Unfortunately, specific risk assessment guidelines based on justifiable standards are lacking. Many of the guidelines are vague, which forces the regulatory agency to form its own interpretation based on a subjective analysis. The essentials of a risk assessment methodology—exposure analysis, hazard analysis, and predictive modeling—should be based on a quantitative assessment of the contaminated site and justifiable standards (Table 7).

LEAKING UNDERGROUND FUEL TANK MANUAL

Summary

The State of California Department of Health Services and State Water Resources Control Board has established procedures for determining whether an underground storage fuel tank site poses a risk to human health. The field manual is a practical extension of the "California Site Mitigation Decision Tree" document produced by the California Department of Health Services, Toxic Substances Control Division.

The guidelines presented in the Leaking Underground Fuel Tank (LUFT) Field Manual apply only to soil and groundwater contamination, rather than surface water contamination or air pollution. In addition, the guidelines only deal with gasoline and diesel fuel products and do not consider waste oil or solvents. The main purpose of the manual is to provide practical guidance to field personnel responsible for dealing with leaking fuel tank problems.

For situations in which gasoline contamination of soil may have occurred, the manual suggests analyzing for benzene, toluene, xylene, and ethylbenzene (BTX and E) and total petroleum hydrocarbons (TPH). For situations in which diesel fuel may have contaminated a site, only TPH is measured.

The manual cautions against measurement of ethylene dibromide (EDB) and organolead for the following reasons. EDB has been in such widespread use that

Table 7. Summary of the Washington Cleanup Methodology with Respect to the Evaluation Criteria

HAZARD ANALYSIS	
Toxicity value	Multiples (10 × or 100 ×) of water standards, air standards or soil background
Noncarcinogenic acute toxicity	Yes
Noncarcinogenic chronic toxicity	Yes
Carcinogenic chronic toxicity	Yes
Assessment of mixtures	No
Body weight factor	No
Pharmacokinetic factor	No
EXPOSURE ANALYSIS	
Air inhalation	Yes
Dust inhalation	No
Water ingestion	Yes
Soil ingestion	No
Dermal absorption	No
Crop uptake	No
Livestock uptake	No
Fish uptake	No
Environmental fate	No
Half-life factor	No
Site-specific factors	Yes
RISK ANALYSIS	
Additivity of toxic effect	No
Multi-media exposure	No
Synergistic effects	No

its detection may not be due to gasoline. Concerning organolead, most laboratories have the ability to only analyze for total lead, and cannot distinguish between inorganic and organic lead. In addition, inorganic lead is native to California soil, which may lead to false positive readings unless background levels are known.

The reasons offered by the manual for choosing BTX and E as indicators of gasoline contamination include the following:

1. They are readily adaptable to gas chromatograph detection.
2. They pose a serious threat to human health (i.e., benzene as a carcinogen).
3. They have the potential to move through the soil and contaminate groundwater.
4. Their vapors can be highly flammable and explosive.

Because BTX and E are highly mobile and can migrate from the site, it is also important to analyze for TPH. TPH detection is reported as the total of all hydrocarbons in the samples.

Tank sites are initially classified as one of the following:

- Category I — No Suspected Soil Contamination, i.e., sites in which tanks are being closed for reasons other than a leak.
- Category II — Suspected or Known Soil Contamination, i.e., sites where tanks or lines have failed precision tests, show discrepancies in monitoring records, or show visual evidence of leakage.
- Category III — Known Groundwater Contamination, i.e., sites where tanks or piping have shown a significant loss of product, especially in areas of high groundwater.

When a leaking tank is discovered, immediate safety issues are assessed and information is collected for site categorization, such as inventory records, precision testing records, and repair histories.

To categorize a site, a field TPH test is conducted in which TPH levels are measured in the ambient air, or air drawn from the soil. The result of the TPH test is compared to background levels. Background levels are determined by taking three soil samples from nearby or adjacent properties. If background TPH levels are exceeded, the site is placed in Category 2 for further testing.

Under Category 2, quantitative lab analyses of soil are conducted for BTX and E, and TPH. Samples are collected from the bottom of the excavation at worst-case locations. The initial trigger levels for BTX and E are the detection limits of the laboratory procedure. According to the manual, the trigger levels should be in the order of 0.3 mg/kg. Cleanup levels for BTX and E are derived from precipitation rates and depth to groundwater, using tables detailed in the manual.

For TPH measurements in Category 2, a leaching potential analysis was developed to determine the levels of TPH that can be safely left in the soil. A leaching potential analysis is based on the tendency of TPH to migrate down to groundwater, depending on the features of the site. Four site characteristics which the manual considers important influences on migration of TPH include depth to groundwater, subsurface fractures, precipitation, and man-made conduits. Each characteristic is rated on a scale of high, medium, and low potential for leaching. These three degrees of sensitivity are expressed in terms of TPH that can be safely left in the soil, i.e., high (10 ppm), medium (100 ppm), and low (1000 ppm). The lowest sensitivity level determined for the four characteristics is used as a cleanup level. If a characteristic cannot be rated due to insufficient data, the lowest value (10 ppm) is used as a cleanup level. In addition, other site features may be considered, such as unique site characteristics, actual use of groundwater, and future land use.

If either the TPH levels or the BTX and E levels exceed the allowable limits, additional site analysis is needed. At this point in the evaluation process, the services of a registered geologist, engineer, or environmental chemist are recommended by the manual. In addition, the manual suggests that a general risk appraisal be conducted using environmental fate and chemistry data, and site-specific information. For this stage, computer modeling is used to estimate the concentrations of BTX and E that can be left in place without risking groundwater pollution. The two models recommended for this step are the SESOIL model and the AT123D model. The SESOIL model involves long-term environmental fate simulation of pollutants in the vadose zone, and predicts the amount of pollutants which will enter groundwater. The AT123D model estimates the rate of pollutant transport and transformation in a groundwater system, and predicts groundwater contamination.

For Category III sites having groundwater contamination, decisions of site investigations and cleanup measures must be made on a case-by-case basis. This step involves assessing groundwater use, and collecting and analyzing groundwater samples. The appendix of the field manual contains numerous procedures for field measurements of this type. Risk analyses for Category III sites also require consultation with a regulatory agency or professionals in the field. The field manual does not attempt to offer guidelines for an in-depth risk assessment, and instead focuses on site categorization and laboratory analyses of indicator chemicals.

Strengths and Weaknesses

The State of California Leaking Underground Fuel Tank (LUFT) Field Manual is a practical extension of the risk assessment process detailed in the California Site Mitigation Decision Tree. The Field Manual describes the steps to be taken for categorizing sites contaminated from leaking underground storage tanks. The guidelines apply only to sites contaminated with gasoline and diesel fuel, and the risk appraisal focuses only on soil and groundwater contamination.

A site is categorized according to extent of contamination determined by field analysis of indicator chemicals. Levels of total petroleum hydrocarbons (TPH) and benzene, toluene, xylene, and ethylbenzene (BTX and E) are measured to determine extent of contamination. The site is classified based on these findings;

however, it is unclear how the trigger levels and the cleanup levels for TPH and BTX and E were derived.

Conclusions

The State of California LUFT Field Manual outlines the step-by-step procedures necessary for field workers to classify sites which may be contaminated by leaking underground fuel tanks. The manual details a classification system which reflects the extent of contamination. Specific testing procedures are explained for the indicator chemicals: benzene, toluene, xylene, ethylbenzene, and total petroleum hydrocarbons. The procedures apply only to leaking underground fuel tanks containing gasoline or diesel fuel products. The risk to human health is evaluated for potential contamination of soil and groundwater.

Environmental fate modeling is encouraged, and several computer programs are detailed for this step. For risk assessment of groundwater contamination, the manual recommends that a regulatory agency and/or professionals in the field be consulted for proper determination of human health risk. Overall, the main goal of the field manual is to provide field workers with the information necessary for accurate analyses of soil and groundwater contamination, and subsequent characterization (Tables 8 and 9).

Table 8. Summary of the State of California Leaking Underground Fuel Tank Manual with Respect to the Evaluation Criteria

HAZARD ANALYSIS	
Toxicity value	None
Noncarcinogenic acute toxicity	No
Noncarcinogenic chronic toxicity	No
Carcinogenic chronic toxicity	No
Assessment of mixtures	Yes (gasoline and diesel fuel)
Body weight factor	No
Pharmacokinetic factor	No
EXPOSURE ANALYSIS	
Air inhalation	No
Dust inhalation	No
Water ingestion	Yes (groundwater)
Soil ingestion	No
Dermal absorption	No
Crop uptake	No
Livestock uptake	No
Fish uptake	No
Environmental fate	Yes
Half-life factor	Yes
Site-specific factors	Yes
RISK ANALYSIS	
Additivity of toxic effect	No
Multi-media exposure	No
Synergistic effects	No

Table 9. Comparison of the Seven Soil Assessment Methodologies for Use in Dealing with Petroleum Contaminated Soil

	California Site Mitigation	U.S. Army Rosenblatt	EPA Superfund	Ford and Gurba	New Jersey Stokman and Dime	Washington Cleanup Policy	California LUFT
HAZARD							
Toxicity Value	NOAEL—epidemiologic, chronic, subchronic TLV LD_{50} Structure Activity Analysis	ADI MCL TLV NEL_L NEL_{90} LD_{50}	AIS AIC CPF	ADI CPF	CPF CPF	Multiples (10× or 100×) of water standards, air standards, or soil background	None
Noncarcinogenic acute toxicity	Yes	Yes	Yes	Yes	No	Yes	No
Noncarcinogenic chronic toxicity	Yes	Yes	Yes	Yes	No	Yes	No
Carcinogenic chronic toxicity	Yes	No	Yes	Yes	Yes	Yes	No
Assessment of mixtures	No	No	Yes	Yes	Yes	No	Yes
Body weight factor	Yes	Yes	Yes	Yes	No	No	No
Pharmacokinetic factors	Yes	No	No	No	No	No	No
EXPOSURE							
Air inhalation	Yes	Yes	Yes	No	No	Yes	No
Dust inhalation	Yes	Yes	Yes	No	No	No	No
Water ingestion	Yes	Yes	Yes	No	No	Yes	Yes

continued

Table 9. *Continued*

	California Site Mitigation	U.S. Army Rosenblatt	EPA Superfund	Ford and Gurba	New Jersey Stokman and Dime	Washington Cleanup Policy	California LUFT
Soil ingestion	No	No	No	Yes	Yes	No	No
Dermal absorption	No	No	No	No	No	No	No
Crop uptake	Yes	Yes	No	No	No	No	No
Livestock uptake	Yes	Yes	No	No	No	No	No
Fish uptake	Yes	Yes	Yes	No	No	No	No
Environmental fate	Yes —soil adsorption —bioconcentration —air transport —water transport	Yes —partition coefficients —bioconcentration —air transport —water transport	Yes —air transport —water transport	No	No	No	Yes
Half-life factor	No	No	Yes	Yes	No	No	Yes
Site-specific factors	Yes	Yes	Yes	No	No	Yes	Yes
RISK ANALYSIS							
Additivity of toxic effect	Yes	No	Yes	No	No	No	No
Multi-media exposure	Yes	Yes	Yes	No	No	No	No
Synergistic effects	No	No	No	No	No	No	No

ACKNOWLEDGMENTS

This work was supported by a grant from the Office of Research and Standards, Massachusetts Department of Environmental Quality Engineering, Boston, MA and The Environmental Institute, University of Massachusetts, Amherst, MA.

REFERENCES

1. California Department of Health Services, Toxic Substances Control Division, Alternative Technology and Policy Development Section, "California Site Mitigation Decision Tree," Draft Working Document, June 1985.
2. Rosenblatt, D. H., J. C. Dacre, and D. R. Cogley. "An Environmental Fate Model Leading to Preliminary Pollutant Limit Values for Human Health Effects," in *Environmental Risk Analysis for Chemicals,* R. A. Conway, Ed. (New York: Van Nostrand Reinhold Co., 1982).
3. *Superfund Health Assessment Manual,* Office of Emergency and Remedial Response, U. S. Environmental Protection Agency, December 1985.
4. Ford, K. L., and P. Gurba. "Health Risk Assessments for Contaminated Soils," in *Proceedings of the 5th National Conference on Management of Uncontrolled Hazardous Waste Sites,* Washington, D.C., November 1984.
5. Stokman, S. K., and R. Dime. "Soil Cleanup Criteria for Selected Petroleum Products," *Risk Assessment,* 342–345.
6. Department of Ecology, State of Washington, "Final Cleanup Policy—Technical," July 10, 1984.
7. "Leaking Underground Fuel Tank Field Manual: Guidelines for Site Assessment, Cleanup and Underground Storage Tank Closure," State of California, Leaking Underground Fuel Tank Task Force, State of California-Department of Health Services, LUFT Task Force, Draft, May 1987, 96 pp.

CHAPTER 25

Creative Approaches in the Study of Complex Mixtures: Evaluating Comparative Potencies

Rita S. Schoeny, Environmental Criteria and Assessment Office, U.S. Environmental Protection Agency, Cincinnati, Ohio

INTRODUCTION

An ongoing problem for agencies such as the U.S. Environmental Protection Agency (EPA) is that of providing guidance as to health risk assessment of mixtures from a variety of sources such as wastewaters, hazardous waste sites, air particulates, or materials spilled in waters or soils. In the absence of data on the mixture in question and its particular matrix, some approach using surrogates must be taken.

One such approach, which is discussed in the U.S. EPA 1986 Guidelines for Health Risk Assessment of Chemical Mixtures,[1] is to obtain an index of toxicity by summing the results of the risk assessments for individual components that have been identified as part of the mixture, after consideration of the potential for interaction among those components. There are generally considerable difficulties with this method. In the case of complex mixtures, such as organic extracts of drinking water or combustion products, the majority of compounds may not have been identified, much less quantified. If identified, the biological activity may not be known for a particular agent. And lastly, the current state of knowledge regarding interactions is such that, with few exceptions, nothing is known as to the type or extent of interactions that could be predicted to occur in any given combination of compounds. Currently, the recommendation in the Guidelines is for the development of a toxicity index that can assist in assigning priorities for action, but does not give a likelihood of adverse effect or an individual risk attributable to mixture exposure.

In the case of a homolog mixture for which the components are known or can be reasonably conjectured, the U.S. EPA Risk Assessment Forum has recommended that a series of toxicity equivalency factors (TEF), be developed. An interim procedure was proposed for estimating risks associated with chlorinated dibenzo-dioxins (CDDs) and dibenzofurans (CDFs) (U.S. EPA[2]). This was revised in 1989. TEF approaches are also under consideration for individual congeners of polychlorinated biphenyl (PCB) mixtures and for a set of polycyclic aromatic hydrocarbons (PAH).

In this risk estimation procedure, information is obtained on the concentrations of the various congeners or homologs in the mixture of interest. Next, available toxicological data and structural activity relationship judgments are used to estimate the potential human hazard associated with exposure to each mixture component. For the CDD and CDF, this potential is expressed in terms of "equivalent amounts of 2,3,7,8-tetrachlorodibenzo-*p*-dioxin," (TCDD), an isomer for which a reasonable health effects data base exists. In other words, exposures are translated into a common scale from which risk can be derived. The assessment outcome is, in this case, an estimate of risk rather than a priority ranking. For the CDD and CDF, TCDD was assigned a value of 1 and the other CDDs and CDFs ranked according to the following criteria. In the absence of human carcinogenicity data, results from long-term animal tests for carcinogenic activity were used. In the absence of any carcinogenicity data, data on reproductive effects were considered. Lacking such data, information from in vitro assays was used, in particular the ability of a specific congener to bind to a cellular receptor which controls induction of aryl hydrocarbon hydroxylase enzymes. This last activity is associated with expression of several manifestations of TCDD toxicity (e.g., cleft palate in the mouse, and thymic involution and body weight loss). Thus, based on available data, the various CDD and CDF components were assigned a toxicity equivalency factor in relation to a signature, well-studied compound. For the PAH, benzo[a]pyrene may serve as the signature compound, with the others to be ranked according to their activities in tests such as skin painting bioassays. For a given mixture, the estimated risks for each component are summed, and information on exposure incorporated to determine the potential human hazard.

For mixtures of components that are not readily identified or separated, another method, that of comparative potency determination, has been suggested, both in the NRC report on *Methods for In Vivo Toxicity Testing of Complex Mixtures*,[3] as well as in the U.S. EPA *Technical Support Document on Risk Assessment of Chemical Mixtures*.[4] Both the TEF and comparative potency approaches utilize types of data not generally considered to be suitable for risk assessment. One point of difference is that the TEF method assumes that either doses or responses can be added (so-called dose or response additivity). This assumption is most suitably applied to agents that share a mode of action or are believed not to interact in producing a biological effect. Comparative potency judgments, by contrast, are validated by comparison to in vivo data on a mixture presumed to be similar to the one in question. Mixture is compared to mixture; the possibility of accounting for interactions is, therefore, not excluded.

DESCRIPTION OF COMPARATIVE POTENCY APPROACH

In its simplest form a comparative potency estimate is done on a mixture for which data are incomplete by carrying out a comparison to a similar mixture on which a human risk assessment has been done. For example, if there are human dose-response data on mixture A and rodent skin painting bioassays for both mixtures A and B, then a comparative potency approach may be used to determine a human risk estimate for mixture B. The underlying hypothesis is that in different assays the potency of the agents relative to one another is the same; if mixture A is twice as potent a human carcinogen as B, then it will be twice as potent a rodent skin carcinogen, twice as active in a *Salmonella* mutagenicity assay, and so on. This can be expressed mathematically as:

$$RP_1 = k(RP_2)$$

where RP_1 and RP_2 are the relative potencies of the agent in assays 1 and 2, and k is a constant.

Albert et al.[5] and Lewtas[6] applied this method in the estimation of human lung cancer risks from a number of combustion products including coke oven emissions, roofing tar, cigarette smoke condensate (CSC), and particulates from diesel- and gasoline-fueled engines. Epidemiologic data were available for the first three materials; namely, dose-related increases in the observation of lung cancer in exposed humans. These data permitted calculation of the lung cancer unit risks presented in Table 1. If the most potent of these three mixtures, coke oven emissions, is given the relative potency of 1, then the RPs for roofing tar and CSC for human lung cancer are those in column two of Table 2. There were no epidemiologic data suitable for deviation of a unit risk for Nissan diesel engine emission samples. An extract of a particulate sample from this engine type, however, had been tested in vivo in a mouse skin initiation/promotion protocol, as well as in vitro for genotoxicity. Likewise, samples of the other three combustion products had been assayed in a similar fashion. The relative potencies of all four agents based on these short-term bioassays are presented in columns 3–5 of Table 2. Examination of the comparative potencies shows good agreement among the assays, with the exception of the relatively high activity of CSC and Nissan diesel in the *Salmonella* mutagenicity assay. Having thus shown that for at least some assays the relative potencies of these mixtures were constant, a determination of Nissan diesel unit risk for humans was made by Albert et al.[5] and Lewtas,[6] by using the relative potency based on the mouse initiation/promotion data (Table 3). The unit risks thus derived (column 4, Table 3) were within a factor of 2 of one another; the final estimate was prepared by taking a simple mean.

Table 1. Human Lung Cancer Risks for Combustion Products[a]

Combustion Product	Cancer Unit Risk/$\mu g/m^3$
Coke oven emission	9.3×10^{-4}
Roofing tar	3.6×10^{-4}
Cigarette smoke condensate	2.2×10^{-6}

[a]Values from Albert et al.[5]

Table 2. Relative Potencies of Emission Extracts in Bioassay Systems[a]

	Relative Potency			
Combustion Product	Human Lung Cancer	Mouse Skin Tumor Initiation, Promotion Assay[c]	Mutation for Mouse Lymphoma Cells[c]	Mutation for *Salmonella* Strain TA98[c]
Coke oven emission	1.0	1.0	1.0	1.0
Roofing tar	0.39	0.20	1.4	0.78
CSC[b]	0.0024	0.0011	0.066	0.52
Nissan diesel	—	0.28	0.24	12.0

[a]Values from Albert et al.[5]
[b]CSC = Cigarette smoke condensate.
[c]In vitro tests employed hepatic homogenates (S9) to provide metabolic activation.

Table 3. Derivation of Nissan Unit Risk[a]

Combustion Product	Comparative Potency Mouse Skin Tumor Initiation[b] A	Cancer Unit Risk/$\mu g/m^3$ B	Nissan Diesel Unit Risk /$\mu g/m^3$ A × B
Coke oven	0.28/1	9.3×10^{-4}	2.6×10^{-4}
Roofing tar	0.28/0.20	3.6×10^{-4}	5.2×10^{-4}
CSC	0.28/0.0011	2.2×10^{-6}	5.4×10^{-4}
		Average	4.4×10^{-4}

[a]Values are from Albert et al.[5]
[b]Nissan relative potency based on mouse skin tumor data ÷ combustion mixture relative potency based on mouse skin tumor data.

This method was further applied to an extended set of diesel emission samples for which there were data in mouse skin tumor and short-term tests for genotoxicity (Table 4) but no epidemiology data. With the exception of the Nissan sample, the engine emission samples were only weakly active in the mouse skin tumor initiation/promotion assay. The correlations between relative potency in that assay and those from the genotoxicity assays were, however, very strong, as can be seen from Table 4. Unit risks for the diesel sample were derived by using the Nissan unit risk and the average relative potencies from *Salmonella* mutagenicity data, mouse lymphoma cell (L5178Y) mutagenicity, and sister chromatid exchanges in CHO cells (Table 5). In a later publication, Lewtas[7] reported an average unit lung cancer risk for the diesel engines to be $2.6 \times 10^{-5}/\mu g/m^3$

particulate matter. The author notes that this compares very well with a unit risk of $1.2 \times 10^{-5}/\mu g/m^3$ derived from data on incidence of adenomas and adenocarcinomas observed in several studies of rats exposed to diesel particulate for their lifetimes. Lewtas[6,7] has published unit risks based on relative potency determination for several emission extracts from several automotive engines and for combustion sources including woodstoves, oil furnaces, and utility power plants. Overall, the author reported a range of only about 150-fold in cancer unit risks calculated using comparative potency based on mutagenicity and tumor initiation data for particulate material obtained from these combustion processes.

Table 4. Relative Potencies of Diesel Emission Extracts Based on Four Bioassays

	Relative Potencies[a]				
	Mouse Skin Tumor Initiation	Mutation in L5178Y Cells[b]	Sister Chromatid Exchange in CHO Cells[b]	Mutation for *Salmonella* Strain TA98[b]	Average
Nissan	1.0	1.0	1.0	1.0	1.0
Volkswagen	0.41	0.25	0.42	0.23	0.30
Oldsmobile	0.53	0.45	0.24	0.11	0.27
Caterpillar	NEG[c]	0.022	NEG	0.023	0.015
Correlation (r^2) with mouse skin tumor initiation		0.95	0.83	0.72	

[a]Values from Albert et al.[5]
[b]In vitro tests incorporated rat hepatic extracts (S9) to provide metabolic activation.
[c]NEG = Negative, no increase in response by comparison to controls on assay.

Table 5. Unit Lung Cancer Risk Estimates for Diesel Emissions[a]

		Unit Risk Estimates (Lifetime Risk/$\mu g/m^3$)		
	Comparative Potency Average	Organics[b]	Organic Extractables %	Particulates[c]
Nissan	1	4.4×10^{-4}	8	3.5×10^{-5}
Volkswagen	0.30	1.3×10^{-4}	18	2.3×10^{-5}
Oldsmobile	0.27	1.2×10^{-4}	17	2.0×10^{-5}
Caterpillar	0.015	6.6×10^{-6}	27	1.8×10^{-6}

[a]Values from Albert et al.[5]
[b]Unit risk estimates were obtained by multiplying the Nissan unit risk by the average comparative potency.
[c]These values were based on percent organic extractables in preceding column.

ASSUMPTIONS, USES, AND LIMITATIONS OF COMPARATIVE POTENCY APPROACHES

The example provided by the comparative potency and cancer risk estimation for combustion products, especially the diesel particulates, indicates that this method could be used to predict the potential human risk of similar mixtures based on short-term in vivo or in vitro tests. If such a method were validated, considerable resources could be saved that otherwise would be used to conduct animal bioassays. This promise of the relative potency approach, however, is tempered by a consideration of the limitations of this approach and an investigation of its underlying assumptions.

Models and Methods of Validation

Although there are references to the use of comparative potency to predict human health risks for endpoints other than cancer, the greatest use of the approach has been for cancer risk prediction. Specifically, methods of deriving the estimates have been tied to use of a linear, nonthreshold model for the low dose region of the dose-response curve for carcinogens. This model is based on the concept of a carcinogen as an agent that produces an irreversible change in a critical site, which initiates the process of neoplasia. Often a carcinogen is thought to produce a mutation (or mutations) as a first step in carcinogenesis. Thus, the tie to in vitro tests that measure mutagenicity or changes (such as damage or repair) in DNA and which generally show low-dose linearity of response, is on a reasonable conceptual basis. Chemical and biological characterization of the combustion materials indicates that these mixtures do contain components which have initiating activity. There are certainly promoting agents present as well, and some of the materials known to be initiators may also serve as promoters for other components of the whole mixture. The contribution of these to the overall carcinogenic potential of the mixture may not be accounted for in a comparative potency judgment based on in vitro genotoxicity, or one based on an initiation/promotion assay. Recently more attention has been turned to the subject of nongenotoxic carcinogens. Short-term assays designed specifically for their detection are now being developed; however, these are not as well characterized or validated as the better-established mutagenicity assays and are not readily applicable to comparative potency determinations at this time.

It seems reasonable to assume that greater confidence can be attached to a comparative potency approach which relies at some point on human data. The agents, both mixtures and single compounds for which such dose-response data exist, are limited. Albert[8] listed 13 human carcinogens considered suitable for comparative potency derivation and suggested animal models that reflect the carcinogenic specificity of these agents. Compounds for which there are no quantitative human data could be used in the process if they are known to have a well-characterized response in one of the animal models.

Another method of preparing estimates of potency for purposes of comparison has been presented by Peto et al.[9] They propose a numerical measure of carcinogenic potency (analogous to an LD_{50}) which they term a TD_{50}, or tumorigenic dose rate for 50% of the test animals. In other words "for any particular sex, strain, species, and set of experimental conditions, the TD_{50} is the dose rate (in mg/kg body weight/day) that if administered chronically for a standard period—the 'standard lifespan' of the species—will halve the mortality-corrected estimate of the probability of remaining tumorless throughout that period." The TD_{50} as an index of carcinogenic potency has been used as the basis for comparison for a Carcinogenic Potency Database (Gold et al.[10]) comprising data from about 3000 long-term animal bioassays on some 770 compounds. One advantage of this method for use in comparisons is that the actual experimental range of doses tested will generally bracket the TD_{50}; this allows for accurate estimation and eliminates the necessity of modeling the data. A disadvantage is the requirement for many interim sacrifices to identify nonlethal tumors.

Decision Points

At many points in the course of preparing a comparative potency determination there are important choices to be made, some of which may not be obvious. The units in which carcinogenic potency will be compared is one such choice. Unit risk can be compared on the basis of risk/μg/m^3 for materials emitted to the air, risk/μg/L of water, risk/mg ingested material/kg body weight, and so on. For particulate samples, one could evaluate the risk per unit of extracted organic material from collected particles (which is generally what is tested in vitro or by skin painting), or risk per unextracted particle (which is the material tested in long-term inhalation studies and which represents the likely exposure to humans). Lewtas[6] suggests normalizing data to responses per kg fuel, km traveled (for engine emissions), or megajoule of heat produced. For spilled materials it will be particularly important to consider the matrix wherein environmental exposure takes place. For example, unit risk might be expressed in terms of risk/kg soil.

Another series of choices particularly germane to comparative potencies of mixtures is the form, source, and preparation of the environmental mixture sample. Optimally one wishes to use data on a sample most relevant to human exposure. This is not always practical in terms of in vitro or short-term testing. For example, diesel particulate is not amenable to assay in *Salmonella typhimurium*. The bacteria have no capacity for phagocytosis, and exposure to potentially mutagenic nonwater-soluble materials associated with particles embedded in agar would be minimal. Likewise, particles would be an inappropriate vehicle for a mouse skin tumor induction assay. When the sample has been modified to facilitate testing (by extraction, fractionation, or in some other way) the validity of its use in the context of potential human exposure must be assured. This is also true for locating points of sampling (e.g., samples taken from the top of a coke oven battery vs samples that could be obtained in greater quantity from within the coke oven battery). An example of this decision-making process and the validation of such decisions is found in the Albert et al.[5] paper.

The type of assay to be used for comparison with the human data is another point of decision. It is clear that no single assay will be suitable for risk estimation of every sort of mixture; thus, those materials that have undergone a battery of assays will be most amenable to comparative potency determination. An instructive illustration of this can be seen in Table 2. The relative potencies of both CSC and the Nissan diesel particulate extract relative to coke oven emissions were a great deal higher when based on the *Salmonella* mutagenicity data than when based on data from the other assays. Chemical composition data for diesel particulates show the presence of nitropyrenes as well as other nitroaromatics. Due to the presence of an endogenous nitroreductase enzyme system, *Salmonella* is particularly sensitive to the mutagenic activity of nitropyrenes (Mermelstein et al.[11]). It was shown that for one diesel sample, a total of 23 nitroaromatics was responsible for about 40% of the TA98 mutagenicity of the complete sample. Use of a congenic *Salmonella* strain which lacked the nitroreductase enzyme system resulted in a decrease of about 50% of the mutagenic activity of the sample.[7] Thus, it would seem that use of the *Salmonella* assay as a basis for comparative potency determinations of nitropyrene-containing materials would have resulted in over-estimation of potential human risk.

The use of a tiered approach to testing agents for potential carcinogenicity has been recommended on many occasions (e.g., NRC[4]) as a means of prioritizing testing and conserving resources. The U.S. EPA Health Effects Research Laboratory has proposed a three-tiered system. The first tier consists of screening bioassays determined to produce a minimum of false negatives (nonpositive agents which are, in fact, carcinogens); this tier is comprised of *Salmonella* mutagenicity and mammalian cytogenicity tests. Tier 3 consists of lifetime bioassays in rodents. The intermediate tier (also called the carcinogen testing matrix or CTM) would be made up of short-term in vivo assays designed to eliminate false positives and to predict the relative carcinogenic potency of agents. The skin tumor initiation/promotion assay in SENCAR mice, the lung adenoma assay in Strain A mice, and the rat liver altered enzyme focus bioassay, have been evaluated as to their ability to identify as positive 18 compounds of varying structure and target organ specificity which are known to be carcinogenic in long-term animal studies.

The short-term in vivo tests were also assessed for their ability to predict the relative potency of carcinogenic compounds or mixtures. Statistical methods for use with the CTM are being developed and validated in the following areas: (1) robust sensitive statistical methods to evaluate each of the three tests for a positive dose-related response; (2) statistical methods for estimation of carcinogenic quantitative risk from the long-term animal tests; (3) statistical methods for estimation of dose-response rate from the short-term tests; (4) and finally, methods for evaluation of the ability of the CTM to predict relative potency in a reliable fashion.

Taken as a battery of three tests, the CTM was able to identify 17 of 18 animal carcinogens, using the criterion of a positive response in at least one of the tests. In this phase of the validation, the lung adenoma assay was most reliable, and the liver focus assay did not identify any compound as positive that was not also detected by the other two.

Relative potencies were also done using data from a standard long-term and from the CTM short-term assays on the basis of TD_{50} determinations. In terms of prediction of relative potency for the long-term cancer bioassays, it is noteworthy that the initiation/promotion test provided significantly poorer predictions than either the liver focus assay or the lung adenoma assays. Future work with the CTM will probably encompass tests with additional chemicals to assist in ascertaining which assays will be most useful for particular chemical types, both for identification of potential carcinogens and for derivation of comparative potencies. The data from these efforts should provide not only some useful bases for relative potency development, but also guidance as to the types of short-term tests most applicable to various classes of agents and ultimately to various types of mixtures.

Many short-term tests have more than one endpoint available for measurement and evaluation. For example, the CTM evaluation effort determined that the most sensitive response measure for the skin tumor assay was time-to-tumor; for lung adenoma, the number of tumors; and for the liver focus assay, the number of foci per unit area. The selection of test endpoint or measurement is, thus, another decision to be made in the choice of assay data for comparative potency. Choice of measurement and the way in which data were handled is discussed at some length in one published description of use of mouse skin tumor data in the calculation of relative potencies of coke oven emission, roofing tar, and CSC (Albert et al.[5]).

CONCLUSIONS

Two requirements for application of comparative potency are (1) information on the composition of the mixture, and (2) data on a mixture which is sufficiently similar, both as to types of components present and as to the biological activity of those components. Thus, this procedure would seem to be less useful in the evaluation of highly complex, very changeable mixtures, such as those associated with toxic waste sites. It should be noted that developing test designs to generate data for comparative potency determinations may involve controlling factors not necessary in other experimental designs. These (as described in Reference #3) include the following: simultaneous testing of all agents used for comparison in one experiment for those bioassays characterized by large variability between experiments; use of identical measures of dose or overlapping dose ranges to facilitate statistical analysis; and a thorough consideration of the specific objectives of the comparative potency study and expected use of the data.

Despite its reliance on some untested assumptions, the comparative potency method offers great promise as a means of evaluating the hazard potential of complex environmental mixtures. It offers the advantage of treating the mixture as a mixture, rather than as a sum of its known components. As in vitro and short-term bioassay data on various mixtures become increasingly available, this procedure should become a focus for research on improving the methods employed and validation of its hypothesis.

REFERENCES

1. U.S. Environmental Protection Agency. "The Risk Assessment Guidelines of 1986," EPA/600/8-87/045, 1986.
2. U.S. EPA. "Interim Procedures for Estimating Risks Associated with Exposures to Mixtures of Chlorinated Dibenzo-p-Dioxins and -Dibenzofurans (CDDs and CDFs)," EPA Risk Assessment Forum, EPA/625/3-87/012, March 1989.
3. National Research Council. *Complex Mixtures. Methods for In Vivo Toxicity Testing* (Washington DC: National Academy Press, 1988).
4. U.S. Environmental Protection Agency. "Technical Support Document on Risk Assessment of Chemical Mixtures." EPA/600/8-90/064, November 1988.
5. Albert, R. E., J. Lewtas, S. Nesnow, T. Thorslund, and E. Anderson. "Comparative Potency Method for Cancer Risk Assessment: Application to Diesel Particulate Emissions," *Risk. Anal.* 3:101–117 (1983).
6. Lewtas, J. "Development of a Comparative Potency Method for Cancer Risk Assessment of Complex Mixtures Using Short-term In Vivo and In Vitro Bioassays," *Toxicol and Indus. Health.* 1:193–203 (1985).
7. Lewtas, J. "Genotoxicity of Complex Mixtures: Strategies for the Identification and Comparative Assessment of Airborne Mutagens and Carcinogens from Combustion Sources," *Fund. Appl. Toxicol.* 10:571–589 (1988).
8. Albert, R. E. "The Comparative Potency Method: An Approach to Quantitive Cancer Risk Assessment," in *Methods for Estimating Risk of Chemical Injury: Humans and Non-human Biota and Ecosystems,* V. B. Vouk, G. C. Butler, D. G. Hoel, and D. B. Peakall, Eds., pp. 281–287. SCOPE. 26, SGOMSEC Conference, 1985.
9. Peto, R., M. C. Pike, L. Bernstein, L. S. Gold, and B. N. Ames. "The TD_{50}: A Proposed General Convention for the Numerical Description of the Carcinogenic Potency of Chemicals in Chronic-Exposure Animal Experiments," *Environ. Health Persp.* 58:1–8 (1984).
10. Gold, L. S., C. B. Sawyer, R. Magaw, G. M. Backman, M. De Veciana, R. Levinson, N. K. Hooper, W. R. Havender, L. Bernstein, R. Peto, M. C. Pike, and B. N. Ames. "A Carcinogenic Potency Database of the Standardized Results of Animal Bioassays," *Environ. Health Perspect.* 58:9–319 (1984).
11. Mermelstein, R., D. K. Kiriazides, M. Butler, E. McCoy, and H. S. Rosenkranz. "The Extraordinary Mutagenicity of Nitropyrenes in Bacteria," *Mutat. Res.* 89:187–196 (1981).

CHAPTER 26

The Effect of Soil Type on Absorption of Toluene and Its Bioavailability

Rita M. Turkall,[*,**] **Gloria A. Skowronski,**[*] and **Mohamed S. Abdel-Rahman,**[*]
Department of Pharmacology, New Jersey Medical School,[*] and Clinical Laboratory Sciences Department, School of Health Related Professions,[**] University of Medicine and Dentistry of New Jersey, Newark

Exposures to soil contaminated by petrochemicals released during manufacture, use, storage, transport or disposal are widespread. Evaluation of the potential health risk to workers and communities exposed to petroleum contaminated soils is of growing concern, and, moreover, is a complex problem for which methods are being developed. The United States Environmental Protection Agency (EPA), the Center for Disease Control (CDC) and numerous independent scientists have shown that inhalation of polluted soils is minimal and is unlikely to cause significant health hazards.[1-4] On the other hand, ingestion and dermal contact represent significant routes of exposure. Children (ages 1.5 to 3.5 years) are especially susceptible to exposure by ingestion.[1] When all of the published information on soil ingestion is considered, the data indicate that the best estimate of soil ingestion by children is about 50 mg/day.[1] The EPA in its risk assessment of 2,3,7,8-tetra-chlorodibenzo-p-dioxin (TCDD) based its estimates of dermal exposures to TCDD polluted soils on actual field investigations.[5] The EPA cited the work of Roels et al.[6] which showed that about 0.5 mg of soil adheres per cm^2 of exposed skin.

The percentage of the total chemical in the soil which subsequently enters the body, i.e., bioavailability, can be influenced by the soil's composition, the chemical's characteristics, as well as the biological environment provided at the site of the body where exposure takes place. Previous studies from this laboratory have demonstrated that route of exposure, in addition to soil-chemical interactions, alter the bioavailability of chemicals from contaminated soils. Adsorption to either of two New Jersey soils increased the amount of benzene available to orally exposed rats, as well as produced changes in plasma kinetics and excretion patterns versus benzene in the absence of soil.[7,8] However, decreased bioavailability resulted when exposure to soil-adsorbed benzene occurred via the dermal route. Altered tissue distribution and excretion patterns were also seen.[8,9]

Similar studies were conducted on soil-adsorbed toluene to evaluate the influence of substituent addition to the benzene ring. The EPA in its 1985 Priority List for cleanup cited toluene as the third most frequent out of 465 substances recorded in 818 abandoned dump sites. Toluene together with benzene and xylene represent the major aromatic compounds of gasoline. Toluene is also used as a solvent for paints, plastics, varnishes, and resins, and in the manufacture of explosives, adhesives, drugs, perfumes, and other synthetic chemicals.[10] Although toluene lacks benzene's chronic hematopoietic effects, animal experiments indicate that toluene is more acutely toxic than benzene. High concentrations of toluene cause symptoms of central nervous system (CNS) depression such as headache, dizziness, weakness, and loss of coordination. Furthermore, chronic exposure to toluene may cause permanent CNS damage.[10-12]

Oral administration of ^{3}H-toluene in peanut oil to rats was followed by rapid absorption of radioactivity. Maximum blood levels of radioactivity were reached within 2 hr after gastric intubation.[13] Sato and Nakajima[14] showed that toluene is poorly absorbed through the skin while Dutkiewicz and Tyras[15] reported that the absorption rate of toluene through human skin is very high (14–23 mg/cm^2/hr). The aim of the present study is to apply pharmacokinetic techniques to assess the effect that soil adsorption has on the bioavailability of toluene by the oral and dermal routes.

MATERIALS AND METHODS

Radioisotopes

All studies were conducted using ring uniformly labeled ^{14}C-toluene with a specific activity of 16.4 mCi/mmole and radiochemical purity of 95% (Amersham Corp., Arlington Heights, IL). Prior to use, the radioisotope was diluted with unlabeled HPLC-grade toluene (Aldrich Chemical Co., Milwaukee, WI) to reduce specific activity to a workable range.

Soils

Two soils were utilized: (1) an Atsion sandy soil (90% sand, 2% clay, 4.4% organic matter) collected from an outcrop site of the Cohansey sand formation near Chatsworth in southcentral New Jersey; and (2) a Keyport clay soil (50% sand, 22% clay, 1.6% organic matter) collected from the Woodbury formation near Moorestown in southwestern New Jersey. The Atsion sandy soil is a deep, poorly drained soil formed in Atlantic coastal plain sediments of New Jersey and New York. Warmer thermic equivalent Leon soil occurs from Maryland to Florida.[16] The Keyport clay soil is a moderately well drained soil formed on clay bed marine deposits of the Atlantic inner-coast plain. In addition to New Jersey, the Keyport soil is found in Delaware, Maryland, and Virginia with similar soils occurring as far southwest as Texas.[17] Gas chromatography/mass spectrometry analysis of extracts of the soil did not detect contamination by toluene or halogenated hydrocarbons.

Animals

Male Sprague-Dawley rats (275–300 g) were purchased from Taconic Farms, Germantown, NY, and quarantined for at least one week prior to administration of the chemical. Animals were housed three per cage and were maintained on a 12 hr light/dark cycle at constant temperature (25°C) and humidity (50%). Ralston Purina rodent lab chow (St. Louis, MO) and tap water were provided ad libitum.

Toluene Administration

The oral administration of toluene was performed as follows: either 150 μL of ^{14}C-toluene solution (5 μCi) alone, or the same volume of radioactivity added to 0.5 g of soil, was combined with 2.85 mL of aqueous 5% gum acacia and a suspension formed by vortexing. This volume of toluene or toluene soil suspension was immediately administered by gavage to groups of rats which had been fasted overnight. Heparinized blood samples were collected at 5, 10, 15, 20, 30, 45, 60, 90, 120, and 180 minutes by cardiac puncture of lightly ether-anesthetized rats.

For the dermal application, a shallow glass cap (Q Glass Co., Towaco, NJ) circumscribing a 13 cm^2 area was tightly fixed with Lang's jet liquid acrylic and powder (Lang Dental Manufacturing Co., Inc., Chicago, IL) on the lightly-shaved right costo-abdominal region of each ether-anesthetized animal one-half hour prior to the administration of toluene. Then 225 μL of ^{14}C-toluene (30 μCi) alone or after the addition of 750 mg of soil was introduced by syringe through a small opening in the cap, which was immediately sealed. This volume of toluene coated the soil with no excess chemical remaining. Rats were rotated from side to side so that the soil-chemical mixture covered the entire circumscribed area.

Heparinized blood samples were collected by cardiac puncture under light ether anesthesia at 0.25, 0.5, 1, 2, 4, 5, 7, 9, 10, 11, 12, 24, 30, 36, and 48 hr. Samples from both routes of administration were processed and radioactivity was measured by liquid scintillation spectrometry as in previous studies.[7,9] Immediately after the collection of the 180 min blood sample in the oral study, rats were sacrificed by an overdose of ether; whole organs or samples of brain, thymus, thyroid, esophagus, stomach, duodenum, ileum, lung, pancreas, adrenal, testes, skin, fat, carcass, bone marrow, liver, kidney, spleen, and heart were collected and stored at $-75°C$. Thawed tissue samples of 300 mg or smaller were used to determine the distribution of radioactivity as previously reported.[7]

Excretion and Metabolism Studies

In the excretion studies, groups of six rats each were administered toluene or toluene adsorbed to the soil, as described above. Animals were housed in all-glass metabolism chambers (Bio Serve Inc., Frenchtown, NJ) for the collection of expired air, urine, and fecal samples. Expired air was passed through activated charcoal tubes (SKC Inc., Eighty-Four, PA) for the collection of ^{14}C-toluene, then bubbled through traps filled with ethanolamine: ethylene glycol monomethyl ether (1:2 v/v) for the collection of $^{14}CO_2$. Charcoal tubes and trap mixtures were collected at 1, 2, 6, 12, 24, and 48 hr after administration of compound. Urine samples were collected at 12, 24, and 48 hr, and fecal samples were collected at 24 and 48 hr. Samples were processed and radioactivity was measured as previously described.[7]

At the conclusion of the dermal excretion studies, rats were sacrificed by an overdose of ether. One to 1.2 mL of ethyl alcohol were introduced into the glass cap and the animals were rotated from side to side. Aliquots of ethanol wash (100 μL) were removed for counting to determine the percent of toluene dose remaining on the skin application sites. Then the glass caps were removed from the rats and tissue specimens were collected for the distribution determination.

Toluene metabolites were determined in n-butanol extracts (>95% efficiency) of urine, utilizing high performance liquid chromatography with a 3.8% phosphoric acid-acetonitrile mobile phase and a C-18 column.

Data Analysis

Exploratory data analysis was used to summarize replicate data in the plasma time course study.[18,19] The curve-fitting procedure which was utilized is called smoothing. For these studies, a "4235EH" smoother (a statistical procedure for treating the data) was utilized.[19] Each replicate was smoothed over all time points, a median value was calculated for all smoothed replicates at each time point, and a second smooth was applied to these median values. The final smoothed data was used to calculate rate constants, $t_{1/2}$ of absorption and elimination from plasma by regression analysis and the method of residuals[20] as well as to determine a maximum concentration and a time at which the maximum concentration was achieved. Plasma concentrations from 0 min to the time at which maximum concentration was achieved were used for absorption calculations.

For the calculation of toluene elimination from plasma, 60 through 180 min were used in oral route studies, while 30 through 48 hr were used in dermal route studies. Since the rate constants and half-lives were calculated from smoothed data, the standard errors (SE) of the rate constants were determined by the bootstrap method.[21,22] The area under the plasma concentration time curve (AUC) was calculated by the trapezoidal rule using individual replicate data and is reported as the mean ± standard error of the mean (SEM). Comparison of slopes was determined by analysis of covariance. Excretion, tissue distribution, and metabolite data are reported as a mean ± SEM. Statistical differences between treatment groups were determined by analysis of variance (ANOVA), F test, and Scheffe's multiple range test.

RESULTS

Data showing the absorption and elimination half-lives following oral and dermal administration of ^{14}C-toluene to male rats are presented in Table 1. No statistically significant differences in the half-life of absorption into plasma were determined in the presence of soil after oral treatment compared to toluene alone. However, adsorption to sandy soil produced a statistically significant decrease (two-fold) in $t_{1/2}$ of absorption after dermal exposure. Absorption half-lives were longer by about 45- to 70-fold after dermal administration compared to oral exposures. Clay soil significantly shortened the $t_{1/2}$ of ^{14}C elimination from plasma compared to control after oral treatment. Neither of the soils altered the half-life of elimination compared to pure toluene following dermal exposure. In the pure and sandy groups, the elimination half-lives after dermal treatment were decreased to approximately one-half of oral treatment, while in the clay group, the dermal elimination was about 2.5-fold of oral elimination. AUC calculations did not reveal any significant differences among the groups during the 3 hr oral period studied or the 48 hr dermal exposure period (Table 2). It is worth noting that there were 270- to 300-fold reductions in dermal AUC's compared to the oral values.

Excretion of radioactivity after oral or percutaneous absorption of ^{14}C-toluene occurs primarily through the kidneys (Table 3). After the oral treatments, radioactivity was excreted rapidly in urine (greater than 50% of the initial dose) for all groups in the first 12 hr of collection. An additional 10% to 20% of the dose was excreted during the 12 to 24 hr time interval. Only about 2% of the administered dose was excreted in urine during the 24–48 hr collection period.

Topical administration of toluene delayed the urinary excretion of radioisotope. About 20% of the applied dose was eliminated between 0–12 hr, while the majority of the dose (approximately 35%) was collected during the 12–24 hr time interval following application of the chemical. The total fractions of the initial dose eliminated in urine by the two routes of administration were comparable for the pure

Table 1. Absorption and Elimination Half-Lives of Radioactivity in Plasma Following Oral or Dermal Administration of ^{14}C-Toluene[a]

Treatment	$t_{1/2}$ (hr)			
	Absorption		Elimination	
	Oral	Dermal	Oral	Dermal
Pure[b]	0.14	6.4	18.1	10.3
Sandy[c]	—[e]	3.2[f]	23.2	10.8
Clay[d]	0.08	5.6	4.2[f]	10.8

[a]Values calculated from 4 to 7 rats per group.
[b]^{14}C-toluene alone.
[c]^{14}C-toluene adsorbed to sandy soil.
[d]^{14}C-toluene adsorbed to clay soil.
[e]Occurence of peak concentration in plasma at the first sampling point did not allow calculation of absorption $t_{1/2}$.
[f]Significantly different than treatment with ^{14}C-toluene alone ($p<0.05$).

Table 2. Area Under Plasma Concentration Time Curve (AUC) Following Oral or Dermal Administration of ^{14}C-Toluene[a]

Treatment	Percent Initial Dose/mL†min	
	Oral	Dermal
Pure[b]	1.13±0.12	0.0038±0.0017
Sandy[c]	1.00±0.07	0.0037±0.0017
Clay[d]	1.27±0.13	0.0042±0.0018

[a]Values calculated from 4 to 7 rats per group.
[b]^{14}C-toluene alone.
[c]^{14}C-toluene adsorbed to sandy soil.
[d]^{14}C-toluene adsorbed to clay soil.

Table 3. Urinary Recovery of Radioactivity Following Oral or Dermal Administration of ^{14}C-Toluene[a]

Time (hr)	Oral			Dermal		
	Pure	Sandy	Clay	Pure	Sandy	Clay
0–12	53.7 ± 10.2	56.0 ± 5.5	61.2 ± 11.8	18.4 ± 2.8	16.5 ± 1.6	18.4 ± 1.9
12–24	20.9 ± 7.3	10.1 ± 1.7	18.6 ± 3.0	36.0 ± 7.5	37.0 ± 3.7	31.4 ± 3.2
0–24	74.5 ± 2.9	66.1 ± 5.4	79.9 ± 8.9	54.3 ± 10.2	53.6 ± 2.5	49.8 ± 4.7
24–48	2.2 ± 0.6	1.6 ± 0.3	1.9 ± 0.3	21.2 ± 4.2	31.3 ± 8.1	30.2 ± 4.4
0–48	76.8 ± 2.4	67.8 ± 5.3	81.7 ± 8.7	75.5 ± 14.2	84.8 ± 10.4	80.1 ± 1.2

[a]Values represent percentage of initial dose (mean ± SEM) for six animals per group.

Table 4. Expired Air Recovery of Radioactivity Following Oral or Dermal Administration of C-Toluene[a]

Time (hr)	Oral			Dermal		
	Pure	Sandy	Clay	Pure	Sandy	Clay
0–12	20.2 ± 1.1	23.5 ± 1.7	13.8 ± 0.7[b]	2.7 ± 1.4	2.6 ± 0.3	0.2 ± 0.1
12–24	0.2 ± 0.0	0.4 ± 0.1	0.5 ± 0.1	0.8 ± 0.3	1.3 ± 0.4	0.5 ± 0.3
0–24	22.4 ± 1.1	23.9 ± 1.8	14.3 ± 0.8[b]	3.5 ± 1.3	4.0 ± 0.7	0.8 ± 0.4
24–48	0.1 ± 0.0	0.1 ± 0.0	0.1 ± 0.0	0.2 ± 0.1	1.4 ± 0.7	0.1 ± 0.0
0–48	22.4 ± 1.1	23.9 ± 1.8	14.4 ± 0.8[b]	3.8 ± 1.2	5.4 ± 1.3	0.9 ± 0.4

[a]Values represent percentage of initial dose (mean ± SEM) for six animals per group.
[b]Significantly different than treatment with ^{14}C-toluene alone ($p<0.01$).

and clay soil-adsorbed groups; however, total urinary recovery in the dermal sandy soil-adsorbed group (85%) exceeded the value of the oral sandy soil group (68%).

The excretion of radioactivity as parent compound in expired air was greater in the oral studies than in the dermal studies (Table 4). Most of the radioactivity (14–24%) was expired in the first 12 hr after oral treatment. Oral clay soil-adsorbed toluene produced significantly smaller amounts of total activity (14% of the initial dose versus 22% and 24%, respectively, for the pure and sandy soil-adsorbed chemicals). The amount of radioactivity eliminated by the lungs was minor after dermal application: 4%, 5%, and 1% for pure, sandy, and clay groups, respectively. Carbon dioxide comprised less than 2% of the total radioactivity in the expired air of all oral treatment groups (data not shown). None of the dose was expired as $^{14}CO_2$ from the dermal route of administration. During the 48-hr period, the radioactivity in feces after oral treatment was 0.8%, 1.2%, and 0.6% of initial dose in pure, sandy, and clay groups, respectively. Negligible amounts of radioactivity (less than 0.5%) were also recovered in all dermal groups during the same time period (data not shown).

The tissue distribution patterns of ^{14}C-activity 3 hr following oral treatment are presented in Table 5. Stomach and fat contained the highest tissue concentrations of radioactivity in the pure toluene group, followed by duodenum, pancreas, and kidney. Radioactivity in the sandy soil-adsorbed group was also greatest in the stomach, followed by the fat, and to a lesser extent, in pancreas and liver. In the clay soil-adsorbed group, fat and stomach contained larger amounts of radioactivity than duodenum, kidney, and ileum. No statistically significant differences in the tissue concentrations of radioactivity were observed between oral treatment groups.

Table 5. Tissue Distribution of Radioactivity in Male Rat Following Oral Administration of ^{14}C-Toluene[a]

Tissue distribution
Pure Toluene Treatment Group:
Stomach = Fat > Duodenum = Pancreas = Kidney
Sandy Treatment Group:
Stomach > Fat > Pancreas = Liver
Clay Treatment Group:
Fat = Stomach > Duodenum = Kidney = Ileum

[a]Data obtained from five rats per group, 3 hr following oral administration.

Although the sequence of ^{14}C-distribution is the same in all treatment groups after dermal exposure, soil-related differences were found in tissue distribution 48 hr post-administration (Table 6). Clay soil treatment significantly decreased radioactivity in treated skin compared to toluene alone. Statistically significant decreases were also found in the untreated skin of both soil groups compared to the control group. High concentrations of activity appeared in all areas of fat examined; namely, fat beneath the treated and untreated skin and gut fat. Ethanol washes of treated sites at necropsy established that only about 0.5% of the initial dose was loosely retained on the skin application sites.

Table 6. Tissue Distribution of Radioactivity in Male Rat Following Dermal Administration of ^{14}C-Toluene[a]

In All Treatment Groups:
Treated Skin[b]>>Untreated Fat = Treated Fat = Gut Fat>Untreated Skin[c]

[a]Data obtained from five rats per group, 48 hr following dermal administration.
[b]Significantly decreased in the clay group compared to the pure toluene group ($p<0.03$).
[c]Significantly decreased in the sandy and clay groups compared to the pure toluene group ($p<0.003$).

Table 7. Urinary Metabolites of ^{14}C-Toluene in the Male Rat[a]

Metabolite	Oral			Dermal		
	Pure	Sandy	Clay	Pure	Sandy	Clay
Hippuric Acid	71.4 ± 6.9	63.5 ± 3.6	72.9 ± 2.0	71.8 ± 4.6	64.5 ± 2.5	63.3 ± 2.3
Undetermined	22.8 ± 4.7	23.6 ± 2.7	24.8 ± 2.2	19.4 ± 1.6	25.4 ± 0.5	27.4 ± 1.2

[a]Values represent % of total radioactivity in the 0–12 hr collection period for six animals per group (mean ± SEM).

The urinary metabolites of ^{14}C-toluene in the male rat after oral and dermal treatments are given in Table 7. Hippuric acid was the major metabolite (>60%) detected in the 0–12 hr samples of all groups in both routes of administration. Smaller quantities of an undetermined metabolite were also found. The type and percentage of toluene metabolites were not significantly altered in the presence of soil after either route of exposure. Similar metabolite percentages were detected in the 12–24 hr urines of all treated groups (data not shown). The parent compound was not detected in the urine of any treatment group.

DISCUSSION

The results of this study indicate that the presence of either sandy or clay soil produced alterations in the bioavailability of toluene following oral or dermal treatments. The absorption of radioactivity into plasma was faster after oral administration of toluene than after topical application. Adsorption to sandy soil, however, significantly reduced the dermal absorption half-life compared to pure compound. Although soil significantly decreased the half-life of elimination of toluene after oral administration, the pure and sandy soil-adsorbed toluene half-lives were longer than by the dermal route. No change in the amount of chemical absorbed was produced by the soils, as evidenced by similar AUCs within the groups exposed by a particular route. Thus, the soils altered the time course of toluene absorption or elimination, but not the amount absorbed. The amount of radioisotope absorbed by the oral route, however, was much greater than by the dermal route.

The oral kinetic data in the current study agree with the previous studies of Pyykko et al.[13] which showed that toluene is rapidly absorbed through the gastrointestinal tract. Approximately 80% of the toluene dose was absorbed at a rate of 1.2 mg/cm²/hr over 48 hr in the present dermal studies. Sato and Nakajima[14] reported low absorption of toluene through skin. Volatilization losses in their study were not determined. Volatilization losses from the application sites in the present experiment were minimized by employing the glass cap arrangement. The high percutaneous absorption rate found by Dutkiewicz and Tyras[15] is based on the dose of applied and remaining toluene on skin. Because no data are available on the quantity of parent compound and/or metabolites in the tissues and excreta of Dutkiewicz and Tyras'[15] study, an accurate comparison of results cannot be made with their work.

Urine as the primary route of excretion in all oral and dermal treatment groups is consistent with the metabolism of toluene to water soluble products which are eliminated via the kidney. The appearance of radioactivity in urine was delayed by the dermal treatments, while most of the oral urinary excretion occurred within 12 hr of exposure. Smaller percentages of toluene were excreted unmetabolized via the expired air by both routes. Furthermore, clay soil significantly decreased the amount of radioactivity excreted through the lungs by the oral route.

The tissue distribution pattern of oral toluene-derived radioactivity was unaffected by the presence of either soil. In all treatment groups, stomach (the site of administration) and fat exhibited the highest percentage of radioactivity. Pyykko et al.[13] also reported high concentrations of radioactivity in the same organs following gastric intubation of rats with ^{3}H-toluene. The high concentration of radioactivity in fat is most likely unmetabolized ^{14}C-toluene which has a high octanol/water partition coefficient. The higher a compound's octanol/water partition coefficient, the more readily it would be bioaccumulated into biological tissues.[23,24] Compared to the oral route, a smaller percentage of radioactivity distributed to fat when ^{14}C-toluene was administered by the dermal route. The difference may be a consequence of slower absorption of radioactivity into blood following dermal exposure, resulting in more extensive metabolism of toluene and less unmetabolized toluene in blood for distribution to tissues including fat.

The metabolic profile of toluene was not changed by the soils in either route of exposure. Hippuric acid appeared as the major urinary metabolite in all treatment groups during all collection periods. The oxidation of toluene to benzoic acid followed by conjugation with glycine is consistent with the findings of Bray et al.[25] in rabbits administered toluene orally. Pathiratne et al.[26] also reported an unidentified ^{14}C-metabolite after incubating ^{14}C-+toluene with rat liver microsomes in the presence of an NADPH-generating system.

The oral and dermal bioavailability of toluene in the presence of either soil is different from that reported for benzene. While the time course of each chemical in the body was altered by the soils, only oral benzene showed an increase in the amount absorbed. The relatively stronger adsorbance of benzene to clay soil resulted in a significantly decreased dermal penetration of chemical. On the other hand, the quantity of oral and dermal toluene absorbed was unaffected by the soils.

ACKNOWLEDGMENT

This research was supported as a project of the National Science Foundation/Industry/University Cooperative Center for Research in Hazardous and Toxic Substances at New Jersey Institute of Technology, an Advanced Technology Center of the New Jersey Commission on Science and Technology.

REFERENCES

1. Paustenbach, D. J. "A Methodology for Evaluating the Environmental and Public Health Risks of Contaminated Soil," in *Petroleum Contaminated Soils, Volume I: Remediation Techniques, Environmental Fate, Risk Assessment,* P. T. Kostecki and E. J. Calabrese, Eds. (Chelsea, MI: Lewis Publishers, Inc., 1989), pp. 225–261.
2. Kimbrough, R., H. Falk, P. Stehr, and G. Fries. "Health Implications of 2,3,7,8-tetrachlorodibenzo-p-dioxin (TCDD) Contamination of Residential Soil," *J. Toxicol. Environ. Health.* 14:47–93 (1984).
3. Paustenbach, D. J., H. P. Shu, and T. J. Murray. "A Critical Analysis of Risk Assessments of TCDD Contaminated Soil," *Regul. Toxicol. Pharmacol.* 6:284–304 (1986).
4. Eschenroeder, A., R. J. Jaeger, J. J. Ospital, and C. P. Doyle. "Health Risk Assessment of Human Exposures to Soil Amended with Sewage Sludge Contaminated with Polychlorinated Dibenzodioxins and Dibenzofurans," *Vet. Hum. Toxicol.* 28:435–442 (1986).
5. Schaum, J. "Risk Analysis of TCDD Contaminated Soil," (Washington, DC, Office of Health and Environmental Assessment, U.S. Environmental Protection Agency, 1983).
6. Roels, H., J. P. Buchet and R. R. Lauwerys. "Exposure to Lead by the Oral and Pulmonary Routes of Children Living in the Vicinity of a Primary Lead Smelter," *Environment.* 22:81–94 (1980).
7. Turkall, R. M., G. Skowronski, S. Gerges, S. Von Hagen, and M. S. Abdel-Rahman. "Soil Adsorption Alters Kinetics and Bioavailability of Benzene in Orally Exposed Male Rats," *Arch. Environ. Contam. Toxicol.* 17:159–164 (1988).
8. Abdel-Rahman, M. S., and R. M. Turkall, "Determination of Exposure of Oral and Dermal Benzene from Contaminated Soils," in *Petroleum Contaminated Soils, Volume I: Remediation Techniques, Environmental Fate, Risk Assessment,* P. T. Kostecki and E. J. Calabrese, Eds. (Chelsea, MI: Lewis Publishers, Inc., 1989) pp. 301–311.
9. Skowronski, G. A., R. M. Turkall, and M. S. Abdel-Rahman. "Soil Adsorption Alters Bioavailability of Benzene in Dermally Exposed Male Rats," *Am. Ind. Hyg. Assoc. J.* 49:506–511 (1988).
10. Von Burg, R. "Toxicology Update, Toluene," *J. Appl. Toxicol.* 1:140 (1981).
11. Bergman, K. "Whole-Body Autoradiography and Allied Tracer Techniques in Distribution and Elimination Studies of Some Organic Solvents," *Scand. J. Work Environ. Health.* 5 (suppl. 1): 1–263 (1979).
12. Sandmeyer, E. E., "Aromatic Hydrocarbons," in *Patty's Industrial Hygiene* and *Toxicology,* Vol. II B. G. D. Clayton and F. E. Clayton, Eds. (New York: John Wiley & Sons, 1981), pp. 3253–3431.

13. Pyykko, K., H. Tahti, and H. Vapaatalo. "Toluene Concentrations in Various Tissues of Rats after Inhalation and Oral Administration," *Arch. Toxicol.* 38:169–176 (1977).
14. Sato, A. and T. Nakajima. "Differences Following Skin or Inhalation Exposure in the Absorption and Excretion Kinetics of Trichloroethylene and Toluene," *Br. J. Ind. Med.* 35:43–49 (1978).
15. Dutkiewicz, T., and H. Tyras. "Skin Absorption of Toluene, Styrene, and Xylene by Man," *Br. J. Ind. Med.* 25:243 (1968).
16. National Cooperative Soil Survey: Official Series Description—Atsion Series. (Washington, DC: United States Department of Agriculture—Soil Conservation Service, 1977).
17. National Cooperative Soil Survey: Official Series Description—Keyport Series. (Washington, DC: United States Department of Agriculture—Soil Conservation Service, 1972).
18. Tukey, J. W. *Exploratory Data Analysis* (Reading, MA: Addison Wesley, 1977), pp. 205–235.
19. Velleman, P. F., and D. C. Hoaglin. *Applications, Basics and Computing* of *Exploratory Data Analysis* (*ABC's of EDA*) (Boston, MA: Duxbury Press, 1981), pp. 159–200.
20. Gibaldi, M., and D. Perrier. *Pharmacokinetics* (New York: Marcel Dekker, 1975), pp. 281–292.
21. Efron, B. *The Jacknife*, the *Bootstrap*, and *Other Resampling Plans* (Philadelphia, PA: Society of Industrial Applied Math, 1982).
22. Efron, B., and R. Tibshirani. "Bootstrap Method for Assessing Statistical Accuracy," Technical Report 101 (Stanford, CA: Division of Biostatistics, 1985).
23. Grisham, J. W., Ed. "Factors Influencing Human Exposure," in *Health Aspects of the Disposal of Waste Chemicals.* (New York: Pergamon Press, 1986), pp. 40–64.
24. Freed, V. H., C. T. Chiou, and R. Hague. "Chemodynamics: Transport and Behavior of Chemicals in the Environment—A Problem in Environmental Health," *Environ. Health Perspect.* 20:55–70 (1977).
25. Bray, H. G., W. V. Thrope, and K. White. "Kinetic Studies of the Metabolism of Foreign Compounds," *Biochem. J.* 48:88–96 (1951).
26. Pathiratne, A., R. L. Puyear, and J. D. Brammer. "Activation of ^{14}C-toluene to Covalently Binding Metabolites by Rat Liver Microsomes," *Drug Metab. Dispos.* 14:386–391 (1986).

CHAPTER 27

A Preliminary Decision Framework for Deriving Soil Ingestion Rate

Edward J. Calabrese, Environmental Health Sciences Program, School of Public Health, University of Massachusetts, Amherst
Edward S. Stanek, Biostatistics and Epidemiology Program, School of Public Health, University of Massachusetts, Amherst
Charles E. Gilbert, Environmental Health Sciences Program, School of Public Health, University of Massachusetts, Amherst

INTRODUCTION

The issue of selecting an appropriate soil ingestion value for children and more recently for adults[1] has become a significant challenge to public health and regulatory agencies as well as consultants performing site-specific risk assessments. The public health implications, along with enormous cost considerations, has played a major role in shaping the debate over what the level(s) should be. This chapter will offer guidance on how to approach the problem of selecting an appropriate daily soil ingestion rate in light of regulatory/public health needs within the context of the quality of the present soil ingestion database.

QUALITY OF THE PRESENT SOIL INGESTION DATABASE

It was shown in Calabrese and Stanek[2] that four studies (i.e., Binder et al.[3]; Calabrese et al.[4]; Van Wijnen et al.[5]; Davis et al.[6]) have been published that were designed to provide quantitative evidence for soil ingestion rates among children. Of those four studies, only the Calabrese et al.[4] and Davis et al.[6] reports provided convincing qualitative evidence that the children in their studies actually eat soil. However, only the Calabrese et al.[4] report was able to provide quantitative estimates of soil ingestion based on acceptable precision of the recovery of tracers ingested with soil. Of the eight tracers employed in the Calabrese et al.[4] study, only two (i.e., Ti and Zr) were recovered with acceptable limits of precision (i.e., 100% + 20% for 2 SD). The current database, therefore, on how much soil is ingested by children is limited to one study. Prior to the development of the model[7] to estimate the detection capacities of soil ingestion studies, it was believed that these studies offered quantitative evidence of soil ingestion, and that their general similarity in estimates provided a complementary and stable database upon which reliable estimates of soil ingestion could be made. However, the quantitative foundations for soil ingestion values were challenged with the recent findings reported in this series of papers,[2,7] leaving a seriously compromised soil ingestion database for children, with only one study providing quantitative estimates. Even for that study,[4] six of the eight tracers employed could not be determined with adequate precision of recovery. Given this situation, how do regulatory/public health officials estimate soil ingestion among children and possibly adults?

The current limited database places great emphasis on the Calabrese et al.[4] study, since it offers the only quantitative estimates of soil ingestion with acceptable precision. Within this context one must recognize the strengths and limitations that this study offers. Based on observations of a mass-balance study of 64 children over a two-week period, this study offered several methodological advantages including:

- the use of multiple (i.e., 8) soil tracers
- following the subjects over two weeks
- obtaining daily tracer ingestion/excretory values
- validating tracer recovery efficiency with a pilot study in adults

While this study contributed to helping assure a high reliability in soil ingestion estimates of the study participants, it is important to emphasize that the study has significant limitations with respect to its generalizability to other populations of children, especially those residing in urban areas and non-Caucasian children. In addition, the nonrandom nature of the selected population affects its capacity to be generalized from an academic community in western Massachusetts to children in other communities. In addition, the study population was observed only between Monday and Thursday for two consecutive weeks. While suggesting a possible magnitude of intra-subject soil ingestion variation, the study provides no insight for seasonal variation, which may be a significant variable, especially in colder climates. Another factor inadequately considered in the study was the relationship of the extent of grass cover, and how that was quantitatively related to soil ingestion.

These collective limitations of the Calabrese et al.[4] study to offer generalizations to other populations of children present a serious challenge to regulatory/public health officials performing soil-based risk assessments. In the absence of an adequate database it becomes even more imperative to establish rational procedures for deciding what is (are) the most appropriate soil ingestion value(s). Ideally, given the limitations of the current database, a national numerical value for children and adults should not be derived until additional appropriate studies are completed. However, given that soil cleanup activities are rapidly proceeding, and that risk-based criteria are needed to assist in the determination of cleanup levels, it appears that soil ingestion values will be selected regardless of whether there is an adequate database or not. In light of this reality and given the current database, how should a daily soil ingestion rate be selected? Five critical issues need to be addressed in the analysis of selecting an interim soil ingestion rate.

1. Selection of Tracer. Which tracer should be selected? Since the two quantitatively acceptable tracer elements in the Calabrese et al.[4] study differ in their median values by a factor of 3.4 (16 for Zr vs 55 for Ti), should they be averaged or should the highest (i.e., Ti) be selected?
2. Statistic Selection. Should the population-based median, geometric mean, or arithmetic mean be selected? Or should a specific percentage of the population, such as the value for the upper 75%, 90%, 95%, or 99% be selected?
3. How do soil ingestion rates change with age? Since soil ingestion is likely to be a function of age, how should these values be estimated and inter-age variations be determined for public health/regulatory purposes?
4. Is it possible to differentiate the urban child from the suburban and rural child with respect to age-dependent and seasonal soil ingestion rates?
5. How should the so-called soil pica child be handled?

SELECTION OF TRACER

As indicated in Calabrese and Stanek,[2] those tracer elements with the smallest confidence interval estimates (expressed as a percent of the estimated median) are considered superior to those with larger confidence intervals. However, it is important that the distribution of respective tracers display a high degree of overlap since this supports the assumption of inter-tracer reliability. Both of the above factors must then be seen within the context of the precision of tracer recovery efficiency.

At the present time, estimations of soil ingestion have been tracer-element specific. However, multiple tracer elements lead to a proliferation of soil ingestion estimates, and possible confusion in interpretation. Novel statistical models (e.g., generally seemingly unrelated regression models, using mixed model variance structures) are possible to be fit that simultaneously account for prediction of soil ingestion based on several elements. Such models have a clear appeal, since they may produce a single soil ingestion estimate for a study. These models may produce better estimates, in part, since they take advantage of the co-variance of soil ingestion estimates from different elements, and in part because they weight different elements proportionally to their precision of recovery. Such models provide perhaps the most economical way of gaining additional insight into soil ingestion. This approach has yet to be implemented, but should be explored with the eight tracers of the Calabrese et al.[4] study as a feasible technique to selecting soil ingestion values.

Since the above analysis has yet to be completed and evaluated, which tracer(s) should be selected for use? The present evidence indicates that only Ti and Zr in the Calabrese et al.[4] study were seen with a reliable degree of precision. For these two tracers, acceptable estimates of soil ingestion exist. For the remaining six tracers in the Calabrese et al.[4] and Davis et al.[6] studies, upper bound values of median soil ingestion per tracer element are provided. Consequently, the estimates offered for Ti and Zr are the most appropriate for further consideration. Statistical analyses[2] revealed that the distribution of median values were tighter for Zr than Ti. In addition, Zr displayed a more favorable (i.e., lower) food to soil ratio than Ti and a more sensitive (i.e., lower) level of soil ingestion detection capacity in the Calabrese et al.[4] study. These factors collectively suggest that Zr is more likely to be a better choice than Ti. However, during the preliminary laboratory analytical development phase of the Calabrese et al.[4] study, difficulties were encountered with Zr recovery. While it was believed that the problem(s) leading to Zr recovery were resolved prior to the actual study, several observations suggest otherwise. First, soil ingestion estimates of a soil-pica child were extremely consistent for seven tracers, with values ranging from 8–13 g/day. However, the estimated soil-ingestion rate for the soil-pica child based on Zr was 2.6 g/day. This reduced estimate is also consistent with the lower median value reported above for Zr compared to Ti. Therefore, it is a reasonable assumption that the Zr value for soil ingestion may represent an underestimate of soil ingestion. Consequently, we tentatively recommend that Ti be viewed as the preferred tracer upon which soil ingestion estimations for the average child in the Calabrese et al.[4] study be based, until this area of uncertainty is more confidently resolved. We caution that selection of the preferred tracer will be study-specific and may vary according to the population statistic selected.

STATISTIC SELECTION

For regulatory/risk assessment purposes, some have recommended using a mean estimate rather than median. The mean has the advantage of being simpler and more easily understood. We feel that this is a weak argument for using the mean, since the mean will be strongly influenced by extreme values. In contrast, the median has been shown to be much more robust than the mean in the analysis of the Calabrese et al.[4] data, using a variety of assumptions in the formulation of different population estimates. However, the use of other percentiles (such as the 75%) would be a sound measure if some estimate larger than the median were desired. If some estimate of the mean is insisted upon, a better estimate will be the geometric mean, since this estimate will account for the skewness of daily soil ingestion values observed among children. In fact, in forming confidence intervals for soil ingestion, better coverage can be obtained by forming confidence intervals based on the geometric mean (using log base e), rather than the simple arithmetic mean. However, the use of the geometric mean argues against simplicity.

If it were necessary to select a single estimate for soil ingestion, we would recommend an estimate on a combined soil and dust element concentration, and consider the best estimate to be based on the median. The rejection of the arithmetic mean is based not only on its instability as a measure of central tendency of the population, but that it does not have any precise meaning in the assessed population. For example, with variables that have skewed distributions, the mean does not represent any benchmark in terms of a percentile. This is in contrast with the median, which represents the 50th percentile.

AGE RELATED CHANGES

It has been generally assumed by various state/federal regulatory and public health agencies that all human age groups ingest soil. It has been concluded, based principally on professional judgment, that children ingest more soil than adults, and that children with high hand-to-mouth activity (i.e., ages 1–4) ingest more soil than children of other ages.

Analysis of the Calabrese et al.[4] data revealed that soil ingestion increased linearly with age for all tracers. This was particularly evident for Ti, while much less for Zr.[7] The slopes of these two most reliable tracers differ to such an extent (slope 7.36-Ti, 0.49-Zr) that it cannot be determined for the Calabrese et al.[4] study population that soil ingestion increases as children increase in age from 1 to 4.

While it is believed that children aged 1–4 years ingest more soil than other age groups, this assumption is not based on empirical data. Incidental or intentional soil ingestion in children 5–10 years may be more or less than 1–4 year-olds, but this study and others provide no quantitative information on the soil ingestion rates that answer this question. An attempt to estimate soil ingestion in the six adults used in the three-week tracer recovery validation study indicated an average soil ingestion rate of approximately 40 mg/day.[8] However, the estimated rate of soil ingestion in this pilot study was considerably below the actual estimated detection limit of that particular study because of the small sample size and high food tracer to soil tracer ratio (see Reference 9). Thus, the reported soil ingestion estimates for adults in Calabrese et al.[8] could not be seen with sufficient recovery precision to derive accurate judgments about soil ingestion rates in adults, while no clear evidence exists that age-related changes in soil ingestion occurred in the Calabrese et al.[4] study. The age analysis for children is also further compromised because the sample size for each age group was small (i.e., ca. 10 for six-month age intervals). Because of the small sample size for age specific soil ingestion values, the soil ingestion detection levels are far higher than values that were estimated.

In light of the inadequacies of the soil ingestion database, how are age adjustments in soil ingestion to be made? It would appear logical that adults should ingest significantly less soil than young children. It would seem reasonable, in the absence of reliable quantitative data, to assume that an "average" adult ingests from 25% to 10% of the "average" child (1–6 years old), based on diminished hand-to-mouth activity and other maturational and social factors.

Based on this rationale, it is recommended that children 6–12 years of age be assumed to ingest 25% of the soil ingestion value of a 1–6 year old child, while those >12 years of age be assumed to ingest 10% of the 1–6 year old child.

RURAL VS URBAN/SUBURBAN CHILDREN

No quantitative data exist on the comparative soil ingestion rates of children from rural, urban, and suburban areas. This remains an important data gap to be filled. At present, any attempt to make a distinction in soil ingestion rates would be speculative. There may be a number of potential factors affecting the differential rate of soil ingestion among rural, urban, and suburban children such as time spent outdoors, degree of grass cover of outdoor play areas, quantity of dust in home, and others. However, in the absence of adequate information on these variables, the present emphasis will focus on the extent of grass cover because of the obvious direct access to contact with soil. Under such circumstances it is not unreasonable to suggest the incorporation of an uncertainty factor analogous to those used in risk assessment activities for noncarcinogens. Since this represents concern with inter-individual variation, an uncertainty factor (UF) of approximately 1–10 is selected, depending on the degree of grass cover in areas where children play. If grass cover were extensive, (>90%) then an UF of 1 would be appropriate. However, if grass cover were more limited (<50–90%) in areas of access, then a 5-fold factor would be recommended, while a 10-fold factor would be used if grass cover were <50%. This approach may have site-specific application, but it is not recommended for national or statewide guidance.

SEASONALITY

The Calabrese et al.[4] study was conducted in the fall (Sept./Oct) in Massachusetts. It may be speculated that ingestion of soil may be highest in the summertime and lowest in the winter in Massachusetts, based on the premise that children play longer hours outdoors in the summer, with greater direct contact with soil. However, it may be argued that soil contact may actually be greater in the spring before the growth of grass becomes significantly thickened or during more rainy seasons such as spring and winter, depending on geographical locations. Thus, it is possible that seasonal effects may markedly vary according to a variety of factors, and that soil ingestion may not be highest in the summer months in all locations. In addition, there may be seasonal variation in the tracking in of dust within the home, with perhaps more mud being tracked into the home during the more rainy seasons. In the absence of information to clarify these uncertainties in the database, no seasonal effect is recommended at this time.

IDENTIFICATION OF PICA CHILDREN

The consumption of nonfood items, especially by young children, is a very common activity; when this activity is excessively performed it becomes characterized as pica. The range of nonfood items that such children may ingest is extremely variable, including: clothing, books/paper, crayons, soil, cigarettes, household furnishings, and other items.

The prevalence of pica behavior appears to be highly variable, being contingent on the definition of pica and the population assessed, among other factors. Table 1 reveals that the prevalence of pica behavior can range from 10% in Caucasian children to 66% in institutionalized psychotic children. It appears, therefore, that children 1 to 6 years old display a pica prevalence that is between 15% and 30%, with no obvious significant variation between males and females.

The identification of pica children presents a major initial stumbling block, since there are no definitive criteria for this behavior. The present literature often represents subjective judgments based on individual perceptions of what comprises pica behavior, with limited standardized behavioral norms concerning whether children display pica behavior.

Some evidence exists suggesting that there is considerable variation among pica behaviors, especially concerning what items are preferentially ingested.[10,11] More specifically, a child with pica behavior may ingest only selective items to the exclusion of others. For example, a pica child ingesting books or paper may not ingest other items such as cigarettes or soil. In contrast, a child with a preference for soil may not ingest other nonfood items. In fact, Harvey et al.[11] has observed that soil pica behavior comprises but a small subset of childhood pica behaviors. Their data suggest that there is an age-dependence in object selectivity and that such selectivity increases with age (Table 1). The potential implications of the Harvey et al.[11] (1986) data are that not all children with pica ingest soil, and in fact only a subset of children with pica ingest greater than average amounts of soil. If one-quarter (25%) of children display pica behavior, it may be reasonable to assume that about 25% of those children are soil pica children.[11] This would result in about 6.25% of the population aged 1–6 years displaying soil pica behavior. Given this theoretical estimate of 6.25% for an estimated soil pica prevalence, it would be important to compare this value with estimates based on soil tracer ingestion studies for children in the 1–6 year age range. The four

Table 1. Range of Pica Behavior Prevalence

Group Description	# Subjects	% of Pica	Reference
Retarded children	30	50	Kanner[12]
Black children >6 mo.	386	27	Cooper[13]
White children >6 mo.	398	17	
Black, 1–6 years	486	32	Millican et al.[14]
White, 1–6 years	294	10	Millican et al.[14]
Children, low income	859	55	Lourie et al.[15]
Children, high income		30	Lourie et al.[15]
Children, 1–6 (interview)	439	15	Barltrop[10]
Children, 1–6 (mail survey)	227	50	
Institutionalized, psychotic 3–13 years	40	66	Oliver[16]
Spanish-American children (California)	21	32	Bruhn and Pangborn[17]
Children (Mississippi)	115	16	Vermeer and Frate[18]

available soil ingestion studies (Binder et al.,[3] Calabrese et al.,[4] Davis et al.,[6] and Van Wijnen et al.[5]) were examined for soil pica evidence. These collective studies have provided daily soil ingestion on 517 children. If soil pica were subjectively defined in quantitative terms as consumption of greater than 1 gm of soil per day, then 10 individuals would be identified from these four studies as having displayed this behavior. This would amount to a soil pica prevalence of 1.9% from the four available soil ingestion studies. This would be lower than the earlier noted theoretical estimate of 6.25% for soil pica in 1–6 year olds.

This soil tracer estimation of the prevalence of soil pica children of course rests on very limited data. The nine individuals in the Van Wijnen et al. study displayed the pica-like behavior (> 1000 mg/day) on only a single observation day over a 2–5 day period. The soil ingestion values of these subjects was not adjusted downward for food ingestion of the tracer elements, thus leading to variable overestimates of soil ingestion. The one child pica subject in the Calabrese et al. study was observed over two separate 4-day periods and displayed soil pica behavior only in the second of the two-week period of observation. These data suggest that those displaying soil pica behavior do so irregularly, and thus would not be predicted to consistently ingest >1 gram of soil per day. If one accepted that about 2% of children 1–6 years of age occasionally ingest >1.0 gram of soil per day, what percentage would ingest >10 grams per day? The Calabrese et al.[4] data suggest that only one child of the 517 (0.2%) ingested greater than 10 grams of soil per day. It should be noted that soil ingestion estimates for this soil pica child were seen with a precision of 100 ± 20 or less for Al, Si, and Ti.

Over what duration of normal life span would one display soil pica? While there are no adequate data to resolve this issue, it is generally believed that pica behavior is of limited duration, with the prevalence in the population being highest over 1–3 years of age, but declining to 1% to 5% by the age of 6.[10] If soil pica declined accordingly, then the soil pica prevalence at age 6 might be predicted to be 0.02 to 1.0%.

Despite the inadequate database concerning soil ingestion about soil pica children, it is becoming necessary to offer tentative guidance in this area. Soil pica is a subset of pica behavior, and has a prevalence in the 1–6 age population of under 8%, based on survey methods.[11] The quantitative tracer methodology for soil ingestion is believed superior to the qualitative survey information of soil pica prevalence, since it offers definably precise estimates of exposure. It is, therefore, more likely that the actual prevalence of soil pica in children is far below the 8% figure given above. It should be noted that the soil tracer methodology estimates that 2% will occasionally ingest at least 1 gram of soil on a given day. The 2% figure is likely to have been considerably lower if Van Wijnen et al.[5] had adjusted for food intake. The soil tracer methodology estimates that 0.19% ingest up to 10–13 grams of soil per day. The duration of exposure for these estimates is most likely restricted to ages 1–6 (i.e., 5 years). While it is possible that soil pica may be observed in some children beyond that age, the prevalence of this behavior is expected to rapidly decrease as one ages. Note that the prevalence calculations employed above (i.e., arriving at the 0.19% value) used the ages of greatest prevalence and would be markedly lower for ages 5 and 6.

It is, therefore, generally recognized that an inadequate database exists with respect to the prevalence of soil pica, the amount of soil such children ingest, and over what duration soil pica behavior occurs. However, the limited data that do exist suggest that soil pica, as defined by approximately 1000 mg/day, may exist in an upper bound of 2% of children aged 1–6. A small subset of this population (ca. 0.2%) is speculated to ingest up to 10 gm/day.

With the notable exception of the one child in the Calabrese et al. study, no conclusive data exist that any other pica children (> 1000 mg/day) were observed in the above four cited studies, since food ingestion was not adjusted by Van Wijnen et al.,[5] nor were these behaviors seen on repeated days.

If one were to err on the side of safety with a speculative upper-bound daily soil ingestion rate, one possible course of action may be the following:

1. Assume that 2% of children aged 1–6 years exhibit soil pica of 1 gram of soil per day.
2. Assume that 0.2% children ingest 10 grams of soil from 1–6 years of age.

If, however, a more realistic estimate of soil ingestion in soil pica children were desired, then the following suggestions may be followed:

1. Assume that 1% of children exhibit soil pica of 1 gram of soil per day for 4 days per week, and 500 mg/day for 3 days per week for 4 years.
2. Assume that 0.2% children ingest 10 grams of soil for 3 days per week, and 200 mg of soil for 4 days per week for four years.

DUST VS SOIL

Several recent reports[19,20] have indicated that the source of residual fecal tracers in the Calabrese et al.[4] study is of both outdoor soil and indoor dust origin. These papers present techniques for how the source of tracers can be quantitatively differentiated. While considerable interspecies variation existed in the source of residual fecal tracers, approximately 50% of the tracers were of indoor origin. A subsequent study by Calabrese et al.[19] estimated that 30% of the indoor dust was comprised of tracked-in soil. Taken collectively, the median child in the Calabrese et al.[19] study had approximately 65% of their residue fecal tracers of outdoor soil origin.

DISCUSSION

Selection of the "correct" soil ingestion number is not a very wise goal, as much as simple solutions are desired. Soil ingestion is likely to be influenced by a variety of factors that need to be assessed and then quantitatively incorporated within a soil ingestion derivation procedure. This chapter provides a guide for how to proceed along such a soil ingestion derivation process. It attempts to identify the critical issues and to show how such factors may affect soil ingestion values, and how they may be quantitatively dealt with, within the context of available data or in default values. Table 2 presents the structure of the decision

Table 2. Decision Framework for Deriving Soil Ingestion Rate

Tracer	1. Needs to have acceptable precision in recovery. 2. Narrow C.I. for distribution of median. 3. High inter-tracer reliability.
Statistic	1. Stable measure of central tendency (median, geometric mean). 2. Select percentile of choice.
Age	1. Data are inadequate to differentiate age. 2. Age-related behavioral changes suggest that older children and adults ingest from ¼ to $^1/_{10}$ that of children.
Urban/suburban/rural	Use UF approach—use UF of from 5–10-fold if grass cover is limited (e.g., <90% covered).
Seasonality	This is also unknown; no adjustment is recommended at this time since reasonable cases can be made for different seasons providing greater risk of soil ingestion.
Dust/soil	Recommend using a combined soil/dust measurement.
Pica	Assume 1/200 children ingest about 1 gm soil 4 days/wk for 4 years during the 1–6 age span.
Soil vs dust	Limited data indicate that up to 50% of residual fecal tracers can be of indoor origin.

framework recommended to lead to a rational and defensible soil ingestion rate. This approach is designed to assist, but not replace, professional judgment by public health/regulatory risk assessment specialists in this site-specific approach to assessing soil ingestion by children.

REFERENCES

1. Porter, J. W. U.S. EPA Office of Solid Waste and Emergency Response, Memorandum to regional administrator, Region 1-X, regarding interim final guidance on soil ingestion rates, Jan. 27, 1989.
2. Calabrese, E. J., and E. J. Stanek, III. "A Guide to Interpreting Soil Ingestion Studies. II. Qualitative and Quantitative Evidence of Soil Ingestion," *Reg. Toxicol. Pharm.* 13:278–292 (1991).
3. Binder, S., D. Sokal, and D. Maughan. "Estimating the Amount of Soil Ingested by Young Children Through Tracer Elements," *Arch. Environ. Health* 41:341–345 (1986).
4. Calabrese, E. J., R. Barnes, E. J. Stanek, H. Pastides, C. E. Gilbert, P. Veneman, X. Wang, A. Lasztity, and P. T. Kostecki. "How Much Soil Do Young Children Ingest: An Epidemiologic Study," *Reg. Toxic. Pharm.* 10:123–137 (1989).
5. Van Wijnen, J. H., P. Clausing, and B. Brunekreef. "Estimated Soil Ingestion By Children," *Env. Res.* 51:147-162 (1989).
6. Davis, S., P. Waller, R. Buschbom, J. Ballou, and P. White. "Quantitative Estimates of Soil Ingestion in Normal Children Between the Ages of 2 and 7 Years: Population-based Estimates Using Aluminum, Silicon, and Titanium as Soil Tracer Elements. *Arch Env. Health* 45:112–122 (1989).
7. Stanek, E. J., III, and E. J. Calabrese. "A Guide to Interpreting Soil Ingestion Studies. I. Development of a Model to Estimate the Soil Ingestion Detection Level of Soil Ingestion Studies," *Reg. Toxicol. Pharm.* 13:253–277 (1991).
8. Calabrese, E. J., and E. J. Stanek, III, C. E. Gilbert, and R. M. Barnes. "Preliminary Adult Soil Ingestion Estimates; Results of a Pilot Study," *Reg. Toxicol. Pharm.* 12:88–95 (1990).
9. Stanek, E. J., III, E. J. Calabrese, and C. E. Gilbert. "Choosing a Best Estimate of Children's Daily Soil Ingestion," in *Petroleum Contaminated Soil,* Vol. 3. P. T. Kostecki and E. J. Calabrese, Eds. (Chelsea, MI: Lewis Publishers, Inc., 1990), pp. 341-348.
10. Barltrop, D. "The Prevalence of Pica," *Am. J. Dis. Child.* 112:116–123 (1966).
11. Harvey, P. G., A. Spurgeon, G. Morgan, J. Chance, and E. Moss. "A Method for Quantifying Hand-to-Mouth Activity in Young Children," *J. Child. Psychol.* (1986).
12. Kanner, L. *Child Psychiatry* (Springfield,IL: Charles C. Thomas, 1937), pp. 340–353.
13. Cooper, M. *Pica* (Springfield, IL: Charles C. Thomas, 1957).
14. Millican, F. K., E. M. Layman, R. S. Lourie, L. Y. Rakahashi, and C. C. Dublin. "The Prevalence of Ingestion and Mouthing of Non-Edible Substances by Children," *Clin. Proc. Child. Hosp.* (Wash.) 18:207–214 (1962).
15. Lourie, R. S., E. M. Layman, and F. K. Millican. "The Epidemiology of Lead Poisoning and Children," *Arch. Pediat.* 79:72–76 (1963).
16. Oliver, B. E., and G. O'Gorman. *Develop. Med. Child. Neurol.* 8:704–706 (1966).

17. Bruhn, C. M., and R. M. Pangborn. *J. Am. Diet. Assoc.* 58:417–420 (1971).
18. Vermeer, D. E., and D. A. Frate. "Geophagia on Rural Mississippi: Environmental and Cultural Contexts and Nutritional Implications," *Am. J. Clin. Nutr.* 32:2129–2135 (1979).
19. Calabrese, E. J., and E. J. Stanek, III. "Distinguishing Outdoor Soil Ingestion from Indoor Dust Ingestion in a Soil Pica Child," *Regulatory Toxicol. Pharm.* 15:83–85 (1992).
20. Stanek, E. J., III, and E. J. Calabrese. "Soil Ingestion in Children: Outdoor Soil or Indoor Dust," *J. Soil Contamination* 1(1):1–28 (1992).

List of Contributors

Mohamed S. Abdel-Rahman, Department of Pharmacology, New Jersey Medical School, University of Medicine and Dentistry of New Jersey, Newark, NJ 07103

John H. Barkach, The Dragun Corporation, 30445 Northwestern Highway, Suite 260, Farmington Hills, MI 48334

Rena Bass, ChemRisk, McLaren/Hart Environmental Engineering, 901 St. Louis Street, Springfield, MO 65806

Ann L. Baugh, Unocal Corporation, 1201 W. Fifth Street, Los Angeles, CA 90017

Marc Bonazountas, Epsilon International SA, 16 Kifisias Avenue, 151-25 Marousi, Greece

Marco D. Boscardin, GEI Consultants, Inc., 1021 Main Street, Winchester, MA 01890-1943

S. R. Boyes, GeoSolutions Inc., P.O. Box 2127, Gainesville, FL 32602

R. Mark Bricka, U.S. Army Engineer Waterways Experiment Station, 3901 Halls Ferry Road, Vicksburg, MS 39180-6911

Dallas L. Byers, Shell Development, Westhollow Research Center, P.O. Box 1380, Houston, TX 77251

Edward J. Calabrese, Northeast Regional Environmental and Public Health Center, Morrill Science Center, University of Massachusetts, Amherst, MA 01003

Michael A. Callahan, Office of Health and Environmental Assessment, Room #M-3817G, Mail Code RD689, U.S. Environmental Protection Agency, 401 M Street, S.W., Washington, DC 20460

L. Denise Clendending, Chevron Oil Field Research Company, P.O. Box 446, La Habra, CA 90633-0446

M. John Cullinane, Jr., U.S. Army Engineer Waterways Experiment Station, 3901 Halls Ferry Road, Vicksburg, MS 39180-6911

Barbara J. Denahan, Denahan and Associates, 1501 SW 96th Street, Gainesville, FL 32607

Steve A. Denahan, Environmental Science & Engineering, Inc., P.O. Box 1703, Gainesville, FL 32602

Dennis Dineen, McLaren/Hart Environmental Engineering, 16755 Von Karman Avenue, Irvine, CA 92714

Brian G. Dixon, Cape Cod Research, Inc., 19 Research Road, East Falmouth, MA 02536

James Dragun, The Dragun Corporation, 30445 Northwestern Highway, Suite 260, Farmington Hills, MI 48334

Linda Eastcott, Department of Chemical Engineering, University of Toronto, 200 College Street, Toronto, Canada M5S, 1A4

Karl Eklund, Eklund Associates, 76 Myricks Street, Berkley, MA 02779

W. G. Elliott, Environmental Science & Engineering, Inc., Gainesville, FL 32602

Christopher J. Englert, The Dragun Corporation, 30445 Northwestern Highway, Suite 260, Farmington Hills, MI 48334

John Fitzgerald, Division of Hazardous Waste, Massachusetts Department of Environmental Protection, NE Region, 5 Commonwealth Avenue, Woburn, MA 01801

Benjamin R. Genes, Remediation Technologies, Inc., 9 Pond Lane, West Concord, MA 01742

Charles E. Gilbert, Environmental Health Sciences Program, School of Public Health, University of Massachusetts, Amherst, MA 01003

James E. Hansen, Geosafe Corporation, 2950 George Washington Way, Richland, WA 99352

Michael E. F. Hansen, Bank of America, Environmental Services Unit, Second Floor, One City Boulevard West, Orange, CA 92668

Evan C. Henry, Bank of America, Environmental Services Unit, Second Floor, One City Boulevard West, Orange, CA 92668

Marvin B. Hertz, Shell Development, Westhollow Research Center, P.O. Box 1380, Houston, TX 77251

Patrick Hicks, McLaren/Hart Environmental Engineering, Santa Ana, CA

James Holland, McLaren/Hart Environmental Engineering, Santa Ana, CA

Holly M. Horton, Northeast Regional Environmental and Public Health Center, University of Massachusetts, Amherst, MA 01003

James D. Jernigan, Amoco Corporation, Mail Code 4905, 200 E. Randolph Drive, Chicago, IL 60601

Paul C. Johnson, Shell Development, Westhollow Research Center, P.O. Box 1380, Houston, TX 77251

Despina Kallidromitou, National Technical University of Athens, Department of Civil Engineering, Athens, Greece

Renee Kalmes, ChemRisk Division, McLaren/Hart Environmental Engineering, 1135 Atlantic Avenue, Alameda, CA 94501

Earl J. Kenzie, The Dragun Corporation, 30445 Northwestern Highway, Suite 260, Farmington Hills, MI 48334

Paul T. Kostecki, Northeast Regional Environmental and Public Health Center, Morrill Science Center, University of Massachusetts, Amherst, MA 01003

Sum Chi Lee, Department of Chemical Engineering, University of Toronto, 200 College Street, Toronto, Canada M5S, 1A4

Jon R. Lovegreen, Applied Geosciences Inc., 29B Technology Drive, Suite 100, Irvine, CA 92718

John Lynch, Remediation Technologies, Inc., 9 Pond Lane, West Concord, MA 01742

Donald Mackay, Department of Chemical Engineering, University of Toronto, 200 College Street, Toronto, Canada M5S, 1A4

Sharon A. Mason, The Dragun Corporation, 30445 Northwestern Highway, Suite 260, Farmington Hills, MI 48334

James H. Nash, Chapman, Inc., P.O. Box 608, Atlantic Highlands, NJ 07716

John W. Noland, Roy F. Weston, Inc., 1 Weston Way, West Chester, PA 19380

David W. Ostendorf, Department of Civil Engineering, University of Massachusetts, Amherst, MA 01003

Sibel Pamukcu, Department of Civil Engineering, Chandler-Ullmann Building #17, Lehigh University, Bethlehem, PA 18015

Dennis J. Paustenbach, ChemRisk Division, McLaren/Hart Environmental Engineering, 1135 Atlantic Avenue, Alameda, CA 94501

Thomas L. Potter, Mass Spectrometry Facility, Massachusetts Agricultural Experiment Station, Chenoweth Laboratory, University of Massachusetts, Amherst, MA 01003

John Sanford, Cape Cod Research, Inc., 19 Research Road, East Falmouth, MA 02536

Rita S. Schoeny, Environmental Criteria and Assessment Office, U.S. Environmental Protection Agency, 26 West Martin Luther King Drive, Cincinnati, OH 45268

Paul Scott, ChemRisk, McLaren/Hart Environmental Engineering, 901 St. Louis Street, Springfield, MO 65806

Wan Ying Shiu, Department of Chemical Engineering, University of Toronto, 200 College Street, Toronto, Canada M5S, 1A4

Gloria A. Skowronski, Department of Pharmacology, New Jersey Medical School, University of Medicine and Dentistry of New Jersey, Newark, NJ 07103

Jill P. Slater, McLaren/Hart Environmental Engineering, Rancho Cordova, CA 95741

Edward S. Stanek, Biostatistics and Epidemiology Program, School of Public Health, University of Massachusetts, Amherst, MA 01003

Brian W. Swift, Cape Cod Research, Inc., 19 Research Road, East Falmouth, MA 02536

Craig L. Timmerman, Geosafe Corporation, 2950 George Washington Way, Richland, WA 99352

Richard P. Traver, Chapman, Inc., P.O. Box 608, Atlantic Highlands, NJ 07716

W. A. Tucker, Environmental Science & Engineering, Inc., Gainesville, FL 32602

Rita M. Turkall, Department of Pharmacology, New Jersey Medical, and Clinical Laboratory Sciences Department, School of Health Related Professions, University of Medicine and Dentistry of New Jersey, Newark, NJ 07103

Luis A. Velazquez, Roy F. Weston, Inc., 1 Weston Way, West Chester, PA 19380

M. G. Winslow, Environmental Science & Engineering, Inc., Gainesville, FL 32602

Index